MEASURE, PROBABILITY, AND MATHEMATICAL FINANCE

MEASURE, PROBABILITY, AND MATHEMATICAL FINANCE

A Problem-Oriented Approach

Guojun Gan
Manulife Financial
Toronto, ON, Canada

Chaoqun Ma
School of Business Administration
Hunan University
Changsha, Hunan, P.R. China

Hong Xie
Manulife Financial
Toronto, ON, Canada

WILEY

Published by John Wiley & Sons, Inc., Hoboken, New Jersey.
Published simultaneously in Canada.

Library of Congress Cataloging-in-Publication Data is available.

Measure, Probability, and Mathematical Finance: A Problem-Oriented Approach
Guojun Gan, Chaoqun Ma, and Hong Xie
ISBN 978-1-118-83196-0

Printed in the United States of America.

10 9 8 7 6 5 4 3 2 1

To my parents
–Guojun Gan

To my wife and my daughter
–Chaoqun Ma

To my family and friends
–Hong Xie

CONTENTS

PREFACE

Mathematical finance, a new branch of mathematics concerned with financial markets, is experiencing rapid growth. During the last three decades, many books and papers in the area of mathematical finance have been published. However, understanding the literature requires that the reader have a good background in measure-theoretic probability, stochastic processes, and stochastic calculus. The purpose of this book is to provide the reader with an introduction to the mathematical theory underlying the financial models being used and developed on Wall Street. To this end, this book covers important concepts and results in measure theory, probability theory, stochastic processes, and stochastic calculus so that the reader will be in a position to understand these financial models. Problems as well as solutions are included to help the reader learn the concepts and results quickly.

In this book, we adopted the definitions and theorems from various books and presented them in a mathematically rigorous way. We tried to cover the most of the basic concepts and the important theorems. We selected the problems in this book in such a way that the problems will help readers understand and know how to apply the concepts and theorems. This book includes 516 problems, most of which are not difficult and can be solved by applying the definitions, theorems, and the results of previous problems.

This book is organized into five parts, each of which is further organized into several chapters. Each chapter is divided into five sections. The first section presents the definitions of important concepts and theorems. The second, third, and fourth sections present the problems, hints on how to solve the problems, and the full solutions to the problems, respectively. The last section contains bibliographic notes. Interdependencies between all chapters are shown in Table 0.1.

Table 0.1: Interdependencies between Chapters.

Chapter	Related to Chapter(s)
1. Sets and Sequences	
2. Measures	1
3. Extension of Measures	1;2
4. Lebesgue-Stieltjes Measures	2;3
5. Measurable Functions	2
6. Lebesgue Integration	1;2;5
7. The Radon-Nikodym Theorem	2;6
8. L^p Spaces	2;6
9. Convergence	1;2;6;8
10. Product Measures	2;3;5;6
11. Events and Random Variables	1;2;4;5
12. Independence	2;3;5;11
13. Expectation	2;6;8;10;11;12
14. Conditional Expectation	1;2;5;6;7;8;10;11;12;13
15. Inequalities	8;11;14
16. Law of Large Numbers	2;8;9;10;12;13;15
17. Characteristic Functions	5;6;8;11;12;13;15
18. Discrete Distributions	12;14;17
19. Continuous Distributions	6;10;12;13;17
20. Central Limit Theorems	6;9;11
21. Stochastic Processes	2;5;10;11;12;19
22. Martingales	2;5;11;13;14;15
23. Stopping Times	2;5;9;11;14;21;22
24. Martingale Inequalities	2;6;8;13;14;15;23
25. Martingale Convergence Theorems	1;6;9;11;14;15;22
26. Random Walks	8;9;13;14;15;19;20;22;23;24
27. Poisson Processes	11;12;14;17;21;22
28. Brownian Motion	8;9;11;12;14;15;16;17;19

29. Markov Processes	2;6;11;14;21
30. Lévy Processes	1;5;6;11;12;14;17;19;22;27;28;29
31. The Wiener Integral	6;9;15;19;28
32. The Itô Integral	5;6;8;10;14;15;22;24;28
33. Extension of the Itô Integrals	9;10;14;22;23;32
34. Martingale Stochastic Integrals	14;15;19;27;32
35. The Itô Formula	6;8;9;22;24;32;34
36. Martingale Representation Theorem	9;14;25;28;32;33;35
37. Change of Measure	7;14;32;34;35
38. Stochastic Differential Equations	8;11;13;32;34;35
39. Diffusion	6;9;11;14;19;21;24;32;35;38
40. The Feynman-Kac Formula	6;14;32;35;38;39
41. Discrete-Time Models	7;12;14;22;23
42. Black-Scholes Option Pricing Models	9;14;19;24;32;33;35;36;37;38;41
43. Path-Dependent Options	10;14;19;28;37;38;42
44. American Options	14;15;21;22;23;32;35;36;37;42;43
45. Short Rate Models	11;14;19;29;32;35;37;38;39;40
46. Instantaneous Forward Rate Models	10;14;19;32;34;35;37;38;40;45
47. LIBOR Market Models	14;32;37;45;46

In Part I, we present measure theory, which is indispensable to the rigorous development of probability theory. Measure theory is also necessary for us to discuss recently developed theories and models in finance, such as the martingale measures, the change of numeraire theory, and the London interbank offered rate (LIBOR) market models.

In Part II, we present probability theory in a measure-theoretic mathematical framework, which was introduced by A.N. Kolmogorov in 1937 in order to deal with David Hilbert's sixth problem. The material presented in this part was selected to facilitate the development of stochastic processes in Part III.

In Part III, we present stochastic processes, which include martingales and Brownian motion. In Part IV, we discuss stochastic calculus. Both stochastic processes and stochastic calculus are important to modern mathematical finance as they are used to model asset prices and develop derivative pricing models.

In Part V, we present some classic models in mathematical finance. Many pricing models have been developed and published since the seminal work of Black and Scholes. This part covers only a small portion of many models.

In this book, we tried to use a uniform set of symbols and notation. For example, we used $\mathbf{N}$, $\mathbf{R}$, and $\emptyset$ to denote the set of natural numbers (i.e., nonnegative integers), the set of real numbers, and the empty set, respectively. A comprehensive list of symbols is also provided at the end of this book.

We have taken great pains to ensure the accuracy of the formulas and statements in this book. However, a few errors are inevitable in almost every book of this size. Please feel free to contact us if you spot errors or have any other constructive suggestions.

How to Use This Book

This book can be used by individuals in various ways:

(a) It can be used as a self-study book on mathematical finance. The prerequisite is linear algebra and calculus at the undergraduate level. This book will provide you with a series of concepts, facts, and problems. You should explore each problem and write out your solution in such a way that it can be shared with others. By doing this you will be able to actively develop an in-depth and comprehensive understanding of the concepts and principles that cannot be archived by passively reading or listening to comments of others.

(b) It can be used as a reference book. This book contains the most important concepts and theorems from mathematical finance. The reader can find the definition of a concept or the statement of a theorem in the book through the index at the end of this book.

(c) It can be used as a supplementary book for individuals who take advanced courses in mathematical finance. This book starts with measure theory and builds up to stochastic financial models. It provides necessary prerequisites for students who take advanced courses in mathematical finance without completing background courses.

Acknowledgments

We would like to thank all the academics and practitioners who have contributed to the knowledge of mathematical finance. In particular, we would like to thank the following academics and practitioners whose work constitutes the backbone of this book: Robert B. Ash, Krishna B. Athreya, Rabi Bhattacharya, Patrick Billingsley, Tomas Björk, Fischer Sheffey Black, Kai Lai Chung, Erhan Çinlar, Catherine A. Doléans-Dade, Darrell Duffie, Richard Durrett, Robert J. Elliott, Damir Filipović, Allan Gut, John Hull, Ioannis Karatzas, Fima C. Klebaner, P. Ekkehard Kopp, Hui-Hsiung Kuo, Soumendra N. Lahiri, Damien Lamberton, Bernard Lapeyre, Gregory F. Lawler, Robert C. Merton, Marek Musiela, Bernt Oksendal, Andrea Pascucci, Jeffrey S. Rosenthal, Sheldon M. Ross, Marek Rutkowski, Myron Scholes, Steven Shreve, J. Michael Steele, and Edward C. Waymire.

We are grateful to Roman Naryshkin and several anonymous reviewers for their helpful comments. Guojun Gan and Hong Xie would like to thank their friends and colleagues at the Global Variable Annuity Hedging Department of Manulife Financial for the pleasant cooperation over the last 4 years.

Guojun Gan gratefully acknowledges support from the CAIS (Canadian Academy of Independent Scholars) grant and thanks Simon Fraser University for giving him full access to its libraries. Guojun Gan wants to thank his parents and parents-in-law for all their love and support. He wants to thank his wife, Xiaoying, for taking care of their children.

This work was supported in part by the National Science Foundation for Distinguished Young Scholars of China (grant 70825006), Program for Changjiang Scholars and Innovative Research Team in University of Ministry of Education of China (grant IRT0916), the Foundation for Innovative Research Groups of the National Natural Science Foundation of China (grant 71221001), and the Furong Scholar Program.

GUOJUN GAN, CHAOQUN MA, AND HONG XIE

Toronto, ON, Canada and Changsha, Hunan, P.R. China, February 28, 2014

Financial Glossary

American option An option that can be exercised at any time prior to the expiration date.

Asia option An option whose payoff is dependent on the average price of the underlying asset during a certain period.

barrier option An option whose payoff is dependent on whether the path of the underlying asset has reached a barrier, which is a certain predetermined level.

call option An option that gives the holder the right to buy an asset.

derivative A financial instrument whose price depends on the price of another asset (called the *underlying asset*); also referred to as *derivative security* or *financial derivative*.

down-and-in option A barrier option that comes into existence when the price of the underlying asset declines to the barrier.

down-and-out option A barrier option that ceases to exist when the price of the underlying asset declines to the barrier.

European option An option that can be exercised only on the expiration date. Let K be the strike price of an option. Let S_T be the price of the underlying asset at maturity. The terminal payoff of a long position (the holder's position) of a European call is given by $\max(S_T - K, 0)$. The terminal payoff of a long position (the holder's position) of a European put is given by $\max(K - S_T, 0)$.

forward contract A nonstandardized agreement between two parties to buy or sell an asset at a certain future time for a certain price.

futures contract A standardized agreement between two parties to buy or sell an asset at a certain future time for a certain price.

LIBOR London interbank offered rate.

lookback option An option whose payoff is dependent on the maximum or minimum price of the underlying asset in a certain period.

option A derivative that gives the holder the right (not the obligation) to buy or sell an asset by a certain date for a predetermined price. The date is called the *expiration date* and the predetermined price is called the *strike price* or *exercise price*. An option is said to be exercised if the holder chooses to buy or sell the underlying asset.

put option An option that gives the holder the right to sell an asset.

term structure The relationship between interest rates and their maturities.

up-and-in option A barrier option that comes into existence when the price of the underlying asset increases to the barrier.

up-and-out option A barrier option that ceases to exist when the price of the underlying asset increases to the barrier.

zero-coupon bond A bond that does not pay coupons.

PART I

MEASURE THEORY

CHAPTER 1

SETS AND SEQUENCES

Sets are the most basic concepts in measure theory as well as in mathematics. In fact, set theory is a foundation of mathematics (Moschovakis, 2006). The algebra of sets develops the fundamental properties of set operations and relations. In this chapter, we shall introduce basic concepts about sets and some set operations such as union, intersection, and complementation. We will also introduce some set relations such as De Morgan's laws.

1.1 Basic Concepts and Facts

Definition 1.1 (Set, Subset, and Empty Set)**.** A set is a collection of objects, which are called *elements*. A set B is said to be a subset of a set A, written as $B \subseteq A$, if the elements of B are also elements of A. A set A is called an *empty set*, denoted by $\emptyset$, if A contains no elements.

Definition 1.2 (Countable Set)**.** A set A is said to be countable if either A contains a finite number of elements or every element of A appears in an infinite sequence $x_1, x_2, \ldots$. A set A is said to be uncountable if it is not countable.

Definition 1.3 (Equality of Sets). Two sets A and B are said to be equal, written as $A = B$, if $A \subseteq B$ and $B \subseteq A$.

Definition 1.4 (Union, Intersection, and Complement of Sets). Let A and B be two subsets of a set S. The union of A and B is defined as

$$A \cup B = \{s \in S : s \in A \text{ or } s \in B\}.$$

The intersection of A and B is defined as

$$A \cap B = \{s \in S : s \in A \text{ and } s \in B\}.$$

The complement of A relative to S is defined as

$$A^c = \{s \in S : s \notin A\}.$$

Definition 1.5 (Difference and Symmetric Difference of Sets). Let A and B be two sets. The difference between A and B, denoted by $A - B$ or $A \backslash B$, is defined as

$$A \backslash B = A \cap B^c.$$

The symmetric difference between A and B is defined as

$$A \Delta B = (A \backslash B) \cup (B \backslash A).$$

Definition 1.6 (Increasing and Decreasing Sequence of Sets). Let $\{A_n\}_{n \geq 1}$ be a sequence of sets. We say that $\{A_n\}_{n \geq 1}$ is an increasing sequence of sets with limit A, written as $A_n \uparrow A$, if

(a) $A_n \subseteq A_{n+1}$ for all $n \geq 1$.

(b) $\bigcup_{n=1}^{\infty} A_n = A$.

We say that $\{A_n\}_{n \geq 1}$ is a decreasing sequence of sets with limit A, written as $A_n \downarrow A$, if

(a) $A_n \supseteq A_{n+1}$ for all $n \geq 1$.

(b) $\bigcap_{n=1}^{\infty} A_n = A$.

Definition 1.7 (Indicator Function). Let A be a set. Then the indicator function of A is defined as

$$I_A(s) = \begin{cases} 1, & \text{if } s \in A; \\ 0, & \text{if } s \notin A. \end{cases}$$

Definition 1.8 (Upper Limit and Lower Limit of Sequences of Sets). Let $\{E_n\}_{n\geq 1}$ be a sequence of subsets of S. Then $\limsup E_n$ and $\liminf E_n$ are defined as

$$\limsup E_n = \bigcap_{j\geq 1}\left(\bigcup_{i\geq j} E_i\right)$$

and

$$\liminf E_n = \bigcup_{j\geq 1}\left(\bigcap_{i\geq j} E_i\right),$$

respectively.

Definition 1.9 (Upper Limit and Lower Limit of Sequences of Real Numbers). Let $\{x_n\}_{n\geq 1}$ be a sequence of real numbers. Then $\limsup x_n$ and $\liminf x_n$ are defined as

$$\limsup x_n = \inf_{m\geq 1}\left\{\sup_{n\geq m} x_n\right\} = \lim_{m\to\infty}\left\{\sup_{n\geq m} x_n\right\}$$

and

$$\liminf x_n = \sup_{m\geq 1}\left\{\inf_{n\geq m} x_n\right\} = \lim_{m\to\infty}\left\{\inf_{n\geq m} x_n\right\},$$

respectively.

Definition 1.10 (Convergence of Sequences). A sequence $\{x_n\}_{n\in\mathbf{N}}$ of real numbers is said to be convergent if and only if

$$\limsup x_n = \liminf x_n.$$

The sequence is said to be convergent to x, written as $x_n \to x$ or $\lim_{n\to\infty} x_n = x$, if and only if x_n is convergent and $\limsup x_n = \liminf x_n = x$.

Definition 1.11 (Convergence of Sets). A sequence $\{A_n\}_{n\in\mathbf{N}}$ of sets is said to converge to A, written as $A_n \to A$ or $\lim_{n\to\infty} A_n = A$, if

$$\lim_{n\to\infty} I_{A_n}(s) = I_A(s)$$

for all $s \in S$.

Definition 1.12 (Partial Ordering, Totally Ordered Sets, and Chains). A partial ordering $\leq$ on a set S is a relation that satisfies the following conditions, where a, b, and c are arbitrary elements of S:

(a) $a \leq a$ (reflexivity).

(b) If $a \leq b$ and $b \leq a$, then $a = b$ (antisymmetricity).

(c) If $a \leq b$ and $b \leq c$, then $a \leq c$ (transitivity).

Let S be a set with a partial ordering $\leq$. A subset C of S is said to be a totally ordered subset of S if and only if for all $a, b \in C$, we have either $a \leq b$ or $b \leq a$. A chain in S is a totally ordered subset of S.

Theorem 1.1 (De Morgan's Laws). *Let $\{A_n\}_{n\geq 1}$ be a sequence of sets. Then*

$$\left(\bigcup_{n=1}^{\infty} A_n\right)^c = \bigcap_{n=1}^{\infty} A_n^c$$

and

$$\left(\bigcap_{n=1}^{\infty} A_n\right)^c = \bigcup_{n=1}^{\infty} A_n^c.$$

Theorem 1.2 (Zorn's Lemma). *Let S be a nonempty set with a partial ordering "$\leq$". Assume that every nonempty chain C in S has an upper bound, that is, there exists an element $x \in S$ such that $a \leq x$ for all $a \in C$. Then S has a maximal element; in other words, there exists an element $m \in S$ such that $a \leq m$ for all $a \in S$.*

1.2 Problems

1.1. Let I be a countable set. For each $i \in I$, let A_i be a countable set. Show that the union

$$A = \bigcup_{i \in I} A_i$$

is again countable.

1.2. Let $\mathbf{Q}$ be the set of all rational numbers, which have the form of a/b, where a and $b (b \neq 0)$ are integers. Let $\mathbf{R}$ be the set of all real numbers. Show that

(a) $\mathbf{Q}$ is countable.

(b) $\mathbf{R}$ is uncountable.

1.3. Let $\{E_n\}_{n\geq 1}$ be a sequence of sets. Show that

$$\liminf E_n \subseteq \limsup E_n.$$

1.4. Let $\{E_n\}_{n\geq 1}$ be a sequence of subsets of S. Show that

$$(\limsup E_n)^c = \liminf E_n^c$$

and

$$(\liminf E_n)^c = \limsup E_n^c,$$

where $A^c = S \backslash A$ for set A.

1.5. Let $\{E_n\}_{n\geq 1}$ be a sequence of subsets of a set S. Show that

$$I_{\limsup E_n}(s) = \limsup I_{E_n}(s), \quad \forall s \in S \tag{1.1}$$

and

$$I_{\liminf E_n}(s) = \liminf I_{E_n}(s), \quad \forall s \in S, \tag{1.2}$$

where I is the indicator function.

1.6. Let $\{A_n\}_{n\geq 1}$ be a sequence of sets of real numbers defined as follows:

$$A_n = \begin{cases} (-\frac{1}{n}, 1], & \text{if } n \text{ is odd;} \\ (-1, \frac{1}{n}], & \text{if } n \text{ is even.} \end{cases}$$

Calculate $\liminf A_n$ and $\limsup A_n$.

1.7. Let $\{x_n\}_{n\geq 1}$ a sequence of real numbers. Show that x_n converges (i.e., the limit $\lim_{n\to\infty} x_n$ exists) in $[-\infty, \infty]$ if and only if

$$\limsup x_n = \liminf x_n.$$

1.8. Let $\{x_n\}_{n\geq 1}$ and $\{y_n\}_{n\geq 1}$ be two sequences of real numbers. Let c be a constant in $(-\infty, \infty)$. Show that

(a) $\limsup(-x_n) = -\liminf x_n$.

(b) $\limsup x_n \geq \liminf x_n$.

(c) $\liminf x_n + \liminf y_n \leq \liminf(x_n + y_n)$.

(d) $\limsup x_n + \limsup y_n \geq \limsup(x_n + y_n)$.

(e) $\limsup x_n + \liminf y_n \leq \limsup(x_n + y_n)$.

(f) $\liminf(c + x_n) = c + \liminf x_n$.

(g) $\liminf(c - x_n) = c - \limsup x_n$.

1.9. Let $\{A_n\}_{n\geq 1}$ be a sequence of subsets of a set S. Show that $\lim_{n\to\infty} A_n$ exists if and only if $\limsup A_n = \liminf A_n$. In addition, if $A = \lim_{n\to\infty} A_n$ exists, then $A = \limsup A_n = \liminf A_n$.

1.10. Let $\{x_n\}_{n\geq 1}$ and $\{y_n\}_{n\geq 1}$ be two sequences of real numbers. Suppose that

$$\lim_{n\to\infty} x_n \quad \text{and} \quad \lim_{n\to\infty} y_n$$

exist. Show that $\lim_{n\to\infty}(x_n + y_n)$ exists and

$$\lim_{n\to\infty}(x_n + y_n) = \lim_{n\to\infty} x_n + \lim_{n\to\infty} y_n.$$

1.3 Hints

1.1. Try to construct a sequence in which every element in A appears.

1.2. To prove part (a), show that all rational numbers can be written as a sequence. Part (b) can be proved by the method of contradiction, that is, by assuming that $\mathbf{R}$ is countable and can be written as a sequence $(x_n)_{n\geq 1}$. Then represent every x_n as a decimal of finite digits and find a new number, which is not in the sequence.

1.3. This problem can be proved by using Definition 1.8.

1.4. This problem can be proved by using Definition 1.8 and Theorem 1.1.

1.5. An indicator function has only two possible values: 0 and 1. Hence the first equality of the problem can be proved by considering two cases: $s \in \limsup E_n$ and $s \notin \limsup E_n$. The second equality of the problem can be proved using the result of Problem 1.4.

1.6. The lower and upper limits of the sequence can be calculated by using Definition 1.8.

1.7. Use the definition of limits. For example, $\lim_{n\to\infty} x_n = \infty$ if and only if, given any $\epsilon > 0$, there exists an integer N_ϵ such that $x_n > \epsilon$ for all $n \geq N_\epsilon$. Similarly, $\lim_{n\to\infty} x_n = L$ for some $L \in (-\infty, \infty)$ if and only if, given any $\epsilon > 0$, there exists an integer N_ϵ such that $L - \epsilon < x_n < L + \epsilon$ for all $n \geq N_\epsilon$.

1.8. Use Definition 1.9 and the fact that $\sup_{n\geq m}(-x_n) = -\inf_{n\geq m} x_n$ to prove part (a). To prove part (b), try to establish

$$\sup_{i\geq j} x_i \geq \inf_{n\geq m} x_n, \quad j, m \geq 1.$$

To prove part (c), try to establish the following inequality

$$\inf_{n\geq m} x_n + \inf_{i\geq j} y_i \leq \sup_{s\geq 1} \inf_{r\geq s} (x_r + y_r).$$

Use parts (a) and (c) to prove part (d). Use parts (a) and (d) to prove part (e). Use part (c) to prove part (f). Use parts (a) and (f) to prove part (g).

1.9. Use Definition 1.11 and the results of Problems 1.5 and 1.7.

1.10. Use the results of Problems 1.7 and 1.8.

1.4 Solutions

1.1. Since I is countable, then $\{A_i : i \in I\}$ is countable. Note that A_i is countable for each $i \in I$. There exists a sequence $(B_n)_{n\geq 1}$ of countable sets such that every

A_i appears in the sequence. For each integer $n \geq 1$, as B_n is countable, there exists a sequence $(x_{n,m})_{m\geq 1}$ such that every element of B_n appears in the sequence. Now let $(y_i)_{i\geq 1}$ be a sequence given by

$$x_{1,1}, x_{2,1}, x_{1,2}, x_{3,1}, x_{2,2}, x_{1,3}, \ldots.$$

Then every element in A appears in the sequence $(y_i)_{i\geq 1}$. Hence A is countable. This completes the proof.

1.2.

(a) For each integer $n \geq 0$, let

$$A_n = \{r \in \mathbf{Q} : n \leq r \leq n+1\}, \quad B_n = \{r \in \mathbf{Q} : -n-1 \leq r \leq -n\}.$$

Then

$$\mathbf{Q} = \bigcup_{n=0}^{\infty} (A_n \cup B_n).$$

By Problem 1.1, it is sufficient to show that A_n and B_n are countable for each integer $n \geq 0$. Note that $A_n = \{r+n : r \in A_0\}$ and $B_n = \{r-n-1 : r \in A_0\}$. We only need to show that A_0 is countable.

Let $(x_n)_{n\geq 1}$ be a sequence given by

$$0, 1, \frac{1}{2}, \frac{1}{3}, \frac{2}{3}, \frac{1}{4}, \frac{2}{4}, \frac{3}{4}, \ldots.$$

Then every number in A_0 appears in the sequence. Hence $\mathbf{Q}$ is countable.

(b) Assume that $\mathbf{R}$ is countable. Then the subset $(0, 1]$ of $\mathbf{R}$ is also countable. Let the numbers in $(0, 1]$ be written as a sequence $(x_n)_{n\geq 1}$. Since $x_n \in (0, 1]$, we can represent x_n as a decimal

$$0.z_{n,1}z_{n,2}\cdots,$$

where $z_{n,j}$ is one of the 10 digits. The decimal representation is not unique for some numbers. For example, we have

$$0.1 = 0.0999\ldots = \sum_{j=1}^{\infty} \frac{9}{10^{j+1}}.$$

In such cases, we choose the representation with 9s at the end. Now let $y = w_1 w_2 \cdots$ be a decimal given by

$$w_j = \begin{cases} 7, & \text{if } z_{j,j} \leq 5; \\ 2, & \text{if } z_{j,j} > 5. \end{cases}$$

Since $w_j \neq 9$ and $w_j \neq z_{j,j}$ for all $j \geq 1$, we have $y \neq x_n$ for all $n \geq 1$. But the sequence $(x_n)_{n\geq 1}$ includes all numbers in $(0, 1]$. Hence $y = x_{n_0}$ for some n_0. This is a contradiction. Hence $\mathbf{R}$ is uncountable.

This completes the proof.

1.3. Let $s \in \liminf E_n$. Then by Definition 1.8, we have $s \in \bigcap_{i\geq j_0} E_i$ for some $j_0 \geq 1$. It follows that $s \in E_i$ for all $i \geq j_0$. Hence we have

$$s \in E_{\max\{j,j_0\}} \subseteq \bigcup_{i\geq j} E_i$$

for all $j \geq 1$. Consequently, $s \in \limsup E_n$. Therefore, $\liminf E_n \subseteq \limsup E_n$.

1.4. By Definition 1.8 and Theorem 1.1, we have

$$\begin{aligned}(\limsup E_n)^c &= \left(\bigcap_{j\geq 1}\left(\bigcup_{i\geq j} E_i\right)\right)^c = \bigcup_{j\geq 1}\left(\bigcup_{i\geq j} E_i\right)^c \\ &= \bigcup_{j\geq 1}\left(\bigcap_{i\geq j} E_i^c\right) = \liminf E_n^c.\end{aligned}$$

Similarly, we can show that $(\liminf E_n)^c = \limsup E_n^c$.

1.5. To prove (1.1), we consider two cases: $s \in \limsup E_n$ and $s \notin \limsup E_n$. If $s \in \limsup E_n$, then

$$s \in \bigcap_{m\geq 1}\bigcup_{n\geq m} E_n,$$

which implies

$$s \in \bigcup_{n\geq m} E_n, \quad \forall m \geq 1.$$

Thus $s \in E_n$ for infinitely many n. Hence we have

$$\sup_{n\geq m} I_{E_n}(s) = 1, \quad \forall m \geq 1,$$

which gives

$$\inf_{m\geq 1}\sup_{n\geq m} I_{E_n}(s) = 1.$$

If $s \notin \limsup E_n$, then $s \in E_n$ for only finitely many n. Thus we have

$$\sup_{n\geq m} I_{E_n}(s) = 0, \quad \forall m \geq M_0,$$

where M_0 is a sufficient large number. Hence we have

$$\inf_{m\geq 1}\sup_{n\geq m} I_{E_n}(s) = 0.$$

Therefore (1.1) is true. From (1.1) and noting that $I_E(s) = 1 - I_{E^c}(s)$, we have

$$I_{\liminf E_n}(s) = 1 - I_{\limsup E_n^c}(s) = 1 - \limsup I_{E_n^c}(s) = \liminf I_{E_n}(s).$$

Thus (1.2) holds.

1.6. To find $\liminf A_n$, we first need to calculate $\bigcap_{i \geq j} A_i$ for all $j \geq 1$. To do that, we let $B_n = A_{2n-1}$ and $C_n = A_{2n}$ for $n = 1, 2, \ldots$. Then B_n and C_n are decreasing, and we have

$$\bigcap_{n \geq k} B_n = [0, 1]$$

and

$$\bigcap_{n \geq k} C_n = (-1, 0]$$

for all $k \geq 1$. Thus $\bigcap_{i \geq j} A_i = \{0\}$ for all $j \geq 1$. Hence $\liminf A_n = \{0\}$. Similarly, we have $\limsup A_n = (-1, 1]$.

1.7. We first prove the "only if" part. Suppose that $\lim_{n \to \infty} x_n$ exists. Let $L = \lim_{n \to \infty} x_n$ and $\epsilon > 0$. If $L = -\infty$, then there exists an $N_\epsilon \geq 1$ such that $x_n < -\epsilon$ for all $n \geq N_\epsilon$. Hence

$$\inf_{m \geq 1} \sup_{n \geq m} x_n \leq \sup_{n \geq N_\epsilon} \leq -\epsilon.$$

Letting $\epsilon \to \infty$ in the above equation gives $\limsup x_n = -\infty$, which implies $\limsup x_n = \liminf x_n = -\infty$. Similarly, we can show that

$$\limsup x_n = \liminf x_n$$

for $-\infty < L < \infty$ and $L = \infty$. Now we prove the "if" part. Suppose that $\limsup x_n = \liminf x_n = L$ and let $\epsilon > 0$. If $-\infty < L < \infty$, then $\sup_{n \geq m_1} x_n \leq L + \epsilon$ for some $m_1 \geq 1$ and $\inf_{n \geq m_2} x_n \geq L - \epsilon$ for some $m_2 \geq 1$. Therefore, we have

$$L - \epsilon \leq x_n \leq L + \epsilon, \quad n \geq \max(m_1, m_2).$$

Since this is true for every $\epsilon > 0$, $\lim_{n \to \infty} x_n$ exists and is equal to L. Similarly, we can show that $\lim_{n \to \infty} x_n$ exists for $L = -\infty$ and $L = \infty$.

1.8.

(a) By Definition 1.9, we have

$$\begin{aligned}
\limsup(-x_n) &= \inf_{m \geq 1} \sup_{n \geq m} (-x_n) \\
&= \inf_{m \geq 1} (-\inf_{n \geq m} x_n) \\
&= -\sup_{m \geq 1} \inf_{n \geq m} x_n \\
&= -\liminf x_n.
\end{aligned}$$

(b) For $j, m \geq 1$, we have

$$\sup_{i \geq j} x_i \geq x_r \geq \inf_{n \geq m} x_n,$$

where $r = \max(j, m)$. Since this is true for all $j, m \geq 1$, we have

$$\inf_{j \geq 1} \sup_{i \geq j} x_i \geq \inf_{n \geq m} x_n, \quad m \geq 1,$$

which implies

$$\inf_{j \geq 1} \sup_{i \geq j} x_i \geq \sup_{m \geq 1} \inf_{n \geq m} x_n.$$

(c) For fixed $m, j \geq 1$,

$$\inf_{n \geq m} x_n + \inf_{i \geq j} y_i \leq x_r + y_r, \quad r \geq \max(m, j),$$

we have

$$\inf_{n \geq m} x_n + \inf_{i \geq j} y_i \leq \inf_{r \geq \max(m,j)} (x_r + y_r) \leq \sup_{s \geq 1} \inf_{r \geq s} (x_r + y_r) \quad m, j \geq 1.$$

Since this is true for every $m, j \geq 1$, we have

$$\sup_{m \geq 1} \inf_{n \geq m} x_n + \sup_{j \geq 1} \inf_{i \geq j} y_i \leq \sup_{s \geq 1} \inf_{r \geq s} (x_r + y_r);$$

that is, $\liminf x_n + \liminf y_n \leq \liminf (x_n + y_n)$.

(d) From parts (a) and (c) of this proof, we have

$$\begin{aligned}
\limsup x_n + \limsup y_n &= -\liminf(-x_n) - \liminf(-y_n) \\
&\geq -\liminf(-x_n - y_n) \\
&= \limsup(x_n + y_n).
\end{aligned}$$

(e) From parts (a) and (d) of this proof, we have

$$\begin{aligned}
\limsup x_n + \liminf y_n &= \limsup(x_n + y_n - y_n) + \liminf y_n \\
&\leq \limsup(x_n + y_n) + \limsup(-y_n) + \liminf y_n \\
&= \limsup(x_n + y_n).
\end{aligned}$$

(f) Since $c = \liminf c$ and $-c = \liminf(-c)$, by part (c), we have

$$c + \liminf x_n = \liminf c + \liminf x_n \leq \liminf(c + x_n),$$

and

$$\liminf(c + x_n) - c = \liminf(c + x_n) + \liminf(-c) \leq \liminf x_n.$$

It follows that $c + \liminf x_n = \liminf(c + x_n)$.

(g) From parts (a) and (f) of this proof, we have

$$\liminf(c - x_n) = c + \liminf(-x_n) = c - \limsup x_n.$$

This finishes the proof.

1.9. First, we prove the "if" part. Suppose that $\limsup A_n = \liminf A_n$. Then by Problem 1.5, we have $\liminf I_{A_n}(s) = \limsup I_{A_n}(s)$ for all $s \in S$. It follows that $\lim I_{A_n}(s)$ exists for all $s \in S$. Hence $\lim_{n\to\infty} A_n$ exists.

Next, we prove the "only if" part. Suppose that $\lim_{n\to\infty} A_n$ exists. Then by definition, we have $\lim_{n\to\infty} I_{A_n}(s)$ exists for all $s \in S$. It follows that

$$\limsup I_{A_n}(s) = \liminf I_{A_n}(s)$$

for all $s \in S$. By Problem 1.5, we have $I_{\limsup A_n}(s) = I_{\liminf A_n}(s)$ for all $s \in S$. Therefore, $\limsup A_n = \liminf A_n$. By Problem 1.5, we have

$$\begin{aligned} I_A(s) &= \lim_{n\to\infty} I_{A_n}(s) = \limsup I_{A_n}(s) = \liminf I_{A_n}(s) \\ &= I_{\limsup A_n}(s) = I_{\liminf A_n}(s), \quad \forall s \in S. \end{aligned}$$

Hence $A = \limsup A_n = \liminf A_n$.

1.10. Since $\lim_{n\to\infty} x_n$ and $\lim_{n\to\infty} y_n$ exist, it follows from Problem 1.7 that

$$\lim_{n\to\infty} x_n = \liminf x_n = \limsup x_n$$

and

$$\lim_{n\to\infty} y_n = \liminf y_n = \limsup y_n.$$

Then by parts (c) and (d) of Problem 1.8, we have

$$\begin{aligned} \liminf(x_n + y_n) &\geq \liminf x_n + \liminf y_n \\ &= \limsup x_n + \limsup y_n \\ &\geq \limsup(x_n + y_n), \end{aligned}$$

which shows that

$$\limsup(x_n + y_n) = \liminf(x_n + y_n) = \lim_{n\to\infty} x_n + \lim_{n\to\infty} y_n.$$

This completes the proof.

1.5 Bibliographic Notes

In this chapter, we introduced some concepts in set theory as well as some set operations and relations. For further information about these concepts, readers are referred to Papoulis (1991), Williams (1991), Ash and Doleans-Dade (1999), Jacod and Protter (2004), and Reitano (2010).

We also introduced some concepts related to sequences of real numbers, which are connected to sequences of sets via indicator functions. The properties of sequences of real numbers and sets are frequently used in later chapters.

Zorn's lemma is an axiom of set theory and is equivalent to the axiom of choice. For a proof of the equivalence, readers are referred to Vaught (1995, p80), Dudley (2002, p20), and Moschovakis (2006, p114).

CHAPTER 2

MEASURES

Measurable sets are to measure theory as open sets are to topology (Williams, 1991). Measures are set functions defined on measurable sets. These concepts are used later to define integration. In this chapter, we shall introduce measurable sets, measures, and other relevant concepts such as algebras and σ-algebras.

2.1 Basic Concepts and Facts

Definition 2.1 (Algebra)**.** An algebra or field Σ_0 on S is a collection of subsets of S that satisfies the following conditions:

(a) $S \in \Sigma_0$.

(b) If $F \in \Sigma_0$, then $F^c \in \Sigma_0$, where $F^c = S \backslash F$.

(c) If $F \in \Sigma_0$ and $G \in \Sigma_0$, then $F \cup G \in \Sigma_0$.

Definition 2.2 (σ-Algebra)**.** A collection Σ of subsets of S is called a *σ-algebra* or *σ-field* if Σ is an algebra on S and is closed under countably infinite unions; that is, if $F_n \in \Sigma$ for $n = 1, 2, \ldots$, then

$$\bigcup_{n=1}^{\infty} F_n \in \Sigma.$$

Definition 2.3 (Measurable Space and Measurable Set)**.** A measurable space is a pair (S, Σ), where S is a set and Σ is a σ-algebra on S. A subset C of S is said to be Σ-measurable if $C \in \Sigma$.

Definition 2.4 (Generated σ-Algebra)**.** Let $\mathcal{C}$ be a collection of subsets of a set S. The σ-algebra $\sigma(\mathcal{C})$ generated by $\mathcal{C}$ is the intersection of all the σ-algebras containing $\mathcal{C}$.

Definition 2.5 (Topology, Topological Space, Open Set)**.** A topology $\mathcal{T}$ in a set S is a a collection of subsets of S that satisfies the following conditions:

(a) $S \in \mathcal{T}$ and $\emptyset \in \mathcal{T}$.

(b) If $F, G \in \mathcal{T}$, then $F \cap G \in \mathcal{T}$.

(c) If $F_i \in \mathcal{T}$ for all $i \in I$, then $\bigcup_{i \in I} F_i \in \mathcal{T}$, where I is an index set (finite, countable, or uncountable).

A topological space is a pair $(S, \mathcal{T})$ where S is a nonempty set and $\mathcal{T}$ is a topology in S. For simplicity, we classify S as a topological space if $\mathcal{T}$ is evident or does not need an explicit name. The elements of $\mathcal{T}$ are called *open sets* in S.

Definition 2.6 (Borel σ-Algebra)**.** The Borel σ-algebra $\mathcal{B}(S)$ on a topological space S is the σ-algebra generated by the collection of open sets in S. The Borel σ-algebra on the real line $\mathbf{R}$ is denoted by $\mathcal{B}$, and the sets in $\mathcal{B}$ are called *Borel sets*. If $E \in \mathcal{B}$, then $\mathcal{B}(E)$ denotes the σ-algebra $\{A \cap E : A \in \mathcal{B}\}$.

Definition 2.7 (Borel Measure)**.** A Borel measure is a measure defined on a Borel σ-algebra.

Definition 2.8 (Finitely Additive and Countably Additive Set Function)**.** Let S be a set and Σ_0 be an algebra on S. Let μ_0 be a nonnegative set function

$$\mu_0 : \Sigma_0 \to [0, \infty].$$

The set function μ_0 is called *finitely additive* if it satisfies the following conditions:

(a) $\mu_0(\emptyset) = 0$, where $\emptyset$ denotes an empty set.

(b) If $F, G \in \Sigma_0$ and $F \cap G = \emptyset$, then $\mu_0(F \cup G) = \mu_0(F) + \mu_0(G)$.

The set function μ_0 is called *countably additive* if it satisfies the following conditions:

(a) $\mu_0(\emptyset) = 0$.

(b) If $F_1, F_2, \ldots$ is a disjoint sequence of elements of Σ_0 and $\bigcup_{n=1}^{\infty} F_n \in \Sigma_0$, then

$$\mu_0\left(\bigcup_{n=1}^{\infty} F_n\right) = \sum_{n=1}^{\infty} \mu_0(F_n).$$

Definition 2.9 (Measure). Let Σ_0 be an algebra of subsets of a set S. A measure on the algebra Σ_0 is a nonnegative, countably additive set function on Σ_0.

Definition 2.10 (Measure Space). A measure space is a triple (S, Σ, μ) where (S, Σ) is a measurable space and μ is a measure on (S, Σ).

Definition 2.11 (Finite Measure and σ-Finite Measure). Let (S, Σ, μ) be a measure space. The measure μ is considered finite if $\mu(S) < \infty$. The measure μ is deemed σ-finite if there exists a sequence $\{S_n\}_{n\in\mathbf{N}}$ of elements of Σ such that $\mu(S_n) < \infty$ for all $n \in \mathbf{N}$ and $\bigcup_{n=1}^{\infty} S_n = S$, where $\mathbf{N}$ is the set of natural numbers, that is, $\mathbf{N} = \{1, 2, 3, \ldots\}$.

Definition 2.12 (Probability Measure and Probability Space). Let $(\Omega, \mathscr{F}, P)$ be a measure space. Then $(\Omega, \mathscr{F}, P)$ is called a *probability measure space*, or simply a *probability space*, if P is a probability measure, that is, $P(\Omega) = 1$.

Definition 2.13 (Counting Measure). Let (S, Σ) be a measurable space. The counting measure μ on (S, Σ) is defined as

$$\mu(A) = \begin{cases} |A|, & \text{if } A \text{ is finite subset of } S; \\ \infty, & \text{if } A \text{ is infinite subset of } S. \end{cases}$$

Definition 2.14 (μ-Null and Almost Everywhere (a.e.)). Let (S, Σ, μ) be a measure space. An element F of Σ is called *μ-null* if its measure is zero (i.e., $\mu(F) = 0$). A statement about points in S is said to be true a.e. if the measure of the set of points for which the statement is false is zero, that is

$$F = \{s : \text{ the statement about } s \text{ is false}\} \in \Sigma$$

and $\mu(F) = 0$.

Definition 2.15 (π-System). A π-system $\mathcal{I}$ on a set S is a collection of subsets of S that is closed under finite intersection:

$$I_1, I_2 \in \mathcal{I} \text{ implies } I_1 \cap I_2 \in \mathcal{I}.$$

Definition 2.16 (d-System). A d-system $\mathcal{D}$ on a set S is a collection of subsets of S such that

(a) $S \in \mathcal{D}$.

(b) If $A, B \in \mathcal{D}$ and $A \subseteq B$, then $B \backslash A \in \mathcal{D}$.

(c) If $A_n \in \mathcal{D}$ for $n \in \mathbf{N}$ and $A_n \uparrow A$, then $A \in \mathcal{D}$.

Theorem 2.1 (Monotone Convergence of Measures). *Let Σ_0 be an algebra of subsets of a set S and μ a measure on Σ_0. Let $F_1, F_2, \ldots, \in \Sigma_0$. If $F_n \uparrow F$ and $F \in \Sigma_0$, then $\mu(F_n) \uparrow \mu(F)$.*

Theorem 2.2 (First Borel-Cantelli Lemma). *Let (S, Σ, μ) be a measure space and $\{E_n\}_{n\geq 1}$ be a sequence of elements of Σ such that*

$$\sum_{n=1}^{\infty} \mu(E_n) < \infty.$$

Then $\mu\left(\limsup E_n\right) = 0$.

2.2 Problems

2.1. Let S be a finite set and 2^S denote the collection of all subsets of S. Show that 2^S is finite and that it is a σ-algebra.

2.2. Let $\mathcal{C}$ be a collection of subsets C of $\mathbf{N} = \{1, 2, 3, \ldots\}$ for which the following limit

$$\lim_{m\to\infty} \frac{|\{k : 1 \leq k \leq m; \quad k \in C\}|}{m}$$

exists, where $|A|$ denotes the number of elements in A. Show that $\mathcal{C}$ is not an algebra.

2.3. Let $A_1, A_2, \ldots, A_n$ be arbitrary subsets of a set S, and let $\Sigma = \sigma(A_1, A_2, \ldots, A_n)$. Give an upper bound of $|\Sigma|$ that is attainable under certain conditions, where $|\Sigma|$ denotes the number of sets in Σ.

2.4. Let $\mathcal{C} = \{(-\infty, c] : c \in \mathbf{R}\}$, where $\mathbf{R} = (-\infty, \infty)$. Show that $\sigma(\mathcal{C}) = \mathcal{B}$. In addition, $\mathcal{C}$ can be replaced by other classes of intervals, for example, all intervals $(-\infty, c)$, $c \in \mathbb{R}$, all intervals $[c, \infty)$, $c \in \mathbb{R}$, all intervals (c, ∞), $c \in \mathbb{R}$, all intervals $[a, b)$, $a, b \in \mathbb{R}$, all intervals $(a, b]$, $a, b \in \mathbb{R}$, or all intervals $[a, b]$, $a, b \in \mathbb{R}$.

2.5. Let Σ be a σ-algebra of subsets of a set S, and let $C \in \Sigma$. Show that $\mathcal{C} = \{A \cap C : A \in \Sigma\}$ is a σ-algebra of subsets of C.

2.6. Let Σ be a collection of subsets of a set S. Show that Σ is a σ-algebra if and only if Σ is both a π-system and a d-system.

2.7 (Dynkin's Lemma). Show that if $\mathcal{I}$ is a π-system, then

$$d(\mathcal{I}) = \sigma(\mathcal{I}),$$

where $d(\mathcal{I})$ is the intersection of all d-systems that contain $\mathcal{I}$.

2.8. Let $\mathcal{I}$ be a π-system on S and $\Sigma = \sigma(\mathcal{I})$. Let μ_1 and μ_2 be measures on the measurable space (S, Σ) such that $\mu_1(S) = \mu_2(S) < \infty$ and $\mu_1(I) = \mu_2(I)$ for all $I \in \mathcal{I}$. Show that

$$\mathcal{D} = \{F \in \Sigma : \mu_1(F) = \mu_2(F)\}$$

is a d-system.

2.9. Let (S, Σ) be a measurable space. Suppose that $\{A_n\}_{n\geq 1}$ is a sequence of Σ-measurable sets. Show that $\liminf A_n \in \Sigma$ and $\limsup A_n \in \Sigma$.

2.10. Let S be a countably infinite set, and let $\Sigma_0 = \{A \subseteq S : |A| < \infty \text{ or } |A^c| < \infty\}$. Let μ_0 be a set function on S be defined as

$$\mu_0(A) = \begin{cases} 0, & \text{if } A \text{ is finite;} \\ 1, & \text{if } A^c \text{ is finite.} \end{cases}$$

(a) Show that Σ_0 is an algebra but not a σ-algebra.

(b) Show that μ_0 is finitely additive but not countably additive on Σ_0.

(c) Show that there exists a sequence $\{A_n\}_{n\geq 1}$ of elements of Σ_0 such that $A_n \uparrow S$ and $\mu_0(A_n) = 0$ for all $n \geq 1$.

2.11. Let Σ_0 be an algebra of subsets of a set S and μ be a measure on Σ_0. Let $A, B \in \Sigma_0$. Show that

(a) $\mu(A \cup B) = \mu(A) + \mu(B) - \mu(A \cap B)$;

(b) $\mu(A \cup B) \leq \mu(A) + \mu(B)$;

(c) $\mu(A \cap B) \leq \mu(A)$.

2.12. Let (S, Σ, μ) be a measure space and $A_i \in \Sigma$ for $i = 1, 2, \ldots, n$. Suppose that μ is finite. Show that

$$\begin{aligned} \mu\left(\bigcup_{i=1}^{n} A_i\right) &= \sum_{1\leq i\leq n} \mu(A_i) - \sum_{1\leq i<j\leq n} \mu(A_i \cap A_j) \\ &\quad + \sum_{1\leq i<j<k\leq n} \mu(A_i \cap A_j \cap A_k) - \cdots \\ &\quad +(-1)^{n+1}\mu(A_1 \cap A_2 \cap \cdots \cap A_n). \end{aligned}$$

2.13. Let (S, Σ, μ) be a measure space and $A_n \in \Sigma$ for $n \geq 1$. Suppose that μ is finite. Show that

$$\mu\left(\bigcup_{n=1}^{\infty} A_n\right) \leq \sum_{n=1}^{\infty} \mu(A_n).$$

2.14. Let μ_0 be a nonnegative, finitely additive set function on an algebra Σ_0. Let $A_1, A_2, \ldots$ be pairwise disjoint sets in Σ_0 and $\bigcup_{n=1}^{\infty} A_n \in \Sigma_0$. Show that

$$\mu_0\left(\bigcup_{n=1}^{\infty} A_n\right) \geq \sum_{n=1}^{\infty} \mu_0(A_n).$$

2.15. Let S be an infinite set and μ the counting measure on $(S, 2^S)$, where 2^S denotes the collection of all subsets of S. Show that there exists a sequence $\{A_n\}_{n\geq 1}$ such that $A_n \downarrow \emptyset$ and $\lim_{n\to\infty} \mu(A_n) \neq 0$.

2.16. Let S be a set, and let $\mathcal{G}$ and $\mathcal{H}$ be two σ-algebras on S. Let

$$\mathcal{I} = \{G \cap H : G \in \mathcal{G}, H \in \mathcal{H}\}.$$

Show that $\mathcal{I}$ is a π-system and $\sigma(\mathcal{I}) = \sigma(\mathcal{G}, \mathcal{H})$, where $\sigma(\mathcal{G}, \mathcal{H})$ is the σ-algebra generated by $\mathcal{G} \cup \mathcal{H}$, that is, $\sigma(\mathcal{G}, \mathcal{H}) = \sigma(\mathcal{G} \cup \mathcal{H})$.

2.17. Let (S, Σ) be a measurable space and $\{A_n\}_{n\geq 1}$ be a sequence of Σ-measurable sets. Show that if $\lim_{n\to\infty} A_n = A$, then $A \in \Sigma$.

2.18. Let (S, Σ, μ) be a measure space. Let $G_1, G_2, \ldots \in \Sigma$ such that $G_n \downarrow G$ and $\mu(G_k) < \infty$ for some k. Show that $\mu(G_n) \downarrow \mu(G)$.

2.19. Let (S, Σ, μ) be a measure space, and let $G_n \in \Sigma$ $(n \geq 1)$ such that $\mu(G_n) = 0$. Show that

$$\mu\left(\bigcup_{i=1}^{\infty} G_i\right) = 0.$$

2.20. Let μ be a finite measure on a measurable space (S, Σ). Show that there cannot be uncountably many disjoint sets $A \in \Sigma$ such that $\mu(A) > 0$.

2.3 Hints

2.1. Consider how many subsets contain zero elements, how many subsets contain one elements, and so forth. The sum of those numbers is the number of elements in 2^S. Use Definition 2.2 to show that 2^S is a σ-algebra.

2.2. To show that $\mathcal{C}$ is not an algebra, one can try to find two sets $A, B \in \mathcal{C}$ but $A \cap B \notin \mathcal{C}$. If $\mathcal{C}$ is indeed an algebra, then $A \cap B = (A^c \cup B^c)^c \in \mathcal{C}$. To make sure that $A \cap B \notin \mathcal{C}$, one needs to show that the sequence

$$r_m = \frac{|\{k : 1 \leq k \leq m; \quad k \in A \cap B\}|}{m}$$

has no limits; that is, $\limsup r_m \neq \liminf r_m$.

2.3. Consider simple cases when $n = 1, 2, 3$, and then consider the general case.

2.4. Use the standard technique to show that $\sigma(\mathcal{C}) = \mathcal{B}$; that is, show that $\sigma(\mathcal{C}) \subseteq \mathcal{B}$ and $\mathcal{B} \subseteq \sigma(\mathcal{C})$.

2.5. Use Definition 2.2.

2.6. Use Definitions 2.2, 2.16, and 2.15.

2.7. Use the result of Problem 2.6.

2.8. Use Definitions 2.16 and 2.9.

2.9. Use Definitions 1.8 and 2.3.

2.10. Since S is a countably infinite set, the elements of S can be enumerated as $a_1, a_2, \ldots$.

2.11. Use Definitions 2.8 and 2.9.

2.12. Consider the cases when $n = 1, 2$, and then use the mathematical induction method.

2.13. Write $\bigcup_{n=1}^{\infty} A_n$ as a union of n mutually disjoint sets.

2.14. Since $A_1, A_2, \ldots$ are pairwise disjoint, we have

$$\mu_0\left(\bigcup_{n=1}^{\infty} A_n\right) = \mu_0\left(\bigcup_{n=1}^{N} A_n\right) + \mu_0\left(\bigcup_{n=N+1}^{\infty} A_n\right) \geq \mu_0\left(\bigcup_{n=1}^{N} A_n\right).$$

Use the above inequality to prove the inequality in question.

2.15. Try to construct a sequence of sets $A_1, A_2, \ldots$ such that $\limsup A_n = \emptyset$ and $\mu(A_n) \neq 0$ for all $n \geq 1$.

2.16. Use Definition 2.15 to show that $\mathcal{I}$ is a π-system and use the standard technique to show that $\sigma(\mathcal{I}) = \sigma(\mathcal{G}, \mathcal{H})$[(i.e., $\sigma(\mathcal{I}) \supseteq \sigma(\mathcal{G}, \mathcal{H})$ and $\sigma(\mathcal{I}) \subseteq \sigma(\mathcal{G}, \mathcal{H})$].

2.17. Note that if $\lim_{n\to\infty} A_n = A$, then $\liminf A_n = \limsup A_n = A$.

2.18. Use Theorem 2.1.

2.19. Write $\bigcup_{i=1}^{\infty} G_i$ as a union of pairwise disjoint sets and utilize the fact that μ is countably additive.

2.20. Use the method of contradiction; that is, suppose that there are uncountably many disjoint sets $A \in \Sigma$ such that $\mu(A) > 0$. Then show that μ is infinity.

2.4 Solutions

2.1. Let $n = |S|$. Since S is finite, we have $n < \infty$. In addition, we can classify the subsets of S into $n + 1$ groups $\mathcal{G}_0, \mathcal{G}_1, \ldots, \mathcal{G}_n$, where $\mathcal{G}_i$ $(0 \leq i \leq n)$ is the group of subsets containing i elements. Then $\bigcup_{i=1}^{n} \mathcal{G}_i = 2^S$, and $\mathcal{G}_i \cap \mathcal{G}_j = \emptyset$ for all $i \neq j$. Hence we have $|2^S| = \sum_{i=0}^{n} |\mathcal{G}_i|$. But $|\mathcal{G}_i| = \binom{n}{i}$, so we have $|2^S| = 2^n < \infty$. By definition of 2^S, we have $S \in 2^S$. If $F \in 2^S$, then $F \subseteq S$, which leads to $F^c = S \backslash F \subseteq S$ and $F^c \in 2^S$. If $F_n \in 2^S$ for $n \geq 1$, then $F_n \subseteq S$ for $n \geq 1$. Let

$x \in \bigcup_{n=1}^{\infty} F_n$. Then $x \in F_{n_0}$ for some $n_0 \geq 1$. But $F_{n_0} \subseteq S$, we have $x \in S$. Thus $\bigcup_{n=1}^{\infty} F_n \subseteq S$, which gives $\bigcup_{n=1}^{\infty} F_n \in 2^S$. Therefore, 2^S is a σ-algebra.

2.2. To show that $\mathcal{C}$ is not an algebra, we only need to find two sets $A, B \in \mathcal{C}$ but $A \cap B \notin \mathcal{C}$. To do this, we divide $\mathbf{N}$ into the following mutually disjoint intervals I_i and I_j for $i = 0, 1, 2, \ldots$, where $I_0 = \{1, 2\}$, $J_0 = \{3, 4\}$, and

$$I_k = \left(2 \cdot 2^k, 3 \cdot 2^k\right] \cap \mathbf{N}, \quad J_k = \left(3 \cdot 2^k, 2 \cdot 2^{k+1}\right] \cap \mathbf{N}, \quad \forall k \geq 1.$$

Then $|I_k| = 2^k$ and $|J_k| = 2^k$ for all $k \geq 1$. We define A and B as

$$A = \bigcup_{k=0}^{\infty} (I_k^e \cup J_k^e)$$

and

$$B = \bigcup_{k=0}^{\infty} (I_k^e \cup J_k^o),$$

respectively, where I_k^e is the set of all even numbers in I_k and I_k^o is the set of all odd numbers in I_k for all $k \geq 0$, J_k^e and J_k^o are defined similarly. Then we have

$$A \cap B = \bigcup_{k=0}^{\infty} I_k^e.$$

Now we claim that $A, B \in \mathcal{C}$ and $A \cap B \notin \mathcal{C}$. Since

$$\frac{|\{k : 1 \leq k \leq m; \quad k \in A\}|}{m} = \begin{cases} \frac{1}{2}, & \text{if } m \text{ is even;} \\ \frac{m-1}{2m}, & \text{if } m \text{ is odd,} \end{cases}$$

and

$$\frac{|\{k : 1 \leq k \leq m; \quad k \in B\}|}{m} = \begin{cases} \frac{1}{2}, & \text{if } m \text{ is even;} \\ \frac{m-1}{2m}, & \text{if } m \text{ is odd and } m \in I_k \text{ for some } k; \\ \frac{m+1}{2m}, & \text{if } m \text{ is odd and } m \in J_k \text{ for some } k, \end{cases}$$

we have

$$\lim_{m \to \infty} \frac{|\{k : 1 \leq k \leq m; \quad k \in A\}|}{m} = \frac{1}{2},$$

and

$$\lim_{m \to \infty} \frac{|\{k : 1 \leq k \leq m; \quad k \in B\}|}{m} = \frac{1}{2}.$$

Thus $A, B \in \mathcal{C}$. As for $A \cap B$, we have

$$\frac{|\{k : 1 \leq k \leq m; \quad k \in A \cap B\}|}{m} = \begin{cases} \frac{1}{3}, & \text{if } m = 3 \cdot 2^k; \\ \frac{1}{4}, & \text{if } m = 2 \cdot 2^{k+1}, \end{cases} \quad \forall k \geq 1$$

which gives

$$\limsup_{m\to\infty} \frac{|\{k : 1 \leq k \leq m; \quad k \in A \cap B\}|}{m} \geq \frac{1}{3},$$

and

$$\liminf_{m\to\infty} \frac{|\{k : 1 \leq k \leq m; \quad k \in A \cap B\}|}{m} \leq \frac{1}{4}.$$

Thus the limit does not exist and $A \cap B \notin \mathcal{C}$.

2.3. The upper bound of $|\Sigma|$ is 2^N, where $N = 2^n - 1$. The upper bound can be attained when $B_1, B_2, \ldots, B_N$ are all nonempty, where

$$B_i = \left(\bigcap_{j \in I_i} A_j \right) \setminus \left(\bigcup_{j \in I_i^c} A_j \right)$$

for $i = 1, 2, \ldots, N$ and $I_1, I_2, \ldots, I_N$ are nonempty, distinct subsets of $\{1, 2, \ldots, n\}$.

The sets $B_1, B_2, \ldots, B_N$ are pairwise disjoint but may contain empty sets. One can use the standard technique (i.e., $A \subseteq B$ and $B \subseteq A$ imply $A = B$) to show that

$$\Sigma = \sigma(B_1, B_2, \ldots, B_N).$$

When $B_1, B_2, \ldots, B_N$ are all nonempty, we have $\Sigma = 2^{\{B_1, B_2, \ldots, B_N\}}$ and $|\Sigma| = 2^N$.

Now we show that such upper bound is attainable. Let $S = \{0, 1, 2, \ldots, 2^n - 1\}$ and $A_i = \{s \in S : s = \beta_1 + \beta_2 2 + \cdots + \beta_n 2^{n-1},\ \beta_k \in \{0,1\} \text{ for } k \neq i,\ \beta_i = 0\}$ for $i = 1, 2, \ldots, n$. Then $B_i = \{\sum_{j \in I_i^c} 2^{j-1}\}$ for $I_i^c \neq \emptyset$ and $B_i = \{0\}$ for $I_i^c = \emptyset$. Hence $B_1, B_2, \ldots, B_N$ are all nonempty. In this case, we have $|\Sigma| = 2^N$.

2.4. Since

$$(-\infty, c] = \bigcap_{n=1}^{\infty} \left(-\infty, c + \frac{1}{n} \right), \quad \forall c \in \mathbf{R},$$

we have $\mathcal{C} \subseteq \mathcal{B}$. Note that $\mathcal{B}$ is a σ-algebra, we have $\sigma(\mathcal{C}) \subseteq \mathcal{B}$.

Now we show that every open subset O of $\mathbf{R}$ is in $\sigma(\mathcal{C})$. Note that O is a finite or countable union of open intervals, we only need to show that

$$(a, b) \in \sigma(\mathcal{C}), \quad \forall a, b \in \mathbf{R} \text{ with } a < b.$$

But

$$(a, b) = (-\infty, b) \cap (-\infty, a]^c = \left(\bigcup_{n=1}^{\infty} \left(-\infty, b - \frac{1}{n} \right] \right) \bigcap (-\infty, a]^c,$$

we have $(a, b) \in \sigma(\mathcal{C})$. Thus we have $\mathcal{B} \subseteq \sigma(\mathcal{C})$. Therefore, $\mathcal{B} = \sigma(\mathcal{C})$.

Other implications can be proved by similar arguments.

2.5. Since $C \in \Sigma$, we have $C \in \mathcal{C}$. Let $B \in \mathcal{C}$. Then $B = A \cap C$ for some $A \in \Sigma$. Thus we have $C \backslash B = C \cap B^c = C \cap A^c \in \mathcal{C}$, where the complement is taken relative to S. Now let $B_n \in \mathcal{C}$ for $n \geq 1$. Then $B_n = A_n \cap C$ for some $A_n \in \Sigma$. But $\bigcup_{n=1}^{\infty} B_n = \bigcup_{n=1}^{\infty}(A_n \cap C) = (\bigcup_{n=1}^{\infty} A_n) \cap C$, so we have $\bigcup_{n=1}^{\infty} B_n \in \mathcal{C}$. Hence $\mathcal{C}$ is a σ-algebra over C.

2.6. The necessity part is trivial. To prove the sufficiency, we assume that Σ is both a π-system and a d-system. From the assumption Σ is a d-system, we have $S \in \Sigma$. If $F \in \Sigma$, then $F^c = S \backslash F \in \Sigma$, since $F, S \in \Sigma$ and $F \subseteq S$. If $F, G \in \Sigma$, then $F \cup G = (F^c \cap G^c)^c \in \Sigma$, since Σ is a π-system and d-system. Let $F_n \in \Sigma\,(n \in \mathbf{N})$ and $A_n = \bigcup_{i=1}^{n} F_i$. Then $A_n \subseteq A_{n+1}$ and $A_n \uparrow \bigcup_{i=1}^{\infty} F_i$. By assumption, we have $\bigcup_{i=1}^{\infty} F_i \in \Sigma$. Thus, Σ is a σ-algebra.

2.7. From Problem 2.6, we know that $\sigma(\mathcal{I})$ is a d-system. Note that as $d(\mathcal{I})$ is the intersection of all d-systems containing $\mathcal{I}$, we have $d(\mathcal{I}) \subseteq \sigma(\mathcal{I})$. To prove $\sigma(\mathcal{I}) \subseteq d(\mathcal{I})$, we only need to show that $d(\mathcal{I})$ is a σ-algebra. From Problem 2.6, we only need to show that $d(\mathcal{I})$ is a π-system.

Let $\mathcal{D}_1 = \{D \in d(\mathcal{I}) : D \cap I \in d(\mathcal{I}), \forall I \in \mathcal{I}\}$, then $\mathcal{D}_1 \subseteq d(\mathcal{I})$. We claim that $\mathcal{D}_1$ is a d-system on $\mathcal{I}$. Since $\mathcal{I}$ is a π-system, we have $\mathcal{I} \subseteq \mathcal{D}_1$. Let $A, B \in \mathcal{D}_1$ and $A \subseteq B$; then

$$(B \backslash A) \cap I = (B \cap I) \cap (A^c \cup I^c) = (B \cap I) \backslash (A \cap I) \in d(\mathcal{I}), \quad \forall I \in \mathcal{I},$$

since $A \cap I, B \cap I \in d(\mathcal{I})$ for all $I \in \mathcal{I}$ and $d(\mathcal{I})$ is a d-system. Thus $B \backslash A \in \mathcal{D}_1$. Let $A_n \in \mathcal{D}_1$ and $A_n \uparrow A$, then $A_n \cap I \in d(\mathcal{I})$ and $A_n \cap I \uparrow A \cap I$ for all $I \in \mathcal{I}$. Since $d(\mathcal{I})$ is a d-system, we have $A \cap I \in d(\mathcal{I})$ $(\forall I \in \mathcal{I})$, which leads to $A \in \mathcal{D}_1$. Thus $\mathcal{D}_1$ is a d-system on $\mathcal{I}$. Note that $d(\mathcal{I})$ is the intersection of all d-systems containing $\mathcal{I}$, so we have $d(\mathcal{I}) \subseteq \mathcal{D}_1$. Hence we have $\mathcal{D}_1 = d(\mathcal{I})$.

Let $\mathcal{D}_2 = \{D \in d(\mathcal{I}) : D \cap J \in d(\mathcal{I}), \forall J \in d(\mathcal{I})\}$. We claim that $\mathcal{D}_2$ is also a d-system on $\mathcal{I}$. Since $\mathcal{D}_1 = d(\mathcal{I})$, we have $\mathcal{I} \subseteq \mathcal{D}_2$. Let $A, B \in \mathcal{D}_2$ and $A \subseteq B$, then $(B \backslash A) \cap J = (B \cap J) \backslash (A \cap J) \in d(\mathcal{I})$ for all $J \in d(\mathcal{I})$, since $B \cap J, A \cap J \in d(\mathcal{I})$ for all $J \in d(\mathcal{I})$ and $d(\mathcal{I})$ is a d-system. Thus we have $B \backslash A \in \mathcal{D}_2$. Let $A_n \in \mathcal{D}_2$ and $A_n \uparrow A$, then $A_n \cap J \in d(\mathcal{I})$ and $A_n \cap J \uparrow A \cap J$ for all $J \in d(\mathcal{I})$. Since $d(\mathcal{I})$ is a d-system, we have $A \cap J \in d(\mathcal{I})$ for all $J \in d(\mathcal{I})$, which gives $A \in \mathcal{D}_2$. Therefore, $\mathcal{D}_2$ is a d-system on $\mathcal{I}$. Hence we have $d(\mathcal{I}) \subseteq \mathcal{D}_2$. By definition, we have $\mathcal{D}_2 \subseteq d(\mathcal{I})$. So we have $\mathcal{D}_2 = d(\mathcal{I})$, which means that $I \cap J \in d(\mathcal{I})$ for all $I, J \in d(\mathcal{I})$. Therefore $d(\mathcal{I})$ is a π-system.

2.8. We only need to show that $\mathcal{D}$ satisfies the conditions in Definition 2.16. Since $\mu_1(S) = \mu_2(S)$, we have $S \in \mathcal{D}$. Let $A, B \in \mathcal{D}$ and $A \subseteq B$; then we have $A, B \in \Sigma$, $\mu_1(A) = \mu_2(A) < \infty$, and $\mu_1(B) = \mu_2(B) < \infty$, which lead to

$$\mu_1(B \backslash A) = \mu_1(B) - \mu_1(A) = \mu_2(B) - \mu_2(A) = \mu_2(B \backslash A).$$

Thus $B \backslash A \in \mathcal{D}$. To check the last item, let $F_n \in \mathcal{D}$ and $F_n \uparrow F$. Then by Theorem 2.1, we have

$$\mu_1(F) = \lim_{n \to \infty} \mu_1(F_n) = \lim_{n \to \infty} \mu_2(F_n) = \mu_2(F).$$

Hence, $\mathcal{D}$ is a d-system.

2.9. Let $B_j = \bigcap_{i\geq j} A_i$ for $j = 1, 2, \ldots$. Since Σ is a σ-algebra and $A_i \in \Sigma$ for all $i \geq 1$, we have $A_i^c \in \Sigma$ for all $i \geq 1$ and thus $B_j = \left(\bigcup_{i\geq j} A_i^c\right)^c \in \Sigma$ for all $j \geq 1$. Note that $\liminf A_n = \bigcup_{j\geq 1} B_j$, we have $\liminf A_n \in \Sigma$. Similarly, we can show that $\limsup A_n \in \Sigma$.

Now we show that $\liminf A_n \subseteq \limsup A_n$. Let $s \in \liminf A_n$. Then $s \in \bigcap_{i\geq j_0} A_i$ for some $j_0 \in \mathbf{N}$; that is, $s \in A_i$ for all $i \geq j_0$. Consequently, we have $s \in \bigcup_{i\geq j} A_i$ for all $j \geq 1$. By definition, we have $s \in \limsup A_n$. Therefore, $\liminf A_n \subseteq \limsup A_n$.

2.10. Since $|S^c| = |\emptyset| = 0 < \infty$, we have $S \in \Sigma_0$. Let $A \in \Sigma_0$. Then A is finite or A^c is finite. In both cases, we have $A^c \in \Sigma_0$. Now let $A, B \in \Sigma_0$. If both A and B are finite, then $A \cup B$ is finite and thus $A \cup B \in \Sigma_0$. If at least one of A and B is infinite, then $(A \cup B)^c = A^c \cap B^c$ is finite and thus $A \cup B \in \Sigma_0$. Hence Σ_0 is an algebra.

Now we show that Σ_0 is not a σ-algebra. Since S is infinite, we can find an infinite subset $B = \{b_1, b_2, \ldots\}$ of S. Let $A_n = \{b_{2n-1}\}$ for all $n \geq 1$. Then $A_n \in \Sigma_0$ for all $n \geq 1$. But $A = \bigcup_{n=1}^{\infty} A_n$ is infinite and A^c is also infinite. Therefore $A \notin \Sigma_0$.

Let $A, B \in \Sigma_0$ and $A \cap B = \emptyset$. We claim that at most one of A and B can be infinite. If both A and B are infinite, then A^c and B^c are finite, which contradicts $A^c \cup B^c = S$. If both A and B are finite, then $\mu_0(A \cup B) = 0 = \mu_0(A) + \mu_0(B)$. If one of A and B is infinite and the other is finite, then $\mu_0(A \cup B) = 1 = \mu_0(A) + \mu_0(B)$. Hence μ_0 is finitely additive.

Since S is countably infinite, we can enumerate the elements of S as $a_1, a_2, \ldots$. But $\sum_{n=1}^{\infty} \mu_0(\{a_n\}) = 0$ and $\mu_0\left(\bigcup_{n=1}^{\infty}\{a_n\}\right) = \mu_0(S) = 1$. Therefore μ_0 is not countably additive.

Let $S = \{a_1, a_2, \ldots\}$ and $A_n = \{a_i : i = 1, 2, \ldots, n\}$ for all $n \geq 1$. Then $A_n \uparrow S$ and $\mu_0(A_n) = 0$ for all $n \geq 1$.

2.11.

(a) Since $A \cup B = A \cup (B\backslash A)$ and $A \cap (B\backslash A) = \emptyset$, we have

$$\mu(A \cup B) = \mu(A) + \mu(B\backslash A).$$

Note that since $(B\backslash A) \cup (A \cap B) = B$ and $(B\backslash A) \cap (A \cap B) = \emptyset$, we have

$$\mu(B) = \mu(B\backslash A) + \mu(A \cap B).$$

Combining the two equation gives $\mu(A \cup B) = \mu(A) + \mu(B) - \mu(A \cap B)$.

(b) Since μ is nonnegative, this implication follows immediately from part (a).

(c) Since $A = (A \cap B) \cup (B\backslash A)$ and $(A \cap B) \cap (B\backslash A) = \emptyset$, we have $\mu(A) = \mu(A \cap B) + \mu(B\backslash A) \geq \mu(A \cap B)$.

2.12. We use the mathematical induction method. When $n = 1$, the equality is trivial. Suppose that the equality holds for $n = m$. Then from problem 2.11 and the induction hypothesis, we have

$$\begin{aligned}
\mu\left(\bigcup_{i=1}^{m+1} A_i\right) &= \mu\left(\bigcup_{i=1}^{m} A_i\right) + \mu(A_{m+1}) - \mu\left(\left(\bigcup_{i=1}^{m} A_i\right) \cap A_{m+1}\right) \\
&= \sum_{1\le i\le m} \mu(A_i) - \sum_{1\le i<j\le m} \mu(A_i \cap A_j) \\
&\quad + \sum_{1\le i<j<k\le m} \mu(A_i \cap A_j \cap A_k) - \cdots \\
&\quad +(-1)^{m+1}\mu(A_1 \cap A_2 \cap \cdots \cap A_m) \\
&\quad +\mu(A_{m+1}) - \mu\left(\left(\bigcup_{i=1}^{m} A_i\right) \cap A_{m+1}\right). \qquad (2.1)
\end{aligned}$$

Also from the induction hypothesis, we have

$$\begin{aligned}
&\mu\left(\left(\bigcup_{i=1}^{m} A_i\right) \cap A_{m+1}\right) \\
&= \mu\left(\bigcup_{i=1}^{m} (A_i \cap A_{m+1})\right) \\
&= \sum_{1\le i\le m} \mu(A_i \cap A_{m+1}) \qquad (2.2) \\
&\quad - \sum_{1\le i<j\le m} \mu(A_i \cap A_j \cap A_{m+1}) \\
&\quad + \sum_{1\le i<j<k\le m} \mu(A_i \cap A_j \cap A_k \cap A_{m+1}) - \cdots \\
&\quad +(-1)^{m+1}\mu(A_1 \cap A_2 \cap \cdots \cap A_m \cap A_{m+1}). \qquad (2.3)
\end{aligned}$$

It follows from Equations (2.1) and (2.3) that the equality holds for $n = m + 1$. Therefore, the equality holds for all $n \in \mathbf{N}$.

2.13. Let $B_1 = A_1$ and $B_n = A_n \backslash \left(\bigcup_{i=1}^{n-1} A_i\right)$ for $n \ge 2$. Then $\bigcup_{n=1}^{\infty} B_n = \bigcup_{n=1}^{\infty} A_n$, $B_i \cap B_j = \emptyset$ for all $1 \le i < j$, and $\mu(B_n) \le \mu(A_n)$ for all $n \ge 1$. Consequently, we have

$$\mu\left(\bigcup_{n=1}^{\infty} A_n\right) = \mu\left(\bigcup_{n=1}^{\infty} B_n\right) = \sum_{n=1}^{\infty} \mu(B_n) \le \sum_{n=1}^{\infty} \mu(A_n).$$

This completes the proof.

2.14. Since μ_0 is finitely additive and $A_1, A_2, \ldots$ are disjoint, we have

$$\mu_0\left(\bigcup_{n=1}^{N} A_n\right) = \sum_{n=1}^{N} \mu_0(A_n)$$

for all $N \geq 1$. Since μ_0 is nonnegative, we have $\mu_0\left(\bigcup_{n=1}^{\infty} A_n\right) = \mu_0\left(\bigcup_{n=1}^{N} A_n\right) + \mu_0\left(\bigcup_{n=N+1}^{\infty} A_n\right) \geq \sum_{n=1}^{N} \mu_0(A_n)$ for all $N \geq 1$. Now the result follows by letting $N \to \infty$.

2.15. Since S is infinite, we can find an infinite subset $B = \{b_1, b_2, \ldots\}$ of S. Let $A_n = \{b_i : i \geq n\}$ for all $n \geq 1$. Then $A_n \downarrow \emptyset$. Note that $\mu(A_n) = \infty$ for all $n \geq 1$, we have $\lim_{n\to\infty} \mu(A_n) = \infty$.

2.16. First, we show that $\mathcal{I}$ is a π-system. Let $I_1 \in \mathcal{I}$ and $I_2 \in \mathcal{I}$. Then, by definition of $\mathcal{I}$, we have $I_1 = G_1 \cap H_1$ and $I_2 = G_2 \cap H_2$ for some $G_1, G_2 \in \mathcal{G}$, and $H_1, H_2 \in \mathcal{H}$. Since $\mathcal{G}$ and $\mathcal{H}$ are σ-algebras, we have $I_1 \cap I_2 = (G_1 \cap G_2) \cap (H_1 \cap H_2) \in \mathcal{I}$. Thus, $\mathcal{I}$ is a π-system.

Next, we show that $\sigma(\mathcal{I}) = \sigma(\mathcal{G}, \mathcal{H})$. Since $\mathcal{I} \subseteq \sigma(\mathcal{G}, \mathcal{H})$, we have $\sigma(\mathcal{I}) \subseteq \sigma(\mathcal{G}, \mathcal{H})$. Note that $\mathcal{G}$ and $\mathcal{H}$ are σ-algebras on S, we have $S \in \mathcal{G}$ and $S \in \mathcal{H}$. Consequently, $\mathcal{G} \subseteq \mathcal{I}$ and $\mathcal{H} \subseteq \mathcal{I}$. Hence, $\mathcal{G} \cup \mathcal{H} \subseteq \sigma(\mathcal{I})$ and $\sigma(\mathcal{G}, \mathcal{H}) \subseteq \sigma(\mathcal{I})$. Therefore, $\sigma(\mathcal{I}) = \sigma(\mathcal{G}, \mathcal{H})$.

2.17. This implication $A \in \Sigma$ follows from Problems 1.9 and 2.9.

2.18. By Problem 2.17, we have $G \in \Sigma$. Let $F_n = G_k \backslash G_n$ for all $n \geq k$. Since $G_n \supseteq G_{n+1}$ for all n and $G_n \downarrow G$, we have $F_n \subseteq F_{n+1}$ for all $n \geq k$ and $F_n \uparrow G_k \backslash G$. Using Theorem 2.1, we have

$$\mu(G_k) - \mu(G_n) = \mu(F_n) \uparrow \mu(G_k \backslash G) = \mu(G_k) - \mu(G).$$

Since $\mu(G_k)$ is finite by hypothesis, we have $\mu(G_n) \downarrow \mu(G)$.

2.19. Let $F_1 = G_1$ and $F_n = G_n \backslash (\bigcup_{i=1}^{n-1} G_i)$ for $n \geq 2$. Then F_n $(n \in \mathbf{N})$ are mutually disjoint and

$$\bigcup_{i=1}^{\infty} F_i = \bigcup_{i=1}^{\infty} G_i, \quad \bigcup_{i=1}^{n} F_i = \bigcup_{i=1}^{n} G_i, \quad \forall n \geq 1.$$

Since μ is countably additive, we have

$$\mu\left(\bigcup_{i=1}^{\infty} G_i\right) = \mu\left(\bigcup_{i=1}^{\infty} F_i\right) = \sum_{i=1}^{\infty} \mu(F_i).$$

Note that μ is nonnegative and $F_i \subseteq G_i$ for all i, we have $0 \leq \mu(F_i) \leq \mu(G_i) = 0$ for all i, which give $\mu(F_i) = 0$ for all i. From the above equation, we have

$$\mu\left(\bigcup_{i=1}^{\infty} G_i\right) = 0.$$

2.20. Suppose that there are uncountably many disjoint sets $A \in \Sigma$ such that $\mu(A) > 0$ and let $\mathcal{C}$ be the collection of such sets. Let $\mathcal{C}_n = \{A \in \mathcal{C} : \mu(A) > 1/n\}$ for all $n \geq 1$. Then $\mathcal{C} = \bigcup_{n=1}^{\infty} \mathcal{C}_n$. Since $\mathcal{C}$ is uncountable, there exists at least one $\mathcal{C}_{i_0}$ for some $i_0 \geq 1$ that has uncountably many elements. It follows that $\mu(S) \geq \sum_{A \in \mathcal{C}_{i_0}} \mu(A) = \infty$, contradicting the assumption that μ is finite. Therefore, there cannot be uncountably many disjoint sets $A \in \Sigma$ such that $\mu(A) > 0$.

2.5 Bibliographic Notes

A σ-algebra is a structure imposed on a collection of sets on which a measure can be defined. Recall that a measure on a set S is a function that assigns a real number to subsets of S. We can think of the measure as a notion of *size* or *volume* for the subsets. However, it might be not possible to assign such a size to every subset of S. For example, it is not possible to assign a size to a Vitali set (Herrlich, 2006, p120). As a result, we consider only a smaller collection of subsets of S when defining measures. This smaller collection of subsets are called *measurable sets*.

The notions of π-system and λ-system are adopted from Williams (1991). Those concepts are useful for dealing with σ-algebras. For more information about σ-algebras and measures, readers are referred to Billingsley (1986), Williams (1991), Ash and Doleans-Dade (1999), Vestrup (2003), and Athreya and Lahiri (2006). Topology space and continuous functions are presented in books on real analysis such as Rudin (1970). For a history of measure theory, readers are referred to Pap (2002).

Theorems 2.1 and 2.2 are very useful. For proofs of these theorems, readers can consult Williams (1991, pp21,27).

CHAPTER 3

EXTENSION OF MEASURES

As defined in the previous chapter, a measure is a set function on an algebra. To extend a measure from an algebra to a σ-algebra, we need some tools. In this chapter, we will introduce concepts and theorems related to measure extension.

3.1 Basic Concepts and Facts

Definition 3.1 (Outer Measure). An outer measure on a set S is a nonnegative, extended real-valued set function $\mu^* : 2^S \to [0, \infty]$ that satisfies the following conditions:

(a) $\mu^*(\emptyset) = 0$.

(b) (Monotonicity) If $A \subseteq B \subseteq S$, then $\mu^*(A) \leq \mu^*(B)$.

(c) (Countable subadditivity) If $A_n \subseteq S$ for all $n \geq 1$, then

$$\mu^*\left(\bigcup_{n=1}^{\infty} A_n\right) \leq \sum_{n=1}^{\infty} \mu^*(A_n).$$

Definition 3.2 (Complete Measure). Let μ be a measure on a σ-algebra Σ. The measure μ is said to be complete if and only if all subsets of a zero measure set in Σ are in Σ. Thus, whenever $A \in \Sigma$ with $\mu(A) = 0$, we have $B \in \Sigma$ for all $B \subseteq A$.

Definition 3.3 (Completion of Measure Spaces). Let (S, Σ, μ) be a measure space. The completion of (S, Σ, μ) is a measure space (S, Σ_μ, μ), where

$$\Sigma_\mu = \{A \cup B : A \in \Sigma, B \subseteq N \in \Sigma, \mu(N) = 0\}.$$

The measure μ is extended to Σ_μ as follows. Let $E \in \Sigma_\mu$. Then $E = A \cup B$, where $A \in \Sigma$ and $B \subseteq N$ for some $N \in \Sigma$ with $\mu(N) = 0$. We define $\mu(E) = \mu(A \cup B) = \mu(A)$.

Theorem 3.1 (Extension of Finite Measures). *If μ is a finite measure on an algebra Σ_0, then μ can be extended to a measure on $\sigma(\Sigma_0)$.*

Theorem 3.2 (Uniqueness of Extension). *Let S be a set, let $\mathcal{I}$ be a π-system on S, and let $\Sigma = \sigma(\mathcal{I})$ be the σ-algebra generated by $\mathcal{I}$. If μ_1 and μ_2 are measures on (S, Σ) such that $\mu_1(S) = \mu_2(S) < \infty$ and $\mu_1(I) = \mu_2(I)$ for all $I \in \mathcal{I}$, then*

$$\mu_1(I) = \mu_2(I) \quad \forall I \in \Sigma.$$

Theorem 3.3 (Monotone Class Theorem). *Let Σ_0 be an algebra of subsets of a set S. Let Σ be a monotone class of subsets of S. (i.e., if $A_1, A_2, \ldots \in \Sigma$ and $A_n \uparrow A$ or $A_n \downarrow A$, then $A \in \Sigma$). If $\Sigma_0 \subseteq \Sigma$, then $\sigma(\Sigma_0) \subseteq \Sigma$.*

Theorem 3.4 (Carathéodory's Extension Theorem). *Let μ be a σ-finite measure on an algebra Σ_0 of subsets of a set S. Then μ has a unique extension to a measure on $\sigma(\Sigma_0)$.*

Theorem 3.5 (Approximation Theorem). *Let (S, Σ, μ) be a measure space. Let Σ_0 be an algebra of subsets of S such that $\sigma(\Sigma_0) = \Sigma$. Assume that μ is σ-finite on Σ_0. If $A \in \Sigma$ and $\mu(A) < \infty$, then, for each $\epsilon > 0$, there exists a set $B \in \Sigma_0$ such that $\mu(A \Delta B) < \epsilon$, where $A \Delta B$ is the symmetric difference of A and B (see Definition 1.5).*

3.2 Problems

3.1. Let Σ_0 be an algebra of subsets of a set S, and let μ be a finite measure on Σ_0 (see Definition 2.9). Let $\{A_n\}_{n \geq 1}$ and $\{B_n\}_{n \geq 1}$ be two sequences of elements of Σ_0 such that $A_n \uparrow A$ and $B_n \uparrow B$ (A and B may not belong to Σ_0). Show that if $A \subseteq B$, then

$$\lim_{n \to \infty} \mu(A_n) \leq \lim_{n \to \infty} \mu(B_n); \tag{3.1}$$

if $A = B$, then

$$\lim_{n \to \infty} \mu(A_n) = \lim_{n \to \infty} \mu(B_n). \tag{3.2}$$

3.2. Let Σ_0 be an algebra of subsets of a set S, and let μ be a finite measure on Σ_0. Let $\mathcal{C}$ be the collection of all limits of increasing sequence of elements of Σ_0. Thus, $A \in \mathcal{C}$ if and only if there exists a sequence $A_1, A_2, \ldots \in \Sigma_0$ such that $A_n \uparrow A$. Let λ be a set function on $\mathcal{C}$ defined as follows. If $A \in \mathcal{C}$ and $A \notin \Sigma_0$, then $\lambda(A) = \lim_{n\to\infty} \mu(A_n)$, where $A_1, A_2, \ldots \in \Sigma_0$ and $A_n \uparrow A \in \mathcal{C}$. If $A \in \mathcal{C}$ and $A \in \Sigma_0$, then $\lambda(A) = \mu(A)$. Show that

(a) λ is well defined.

(b) $\emptyset \in \mathcal{C}$ and $\lambda(\emptyset) = 0$; $S \in \mathcal{C}$ and $\lambda(S) < \infty$.

(c) If $A, B \in \mathcal{C}$, then $A \cup B, A \cap B \in \mathcal{C}$ and $\lambda(A \cup B) = \lambda(A) + \lambda(B) - \lambda(A \cap B)$.

(d) If $A, B \in \mathcal{C}$ and $A \subseteq B$, then $\lambda(A) \leq \lambda(B)$.

(e) If $A_1, A_2, \ldots \in \mathcal{C}$ and $A_n \uparrow A$, then $A \in \mathcal{C}$ and $\lambda(A_n) \to \lambda(A)$.

3.3. Let $\mathcal{C}$ be a collection of subsets of a set S and λ be a nonnegative, real-valued set function on $\mathcal{C}$ such that $\mathcal{C}$ and λ satisfy the four conditions (b)-(e) in Problem 3.2; that is, $\mathcal{C}$ and λ satisfy:

(a) $\emptyset \in \mathcal{C}$ and $\lambda(\emptyset) = 0$; $S \in \mathcal{C}$ and $\lambda(S) < \infty$.

(b) If $A, B \in \mathcal{C}$, then $A \cup B, A \cap B \in \mathcal{C}$ and $\lambda(A \cup B) = \lambda(A) + \lambda(B) - \lambda(A \cap B)$.

(c) If $A, B \in \mathcal{C}$ and $A \subseteq B$, then $\lambda(A) \leq \lambda(B)$.

(d) If $A_1, A_2, \ldots \in \mathcal{C}$ and $A_n \uparrow A$, then $A \in \mathcal{C}$ and $\lambda(A_n) \to \lambda(A)$.

For each $A \subseteq S$, define

$$\mu^*(A) = \inf\{\lambda(G) : G \in \mathcal{C},\ A \subseteq G\}.$$

Show that

(a) $\mu^* = \lambda$ on $\mathcal{C}$, and $0 \leq \mu^*(A) \leq \lambda(S)$ for all $A \subseteq S$.

(b) $\mu^*(A \cup B) + \mu^*(A \cap B) \leq \mu^*(A) + \mu^*(B)$ for all $A, B \subseteq S$.

(c) If $A, B \subseteq S$ and $A \subseteq B$, then $\mu^*(A) \leq \mu^*(B)$.

(d) If $A_1, A_2, \ldots \subseteq S$ and $A_n \uparrow A$, then $\mu^*(A_n) \uparrow \mu^*(A)$.

(e) μ^* is an outer measure on S.

(f) If $\mathcal{H} = \{H \subseteq S : \mu^*(H) + \mu^*(H^c) = \mu^*(S)\}$, then $\mathcal{H}$ is a σ-algebra of subsets of S and μ^* is a finite measure on $\mathcal{H}$.

3.4. Let (S, Σ, μ) be a measure space and Σ_μ be the completion of Σ relative to μ. Show that Σ_μ is a σ-algebra.

3.5. Let (S, Σ, μ) be a measure space, and let Σ_μ be the completion of Σ relative to μ. Let μ_1 and μ_2 be two set functions on S defined as

$$\begin{aligned} \mu_1(A) &= \sup\{\mu(B) : B \in \Sigma, B \subseteq A\}, \quad A \subseteq S, \\ \mu_2(A) &= \inf\{\mu(B) : B \in \Sigma, A \subseteq B\}, \quad A \subseteq S. \end{aligned}$$

Show that

(a) If $A \in \Sigma_\mu$, then $\mu_1(A) = \mu_2(A) = \mu(A)$.

(b) If $A \subseteq S$ and $\mu_1(A) = \mu_2(A) < \infty$, then $A \in \Sigma_\mu$.

3.3 Hints

3.1. Consider the sequence $A_m \cap B_n$ and use Theorem 2.1.

3.2. The parts (a)-(d) should be straightforward according to the definition of λ, Problem 3.1, and Problem 2.11. For part (e), we need to construct an increasing sequence of sets in Σ_0 from the increasing sequences that converge to $A_1, A_2, \ldots, A_n, \ldots$ such that the constructed increasing sequence of sets converges to A. For example, let $B_{nm} \uparrow A_n$ as $m \to \infty$ for each $n \geq 1$. Then we let $C_m = \bigcup_{i=1}^m B_{i,m+1-i}$ for $m \geq 1$. The sequence $\{C_m\}_{m \geq 1}$ is an increasing sequence and converges to A.

3.3. Part (a) follows from the definition of μ^* directly. For parts (b)-(d), we need to use the definition of μ^* in another form. For example, for each $n \geq 1$, there exists a set $A_n \in \mathcal{C}$ such that $A \subseteq A_n$ and $\lambda(A_n) < \mu^*(A) + 1/n$. Part (e) can be proved by using the definition of outer measures and parts (a), (b), and(d). For part (f), we can use the concepts of π-system and λ-system to prove that $\mathcal{H}$ is a σ-algebra. The method of mathematical induction can be used to prove that μ^* is a measure on $\mathcal{H}$.

3.4. Follow the definition of σ-algebra (see Definition 2.2).

3.5. Part (a) can be proved by using the definition of μ_1 and μ_2. For part (b), we need to use the ϵ technique; that is, for any $\epsilon > 0$, there exists a set $B_\epsilon \in \Sigma$ and $B_\epsilon \subseteq A$ such that $\mu_1(A) < \mu(B_\epsilon) + \epsilon$.

3.4 Solutions

3.1. Let $m \geq 1$ be fixed. Then $A_m \cap B_n \uparrow A_m \cap B = A_m$ as $n \to \infty$. By Theorem 2.1, we have $\mu(A_m \cap B_n) \uparrow \mu(A_m)$ as $n \to \infty$. But by Problem 2.11, we have $\mu(A_m \cap B_n) \leq \mu(B_n)$ for all $n \geq 1$. Hence we have $\mu(A_m) = \lim_{n\to\infty} \mu(A_m \cap B_n) \leq \lim_{n\to\infty} \mu(B_n)$. Inequality (3.1) follows by letting $m \to \infty$. Equality (3.2) follows from inequality (3.1).

3.2.

(a) This follows from Problem 3.1.

(b) This is trivial since Σ_0 is an algebra, $\Sigma_0 \subseteq \mathcal{C}$, and $\lambda = \mu$ on Σ_0.

(c) Let $A_n, B_n \in \Sigma_0$ for $n \geq 1$ such that $A_n \uparrow A$ and $B_n \uparrow B$. Then we have $A_n \cup B_n \uparrow A \cup B \in \mathcal{C}$ and $A_n \cap B_n \uparrow A \cap B \in \mathcal{C}$. By Problem 2.11, $\mu(A_n \cup B_n) = \mu(A_n) + \mu(B_n) - \mu(A_n \cap B_n)$ for all $n \geq 1$. Letting $n \to \infty$, we get $\lambda(A \cup B) = \lambda(A) + \lambda(B) - \lambda(A \cap B)$.

(d) This follows from Problem 3.1.

(e) For each $n \geq 1$, since $A_n \in \mathcal{C}$, there exists a sequence $\{B_{nm}\}_{m\geq 1}$ in Σ_0 such that $B_{nm} \uparrow A_n$ as $m \to \infty$. Let $C_n = \bigcup_{i=1}^{n} B_{in}$ for all $n \geq 1$. Then $B_{in} \subseteq C_n \subseteq A_n$ and $\mu(B_{in}) \leq \mu(C_n) \leq \lambda(A_n)$ for all $i \leq n$. Letting $n \to \infty$, we get $A_i \subseteq \bigcup_{n=1}^{\infty} C_n \subseteq A$ and $\lambda(A_i) \leq \lim_{n\to\infty} \mu(C_n) \leq \lim_{n\to\infty} \lambda(A_n)$. Letting $i \to \infty$, we get $A = \bigcup_{n=1}^{\infty} C_n$ (or $C_n \uparrow A$) and $\lim_{n\to\infty} \lambda(A_n) = \lim_{n\to\infty} \mu(C_n) = \lambda(A)$.

3.3.

(a) This follows from the definition of μ^*.

(b) Let $\epsilon > 0$. By the definition of μ^*, there exists a set $E \in \mathcal{C}$ such that $A \subseteq E$ and $\lambda(E) \leq \mu^*(A) + \epsilon$. Similarly, there exists a set $F \in \mathcal{C}$ and $\lambda(F) \leq \mu^*(B) + \epsilon$. Note that $A \cup B \subseteq E \cup F \in \mathcal{C}$ and $A \cap B \subseteq E \cap F \in \mathcal{C}$, we have $\mu^*(A \cup B) \leq \lambda(E \cup F)$ and $\mu^*(A \cap B) \leq \lambda(E \cap F)$ by the definition of μ^*. Since $\lambda(E \cup F) + \lambda(E \cap F) = \lambda(E) + \lambda(F)$, we have

$$\mu^*(A \cup B) + \mu^*(A \cap B) \leq \mu^*(A) + \mu^*(B) + 2\epsilon.$$

Since $\epsilon > 0$ is arbitrary, the result follows.

(c) By the definition of μ^*, there exists a set $B_n \in \mathcal{C}$ such that $B \subseteq B_n$ and $\lambda(B_n) < \mu^*(B) + 1/n$. Since $A \subseteq B$, we have $A \subseteq B_n$ and thus $\mu^*(A) \leq \lambda(B_n)$. Hence we have $\mu^*(A) < \mu^*(B) + 1/n$. The result follows by letting $n \to \infty$.

(d) Since $A_n \uparrow A$, then $\mu^*(A_n)$ is increasing and $\mu^*(A_n) \leq \mu^*(A) \leq \lambda(S) < \infty$ for all $n \geq 1$. Hence the limit of $\mu^*(A_n)$ exists and $\lim_{n\to\infty} \mu^*(A_n) \leq \mu^*(A)$. Thus we only need to prove $\mu^*(A) \leq \lim_{n\to\infty} \mu^*(A_n)$.

Let $\epsilon > 0$. For each $n \geq 1$, by the definition of μ^*, there exists a set $E_n \in \mathcal{C}$ such that $A_n \subseteq E_n$ and $\lambda(E_n) \leq \mu^*(A_n) + \epsilon/2^n$. Then $A = \bigcup_{n=1}^{\infty} A_n \subseteq \bigcup_{n=1}^{\infty} E_n$. It follows that

$$\mu^*(A) \leq \lambda\left(\bigcup_{n=1}^{\infty} E_n\right) = \lim_{N\to\infty} \lambda\left(\bigcup_{n=1}^{N} E_n\right). \tag{3.3}$$

We claim that

$$\lambda\left(\bigcup_{n=1}^{N} E_n\right) \le \mu^*(A_N) + \epsilon \sum_{n=1}^{N} \frac{1}{2^n}. \tag{3.4}$$

To prove inequality (3.4), we use mathematical induction. When $N = 1$, inequality (3.4) is true by the choice of E_1. Assume that inequality (3.4) is true for $N = m$. Then by the property of λ, we have

$$\begin{aligned}
\lambda\left(\bigcup_{n=1}^{m+1} E_n\right) &= \lambda\left(\bigcup_{n=1}^{m} E_n\right) + \lambda(E_{m+1}) - \lambda\left(\left(\bigcup_{n=1}^{m} E_n\right) \cap E_{m+1}\right) \\
&\le \lambda\left(\bigcup_{n=1}^{m} E_n\right) + \lambda(E_{m+1}) - \lambda(E_m \cap E_{m+1}) \\
&\le \mu^*(A_m) + \epsilon \sum_{n=1}^{m} \frac{1}{2^n} + \mu^*(A_{m+1}) + \frac{\epsilon}{2^{m+1}} - \mu^*(A_m) \\
&\le \mu^*(A_{m+1}) + \epsilon \sum_{n=1}^{m+1} \frac{1}{2^n}.
\end{aligned}$$

Hence inequality (3.4) is true. Letting $N \to \infty$ in inequality (3.4) gives

$$\lim_{N\to\infty} \lambda\left(\bigcup_{n=1}^{N}\right) \le \lim_{N\to\infty} \mu^*(A_N) + \epsilon.$$

By inequality (3.3), we get $\mu^*(A) \le \lim_{N\to\infty} \mu^*(A_N) + \epsilon$. Since ϵ is arbitrary, we have $\mu^*(A) \le \lim_{n\to\infty} \mu^*(A_n)$. This completes the proof.

(e) It is clear that $\mu^*(\emptyset) = 0$ and $\mu^*(A) \le \mu^*(B)$ for $A \subseteq B \subseteq S$. Now let $A_n \subseteq S$ for $n \ge 1$. Then by parts (d) and (b) of this proof, we have

$$\mu^*\left(\bigcup_{n=1}^{\infty} A_n\right) = \lim_{n\to\infty} \mu^*\left(\bigcup_{i=1}^{n} A_i\right) \le \lim_{n\to\infty} \sum_{i=1}^{n} \mu^*(A_i) = \sum_{n\to\infty} \mu^*(A_n).$$

Hence μ^* is an outer measure on S.

(f) By Problem 2.6, to show that $\mathcal{H}$ is a σ-algebra, we only need to show that $\mathcal{H}$ is both a π-system and a d-system. Let $A, B \in \mathcal{H}$. By part (b) of this proof, we have $\mu^*(A \cap B) + \mu^*((A \cap B)^c) \ge \mu^*(S)$ and $\mu^*(A \cup B) + \mu^*((A \cup B)^c) \ge \mu^*(S)$. Hence, we have $\mu^*(A \cap B) + \mu^*((A \cap B)^c) \le \mu^*(A) + \mu^*(B) - \mu^*(A \cup B) + \mu^*(A^c) + \mu^*(B^c) - \mu^*(A^c \cap B^c) = 2\mu^*(S) - [\mu^*(A \cup B) + \mu^*(A^c \cap B^c)] \le \mu^*(S)$. Hence $\mu^*(A \cap B) + \mu^*((A \cap B)^c) = \mu^*(S)$ and then $A \cap B \in \mathcal{H}$. Therefore, $\mathcal{H}$ is a π-system.

Now we show that $\mathcal{H}$ is a d-system. Since $\mu^*(\emptyset) = 0$, we have $S \in \mathcal{H}$. Let $A, B \in \mathcal{H}$ and $A \subseteq B$. Then $A^c \in \mathcal{H}$. Since $\mathcal{H}$ is a π-system, we have $B \cap A^c \in \mathcal{H}$. Now let $A_n \in \mathcal{H}$ for $n \ge 1$ and $A_n \uparrow A$. By part (d) of this proof, we have $\mu^*(A_n) \uparrow \mu^*(A)$. Since $A_n^c \downarrow A^c$, $\mu^*(A^c) \le \mu^*(A_n^c) = \mu^*(S) - \mu^*(A_n)$ or $\mu^*(A_n) \le \mu^*(S) - \mu^*(A^c)$ for all $n \ge 1$. Letting $n \to \infty$ gives $\mu^*(A) \le \mu^*(S) - \mu^*(A^c)$. But since $\mu^*(A) + \mu^*(A^c) \ge \mu^*(A \cup A^c) = \mu^*(S)$, we have $\mu^*(A) + \mu^*(A^c) = \mu^*(S)$. Therefore, $A \in \mathcal{H}$. Hence $\mathcal{H}$ is a d-system.

Finally, we show that μ^* is a finite measure on $\mathcal{H}$. Clearly, $\mu^*(\emptyset) = 0$ and $\mu^*(S) < \infty$. To show that μ^* is countably additive, let $A_1, A_2, \ldots$ be disjoint sets in $\mathcal{H}$. We claim that

$$\mu^*\left(\bigcup_{n=1}^{N} A_n\right) = \sum_{n=1}^{N} \mu^*(A_n), \quad \forall N \ge 1. \tag{3.5}$$

To show this, we use mathematical induction. When $N = 1$, Equation (3.5) is trivial. Assume that Equation (3.5) holds for $N = m$. Then by part (b) of this proof, we obtain

$$\begin{aligned}
\mu^*\left(\bigcup_{n=1}^{m+1} A_n\right) &= \mu^*(S) - \mu^*\left(\bigcap_{n=1}^{m+1} A_n^c\right) \\
&\ge \mu^*(S) - \mu^*\left(\bigcap_{n=1}^{m} A_n^c\right) - \mu^*(A_{m+1}^c) \\
&\quad + \mu^*\left(\left(\bigcap_{n=1}^{m} A_n^c\right) \cup A_{m+1}^c\right) \\
&= 2\mu^*(S) - \mu^*\left(\bigcap_{n=1}^{m} A_n^c\right) - \mu^*(A_{m+1}^c) \\
&= \mu^*\left(\bigcup_{n=1}^{m} A_n\right) + \mu^*(A_{m+1}) \\
&= \sum_{n=1}^{m+1} \mu^*(A_n).
\end{aligned}$$

Therefore, equation (3.5) holds for all $N \ge 1$. The result follows by letting $N \to \infty$ in Equation (3.5) and part (d) of this proof.

3.4. Since $S \in \Sigma$ and $\mu(\emptyset) = 0$, we have $S = S \cup \emptyset \in \Sigma_\mu$. Let $E \in \Sigma_\mu$. Then $E = A \cup B$, where $A \in \Sigma$ and $B \subseteq N$ for some $N \in \Sigma$ with $\mu(N) = 0$. Since $N^c \subseteq B^c$, we have $E^c = A^c \cap B^c = (A^c \cap N^c) \cup (A^c \cap B^c \cap N)$. Note that since $\mu(A^c \cap B^c \cap N) \le \mu(N) = 0$, we have $\mu(A^c \cap B^c \cap N) = 0$. Thus $E^c \in \Sigma_\mu$. Now let $E_n \in \Sigma_\mu$ for $n \ge 1$. Then $E_n = A_n \cup B_n$, where A_n and B_n are defined similarly. Since $\bigcup_{n=1}^\infty E_n = \bigcup_{n=1}^\infty A_n \cup B_n = (\bigcup_{n=1}^\infty A_n) \cup (\bigcup_{n=1}^\infty B_n)$ and $\mu\left(\bigcup_{n=1}^\infty B_n\right) = 0$ (see Problem 2.19), we have $\bigcup_{n=1}^\infty E_n \in \Sigma_\mu$. Therefore, Σ_μ is a σ-algebra.

3.5.

(a) Since $A \in \Sigma_\mu$, we can express $A = A_1 \cup A_2$, where $A_1 \in \Sigma$ and $A_2 \subseteq N$ for some set $N \in \Sigma$ with $\mu(N) = 0$. Note that $A_1 \in \Sigma$ and $A_1 \subseteq A$, we have $\mu_1(A) \geq \mu(A_1) = \mu(A)$. On the other hand, we have $A \subseteq A_1 \cup N$. We have $\mu_1(A) \leq \mu(A_1 \cup N) = \mu(A_1) = \mu(A)$. Thus $\mu_1(A) = \mu(A)$. Similarly, we have $\mu_2(A) = \mu(A)$.

(b) Suppose that $A \subseteq S$ and $\mu_1(A) = \mu_2(A) < \infty$. For each $n \geq 1$, by definition of μ_1, there exists a set $B_n \in \Sigma$ such that $B_n \subseteq A$ and $\mu_1(A) < \mu(B_n) + 1/n$. Similarly, there exists a sequence $\{C_n\}_{n\geq 1}$ of sets in Σ such that $C_n \supseteq A$ for all $n \geq 1$ and $\mu_2(A) > \mu(C_n) - 1/n$. Then $\bigcup_{i=1}^\infty B_i \subseteq A \subseteq \bigcap_{i=1}^\infty C_i$. Now we let $B = \bigcup_{n=1}^\infty B_n$ and $C = \bigcap_{n=1}^\infty C_n$ and write $A = B \cup (A \backslash B)$. Note that $\mu(B_n) \leq \mu\left(\bigcup_{i=1}^\infty B_i\right)$, $\mu(C_n) \geq \mu(\bigcap_{n=1}^\infty C_n)$, and $\mu(B_n) + 1/n > \mu_1(A) = \mu_2(A) > \mu(C_n) - 1/n$ for all $n \geq 1$. We have $\mu(C \backslash B) = \mu(C) - \mu(B) \leq \mu(C_n) - \mu(B_n) < 2/n$. Letting $n \to \infty$ gives $\mu(C \backslash B) = 0$. This completes the proof.

3.5 Bibliographic Notes

Usually we define a set function on a field Σ_0 of subsets of a set S and then extend the set function to a measure on $\sigma(\Sigma_0)$, the smallest σ-field containing Σ_0. The theorems presented in this chapter provide a theoretical foundation for measure extension. For example, we can extend the set function $\mu(a, b] = b - a$ defined on the set of all right-semiclosed intervals to $\mathcal{B}$ (Ash and Doleans-Dade, 1999, p12).

In a general case, we consider extending a finite measure μ_0 from a field Σ_0 of subsets of a set S to the minimal σ-field $\Sigma = \sigma(\Sigma_0)$. To do this, we first consider extending the measure to the set of all limits of increasing sequences of sets in Σ_0, as follows:

$$\Sigma_1 = \{A \subseteq S : \text{ there are sets } A_n \in \Sigma_0 \text{ such that } A_n \uparrow A\}.$$

We can do this by defining μ_1 on Σ_1 as follows:

$$\mu_1(A) = \lim_{n\to\infty} \mu_0(A_n),$$

where $A \in \Sigma_1$ and $A_n \in \Sigma_0$ is a sequence such that $A_n \uparrow A$.

Then we define an extended set function μ^* on S as

$$\mu^*(A) = \inf\{\mu_1(G) : G \in \Sigma_1, G \supseteq A\}.$$

It can be shown that μ^* defined in the above equation is an outer measure on S (see Problem 3.3). Using the outer measure, we define a set $\mathcal{H} = \{H \subseteq S : \mu^*(H) + \mu^*(H^c) = \mu^*(S)\}$. It can be shown that $\mathcal{H}$ is a σ-field and μ^* is a finite measure on $\mathcal{H}$. The set $\mathcal{H}$ is a superset of $\sigma(\Sigma_0)$. In fact, if a set $B \in \mathcal{H}$, then we can write $B = A \cup N$ for some $A \in \sigma(\Sigma_0)$ and $N \subseteq M$, where $M \in \sigma(\Sigma_0)$ and $\mu^*(M) = 0$. The measure space $(S, \mathcal{H}, \mu^*)$ is also called the *completion* of the measure space $(S, \sigma(\Sigma_0), \mu^*)$.

Rosenthal (2006, Chapter 2) presented some examples of constructing probability measures using the extension theorem.

CHAPTER 4

LEBESGUE-STIELTJES MEASURES

The Lebesgue measure is a special measure defined on subsets of Euclidean spaces and is the standard way of measuring the length, area, and volume of these subsets. In this chapter, we define the Lebesgure measure and present some relevant theorems.

4.1 Basic Concepts and Facts

Definition 4.1 (Lebesgue-Stieltjes Measure). A Lebesgue-Stieltjes measure on $\mathbf{R} = (-\infty, \infty)$ is a measure μ on $\mathcal{B}$ such that $\mu(I) < \infty$ for each bounded interval $I \subset \mathbf{R}$.

Definition 4.2 (Distribution Function). A distribution function on $\mathbf{R}$ is a map $F : \mathbf{R} \to \mathbf{R}$ that satisfies the following conditions:

(a) F is increasing; that is, $a < b$ implies $F(a) \le F(b)$.

(b) F is right-continuous:

$$\lim_{x \to x_0^+} = F(x_0),$$

where $x \to x_0^+$ means that $x > x_0$ and x converges to x_0.

Definition 4.3 (Lebesgue Measure). The Lebesgue-Stieltjes measure μ on $\mathbf{R}$ defined by

$$\mu(a, b] = b - a, \quad a < b$$

is called the *Lebesgue measure* on $\mathcal{B}$ (see Theorem 4.1). The completion $\bar{\mathcal{B}}$ of $\mathcal{B}$ relative to the Lebesgue measure is called the *class of Lebesgue measurable sets*.

Definition 4.4 (Lebesgue-Stieltjes Measure on $\mathbf{R}^d$). A Lebesgue-Stieltjes measure on $\mathbf{R}^d$ is a measure μ on $\mathcal{B}(\mathbf{R}^d)$ such that $\mu(\mathbf{I}) < \infty$ for each bounded interval $\mathbf{I} \subset \mathbf{R}^d$, where $d \in \mathbf{N}$ and $\mathcal{B}(\mathbf{R}^d)$ is the collections of Borel sets of $\mathbf{R}^d$, that is, the σ-algebra generated by all intervals $(\mathbf{a}, \mathbf{b}]$, $\mathbf{a}, \mathbf{b} \in \mathbf{R}^d$.

Definition 4.5 (Difference Operator). Let $G : \mathbf{R}^d \to \mathbf{R}$ be a function. The difference operator $\Delta_{i,b,a}G(x_1, x_2, \ldots, x_d)$, $1 \leq i \leq d$, is defined as

$$\begin{aligned} \Delta_{i,b,a}G(x_1, x_2, \ldots, x_d) &= G(x_1, \ldots, x_{i-1}, b, x_{i+1}, \ldots, x_d) \\ &\quad -G(x_1, \ldots, x_{i-1}, a, x_{i+1}, \ldots, x_d). \end{aligned} \tag{4.1}$$

Definition 4.6 (Distribution Function on $\mathbf{R}^d$). A distribution function on $\mathbf{R}^d$ is a map $F : \mathbf{R}^d \to \mathbf{R}$ that satisfies the following conditions:

(a) F is increasing; that is, $\mathbf{a} \leq \mathbf{b}$ (i.e., $a_i \leq b_i$ for $i = 1, 2, \ldots, d$, where a_i and b_i are the ith components of $\mathbf{a}$ and $\mathbf{b}$, respectively), implies $F(\mathbf{a}, \mathbf{b}] \geq 0$, where

$$F(\mathbf{a}, \mathbf{b}] = \Delta_{1,b_1,a_1} \cdots \Delta_{d,b_d,a_d} F(x_1, x_2, \ldots, x_d); \tag{4.2}$$

(b) F is right-continuous; that is, for any sequence $\mathbf{x}_1 \geq \mathbf{x}_2 \geq \cdots \to \mathbf{x}$, we have $F(\mathbf{x}_k) \to F(\mathbf{x})$.

Definition 4.7 (Lebesgue Measure on $\mathbf{R}^d$). The Lebesgue-Stieltjes measure μ on $\mathbf{R}^d$ defined by

$$\mu(\mathbf{a}, \mathbf{b}] = F(\mathbf{a}, \mathbf{b}], \quad \mathbf{a} \leq \mathbf{b}$$

is called the *Lebesgue measure* on $\mathcal{B}(\mathbf{R}^d)$ if $F(\mathbf{x}) = x_1 x_2 \cdots x_d$ for $\mathbf{x} \in \mathbf{R}^d$. The completion $\bar{\mathcal{B}}(\mathbf{R}^d)$ of $\mathcal{B}(\mathbf{R}^d)$ relative to the Lebesgue measure is called the *class of Lebesgue measurable sets* in $\mathbf{R}^d$.

Theorem 4.1. *Let F be a distribution function on $\mathbf{R}$, and let $\mathcal{B}_0(\bar{\mathbf{R}})$ be the algebra of all finite disjoint unions of right-semiclosed intervals of $\bar{\mathbf{R}}$ (counting $[-\infty, b]$ as right-semiclosed). Let $\mu : \mathcal{B}_0(\bar{\mathbf{R}}) \to \bar{\mathbf{R}}$ be a set function defined as*

$$\mu(a, b] = F(b) - F(a), \quad a < b.$$

Then μ has a unique extension to a Lebesgue-Stieltjes measure on $\mathbf{R}$.

Theorem 4.2. *Let F be a distribution function on $\mathbf{R}^d$, and let $\mathcal{B}_0(\bar{\mathbf{R}}^d)$ be the algebra of all finite disjoint unions of right-semiclosed intervals $(\mathbf{a}, \mathbf{b}]$ of $\bar{\mathbf{R}}^d$ (counting $[-\infty, \mathbf{b}]$ as right-semiclosed), where*

$$(\mathbf{a}, \mathbf{b}] = \{\mathbf{x} \in \mathbf{R}^d : a_i < x_i \leq b_i, i = 1, 2, \ldots, d\}.$$

Let $\mu : \mathcal{B}_0(\bar{\mathbf{R}}^d) \to \bar{\mathbf{R}}$ be a set function defined as

$$\mu(\mathbf{a}, \mathbf{b}] = F(\mathbf{a}, \mathbf{b}], \quad \mathbf{a} \leq \mathbf{b},$$

where $F(\mathbf{a}, \mathbf{b}]$ is defined in Equation (4.2). *Then μ has a unique extension to a Lebesgue-Stieltjes measure on $\mathbf{R}^d$.*

4.2 Problems

4.1. Let μ be a Lebesgue-Stieltjes measure on $\mathbf{R}$. Let $F : \mathbf{R} \to \mathbf{R}$ be a function defined by $F(b) - F(a) = \mu(a, b]$ for $b \geq a$ and $F(0) = c$ for some constant $c \in \mathbf{R}$. Show that F is a distribution function on $\mathbf{R}$.

4.2. Find a measure μ on $\mathbf{R}$ such that the distribution function F defined by $F(b) - F(a) = \mu(a, b]$ for $b \geq a$ is not left-continuous at some point x_0, where $x \to x_0^-$ means $x < x_0$ and x converges to x_0:

$$\lim_{x \to x_0^-} F(x) \neq F(x_0),$$

4.3. Let F be a distribution function, and let μ be the corresponding Lebesgue-Stieltjes measure:

$$\mu(a, b] = b - a, \quad a < b.$$

Let $F(x_0^-)$ denote $\lim_{x \to x_0^-} F(x)$. Show that

(a) $\mu[a, b] = F(b) - F(a^-)$.

(b) $\mu(a, b) = F(b^-) - F(a)$.

(c) $\mu[a, b) = F(b^-) - F(a^-)$.

(d) $\mu(-\infty, x] = F(x) - F(-\infty)$.

(e) $\mu(-\infty, x) = F(x^-) - F(-\infty)$.

(f) $\mu[x, \infty) = F(\infty) - F(x^-)$.

(g) $\mu(x, \infty) = F(\infty) - F(x)$.

(h) $\mu(\mathbf{R}) = F(\infty) - F(-\infty)$.

4.4. Let $F : \mathbf{R}^d \to \mathbf{R}$ be a function. Show that

$$\Delta_{1,b_1,a_1} \cdots \Delta_{d,b_d,a_d} F(\mathbf{x}) = \sum_{i=0}^{d} (-1)^i S_i, \tag{4.3}$$

where

$$S_i = \sum_{A \subseteq Q_d, |A|=i} F(\mathbf{x})|_{x_j = a_j \text{ for } j \in A, x_j = b_j \text{ for } j \in Q_d \backslash A},$$

with $Q_d = \{1, 2, \ldots, d\}$. For example, $S_0 = F(b_1, b_2, \ldots, b_d)$ and

$$S_1 = \sum_{i=1}^{d} F(b_1, \ldots, b_{i-1}, a_i, b_{i+1}, \ldots, b_d).$$

4.5. Let μ be a finite measure on $\mathcal{B}(\mathbf{R}^d)$, and let $F : \mathbf{R}^d \to \mathbf{R}$ be defined as

$$F(\mathbf{x}) = \mu(-\infty, \mathbf{x}] = \mu\{\mathbf{y} \in \mathbf{R}^d : y_i \le x_i, i = 1, 2, \ldots, d\}.$$

Let $\mathbf{a}, \mathbf{b} \in \mathbf{R}^d$ and $\mathbf{a} \le \mathbf{b}$. Show that

$$\mu(\mathbf{a}, \mathbf{b}] = \Delta_{1,b_1,a_1} \cdots \Delta_{d,b_d,a_d} F(\mathbf{x}).$$

4.6. Let $F_1, F_2, \ldots, F_d$ be distribution functions on $\mathbf{R}$, and define $F : \mathbf{R}^d \to \mathbf{R}$ as

$$F(\mathbf{x}) = \prod_{i=1}^{d} F_i(x_i).$$

Show that

(a)

$$F(\mathbf{a}, \mathbf{b}] = \prod_{i=1}^{d} (F_i(b_i) - F_i(a_i)), \quad \mathbf{a} \le \mathbf{b},$$

where $F(\mathbf{a}, \mathbf{b}]$ is as defined in Equation (4.2).

(b) F is a distribution function on $\mathbf{R}^d$.

4.7. Let F be a continuous distribution function on $\mathbf{R}$. Let μ be the Lebesgue-Stieltjes measure defined by $\mu(a, b] = F(b) - F(a)$. Show that $\mu(A) = 0$ if A is a countable subset of $\mathbf{R}$.

4.8 (Cantor Ternary Set)**.** The Cantor ternary set is constructed by removing the middle thirds of line segments in $[0, 1]$ as follows. Let M_1 be the middle third of $[0, 1]$, that is, $M_1 = (\frac{1}{3}, \frac{2}{3})$. Let M_2 be the union of the middle thirds of the two intervals in $[0, 1]\backslash M_1$, i.e., $M_2 = (\frac{1}{9}, \frac{2}{9}) \cup (\frac{7}{9}, \frac{8}{9})$. Sets $M_3, M_4, \cdots$ are constructed similarly. The Cantor ternary set C is defined to be $[0, 1]\backslash \bigcup_{n=1}^{\infty} M_n$.

(a) Show that C has Lebesgue measure 0.

(b) Let $x = (0.a_1 a_2 \cdots)_3 = \sum_{n=1}^{\infty} a_n/3^n$ be the ternary form (i.e., the base 3 expansion) of $x \in [0, 1]$. For example, $\frac{1}{3} = (0.1)_3 = (0.0222\cdots)_3$. Show that $x \in C$ if and only if $a_n \in \{0, 2\}$ for all $n \ge 1$.

(c) Show that C is uncountable.

4.9. Let μ be the Lebesgue measure on $\mathbf{R}$. Give an example set $A \subseteq \mathbf{R}$ such that $\mu(A) > 0$ and A does not contain any open interval.

4.3 Hints

4.1. Use the definition of distribution functions (see Definition 4.2) and the result of Problem 2.18.

4.2. It might not be easy to find such a measure directly. But one can easily find a distribution function that is not left-continuous at some point. Then one can define a measure based on the distribution function.

4.3. To prove those equalities, we need to write the intervals in terms of right-semiclosed intervals.

4.4. Consider simple cases when $d = 1, 2$ and then use the method of mathematical induction to deal with the general case.

4.5. First consider simple cases when $d = 1, 2$ and then consider the general case.

4.6. Part (a) can be proved by using the definition of $F(\mathbf{a}, \mathbf{b}]$. Part (b) can be proved by using the proof of part (a) and the fact that $F_1, F_2, \ldots, F_d$ are distribution functions on $\mathbf{R}$.

4.7. Use the result of problem 4.3 and the fact that a countable set can be expressed as $\{a_1, a_2, \ldots\}$.

4.8. For part (a), we can show that C has Lebesgue measure 0 by calculating the Lebesgue measure of $\bigcup_{n=1}^{\infty} M_n$. Part (b) can be proved by examining the construction procedure of C. For part (c), we can use the method of contradiction to prove that C is uncountable; that is, we assume that C is countable and try to find another point in C using the ternary forms (Cantor's diagonalizing method).

4.9. Use the idea of constructing the Cantor ternary set in Problem 4.8.

4.4 Solutions

4.1. Clearly, F is increasing. To show that F is right-continuous, let $x_n \downarrow x_0$, that is, $x_1 \geq x_2 \geq \cdots > x_0$ and $x_n \to x_0$. Then we have $(x_0, x_n] \downarrow \emptyset$. Then by Problem 2.18, we have $F(x_n) - F(x_0) \downarrow 0$. Thus F is right-continuous and hence a distribution function.

4.2. Let μ be a set function defined on all right-semiclosed intervals as follows:

$$\mu(a, b] = \begin{cases} 0, & \text{if } a < b < 0; \\ 2, & \text{if } a < 0 \text{ and } 0 \leq b; \\ 0, & \text{if } 0 \leq a < b. \end{cases}$$

Then μ can be extended to a measure on $\mathcal{B}$ according to Theorem 3.1. The corresponding distribution function F is

$$F(x) = \begin{cases} F(0) - 2, & \text{if } x < 0; \\ F(0), & \text{if } 0 \leq x. \end{cases}$$

Clearly, the distribution function is not left-continuous at 0.

4.3. Parts (d)–(h) are obtained from the definition and items (a)–(c) by letting $a \to -\infty$ or $b \to \infty$. In the following we prove parts (a)–(c).

(a) By definition and Problem 2.18, we have

$$\begin{aligned} \mu[a, b] &= \lim_{n\to\infty} \mu\left(a - \frac{1}{n}, b\right] \\ &= \lim_{n\to\infty} \left(F(b) - F\left(a - \frac{1}{n}\right)\right) \\ &= F(b) - F(a^-). \end{aligned}$$

(b) By definition and Theorem 2.1, we get

$$\begin{aligned} \mu(a, b) &= \lim_{n\to\infty} \mu\left(a, b - \frac{1}{n}\right] \\ &= \lim_{n\to\infty} \left(F\left(b - \frac{1}{n}\right) - F(a)\right) \\ &= F(b^-) - F(a). \end{aligned}$$

(c) Similarly, we have

$$\mu[a, b) = \lim_{n\to\infty} \mu\left(a - \frac{1}{n}, b - \frac{1}{n}\right] = F(b^-) - F(a^-).$$

4.4. We use mathematical induction on d. When $d = 1$, Equation (4.3) reduces to $\Delta_{1,b_1,a_1} = F(b_1) - F(a_1)$, which is true. Suppose that the equation holds when $d = m$. When $d = m + 1$, let

$$\begin{aligned} E(x_1, \ldots, x_m) &= \Delta_{m+1,b_{m+1},a_{m+1}} F(x_1, \ldots, x_{m+1}) \\ &= F(x_1, \ldots, x_m, b_{m+1}) - F(x_1, \ldots, x_m, a_{m+1}). \end{aligned}$$

Then by the induction hypothesis, we obtain

$$\begin{aligned}
& \Delta_{1,b_1,a_1} \cdots \Delta_{m+1,b_{m+1},a_{m+1}} F(\mathbf{x}) \\
= \; & \Delta_{1,b_1,a_1} \cdots \Delta_{m,b_m,a_m} E(x_1, \ldots, x_m) \\
= \; & \sum_{j=0}^{m} (-1)^j \sum_{A \subseteq Q_m, |A|=j} E(x_1, \ldots, x_m)|_{x_j=a_j \text{ for } j \in A, x_j=b_j \text{ for } j \in Q_m \backslash A} \\
= \; & \sum_{j=0}^{m} (-1)^j \sum_{A \subseteq Q_m, |A|=j} F(x_1, \ldots, x_m, b_{m+1})|_{x_j=a_j \text{ for } j \in A, x_j=b_j \text{ for } j \in Q_m \backslash A} \\
& - \sum_{j=0}^{m} (-1)^j \sum_{A \subseteq Q_m, |A|=j} F(x_1, \ldots, x_m, a_{m+1})|_{x_j=a_j \text{ for } j \in A, x_j=b_j \text{ for } j \in Q_m \backslash A} \\
= \; & \sum_{j=0}^{m+1} (-1)^j \sum_{A \subseteq Q_{m+1}, |A|=j} F(x_1, \ldots, x_{m+1})|_{x_j=a_j \text{ for } j \in A, x_j=b_j \text{ for } j \in Q_{m+1} \backslash A}.
\end{aligned}$$

Hence Equation (4.3) holds for all $d \geq 1$.

4.5. By definition, we have

$$\begin{aligned}
\Delta_{d,b_d,a_d} F(x_1, \ldots, x_d) &= F(x_1, \ldots, x_{d-1}, b_d) - F(x_1, \ldots, x_{d-1}, a_d) \\
&= \mu\{\mathbf{y} : y_i \leq x_i, i = 1, 2, \ldots, d-1, y_d \leq b_d\} \\
&\quad - \mu\{\mathbf{y} : y_i \leq x_i, i = 1, 2, \ldots, d-1, y_d \leq a_d\} \\
&= \mu\{\mathbf{y} : y_i \leq x_i, i = 1, 2, \ldots, d-1, a_d \leq y_d \leq a_d\}
\end{aligned}$$

and

$$\begin{aligned}
& \Delta_{d-1,b_{d-1},a_{d-1}} \Delta_{b_d a_d} F(x_1, \ldots, x_d) \\
= \; & F(x_1, \ldots, x_{d-2}, b_{d-1}, b_d) - F(x_1, \ldots, x_{d-2}, b_{d-1}, a_d) \\
& - F(x_1, \ldots, x_{d-2}, a_{d-1}, b_d) + F(x_1, \ldots, x_{d-2}, a_{d-1}, a_d) \\
= \; & \mu\{\mathbf{y} : y_i \leq x_i, i = 1, 2, \ldots, d-2, y_{d-1} \leq b_{d-1}, a_d \leq y_d \leq a_d\} \\
& - \mu\{\mathbf{y} : y_i \leq x_i, i = 1, 2, \ldots, d-2, y_{d-1} \leq a_{d-1}, a_d \leq y_d \leq a_d\} \\
= \; & \mu\{\mathbf{y} : y_i \leq x_i, i = 1, 2, \ldots, d-2, a_{d-1} \leq y_{d-1} \leq b_{d-1}, a_d \leq y_d \leq a_d\}.
\end{aligned}$$

Repeating the above procedure d times gives the result.

4.6. By definition, we have

$$\begin{aligned}
& \Delta_{d,b_d,a_d} F(x_1, \ldots, x_d) \\
= \; & F(x_1, \ldots, x_{d-1}, b_d) - F(x_1, \ldots, x_{d-1}, a_d) \\
= \; & F(x_1) \cdots F(x_{d-1}) F(b_d) - F(x_1) \cdots F(x_{d-1}) F(a_d) \\
= \; & F(x_1) \cdots F(x_{d-1}) (F(b_d) - F(a_d)),
\end{aligned}$$

and

$$\begin{aligned}
&\Delta_{d-1,b_{d-1},a_{d-1}}\Delta_{d,b_d,a_d}F(x_1,\ldots,x_d)\\
=\ &\Delta_{b_{d-1}a_{d-1}}F(x_1)\cdots F(x_{d-1})(F(b_d)-F(a_d))\\
=\ &F(x_1)\cdots F(x_{d-2})(F(b_{d-1})-F(a_{d-1}))(F(b_d)-F(a_d)).
\end{aligned}$$

The result follows by repeating the above procedure d times.

Now we show that F is a distribution function on $\mathbf{R}^d$. Let $\mathbf{a} \le \mathbf{b}$. Then by the first part of this proof, we have

$$F(\mathbf{a},\mathbf{b}] = \prod_{i=1}^{d}(F_i(b_i) - F_i(a_i)) \ge 0.$$

Hence F is an increasing function. The right-continuity of F follows from the right-continuity of $F_1, F_2, \ldots, F_d$. This completes the proof.

4.7. Since A is a countable set, we can enumerate the elements of A and write $A = \{a_1, a_2, \ldots\}$. Since μ is countably additive, we have $\mu(A) = \sum_{n=1}^{\infty}\mu(a_n)$. By Problem 4.3 and the assumption that F is continuous, we have $\mu(a_n) = \mu[a_n, a_n] = F(a_n) - F(a_n^-) = 0$. Hence $\mu(A) = 0$.

4.8.

(a) Let μ be the Lebesgue measure. Then $\mu(M_n) = \frac{1}{3}\cdot\left(\frac{2}{3}\right)^{n-1}$ for $n = 1, 2, \ldots$. Noting that $M_1, M_2, \cdots$ are pairwise disjoint, we have

$$\mu(C) = 1 - \mu\left(\bigcup_{n=1}^{\infty} M_n\right) = 1 - \sum_{n=1}^{\infty}\mu(M_n) = 0.$$

(b) Suppose that $x \in C$. Then in the first step of constructing C we have $x \in [0, \frac{1}{3}]$ or $x \in [\frac{2}{3}, 1]$. If $x \in [0, \frac{1}{3}]$, then $a_1 = 0$ [$\frac{1}{3}$ can be written as $(0.0222\cdots)_3$]. If $x \in [\frac{2}{3}, 1]$, then $a_1 = 2$. In this second step, we have $x \in [0, \frac{1}{9}] \cup [\frac{2}{9}, \frac{1}{3}] \cup [\frac{2}{3}, \frac{7}{9}] \cup [\frac{8}{9}, 1]$, which implies that $a_1, a_2 \in \{0, 2\}$. Repeating the above process shows that $a_n \in \{0, 2\}$ for all $n \ge 1$.

Suppose that $a_n \in \{0, 2\}$ for all $n \ge 1$. Then $x \in [0, \frac{1}{3}] \cup [\frac{2}{3}, 1]$ in the first step of constructing C as $a_1 \in \{0, 2\}$. Similarly, x is not in the removed intervals in all steps. Hence $x \in C$.

(c) Suppose that C is countable. Then we can write $C = \{c_1, c_2, \ldots\}$. According to the second item of this problem, we can write $c_n = (0.a_{n1}a_{n2}\cdots)_3$ with $a_{nm} \in \{0, 2\}$ for all $n, m \ge 1$. Now we let $b = (0.b_1b_2\cdots)_3$, where

$$b_n = \begin{cases} 0, & \text{if } a_{nn} = 2; \\ 2, & \text{if } a_{nn} = 0. \end{cases}$$

Then $b \neq c_n$ for all $n \geq 1$ since the difference of b_n and a_{nn} is 2. Hence $b \notin C$. But $b_n \in \{0, 2\}$ for all $n \geq 1$, we have $b \in C$, which contradicts $b \notin C$. Thus C is uncountable.

4.9. We use a similar approach used to construct the Cantor ternary set to construct A. Let $\alpha \in (0, 1)$ be a constant. First we remove the middle $\alpha/2$ from interval $[0, 1]$. That is, we remove $M_1 = (\alpha/4, 3\alpha/4)$ from $[0, 1]$. Then we remove the middle $\alpha/8$ of the remaining two intervals; that is, we remove $M_2 = (\alpha/16, 3\alpha/16) \cup (8 + 5\alpha/16, 8 + 7\alpha/16)$. We continue the above process to remove $M_3, M_4, \ldots$. Finally, we let $A = [0, 1] \backslash \bigcup_{n=1}^{\infty} M_n$. Note that $\mu(M_n) = \alpha/2^n$ and $M_1, M_2, \ldots$ are disjoint, we have $\mu(A) = 1 - \sum_{n=1}^{\infty} \mu(M_n) = 1 - \alpha > 0$.

We claim that A does not contain any open interval. In fact, suppose that $(a, b) \subseteq A$ for some $a < b$. Then (a, b) must be in some remaining interval after M_n is removed for all $n \geq 1$. However, for sufficient large n, the length of each remaining interval after M_n is removed is less than $b - a$. Hence (a, b) must not be in any of those remaining intervals. Thus, A does not contain any open interval.

4.5 Bibliographic Notes

In this chapter, we introduced the Lebesgue measure and the Lebesgue-Stieltjes measure, which is more general than the Lebesgue measure. To define the Lebesgue measure on the real line, we first define the measure on the collection of left-open and right-closed intervals. Then we extend this measure to the collection of Borel sets. For more information about Lebesgue measures, readers are referred to Ash and Doleans-Dade (1999).

Theorems 4.1 and 4.2 state that we can define measures based on distribution functions. The proofs of these theorems can be found in Ash and Doleans-Dade (1999, pp24,30).

CHAPTER 5

MEASURABLE FUNCTIONS

In measure theory, measurable functions are defined by the property of their preimages. The definition of measurable functions is similar to the definition of continuous functions in topology. In this chapter, we introduce measurable functions and relevant theorems.

5.1 Basic Concepts and Facts

Definition 5.1 (Σ-Measurable Function)**.** Let (S, Σ) be a measurable space. A function $h : S \to \mathbf{R}$ is called Σ-*measurable*, or measurable relative to the σ-algebra Σ, if and only if

$$h^{-1}(A) \in \Sigma, \quad \forall A \in \mathcal{B},$$

where $\mathcal{B}$ is the Borel σ-algebra on $\mathbf{R}$ (see Definition 2.6) and $h^{-1}(A)$ is defined as

$$h^{-1}(A) = \{s \in S : h(s) \in A\}.$$

The set of all Σ-measurable functions is denoted by $m\Sigma$.

Definition 5.2 (Borel Function)**.** Let S be a topological space (see Definition 2.5). A function $h : S \to \mathbf{R}$ is called a *Borel function* if h is $\mathcal{B}(S)$-measurable or Borel measurable, where $\mathcal{B}(S)$ is the σ-algebra generated by the collection of all open sets in S.

Definition 5.3 (Σ_1/Σ_2-measurable Function)**.** Let (S_1, Σ_1) and (S_2, Σ_2) be two measurable spaces and $h : S_1 \to S_2$. h is considered Σ_1/Σ_2-measurable if and only if

$$h^{-1}(A) \in \Sigma_1, \quad \forall A \in \Sigma_2,$$

where $h^{-1}(A) = \{s \in S_1 : h(s) \in A\}$.

Definition 5.4 (Monotone Class of Functions)**.** Let (S, Σ) be a measurable space and $\mathcal{M}$ be a collection of functions from S to $\mathbf{R}$. The collection $\mathcal{M}$ is called a *monotone class* if it satisfies the following properties:

(a) $\mathcal{M}$ includes the constant function 1.

(b) If f and g are bounded functions in $\mathcal{M}$ and $a, b \in \mathbf{R}$, then $af + bg \in \mathcal{M}$.

(c) If f is the limit of an increasing sequence $\{f_n : n \geq 1\}$ of positive functions in $\mathcal{M}$ (i.e., $f_n \uparrow f$), then $f \in \mathcal{M}$.

Theorem 5.1. *Let (S, Σ) be a measurable space and $m\Sigma$ denote the set of all Σ-measurable functions from S to $\mathbf{R}$. If $\lambda \in \mathbf{R}$ and $h_1, h_2 \in m\Sigma$, then*

$$h_1 + h_2 \in m\Sigma, \quad h_1^2 \in m\Sigma, \quad h_1 h_2 \in m\Sigma, \quad \lambda h_1 \in m\Sigma.$$

Theorem 5.2. *Suppose that $f : \mathbf{R} \to \mathbf{R}$ is a continuous function or a piecewise-continuous function. Then f is Borel measurable.*

5.2 Problems

5.1. Let (S_1, Σ_1) and (S_2, Σ_2) be measurable spaces, and let $h : S_1 \to S_2$ be a function. Show that

(a) If $A_i \in \Sigma_2$ for all $i \in \mathbf{N}$, then

$$h^{-1}\left(\bigcup_{i=1}^{\infty} A_i\right) = \bigcup_{i=1}^{\infty} h^{-1}(A_i);$$

(b) $h^{-1}(A^c) = (h^{-1}(A))^c$ for all $A \in \Sigma_2$.

5.2. Let (S, Σ) be a measurable space. Show that a subset A of S is Σ-measurable if and only if I_A is Σ-measurable; that is, $A \in \Sigma$ if and only if $I_A \in m\Sigma$.

5.3. Let (S, Σ) be a measurable space, and let $h : S \to \mathbf{R}$ be a function. Suppose that $\mathcal{C} \subseteq \mathcal{B}$ and $\sigma(\mathcal{C}) = \mathcal{B}$. Show that if $h^{-1}(A) \in \Sigma$ for every $A \in \mathcal{C}$, then h is Σ-measurable.

5.4. Let (S_1, Σ_1) and (S_2, Σ_2) be measurable spaces, and let $h : S_1 \to S_2$ be a function. Suppose that $\mathcal{C} \subseteq \Sigma_2$ and $\sigma(\mathcal{C}) = \Sigma_2$. Show that if $h^{-1}(A) \in \Sigma_1$ for every $A \in \mathcal{C}$, then h is Σ_1/Σ_2-measurable.

5.5. Let (S, Σ) be a measurable space, and let $h : S \to \mathbf{R}$ be a Σ-measurable function. Let $\mathcal{C} \subseteq \mathcal{B}$. Show that

$$h^{-1}(\sigma(\mathcal{C})) = \sigma(h^{-1}(\mathcal{C})),$$

where

$$\begin{aligned} h^{-1}(\sigma(\mathcal{C})) &= \{h^{-1}(A) : A \in \sigma(\mathcal{C})\}, \\ h^{-1}(\mathcal{C}) &= \{h^{-1}(A) : A \in \mathcal{C}\}. \end{aligned}$$

5.6. Let S be a topological space. If $h : S \to \mathbf{R}$ is continuous (i.e., $h^{-1}(O)$ is open for any open set $O \in \mathcal{B}$), then h is a Borel function.

5.7. Let (S, Σ) be a measurable space. Show that $h : S \to \mathbf{R}$ is Σ-measurable if

$$\{h \leq c\} \in \Sigma, \quad \forall c \in \mathbf{R},$$

where $\{h \leq c\} = \{s \in S : h(s) \leq c\}$.

5.8. Let (S, Σ) be a measurable space. If $h \in m\Sigma$ and $f \in m\mathcal{B}$, then $f \circ h \in m\Sigma$, where $f \circ h : S \to \mathbf{R}$ is defined as

$$(f \circ h)(s) = f(h(s)), \quad \forall s \in S.$$

5.9. Let (S, Σ) be a measurable space and $\{h_n\}_{n \in \mathbf{N}}$ be a sequence of Σ-measurable functions. Show that

(a) $\inf_{n \geq m} h_n$ is Σ-measurable for all $m \in \mathbf{N}$.

(b) $\sup_{n \geq m} h_n$ is Σ-measurable for all $m \in \mathbf{N}$.

(c) $\liminf h_n$ is Σ-measurable.

(d) $\limsup h_n$ is Σ-measurable.

(e) $\{s : \lim_{n \to \infty} h_n(s) \text{ exists in } \mathbf{R}\} \in \Sigma$.

5.10. Let (S, Σ) be a measurable space. Assume that $h_1 \in m\Sigma$ and $h_2 \in m\Sigma$. Show that $\max(h_1, h_2) \in m\Sigma$.

5.11. Let (S, Σ) be a measure space and f a positive Σ-measurable function. For $n \in \mathbf{N}$, let $\alpha_n : [0, \infty] \to [0, \infty]$ be defined as

$$\alpha_n(x) = \begin{cases} 0, & \text{if } x = 0; \\ (i-1)2^{-n}, & \text{if } (i-1)2^{-n} < x \leq i2^{-n}, i = 1, 2, \ldots, n2^n; \\ n, & \text{if } n < x. \end{cases} \tag{5.1}$$

Show that

$$f_n \uparrow f,$$

where $f_n = \alpha_n \circ f$, i.e., $f_n(s) = \alpha_n(f(s))$ for $s \in S$.

5.12. Let (S, Σ) be a measurable space and $\mathcal{M}$ be a monotone class of functions on S. Let $\mathcal{I}$ be a π-system such that $\Sigma = \sigma(\mathcal{I})$. Suppose that $I_A \in \mathcal{M}$ for every $A \in \mathcal{I}$, where I_A is the indicator function of A. Show that $\mathcal{M}$ includes all positive Σ-measurable functions and all bounded Σ-measurable functions.

5.3 Hints

5.1. Use the definition of h^{-1} (Definition 5.1) and the standard technique to prove that two sets are equal. In other words, we can prove $A = B$ by establishing $A \subseteq B$ and $B \subseteq A$.

5.2. Use the definition of Σ-measurable functions (see Definition 5.1).

5.3. A function h is Σ-measurable if and only if $h^{-1}(A) \in \Sigma$ for all $A \in \mathcal{B}$ (see Definition 5.1). Hence we can show that $\mathcal{A} = \left\{A \in \mathcal{B} : h^{-1}(A) \in \Sigma\right\}$ is a σ-algebra in order to establish $\mathcal{A} = \mathcal{B}$.

5.4. Use the same idea used to prove Problem 5.3.

5.5. First show that $h^{-1}(\sigma(\mathcal{C}))$ is a σ-algebra and then use the standard technique to show that $h^{-1}(\sigma(\mathcal{C})) = \sigma(h^{-1}(\mathcal{C}))$; that is, show that $h^{-1}(\sigma(\mathcal{C})) \supseteq \sigma(h^{-1}(\mathcal{C}))$ and $h^{-1}(\sigma(\mathcal{C})) \subseteq \sigma(h^{-1}(\mathcal{C}))$.

5.6. Compare this problem with Problem 5.3 and note that $\mathcal{B} = \sigma(\mathcal{C})$ (see Definition 2.6), where $\mathcal{C}$ is the set of all open sets.

5.7. Compare this problem with Problem 5.3 and note that $\mathcal{B} = \sigma(\mathcal{C})$ (see Problem 2.4), where $\mathcal{C} = \{(-\infty, c] : c \in \mathbf{R}\}$ is the set of all open sets.

5.8. Use the result of Problem 5.7.

5.9. Use the result of Problem 5.7 to prove parts (a) and (b). Parts (c)-(e) follow from parts (a) and (b).

5.10. Use the result of Problem 5.8 and Theorem 5.1.

5.11. Use the ϵ technique to show that $\liminf f_n \geq f$; that is, show that $\liminf f_n \geq f - \epsilon$ for an arbitrary $\epsilon > 0$.

5.12. Use Dynkin's lemma (see Problem 2.7) and the result of Problem 5.11.

5.4 Solutions

5.1.

(a) First, let $x \in h^{-1}\left(\bigcup_{i=1}^{\infty} A_i\right)$. By definition, we have

$$h^{-1}\left(\bigcup_{i=1}^{\infty} A_i\right) = \left\{s \in S : h(s) \in \bigcup_{i=1}^{\infty} A_i\right\}.$$

Thus we have $h(x) \in \bigcup_{i=1}^{\infty} A_i$, which implies that $h(x) \in A_{i_x}$ for some $i_x \in \mathbf{N}$. Therefore

$$x \in h^{-1}(A_{i_x}) \subseteq \bigcup_{i=1}^{\infty} h^{-1}(A_i).$$

Since x is arbitrary, we have

$$h^{-1}\left(\bigcup_{i=1}^{\infty} A_i\right) \subseteq \bigcup_{i=1}^{\infty} h^{-1}(A_i). \tag{5.2}$$

Next, let $x \in \bigcup_{i=1}^{\infty} h^{-1}(A_i)$, then $x \in h^{-1}(A_{i_x})$ for some $i_x \in \mathbf{N}$. By definition, we have

$$h(x) \in A_{i_x} \subseteq \bigcup_{i=1}^{\infty} A_i,$$

which gives $x \in h^{-1}\left(\bigcup_{i=1}^{\infty} A_i\right)$. Since x is arbitrary, we have

$$\bigcup_{i=1}^{\infty} h^{-1}(A_i) \subseteq h^{-1}\left(\bigcup_{i=1}^{\infty} A_i\right). \tag{5.3}$$

Combining (5.2) and (5.3) gives

$$h^{-1}\left(\bigcup_{i=1}^{\infty} A_i\right) = \bigcup_{i=1}^{\infty} h^{-1}(A_i).$$

(b) Since $x \in h^{-1}(A^c)$ is equivalent to $h(x) \in A^c$, which is equivalent to $h(x) \notin A$, and $h(x) \notin A$ is equivalent to $x \notin h^{-1}(A)$, which is equivalent to $x \in (h^{-1}(A))^c$, we have $h^{-1}(A^c) = (h^{-1}(A))^c$.

5.2. First, we prove necessity. Suppose that $I_A \in m\Sigma$. Then, by definition of Σ-measurable functions, we have $A = I_A^{-1}(1) \in \Sigma$.

Next, we prove sufficiency. Suppose that $A \in \Sigma$. Let $B \in \mathcal{B}$. Then we have

$$I_A^{-1}(B) = \begin{cases} A, & \text{if } 1 \in B \text{ and } 0 \notin B; \\ A^c, & \text{if } 1 \notin B \text{ and } 0 \in B; \\ S, & \text{if } 1 \in B \text{ and } 0 \in B; \\ \emptyset, & \text{if } 1 \notin B \text{ and } 0 \notin B. \end{cases}$$

Thus we have $I_A^{-1}(B) \in \Sigma$. Hence I_A is Σ-measurable.

5.3. Let $\mathcal{A}$ be the collection of elements A of $\mathcal{B}$ such that $h^{-1}(A) \in \Sigma$:

$$\mathcal{A} = \left\{A \in \mathcal{B} : h^{-1}(A) \in \Sigma\right\}.$$

We claim that $\mathcal{A}$ is a σ-algebra on $\mathbf{R}$. Since $h^{-1}(\mathbf{R}) = \{s \in S : h(s) \in \mathbf{R}\} = S \in \Sigma$, we have $\mathbf{R} \in \mathcal{A}$. Let $A \in \mathcal{A}$, then $h^{-1}(A) \in \Sigma$. From Problem 5.1, we have $h^{-1}(A^c) = (h^{-1}(A))^c$. Note that Σ is a σ-algebra on S, so we have $h^{-1}(A^c) \in \Sigma$ which implies $A^c \in \mathcal{A}$. Let $A_i \in \mathcal{A}$ $(i \in \mathbf{N})$, then we have $h^{-1}(A_i) \in \Sigma$ for all $i \in \mathbf{N}$. From Problem 5.1, we have

$$h^{-1}\left(\bigcup_{i=1}^{\infty} A_i\right) = \bigcup_{i=1}^{\infty} h^{-1}(A_i) \in \Sigma,$$

which gives

$$\bigcup_{i=1}^{\infty} A_i \in \mathcal{A}.$$

Therefore, $\mathcal{A}$ is a σ-algebra on $\mathbf{R}$. From the definition of $\mathcal{A}$, we have $\mathcal{A} \subseteq \mathcal{B}$. Since $\mathcal{B} = \sigma(\mathcal{C})$ and $\mathcal{C} \subseteq \mathcal{A}$ (by hypothesis), we have $\mathcal{B} \subseteq \mathcal{A}$. Thus we have $\mathcal{A} = \mathcal{B}$, which implies that h is Σ-measurable.

5.4. This proof is similar to that of Problem 5.3. Let $\mathcal{A}$ be defined as

$$\mathcal{A} = \left\{A \in \Sigma_2 : h^{-1}(A) \in \Sigma_1\right\}.$$

We show that $\mathcal{A}$ is a σ-algebra on S_2. Since $h^{-1}(S_2) = \{x \in S_1 : h(x) \in S_2\} = S_1 \in \Sigma_1$, we have $S_2 \in \mathcal{A}$. Let $A \in \mathcal{A}$, then $h^{-1}(A) \in \Sigma_1$. From Problem 5.1, we have $h^{-1}(A^c) = (h^{-1}(A))^c$, and so $h^{-1}(A^c) \in \Sigma_1$. Hence $A^c \in \mathcal{A}$. Let $A_i \in \mathcal{A}$ $(i \in \mathbf{N})$; then we have $h^{-1}(A_i) \in \Sigma_1$ for every $i \in \mathbf{N}$. From Problem 5.1, we have

$$h^{-1}\left(\bigcup_{i=1}^{\infty} A_i\right) = \bigcup_{i=1}^{\infty} h^{-1}(A_i) \in \Sigma_1,$$

which gives

$$\bigcup_{i=1}^{\infty} A_i \in \mathcal{A}.$$

Therefore, $\mathcal{A}$ is a σ-algebra on S_2. From the definition of $\mathcal{A}$, we have $\mathcal{A} \subseteq \Sigma_2$. Since $\Sigma_2 = \sigma(\mathcal{C})$ and $\mathcal{C} \subseteq \mathcal{A}$ (by hypothesis), we have $\Sigma_2 \subseteq \mathcal{A}$. Thus we have $\mathcal{A} = \Sigma_2$, which implies that h is Σ_1/Σ_2-measurable.

5.5. First, we show that $h^{-1}(\sigma(\mathcal{C}))$ is a σ-algebra on S. In fact, since $\sigma(\mathcal{C})$ is a σ-algebra on $\mathbf{R}$, we have $S = h^{-1}(\mathbf{R}) \in h^{-1}(\sigma(\mathcal{C}))$. Let $F \in h^{-1}(\sigma(\mathcal{C}))$; then we have $F = h^{-1}(A)$ for some $A \in \sigma(\mathcal{C})$. From Problem 5.1, we have $F^c =$

$(h^{-1}(A))^c = h^{-1}(A^c) \in h^{-1}(\sigma(\mathcal{C}))$. Let $F_n \in h^{-1}(\sigma(\mathcal{C}))$ $(n \in \mathbf{N})$; then $F_n = h^{-1}(A_n)$ for some $A_n \in \sigma(\mathcal{C})$. Again from Problem 5.1, we have

$$\bigcup_{n=1}^{\infty} F_n = \bigcup_{n=1}^{\infty} h^{-1}(A_n) = h^{-1}\left(\bigcup_{n=1}^{\infty} A_n\right) \in h^{-1}(\sigma(\mathcal{C})).$$

Thus, $h^{-1}(\sigma(\mathcal{C}))$ is a σ-algebra on S. Note that as $h^{-1}(\mathcal{C}) \subseteq h^{-1}(\sigma(\mathcal{C}))$, we have

$$\sigma(h^{-1}(\mathcal{C})) \subseteq h^{-1}(\sigma(\mathcal{C})). \tag{5.4}$$

Next, we show that

$$h^{-1}(\sigma(\mathcal{C})) \subseteq \sigma(h^{-1}(\mathcal{C})). \tag{5.5}$$

To do this, we define $\mathcal{A}$ as

$$\mathcal{A} = \{A \in \sigma(\mathcal{C}) : h^{-1}(A) \in \sigma(h^{-1}(\mathcal{C}))\}.$$

We claim that $\mathcal{A}$ is a σ-algebra on $\mathbf{R}$. In fact, since $\mathbf{R} \in \sigma(\mathcal{C})$ and $h^{-1}(\mathbf{R}) = S \in \sigma(h^{-1}(\mathcal{C}))$, we have $\mathbf{R} \in \mathcal{A}$. Let $A \in \mathcal{A}$, then $A \in \sigma(\mathcal{C})$ and $h^{-1}(A) \in \sigma(h^{-1}(\mathcal{C}))$. From Problem 5.1, we have $h^{-1}(A^c) = (h^{-1}(A))^c \in \sigma(h^{-1}(\mathcal{C}))$. Thus, $A^c \in \mathcal{A}$. Let $A_n \in \mathcal{A}$ $(n \in \mathbf{N})$, then $A_n \in \sigma(\mathcal{C})$ and $h^{-1}(A_n) \in \sigma(h^{-1}(\mathcal{C}))$. From Problem 5.1, we have

$$h^{-1}\left(\bigcup_{n=1}^{\infty} A_n\right) = \bigcup_{n=1}^{\infty} h^{-1}(A_n) \in \sigma(h^{-1}(\mathcal{C})),$$

which implies $\bigcup_{n=1}^{\infty} A_n \in \mathcal{A}$. Therefore, $\mathcal{A}$ is a σ-algebra on $\mathbf{R}$. Note that with $\mathcal{C} \subseteq \mathcal{A}$, we have $\sigma(\mathcal{C}) \subseteq \mathcal{A}$. By definition of $\mathcal{A}$, we have $\mathcal{A} \subseteq \sigma(\mathcal{C})$. Therefore, $\mathcal{A} = \sigma(\mathcal{C})$, which implies (5.5).

Combining (5.4) and (5.5) gives the result.

5.6. Let $\mathcal{C} = \{O \subseteq \mathbf{R} : O \text{ is open}\}$, then $\mathcal{B} = \sigma(\mathcal{C})$. By hypothesis, we have $h^{-1}(O) \in \mathcal{B}(S), \forall O \in \mathcal{C}$. From Problem 5.3, we know that h is $\mathcal{B}(S)$-measurable; that is, h is Borel (see Definition 5.2).

5.7. Let $\mathcal{C} = \{(-\infty, c] : c \in \mathbf{R}\}$, then we have $\mathcal{B} = \sigma(\mathcal{C})$ (see Problem 2.4). By hypothesis, $h^{-1}(A) \in \Sigma$, $\forall A \in \mathcal{C}$. From Problem 5.3, we know that h is Σ-measurable. This completes the proof.

5.8. To prove a function $f \circ h \in m\Sigma$, we only need to show that $\{f \circ h > c\} \in \Sigma$, $\forall c \in \mathbf{R}$, where $\{f \circ h > c\} = \{s \in S : f(h(s)) > c\}$ (see Problem 5.7). To do this, let

$$A_c = \{a \in \mathbf{R} : f(a) > c\} = f^{-1}((c, \infty)), \quad \forall c \in \mathbf{R}.$$

Since $f \in m\mathcal{B}$, we have $A_c \in \mathcal{B}$, $\forall c \in \mathbf{R}$. Note that

$$\{f \circ h > c\} = \{s \in S : h(s) \in A_c\} = h^{-1}(A_c), \quad \forall c \in \mathbf{R}$$

and $h \in m\Sigma$, we have $\{f \circ h > c\} \in \Sigma$, which implies that $f \circ h$ is Σ-measurable, that is, $f \circ h \in m\Sigma$.

5.9. To prove a function $h \in m\Sigma$, we only need to show that $\{h > c\} \in \Sigma$, $\forall c \in \mathbf{R}$, where $\{h > c\} = \{s \in S : h(s) > c\}$ (see Problem 5.7).

(a) Note that

$$\left\{\inf_{n\geq m} h_n > c\right\} = \bigcap_{n\geq m} \{h_n > c\}, \quad \forall m \in \mathbf{N},$$

which implies $\inf_{n\geq m} h_n$ is Σ-measurable for all $m \in \mathbf{N}$.

(b) From

$$\left\{\sup_{n\geq m} h_n \leq c\right\} = \bigcap_{n\geq m} \{h_n \leq c\}, \quad \forall m \in \mathbf{N},$$

we know that $\sup_{n\geq m} h_n$ is Σ-measurable for all $m \in \mathbf{N}$.

(c) Since

$$\liminf h_n = \sup_{m\geq 1} \inf_{n\geq m} h_n$$

and $\inf_{n\geq m} h_n$ is Σ-measurable for every $m \in \mathbf{N}$, we have $\liminf h_n \in m\Sigma$.

(d) Similarly, note that

$$\limsup h_n = \inf_{m\geq 1} \sup_{n\geq m} h_n,$$

and $\sup_{n\geq m} h_n$ is Σ-measurable for every $m \in \mathbf{N}$, we have $\limsup h_n \in m\Sigma$.

(e) Note that $\lim_{n\to\infty} h_n(s)$ exists in $\mathbf{R}$ if and only if

$$\limsup h_n(s) = \liminf h_n(s).$$

Thus we have

$$\begin{aligned} & \left\{s : \lim_{n\to\infty} h_n(s) \text{ exists in } \mathbf{R}\right\} \\ = \; & \{\limsup h_n < \infty\} \cap \{\liminf h_n > -\infty\} \cap f^{-1}(\{0\}), \end{aligned}$$

where $f = \limsup h_n - \liminf h_n$. Since $\limsup h_n$ and $\liminf h_n$ are Σ-measurable, we have

$$\left\{s : \lim_{n\to\infty} h_n(s) \text{ exists in } \mathbf{R}\right\} \in \Sigma.$$

5.10. By Theorem 5.1, we know that $h_1 - h_2$ and $h_2 - h_1$ are Σ-measurable functions. Let $f : \mathbf{R} \to \mathbf{R}$ be defined as

$$f(x) = \begin{cases} x & \text{if } x > 0; \\ 0 & \text{if } x \leq 0. \end{cases}$$

Then we have

$$\max(h_1, h_2) = f(h_1 - h_2) + h_2.$$

By Problem 5.8, we only need to show that f is $\mathcal{B}$-measurable in order to show that $\max(h_1, h_2)$ is Σ-measurable. Let $A \in \mathcal{B}$. Then we have

$$f^{-1}(A) = \begin{cases} A \cap (0, \infty) & \text{if } 0 \notin A; \\ A \cup (-\infty, 0] & \text{if } 0 \in A, \end{cases}$$

which implies that $f^{-1}(A) \in \mathcal{B}$. Hence $f \in m\mathcal{B}$. This completes the proof.

5.11. Clearly, f_n $(n \geq 1)$ are nonnegative simple functions and $f_n \leq f$ for all $n \geq 1$.

To show that $f_n \uparrow f$, we first show that

$$f_n \leq f_{n+1}, \quad n \geq 1. \tag{5.6}$$

To do this, let $x \in S$. If $f(x) = 0$, then $f_n(x) = f_{n+1}(x) = 0$. If $(i-1)2^{-n} < f(x) \leq i2^{-n}$ for some $1 \leq i \leq n2^n$, then $f_n(x) = (i-1)2^{-n}$. But $(2i-2)2^{-(n+1)} < f(x) \leq (2i)2^{-(n+1)}$, we have $f_{n+1}(x) \geq (2i-2)2^{-(n+1)} = f_n(x)$. If $f(x) > n$, then $f_n(x) = n$. But $f_{n+1}(x) > n$. Therefore, inequality (5.6) is true.

Next, we show that

$$\lim_{n\to\infty} f_n = f. \tag{5.7}$$

Since $\limsup f_n \leq f$, we only need to prove $\liminf f_n \geq f$. To do this, let $\epsilon > 0$. For fixed $x \in S$, choose $N_\epsilon \in \mathbf{N}$ such that $N_\epsilon \geq \max[f(x), \log_2(1/\epsilon)]$. Then

$$\liminf f_n \geq \inf_{n\geq N_\epsilon} f_n(x) \geq f_{N_\epsilon}(x) > f(x) - 2^{-N_\epsilon} \geq f(x) - \epsilon.$$

Since this is true for every $\epsilon > 0$, we have $\liminf f_n \geq f$. On combining Equation (5.6) and Equation (5.7), we get $f_n \uparrow f$.

5.12. First, we show that $I_A \in \mathcal{M}$ for every $A \in \Sigma$. To do that, let

$$\mathcal{D} = \{A \in \Sigma : I_A \in \mathcal{M}\}.$$

It follows from the definition of monotone classes that $\mathcal{D}$ is a d-system. From the assumption, we have $\mathcal{I} \subseteq \mathcal{D}$. By Problem 2.7, $\Sigma = \sigma(\mathcal{I}) \subseteq \mathcal{D}$. Hence $I_A \in \mathcal{M}$ for every $A \in \Sigma$. Since $\mathcal{M}$ is a monotone class, $\mathcal{M}$ contains all the simple functions on S.

Next, let f be a positive Σ-measurable function on S. Then by Problem 5.11, there exists a sequence $\{f_n : n \geq 1\}$ of positive simple functions such that $f_n \uparrow f$. Since $f_n \in \mathcal{M}$, we have $f \in \mathcal{M}$.

Finally, let f be a bounded Σ-measurable function on S. Then $f^+ = \max(f, 0) \in \mathcal{M}$ and $f^- = \max(-f, 0) \in \mathcal{M}$. Hence we have $f = f^+ - f^- \in \mathcal{M}$.

This completes the proof.

5.5 Bibliographic Notes

Most material in this chapter was taken from Williams (1991), where readers can find proofs of Theorem 5.1. Proving Theorem 5.2 requires the following basic result of point-set topology: If f is continuous, then $f^{-1}(O)$ is an open subset of $\mathbf{R}$ whenever O is an open subset of $\mathbf{R}$ (Rosenthal, 2006, p31).

In measure theory, the measurability of a function $f : (S_1, \Sigma_1) \rightarrow (S_2, \Sigma_2)$ is defined in terms of the inverse image of f rather than the direct image of f. Hence, the function f is Σ_1/Σ_2-measurable does not imply than $f(A) \in \Sigma_2$ for each $A \in \Sigma_1$. In fact, basic set operations are preserved by inverse images but not in general by direct images (Ash and Doleans-Dade, 1999, p36). For example, we have $f^{-1}(A^c) = [f^{-1}(A)]^c$ but not $f(A^c) = [f(A)]^c$ in general.

CHAPTER 6

LEBESGUE INTEGRATION

Lebesgue integration is the general theory of integration of a function with respect to a general measure. In this chapter, we present some concepts and theorems used to develop the Lebesgue integration theory.

6.1 Basic Concepts and Facts

Definition 6.1 (Simple Function)**.** Let (S, Σ) be a measurable space and $(m\Sigma)^+$ denote the set of all nonnegative Σ-measurable functions. An element f of $(m\Sigma)^+$ is considered simple if f can be written as a finite sum

$$f = \sum_{k=1}^{m} a_k I_{A_k}, \tag{6.1}$$

where $a_k \in [0, \infty]$ and $A_k \in \Sigma$ for $k = 1, 2, \ldots, m$. The set of all nonnegative simple functions is denoted by SF^+.

Definition 6.2 (Integral of Nonnegative Simple Functions). Let (S, Σ, μ) be a measure space, and let

$$f = \sum_{k=1}^{m} a_k I_{A_k}$$

be a nonnegative simple function on S. The integral of f with respect to μ is defined as

$$\mu_0(f) = \sum_{k=1}^{m} a_k \mu(A_k). \tag{6.2}$$

Definition 6.3 (Integral of Nonnegative Measurable Functions). Let (S, Σ, μ) be a measure space and $(m\Sigma)^+$ be the set of all nonnegative Σ-measurable functions on S. Let $f \in (m\Sigma)^+$, then the Lebesgue integral of f with respect to μ is defined as

$$\mu(f) = \sup\{\mu_0(h) : h \in SF^+,\ h \leq f\}.$$

The integral $\mu(f)$ is also written as $\int_S f(x)\mu(\mathrm{d}x)$ or $\int f \mathrm{d}\mu$.

Definition 6.4 (Integral of General Measurable Functions). Let (S, Σ, μ) be a measure space and $f \in m\Sigma$, where $m\Sigma$ is the set of all measurable functions on S. The function f is called *μ-integrable* , denoted $f \in L^1(S, \Sigma, \mu)$, if

$$\mu(|f|) = \mu(f^+) + \mu(f^-) < \infty,$$

where $f^+ = \max(f, 0)$ and $f^- = \max(-f, 0)$. If f is μ-integrable, then the integral of f is defined as

$$\int f \mathrm{d}\mu = \int_S f \mathrm{d}\mu = \mu(f) = \mu(f^+) - \mu(f^-).$$

Definition 6.5 (Integral over Subsets). Let (S, Σ, μ) be a measure space and $f \in m\Sigma$. Let $A \in \Sigma$. The integral of f over A is defined as

$$\int_A f \mathrm{d}\mu = \mu(f; A) = \mu(f I_A).$$

Theorem 6.1 (Integration by Substitution). *Let (S_1, Σ_1) and (S_2, Σ_2) be two measurable spaces and $T : S_1 \to S_2$ be a Σ_1/Σ_2-measurable function. Let μ_1 be a measure on Σ_1, and let $\mu_2 = \mu_1 T^{-1}$ be a measure on Σ_2 defined as*

$$\mu_2(A) = \mu_1(T^{-1}(A)), \quad A \in \Sigma_2.$$

If $f : S_2 \to \bar{\mathbf{R}}$ is $\Sigma_2/\mathcal{B}(\bar{\mathbf{R}})$-measurable and $A \in \Sigma_2$, then

$$\int_{T^{-1}A} f(T(s))\mu_1(\mathrm{d}s) = \int_A f(t)\mu_2(\mathrm{d}t),$$

in the sense that if one of the integrals exists (finite or infinite), then the other also exists and the two integrals are equal.

Theorem 6.2 (Monotone Convergence Theorem). *Let (S, Σ, μ) be a measure space, and let $\{f_n\}_{n\geq 1}$ be a sequence of elements of $(m\Sigma)^+$ such that $f_n \uparrow f$. Then*

$$\mu(f_n) \uparrow \mu(f).$$

Theorem 6.3 (Fatou's Lemma). *Let (S, Σ, μ) be a measure space and $f_n \in (m\Sigma)^+$ for all $n \geq 1$. Then*

$$\mu(\liminf f_n) \leq \liminf \mu(f_n).$$

Theorem 6.4 (Dominated Convergence Theorem). *Let (S, Σ, μ) be a measure space. Suppose that $f \in m\Sigma$ and $f_n \in m\Sigma$ for $n \geq 1$, that the sequence $\{f_n\}_{n\geq 1}$ converges pointwise to f, that is,*

$$\lim_{n\to\infty} f_n(s) = f(s), \quad \forall s \in S,$$

and that the sequence $\{f_n\}_{n\geq 1}$ is dominated by some nonnegative μ-integrable function g:

$$|f_n(s)| \leq g(s), \quad s \in S,\ n \geq 1.$$

Then f is μ-integrable, that is,

$$\lim_{n\to\infty} \mu(|f_n - f|) = 0,$$

and

$$\lim_{n\to\infty} \mu(f_n) = \mu(f).$$

6.2 Problems

6.1. Let (S, Σ, μ) be a measure space and

$$f = \sum_{i=1}^{m} a_i I_{A_i}$$

be a nonzero simple function on S, where $a_i \in [0, \infty]$ and $A_i \in \Sigma$. Show that there exists a finite collection of mutually disjoint sets C_H $(H \in \mathcal{H})$ and $u_H > 0$ such that

$$\sum_{i=1}^{m} a_i I_{A_i} = \sum_{H\in\mathcal{H}} u_H I_{C_H}, \quad \sum_{i=1}^{m} a_i \mu(A_i) = \sum_{H\in\mathcal{H}} u_H \mu(C_H).$$

6.2. Show that $\mu_0(f)$ in Equation (6.2) is well defined.

6.3. Let (S, Σ, μ) be a measure space and SF^+ be the set of nonnegative simple functions on S. Let μ_0 be as defined in Equation (6.2). Show that

(a) If $f, g \in SF^+$ and $\mu(f \neq g) = 0$, then $\mu_0(f) = \mu_0(g)$.

(b) (Linearity) If $f, g \in SF^+$ and $c \geq 0$, then $f + g, cf \in SF^+$ and

$$\mu_0(f + g) = \mu_0(f) + \mu_0(g), \quad \mu_0(cf) = c\mu_0(f);$$

(c) (Monotonicity) If $f, g \in SF^+$ and $f \leq g$, then $\mu_0(f) \leq \mu_0(g)$.

(d) If $f, g \in SF^+$, then $f \wedge g \in SF^+$ and $f \vee g \in SF^+$, where $f \wedge g = \min(f, g)$ and $f \vee g = \max(f, g)$.

6.4. Let (S, Σ, μ) be a measure space. Let $A \in \Sigma$ and $\{h_n\}_{n\geq 1}$ be a sequence of nonnegative simple functions such that $h_n \uparrow I_A$. Show that

$$\mu_0(h_n) \uparrow \mu(A),$$

where μ_0 is as defined in Equation (6.2).

6.5. Let (S, Σ, μ) be a measure space, and let $f \in SF^+$. Suppose that $\{f_n\}_{n\geq 1}$ is a sequence of elements of SF^+ such that $f_n \uparrow f$. Show that

$$\mu_0(f_n) \uparrow \mu_0(f), \tag{6.3}$$

where μ_0 is as defined in Equation (6.2).

6.6. Let (S, Σ, μ) be a measure space and $f \in (m\Sigma)^+$, where f is a nonnegative Σ-measurable function on S. Suppose that $\{f_n\}_{n\geq 1}$ and $\{g_m\}_{m\geq 1}$ are two sequences of elements of SF^+ such that

$$f_n \uparrow f, \quad g_m \uparrow f.$$

Show that

$$\lim_{n\to\infty} \mu_0(f_n) = \lim_{m\to\infty} \mu_0(g_m),$$

where μ_0 is as defined in Equation (6.2).

6.7. Let (S, Σ, μ) be a measure space, and $f \in (m\Sigma)^+$. Let $\{y_n\}_{n\geq 1}$ be any sequence of elements of SF^+ such that $y_n \uparrow f$. Show that

$$\mu(y_n) \uparrow \mu(f).$$

6.8. Let (S, Σ, μ) be a measure space, and let $f \in (m\Sigma)^+$ and $\mu(f) = 0$. Show that

$$\mu(\{f > 0\}) = 0.$$

6.9. Let (S, Σ, μ) be a measure space and let $f, g \in (m\Sigma)^+$ such that $\mu(f \neq g) = 0$. Show that $\mu(f) = \mu(g)$.

6.10. Let (S, Σ, μ) be a measure space and $A \in \Sigma$. Let $f \in m\Sigma$. Show that if $\mu(A) = 0$, then

$$\int_A f \mathrm{d}\mu = 0.$$

6.11. Let (S, Σ, μ) be a measure space and $f \in (m\Sigma)^+$. Suppose that $\{f_n\}_{n\geq 1}$ is a sequence of elements of $(m\Sigma)^+$ such that, except on a μ-null set A, $f_n \uparrow f$. Show that $\mu(f_n) \uparrow \mu(f)$.

6.12. Let (S, Σ, μ) be a measure space and $f, g \in (m\Sigma)^+$. Let $\alpha, \beta > 0$. Show that

$$\mu(\alpha f + \beta g) = \alpha\mu(f) + \beta\mu(g).$$

6.13 (Reverse Fatou's Lemma)**.** Let (S, Σ, μ) be a measure space and $\{f_n\}_{n\geq 1}$ be a sequence of elements of $(m\Sigma)^+$ such that $f_n \leq g$ for all $n \geq 1$, where $g \in (m\Sigma)^+$ and $\mu(g) < \infty$. Show that

$$\mu(\limsup f_n) \geq \limsup \mu(f_n).$$

6.14. Let (S, Σ, μ) be a measure space and h be a Σ-measurable function. Show that

(a) If h is μ-integrable, then hI_A is also μ-integrable for each $A \in \Sigma$, where I_A is the indicator function.

(b) If $\int_S h\mathrm{d}\mu$ is finite, then $\int_A h\mathrm{d}\mu$ is also finite for each $A \in \Sigma$.

6.15. Let (S, Σ, μ) be a measure space and h a Σ-measurable function. Show that if h is μ-integrable, then $\mu(\{|h| = \infty\}) = 0$, that is, h is finite a.e.

6.16 (Chebyshev's Inequality)**.** Let (S, Σ, μ) be a measure space and f be a nonnegative, extended-value, Borel measuralbe function on S. Let $p \in (0, \infty)$ and $\epsilon \in (0, \infty)$. Show that

$$\mu\{s : f(s) \geq \epsilon\} \leq \frac{1}{\epsilon^p} \int_S f^p \mathrm{d}\mu.$$

6.17. Let (S, Σ, μ) be a measure space and $f, g \in L^1(S, \Sigma, \mu)$, where $L^1(S, \Sigma, \mu)$ is the set of all μ-integrable functions on S. Let $\alpha, \beta \in \mathbf{R}$. Show that $\alpha f + \beta g \in L^1(S, \Sigma, \mu)$ and

$$\mu(\alpha f + \beta g) = \alpha\mu(f) + \beta\mu(g).$$

6.18 (Scheffé's Lemma A)**.** Let (S, Σ, μ) be a measure space. Suppose that $f_n, f \in L^1(S, \Sigma, \mu)^+$,, that is, f_n, f are nonnegative μ-integrable functions on S, and that $f_n \to f$ a.e. Show that $\mu(|f_n - f|) \to 0$ if and only if $\mu(f_n) \to \mu(f)$.

6.19 (Scheffé's Lemma B)**.** Let (S, Σ, μ) be a measure space. Suppose that $f_n, f \in L^1(S, \Sigma, \mu)$, and that $f_n \to f$ $(a.e.)$. Show that $\mu(|f_n - f|) \to 0$ if and only if $\mu(|f_n|) \to \mu(|f|)$.

6.20. Let (S, Σ, μ) be a measure space and $h : \mathbf{R} \to \mathbf{R}$ be a Borel measurable function. Show that $h \circ f \in L^1(S, \Sigma, \mu)$ if and only if $h \in L^1(\mathbf{R}, \mathcal{B}, \Lambda_f)$, where Λ_f is a measure on $(\mathbf{R}, \mathcal{B})$ and is defined as $\Lambda_f(B) = \mu(f \in B) = \mu(\{s \in S : f(s) \in B\})$, and that

$$\mu(h \circ f) = \int_S h(f(s))\mu(\mathrm{d}s) = \int_{\mathbf{R}} h(x)\Lambda_f(\mathrm{d}x) = \Lambda_f(h).$$

6.21. Let (S, Σ, μ) be a measure space and $\{h_n\}_{n\geq 1}$ be a sequence of elements of $(m\Sigma)^+$. Show that

$$\int_S \left(\sum_{n=1}^{\infty} h_n \right) \mathrm{d}\mu = \sum_{n=1}^{\infty} \int_S h_n \mathrm{d}\mu.$$

6.22 (Extended Monotone Convergence Theorem)**.** Let (S, Σ, μ) be a measure space, and let $g, h, g_1, g_2, \ldots$ be Σ-measurable functions. Show that

(a) If $\int_S h\mathrm{d}\mu > -\infty$, $g_n \geq h$ for all $n \geq 1$, and $g_n \uparrow g$, then

$$\int_S g_n \mathrm{d}\mu \uparrow \int_S g\mathrm{d}\mu;$$

(b) If $\int_S h\mathrm{d}\mu < \infty$, $g_n \leq h$ for all $n \geq 1$, and $g_n \downarrow g$, then

$$\int_S g_n \mathrm{d}\mu \downarrow \int_S g\mathrm{d}\mu.$$

6.23 (Gronwall's Lemma)**.** Let f and g be two nonnegative Borel measurable functions on $[0, \infty)$. Let $T > 0$ and $a \geq 0$ be constants. Suppose that

$$\int_0^T f(s)\mathrm{d}s < \infty$$

and

$$g(t) \leq a + \int_0^t g(s)f(s)\mathrm{d}s, \quad \forall t \in [0, T].$$

Show that

$$g(t) \leq a \exp\left(\int_0^t f(s)\mathrm{d}s \right), \quad \forall t \in [0, T].$$

6.3 Hints

6.1. First find a way to construct a collection of pairwise disjoint sets from $A_1, A_2, \ldots, A_m$. The remaining proof is straightforward.

6.2. Use the result of Problem 6.1.

6.3. For parts (a), (c), and (d), we need to use the result of Problem 6.1. Part (b) can be proved by using the definition of simple functions (see Definition 6.1).

6.4. According to Problem 6.3, $\{\mu_0(H_n)\}_{n\geq 1}$ is increasing. To show that

$$\lim_{n\to\infty} \mu_0(h_n) = \mu(A),$$

we can show that

$$\limsup \mu_0(h_n) \leq \mu(A)$$

and

$$\liminf \mu_0(h_n) \geq \mu(A).$$

The first equation here is straightforward. The second one can be established by showing that $\liminf \mu_0(h_n) \geq (1-\epsilon)\mu(A)$ for all $\epsilon > 0$.

6.5. Since f_n and f are nonnegative simple functions, we can write them as the linear combinations of indicator functions (see Problem 6.1). Then we can use the result of Problem 6.4 to prove this problem.

6.6. Use the ϵ technique to show that the two limits are equal; that is, we can show that $R_\infty > L_\infty - \epsilon$ and $L_\infty > R_\infty - \epsilon$, where $R_\infty = \lim_{m\to\infty} \mu_0(g_m)$ and $L_\infty = \lim_{n\to\infty} \mu_0(f_n)$.

6.7. Use the result of Problem 5.11 to construct an increasing sequence $\{f_n\}_{n\geq 1}$ of nonnegative simple functions such that $f_n \uparrow f$ and $\mu_0(f) \uparrow \mu(f)$. Then the result follows from Problem 6.6.

6.8. Consider the measure of sets $A_n = \{f > 1/n\}$ for $n \geq 1$.

6.9. Construct two sequences $\{f_n\}_{n\geq 1}$ and $\{g_n\}_{n\geq 1}$ of increasing nonnegative simplex functions such that $\mu(f_n \neq g_n) = 0$ for all $n \geq 1$, $f_n \uparrow f$, and $g_n \uparrow g$ (see Problem 5.11).

6.10. Consider the sequence $f_n = \alpha_n \circ f^+$ and $g_n = \alpha_n \circ f^-$, where α_n is defined in Problem 5.11. Then use the monotone convergence theorem (Theorem 6.2).

6.11. Consider functions $f_n I_{S\setminus A}$ and $f I_{S\setminus A}$ and use the result of Problem 6.9 and Theorem 6.2.

6.12. Use the result of Problem 5.11 to choose two sequences $f_n (n \geq 1)$ and $g_n (n \geq 1)$ of nonnegative simple functions on S such that $f_n \uparrow f$ and $g_n \uparrow g$ and use the result of Problem 6.3.

6.13. Consider functions $g - f_n$ and $g - f$ and use Fatou's lemma (see Theorem 6.3).

6.14. Consider the positive and negative parts of h and use the definition of integration of general measurable functions (see Definition 6.4).

6.15. Use the method of contradictino.

6.16. Compare the integration of f^p over S and over $\{s : f(s) \geq \epsilon\}$.

6.17. Consider the positive and negative parts of the functions and the result of Problem 6.12.

6.18. Consider the positive and negative parts of the function $f_n - f$ and use the dominated convergence theorem (see Theorem 6.4).

6.19. Consider the positive and negative parts of the functions f_n and f and use Fatou's lemma (see Theorem 6.3).

6.20. Prove the problem by considering different cases of h. First show that the result holds when h is an indicator function. Then show that the result holds when h is a nonnegative simple function, a nonnegative Borel measurable function, and a general measurable function.

6.21. Consider the partial sum of the sequence of functions (i.e., $g_n = \sum_{i=1}^n h_i$), and use Theorem 6.2 and the result of Problem 6.17.

6.22. Consider the cases $\int_S h\mathrm{d}\mu = \infty$ and $\int_S h\mathrm{d}\mu < \infty$ and use Theorem 6.2.

6.23. Let

$$F(t) = \int_0^t f(s)\mathrm{d}s, \quad G(t) = \int_0^t f(s)g(s)\mathrm{d}s.$$

Then consider the function $e^{-F(t)}G(t)$.

6.4 Solutions

6.1. Let $\mathcal{H}_m = P(\{1, 2, \ldots, m\})$, where $P(\{1, 2, \ldots, m\})$ is the power set of $\{1, 2, \ldots, m\}$. We define

$$C_H = \left(\bigcap_{i \in H} A_i\right) \setminus \left(\bigcup_{i \in H^c} A_i\right), \quad u_H = \sum_{i \in H} a_i, \quad H \in \mathcal{H}_m, \quad H \in \mathcal{H}_m,$$

where $H^c = \{1, 2, \ldots, m\} \backslash H$. Then $C_H \in \Sigma\,(H \in \mathcal{H}_m)$ are mutually disjoint, and

$$A_i = \bigcup_{H \in \mathcal{H}_m, i \in H} C_H, \quad \forall 1 \leq i \leq m.$$

Define $\mathcal{H}_m^+$ as

$$\mathcal{H}_m^+ = \{H \in \mathcal{H}_m : u_H > 0\}.$$

Then we have

$$\begin{aligned}\sum_{i=1}^{m} a_i I_{A_i} &= \sum_{i=1}^{m} a_i \sum_{H\in\mathcal{H}_m, i\in H} I_{C_H}\\ &= \sum_{H\in\mathcal{H}_m} u_H I_{C_H}\\ &= \sum_{H\in\mathcal{H}_m^+} u_H I_{C_H},\end{aligned}$$

and

$$\begin{aligned}\sum_{i=1}^{m} a_i \mu(A_i) &= \sum_{i=1}^{m} a_i \sum_{H\in\mathcal{H}_m, i\in H} \mu(C_H)\\ &= \sum_{H\in\mathcal{H}_m} u_H \mu(C_H)\\ &= \sum_{H\in\mathcal{H}_m^+} u_H \mu(C_H).\end{aligned}$$

Since $|\mathcal{H}_m| = 2^m$, we can show that $\mathcal{H}_m^+$ is a finite set. We let $\mathcal{H} = \mathcal{H}_m^+$. This completes the proof.

6.2. Let $f \in SF^+$. Suppose that f has two different representations:

$$f = \sum_{i=1}^{m} a_i I_{A_i} = \sum_{j=1}^{n} b_j I_{B_j}, \tag{6.4}$$

where $a_i \geq 0$, $b_j \geq 0$, $A_i \in \Sigma$, and $B_j \in \Sigma$ for $i = 1, 2, \ldots, m$ and $j = 1, 2, \ldots, n$.

If $f = 0$, then we have $a_i I_{A_i} = 0$ and $b_j I_{B_j} = 0$ for all $1 \leq i \leq m$ and $1 \leq j \leq n$. In this case, both representations give $\mu_0(f) = 0$. If $f \neq 0$, then from Problem 6.1 we have

$$f = \sum_{H\in\mathcal{H}_m^+} u_H I_{C_H} = \sum_{J\in\mathcal{H}_n^+} v_J I_{D_J}, \tag{6.5}$$

where $C_H\,(H \in \mathcal{H}_m^+)$ are pairwise disjoint sets, $u_H > 0$, $D_J\,(J \in \mathcal{H}_n^+)$ are pairwise disjoint sets, and $v_J > 0$.

Since

$$\{f > 0\} = \bigcup_{H\in\mathcal{H}_m^+} C_H = \bigcup_{J\in\mathcal{H}_n^+} D_J,$$

we have

$$C_H = \bigcup_{J\in\mathcal{H}_n^+} (C_H \cap D_J), \quad D_J = \bigcup_{H\in\mathcal{H}_m^+} (C_H \cap D_J).$$

Then, from (6.4) and (6.5) we have

$$\sum_{i=1}^{m} a_i\mu(A_i) = \sum_{H\in\mathcal{H}_m^+} u_H\mu(C_H) = \sum_{H\in\mathcal{H}_m^+}\sum_{J\in\mathcal{H}_n^+} u_H\mu(C_H\cap D_J), \quad (6.6)$$

$$\sum_{j=1}^{n} b_j\mu(B_j) = \sum_{J\in\mathcal{H}_n^+} v_J\mu(D_J) = \sum_{H\in\mathcal{H}_m^+}\sum_{J\in\mathcal{H}_n^+} v_J\mu(C_H\cap D_J). \quad (6.7)$$

We claim that

$$u_H\mu(C_H\cap D_J) = v_J\mu(C_H\cap D_J), \quad \forall H\in\mathcal{H}_m^+,\ J\in\mathcal{H}_n^+. \quad (6.8)$$

In fact, if $C_H\cap D_J = \phi$, then both sides of (6.8) are zero. If $C_H\cap D_J \neq \phi$, we have, for $\omega\in C_H\cap D_J$,

$$f(\omega) = u_H = v_J.$$

Thus (6.8) is true.

Therefore, (6.6) and (6.7) give the same $\mu_0(f)$. Thus μ_0 is well defined.

6.3. Let

$$f = \sum_{i=1}^{m} a_i I_{A_i}, \quad (6.9)$$

$$g = \sum_{j=1}^{n} b_j I_{B_j}, \quad (6.10)$$

where $a_i, b_j\in[0,\infty]$ and $A_i, B_j\in\Sigma$. Then from Problem 6.1 we can write f, g as

$$f = \sum_{H\in\mathcal{H}_m^+} u_H I_{C_H},$$

$$g = \sum_{J\in\mathcal{H}_n^+} v_J I_{D_J}$$

without changing $\mu_0(f)$, where C_H $(H\in\mathcal{H}_m^+)$ are mutually disjoint, $u_H > 0$, D_J $(J\in\mathcal{H}_n^+)$ are mutually disjoint, and $v_J > 0$. Let

$$C = \bigcup_{H\in\mathcal{H}_m^+} C_H, \quad D = \bigcup_{J\in\mathcal{H}_n^+} D_J,$$

then we have

$$
\begin{aligned}
f &= \sum_{H\in\mathcal{H}_m^+} u_H I_{C_H} \\
&= \sum_{H\in\mathcal{H}_m^+}\sum_{J\in\mathcal{H}_n^+} u_H I_{C_H\cap D_J} + \sum_{H\in\mathcal{H}_m^+} u_H I_{C_H\cap D^c}, \qquad (6.11)\\
g &= \sum_{J\in\mathcal{H}_n^+} v_J I_{D_J} \\
&= \sum_{H\in\mathcal{H}_m^+}\sum_{J\in\mathcal{H}_n^+} v_J I_{C_H\cap D_J} + \sum_{J\in\mathcal{H}_n^+} v_J I_{D_J\cap C^c}. \qquad (6.12)
\end{aligned}
$$

(a) From (6.11) and (6.12), we have

$$
\begin{aligned}
\mu_0(f)-\mu_0(g) &= \sum_{H\in\mathcal{H}_m^+}\sum_{J\in\mathcal{H}_n^+}(u_H-v_J)\mu(C_H\cap D_J)\\
&+ \sum_{H\in\mathcal{H}_m^+} u_H\mu(C_H\cap D^c) - \sum_{J\in\mathcal{H}_n^+} v_J\mu(D_J\cap C^c).
\end{aligned}
$$

Since $\mu(f\neq g)=0$, we can show that $(u_H-v_J)\mu(C_H\cap D_J)=0$, $u_H\mu(C_H\cap D^c)=0$, and $v_J\mu(D_J\cap C^c)=0$. In fact, if $u_H\neq v_J$, then $C_H\cap D_J\subseteq\{f\neq g\}$, which gives $\mu(C_H\cap D_J)\leq\mu(f\neq g)=0$. If $C_H\cap D^c\neq\phi$, then $f(s)=u_H>g(s)=0$ for $s\in C_H\cap D^c$, which implies that $\mu(C_H\cap D^c)=0$. Similarly, $\mu(D_J\cap C^c)=0$. Therefore, $\mu_0(f)=\mu_0(g)$.

(b) From (6.9) and (6.10), we have

$$
\begin{aligned}
\mu_0(f+g) &= \mu_0\left(\sum_{i=1}^m a_i I_{A_i} + \sum_{j=1}^n b_j I_{B_j}\right)\\
&= \sum_{i=1}^m a_i\mu(A_i) + \sum_{j=1}^n b_j\mu(B_j)\\
&= \mu_0(f)+\mu_0(g),
\end{aligned}
$$

and

$$
\mu_0(cf)=\mu_0\left(\sum_{i=1}^m ca_i I_{A_i}\right)=\sum_{i=1}^m ca_i\mu(A_i)=c\mu_0(f).
$$

(c) Let f and g be written as in (6.11) and (6.12), respectively. Since $f\leq g$, we have $f(\omega)=u_H\leq v_J=g(\omega)$ for $\omega\in C_H\cap D_J$ if $C_H\cap D_J\neq\phi$. If

$C_H \cap D^c \neq \phi$, we have $f(\omega) = u_H = 0$ for $\omega \in C_H \cap D^c$. Thus we have

$$\begin{aligned}
& \mu_0(f) - \mu_0(g) \\
= & \sum_{H \in \mathcal{H}_m^+} \sum_{J \in \mathcal{H}_n^+} (u_H - v_J)\mu(C_H \cap D_J) \\
& + \sum_{H \in \mathcal{H}_m^+} u_H \mu(C_H \cap D^c) - \sum_{J \in \mathcal{H}_n^+} v_J \mu(D_J \cap C^c) \\
= & \sum_{H \in \mathcal{H}_m^+} \sum_{J \in \mathcal{H}_n^+} (u_H - v_J)\mu(C_H \cap D_J) - \sum_{J \in \mathcal{H}_n^+} v_J \mu(D_J \cap C^c) \\
\leq & \ 0.
\end{aligned}$$

(d) From (6.11) and (6.12), we have

$$\begin{aligned}
f \wedge g &= \sum_{H \in \mathcal{H}_m^+} \sum_{J \in \mathcal{H}_n^+} \min(u_H, v_J) I_{C_H \cap D_J}, \\
f \vee g &= \sum_{H \in \mathcal{H}_m^+} \sum_{J \in \mathcal{H}_n^+} \max(u_H, v_J) I_{C_H \cap D_J} \\
& \quad + \sum_{H \in \mathcal{H}_m^+} u_H I_{C_H \cap D^c} + \sum_{J \in \mathcal{H}_n^+} v_J I_{D_J \cap C^c},
\end{aligned}$$

which imply that $f \wedge g$ and $f \vee g$ are simple functions.

6.4. Since $h_n \uparrow I_A$, from Problem 6.3, we have $\mu_0(h_n) \leq \mu_0(h_{n+1})$ for $n \geq 1$ and $\mu_0(h_n) \leq \mu_0(I_A) = \mu(A)$, which implies that $\limsup \mu_0(h_n) \leq \mu(A)$. Thus, we only need to show that

$$\liminf \mu_0(h_n) \geq \mu(A). \tag{6.13}$$

Let $\epsilon > 0$, and let $A_n = \{s \in A : h_n(s) > 1 - \epsilon\}$. We claim that $A_n \uparrow A$. In fact, since $h_n \leq h_{n+1}$, we have $A_n \subseteq A_{n+1}$. By definition, we have $\lim_{n\to\infty} A_n \subseteq A$. Let $s \in A$. Since $h_n \uparrow I_A$, then $h_{n_s}(s) > 1 - \epsilon$ for some $n_s \in \mathbf{N}$. Hence we have $s \in A_{n_s}$. Thus, $s \in \lim_{n\to\infty} A_n$. Hence $A_n \uparrow A$. By Theorem 2.1, we have $\mu(A_n) \uparrow \mu(A)$. But

$$(1 - \epsilon) I_{A_n} \leq h_n, \quad n \in \mathbf{N},$$

thus, by Problem 6.3, $(1 - \epsilon)\mu(A_n) \leq \mu_0(h_n)$. Noting that $h_j \leq h_i$ for $j \leq i$, by Problem 6.3, we have

$$\inf_{i \geq j} \mu_0(h_i) = \mu_0(h_j) \geq (1 - \epsilon)\mu(A_j),$$

which gives

$$\sup_{j \geq 1} \inf_{i \geq j} \mu_0(h_i) \geq (1 - \epsilon)\mu(A_m), \quad m \geq 1. \tag{6.14}$$

Letting $m \to \infty$ in (6.14), we have

$$\liminf \mu_0(h_n) \geq (1 - \epsilon)\mu(A). \tag{6.15}$$

Since (6.15) is true for every $\epsilon > 0$, (6.13) is true.

6.5. By hypothesis, f is a nonnegative simple function on S. Hence

$$f = \sum_{k=1}^{r} a_k I_{A_k},$$

where $a_k \geq 0$ and $A_k \in \Sigma$ for $1 \leq k \leq r$.

If $f = 0$, then $f_n = 0$, since, by hypothesis, $f_n \leq f$ and $f_n \geq 0$. By Problem 6.2, $\mu_0(f_n) = \mu_0(f) = 0$. Therefore (6.3) is true.

If $f > 0$, then, by Problem 6.1, f can be written as

$$f = \sum_{i=1}^{s} c_i I_{C_i},$$

where $s \in \mathbf{N}$, $c_i > 0$ for $1 \leq i \leq s$, and C_i $(1 \leq i \leq s)$ are pairwise disjoint. We claim that

$$f_n = \sum_{i=1}^{s} I_{C_i} f_n, \quad n \geq 1. \tag{6.16}$$

To do this, let $x \in S$. If $f_n(x) = 0$, then Equation (6.16) is trivial. If $f_n(x) > 0$, then, by hypothesis, $f(x) \geq f_n(x) > 0$, which implies that $x \in C_{i_0}$ for some $1 \leq i_0 \leq s$. Accordingly, we have

$$f_n(x) = I_{C_{i_0}} f_n(x) = \sum_{i=1}^{s} I_{C_i} f_n(x).$$

Therefore, Equation (6.16) is true. However,

$$\frac{I_{C_i}}{c_i} f_n \uparrow \frac{I_{C_i}}{c_i} f = I_{C_i}, \quad 1 \leq i \leq s,$$

and by Problems 6.4 and 6.3, we have

$$\mu_0\left(I_{C_i} f_n\right) \uparrow c_i \mu(C_i), \quad 1 \leq i \leq s. \tag{6.17}$$

Thus, by Problem 6.3 and Equation (6.17), we have

$$\mu_0(f_n) = \mu_0\left(\sum_{i=1}^{s} I_{C_i} f_n\right) = \sum_{i=1}^{s} \mu_0(I_{C_i} f_n) \uparrow \sum_{i=1}^{s} c_i \mu(C_i) = \mu_0(f).$$

This completes the proof.

6.6. Let

$$L_\infty = \lim_{n \to \infty} L_n, \text{ and } R_\infty = \lim_{m \to \infty} R_m,$$

where $L_n = \mu_0(f_n)$ for $n \geq 1$ and $R_m = \mu_0(g_m)$ for $m \geq 1$. Let $\epsilon > 0$. Then

$$L_{n_0} > L_\infty - \frac{\epsilon}{2} \tag{6.18}$$

for some $n_0 \in \mathbf{N}$. Let

$$h_{n,m} = \min(f_n, g_m), \quad n \geq 1,\ m \geq 1.$$

By Problem 6.3, $h_{n,m}$ is a simple function on S. Noting that, for fixed n,

$$h_{n,m} \uparrow \min(f_n, f) = f_n,$$

we have

$$\mu_0(h_{n_0,m_0}) > L_{n_0} - \frac{\epsilon}{2}, \tag{6.19}$$

for some $m_0 \in \mathbf{N}$. By Problem 6.3, we have

$$R_\infty \geq \mu_0(g_{m_0}) \geq \mu_0(h_{n_0,m_0}). \tag{6.20}$$

On combining Equations (6.18)-(6.20), we get

$$R_\infty > L_\infty - \epsilon,$$

so that $R_\infty \geq L_\infty$. Similarly, $L_\infty \geq R_\infty$. Therefore, $L_\infty = R_\infty$. This completes the proof.

6.7. By Definition 6.3, we may choose a sequence h_n $(n \geq 1)$ of simple functions on S such that $h_n \leq f$ and $\mu_0(h_n) \uparrow \mu(f)$. By Problem 5.11, we may also choose a sequence g_n $(n \geq 1)$ of simple functions on S such that $g_n \uparrow f$.

Let

$$f_n = \max(g_n, h_1, h_2, \ldots, h_n), \quad n \in \mathbf{N}.$$

Then, $f_n \leq f_{n+1} \leq f$ for $n \geq 1$ and, by Problem 6.3, f_n $(n \geq 1)$ are nonnegative simple functions. Since $g_n \uparrow f$ and $f \geq f_n \geq g_n$, we get $f_n \uparrow f$. Since $f_n \leq f$ and f_n is simple, by definition of $\mu(f)$, $\mu_0(f_n) \leq \mu(f)$. Noting that $h_n \leq f_n$ and $\mu_0(h_n) \uparrow \mu(f)$, we have $\mu_0(f_n) \uparrow \mu(f)$.

By hypothesis, $y_n \uparrow f$. Since $f_n \uparrow f$ and $\mu_0(f_n) \uparrow \mu(f)$, by Problem 6.6, we have

$$\mu_0(y_n) \uparrow \mu(f).$$

Note that y_n $(n \geq 1)$ are simple functions, we have $\mu(y_n) = \mu_0(y_n)$. Therefore

$$\mu(y_n) \uparrow \mu(f).$$

6.8. We will show that $\mu(\{f > 0\}) = 0$ by contradiction. To do this, let $A_n = \{f > 1/n\}$ for $n \geq 1$. Then $A_n \uparrow \{f > 0\}$. By Theorem 2.1, $\mu(A_n) \uparrow \mu(\{f > 0\})$. Suppose that $\mu(\{f > 0\}) > 0$. Then $\mu(A_N) > 0$ for some $N \in \mathbf{N}$. By definition of $\mu(f)$, we obtain

$$\mu(f) \geq \mu_0\left(\frac{1}{N} I_{A_N}\right) > 0,$$

which contradicts the hypothesis that $\mu(f) = 0$. Thus $\mu(\{f > 0\}) = 0$.

6.9. Let $f_n = \alpha_n \circ f$ and $g_n = \alpha_n \circ g$, where α_n is as defined in Equation (5.1). Then f_n and g_n are nonnegative simple functions. Since, by hypothesis, $\mu(f \neq g) = 0$, $\mu(f_n \neq g_n) = 0$ and, by Problem 6.3, $\mu(f_n) = \mu(g_n)$. Noting that $f_n \uparrow f$ and $g_n \uparrow g$, by Theorem 6.2, we have $\mu(f) = \mu(g)$.

6.10. Let $f_n = \alpha_n \circ f^+$ and $g_n = \alpha_n \circ f^-$, where α_n is as defined in Problem 5.11. Then we have $f_n \uparrow f^+$ and $g_n \uparrow f^-$. By Theorem 6.2, we have

$$\int_A f_n \mathrm{d}\mu = \int_S f_n I_A \mathrm{d}\mu \uparrow \int_S f I_A \mathrm{d}\mu = \int_A f \mathrm{d}\mu.$$

But

$$0 \leq \int_A f_n \mathrm{d}\mu \leq \int_A n \mathrm{d}\mu = n\mu(A) = 0,$$

which gives $\int_A f_n \mathrm{d}\mu = 0$. Hence we have $\int_A f^+ \mathrm{d}\mu = 0$. Similarly, we have $\int_A f^- \mathrm{d}\mu = 0$. Therefore, $\int_A f \mathrm{d}\mu = 0$. This completes the proof.

6.11. By hypothesis, $\mu(A) = 0$. Hence $\mu(f \neq f I_{S \setminus A}) = 0$ and $\mu(f_n \neq f_n I_{S \setminus A}) = 0$. Consequently, by Problem 6.9, $\mu(f) = \mu(f I_{S \setminus A})$ and $\mu(f_n) = \mu(f_n I_{S \setminus A})$. By Theorem 6.2, $\mu(f_n I_{S \setminus A}) \uparrow \mu(f I_{S \setminus A})$. Therefore, $\mu(f_n) \uparrow \mu(f)$.

6.12. By Problem 5.11, we may choose two sequences f_n $(n \geq 1)$ and g_n $(n \geq 1)$ of nonnegative simple functions on S such that $f_n \uparrow f$ and $g_n \uparrow g$. Then we have $\alpha f_n + \beta g_n \uparrow \alpha f + \beta g$. But, by Problem 6.3,

$$\mu(\alpha f_n + \beta g_n) = \alpha \mu(f_n) + \beta \mu(g_n).$$

Therefore, by Theorem 6.2, we have

$$\mu(\alpha f + \beta g) = \alpha \mu(f) + \beta \mu(g).$$

6.13. By hypothesis, $f_n, g \in (m\Sigma)^+$ and $f_n \leq g$ for $n \geq 1$. Then $g - f_n \in (m\Sigma)^+$. By Theorem 6.3, we have

$$\mu(\liminf (g - f_n)) \leq \liminf \mu(g - f_n).$$

Noting that $\mu(g) < \infty$, by Problem 6.12, we have

$$\mu(g) - \mu(\limsup f_n) \leq \mu(g) - \limsup \mu(f_n),$$

which leads to

$$\mu(\limsup f_n) \geq \limsup \mu(f_n).$$

6.14.

(a) Since h is μ-integrable, we have $\int (h^+ + h^-) \mathrm{d}\mu < \infty$. Noting that $h^+ I_A \leq h^+$ and $h^- I_A \leq h^-$, we have $\int_A (h^+ + h^-) \mathrm{d}\mu = \int (h^+ I_A + h^- I_A) \mathrm{d}\mu \leq \int (h^+ + h^-) \mathrm{d}\mu < \infty$ by definition. Hence $h I_A$ is μ-integrable.

(b) Since

$$\left|\int_A h\mathrm{d}\mu\right| \le \int |h|I_A\mathrm{d}\mu \le \int |h|\mathrm{d}\mu < \infty,$$

the result follows.

6.15. Suppose that $\mu(\{|h| = \infty\}) > 0$. Then $\int_S |h|\mathrm{d}\mu \ge \infty\mu(\{|h| = \infty\}) = \infty$, which contradicts the hypothesis that h is μ-integrable. Hence h is finite a.e.

6.16. Let $A = \{s : f(s) \ge \epsilon\}$. Then by Definition 6.5 and Problem 6.12, we have

$$\begin{aligned}
\int_S f^p \mathrm{d}\mu &= \int_S (f^p I_A + f^p I_{A^c})\mathrm{d}\mu \\
&= \int_S f^p I_A \mathrm{d}\mu + \int_S f^p I_{A^c}\mathrm{d}\mu \\
&\ge \int_A f^p \mathrm{d}\mu \\
&\ge \int_A \epsilon^p \mathrm{d}\mu \\
&= \epsilon^p \mu(A).
\end{aligned}$$

The result follows by dividing ϵ^p on both sides of the above inequality.

6.17. We first show that $\alpha f + \beta g \in L^1(S, \Sigma, \mu)$. Since, by hypothesis, $f, g \in L^1(S, \Sigma, \mu)$, we have

$$\mu(|\alpha f + \beta g|) \le \mu(|\alpha||f| + |\beta||g|) \le |\alpha|\mu(|f|) + |\beta|\mu(|g|) < \infty.$$

Therefore, $\alpha f + \beta g$ is μ-integrable.

Next, we show that

$$\mu(\alpha f + \beta g) = \alpha\mu(f) + \beta\mu(g). \tag{6.21}$$

Let $f_\alpha^+ = \max(\mathrm{sgn}(\alpha)f, 0)$, $f_\alpha^- = \max(-\mathrm{sgn}(\alpha)f, 0)$, $g_\beta^+ = \max(\mathrm{sgn}(\beta)g, 0)$, and $g_\beta^- = \max(-\mathrm{sgn}(\beta)g, 0)$, where

$$\mathrm{sgn}(x) = \begin{cases} 1, & \text{if } x > 0; \\ 0, & \text{if } x = 0; \\ -1, & \text{if } x < 0. \end{cases}$$

Then $\alpha f = |\alpha|\,\mathrm{sgn}(\alpha) f = |\alpha| f_\alpha^+ - |\alpha| f_\alpha^-$, $\beta g = |\beta|\,\mathrm{sgn}(\beta) g = |\beta| g_\beta^+ - |\beta| g_\beta^-$, and so, by Definition 6.4 and Problem 6.12, we obtain

$$\begin{aligned}
\mu(\alpha f + \beta g) &= \mu(|\alpha| f_\alpha^+ - |\alpha| f_\alpha^- + |\beta| g_\beta^+ - |\beta| g_\beta^-) \\
&= \mu(|\alpha| f_\alpha^+ + |\beta| g_\beta^+) - \mu(|\alpha| f_\alpha^- + |\beta| g_\beta^-) \\
&= |\alpha|\mu(f_\alpha^+) + |\beta|\mu(g_\beta^+) - |\alpha|\mu(f_\alpha^-) - |\beta|\mu(g_\beta^-) \\
&= |\alpha|\mu(f_\alpha^+ - f_\alpha^-) + |\beta|\mu(g_\beta^+ - g_\beta^-) \\
&= \mu(\alpha f) + \mu(\beta g).
\end{aligned}$$

Now we show that $\mu(\alpha f) = \alpha\mu(f)$. Let $f^+ = \max(f, 0)$ and $f^- = \max(-f, 0)$. If $\alpha \geq 0$, then $\mu(\alpha f) = \mu(\alpha f^+ - \alpha f^-) = \alpha\mu(f^+) - \alpha\mu(f^-) = \alpha\mu(f)$. If $\alpha < 0$, then $\mu(\alpha f) = \mu(-\alpha f^- - (-\alpha f^+)) = -\alpha\mu(f^-) - (-\alpha)\mu(f^+) = \alpha\mu(f)$. Similarly, we can show that $\mu(\beta g) = \beta\mu(g)$. Therefore, Equation (6.21) is true.

6.18. We first prove the "only if" part. Suppose that $\mu(|f_n - f|) \to 0$. Note that

$$-|f_n - f| \leq f_n - f \leq |f_n - f|,$$

which implies that

$$|\mu(f_n - f)| \leq \mu(|f_n - f|).$$

Therefore, $\mu(f_n) \to \mu(f)$.

Next, we prove the "if" part. Suppose that $\mu(f_n) \to \mu(f)$. Since, by hypothesis, f_n and f are nonnegative μ-integrable, $(f_n - f)^- = \max(-f_n + f, 0) \leq f$. Noting that $f_n \to f$, by Theorem 6.4, we have

$$\lim_{n\to\infty} \mu((f_n - f)^-) = \mu(0) = 0.$$

Since

$$\mu((f_n - f)^+) = \mu(f_n - f) + \mu((f_n - f)^-) = \mu(f_n) - \mu(f) + \mu((f_n - f)^-),$$

we have $\mu((f_n - f)^+) \to 0$. Therefore

$$\lim_{n\to\infty} \mu(|f_n - f|) = \lim_{n\to\infty} \mu((f_n - f)^+) + \lim_{n\to\infty} \mu((f_n - f)^-) = 0.$$

This completes the proof.

6.19. We first prove the "only if" part. Suppose that $\mu(|f_n - f|) \to 0$. Since $|f_n| - |f| = |f_n - f + f| - |f| \leq |f_n - f| + |f| - |f| = |f_n - f|$ and $|f_n| - |f| = |f_n| - |f - f_n + f_n| \geq |f_n| - (|f - f_n| + |f_n|) = -|f_n - f|$, $||f_n| - |f|| \leq |f_n - f|$, which implies that $\mu(||f_n| - |f||) \to 0$. Therefore, $\mu(|f_n|) \to \mu(|f|)$.

Next, we prove the "if" part. Suppose that $\mu(|f_n|) \to \mu(|f|)$. Then, by Fatou's lemma (see Theorem 6.3), we have

$$\mu(f^+) = \mu(\liminf f_n^+) \leq \liminf \mu(f_n^+)$$

and

$$\mu(f^-) = \mu(\liminf f_n^-) \leq \liminf \mu(f_n^-).$$

Therefore, by Problem 1.8, we have

$$\mu(|f|) = \mu(f^+) + \mu(f^-) \leq \liminf(\mu(f_n^+) + \mu(f_n^-)) = \liminf \mu(|f_n|) = \mu(|f|),$$

which implies that $\mu(f^+) = \liminf \mu(f_n^+)$ and $\mu(f^-) = \liminf \mu(f_n^-)$. But, by Problem 1.8, we have

$$\begin{aligned}
\limsup \mu(f_n^+) + \liminf \mu(f_n^-) &\leq \limsup(\mu(f_n^+) + \mu(f_n^-)) \\
&= \limsup \mu(|f_n|) \\
&= \mu(|f|) = \mu(|f^+|) + \mu(|f^-|),
\end{aligned}$$

which implies that $\limsup \mu(f_n^+) = \mu(f^+)$. Therefore, $\mu(f_n^+) \to \mu(f^+)$. Similarly, we can show that $\mu(f_n^-) \to \mu(f^-)$. Thus, by Problem 6.18, we have $\mu(|f_n^+ - f^+|) \to 0$ and $\mu(|f_n^- - f^-|) \to 0$. But

$$\begin{aligned}\mu(|f_n - f|) &= \mu(|f_n^+ - f_n^- - f^+ + f^-|) \\ &\leq \mu(|f_n^+ - f^+|) + \mu(|f_n^- - f^-|).\end{aligned}$$

Therefore, $\mu(|f_n - f|) \to 0$. This completes the proof.

6.20. First, we consider the case when $h = I_B$ for $B \in \mathcal{B}$. In this case, we have

$$\mu(h \circ f) = \int_S h(f(s))\mu(\mathrm{d}s) = \int_{\{f \in B\}} 1\mu(\mathrm{d}s) = \mu(f \in B), \tag{6.22}$$

and

$$\Lambda_f(h) = \int_{\mathbf{R}} h(x)\Lambda_f(\mathrm{d}x) = \int_B 1\Lambda_f(\mathrm{d}x) = \Lambda_f(B). \tag{6.23}$$

Suppose that $h \circ f \in L^1(S, \Sigma, \mu)$. Then, from Equation (6.22), we have $\mu(f \in B) < \infty$. By the definition of Λ_f, $\Lambda_f(B) = \mu(f \in B)$, and so $\mu(h \circ f) = \Lambda_f(h)$. Therefore, $\Lambda_f(h) < \infty$, and so $h \in L^1(\mathbf{R}, \mathcal{B}, \Lambda_f)$. Similarly, we can show that $h \circ f \in L^1(S, \Sigma, \mu)$ if $h \in L^1(\mathbf{R}, \mathcal{B}, \Lambda_f)$.

Next, we consider the case when h is a simple $\mathcal{B}$-measurable function. By Problem 6.1, we can write h as

$$h = \sum_{i=1}^n b_i I_{B_i},$$

where $b_i > 0$ and $B_i \in \mathcal{B}$ $(i = 1, 2, \ldots, n)$ are pairwise disjoint. Then, by Problem 6.3 and Equation (6.22), we have

$$\begin{aligned}\mu(h \circ f) &= \mu\left(\sum_{i=1}^n b_i (I_{B_i} \circ f)\right) \\ &= \sum_{i=1}^n b_i \mu(I_{B_i} \circ f) \\ &= \sum_{i=1}^n b_i \mu(f \in B_i),\end{aligned} \tag{6.24}$$

and

$$\Lambda_f(h) = \Lambda_f\left(\sum_{i=1}^n b_i I_{B_i}\right) = \sum_{i=1}^n b_i \Lambda_f(I_{B_i}) = \sum_{i=1}^n b_i \Lambda_f(B_i). \tag{6.25}$$

Since, by definition of Λ_f, $\mu(f \in B_i) = \Lambda_f(B_i)$, we have $\mu(h \circ f) = \Lambda_f(h)$. Therefore, $h \circ f \in L^1(S, \Sigma, \mu)$ if and only if $h \in L^1(\mathbf{R}, \mathscr{F}, \Lambda_f)$.

Next, we consider the case when h is a nonnegative $\mathcal{B}$-measurable function. Then, by Problem 5.11, there exists a sequence h_n $(n \geq 1)$ of simple functions such that

$h_n \uparrow h$. Since $h_n \circ f \uparrow h \circ f$ and $\mu(h_n \circ f) = \Lambda_f(h_n)$ for $n \geq 1$, we have, by Theorem 6.2, $\mu(h_n \circ f) \uparrow \mu(h \circ f)$ and $\Lambda_f(h_n) \uparrow \Lambda_f(h)$. Therefore, $\mu(h \circ f) = \Lambda_f(h)$ and so that $h \circ f \in L^1(S, \Sigma, \mu)$ if and only if $h \in L^1(\mathbf{R}, \mathscr{F}, \Lambda_f)$.

Finally, we consider the case when h is a $\mathcal{B}$-measurable function. Then $h \circ f = h^+ \circ f - h^- \circ f$, where $h^+ = \max(h, 0)$ and $h^- = \max(-h, 0)$. We have already shown that $\mu(h^+ \circ f) = \Lambda_f(h^+)$ and $\mu(h^- \circ f) = \Lambda_f(h^-)$. If $h \circ f \in L^1(S, \Sigma, \mu)$, then $h^+ \circ f$ and $h^- \circ f$ are in $L^1(S, \Sigma, \mu)$. Accordingly, h^+ and h^- are in $L^1(\mathbf{R}, \mathcal{B}, \Lambda_f)$ and so is h. Similarly, we can show that $h \in L^1(\mathbf{R}, \mathcal{B}, \Lambda_f)$ if $h \circ f \in L^1(S, \Sigma, \mu)$.

This completes the proof.

6.21. Let $g_n = \sum_{i=1}^n h_i$. Since $h_n \in (m\Sigma)^+$ for all $n \geq 1$, we have $g_n \in (m\Sigma)^+$ for all $n \geq 1$. Since $g_n \uparrow \sum_{i=1}^\infty h_i$, the result follows from Theorem 6.2 and Problem 6.17.

6.22.

(a) If $\int_S h \mathrm{d}\mu = \infty$, then $\int_S g_n \mathrm{d}\mu = \infty$ for all $n \geq 1$ and $\int_S g \mathrm{d}\mu = \infty$. If $\int_S h \mathrm{d}\mu < \infty$, then h is finite a.e. by Problem 6.15. Let $A = \{s \in S : |h(s)| = \infty\}$. Then $\mu(A) = 0$. Also we have $g_n - h \geq 0$ on $S \backslash A$ and $g_n - h \uparrow g - h$ on $S \backslash A$. By Theorem 6.2, $\int_{S \backslash A} (g_n - h) \mathrm{d}\mu \uparrow \int_{S \backslash A} (g - h) \mathrm{d}\mu$. Since $\mu(A) = 0$, we have $\int_S (g_n - h) \mathrm{d}\mu \uparrow \int_S (g - h) \mathrm{d}\mu$. Since $\int_S h \mathrm{d}\mu$ is finite, the result follows from Problem 6.17.

(b) From the assumption, we have $\int_S -h \mathrm{d}\mu > -\infty$, $-g_n \geq -h$ for $n \geq 1$, and $-g_n \uparrow -g$. Hence the result follows from part (a) of this proof.

6.23. Let $F(t)$ and $G(t)$ be defined as

$$F(t) = \int_0^t f(s) \mathrm{d}s, \quad G(t) = \int_0^t f(s) g(s) \mathrm{d}s.$$

Then by the assumption, we have

$$\frac{\mathrm{d}}{\mathrm{d}t} \left[e^{-F(t)} G(t) \right] = f(t) e^{-F(t)} (g(t) - G(t)) \leq a f(t) e^{-F(t)}.$$

Integrating the first and the last terms in this equation gives

$$e^{-F(t)} G(t) \leq a \left(1 - e^{-F(t)} \right), \quad \forall t \in [0, T],$$

which leads to

$$G(t) \leq a \left(e^{F(t)} - 1 \right).$$

Again by the assumption, we have

$$g(t) \leq a + G(t) \leq a e^{F(t)}.$$

This completes the proof.

6.5 Bibliographic Notes

In this chapter, we introduced Lebesgue integration. To construct the Lebesgue integral, we first deal with nonnegative simple functions, which are Borel measurable and take on only finitely many distinct values. Then we consider general nonnegative measurable functions. Finally we define the integration for general measurable functions. For more information about Lebesgue integration, readers are referred to Williams (1991) and Ash and Doleans-Dade (1999). Lebesgue integration on the real line is introduced in many books on real analysis such as Wheeden and Zygmund (1977), Royden (1988), and Yeh (2006).

The Lebesgue integration is more general than the Riemann integration (Royden, 1988, Section 4.1). For a comparison of Lebesgue and Riemann integrals, readers are referred to Ash and Doleans-Dade (1999, Section 1.7).

CHAPTER 7

THE RADON-NIKODYM THEOREM

The Radon-Nikodym theorem is a fundamental and important result in measure theory and has important applications in modern probability theory. For example, the Radon-Nikodym theorem can be used to prove the existence of conditional expectations (Klenke, 2006). In this chapter, we present some concepts and results related to this theorem.

7.1 Basic Concepts and Facts

Definition 7.1 (Signed Measure). A signed measure on a σ-algebra is a set function $\mu : \Sigma \to \bar{\mathbf{R}}$ that is countably additive:

(a) $\mu(\emptyset) = 0$.

(b) For any disjoint sets $A_n \in \Sigma$, $n = 1, 2, \ldots$, have

$$\mu\left(\sum_{n=1}^{\infty}\right) = \sum_{n=1}^{\infty} \mu(A_n).$$

Definition 7.2 (Absolute Continuity). Let (S, Σ, μ) be a measure space and λ a signed measure on Σ. The signed measure λ is said to be absolutely continuous with respect to μ, denoted by $\lambda \ll \mu$, if and only if $\mu(A) = 0\,(A \in \Sigma)$ implies $\lambda(A) = 0$.

Definition 7.3 (Equivalent Measures). Let μ and λ be two measures defined on a measurable space (S, Σ). The two measures are said to be equivalent, written as $\mu \sim \lambda$, if and only if $\mu \ll \lambda$ and $\lambda \ll \mu$.

Definition 7.4 (Singularity). Let μ_1 and μ_2 be two measures on a measurable space (S, Σ). Let λ_1 and λ_2 be two signed measures on (S, Σ). Let $|\lambda_1|$ and $|\lambda_2|$ be the total variations of λ_1 and λ_2, respectively (see Theorem 7.2). The two measures μ_1 and μ_2 are said to be mutually singular, denoted by $\mu_1 \perp \mu_2$, if and only if there exists a set $A \in \Sigma$ such that $\mu_1(A) = 0$ and $\mu_2(A^c) = 0$. The two signed measures λ_1 and λ_2 are said to be mutually singular if and only if $|\lambda_1| \perp |\lambda_2|$. The signed measure λ_1 and the measure μ_1 are said to be mutually singular if and only if there exists a set $A \in \Sigma$ such that $\mu_1(A) = 0$ and $|\lambda_1|(A) = 0$.

Theorem 7.1. *Let (S, Σ) be a measurable space and λ a countably additive extended real-valued set function on Σ. Then the maximum and the minimum values of λ are attainable; that is, there exist sets $C, D \in \Sigma$ such that*

$$\lambda(C) = \sup\{\lambda(A) : A \in \Sigma\},$$

and

$$\lambda(D) = \inf\{\lambda(A) : A \in \Sigma\}.$$

Theorem 7.2 (Jordan-Hahn Decomposition). *Let λ be a countably additive extended real-valued set function on a σ-algebra Σ. Let λ^+ and λ^- be set functions on Σ defined as*

$$\lambda^+(A) = \sup\{\lambda(B) : B \in \Sigma,\ B \subseteq A\}, \tag{7.1a}$$

$$\lambda^-(A) = -\inf\{\lambda(B) : B \in \Sigma,\ B \subseteq A\}. \tag{7.1b}$$

Then λ^+ and λ^- are measures on Σ and $\lambda = \lambda^+ - \lambda^-$.

The measure λ^+ is the upper variation or positive part of λ, the measure λ^- is the lower variation or negative part of λ, and the measure $|\lambda| = \lambda^+ + \lambda^-$ is the total variation of λ.

Theorem 7.3 (Lebesgue Decomposition Theorem). *Let (S, Σ) be a measurable space and μ a measure on Σ. If λ is a σ-finite signed measure (i.e., $|\lambda|$ is σ-finite), then λ has a unique decomposition as $\lambda_1 + \lambda_2$, where λ_1 and λ_2 are signed measures on Σ such that $\lambda_1 \ll \mu$ and $\lambda_2 \perp \mu$.*

Theorem 7.4 (Radon-Nikodym Theorem). *Let (S, Σ) be a measurable space and μ a σ-finite measure on Σ. Suppose that λ is a signed measure on Σ that is absolutely continuous with respect to μ. Then there exists a Borel measurable function $g : S \to \bar{\mathbf{R}}$ such that*

$$\lambda(A) = \mu(g; A) = \int_A g \mathrm{d}\mu, \quad \forall A \in \Sigma.$$

If h is another such function, then $\mu(g \neq h) = 0$. The function g is the Radon-Nikodym derivative and is commonly denoted by

$$g = \frac{\mathrm{d}\lambda}{\mathrm{d}\mu}$$

or

$$\mathrm{d}\lambda = g\mathrm{d}\mu.$$

7.2 Problems

7.1. Let (S, Σ) be a measurable space and λ a countably additive extended real-valued set function on Σ. Let $D \in \Sigma$ such that $\lambda(A \cap D) \geq 0$ and $\lambda(A \cap D^c) \leq 0$ for all $A \in \Sigma$. Show that $\lambda^+(A) = \lambda(A \cap D)$ and $\lambda^-(A) = \lambda(A \cap D^c)$ for all $A \in \Sigma$, where λ^+ and λ^- are as defined in Equation (7.1).

7.2. Let μ_1 be an arbitrary finite measure on $\mathcal{B}$. Let μ_2 be a measure on $\mathcal{B}$ defined as follows:

$$\mu_2(A) = \begin{cases} \mu_1(\mathbf{R}), & \text{if } 0 \in A; \\ 0, & \text{if } 0 \notin A. \end{cases}$$

Find the Jordan-Hahn decomposition of the signed measure $\lambda = \mu_1 - \mu_2$.

7.3. Let (S, Σ) be a measurable space and λ be a signed measure on Σ. Suppose that $\lambda(A) < \infty$ for all $A \in \Sigma$ and that $D \in \Sigma$ is a set on which λ attains its maximum (see Theorem 7.1). Show that

(a) $\lambda(A \cap D) \geq 0$ and $\lambda(A \cap D^c) \leq 0$ for all $A \in \Sigma$.

(b) $\lambda^+(A) = \lambda(A \cap D)$ and $\lambda^-(A) = -\lambda(A \cap D^c)$ for all $A \in \Sigma$.

7.4. Let (S, Σ) be a measurable space. Let λ, λ_1, and λ_2 be signed measures on Σ. Show that

(a) $|\lambda|(A) = \sup\{\sum_{i=1}^n |\lambda(B_i)| : B_1, B_2, \ldots, B_n$ are arbitrary disjoint measurable subsets of A, $n = 1, 2, \ldots\}$, where $|\lambda|$ is the total variation of λ (see Theorem 7.2).

(b) $|\lambda_1 + \lambda_2| \leq |\lambda_1| + |\lambda_2|$.

7.5. Let (S, Σ) be a measurable space and μ a measure on Σ. Let λ_1 and λ_2 be signed measures on Σ. Show that

(a) If $\lambda_1 \perp \mu$ and $\lambda_2 \perp \mu$, then $\lambda_1 + \lambda_2 \perp \mu$.

(b) If $\lambda_1 \ll \mu$, then $|\lambda_1| \ll \mu$.

(c) If $|\lambda_1| \ll \mu$, then $\lambda_1 \ll \mu$.

(d) If $\lambda_1 \ll \mu$ and $\lambda_2 \perp \mu$, then $\lambda_1 \perp \lambda 2$.

(e) If $\lambda_1 \ll \mu$ and $\lambda_1 \perp \mu$, then $\lambda_1 = 0$ (i.e., $\lambda(A) = 0$ for all $A \in \Sigma$).

(f) If λ_1 is finite, then $\lambda_1 \ll \mu$ if and only if $\lim_{\mu(A)\to 0} \lambda_1(A) = 0$.

7.6. Let P and Q be probability measures on a measurable space $(\Omega, \mathscr{F})$. Suppose that Q is absolutely continuous relative to P and let $g = \mathrm{d}Q/\mathrm{d}P$ be the Radon-Nikodym derivative. Show that P and Q are equivalent if and only if

$$P\{g > 0\} = 1.$$

7.3 Hints

7.1. Follow the definitions of λ^+ and λ^-.

7.2. Use the definition of λ^+ and λ^- in Equation (7.1).

7.3. Part (a) can be proved by using the method of contradiction. Part (b) can be proved by using the result of part (a) and the definition of upper and lower variations.

7.4. Use the result of Problem 7.3 and the definition of $|\lambda|$ to prove part (a). Then use the result of part (a) to prove part (b).

7.5. Use the definition of singularity (Definition 7.4) and the result of Problem 7.4 to prove part (a). Use the result of Problem 7.3 to prove part (b). Use the definition of $|\lambda_1|$ (see Theorem 7.1) to prove part (c). Use part (b) to prove part (d). Part (e) can be proved by using part (d). For part (f), we need to use the first Borel-Cantelli lemma (Theorem 2.2), the result of Problem 2.18, and the method of contradiction.

7.6. Follow the definition of equivalent measures (Definition 7.3). To prove the necessity part, try to show that $Q\{g \le 0\} = 0$. Use the result of Problem 6.8 to prove the sufficiency part.

7.4 Solutions

7.1. Let $A \in \Sigma$. Since $A \cap D \subseteq A$, we only need to show that $\lambda(B) \le \lambda(A \cap D)$ for all $B \subseteq A$. Using the assumption, we have $\lambda(B) = \lambda(B \cap D) + \lambda(B \cap D^c) \le \lambda(B \cap D) \le \lambda(B \cap D) + \lambda((A \backslash B) \cap D) = \lambda(A \cap D)$. Hence $\lambda^+(A) = \lambda(A \cap D)$. Similarly, we can show that $\lambda^-(A) = \lambda(A \cap D^c)$.

7.2. First let us calculate the measure λ^+. By definition, we have

$$\lambda^+(A) = \sup\{\lambda(B) : B \in \mathcal{B},\ B \subseteq A\}$$

for all $A \in \mathcal{B}$. Note that

$$\lambda(B) = \begin{cases} \mu_1(B) - \mu_1(\mathbf{R}), & \text{if } 0 \in B; \\ \mu_1(B), & \text{if } 0 \notin B. \end{cases}$$

we have $\lambda^+(A) = \mu_1(A \backslash \{0\})$ for all $A \in \mathcal{B}$. Similarly, we have $\lambda^-(A) = \mu_1(\mathbf{R}) - \mu_1(\{0\})$ if $0 \in A$ and $\lambda^-(A) = 0$ if $0 \notin A$.

7.3.

(a) Suppose that $\lambda(A \cap D) < 0$ for some $A \in \Sigma$. Note that $\lambda(D) = \lambda(D \cap A) + \lambda(D \cap A^c)$. By the assumption that λ is bounded above, we have $\lambda(D \cap A^c) = \lambda(D) - \lambda(D \cap A) > \lambda(D)$. However, this contradicts the assumption that D is a set on which λ attains its maximum. Hence $\lambda(A \cap D) \geq 0$ for all $A \in \Sigma$. Similarly, we can show that $\lambda(A \cap D^c) \leq 0$.

(b) Let $A \in \Sigma$. To show that $\lambda^+(A) = \lambda(A \cap D)$, we only need to show that $\lambda(B) \leq \lambda(A \cap D)$ for all $B \subseteq A$ and $B \in \Sigma$. In fact, by the result of part (a), we have $\lambda(B) = \lambda(B \cap D) + \lambda(B \cap D^c) \leq \lambda(B \cap D) \leq \lambda(B \cap D) + \lambda((A \backslash B) \cap D) = \lambda(A \cap D)$. Similarly, we have $\lambda^-(A) = -\lambda(A \cap D^c)$. This completes the proof.

7.4.

(a) Noting that $\lambda = \lambda^+ - \lambda^-$ and $|\lambda|$ is a measure, we have

$$\begin{aligned} \sum_{i=1}^n |\lambda(B_i)| &= \sum_{i=1}^n |\lambda^+(B_i) - \lambda^-(B_i)| \leq \sum_{i=1}^n \lambda^+(B_i) + \lambda^-(B_i) \\ &= \sum_{i=1}^n |\lambda|(B_i) \\ &= |\lambda|(\bigcup_{i=1}^n B_i) \\ &\leq |\lambda|(A). \end{aligned}$$

By Problem 7.3, there exists a set $D \in \Sigma$ such that $\lambda^+(A) = \lambda(A \cap D) \geq 0$ and $\lambda^-(A) = -\lambda(A \cap D^c) \geq 0$. Hence

$$\begin{aligned} |\lambda|(A) &= \lambda^+(A) + \lambda^-(A) \\ &= \lambda(A \cap D) - \lambda(A \cap D^c) \\ &= |\lambda(A \cap D)| + |\lambda(A \cap D^c)|. \end{aligned}$$

Since $\sum_{i=1}^n |\lambda(B_i)|$ is bounded above by $|\lambda|(A)$ and can reach $|\lambda|(A)$ for some B_i, the result follows.

(b) Let $A \in \Sigma$ be an arbitrary set. By part (a) of this problem, we have

$$
\begin{aligned}
& |\lambda_1 + \lambda_2|(A) \\
= \ & \sup\{\sum_{i=1}^{n} |\lambda_1(B_i) + \lambda_2(B_i)| : B_1, B_2, \ldots, B_n \\
& \text{are arbitrary disjoint measurable subsets of } A, n = 1, 2, \ldots\} \\
\leq \ & \sup\{\sum_{i=1}^{n} |\lambda_1(B_i)| : B_1, B_2, \ldots, B_n \\
& \text{are arbitrary disjoint measurable subsets of } A, n = 1, 2, \ldots\} \\
+ \ & \sup\{\sum_{i=1}^{n} |\lambda_2(B_i)| : B_1, B_2, \ldots, B_n \\
& \text{are arbitrary disjoint measurable subsets of } A, n = 1, 2, \ldots\} \\
= \ & |\lambda_1|(A) + |\lambda_2|(A).
\end{aligned}
$$

Since A is arbitrary, the result follows.

This completes the proof.

7.5.

(a) Since $\lambda_1 \perp \mu$ and $\lambda_2 \perp \mu$, we have $\mu(A) = \mu(B) = 0$ and $|\lambda_1|(A^c) = |\lambda_2|(B^c) = 0$ for some $A \in \Sigma$ and $B \in \Sigma$. Noting that μ is a measure, we have $0 \leq \mu(A \cup B) \leq \mu(A) + \mu(B) = 0$, which implies that $\mu(A \cup B) = 0$. By Problem 7.4 and noting that $|\lambda_1 + \lambda_2|$, $|\lambda_1|$, and $|\lambda_2|$ are measures, we have $0 \leq |\lambda_1 + \lambda_2|((A \cup B)^c) = |\lambda_1 + \lambda_2|(A^c \cap B^c) \leq |\lambda_1|(A^c \cap B^c) + |\lambda_2|(A^c \cap B^c) \leq |\lambda_1|(A^c) + |\lambda_2|(B^c) = 0$, which implies that $|\lambda_1 + \lambda_2|((A \cup B)^c) = 0$. Hence $\lambda_1 + \lambda_2 \perp \mu$.

(b) Let $\mu(A) = 0$. By Problem 7.3, we have $\lambda_1^+ = \lambda_1(A \cap D)$ and $\lambda_1^- = \lambda_1(A \cap D^c)$ for some set $D \in \Sigma$, where λ_1^+ and λ_2^- are the upper variation and lower variation of λ_1, respectively. Since $A \cap D \subseteq A$, we have $\mu(A \cap D) = 0$. By the assumption that $\lambda_1 \ll \mu$, we have $\lambda_1(A \cap D) = 0$. Hence $\lambda_1^+ \ll \mu$. Similarly, we have $\lambda_1^- \ll \mu$. Hence $|\lambda_1| \ll \mu$.

(c) Let $\mu(A) = 0$. Then $|\lambda_1|(A) = 0$. Noting that $|\lambda_1| = \lambda_1^+ + \lambda_1^-$, we have $\lambda_1^+(A) = 0$ and $\lambda_1^-(A) = 0$. Thus $\lambda_1(A) = \lambda_1^+(A) - \lambda_1^-(A) = 0$. Therefore, $\lambda_1 \ll \mu$.

(d) By assumption, we have $\mu(A) = |\lambda_2|(A^c) = 0$ for some $A \in \Sigma$. By part (b) of this problem and the assumption that $\lambda_1 \ll \mu$, we have $|\lambda_1| \ll \mu$. Thus $|\lambda_1|(A) = 0$. Hence $\lambda_1 \perp \lambda_2$.

(e) By part (d) of this problem, we have $\lambda_1 \perp \lambda_1$, which implies that $|\lambda_1|(A) = |\lambda_1|(A^c) = 0$ for some $A \in \Sigma$. Therefore, $|\lambda_1|(S) = 0$. It follows that $\lambda_1^+(S) = \lambda_1^-(S) = 0$. Hence $\lambda_1(B) = 0$ for all $B \in \Sigma$.

(f) First we prove the "if" part. Suppose that $\mu(A) = 0$ and

$$\lim_{\mu(B)\to 0} \lambda_1(B) = 0.$$

We need to show that $\lambda_1(A) = 0$. To do this, we let $B_n = A$ for $n = 1, 2, \ldots$. Then by assumption we have $\lambda_1(A) = \lim_{\mu(B_n)\to 0} \lambda_1(B_n) = 0$. Hence $\lambda_1 \ll \mu$.

Then we prove the "only if" part. Suppose that $\lambda_1 \ll \mu$. We need to show that $\lim_{\mu(B)\to 0} \lambda_1(B) = 0$. Since $|\lambda_1(B)| \leq |\lambda_1|(B)$ for all $B \in \Sigma$, we only need to show that $\lim_{n\to\infty} |\lambda_1|(A_n) = 0$, where $\mu(A_n) \to 0$ as $n \to \infty$. To do this, we only need to show that $\limsup |\lambda_1|(A_n) = 0$. If $\limsup |\lambda_1|(A_n) \neq 0$, then there exists some $\epsilon > 0$ such that $\sup_{i\geq j} |\lambda_1|(A_i) \geq 2\epsilon$ for all $j \geq 1$. Hence there are infinitely many A_n such that $|\lambda_1|(A_n) \geq \epsilon$. Noting also that $\mu(A_n) \to 0$ as $n \to \infty$, we can find a subsequence $\{B_n\}_{n\geq 1}$ from $\{A_n\}_{n\geq 1}$ such that $\mu(B_n) \leq 2^{-n}$ and $|\lambda_1|(B_n) \geq \epsilon$. Then $\sum_{n=1}^{\infty} \mu(B_n) < \infty$. By Theorem 2.2, we have $\mu(\limsup B_n) = 0$. By part (b), we have $|\lambda_1|(\limsup B_n) = 0$. However, $|\lambda_1|(\bigcup_{i\geq j}(B_i) \geq |\lambda_1|(B_j) \geq \epsilon$ for all $j \geq 1$. By Problem 2.18 and the assumption that λ_1 is finite, we have $|\lambda_1|(\limsup B_n) \geq \epsilon$, which contradicts $|\lambda_1|(\limsup B_n) = 0$. Thus $\lim_{\mu(B)\to 0} |\lambda_1|(B) = 0$.

This completes the proof.

7.6. First let us prove the necessity ("only if") part. Suppose that $Q \sim P$. Since

$$0 \leq Q\{g \leq 0\} = \int_{\{g\leq 0\}} g\mathrm{d}P \leq 0,$$

we have $Q\{g \leq 0\} = 0$. By the assumption that $Q \sim P$, we have $P\{g \leq 0\} = 0$, which gives $P\{g > 0\} = 1$.

Now we prove the sufficiency part. Suppose that $P\{g > 0\} = 1$. We need to show that P is absolutely continuous relative to Q. To do that, let $A \in \mathscr{F}$ such that $Q(A) = 0$. Note that

$$Q(A) = \int_A g\mathrm{d}P = \int_{A\cap\{g>0\}} g\mathrm{d}P.$$

Then by Problem 6.8, $P(A \cap \{g > 0\}) = 0$. Since $P\{g > 0\} = 1$, we have

$$P(A \cap \{g > 0\}) = P(A) + P\{g > 0\} - P(A \cup \{g > 0\}) \geq P(A).$$

Hence $P(A) = 0$. Since A is arbitrary, P is absolutely continuous relative to Q. This finishes the proof.

7.5 Bibliographic Notes

In this chapter, we introduced the Radon-Nikodym theorem. As a fundamental result in measure theory, the Radon-Nikodym theorem is presented and discussed in many textbooks such as Rudin (1970), Taylor (1973), Royden (1988), Ash and Doleans-Dade (1999), Schilling (2005), Klenke (2006), and Capinski and Kopp (2007). The Jordan-Hahn decomposition (see Theorem 7.2) was used to develop the Radon-Nikodym theorem, the proof of which is not presented in this chapter. One can find a proof in Ash and Doleans-Dade (1999, p65).

CHAPTER 8

L^P SPACES

L^p spaces are spaces of functions whose pth power is integrable. For functions in a L^p space, we can define norms and metrics and study the convergence of sequences of functions. In this chapter, we introduce the concepts of L^p spaces and some important inequalities for functions in the L^p spaces.

8.1 Basic Concepts and Facts

Definition 8.1 (L^p Space). Let (S, Σ, μ) be a measure space and $p \in (0, \infty]$. Then $L^p(S, \Sigma, \mu)$ is defined as

$$\begin{aligned} L^p(S, \Sigma, \mu) &= \{f : |f|^p \text{ is } \mu\text{-integrable}\} \\ &= \left\{ f : \int_S |f|^p \mathrm{d}\mu < \infty \right\}, \quad p \in (0, \infty), \end{aligned}$$

and

$$L^\infty(S, \Sigma, \mu) = \{f : \mu(|f| > K) = 0 \text{ for some } K \in (0, \infty)\}.$$

Definition 8.2 (L^p Norm)**.** Let $p \in (0, \infty]$ and $f \in L^p(S, \Sigma, \mu)$. The norm of f on $L^p(S, \Sigma, \mu)$ is defined as

$$\|f\|_p = \begin{cases} \left(\int_S |f|^p \mathrm{d}\mu\right)^{\min\left(\frac{1}{p}, 1\right)}, & \text{if } 0 < p < \infty; \\ \sup\{k : \mu(|f| > k) > 0\}, & \text{if } p = \infty. \end{cases}$$

When $p = \infty$, the norm is called the *infinite norm* or the *maximum norm*.

Definition 8.3 (Metric Space)**.** Let S be a set and $d : S \times S \to [0, \infty]$ be a function. (S, d) is called a *metric space* if for all $x, y, z \in S$,

(a) (Symmetry) $d(x, y) = d(y, x)$.

(b) (Triangle inequality) $d(x, y) \leq d(x, z) + d(y, z)$.

(c) $d(x, y) = 0$ if and only if $x = y$.

The function $d(\cdot, \cdot)$ is called a *metric* on S.

Definition 8.4 (Cauchy Sequence)**.** Let (S, d) be a metric space and $\{x_n\}_{n\geq 1}$ be a sequence of elements of S. Then the sequence $\{x_n\}_{n\geq 1}$ is called a *Cauchy sequence* in (S, d) if

$$\lim_{k\to\infty} \sup_{r,s\geq k} d(x_r, x_s) = 0.$$

Definition 8.5 (Completeness of Metric Spaces)**.** A metric space (S, d) is considered complete if every Cauchy sequence in (S, d) converges to an element in S; that is, if $\{x_n\}_{n\geq 1}$ is a Cauchy sequence in (S, d), then there exists an element x in (S, d) such that

$$\lim_{n\to\infty} d(x_n, x) = 0.$$

Definition 8.6 (Equivalence Relation)**.** Let S be a set. A relation $\sim$ is called an *equivalence relation* on S if

(a) (Reflexive) $x \sim x$ for all x in S.

(b) (Symmetry) $x \sim y$ implies $y \sim x$.

(c) (Transitive) $x \sim y$ and $y \sim z$ imply $x \sim z$.

Definition 8.7 (Equivalence Relation on L^p Spaces)**.** Let $f, g \in L^p(S, \Sigma, \mu)$. f is considered equivalent to g, written as $f \sim g$, if $f = g$ a.e. The relation $\sim$ is an equivalent relation on the L^p space (see Problem 8.7).

Definition 8.8 (Metric on L^p Spaces)**.** For $f, g \in L^p(S, \Sigma, \mu)$, $1 \leq p \leq \infty$, let $d_p(\cdot, \cdot)$ be defined as

$$d_p([f], [g]) = \|f - g\|_p, \tag{8.1}$$

where $[f]$ and $[g]$ denote the equivalence classes of functions containing f and g, respectively. Then $d_p(\cdot,\cdot)$ is a metric on $L^p(S,\Sigma,\mu)$ and $(L^p(S,\Sigma,\mu), d_p)$ is a metric space (see Problem 8.8).

Definition 8.9 (Real Inner Product Space). A vector space V is called a *real inner product space* if there exists a function $f : V\times V \to \mathbf{R}$, denoted by $f(x,y) = \langle x,y\rangle$, such that

(a) $\langle x,y\rangle = \langle y,x\rangle$ for all $x,y\in V$.

(b) $\langle ax+by,z\rangle = a\langle x,z\rangle + b\langle y,z\rangle$ for all $a,b\in\mathbf{R}$ and $x,y,z\in V$.

(c) $\langle x,x\rangle \geq 0$ for all $x\in V$ and $\langle x,x\rangle = 0$ if and only if $x=\theta$, the zero vector in V.

The function $\langle\cdot,\cdot\rangle$ is called the inner product.

Definition 8.10 (Orthogonal Vector and Collection). Let $(V,\langle\cdot,\cdot\rangle)$ be an inner product space. Two vectors $x,y\in V$ are considered orthogonal, written as $x\perp y$, if $\langle x,y\rangle = 0$. A collection $B\subseteq V$ is considered orthogonal if for $x,y\in B$, $x\neq y$ implies $\langle x,y\rangle = 0$.

Definition 8.11 (Inner product Space L^2). Let (S,Σ,μ) be a measure space. Let the inner product on $L^2(S,\Sigma,\mu)$ be defined as

$$\langle f,g\rangle = \int_S fg\mathrm{d}\mu,\quad f,g\in L^2(S,\Sigma,\mu).$$

Then $(L^2(S,\Sigma,\mu),\langle\cdot,\cdot\rangle)$ is a complete inner product space.

Theorem 8.1 (Hölder's Inequality). *Let (S,Σ,μ) be a measure space. Let $f\in L^p(S,\Sigma,\mu)$ and $g\in L^q(S,\Sigma,\mu)$, where $1<p<\infty$ and $q=p/(p-1)$. Then $\|fg\|_1\leq\|f\|_p\|g\|_q$:*

$$\int_S |fg|\mathrm{d}\mu \leq \left(\int_S |f|^p\mathrm{d}\mu\right)^{\frac{1}{p}} \left(\int_S |g|^q\mathrm{d}\mu\right)^{\frac{1}{q}}. \tag{8.2}$$

If $\|fg\|_1\neq 0$, then the equality in (8.2) holds if and only if $|f|^p = c|g|^q$ (a.e.) for some constant $c\in(0,\infty)$.

Theorem 8.2 (Minkowski's Inequality). *Let $1<p<\infty$ and $f,g\in L^p(S,\Sigma,\mu)$. Then $\|f+g\|_p\leq\|f\|_p+\|g\|_p$, specifically*

$$\left(\int_S |f+g|^p\mathrm{d}\mu\right)^{\frac{1}{p}} \leq \left(\int_S |f|^p\mathrm{d}\mu\right)^{\frac{1}{p}} + \left(\int_S |g|^p\mathrm{d}\mu\right)^{\frac{1}{p}}, \tag{8.3}$$

and the equality in (8.3) holds if and only if $f=\lambda g$ (a.e.) for some constant $\lambda\in(0,\infty)$.

Theorem 8.3 (Schwarz's Inequality). *Let* $f, g \in L^2(S, \Sigma, \mu)$. *Then*

$$fg \in L^1(S, \Sigma, \mu)$$

and

$$|\mu(fg)| \leq \mu(|fg|) \leq ||f||_2 \cdot ||g||_2.$$

Theorem 8.4 (Completeness of L^p Spaces). *For* $1 \leq p \leq \infty$, $(L^p(S, \Sigma, \mu), d_p)$ *is complete, where* d_p *is as defined in Equation* (8.1).

Theorem 8.5 (Orthogonal Projection). *Let* $(L^2(S, \Sigma, \mu), \langle \cdot, \cdot \rangle)$ *be an inner product space and* $\mathcal{K}$ *a subspace of* $(L^2(S, \Sigma, \mu), \langle \cdot, \cdot \rangle)$ *that is complete in that whenever* $\{f_n\}_{n \geq 1}$ *is a Cauchy sequence in* $\mathcal{K}$, *then there exists a function* $f \in \mathcal{K}$ *such that*

$$\|f_n - f\|_2 \to 0.$$

Then, given $X \in L^2(S, \Sigma, \mu)$, *there exists a function* $Y \in \mathcal{K}$ *such that*

$$\|X - Y\|_2 = \inf\{\|X - W\|_2 : W \in \mathcal{K}\}, \tag{8.4}$$

$$X - Y \perp Z, \quad \forall Z \in \mathcal{K}. \tag{8.5}$$

Also, (8.4) *and* (8.5) *are equivalent. If* $\tilde{Y}$ *satisfies* (8.4) *or* (8.5), *then* $\tilde{Y} = Y$ *(a.e.) The function* Y *is considered a version of the orthogonal projection of* X *into* $\mathcal{K}$.

8.2 Problems

8.1. Let $0 < \alpha < 1$ and $\beta = 1 - \alpha$. Let $x > 0$ and $y > 0$. Show that

$$x^\alpha y^\beta < \alpha x + \beta y.$$

8.2. Prove Hölder's inequality in Theorem 8.1.

8.3. Let $a \geq 0$, $b \geq 0$, and $p \geq 1$. Show that

$$(a + b)^p \leq 2^{p-1}(a^p + b^p). \tag{8.6}$$

8.4. Prove Minkowski's inequality in Theorem 8.2.

8.5. Let (S, Σ, μ) be a finite measure space [i.e., $\mu(S) < \infty$] and $1 \leq p < r < \infty$. Show that $L^r(S, \Sigma, \mu) \subseteq L^p(S, \Sigma, \mu)$.

8.6. Let $0 < p \leq \infty$. Let $a, b \in \mathbf{R}$ and $f, g \in L^p(S, \Sigma, \mu)$. Then $af + bg \in L^p(S, \Sigma, \mu)$.

8.7. Let $\sim$ be a relation on $L^p(S, \Sigma, \mu)$ such that $f \sim g$ if $f = g$ a.e. for $f, g \in L^p(S, \Sigma, \mu)$. Show that $\sim$ is an equivalent relation.

8.8. For $f, g \in L^p(S, \Sigma, \mu)$, $0 < p \leq \infty$, let $d_p(\cdot, \cdot)$ be defined as

$$d_p([f], [g]) = \|f - g\|_p,$$

where $[f]$ and $[g]$ denote the equivalence classes of functions containing f and g, respectively. Show that d_p is a metric on $L^p(S, \Sigma, \mu)$.

8.9 (Parallelogram Law). Let $U, V \in (L^2(S, \Sigma, \mu), \langle \cdot, \cdot \rangle)$. Show that

$$\|U + V\|_2^2 + \|U - V\|_2^2 = 2\|U\|_2^2 + 2\|V\|_2^2.$$

8.10. Prove Theorem 8.4.

8.11. Prove Theorem 8.5.

8.3 Hints

8.1. Consider the inequality $z^\alpha - 1 \leq \alpha(z - 1)$ for $\alpha \in (0, 1)$ and $z \geq 1$.

8.2. Use the result of Problem 8.1.

8.3. Consider the derivative of the following function

$$f(x) = (a + x)^p - 2^{p-1}(a^p + x^p).$$

8.4. Consider applying Hölder's inequality (Theorem 8.1) to the functions $\int_S |f + g|^{p-1}|f| \mathrm{d}\mu$ and $\int_S |f + g|^{p-1}|g| \mathrm{d}\mu$.

8.5. Use Hölder's inequality (Theorem 8.1) and the definition of L^p space (Definition 8.1).

8.6. Consider the case when $p \in (0, \infty)$ and the case when $p = \infty$.

8.7. Just follow the definition of an equivalence relation (Definition 8.6).

8.8. Follow the definition of a metric (Definition 8.3).

8.9. Follow the definition of the inner product in L^2 space (Definition 8.9).

8.10. Find a subsequence $\{n_k\}_{k \geq 1}$ such that $\{f_{n_k}\}_{k \geq 1}$ converges a.e. to a limit function f.

8.11. Consider a sequence $\{Y_n\}_{n \geq 1}$ in $\mathcal{K}$ such that

$$\|X - Y_n\|_2 \rightarrow \inf\{\|X - W\|_2 : W \in \mathcal{K}\}$$

and show that $\{Y_n\}_{n \geq 1}$ is a Cauchy sequence.

8.4 Solutions

8.1. We first show, for $\alpha \in (0,1)$ and $z \geq 1$, that

$$z^\alpha - 1 \leq \alpha(z-1), \tag{8.7}$$

and the equality in (8.7) holds if and only if $z = 1$. To do this, consider the function $\varphi(z) = \alpha(z-1) - z^\alpha + 1$. The derivative of this function is $\varphi'(z) = \alpha(1 - z^{\alpha-1})$. Since $\varphi'(z) > 0$ for $z > 1$, $\varphi(z)$ is strictly increasing in $(1, \infty)$. Noting that $\varphi(1) = 0$, we have $\varphi(z) > 0$ for $z > 1$.

Now suppose that $x \geq y$ (since x and y are symmetric, we can do this) and let $z = x/y$. Then

$$\left(\frac{x}{y}\right)^\alpha - 1 \leq \alpha\left(\frac{x}{y} - 1\right),$$

which, after some algebraic manipulation, is

$$x^\alpha y^\beta \leq \alpha x + \beta y,$$

and the equality holds if and only if $x/y = 1$ or $x = y$.

8.2. If $f = 0$ or $g = 0$, then both sides of (8.2) are equal to zero. If $f \neq 0$ and $g \neq 0$, then, by Problem 8.1, we have

$$\begin{aligned}\frac{|fg|}{\left(\int_S |f|^p \mathrm{d}\mu\right)^{\frac{1}{p}} \left(\int_S |g|^q \mathrm{d}\mu\right)^{\frac{1}{q}}} &= \left(\frac{|f|^p}{\int_S |f|^p \mathrm{d}\mu}\right)^{\frac{1}{p}} \left(\frac{|g|^q}{\int_S |g|^q \mathrm{d}\mu}\right)^{\frac{1}{q}} \\ &\leq \frac{1}{p}\left(\frac{|f|^p}{\int_S |f|^p \mathrm{d}\mu}\right) + \frac{1}{q}\left(\frac{|g|^q}{\int_S |g|^q \mathrm{d}\mu}\right).\end{aligned}$$

Therefore (8.2) follows from integrating both sides of the above equation over S and the equality in (8.2) holds if and only if

$$\frac{|f|^p}{\int_S |f|^p \mathrm{d}\mu} = \frac{|g|^q}{\int_S |g|^q \mathrm{d}\mu}, \quad \text{a.e.}$$

or $|f|^p = c|g|^q$ (a.e.), where

$$c = \frac{\int_S |f|^p \mathrm{d}\mu}{\int_S |g|^q \mathrm{d}\mu}.$$

This completes the proof.

8.3. Let

$$f(x) = (a+x)^p - 2^{p-1}(a^p + x^p), \quad x \geq 0.$$

Then

$$f'(x) = \frac{\mathrm{d}f(x)}{\mathrm{d}x} = p(a+x)^{p-1} - 2^{p-1}px^{p-1}, \quad x \geq 0.$$

Note that $f'(x) = p[(a+x)^{p-1} - (2x)^{p-1}]$. Hence

$$f'(x) \begin{cases} > 0 & \text{if } 0 \le x < a; \\ = 0 & \text{if } x = a; \\ < 0 & \text{if } x > a, \end{cases}$$

which implies that $f(x)$ is maximized at $x = a$. Therefore

$$f(b) = (a+b)^p - 2^{p-1}(a^p + b^p) \le f(a) = 0.$$

This completes the proof.

8.4. If $f + g = 0$, then (8.3) is true. Suppose that $f + g \ne 0$ and let $q = p/(p-1)$. Then $(p-1)q = p$ and, by Theorem 8.1, we get

$$\begin{aligned} \int_S |f+g|^{p-1}|f|\mathrm{d}\mu &\le \left(\int_S |f|^p \mathrm{d}\mu\right)^{\frac{1}{p}} \left(\int_S |f+g|^{(p-1)q}\mathrm{d}\mu\right)^{\frac{1}{q}} \\ &= \left(\int_S |f|^p \mathrm{d}\mu\right)^{\frac{1}{p}} \left(\int_S |f+g|^{p}\mathrm{d}\mu\right)^{\frac{1}{q}}, \end{aligned} \tag{8.8}$$

and, similarly,

$$\int_S |f+g|^{p-1}|g|\mathrm{d}\mu \le \left(\int_S |g|^p \mathrm{d}\mu\right)^{\frac{1}{p}} \left(\int_S |f+g|^{p}\mathrm{d}\mu\right)^{\frac{1}{q}}. \tag{8.9}$$

Thus we have

$$\begin{aligned} \int_S |f+g|^p \mathrm{d}\mu &\le \int_S |f+g|^{p-1}(|f|+|g|)\mathrm{d}\mu \\ &\le \left(\int_S |f+g|^p \mathrm{d}\mu\right)^{\frac{1}{q}} (\|f\|_p + \|g\|_p). \end{aligned} \tag{8.10}$$

Dividing (8.10) by $(\int_S |f+g|^p \mathrm{d}\mu)^{\frac{1}{q}}$ gives (8.3). In addition, the equality in (8.3) holds if and only if the equalities in Equations (8.8)-(8.10) hold. Therefore $|f+g| = |f| + |g|$ (a.e.) and , by Theorem 8.1, $|f+g|^p = |f+g|^{(p-1)q} = c_1|f|^p = c_2|g|^p$ (a.e.) for some constants $c_1, c_2 \in (0, \infty)$. Accordingly, $f = \lambda g$ (a.e.) for some $\lambda \in (0, \infty)$.

8.5. Since $p < r$, $r/p > 1$. Let $q = r/(r-p)$. Then $1/(r/p) + 1/q = 1$. Let $f \in L^r(S, \Sigma, \mu)$. Then, by Theorem 8.1, we obtain

$$\int_S |f|^p \mathrm{d}\mu \le \left(\int_S |f|^r \mathrm{d}\mu\right)^{\frac{p}{r}} \left(\int_S 1^q \mathrm{d}\mu\right)^{\frac{1}{q}} = \left(\int_S |f|^r \mathrm{d}\mu\right)^{\frac{p}{r}} \mu(S)^{\frac{1}{q}}.$$

Noting that $\mu(S) < \infty$ and $f \in L^r(S, \Sigma, \mu)$, we have

$$\int_S |f|^p \mathrm{d}\mu < \infty.$$

Thus $f \in L^p(S, \Sigma, \mu)$. This completes the proof.

8.6. First, we consider $p \in (0, \infty)$. Then

$$\begin{aligned}
|af + bg|^p &\le (|af| + |bg|)^p \\
&\le [2\max(|af|, |bg|)]^p \\
&\le 2^p(|af|^p + |bg|^p) \\
&= |2a|^p|f|^p + |2b|^p|g|^p.
\end{aligned}$$

The result follows by integrating the above equation.

Next, we consider $p = \infty$. By definition, there exist constants $K_1 < \infty$ and $K_2 < \infty$ such that $\mu(|f| > K_1) = 0$ and $\mu(|g| > K_2) = 0$. Then $\mu(|af + bg| > k) \le \mu(|af| + |bg| > K) \le \mu(|f| > K_1) + \mu(|g| > K_2) = 0$ for $K > |a|K_1 + |b|K_2$. Hence $af + bg \in L^\infty(S, \Sigma, \mu)$.

8.7. Since $\mu(f \ne f) = \mu(\emptyset) = 0$ for $f \in L^p(S, \Sigma, \mu)$, $f \sim f$. If $f \sim g$, then $\mu(f \ne g) = 0$. But $\mu(g \ne f) = \mu(f \ne g) = 0$. Hence $g \sim f$. Suppose that $f \sim g$ and $g \sim h$ for $f, g, h \in L^p(S, \Sigma, \mu)$. Then $\mu(f \ne g) = 0$ and $\mu(g \ne h) = 0$. Since $\{f = g\} \cap \{g = h\} \subseteq \{f = h\}$, $\{f \ne h\} \subseteq \{f \ne g\} \cup \{g \ne h\}$, and so $\mu(f \ne h) \le \mu(f \ne g) + \mu(g \ne h) = 0$. Hence $f \sim h$. Thus $\sim$ is an equivalent relation.

8.8. Since functions in an equivalence class are equivalent, d_p is well defined. Since $d_p([f], [g]) = \|f - g\|_p = \|g - f\|_p = d_p([g], [f])$, symmetry holds. Let $f, g, h \in L^p(S, \Sigma, \mu)$. Since

$$\|f - g\|_1 = \int_S |f - g| \mathrm{d}\mu \le \int_S (|f - h| + |g - h|) \mathrm{d}\mu = \|f - h\|_1 + \|g - h\|_1,$$

and, by Theorem 8.1,

$$\|f - g\|_p \le \|f - h\|_p + \|g - h\|_p, \quad p > 1,$$

we have $d_p([f], [g]) \le d_p([f], [h]) + d_p([g], [h])$. Suppose that $d_p([f], [g]) = 0$. Then, by Problem 6.8, $f = g$ a.e. and so $[f] = [g]$. Suppose that $[f] = [g]$. Then $f = g$ a.e. and so $d_p([f], [g]) = \|f - g\|_p = 0$. Therefore, d_p is a metric on $L^p(S, \Sigma, \mu)$.

8.9. By definition of the inner product, we obtain

$$\begin{aligned}
&\|U + V\|_2^2 + \|U - V\|_2^2 \\
=&\langle U + V, U + V\rangle + \langle U - V, U - V\rangle \\
=&\langle U, U\rangle + 2\langle U, V\rangle + \langle V, V\rangle + \langle U, U\rangle - 2\langle U, V\rangle + \langle V, V\rangle \\
=&2\|U\|_2^2 + 2\|V\|_2^2.
\end{aligned}$$

8.10. Let $\{f_n\}_{n\ge1}$ be a Cauchy sequence in $(L^p(S, \Sigma, \mu), d_p)$. We need to show that there exists a function $f \in (L^p(S, \Sigma, \mu), d_p)$ such that

$$\lim_{n\to\infty} \|f_n - f\|_p = 0.$$

First, we show that there exists a subsequence $\{n_k\}_{k\geq 1}$ such that $\{f_{n_k}\}_{k\geq 1}$ converges (a.e.) to a limit function f. To do this, let $\epsilon_k = 2^{-(p+1)k}$ and $\delta_k = 2^{-k}$ for $k \geq 1$. Since $\{f_n\}_{n\geq 1}$ is a Cauchy sequence, for each $k \geq 1$, there exists an integer n_k such that

$$\int_S |f_r - f_s|^p \mathrm{d}\mu < \epsilon_k, \quad \text{for all } r, s \geq n_k. \tag{8.11}$$

Let $A_k = \{|f_{n_{k+1}} - f_{n_k}| \geq \delta_k\}$ for $k \geq 1$. Then

$$\begin{aligned}\mu(A_k) &= \int_{A_k} 1 \mathrm{d}\mu \\ &\leq \delta_k^{-p} \int_{A_k} |f_{n_{k+1}} - f_{n_k}|^p \mathrm{d}\mu \\ &\leq \delta_k^{-p} \int_S |f_{n_{k+1}} - f_{n_k}|^p \mathrm{d}\mu \\ &< \delta_k^{-p} \epsilon_k = 2^{-k}, \quad k \geq 1,\end{aligned}$$

which implies $\sum_{k=1}^{\infty} \mu(A_k) < \infty$. By Theorem 2.2, $\mu(A) = 0$, where $A = \limsup A_k$. Then for $s \in A^c$, there exists an integer k_s such that $s \in A_k^c$ for all $k \geq k_s$, and so

$$\sum_{k\geq k_s} |f_{n_{k+1}}(s) - f_{n_k}(s)| < \sum_{k\geq k_s} \delta_k < \infty.$$

Thus $\{f_{n_k}(s)\}_{k\geq 1}$ converges in $\mathbf{R}$. Let

$$f(s) = \begin{cases} \lim_{k\to\infty} f_{n_k}(s), & \text{if } s \in A^c; \\ 0, & \text{if } s \in A. \end{cases}$$

Then we have

$$\lim_{k\to\infty} f_{n_k} = f \quad \text{a.e.}$$

Next, we show that

$$\lim_{k\to\infty} \|f_{n_k} - f\|_p = 0. \tag{8.12}$$

By Theorem 6.3 and (8.11), we have

$$\int_S |f_{n_k} - f|^p \mathrm{d}\mu \leq \liminf \int_S |f_{n_k} - f_{n_{k+j}}|^p \mathrm{d}\mu < \epsilon_k, \quad k \geq 1. \tag{8.13}$$

Thus $f_{n_k} - f \in L^p(S, \Sigma, \mu)$. Since $f_{n_k} \in L^p(S, \Sigma, \mu)$, by Problem 8.6, $f \in L^p(S, \Sigma, \mu)$. The result in (8.12) follows by letting $k \to \infty$ in (8.13).

Next, we show that

$$\lim_{n\to\infty} \|f_n - f\|_p = 0. \tag{8.14}$$

By triangle inequality (see Problem 8.8), Equations (8.11) and (8.13), we have

$$\|f_n - f\|_p \leq |f_n - f_{n_k}\|_p + \|f_{n_k} - f\|_p < 2\epsilon_k^{\frac{1}{p}}, \quad n \geq n_k.$$

The result in (8.14) follows by letting $k \to \infty$ in the above inequality. This completes the proof.

8.11. Let $\{Y_n\}_{n\geq 1}$ be a sequence in $\mathcal{K}$ such that

$$\|X - Y_n\|_2 \to \Delta,$$

where $\Delta = \inf\{\|X - W\|_2 : W \in \mathcal{K}\}$. Then by Problem 8.9, we have

$$\|X - Y_r\|_2^2 + \|X - Y_s\|_2^2 = 2\left\|X - \frac{1}{2}(Y_r + Y_s)\right\|_2^2 + 2\left\|\frac{1}{2}(Y_r - Y_s)\right\|_2^2.$$

Since $\frac{1}{2}(Y_r + Y_s) \in \mathcal{K}$, $\|X - \frac{1}{2}(Y_r + Y_s)\|_2 \geq \Delta$. Let $\epsilon > 0$. Choose $N_\epsilon \in \mathbf{N}$ such that

$$\|X - Y_n\|_2 < \sqrt{\Delta^2 + \epsilon^2}, \quad \forall n \geq N_\epsilon.$$

Then for $r, s \geq N_\epsilon$, we have

$$2\left\|\frac{1}{2}(Y_r - Y_s)\right\|_2^2 < 2(\Delta^2 + \epsilon^2) - 2\Delta^2 = 2\epsilon^2,$$

which gives $\|\frac{1}{2}(Y_r - Y_s)\|_2 < \epsilon$. Thus $\{Y_n\}_{n\geq 1}$ is a Cauchy sequence in $\mathcal{K}$. Since $\mathcal{K}$ is complete, there exists a $Y \in \mathcal{K}$ such that

$$\|Y_n - Y\|_2 \to 0.$$

Since, by Theorem 8.2, $\|X - Y\|_2 \leq \|X - Y_n\|_2 + \|Y_n - Y\|_2$, we have $\|X - Y\|_2 = \Delta$.

Now we show that Equations (8.4) and (8.5) are equivalent. Suppose that $Y \in \mathcal{K}$ satisfies $\|X - Y\|_2 = \Delta$. Then for any $Z \in \mathcal{K}$, we have $Y + tZ \in \mathcal{K}$ for $t \in \mathbf{R}$. Thus $\|X - Y - tZ\|_2 \geq \|X - Y\|_2$. Note that $\|X - Y - tZ\|_2^2 = \|X - Y\|_2^2 - 2t\langle X - Y, Z\rangle + t^2\|Z\|_2^2$, we have

$$-2t\langle X - Y, Z\rangle + t^2\|Z\|_2^2 \geq 0, \quad t \in \mathbf{R}.$$

Therefore $\langle X - Y, Z\rangle = 0$. Suppose that Y satisfies (8.5), that is, $X - Y \perp Z$ for all $Z \in \mathcal{K}$. Then for any $Z \in \mathcal{K}$, we have $Y - Z \in \mathcal{K}$, and so $\langle X - Y, Y - Z\rangle = 0$. Thus

$$\|X - Z\|_2^2 = \|X - Y\|_2^2 + \|Y - Z\|_2^2 \geq \|X - Y\|_2^2, \quad \forall Z \in \mathcal{K},$$

which gives (8.4).

If $Y, \tilde{Y} \in \mathcal{K}$ satisfy (8.5), then

$$\langle X - Y, Z\rangle = 0, \quad \langle X - \tilde{Y}, Z\rangle = 0, \quad \forall Z \in \mathcal{K},$$

and so $\langle Y - \tilde{Y}, Z\rangle = 0$ for all $Z \in \mathcal{K}$. Since $Y - \tilde{Y} \in \mathcal{K}$, we have $\langle Y - \tilde{Y}, Y - \tilde{Y}\rangle = 0$, which implies $Y = \tilde{Y}$ a.e. This completes the proof.

8.5 Bibliographic Notes

In this chapter, we introduced basic concepts of L^p spaces and some inequalities such as the Hölder's inequality, Minkowski's inequality, and Schwarz's inequality. L^p spaces are covered by books on functional analysis, measure theory, and probability theory. For example, readers can find more information on L^p spaces in Williams (1991), Ash and Doleans-Dade (1999), Athreya and Lahiri (2006), and Capinski and Kopp (2007).

CHAPTER 9

CONVERGENCE

In measure theory, there exist several different notions of convergence of measurable functions, for example, convergence almost everywhere, convergence in measure, convergence in L^p, to just name a few. In this chapter, we present definitions of some common notions of convergence.

9.1 Basic Concepts and Facts

Definition 9.1 (Convergence in Measure). Let $f, f_1, f_2, \cdots$ be Borel measurable functions on a measure space (S, Σ, μ). The sequence $\{f_n\}_{n\geq 1}$ is said to converge to f in measure, written as $f_n \xrightarrow{\mu} f$, if and only if for every $\epsilon > 0$, $\mu\{s : |f_n(s) - f(s)| \geq \epsilon\} \to 0$ as $n \to \infty$.

In particular, if μ is a probability measure, then convergence in measure is also referred to as *convergence in probability*.

Definition 9.2 (Convergence in L^p). Let f be a Borel measurable function on a measure space (S, Σ, μ) and $\{f_n\}_{n\geq 1}$ a sequence of of functions in $L^p(S, \Sigma, \mu)$.

The sequence $\{f_n\}_{n\geq 1}$ is said to converge to f in L^p, written as $f_n \xrightarrow{L^p} f$, if and only if $\|f_n - f\|_p \to 0$ as $n \to \infty$, where $\|f_n - f\|_p = \left(\int_S |f_n - f|^p \mathrm{d}\mu\right)^{\frac{1}{p}}$.

Definition 9.3 (Uniform Convergence)**.** Let $f, f_1, f_2, \ldots$ be Borel measurable functions on a measure space (S, Σ, μ) and $B \in \Sigma$. The sequence $\{f_n\}_{n\geq 1}$ is said to converge to f uniformly on B if and only if for every $\epsilon > 0$, there exists an $N \in \mathbf{N}$ such that $|f_n(s) - f(s)| < \epsilon$ for all $s \in B$ and $n \geq N$.

Definition 9.4 (Almost Uniform Convergence)**.** Let $f, f_1, f_2, \ldots$ be Borel measurable functions on a measure space (S, Σ, μ). The sequence $\{f_n\}_{n\geq 1}$ is said to converge to f almost uniformly if and only if for a given $\epsilon > 0$, there is a set $A \in \Sigma$ such that $\mu(A) < \epsilon$ and $f_n \to f$ uniformly on A^c.

Definition 9.5 (Almost Everywhere Convergence)**.** Let $f, f_1, f_2, \ldots$ be Borel measurable functions on a measure space (S, Σ, μ). The sequence $\{f_n\}_{n\geq 1}$ is said to converge to f almost everywhere, written as $f_n \xrightarrow{\text{a.e.}} f$, if and only if there is a set $A \in \Sigma$ such that $f_n \to f$ on A and $\mu(A^c) = \mu(S\backslash A) = 0$.

Definition 9.6 (Uniform Integrability)**.** Let $\{f_i\}_{i\in I}$ be a collection of Borel measurable functions on the measure space (S, Σ, μ), where μ is finite and I is an index set (countable or uncountable). The f_i are said to be uniformly integrable if and only if

$$\lim_{c\to\infty} \sup_{i\in I} \int_{\{|f_i|\geq c\}} |f_i| \mathrm{d}\mu = 0.$$

Theorem 9.1 (Weierstrass M-Test)**.** *Let $\{f_n\}_{n\geq 1}$ be a sequence of Borel measurable functions defined on a set A. Suppose that there exist positive constants M_n such that*

$$|f_n(s)| \leq M_n, \quad \forall s \in A,\ n \geq 1,$$

and the series $\sum_{n=1}^{\infty} M_n$ converges. Then the series $\sum_{n=1}^{\infty} f_n$ converges uniformly on A.

9.2 Problems

9.1. Let $p \in (0, \infty)$. If $f, f_1, f_2, \ldots \in L^p(S, \Sigma, \mu)$, then $f_n \xrightarrow{L^p} f$ implies $f_n \xrightarrow{\mu} f$.

9.2. Let (S, Σ, μ) be a measure space and $f, f_1, f_2, \ldots \in m\Sigma$. Suppose that $f_n \to f$ almost uniformly. Show that $f_n \to f$ in measure and almost everywhere.

9.3. Let (S, Σ, μ) be a measure space and $f, f_1, f_2, \cdots \in m\Sigma$. Given μ finite, show that $f_n \to f$ almost everywhere if and only if for every $\delta > 0$,

$$\mu\left(\bigcup_{k=n}^{\infty} \{s : |f_k(s) - f(s)| \geq \delta\}\right) \to 0 \quad \text{as} \quad n \to \infty.$$

9.4 (Egoroff's Theorem)**.** Let (S, Σ, μ) be a measure space and $f, f_1, f_2, \ldots \in m\Sigma$. Suppose that μ is finite and $f_n \to f$ a.e. Show that $f_n \to f$ almost uniformly.

9.5. Let (S, Σ, μ) be a measure space and $f, g, f_1, f_2, \ldots \in m\Sigma$. Suppose that $f_n \xrightarrow{\mu} f$ and $f_n \xrightarrow{\mu} g$. Show that $f = g$ a.e.[i.e., $\mu(f \neq g) = 0$].

9.6. Let (S, Σ, μ) be a measure space and $f, f_1, f_2, \ldots \in m\Sigma$. Suppose that $f_n \xrightarrow{\mu} f$. Show that

(a) There is a subsequence converging almost uniformly to f.

(b) There is a subsequence converging to f a.e.

9.7. Let (S, Σ, μ) be a measure space and $f_1, f_2, \ldots \in m\Sigma$. Show that

$$\lim_{n\to\infty} f_n$$

exists a.e. if and only if $f_j - f_k \to 0$ as $j, k \to \infty$ a.e, that is, for each $\epsilon > 0$,

$$\mu\left(\bigcup_{j,k\geq n} \{|f_j - f_k| \geq \epsilon\}\right) \to 0 \quad \text{as} \quad n \to \infty.$$

9.8. Let (S, Σ, μ) be a measure space and $f_1, f_2, \cdots \in m\Sigma$. Show that

$$\lim_{n\to\infty} f_n$$

exists a.e. if and only if for each $\epsilon > 0$,

$$\mu\left(\bigcup_{j=1}^{\infty} \{|f_{n+j} - f_n| \geq \epsilon\}\right) \to 0 \quad \text{as} \quad n \to \infty.$$

9.9. Let $\{f_n : n \geq 1\}$ be a sequence of measurable functions on a finite measure space (S, Σ, μ) such that for every $n \geq 1$,

$$\int_S |f_n|^2 \mathrm{d}\mu < C,$$

where $C < \infty$ is a constant. Show that $\{f_n : n \geq 1\}$ is uniformly integrable.

9.10. Let (S, Σ, μ) be a finite measure space and $f_1, f_2, \ldots$ uniformly integrable functions on (S, Σ, μ). Show that

(a)

$$\int_S \liminf f_n \mathrm{d}\mu \leq \liminf \int_S f_n \mathrm{d}\mu \leq \limsup \int_S f_n \mathrm{d}\mu \leq \int_S \limsup f_n \mathrm{d}\mu. \tag{9.1}$$

(b) If $f_n \xrightarrow{\text{a.e.}} f$, then f is μ-integrable and

$$\lim_{n\to\infty} \int_S f_n \mathrm{d}\mu = \int_S f \mathrm{d}\mu. \tag{9.2}$$

(c) If $f_n \xrightarrow{\mu} f$, then f is μ-integrable and

$$\lim_{n\to\infty} \int_S f_n \mathrm{d}\mu = \int_S f \mathrm{d}\mu. \tag{9.3}$$

9.11. Let $\{f_n\}_{n\geq 1}$ be a sequence of Borel measurable functions on the measure space (S, Σ, μ), where μ is finite. Show that $\{f_n\}_{n\geq 1}$ are uniformly integrable if and only if the integrals $\int_S |f_n| \mathrm{d}\mu$ are uniformly bounded, that is,

$$\sup_n \int_S |f_n| \mathrm{d}\mu \leq M$$

for some $M > 0$, and uniformly continuous:

$$\sup_n \int_A |f_n| \mathrm{d}\mu \to 0 \quad \text{as} \quad \mu(A) \to 0.$$

9.12. Let (S, Σ, μ) be a finite measure space and $0 < p < \infty$. Let $\{f_n\}_{n\geq 1}$ be a sequence of Borel measurable functions. Assume that $f_n \xrightarrow{\mu} f$ and the $|f_n|^p$ are uniformly integrable. Show that

$$f_n \xrightarrow{L^p} f \quad \text{as} \quad n \to \infty.$$

9.13. Let f_i, $i \in I$, be integrable functions on a finite measure space (S, Σ, μ). Let $h : [0, \infty) \to [0, \infty)$ be a Borel measurable function such that

$$\frac{h(t)}{t} \to \infty \quad \text{as} \quad t \to \infty.$$

Show that if

$$\sup_{i\in I} \int_S h(|f_i|) \mathrm{d}\mu < \infty,$$

then the f_i are uniformly integrable.

9.14. Let $0 < p < \infty$ and $f, f_1, f_2, \ldots \in L^p(S, \Sigma, \mu)$. Assume that $f_n \xrightarrow{L^p} f$ and μ is finite. Show that the $|f_n|^p$ are uniformly integrable.

9.3 Hints

9.1. Follow the definition of convergence L^p and convergence in measure and use Chebyshev's inequality (see Problem 6.16).

9.2. Follow the definition of almost uniform convergence (see Definition 9.4), convergence in measure (see Definition 9.1), and almost everywhere convergence (see Definition 9.5).

9.3. Follow the definition of convergence almost everywhere (see Definition 9.5).

9.4. Use the result of Problem 9.3.

9.5. Use the definition of convergence in measure (see Definition 9.1) and the fact that

$$\{s : |f(s)-g(s)| \geq \frac{1}{k}\} \subseteq \{s : |f_n(s)-f(s)| \geq \frac{1}{2k}\} \cup \{s : |f_n(s)-g(s)| \geq \frac{1}{2k}\}.$$

9.6. To prove part (a), first show that $\{f_n\}_{n\geq 1}$ is Cauchy in measure; that is, for every $\epsilon > 0$,

$$\lim_{n\to\infty} \sup_{j,k\geq n} \mu\{|f_j - f_k| \geq \epsilon\} = 0.$$

Then use the result of Problem 9.5 to show that the limit of the subsequence is equal to f a.e.

9.7. Use the result of Problem 9.3.

9.8. Use the result of Problem 9.7.

9.9. Follow the definition of uniform integrability (see Definition 9.6).

9.10. Part (a) can be proved by Fatou's lemma (see Theorem 6.3) and the definition of uniformly integrability (see Definition 9.6). Try to establish

$$\int_S \liminf f_n \mathrm{d}\mu \leq \liminf \int_S f_n \mathrm{d}\mu + \epsilon.$$

Parts (b) and (c) can be proved by part (a) and the result of Problem 9.6.

9.11. The "if" part can be proved by Chebyshev's inequality (see Problem 6.16). The "only if" part can be proved by following the definition of uniform integrability (see Definition 9.6).

9.12. First use the result of Problem 9.11 to show that the $|f_n - f|^p$ are uniformly integrable. Then use the result of Problem 9.10 to show that the subsequence converges to f in L^p.

9.13. Use the definition of uniform integrability (see Definition 9.6).

9.14. Use the results of Problems 8.3 and 9.11. First try to establish that $|f_n - f|^p$ is uniformly integrable.

9.4 Solutions

9.1. Let $\epsilon > 0$. Then by Problem 6.16, we have

$$\mu\{s : |f_n(s) - f(s)| \geq \epsilon\} \leq \frac{1}{\epsilon^p} \int_S |f_n(s) - f(s)|^p \mathrm{d}\mu.$$

By assumption, we have $\int_S |f_n(s) - f(s)|^p \mathrm{d}\mu \to 0$ as $n \to \infty$. Thus we have $\mu\{s : |f_n(s) - f(s)| \geq \epsilon\} \to 0$ as $n \to \infty$. This completes the proof.

9.2. First, we show that $f_n \to f$ in measure. Let $\epsilon > 0$. By definition, we only need to show that $\mu\{s : |f_n(s) - f(s)| \geq \epsilon\} \to 0$ as $n \to \infty$. To do this, we only need to show that for every $k > 1$, there exists an integer N_k such that

$$\mu\{s : |f_n(s) - f(s)| \geq \epsilon\} < \frac{1}{k}$$

for all $n \geq N_k$. By assumption that $f_n \to f$ almost uniformly, for every $k > 1$ there exists a set $A_k \in \Sigma$ such that $f_n \to f$ uniformly on A_k^c and $\mu(A_k) < 1/k$. By Definition 9.3, there exists an N_k such that $|f_n(s) - f(s)| < \epsilon$ on A_k^c for all $n \geq N_k$. Hence the result follows since $\mu\{s : |f_n(s) - f(s)| \geq \epsilon\} = \mu(A_k) < 1/k$ for all $n \geq N_k$.

Next, we show that $f_n \to f$ almost everywhere. Let A_k $(k \geq 1)$ be the sets mentioned in the first part of this proof and $B = \bigcup_{k=1}^{\infty} A_k^c$. Then $f_n \to f$ on B. Noting that $\mu(B^c) = \mu(\bigcap_{k=1}^{\infty} A_k) \leq \mu(A_k) < 1/k$, we have $\mu(B^c) =$ if we let $k \to \infty$. Hence $f_n \to f$ almost everywhere.

9.3. First, we prove the "if" part. Suppose that for every $\delta > 0$,

$$\mu\left(\bigcup_{k=n}^{\infty} \{s : |f_k(s) - f(s)| \geq \delta\}\right) \to 0 \quad \text{as} \quad n \to \infty.$$

Let $A_m = \bigcap_{n=1}^{\infty} \bigcup_{k=n}^{\infty} \{s : |f_k(s) - f(s)| \geq 1/m\}$ for $m = 1, 2, \cdots$. Then we have $\mu(A_m) = 0$ for all $m \geq 1$. Now let $A = \bigcup_{m=1}^{\infty} A_m$. Then $0 \leq \mu(A) \leq \sum_{m=1}^{\infty} \mu(A_m) = 0$, implying that $\mu(A) = 0$. The result follows if we show that $f_n \to f$ on A^c. To do this, let $\epsilon > 0$ and $t \in A^c$. Then $t \in A_m^c$ for all $m \geq 1$. Hence $t \in A_r^c$, where $r \in \mathbf{N}$ and $r > 1/\epsilon$. Therefore, $t \in \bigcap_{k=n_r}^{\infty} \{s : |f_k(s) - f(s)| < 1/r < \epsilon\}$ for some $n_r \geq 1$, which means that $|f_k(t) - f(t)| < \epsilon$ for all $k \geq n_r$. Since ϵ and t are arbitrary, we have $f_n \to f$ on A^c.

Now we prove the "only if" part. Suppose that $f_n \to f$ almost everywhere. Then there exists a set $A \in \Sigma$ such that $\mu(A) = 0$ and $f_n \to f$ on A^c. Let $\delta > 0$. Then for every $t \in A^c$, there exists an integer $N_{t,\delta}$ such that $|f_k(t) - f(t)| < \delta$ for all $k \geq N_{t,\delta}$. Hence $t \in \bigcup_{n=1}^{\infty} \bigcap_{k=n}^{\infty} \{s : |f_k(s) - f(s)| < \delta\}$. Since t is arbitrary, we have $A^c \subseteq \bigcup_{n=1}^{\infty} \bigcap_{k=n}^{\infty} \{s : |f_k(s) - f(s)| < \delta\}$. Noting that μ is finite, we $\mu(S) \geq \mu(\bigcup_{n=1}^{\infty} \bigcap_{k=n}^{\infty} \{s : |f_k(s) - f(s)| < \delta\}) \geq \mu(A^c) = \mu(S) - \mu(A) = \mu(S)$, implying that $\mu(\bigcup_{n=1}^{\infty} \bigcap_{k=n}^{\infty} \{s : |f_k(s) - f(s)| < \delta\}) = \mu(S)$. Therefore, $\mu(\bigcap_{n=1}^{\infty} \bigcup_{k=n}^{\infty} \{s : |f_k(s) - f(s)| \geq \delta\}) = 0$. But $\bigcup_{k=n}^{\infty} \{s : |f_k(s) - f(s)| \geq \delta\} \downarrow$

$\bigcap_{n=1}^{\infty}\bigcup_{k=n}^{\infty}\{s: |f_k(s)-f(s)|\geq\delta\}$ as $n\to\infty$. The result follows from Problem 2.18.

9.4. Since μ is finite and $f_n \to f$ a.e., by Problem 9.3 we have for every $\delta > 0$,

$$\mu\left(\bigcup_{k=n}^{\infty}\{s: |f_k(s)-f(s)|\geq\delta\}\right)\to 0 \quad \text{as} \quad n\to\infty.$$

Let $\epsilon > 0$. Then for every $m \geq 1$, there exists an integer N_m such that $\mu(A_m) < \epsilon/2^m$, where

$$A_m = \bigcup_{k=N_m}^{\infty}\{s: |f_k(s)-f(s)|\geq\frac{1}{m}\}.$$

Define $A = \bigcup_{m=1}^{\infty} A_m$. Then $\mu(A) < \epsilon$.

Now we show that $f_n \to f$ uniformly on A^c. To do this, let $\theta > 0$. Choose m_0 such that $(1/m_0) < \theta$. Then for all $m \geq N_{m_0}$ and $s \in A^c$, we have $|f_m(s)-f(s)| < \frac{1}{m_0} < \theta$. Hence $f_n \to f$ uniformly on A^c. This completes the proof.

9.5. Let $\epsilon > 0$. Note that

$$\{s: f(s)\neq g(s)\} = \bigcup_{k=1}^{\infty}\{s: |f(s)-g(s)|\geq\frac{1}{k}\}.$$

Since $|f(s)-g(s)| \leq |f_n(S)-f(s)| + |f_n(s)-g(s)|$, we have

$$\{s: |f(s)-g(s)|\geq\frac{1}{k}\}\subseteq\{s: |f_n(s)-f(s)|\geq\frac{1}{2k}\}\cup\{s: |f_n(s)-g(s)|\geq\frac{1}{2k}\}.$$

By assumption, for every $k \geq 1$, there exists an integer N_k such that $\mu\{s: |f_n(S) - f(s)| \geq \frac{1}{2k}\} < \frac{\epsilon}{2^{k+1}}$ and $\mu\{s: |f_n(S)-g(s)| \geq \frac{1}{2k}\} < \frac{\epsilon}{2^{k+1}}$ for all $n \geq N_k$. Then we have $\mu\{s: |f(s)-g(s)| \geq \frac{1}{k}\} < \frac{\epsilon}{2^k}$. Therefore, $\mu(f\neq g) \leq \sum_{k=1}^{\infty}\mu\{s: |f(s)-g(s)| \geq \frac{1}{k}\} < \epsilon$. Since ϵ is arbitrary, we have $\mu(f\neq g) = 0$. This completes the proof.

9.6.

(a) First we show that $\{f_n\}_{n\geq 1}$ is Cauchy in measure. Let $\epsilon > 0$ and $\eta > 0$. Since $f_n \xrightarrow{\mu} f$, we have

$$\mu\left\{|f_j - f|\geq\frac{\epsilon}{2}\right\} < \frac{\eta}{2}, \quad \forall j\geq N_\eta,$$

where N_η is some integer. Then for all $j, k \geq N_\eta$, we have

$$\mu\{|f_j - f_k|\geq\epsilon\}\leq\mu\left\{|f_j - f|\geq\frac{\epsilon}{2}\right\}+\mu\left\{|f_k - f|\geq\frac{\epsilon}{2}\right\} < \eta.$$

Hence $\{f_n\}_{n\ge 1}$ is Cauchy in measure. Then for each positive integer k, we can choose an integer N_k such that $N_{k+1} > N_k$ and

$$\mu\left\{|f_j - f_k| \ge \frac{1}{2^k}\right\} < \frac{1}{2^k}, \quad \forall j,k \ge N_k.$$

Let $g_k = f_{N_k}$, $k = 1, 2, \ldots$. Then

$$\mu\left\{|g_k - g_{k+1}| \ge \frac{1}{2^k}\right\} < \frac{1}{2^k}, \quad k \ge 1.$$

Let $A = \limsup A_k$, where $A_k = \{|g_k - g_{k+1}| \ge 1/2^k\}$. Then by Theorem 2.2, we have $\mu(A) = 0$.

If $s \notin A$, then $s \in A_k$ for only finitely many k. Hence $s \notin A_k$ for $k \ge M_s$ for some M_s. Therefore, $|g_k(s) - g_{k+1}(s)| < 1/2^k$ for $k \ge M_s$. Thus $\{g_k(s)\}_{k\ge 1}$ converges to a limit $g(s)$. Since $\mu(A) = 0$, we have $g_k \xrightarrow{\text{a.e.}} g$.

Now we show that g_k converges to g almost uniformly. Let $\delta > 0$ and $B_n = \bigcup_{k=n}^{\infty} A_k$. Then we can choose a large integer r such that

$$\mu(B_r) \le \sum_{k=r}^{\infty} \mu(A_k) = \frac{1}{2^{r-1}} < \delta.$$

By Theorem 9.1, g_k converges to g uniformly. Since δ is arbitrary, g_k converges to g almost uniformly.

By assumption, we have $f_n \xrightarrow{\mu} f$. Hence $g_k \xrightarrow{\mu} f$. By Problem 9.2, we have $g_k \xrightarrow{\mu} g$. Thus, by Problem 9.5, we have $f = g$ a.e.

(b) This part follows from part (a) and Problem 9.2.

9.7. First let us prove the "if" part. Suppose that, for each $\epsilon > 0$, we have

$$\mu\left(\bigcup_{j,k\ge n} \{|f_j - f_k| \ge \epsilon\}\right) \to 0 \quad \text{as} \quad n \to \infty.$$

Let $\eta > 0$ and $\delta > 0$. Then there exists an N_η such that

$$\mu\left(\bigcup_{j,k\ge N_\eta} \left\{|f_j - f_k| \ge \frac{\delta}{2}\right\}\right) < \eta.$$

Let $f = \limsup f_n$. Now we consider the following set:

$$A = \bigcup_{k=N_\eta}^{\infty} \{|f_k - f| \ge \delta\}.$$

We claim that

$$A \subseteq \bigcup_{j,k \geq N_\eta} \left\{ |f_j - f_k| \geq \frac{\delta}{2} \right\}.$$

To show that, let $s \in A$. Then there exists an integer $k_0 \geq N_\eta$ such that $|f_{k_0}(s) - f(s)| \geq \delta$. Note that $f = \limsup f_n$. There exists an integer $j_0 \geq N_\eta$ such that $|f_{j_0}(s) - f(s)| < \delta/2$. Thus, we have

$$|f_{k_0}(s) - f_{j_0}(s)| \geq |f_{k_0}(s) - f(s)| - |f_{j_0}(s) - f(s)| \geq \delta/2,$$

which implies that $s \in \bigcup_{j,k \geq N_\eta} \{|f_j - f_k| \geq \delta/2\}$. Since s is arbitrary, we have $A \subseteq \bigcup_{j,k \geq N_\eta} \{|f_j - f_k| \geq \frac{\delta}{2}\}$. Therefore, we have

$$\mu(A) \leq \mu\left(\bigcup_{j,k \geq N_\eta} \left\{ |f_j - f_k| \geq \frac{\delta}{2} \right\} \right) < \eta.$$

Since δ and η are arbitrary positive numbers, $f_n \xrightarrow{\text{a.e}} f$ by Problem 9.3. Hence $\lim f_n$ exists a.e.

Next, let us prove the "only if" part. Suppose that $\lim f_n$ exists a.e. Let $f = \limsup f_n$. Then $f_n \xrightarrow{\text{a.e}} f$. By Problem 9.3, we have for every $\delta > 0$,

$$\lim_{n \to \infty} \mu\left(\bigcup_{k=n}^{\infty} \{|f_k - f| \geq \delta\} \right) = 0.$$

Let $\epsilon > 0$ and $\eta > 0$. Then there exists an integer N_η such that

$$\mu\left(\bigcup_{k=N_\eta}^{\infty} \{|f_k - f| \geq \epsilon\} \right) < \eta.$$

Now we consider the following set

$$B = \bigcup_{j,k \geq N_\eta} \{|f_j - f_k| \geq \epsilon\}.$$

We claim that $B \subseteq \bigcup_{k=N_\eta}^{\infty} \{|f_k - f| \geq \epsilon\}$. To show that, let $s \in B$. Then there exist two integers $j_0, k_0 \geq N_\eta$ such that $|f_{j_0}(s) - f_{k_0}(s)| \geq \epsilon$. Note that $f(s) \geq f_{j_0}(s)$ and $f(s) \geq f_{k_0}(s)$. If $f_{j_0}(s) < f_{k_0}(s)$, then $|f(s) - f_{j_0}(s)| = f(s) - f_{j_0}(s) \geq f_{k_0}(s) - f_{j_0}(s) \geq \epsilon$. If $f_{j_0}(s) \geq f_{k_0}(s)$, then $|f(s) - f_{k_0}(s)| = f(s) - f_{k_0}(s) \geq f_{j_0}(s) - f_{k_0}(s) \geq \epsilon$. Hence $s \in \bigcup_{k=N_\eta}^{\infty} \{|f_k - f| \geq \epsilon\}$. Therefore

$$B \subseteq \bigcup_{k=N_\eta}^{\infty} \{|f_k - f| \geq \epsilon\}.$$

Hence $\mu(B) < \eta$. Since ϵ and η are arbitrary positive numbers, we have

$$\mu\left(\bigcup_{j,k\geq n}\{|f_j - f_k| \geq \epsilon\}\right) \to 0 \quad \text{as} \quad n \to \infty.$$

This completes the proof.

9.8. First let us prove the "if" part. Suppose that for each $\epsilon > 0$, we have

$$\mu\left(\bigcup_{j=1}^{\infty}\{|f_{n+j} - f_n| \geq \epsilon\}\right) \to 0 \quad \text{as} \quad n \to \infty.$$

Let $\delta > 0$ and $\eta > 0$. Then there exists an integer N_η such that

$$\mu\left(\bigcup_{j=1}^{\infty}\left\{|f_{N_\eta+j} - f_{N_\eta}| \geq \frac{\delta}{2}\right\}\right) < \eta.$$

Now we consider the following set:

$$A = \bigcup_{j,k\geq N_\eta}\{|f_j - f_k| \geq \delta\}.$$

We claim that

$$A \subseteq \bigcup_{j=1}^{\infty}\left\{|f_{N_\eta+j} - f_{N_\eta}| \geq \frac{\delta}{2}\right\}.$$

To show that, let $s \in A$. Then $|f_{j_0}(s) - f_{k_0}(s)| \geq \delta$ for some $j_0, k_0 \geq N_\eta$. Hence

$$|f_{j_0}(s) - f_{N_\eta}(s)| + |f_{k_0}(s) - f_{N_\eta}(s)| \geq |f_{j_0}(s) - f_{k_0}(s)| \geq \delta,$$

which implies that at least one of the following two inequalities is true: $|f_{j_0}(s) - f_{N_\eta}(s)| \geq \delta/2$ and $|f_{k_0}(s) - f_{N_\eta}(s)| \geq \delta/2$. Therefore $s \in \bigcup_{j=1}^{\infty}\{|f_{N_\eta+j} - f_{N_\eta}| \geq \delta/2\}$. Hence we have

$$\mu(A) < \eta.$$

Since δ and η are arbitrary positive numbers, it follows from Problem 9.7 that

$$\lim_{n\to\infty} f_n$$

exists a.e.

The "only if" part is straightforward. Suppose that $\lim_{n\to\infty} f_n$ exists a.e. Then by Problem 9.7, for each $\epsilon > 0$, we have

$$\mu\left(\bigcup_{j,k\geq n}\{|f_j - f_k| \geq \epsilon\}\right) \to 0 \quad \text{as} \quad n \to \infty.$$

But

$$\bigcup_{j=1}\{|f_{n+j}-f_n|\geq\epsilon\}\subseteq\bigcup_{j,k\geq n}\{|f_j-f_k|\geq\epsilon\},$$

we have

$$\mu\left(\bigcup_{j=1}\{|f_{j+n}-f_n|\geq\epsilon\}\right)\to 0 \quad \text{as} \quad n\to\infty.$$

This completes the proof.

9.9. Let $\epsilon>0$. Let $N=C/\epsilon$. Then we have

$$\begin{aligned}\int_S|f_n|^2\mathrm{d}\mu &= \int_{\{|f_n|\geq N\}}|f_n|^2\mathrm{d}\mu+\int_{\{|f_n|<N\}}|f_n|^2\mathrm{d}\mu\\ &\geq \int_{\{|f_n|\geq N\}}|f_n|N\mathrm{d}\mu\\ &= \frac{C}{\epsilon}\int_{\{|f_n|\geq N\}}|f_n|\mathrm{d}\mu.\end{aligned}$$

By the assumption that $\int_S|f_n|^2\mathrm{d}\mu<C$, we have

$$\sup_{n\geq 1}\int_{\{|f_n|\geq N\}}|f_n|\mathrm{d}\mu<\epsilon,\quad n\geq 1.$$

Since ϵ is arbitrary, the sequence is uniformly integrable. This completes the proof.

9.10.

(a) Let $\epsilon>0$. Since $f_1,f_2,\ldots$ are uniformly integrable, we have

$$\sup_{n\geq 1}\int_{\{|f_n|\geq c\}}|f_n|\mathrm{d}\mu<\epsilon$$

for some $c>0$. Hence

$$\left|\int_{\{f_n<-c\}}f_n\mathrm{d}\mu\right|\leq\int_{\{f_n<-c\}}|f_n|\mathrm{d}\mu<\epsilon,\quad n\geq 1. \tag{9.4}$$

Note that $f_nI_{\{f_n\geq -c\}}+c\geq 0$. By Theorem 6.3, we have

$$\int_S\liminf(f_nI_{\{f_n\geq -c\}}+c)\mathrm{d}\mu\leq\liminf\int_S(f_nI_{\{f_n\geq -c\}}+c)\mathrm{d}\mu.$$

Since μ is finite and $\liminf f_nI_{\{f_n\geq -c\}}+c=\liminf(f_nI_{\{f_n\geq -c\}}+c)$ (see Problem 1.8), we have

$$\int_S\liminf f_nI_{\{f_n\geq -c\}}\mathrm{d}\mu\leq\liminf\int_S f_nI_{\{f_n\geq -c\}}\mathrm{d}\mu.$$

Note that $f_n I_{\{f_n \geq -c\}} \geq f_n$. From the above inequality, We have

$$\int_S \liminf f_n \mathrm{d}\mu \leq \liminf \int_S f_n I_{\{f_n \geq -c\}} \mathrm{d}\mu. \tag{9.5}$$

By Problem 1.8, we have

$$\liminf \int_S f_n I_{\{f_n \geq -c\}} \mathrm{d}\mu + \liminf \int_S f_n I_{\{f_n < -c\}} \mathrm{d}\mu \leq \liminf \int_S f_n \mathrm{d}\mu. \tag{9.6}$$

Combining inequality (9.4) and inequality (9.6) gives

$$\liminf \int_S f_n I_{\{f_n \geq -c\}} \mathrm{d}\mu \leq \liminf \int_S f_n \mathrm{d}\mu + \epsilon.$$

Combining inequality (9.5) and the above inequality gives

$$\int_S \liminf f_n \mathrm{d}\mu \leq \liminf \int_S f_n \mathrm{d}\mu + \epsilon.$$

Since $\epsilon > 0$ is arbitrary, the first inequality in Equation (9.1) is true. The second inequality is obvious by definition of $\liminf$ and $\limsup$. The third inequality can be proved similarly.

(b) If $f_n \xrightarrow{\text{a.e.}} f$, then $\liminf f_n = \limsup f_n = f$. By part (a), we have

$$\int_S f \mathrm{d}\mu \leq \liminf \int_S f_n \mathrm{d}\mu \leq \limsup \int_S f_n \mathrm{d}\mu \leq \int_S f \mathrm{d}\mu,$$

which implies inequality (9.2).

(c) By Problem 9.6, there exists a subsequence f_{n_k} such that $f_{n_k} \xrightarrow{\text{a.e.}} f$. By part (b), we know that f is integrable and

$$\int_S f_{n_k} \mathrm{d}\mu \to \int_S f \mathrm{d}\mu \quad \text{as} \quad k \to \infty.$$

We claim that

$$\int_S f_n \mathrm{d}\mu \to \int_S f \mathrm{d}\mu \quad \text{as} \quad n \to \infty.$$

If $\int_S f_n \mathrm{d}\mu$ does not converge to $\int_S f \mathrm{d}\mu$, then we have for some $\epsilon > 0$,

$$\left| \int_S f_n \mathrm{d}\mu - \int_S f \mathrm{d}\mu \right| \geq \epsilon$$

for infinitely many n. But we can find a subsequence f_{m_k} from these f_n such that $f_{m_k} \xrightarrow{\text{a.e.}} f$. Using the above argument, we have

$$\int_S f_{m_k} \mathrm{d}\mu \to \int_S f \mathrm{d}\mu \quad \text{as} \quad k \to \infty.$$

This contradicts the hypothesis. Hence our claim is true.

9.11. First let us prove the "if" part. Assume that the integrals $\int_S |f_n| \mathrm{d}\mu$ are uniformly bounded and uniformly continuous. By Problem 6.16, we have

$$\mu\{|f_n| \geq c\} \leq \frac{1}{c} \int_S |f_n| \mathrm{d}\mu \leq \frac{M}{c}.$$

Hence $\mu\{|f_n| \geq c\} \to 0$ uniformly as $c \to \infty$. Since the integrals are also uniformly continuous, we have

$$\sup_n \int_{\{|f_n| \geq c\}} |f_n| \mathrm{d}\mu \to 0 \quad \text{as} \quad c \to \infty.$$

Hence the f_n are uniformly integrable.

Now we prove the "only if" part. Suppose that the f_n are uniformly integrable. Let $\epsilon > 0$. Then there exists a $c > 0$ such that

$$\sup_{n \geq 1} \int_{\{|f_n| \geq c\}} |f_n| \mathrm{d}\mu < \epsilon.$$

Hence we have

$$\begin{aligned} \int_S |f_n| \mathrm{d}\mu &= \int_{\{|f_n| \geq c\}} |f_n| \mathrm{d}\mu + \int_{\{|f_n| < c\}} |f_n| \mathrm{d}\mu \\ &\leq \epsilon + c\mu(S), \quad n \geq 1, \end{aligned}$$

which shows that the integrals are uniformly bounded.

Let $\delta > 0$. Then we can choose d such that

$$\sup_{n \geq 1} \int_{\{|f_n| \geq d\}} |f_n| \mathrm{d}\mu < \frac{\delta}{2}.$$

Let $A \in \Sigma$ such that $\mu(A) < \delta/(2d)$. Then we have

$$\begin{aligned} \int_A |f_n| \mathrm{d}\mu &= \int_{A \cap \{|f_n| \geq d\}} |f_n| \mathrm{d}\mu + \int_{A \cap \{|f_n| < d\}} |f_n| \mathrm{d}\mu \\ &\leq \frac{\delta}{2} + d\mu(A) \\ &\leq \delta, \quad n \geq 1. \end{aligned}$$

Since δ is arbitrary, it follows that the integrals are uniformly continuous. This completes the proof.

9.12. First we show that the $|f_n - f|^p$ are uniformly integrable. By Problem 8.3, we have

$$|f_n - f|^p \leq 2^{p-1}(|f_n|^p + |f|^p), \quad p > 1.$$

Also $|f_n - f|^p \le |f_n|^p + |f|^p$ for $0 < p \le 1$. Note that $f_n \xrightarrow{\mu} f$. By Problem 9.6, there exists a subsequence $\{f_{n_k}\}_{k\ge1}$ converging to f a.e. Hence $|f_{n_k}|^p \xrightarrow{a.e.} |f|^p$. By Problem 9.10, $|f|^p$ is integrable. Note that the $|f_n|^p$ are uniformly integrable. Hence the integrals $\int_S |f_n - f|^p \mathrm{d}\mu$ are uniformly bounded and uniformly continuous. By Problem 9.11, the $|f_n - f|^p$ are uniformly integrable.

Since $f_{n_k} \xrightarrow{\text{a.e.}} f$ and $|f_{n_k} - f|^p$ are uniformly integrable, by Problem 9.10, we have

$$\int_S |f_{n_k} - f|^p \mathrm{d}\mu \to 0 \quad \text{as} \quad k \to \infty.$$

Using the same argument, we can show that any subsequence of $\{f_n\}_{n\ge1}$ has a subsequence converging to f in L^p. It follows that $f_n \xrightarrow{L^p} f$. This completes the proof.

9.13. Let $\epsilon > 0$ and $M = \sup_{i\in I} \int_S h(|f_i|)\mathrm{d}\mu$. By assumption, there is a positive number c such that

$$\frac{h(t)}{t} \ge \frac{M}{\epsilon}, \quad \forall t \ge c.$$

Then

$$\begin{aligned}
\int_{\{|f_i|\ge c\}} |f_i|\mathrm{d}\mu &\le \int_{\{|f_i|\ge c\}} \frac{\epsilon h(|f_i|)}{M}\mathrm{d}\mu \\
&\le \int_S \frac{\epsilon h(|f_i|)}{M}\mathrm{d}\mu \\
&\le \epsilon.
\end{aligned}$$

Since ϵ is arbitrary, the f_i are uniformly integrable by definition. This completes the proof.

9.14. First we show that the $|f_n - f|^p$ are uniformly integrable. Let $\epsilon > 0$. Since $f_n \xrightarrow{L^p} f$, there is a positive integer N_ϵ such that

$$\int_S |f_n - f|^p \mathrm{d}P < \frac{\epsilon}{2}, \quad \forall n > N_\epsilon.$$

Let $M = \max_{1\le n\le N_\epsilon} |f_n - f|^p$. Then

$$\begin{aligned}
\sup_{n\ge1} \int_A |f_n - f|^p \mathrm{d}P &\le \sup_{1\le n\le N_\epsilon} \int_A |f_n - f|^p \mathrm{d}P + \sup_{n>N_\epsilon} \int_A |f_n - f|^p \mathrm{d}P \\
&\le \int_A M\mathrm{d}P + \sup_{n>N_\epsilon} \int_S |f_n - f|^p \mathrm{d}P \\
&\le \mu(A)M + \frac{\epsilon}{2}, \quad A \in \Sigma.
\end{aligned}$$

If $\mu(A) < \epsilon(2M)$, then

$$\sup_{n\ge1} \int_A |f_n - f|^p \mathrm{d}P \le \epsilon.$$

Since ϵ is arbitrary, we have

$$\int_A |f_n - f|^p \mathrm{d}P \to 0 \quad \text{as} \quad \mu(A) \to 0.$$

Hence the integrals of $|f_n - f|^p$ are uniformly continuous. If $A = S$, then

$$\sup_{n\geq 1} \int_S |f_n - f|^p \mathrm{d}P \leq \mu(S)M + \frac{\epsilon}{2}.$$

Since μ is finite, the above inequality shows that the integrals of $|f_n - f|^p$ are uniformly bounded. Hence $|f_n - f|^p$ are uniformly integrable by Problem 9.11.

Now, by Problem 8.3, we have

$$|f_n|^p = |f_n - f + f|^p \leq 2^{p-1}(|f_n - f|^p + |f|^p).$$

It follows that $|f_n|^p$ are uniformly bounded and uniformly continuous. Hence the $|f_n|^p$ are uniformly integrable. This completes the proof.

9.5 Bibliographic Notes

In this chapter, we defined several types of convergence in measure theory. The relationships of those types of convergence are shown in Figure 9.1.

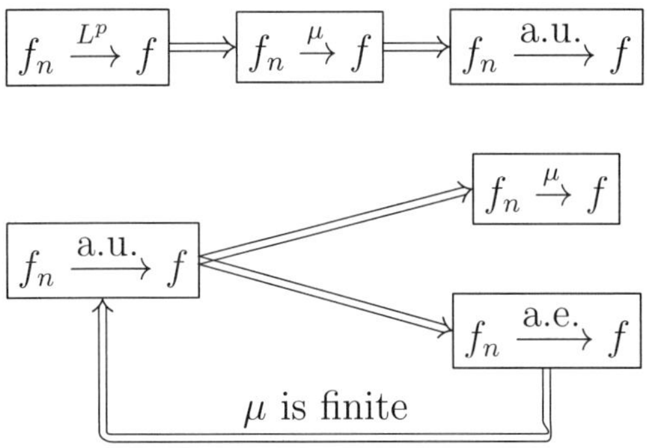

Figure 9.1 Relationships of several types of convergence, where $\xrightarrow{\text{a.u}}$ and $\xrightarrow{\text{a.e.}}$ mean almost uniform convergence and almost everywhere convergence, respectively.

Convergence in L^p is stronger than convergence in measure. Almost uniform convergence is stronger than both convergence in measure and almost everywhere convergence. Convergence in measure does not necessarily lead to almost uniform convergence. However, if a sequence is convergent in measure, then there is a subsequence that is convergent almost uniformly to the same limit function. If the underlying measure is finite, then almost everywhere convergence implies almost uniform convergence. More information about convergence can be found in Ash and Doleans-Dade (1999, Section 2.5) and Lawler (2006).

CHAPTER 10

PRODUCT MEASURES

Product measures are measures defined on the product of several measurable spaces. Fubini's theorem gives conditions under which we can evaluate integrals with respect to product measures by computing iterated integrals. In this chapter, we present the definition of product measures and Fubini's theorem.

10.1 Basic Concepts and Facts

Definition 10.1 (Cartesian Product)**.** Let S_1 and S_2 be two sets. The Cartesian product of S_1 and S_2 is a set defined as

$$S_1 \times S_2 = \{(x, y) : x \in S_1 \text{ and } y \in S_2\}.$$

Definition 10.2 (Measurable Rectangle)**.** Let (S_1, Σ_1) and (S_2, Σ_2) be two measurable spaces. Let $S = S_1 \times S_2$ be the Cartesian product of S_1 and S_2. A measurable rectangle in the product space S is a product $A_1 \times A_2$ for which $A_1 \in \Sigma_1$ and $A_2 \in \Sigma_2$.

Definition 10.3 (Product σ-Algebra)**.** Let (S_1, Σ_1) and (S_2, Σ_2) be two measurable spaces. The product σ-algebra $\Sigma \otimes \Sigma_2$ ($\Sigma_1 \otimes \Sigma_2$ is not a Cartesian product in the usual sense.) is defined as the σ-algebra generated by all measurable rectangles in the product space $S_1 \times S_2$:

$$\Sigma_1 \otimes \Sigma_2 = \sigma(\{A_1 \times A_2 : A_1 \in \Sigma_1, A_2 \in \Sigma_2\}).$$

Definition 10.4 (Infinite Product)**.** Let I be an index set, countable or uncountable. For each $t \in I$, let (S_t, Σ_t) be a measurable space. Let S be the product space of S_t, $t \in I$. A rectangle in S is defined as

$$\underset{t \in I}{\times} A_t = \{(x_t)_{t \in I} \in S : x_t \in A_t,\ t \in I\},$$

where $A_t \subseteq S_t$ and A_t differs from S_t for only a finite number of t. The rectangle is said to be measurable if $A_t \in \Sigma_t$ for every $t \in I$.

The σ-algebra on S generated by the collection of all measurable rectangles is called the product σ-algebra and is denoted by

$$\bigotimes_{t \in I} \Sigma_t.$$

The resulting measurable space is denoted by

$$\left(\underset{t \in I}{\times} S_t, \bigotimes_{t \in I} \Sigma_t \right).$$

Definition 10.5 (Coordinate Maps)**.** Let $S_1 \times S_2$ be the Cartesian product of S_1 and S_2. Then the ith coordinate map $\rho_i : S_1 \times S_2 \to S_i$ is defined as

$$\rho_i(s_1, s_2) = s_i, \quad i = 1, 2.$$

Definition 10.6 (Product Measure and Product Measure Space)**.** Let (S_1, Σ_1, μ_1) and (S_2, Σ_2, μ_2) be measure spaces and suppose that μ_1 and μ_2 are σ-finite. The product measure $\mu = \mu_1 \times \mu_2$ of μ_1 and μ_2 is defined as follows (see Problems 10.6 and 10.7):

$$\begin{aligned}
\mu(E) &= \int_{S_1} \mu_2(\{y \in S_2 : (x, y) \in E\}) \mu_1(\mathrm{d}x) \\
&= \int_{S_2} \mu_1(\{x \in S_1 : (x, y) \in E\}) \mu_2(\mathrm{d}y).
\end{aligned}$$

The measure space $(S, \Sigma, \mu) = (S_1 \times S_2, \Sigma_1 \otimes \Sigma_2, \mu_1 \times \mu_2)$ is called the *product measure space*.

Theorem 10.1 (Tonelli's Theorem)**.** *Let (S_1, Σ_1, μ_1) and (S_2, Σ_2, μ_2) be σ-finite measure spaces, and let $f : S_1 \times S_1 \to \mathbf{R}^+$ be a nonnegative $\Sigma_1 \otimes \Sigma_2$-measurable function, where $\mathbf{R}^+ = [0, \infty)$ denotes the set of all nonnegative real numbers. Let $g_1 : S_1 \to \bar{\mathbf{R}}$ and $g_2 : S_2 \to \bar{\mathbf{R}}$ be defined as*

$$g_1(x) = \int_{S_2} f(x, y) \mu_2(\mathrm{d}y), \quad g_2(y) = \int_{S_1} f(x, y) \mu_1(\mathrm{d}x),$$

where $\bar{\mathbf{R}} = [-\infty, \infty]$ *denotes the set of extended real numbers.*

Then g_1 *is* $\Sigma_1/\mathcal{B}(\bar{\mathbf{R}})$*-measurable and* g_2 *is* $\Sigma_2/\mathcal{B}(\bar{\mathbf{R}})$*-measurable. In addition*

$$\int_{S_1\times S_2} f(z)\mu(\mathrm{d}z) = \int_{S_1} g_1(x)\mu_1(\mathrm{d}x) = \int_{S_2} g_2(y)\mu_2(\mathrm{d}y), \tag{10.1}$$

where $\mu = \mu_1 \times \mu_2$.

Theorem 10.2 (Fubini's Theorem). *Let* (S_1, Σ_1, μ_1) *and* (S_2, Σ_2, μ_2) *be* σ*-finite measure spaces and* $f \in L^1(S_1 \times S_2, \Sigma_1 \otimes \Sigma_2, \mu_1 \times \mu_2)$*; that is,* f *is* $\mu_1 \times \mu_2$*-integrable. Then there exist sets* $A_1 \in \Sigma_1$ *and* $A_2 \in \Sigma_2$ *such that*

(a) $\mu_1(S_1 \backslash A_1) = 0$ *and* $\mu_2(S_2 \backslash A_2) = 0$.

(b) *For fixed* $x \in A_1$, $f_x(y) = f(x, y) \in L^1(S_2, \Sigma_2, \mu_2)$.

(c) *The function*

$$g_1(x) = \begin{cases} \int_{S_2} f(x,y)\mu_2(\mathrm{d}y), & \text{if } x \in A_1; \\ 0, & \text{if } x \notin A_1 \end{cases} \tag{10.2}$$

is Σ_1*-measurable and*

$$\int_{S_1} g_1 \mathrm{d}\mu_1 = \int_{S_1\times S_2} f \mathrm{d}(\mu_1 \times \mu_2).$$

(d) *For fixed* $y \in A_2$, $f_y(x) = f(x, y) \in L^1(S_1, \Sigma_1, \mu_1)$.

(e) *The function*

$$g_2(y) = \begin{cases} \int_{S_1} f(x,y)\mu_1(\mathrm{d}x), & \text{if } y \in A_2; \\ 0, & \text{if } y \notin A_2 \end{cases} \tag{10.3}$$

is Σ_2*-measurable and*

$$\int_{S_2} g_2 \mathrm{d}\mu_2 = \int_{S_1\times S_2} f \mathrm{d}(\mu_1 \times \mu_2).$$

10.2 Problems

10.1. Let (S_1, Σ_1) and (S_2, Σ_2) be measurable spaces, and $\mathcal{I}$ be the collection of all measurable rectangles in the product space $S_1 \times S_2$:

$$\mathcal{I} = \{A_1 \times A_2 : A_1 \in \Sigma_1, A_2 \in \Sigma_2\}.$$

Show that $\mathcal{I}$ is a π-system.

10.2. Let (S_1, Σ_1) and (S_2, Σ_2) be measurable spaces and $\rho_i : S_1 \times S_2 \to S_i$ be the ith coordinate map for $i = 1, 2$. Show that ρ_i is $\Sigma_1 \otimes \Sigma_2/\Sigma_i$-measurable.

10.3. Let (S_1, Σ_1) and (S_2, Σ_2) be measurable spaces and $\rho_i : S_1 \times S_2 \to S_i$ be the ith coordinate map for $i = 1, 2$. Show that

$$\Sigma_1 \otimes \Sigma_2 = \sigma(\rho_1, \rho_2),$$

where $\Sigma_1 \otimes \Sigma_2$ is as defined in Definition 10.3 and $\sigma(\rho_1, \rho_2)$ is defined as

$$\begin{aligned}\sigma(\rho_1, \rho_2) &= \sigma(\rho_i^{-1}(B) : i = 1, 2, B \in \mathcal{B}) \\ &= \sigma(\{(s_1, s_2) \in S_1 \times S_2 : \rho_i(s_1, s_2) \in B\} : i = 1, 2, B \in \mathcal{B}).\end{aligned}$$

10.4. Let (S_1, Σ_1) and (S_2, Σ_2) be measurable spaces, and let $\Sigma_1 \otimes \Sigma_2$ be the product σ-algebra. Show that

(a) If $E \in \Sigma_1 \otimes \Sigma_2$, then for each $x \in S_1$ the set $E_x = \{y : (x, y) \in E\}$ is Σ_2-measurable and for each $y \in S_2$ the set $E_y = \{x : (x, y) \in E\}$ is Σ_1-measurable.

(b) If f is $\Sigma_1 \otimes \Sigma_2$-measurable, then for each fixed $x \in S_1$ the function $f_x(y) = f(x, y)$ is Σ_2-measurable and for each fixed $y \in S_2$ the function $f_y(x) = f(x, y)$ is Σ_1-measurable.

10.5. Let (S_1, Σ_1, μ_1) and (S_2, Σ_2, μ_2) be measure spaces and suppose that μ_1 and μ_2 are finite measures. For $E \in \Sigma_1 \otimes \Sigma_2$, let $g_{E,i} : S_i \to \mathbf{R}$ $(i = 1, 2)$ be defined as

$$\begin{aligned}g_{E,1}(x) &= \mu_2(\{y \in S_2 : (x, y) \in E\}), \\ g_{E,2}(y) &= \mu_1(\{x \in S_1 : (x, y) \in E\}).\end{aligned}$$

Show that

(a) $g_{E,1}$ and $g_{E,2}$ are well defined and bounded for $E \in \Sigma_1 \otimes \Sigma_2$.

(b) If $\mathcal{E}_i (i = 1, 2)$ is the collection of E in $\Sigma_1 \otimes \Sigma_2$ for which $g_{E,i}$ is Σ_i-measurable, then $\mathcal{E}_i = \Sigma_1 \otimes \Sigma_2$.

10.6. Let (S_1, Σ_1, μ_1) and (S_2, Σ_2, μ_2) be measure spaces and suppose that μ_1 and μ_2 are finite measures. Then $g_{E,1}(x) = \mu_2(\{y \in S_2 : (x, y) \in E\})$ is bounded and Σ_1-measurable and $g_{E,2}(y) = \mu_1(\{x \in S_1 : (x, y) \in E\})$ is bounded and Σ_2-measurable for $E \in \Sigma_1 \otimes \Sigma_2$ (see Problem 10.5). Let $\theta_1 : \Sigma_1 \otimes \Sigma_2 \to \mathbf{R}$ and $\theta_2 : \Sigma_1 \otimes \Sigma_2 \to \mathbf{R}$ be defined as

$$\begin{aligned}\theta_1(E) &= \int_{S_1} g_{E,1}(x)\mu_1(\mathrm{d}x) = \int_{S_1} \mu_2(\{y \in S_2 : (x, y) \in E\})\mu_1(\mathrm{d}x), \\ \theta_2(E) &= \int_{S_2} g_{E,2}(y)\mu_2(\mathrm{d}y) = \int_{S_2} \mu_1(\{x \in S_1 : (x, y) \in E\})\mu_2(\mathrm{d}y).\end{aligned}$$

Show that

(a) θ_1 and θ_2 are finite measures on $(S, \Sigma) = (S_1 \times S_2, \Sigma_1 \otimes \Sigma_2)$.

(b) For measurable rectangles $E = A_1 \times A_2$, we have

$$\theta_1(E) = \theta_2(E) = \mu_1(A_1)\mu_2(A_2).$$

(c) For $E \in \Sigma_1 \otimes \Sigma_2$, we have

$$\theta_1(E) = \theta_2(E).$$

10.7. Let (S_1, Σ_1, μ_1) and (S_2, Σ_2, μ_2) be measure spaces and suppose that μ_1 and μ_2 are σ-finite measures. Show that

(a) $g_{E,1}(x) = \mu_2(\{y \in S_2 : (x, y) \in E\})$ is well defined and Σ_1-measurable and $g_{E,2}(y) = \mu_1(\{x \in S_1 : (x, y) \in E\})$ is well defined and Σ_2-measurable for $E \in \Sigma_1 \otimes \Sigma_2$.

(b) Let $\theta_1 : \Sigma_1 \otimes \Sigma_2 \to \mathbf{R}$ and $\theta_2 : \Sigma_1 \otimes \Sigma_2 \to \mathbf{R}$ be defined as

$$\theta_1(E) = \int_{S_1} g_{E,1}(x)\mu_1(\mathrm{d}x) = \int_{S_1} \mu_2(\{y \in S_2 : (x, y) \in E\})\mu_1(\mathrm{d}x),$$
$$\theta_2(E) = \int_{S_2} g_{E,2}(y)\mu_2(\mathrm{d}y) = \int_{S_2} \mu_1(\{x \in S_1 : (x, y) \in E\})\mu_2(\mathrm{d}y).$$

Then θ_1 and θ_2 are measures on $(S, \Sigma) = (S_1 \times S_2, \Sigma_1 \otimes \Sigma_2)$.

(c) For measurable rectangles $E = A_1 \times A_2$, we have

$$\theta_1(E) = \theta_2(E) = \mu_1(A_1)\mu_2(A_2);$$

(d) For $E \in \Sigma_1 \otimes \Sigma_2$, we have

$$\theta_1(E) = \theta_2(E).$$

10.8. Let $(\Omega, \mathscr{F})$ be a measurable space. Let $(E_t, \mathscr{E}_t)$, $t \in I$, be measurable spaces, where I is an index set. Let $(F, \mathcal{G})$ be the product measurable space of $(E_t, \mathscr{E}_t)$, $t \in I$:

$$(F, \mathscr{G}) = \left(\times_{t \in I} E_t, \bigotimes_{t \in I} \mathscr{E}_t \right).$$

For each $t \in I$, let f_t be a mapping from Ω to E_t. For each $\omega \in \Omega$, define $f(\omega) = (f_t(\omega))_{t \in I} \in F$. Show that the mapping $f : \Omega \to F$ is $\mathscr{F}/\mathscr{G}$-measurable if and only if for every $t \in I$, f_t is $\mathscr{F}/\mathscr{E}_t$-measurable.

10.3 Hints

10.1. Use the definition of the Cartesian product (Definition 10.1) and the definition of π-system (Definition 2.15).

10.2. Use the definition of measurable functions (Definition 5.3).

10.3. Show that $\Sigma_1 \otimes \Sigma_2 \subseteq \sigma(\rho_1, \rho_2)$ and $\sigma(\rho_1, \rho_2) \subseteq \Sigma_1 \otimes \Sigma_2$.

10.4. For part (a), consider mapping $T_x : S_2 \to S_1 \times S_2$. For part (b), consider the set $f_x^{-1}(B)$ for $B \in \mathcal{B}$ and use the result of part (a).

10.5. To prove part (a), use the result of Problem 10.4. To prove part (b), use the results of Problems 10.1 and 2.7.

10.6. To prove part (a), follow the definition of measures and use Theorem 6.2 and the result of Problem 6.12. To prove part (b), note that

$$\mu_2(\{y \in S_2 : (x, y) \in A_1 \times A_2\}) = I_{A_1}(x)\mu_2(A_2).$$

To prove part (c), use Theorem 3.2.

10.7. To prove part (a), use the result of Problem 10.5 and Theorem 5.1. Use the result of Problem 10.6 and Theorem 6.2 to prove part (b). Parts (c) and (d) can be proved by using the result of Problem 10.6.

10.8. To prove the "if" part, use the definition of infinite products (Definition 10.4) and the result of 5.4. The "only if" part can be proved by considering special rectangles.

10.4 Solutions

10.1. Let $A_1 \times A_2 \in \mathcal{I}$ and $B_1 \times B_2 \in \mathcal{I}$. Since

$$(A_1 \times A_2) \cap (B_1 \times B_2) = (A_1 \cap B_1) \times (A_2 \cap B_2),$$

we have $(A_1 \times A_2) \cap (B_1 \times B_2) \in \mathcal{I}$. Hence $\mathcal{I}$ is a π-system.

10.2. We show that ρ_1 is $\Sigma_1 \otimes \Sigma_2/\Sigma_1$-measurable. By the definition of ρ_1, we have

$$\rho_1^{-1}(B_1) = B_1 \times S_2, \quad B_1 \in \Sigma_1.$$

By the definition of $\Sigma_1 \otimes \Sigma_2$, we have $B_1 \times S_2 \in \Sigma_1 \otimes \Sigma_2$ for $B_1 \in \Sigma_1$. Therefore, $\rho_1^{-1}(B_1) \in \Sigma_1 \otimes \Sigma_2$ for $B_1 \in \Sigma_1$; that is, ρ_1 is $\Sigma_1 \otimes \Sigma_2/\Sigma_1$-measurable. Similarly, we can show that ρ_2 is $\Sigma_1 \otimes \Sigma_2/\Sigma_2$-measurable.

10.3. We will prove $\Sigma_1 \otimes \Sigma_2 = \sigma(\rho_1, \rho_2)$ by showing that $\Sigma_1 \otimes \Sigma_2 \subseteq \sigma(\rho_1, \rho_2)$ and $\sigma(\rho_1, \rho_2) \subseteq \Sigma_1 \otimes \Sigma_2$. To show that $\Sigma_1 \otimes \Sigma_2 \subseteq \sigma(\rho_1, \rho_2)$, we let $A_1 \in \Sigma_1$ and

$A_2 \in \Sigma_2$. Note that $A_1 \times A_2 = (A_1 \times S_2) \cap (S_1 \times A_2)$ and $A_1 \times S_2 = \rho_1^{-1}(A_1) \in \sigma(\rho_1, \rho_2)$ and $S_1 \times A_2 = \rho_2^{-1}(A_2) \in \sigma(\rho_1, \rho_2)$, we have $A_1 \times A_2 \in \sigma(\rho_1, \rho_2)$. Therefore, $\Sigma_1 \otimes \Sigma_2 \subseteq \sigma(\rho_1, \rho_2)$.

Now we show that $\sigma(\rho_1, \rho_2) \subseteq \Sigma_1 \otimes \Sigma_2$. But given $\rho_1^{-1}(A_1) = A_1 \times S_2 \in \Sigma_1 \otimes \Sigma_2$ and $\rho_2^{-1}(A_2) = S_1 \times A_1 \in \Sigma_1 \otimes \Sigma_2$ for all $A_1 \in \Sigma_1$ and $A_2 \in \Sigma_2$, we have $\sigma(\rho_1, \rho_2) \subseteq \Sigma_1 \otimes \Sigma_2$. This completes the proof.

10.4.

(a) Let $x \in S_1$ be fixed. We consider the mapping $T_x : S_2 \to S_1 \times S_2$ defined by $T_x(y) = (x, y)$. If $E = A_1 \times A_2$ is a measurable rectangle, then

$$T_x^{-1}(E) = \begin{cases} A_2 & \text{if } x \in A_1; \\ \emptyset & \text{if } x \notin A_1, \end{cases}$$

and so $T_x^{-1}(E) \in \Sigma_2$. Noting that $\Sigma_1 \otimes \Sigma_2$ is generated by all measurable rectangles, we have, by Problem 5.4, T_x is $\Sigma_2/\Sigma_1 \otimes \Sigma_2$-measurable. It follows that $E_x = T_x^{-1}(E) \in \Sigma_2$ for $E \in \Sigma_1 \otimes \Sigma_2$; that is, E_x is Σ_2-measurable. Similarly, we can show that for each fixed y the set E_y is Σ_1-measurable.

(b) Now we show that f_x is Σ_2-measurable. To do this, let $B \in \mathcal{B}$. Then $f_x^{-1}(B) = \{y \in S_2 : f_x(y) \in B\} = \{y \in S_2 : f(x, y) \in B\}$. Since f is $\Sigma_1 \otimes \Sigma_2$-measurable, we have $E_B = \{(x, y) \in S_1 \times S_2 : f(x, y) \in B\} = f^{-1}(B) \in \Sigma_1 \otimes \Sigma_2$. Noting that $f_x^{-1}(B) = \{y \in S_2 : (x, y) \in E_B\}$, we have, by the first part of this proof, $f_x^{-1}(B) \in \Sigma_2$. Hence f_x is Σ_2-measurable. Similarly, we can show that f_y is Σ_1-measurable for fixed y.

10.5.

(a) Let $E \in \Sigma_1 \otimes \Sigma_2$. By Problem 10.4, for fixed $x \in S_1$ the set $\{y \in S_2 : (x, y) \in E\}$ is Σ_2-measurable. Therefore, the function $g_{E,1}$ is well defined. By hypothesis, μ_2 is finite. Then $g_{E,1}(x) \le \mu_2(S_2) < \infty$ and so $g_{E,1}$ is bounded. Similarly, $g_{E,2}$ is well defined and bounded.

(b) Now we show that $\mathcal{E}_1 = \Sigma_1 \otimes \Sigma_2$. First, we show that $\mathcal{E}_1$ is a d-system on $S_1 \times S_2$. Since $g_{S_1 \times S_2, 1}(x) = \mu_2(\{y \in S_2 : (x, y) \in S_1 \times S_2\}) = \mu_2(S_2)$ for $x \in S_1$, for $B \in \mathcal{B}$

$$g_{S_1 \times S_2, 1}^{-1}(B) = \begin{cases} S_1, & \text{if } \mu_2(S_2) \in B; \\ \emptyset, & \text{if } \mu_2(S_2) \notin B, \end{cases}$$

and so $g_{S_1 \times S_2, 1}^{-1} \in \Sigma_1$. Hence $g_{S_1 \times S_2, 1}$ is Σ_1-measurable, and so $S_1 \times S_2 \in \mathcal{E}_1$. Let $E_1, E_2 \in \mathcal{E}_1$ and $E_1 \subseteq E_2$. Then $g_{E_1,1}(x)$ and $g_{E_2,1}(x)$ are Σ_1-measurable. Since

$$\begin{aligned} g_{E_2 \backslash E_1, 1}(x) &= \mu_2(\{y \in S_2 : (x, y) \in E_2 \backslash E_1\}) \\ &= \mu_2(\{y \in S_2 : (x, y) \in E_2\} \backslash \{y \in S_2 : (x, y) \in E_1\}) \\ &= \mu_2(\{y \in S_2 : (x, y) \in E_2\}) - \mu_2(\{y \in S_2 : (x, y) \in E_1\}) \\ &= g_{E_2,1}(x) - g_{E_1,1}(x), \end{aligned}$$

$g_{E_2\backslash E_1,1}(x)$ is Σ_1-measurable. Therefore, $E_2\backslash E_1 \in \mathcal{E}_1$. Let $E_n \in \mathcal{E}_1$ $(n \geq 1)$ and $E_n \uparrow E$. Then $\{y \in S_2 : (x,y) \in E_n\} \uparrow \{y \in S_2 : (x,y) \in E\}$. By Theorem 2.1, $g_{E_n,1}(x) = \mu_2(\{y \in S_2 : (x,y) \in E_n\}) \uparrow g_{E,1}(x) = \mu_2(\{y \in S_2 : (x,y) \in E\})$. By Problem 5.9, $g_{E,1}(x)$ is Σ_1-measurable. Hence $E \in \mathcal{E}_1$. Therefore, $\mathcal{E}_1$ is a d-system on $S_1 \times S_2$.

Next, we show that $\mathcal{E}_1$ contains all measurable rectangles in the product space $S_1 \times S_2$. Let $E = A_1 \times A_2$ be a measurable rectangle. Then $g_{E,1}(x) = \mu_2(\{y \in S_2 : (x,y) \in A_1 \times A_2\}) = I_{A_1}(x)\mu_2(A_2)$ and so $g_{E,1}(x)$ is Σ_1-measurable. Hence $E \in \mathcal{E}_1$.

By Problem 10.1, the collection $\mathcal{I}$ of all measurable rectangles is a π-system. By definition, $\Sigma_1 \otimes \Sigma_2 = \sigma(\mathcal{I})$. Recall that $\mathcal{E}_1$ is a d-system. Then, by Problem 2.7, $\Sigma_1 \otimes \Sigma_2 = \sigma(\mathcal{I}) = d(\mathcal{I}) \subseteq \mathcal{E}_1$. By the definition of $\mathcal{E}_1$, we have $\mathcal{E}_1 \subseteq \Sigma_1 \otimes \Sigma_2$. Therefore, $\mathcal{E}_1 = \Sigma_1 \otimes \Sigma_2$. Similarly, we can show that $\mathcal{E}_2 = \Sigma_1 \otimes \Sigma_2$. This completes the proof.

10.6. First, we show that θ_1 is a finite measure. By the definition of θ_1, we have $\theta_1 \geq 0$ and $\theta_1(\emptyset) = \int_{S_1} \mu_2(\emptyset)\mu_1(\mathrm{d}x) = \int_{S_1} 0\mu_1(\mathrm{d}x) = 0$. Let E_n $(n \geq 1)$ be a sequence of pairwise disjoint sets in $\Sigma_1 \otimes \Sigma_2$. Let $f_n : S_1 \to \mathbf{R}$ and $f : S_1 \to \mathbf{R}$ be defined as

$$f_n(x) = \mu_2\left(\left\{y \in S_2 : (x,y) \in \bigcup_{i=1}^{n} E_i\right\}\right), \quad n \in \mathbf{N},$$
$$f(x) = \mu_2\left(\left\{y \in S_2 : (x,y) \in \bigcup_{i=1}^{\infty} E_i\right\}\right), \quad n \in \mathbf{N}.$$

Then $f_n \uparrow f$ and, by Theorem 6.2, we obtain

$$\int_{S_1} f_n(x)\mu_1(\mathrm{d}x) \uparrow \int_{S_1} f(x)\mu_1(\mathrm{d}x) = \theta_1\left(\bigcup_{i=1}^{\infty} E_i\right). \tag{10.4}$$

But, by Problem 6.12, we have

$$\int_{S_1} f_n(x)\mu_1(\mathrm{d}x) = \sum_{i=1}^{n} \int_{S_1} \mu_2(\{y \in S_2 : (x,y) \in E_i\})\mu_1(\mathrm{d}x) = \sum_{i=1}^{n} \theta(E_i).$$

From equation (10.4), we have

$$\sum_{i=1}^{n} \theta(E_i) \uparrow \theta_1\left(\bigcup_{i=1}^{\infty} E_i\right),$$

and so

$$\sum_{i=1}^{\infty} \theta(E_i) = \theta_1\left(\bigcup_{i=1}^{\infty} E_i\right).$$

Therefore, θ_1 is a measure on $(S_1 \times S_2, \Sigma_1 \otimes \Sigma_2)$. Since μ_1 and μ_2 are finite and $\theta_1(S_1 \times S_2) = \mu_1(S_1)\mu_2(S_2) < \infty$, we see that θ_1 is finite. Similarly, we can show that θ_2 is a finite measure on $(S_1 \times S_2, \Sigma_1 \otimes \Sigma_2)$.

Next, let $E = A_1 \times A_2$ be a measurable rectangle. Then

$$\begin{aligned}\theta_1(E) &= \int_{S_1} \mu_2(\{y \in S_2 : (x, y) \in A_1 \times A_2\})\mu_1(\mathrm{d}x) \\ &= \int_{S_1} I_{A_1}(x)\mu_2(A_2)\mu_1(\mathrm{d}x) \\ &= \mu_1(A_1)\mu_2(A_2).\end{aligned}$$

Similarly, we have $\theta_2(E) = \mu_1(A_1)\mu_2(A_2)$.

Finally, we show that $\theta_1(E) = \theta_2(E)$ for $E \in \Sigma_1 \otimes \Sigma_2$. Since $\theta_1(E) = \theta_2(E)$ for every $E \in \mathcal{I}$, where $\mathcal{I}$ is the collection of all measurable rectangles in the product space $S_1 \times S_2$, and $\Sigma_1 \otimes \Sigma_2 = \sigma(\mathcal{I})$, the equality of θ_1 and θ_2 on $\Sigma_1 \otimes \Sigma_2$ follows from Theorem 3.2.

10.7. By hypothesis, μ_1 and μ_2 are σ-finite. Let C_n $(n \geq 1)$ and D_m $(m \geq 1)$ be decompositions of S_1 and S_2 into sets of finite measure, and let $\mu_{n,1}(C) = \mu_1(C \cap C_n)$ for $C \in S_1$ and $\mu_{m,2}(D) = \mu_2(D \cap D_m)$ for $D \in S_2$.

(a) Let $E \in \Sigma_1 \otimes \Sigma_2$. Since

$$\begin{aligned}g_{E,1}(x) &= \mu_2(\{y \in S_2 : (x, y) \in E\}) \\ &= \mu_2\left(\bigcup_{m=1}^{\infty} (\{y \in S_2 : (x, y) \in E\} \cap D_m)\right) \\ &= \sum_{m=1}^{\infty} \mu_2(\{y \in S_2 : (x, y) \in E\} \cap D_m) \\ &= \sum_{m=1}^{\infty} \mu_{m,2}(\{y \in S_2 : (x, y) \in E\}),\end{aligned} \tag{10.5}$$

by Problem 10.5 and Theorem 5.1, $g_{E,1}(x)$ is well defined and Σ_1-measurable. Similarly, $g_{E,2}(y)$ is well defined and Σ_2-measurable.

(b) By Equation (10.5) and Theorem 6.2, we have

$$\begin{aligned}\theta_1(E) &= \int_{S_1} g_{E,1}(x)\mu_1(\mathrm{d}x) \\ &= \int_{S_1} \sum_{m=1}^{\infty} \mu_{m,2}(\{y \in S_2 : (x,y) \in E\})\mu_1(\mathrm{d}x) \\ &= \sum_{m=1}^{\infty} \int_{S_1} \mu_{m,2}(\{y \in S_2 : (x,y) \in E\})\mu_1(\mathrm{d}x) \\ &= \sum_{m=1}^{\infty}\sum_{n=1}^{\infty} \int_{S_1\cap C_n} \mu_{m,2}(\{y \in S_2 : (x,y) \in E\})\mu_{n,1}(\mathrm{d}x) \\ &= \sum_{m=1}^{\infty}\sum_{n=1}^{\infty} \theta_{m,n,1}(E),\end{aligned} \tag{10.6}$$

where

$$\theta_{m,n,1}(E) = \int_{S_1\cap C_n} \mu_{m,2}(\{y \in S_2 : (x,y) \in E\})\mu_{n,1}(\mathrm{d}x), \quad m,n \geq 1.$$

Similarly, we have

$$\theta_2(E) = \sum_{m=1}^{\infty}\sum_{n=1}^{\infty} \theta_{m,n,2}(E), \tag{10.7}$$

where

$$\theta_{m,n,2}(E) = \int_{S_2\cap D_m} \mu_{n,1}(\{x \in S_1 : (x,y) \in E\})\mu_{m,2}(\mathrm{d}y), \quad m,n \geq 1.$$

By Problem 10.6, $\theta_{m,n,1}$ and $\theta_{m,n2}$ are measures on $S_1 \times S_2$. Therefore, $\theta_1(E)$ and $\theta_2(E)$ are measures on $S_1 \times S_2$.

(c) Let $E = A_1 \times A_2$ be a measurable rectangle.By Problem 10.6, we have

$$\theta_{m,n,1}(E) = \theta_{m,n,2}(E) = \mu_{n,1}(A_1)\mu_{m,2}(A_2).$$

Hence

$$\begin{aligned}\theta_1(E) &= \sum_{m=1}^{\infty}\sum_{n=1}^{\infty} \theta_{m,n,1}(E) \\ &= \sum_{m=1}^{\infty}\sum_{n=1}^{\infty} \mu_{n,1}(A_1)\mu_{m,2}(A_2) = \mu_1(A_1)\mu_2(A_2).\end{aligned}$$

Similarly, we have $\theta_2(E) = \mu_1(A_1)\mu_2(A_2)$.

(d) By Problem 10.6, we have $\theta_{m,n,1}(E) = \theta_{m,n,2}(E)$ for $E \in \Sigma_1 \otimes \Sigma_2$. Therefore, by Equations (10.6) and (10.7), $\theta_1(E) = \theta_2(E)$ for $E \in \Sigma_1 \otimes \Sigma_2$.

This completes the proof.

10.8. First, let us prove the "if" part. Suppose that for every $t \in I$, f_t is $\mathscr{F}/\mathscr{E}_t$-measurable. Let $\mathcal{C}$ be the collection of all measurable rectangles on F. Let $A \in \mathcal{C}$. Then

$$A = \times_{t\in I} A_t,$$

where $A_t \in \mathscr{E}_t$ and A_t differs from E_t for only a finite number of t. Note that

$$f^{-1}(A) = \{\omega \in \Omega : f_t(\omega) \in A_t\} = \bigcap_{t\in J} \{f_t \in A_t\},$$

where J is the finite index set of t where A_t differs from E_t. Hence we have $f^{-1}(A) \in \mathscr{F}$. Since $\mathscr{G} = \sigma(\mathcal{C})$, by Problem 5.4, f is $\mathscr{F}/\mathscr{G}$-measurable.

Second, let us prove the "only if" part. Suppose that f is $\mathscr{F}/\mathscr{G}$-measurable. Let $s \in I$ be fixed and $B \in \mathscr{E}_s$. Let A be a measurable rectangle on F given by

$$A = \times_{t\in I} A_t,$$

where $A_t = E_t$ for $t \neq s$ and $A_t = B$ for $t = s$. By the assumption, we have

$$f^{-1}(A) \in \mathscr{F}.$$

But $f^{-1}(A) = f_s^{-1}(B)$. Thus we have $f_s^{-1}(B) \in \mathscr{F}$.

Hence f_s is $\mathscr{F}/\mathscr{E}_t$-measurable. This completes the proof.

10.5 Bibliographic Notes

In this chapter, we introduced product measures and Fubini's theorem. We presented the definitions using the product of two measurable spaces. In fact, these definition and theorems can be extended to the product of n measurable spaces. For more information on product measures and Fubini's theorem, readers are referred to Weir (1979), Billingsley (1986), Williams (1991), Vestrup (2003), and Athreya and Lahiri (2006).

PART II

PROBABILITY THEORY

CHAPTER 11

EVENTS AND RANDOM VARIABLES

Samples, events, and random variables are the fundamental concepts in probability theory. In this chapter, we shall introduce the definitions of these concepts based on measure theory.

11.1 Basic Concepts and Facts

Definition 11.1 (Sample Space, Sample Point, and Event)**.** Let $(\Omega, \mathscr{F}, P)$ be a probability space (see Definition 2.12). The set Ω is called the *sample space*, a point in Ω is called a *sample point*, and a subset of $\mathscr{F}$ is called an *event*.

Definition 11.2 (Discrete Probability Space)**.** Let $\Omega = \{\omega_1, \omega_2, \ldots\}$ be a finite or countably infinite set. Let P be a set function defined as

$$P\{\omega_i\} = p_i, \quad i = 1, 2, \ldots$$

and

$$P(A) = \sum_{\omega_i \in A} p_i, \quad A \subseteq \Omega,$$

Measure, Probability, and Mathematical Finance.
By Guojun Gan, Chaoqun Ma, and Hong Xie

where $p_1, p_2, \ldots$ are nonnegative numbers whose sum is 1. Then P is a probability measure on 2^{Ω}, which is the collection of all subsets of Ω. The probability space $(\Omega, 2^{\Omega}, P)$ is called a *discrete probability space*.

Definition 11.3 (Almost Surely)**.** Let $(\Omega, \mathscr{F}, P)$ be a probability space. A statement about outcomes is said to be true almost surely (a.s.), or with probability 1, if the set of sample points for which the statement is true has probability 1:

$$F = \{\omega \in \Omega : \text{the statement about } \omega \text{ is true}\} \in \mathscr{F} \text{ and } P(F) = 1.$$

Definition 11.4 (Infinitely Often Event)**.** Let $(\Omega, \mathscr{F}, P)$ be a probability space, and let $\{E_n\}_{n\in\mathbf{N}}$ be a sequence of events. Infinitely often event $\limsup E_n$ is defined as

$$\limsup E_n = \bigcap_{m\geq 1}\left(\bigcup_{n\geq m} E_n\right).$$

Definition 11.5 (Eventual Event)**.** Let $(\Omega, \mathscr{F}, P)$ be a probability space and let $\{E_n\}_{n\in\mathbf{N}}$ be a sequence of events. Eventual event $\liminf E_n$ is defined as

$$\liminf E_n = \bigcup_{m\geq 1}\left(\bigcap_{n\geq m} E_n\right).$$

Definition 11.6 (Random Variable)**.** Let $(\Omega, \mathscr{F})$ be a measurable space. A random variable X is a $\mathscr{F}$-measurable function; that is,

$$X : \Omega \to \mathbf{R},$$

and $X^{-1}(A) \in \mathscr{F}$, $\forall A \in \mathcal{B}$, where

$$X^{-1}(A) = \{\omega \in \Omega : X(\omega) \in A\}.$$

Definition 11.7 (E-Valued Random Variable)**.** Let $(\Omega, \mathscr{F})$ and $(E, \mathscr{E})$ be two measurable spaces. An E-valued random variable X is a $\mathscr{F}/\mathscr{E}$-measurable function; that is,

$$X : \Omega \to E,$$

and $X^{-1}(A) \in \mathscr{F}$, $\forall A \in \mathscr{E}$, where

$$X^{-1}(A) = \{\omega \in \Omega : X(\omega) \in A\}.$$

Definition 11.8 (Random Vectors)**.** Let $(\Omega, \mathscr{F})$ be a measurable space. An n-dimensional random vector X is a $\mathscr{F}/\mathcal{B}(\mathbf{R}^n)$-measurable function. That is,

$$X : \Omega \to \mathbf{R}^n,$$

and $X^{-1}(A) \in \mathscr{F}$, $\forall A \in \mathcal{B}(\mathbf{R}^n)$, where

$$X^{-1}(A) = \{\omega \in \Omega : X(\omega) \in A\}.$$

A random vector X can be regarded as an n-tuple $(X_1, X_2, \ldots, X_n)$ of random variables defined on a common probability space.

Definition 11.9 (σ-algebra Generated by a Collection of Functions)**.** Let $(\Omega, \mathscr{F})$ be a measurable space and $(Y_\gamma : \gamma \in C)$ be a collection of functions $Y_\gamma : \Omega \to \mathbf{R}$. The σ-algebra $\mathcal{Y}$ generated by the collection $(Y_\gamma : \gamma \in C)$ of functions is defined as

$$\mathcal{Y} = \sigma(Y_\gamma : \gamma \in C) = \sigma\left(\{\omega \in \Omega : Y_\gamma(\omega) \in B\} : \gamma \in C, B \in \mathcal{B}\right).$$

In other words, $\mathcal{Y}$ is the smallest σ-algebra on Ω such that each function Y_γ is $\mathcal{Y}$-measurable.

Definition 11.10 (Probability Law and Distribution Function)**.** Let $(\Omega, \mathscr{F}, P)$ be a probability space and X be a random variable on Ω. The probability law of X is a function $\mathcal{L}_X : \mathcal{B} \to [0, 1]$ defined as

$$\mathcal{L}_X = P \circ X^{-1}.$$

The distribution function $F_X : \mathbf{R} \to [0, 1]$ of X is defined as

$$F_X(c) = \mathcal{L}_X(-\infty, c] = (P \circ X^{-1})(-\infty, c] = P(X \leq c) = P\{\omega : X(\omega) \leq c\}.$$

Definition 11.11 (Joint Law and Joint Distribution Function)**.** Let $X = (X_1, X_2, \ldots, X_n)$ be a random vector on a probability space $(\Omega, \mathscr{F}, P)$, where n is a natural number. Then the distribution function of X is the function $F_X : \mathbf{R}^d \to \mathbf{R}$ given by (see Definition 4.2)

$$F_X(x_1, x_2, \ldots, x_n) = P\left\{\omega : X_1(\omega) \leq x_1, X_2(\omega) \leq x_2, \ldots, X_n(\omega) \leq x_n\right\}.$$

Definition 11.12 (Absolutely Continuous Random Variable and Density Function)**.** A random variable X is said to be absolutely continuous if and only if there exists a nonnegative Borel function f on $\mathbf{R}$ such that

$$F(x) = \int_{-\infty}^{x} f(t)\mathrm{d}t, \quad x \in \mathbf{R},$$

where F is the distribution function of X. The function f is called the *density function* of X.

A random vector X is said to be absolutely continuous if and only if there exists a nonnegative Borel function f on $\mathbf{R}^n$ such that

$$F(x) = \int_{-\infty}^{x_1} \int_{-\infty}^{x_2} \cdots \int_{-\infty}^{x_n} f(t_1, t_2, \ldots, t_n)\mathrm{d}t_1\mathrm{d}t_2 \cdots \mathrm{d}t_n, \quad x \in \mathbf{R}^n,$$

where F is the joint distribution function of X.

Definition 11.13 (Degenerate and Nondegenerate Random Variable)**.** A random variable X on a probability space $(\Omega, \mathscr{F}, P)$ is said to be degenerate if its law is a Dirac measure δ_c, where

$$\delta_c(A) = \begin{cases} 1, & \text{if } c \in A; \\ 0, & \text{if } c \notin A, \end{cases} \qquad A \subset \mathcal{B}.$$

A random variable is said to be nondegenerate if it is not degenerate.

Theorem 11.1 (Fatou's Lemma). *Let (S, Σ, μ) be a measure space and let $\{E_n\}_{n\in\mathbf{N}}$ be a sequence of Σ-measurable subsets of S. Then we have*

$$\mu(\liminf E_n) \le \liminf \mu(E_n).$$

Theorem 11.2. *Let $(\Omega, \mathscr{F})$ be a measurable space. Let $X_1, X_2, \ldots, X_n$ be functions from Ω to $\mathbf{R}$. Then a function $Z : \Omega \to \mathbf{R}$ is $\sigma(X_1, X_2, \ldots, X_n)$-measurable if and only if there exists a Borel function $f : \mathbf{R}^n \to \mathbf{R}$ such that*

$$Z = f(X_1, X_2, \ldots, X_n).$$

Theorem 11.3. *Let $(\Omega, \mathscr{F})$ be a measurable space. Let $(X_\gamma : \gamma \in C)$ be a collection of functions from Ω to $\mathbf{R}$, where C is an infinite index set. Then a function $Z : \Omega \to \mathbf{R}$ is $\sigma(X_\gamma : \gamma \in C)$-measurable if and only if there exists a countable sequence $(\gamma_i : i \in \mathbf{N})$ of elements of C and a Borel function $f : \mathbf{R}^{\mathbf{N}} \to \mathbf{R}$ such that*

$$Z = f(X_{\gamma_1}, X_{\gamma_2}, \ldots).$$

11.2 Problems

11.1. Let $(\Omega, \mathscr{F}, P)$ be a probability space, and let $\{F_n\}_{n\in\mathbf{N}}$ be a sequence of events such that $P(F_n) = 1$ for all $n \in \mathbf{N}$. Show that

$$P\left(\bigcap_{n=1}^{\infty} F_n\right) = 1.$$

11.2. Prove Fatou's lemma in Theorem 11.1.

11.3 (Reverse Fatou's Lemma). Let $(\Omega, \mathscr{F}, P)$ be a probability space and $\{E_n\}_{n\in\mathbf{N}}$ a sequence of events. Show that

$$P(\limsup E_n) \ge \limsup P(E_n).$$

11.4. Let $(\Omega, \mathscr{F})$ be a measurable space, where $\Omega = \{H, T\}$ and

$$\mathscr{F} = \{\emptyset, \{H\}, \{T\}, \{H, T\}\}.$$

Let $X : \Omega \to \mathbf{R}$ be defined as

$$X(\omega) = \begin{cases} p, & \text{if } \omega = H; \\ 1 - p, & \text{if } \omega = T, \end{cases}$$

where $0 < p < 1$. Show that X is a random variable.

11.5. Let $(\Omega, \mathscr{F}, P)$ be a probability space and $E \in \mathscr{F}$ be an event. Show that

$$\{\emptyset, E, \Omega\backslash E, \Omega\} = \sigma(I_E),$$

where I_E is the indicator function.

11.6. Let $(\Omega, \mathscr{F}, P)$ be a probability space and $\{E_n\}_{n\geq 1}$ be a sequence of events. Show that

(a) If $E_n \uparrow E$, then $P(E_n) \uparrow P(E)$.

(b) If $E_n \downarrow E$, then $P(E_n) \downarrow P(E)$.

11.7. Let $(\Omega, \mathscr{F}, P)$ be a probability space and $\{E_n\}_{n\geq 1}$ be a sequence of events such that $E_n \to E$. Show that $P(E_n) \to P(E)$.

11.8. Let $(\Omega, \mathscr{F}, P)$ be a probability space and $\{E_n\}_{n\geq 1}$ be a sequence of events. Show that

$$P\left(\bigcup_{n=1}^{\infty} E_n\right) \leq \sum_{n=1}^{\infty} P(E_n).$$

11.9 (Bonferroni Inequalities)**.** Let $(\Omega, \mathscr{F}, P)$ be a probability space and $\{E_n\}_{n\geq 1}$ be a sequence of events. Show that

$$P\left(\bigcup_{i=1}^{n} E_i\right) \geq \sum_{i=1}^{n} P(E_i) - \sum_{1\leq i<j\leq n} P(E_i \cap E_j), \tag{11.1}$$

and

$$P\left(\bigcup_{i=1}^{n} E_i\right) \leq \sum_{i=1}^{n} P(E_i) - \sum_{1\leq i<j\leq n} P(E_i \cap E_j) + \sum_{1\leq i<j<k\leq n} P(E_i \cap E_j \cap E_k). \tag{11.2}$$

11.10. Let $(\Omega, \mathscr{F}, P)$ be a probability space and X be a random variable on Ω. Let $\pi(X)$ be defined as

$$\pi(X) = \{\{\omega \in \Omega : X(\omega) \leq c\} : c \in \mathbf{R}\}.$$

Show that $\pi(X)$ is a π-system and

$$\sigma(X) = \sigma(\pi(X)).$$

11.11. Let $(\Omega, \mathscr{F}, P)$ be a probability space and $\{X_n\}_{n\in\mathbf{N}}$ be a sequence of random variables on Ω. Define $\pi(X_n : n \in \mathbf{N})$ as

$$\pi(X_n : n \in \mathbf{N}) = \{\{\omega \in \Omega : X_i(\omega) \leq c_i,\ i \in \mathbf{N}\} :\ c_i \in \mathbf{R} \cup \{\infty\}\}.$$

Show that $\pi(X_n : n \in \mathbf{N})$ is a π-system and

$$\sigma(X_n : n \in \mathbf{N}) = \sigma(\pi(X_n : n \in \mathbf{N})).$$

11.12. Let $(\Omega, \mathscr{F}, P)$ be a probability space and $\{X_n\}_{n\geq 1}$ be a sequence of random variables on Ω. Show that

$$\{\limsup X_n > a\} = \limsup\{X_n > a\}, \quad a \in \mathbf{R}.$$

11.13. Let F_X be a distribution function of a random variable X. Show that F has only countably many discontinuities, that is, the set

$$D = \{a \in \mathbf{R} : \ F(x) \text{ is not continuous at } a\}$$

is countable.

11.14. Let X be a random variable with continuous distribution function F. Find the distribution function of the random variable $Y = F(X)$.

11.15. Let $\{E_n\}_{n\geq 1}$ be a sequence of events in a probability space $(\Omega, \mathscr{F}, P)$ such that

$$\lim_{n\to\infty} P(E_n) = 0$$

and

$$\sum_{n=1}^{\infty} P\left(E_{n+1} \cap E_n^c\right) = 0.$$

Show that

$$P\left(\limsup E_n\right) = 0.$$

11.3 Hints

11.1. Apply the result of Problem 2.19.

11.2. Establish $\mu\left(\bigcap_{i\geq m} E_i\right) \leq \sup_{r\geq 1}\left(\inf_{n\geq r} \mu(E_n)\right)$ for all $m \geq 1$ first and then apply Theorem 2.1.

11.3. First establish $P\left(\bigcup_{n\geq m} E_n\right) \geq \inf_{r\geq 1}\left(\sup_{n\geq r} P(E_n)\right)$ for all $m \geq 1$ and then use the result of Problem 2.18.

11.4. By definition, a random variable is a measurable function. To show that X is $\mathscr{F}$-measurable, one can use the result of Problem 5.7.

11.5. Follow the definition of σ-algebra generated by functions (see Definition 11.9).

11.6. Look at Theorem 2.1 and Problem 2.18.

11.7. First get familiar with what $E_n \to E$ means (see Problem 1.9) and then apply the result of Problem 11.6.

11.8. Write $\bigcup_{n=1}^{\infty} E_n$ as an infinite union of disjoint sets and use properties of measures.

11.9. Use the mathematical induction method.

11.10. Use the results of Problems 2.4 and 2.18 to show that $\sigma(X) = \sigma(\pi(X))$.

11.11. To show that $\sigma(X_n : n \in \mathbf{N}) = \sigma(\pi(X_n : n \in \mathbf{N}))$, one can use the standard technique, that is, show that $\sigma(X_n : n \in \mathbf{N}) \supseteq \sigma(\pi(X_n : n \in \mathbf{N}))$ and $\sigma(X_n : n \in \mathbf{N}) \subseteq \sigma(\pi(X_n : n \in \mathbf{N}))$.

11.12. Use the standard technique to show that two sets are equal (see Hint 11.11).

11.13. Use the result of Problem 2.20.

11.14. Note that if F is continuous, then $F^{-1}(y)$ is nonempty for any $y \in [0, 1]$. Then use the definition of distribution functions (Definition 11.10) to calculate the distribution function of Y.

11.15. First try to establish that

$$\limsup E_n \subseteq [\limsup (E_{n+1} \cap E_n^c)] \cup \liminf E_n.$$

and then use the first Borel-Cantelli lemma (Theorem 2.2) and Fatou's lemma (Theorem 11.1).

11.4 Solutions

11.1. Since P is a probability measure, we have $P(F_n^c) = 1 - P(F_n) = 0$ for all $n \geq 1$. From Problem 2.19, we have

$$P\left(\bigcap_{n=1}^{\infty} F_n\right) = 1 - P\left(\bigcup_{n=1}^{\infty} F_n^c\right) = 1.$$

This finishes the proof.

11.2. Since

$$\mu\left(\bigcap_{i\geq m} E_i\right) \leq \mu(E_n), \quad \forall n \geq m,\ m \geq 1,$$

we have

$$\mu\left(\bigcap_{i\geq m} E_i\right) \leq \inf_{i\geq m} \mu(E_i) \leq \sup_{r\geq 1}\left(\inf_{n\geq r} \mu(E_n)\right), \quad \forall m \geq 1. \tag{11.3}$$

From Theorem 2.1, we have

$$\mu\left(\bigcap_{n\geq m} E_n\right) \uparrow \mu\left(\bigcup_{m=1}^{\infty}\bigcap_{n\geq m} E_n\right) = \mu(\liminf E_n) \tag{11.4}$$

as $m \to \infty$. Combining (11.3) and (11.4) gives

$$\mu(\liminf E_n) \leq \liminf \mu(E_n).$$

This proves the theorem.

11.3. Since

$$P\left(\bigcup_{i\geq m} E_i\right) \geq P(E_n), \quad \forall n \geq m,\ m \in \mathbf{N},$$

we have

$$P\left(\bigcup_{n\geq m} E_n\right) \geq \sup_{n\geq m} P(E_n) \geq \inf_{r\geq 1}\left(\sup_{n\geq r} P(E_n)\right), \quad \forall m \in \mathbf{N}. \tag{11.5}$$

From Problem 2.18, we have

$$P\left(\bigcup_{n\geq m} E_n\right) \downarrow P\left(\bigcap_{n\geq 1}\bigcup_{n\geq m} E_n\right) = P\left(\limsup E_n\right) \tag{11.6}$$

as $m \to \infty$. Combining (11.5) and (11.6), we have

$$P\left(\limsup E_n\right) \geq \limsup P(E_n).$$

11.4. To show that X is a random variable, we only need to show that X is $\mathscr{F}$-measurable. To do this, we show that $\{X > c\} \in \mathscr{F}$, $\forall c \in \mathbf{R}$ (see Problem 5.7), where $\{X > c\} = \{\omega \in \Omega : X(\omega) > c\}$.

If $p < \frac{1}{2}$, then $p < 1 - p$ and

$$\{X > c\} = \begin{cases} \{H, T\}, & \text{if } c < p; \\ \{T\}, & \text{if } p \leq c < 1-p; \\ \emptyset, & \text{if } 1-p \leq c. \end{cases}$$

If $p = \frac{1}{2}$, then $p = 1 - p$ and

$$\{X > c\} = \begin{cases} \{H, T\}, & \text{if } c < p; \\ \emptyset, & \text{if } 1-p \leq c. \end{cases}$$

If $p > \frac{1}{2}$, then $p > 1 - p$ and

$$\{X > c\} = \begin{cases} \{H, T\}, & \text{if } c < 1 - p; \\ \{H\}, & \text{if } 1 - p \le c < p; \\ \emptyset, & \text{if } p \le c. \end{cases}$$

Thus we have $\{X > c\} \in \mathscr{F}$, $\forall c \in \mathbf{R}$.

11.5. By definition, we have

$$\sigma(I_E) = \sigma(\{\omega \in \Omega : I_E(\omega) \in B\} : B \in \mathcal{B}).$$

But

$$\{\omega \in \Omega : I_E(\omega) \in B\} = \begin{cases} \emptyset, & \text{if } 0 \notin B \text{ and } 1 \notin B; \\ \Omega \backslash E, & \text{if } 0 \in B \text{ and } 1 \notin B; \\ E, & \text{if } 0 \notin B \text{ and } 1 \in B; \\ \Omega, & \text{if } 0 \in B \text{ and } 1 \in B, \end{cases}$$

Thus we have

$$\sigma(I_E) = \sigma(\emptyset, E, \Omega\backslash E, \Omega).$$

Noting that $\{\emptyset, E, \Omega\backslash E, \Omega\}$ is the smallest σ-algebra containing $\emptyset$, E, $\Omega\backslash E$, and Ω, we have $\sigma(\emptyset, E, \Omega\backslash E, \Omega) = \{\emptyset, E, \Omega\backslash E, \Omega\}$, which gives the result.

11.6. The results follow from Theorem 2.1 and Problem 2.18 immediately.

11.7. Since $E_n \to E$, we have, by Problem 1.9, $E \in \mathscr{F}$ and $\limsup E_n = \liminf E_n = E$. Let $A_n = \bigcup_{i \ge n} E_i$ and $B_n = \bigcap_{i \ge n} E_i$ for all $n \ge 1$. Then $A_n \downarrow E$ and $B_n \uparrow E$. By Problem 11.6, we have $P(A_n) \downarrow P(E)$ and $P(B_n) \uparrow P(E)$. But $P(A_n) \ge P(E_n) \ge P(B_n)$ for all $n \ge 1$, we have $P(E_n) \to P(E)$.

11.8. Let $A_1 = E_1$ and $A_n = E_n \backslash \bigcup_{i=1}^{n-1} E_i$ for $n \ge 2$. Then $A_1, A_2, \ldots$ are pairwise disjoint and $\bigcup_{n=1}^{\infty} A_n = \bigcup_{n=1}^{\infty} E_n$. Noting that P is a probability measure and $P(A_n) \le P(E_n)$ for $n \ge 1$, we have

$$P\left(\bigcup_{n=1}^{\infty} E_n\right) = P\left(\bigcup_{n=1}^{\infty} A_n\right) = \sum_{n=1}^{\infty} P(A_n) \le \sum_{n=1}^{\infty} P(E_n).$$

11.9. We use the mathematical induction method. When $n = 1$, inequality (11.1) reduces to $P(E_1) \ge P(E_1)$, which is true. Suppose that inequality (11.1) is true for

$n = m$. Then by Problem 2.11, the induction hypothesis, and Problem 11.8, we have

$$\begin{aligned} P\left(\bigcup_{i=1}^{m+1} E_i\right) &= P\left(\bigcup_{i=1}^{m} E_i\right) + P(E_{m+1}) - P\left(\left(\bigcup_{i=1}^{m} E_i\right) \cap E_{m+1}\right) \\ &\geq \sum_{i=1}^{m} P(E_i) - \sum_{1\leq i<j\leq m} P(E_i \cap E_j) \\ &\quad + P(E_{m+1}) - P\left(\bigcup_{i=1}^{m} (E_i \cap E_{m+1})\right) \\ &\geq \sum_{i=1}^{m+1} P(E_i) - \sum_{1\leq i<j\leq m+1} P(E_i \cap E_j). \end{aligned}$$

Therefore, inequality (11.1) is true for all $n \geq 1$.

We can prove inequality (11.2) by mathematical induction and inequality (11.1).

11.10. We first show that $\pi(X)$ is a π-system. Let $I_1 \in \pi(X)$ and $I_2 \in \pi(X)$; then

$$\begin{aligned} I_1 &= \{\omega \in \Omega : X(\omega) \leq c_1\}, \\ I_2 &= \{\omega \in \Omega : X(\omega) \leq c_2\} \end{aligned}$$

for some $c_1, c_2 \in \mathbf{R}$. But

$$I_1 \cap I_2 = \{\omega \in \Omega : X(\omega) \leq \min(c_1, c_2)\} \in \pi(X)$$

which implies that $\pi(X)$ is a π-system.

Note that

$$\sigma(X) = \{\{\omega \in \Omega : X(\omega) \in B\} : B \in \mathcal{B}\} = X^{-1}(\mathcal{B}),$$

and

$$\pi(X) = \{\{\omega \in \Omega : X(\omega) \leq c\} : c \in \mathbf{R}\} = X^{-1}(\pi(\mathbf{R})),$$

where $\pi(\mathbf{R}) = \{(-\infty, c] : c \in \mathbf{R}\}$. By Problem 2.4 and Problem 5.5, we have

$$\sigma(X) = X^{-1}(\mathcal{B}) = \sigma(X^{-1}(\pi(\mathbf{R}))) = \sigma(\pi(X)).$$

This completes the proof.

11.11. We first show that $\pi(X_n : n \in \mathbf{N})$ is a π-system. To do this, let $I_1, I_2 \in \pi(X_n : n \in \mathbf{N})$; then we have

$$\begin{aligned} I_1 &= \{\omega \in \Omega : X_i(\omega) \leq c_i,\ i \in \mathbf{N}\}, \\ I_2 &= \{\omega \in \Omega : X_i(\omega) \leq d_i,\ i \in \mathbf{N}\}, \end{aligned}$$

for some $c_i, d_i \in \mathbf{R} \cup \{\infty\}$. But

$$I_1 \cap I_2 = \{\omega \in \Omega : X_i(\omega) \leq \min(c_i, d_i),\ i \in \mathbf{N}\} \in \pi(X_n : n \in \mathbf{N}).$$

Thus, $\pi(X_n : n \in \mathbf{N})$ is a π-system.

Noting that, for any $i \in \mathbf{N}$, we obtain

$$\begin{aligned} \pi(X_i) &\in \{\omega : X_i(\omega) \leq c,\ X_j(\omega) \leq \infty, j \in \mathbf{N}, j \neq i\} \\ &= \{\omega : X_i(\omega) \leq c\} \\ &\in \pi(X_n : n \in \mathbf{N}), \quad \forall c \in \mathbf{R} \cup \{\infty\}, \end{aligned}$$

which implies

$$\pi(X_i) \subseteq \pi(X_n : n \in \mathbf{N}).$$

From Problem 11.10, we have

$$\sigma(X_i) = \sigma(\pi(X_i)) \subseteq \sigma(\pi(X_n : n \in \mathbf{N})).$$

Therefore

$$\bigcup_{i=1}^{\infty} \sigma(X_i) \subseteq \sigma(\pi(X_n : n \in \mathbf{N})). \tag{11.7}$$

But

$$\begin{aligned} \sigma(X_n : n \in \mathbf{N}) &= \sigma(\{\omega \in \Omega : X_n(\omega) \in B\} :\ n \in \mathbf{N}, B \in \mathcal{B}) \\ &= \sigma\left(\bigcup_{n=1}^{\infty} \sigma(X_n)\right). \end{aligned}$$

Thus, from (11.7) we have

$$\sigma(X_n : n \in \mathbf{N}) \subseteq \sigma(\pi(X_n : n \in \mathbf{N})). \tag{11.8}$$

Now we show that

$$\sigma(\pi(X_n : n \in \mathbf{N})) \subseteq \sigma(X_n : n \in \mathbf{N}). \tag{11.9}$$

To do this, let $I \in \pi(X_n : n \in \mathbf{N})$; then

$$I = \{\omega \in \Omega : X_i(\omega) \leq c_i,\ i \in \mathbf{N}\}.$$

But

$$\{\omega \in \Omega : X_i(\omega) \leq c_i,\ i \in \mathbf{N}\} = \bigcap_{i=1}^{\infty} \{\omega \in \Omega : X_i(\omega) \leq c_i\} \in \sigma(X_n : n \in \mathbf{N}).$$

We have $I \in \sigma(X_n : n \in \mathbf{N})$. Since I is arbitrary, we have

$$\pi(X_n : n \in \mathbf{N}) \subseteq \sigma(X_n : n \in \mathbf{N}),$$

which gives (11.9). Combining (11.8) and (11.9) gives the result.

11.12. First we show that

$$\{\limsup X_n > a\} \subseteq \limsup\{X_n > a\}.$$

To do this, let $\omega \in \{\limsup X_n > a\}$. Then $\limsup X_n(\omega) > a$. By Definition 1.9, we have $\sup_{n\geq m} X_n(\omega) > a$ for all $m \geq 1$, which implies that $\omega \in \bigcup_{n\geq m}\{X_n > a\}$ for all $m \geq 1$. Hence $\omega \in \bigcap_{m\geq 1}\bigcup_{n\geq m}\{X_n > a\}$. Thus $\{\limsup X_n > a\} \subseteq \limsup\{X_n > a\}$.

Similarly, we can show that $\{\limsup X_n > a\} \supseteq \limsup\{X_n > a\}$. This completes the proof.

11.13. Suppose that X is a random variable on the probability space $(\Omega, \mathscr{F}, P)$. For each $a \in D$, define an event

$$E_a = \{\omega \in \Omega : X(\omega) = a\}.$$

Then the E_a, $a \in D$, are disjoint sets in $\mathscr{F}$. By the definition of D, for $a \in D$, we have $P(E_a) = F(a) - F(a-) > 0$, where

$$F(a-) = \lim_{x\to a-} F(x).$$

It follows from Problem 2.20 that D is a countable set. This completes the proof.

11.14. Let F_Y be the distribution function of Y. Then by Definition 11.10, we have

$$F_Y(y) = P\{Y \leq y\} = P\{F(X) \leq y\}, \quad y \in \mathbf{R}.$$

Note that $F(X) \in [0, 1]$. We have $F_Y(y) = 0$ for $y < 0$ and $F_Y(y) = 1$ for $y > 1$. Now let us consider the case when $y \in [0, 1]$. By the assumption that the distribution function F is continuous, we know that $F^{-1}(y)$ is nonempty. Hence we can find an $x_0 \in F^{-1}(y)$; that is, we can find an x_0 such that $F(x_0) = y$. Thus

$$F_Y(y) = P\{F(X) \leq y\} = P\{F(X) \leq F(x_0)\} = P\{X \leq x_0\} = F(x_0) = y.$$

Therefore, Y has a distribution function given by

$$F_Y(y) = \begin{cases} 0, & \text{if } y < 0; \\ y, & \text{if } 0 \leq y \leq 1; \\ 1, & \text{if } y > 1. \end{cases}$$

11.15. To prove the result, we first show that

$$\limsup E_n \subseteq [\limsup (E_{n+1} \cap E_n^c)] \cup \liminf E_n. \tag{11.10}$$

To do that, let $\omega \in \limsup E_n$. Then there are infinitely many events containing ω. That is, $\omega \in E_{n_i}$ for $n_1 < n_2 < \ldots$.

Now if there exists an integer k_0 such that $n_k = n_{k_0} + k - k_0$ all $k \geq k_0$, then $\omega \in \liminf E_n$; that is, if ω is contained in all E_i for $i \geq n_{k_0}$, then $\omega \in \liminf E_n$. In this case, we have

$$\omega \in [\limsup (E_{n+1} \cap E_n^c)] \cup \liminf E_n.$$

Now suppose that no such k_0 exists. Then for any $i \in \mathbf{N}$, we can find an integer $k > i$ such that $\omega \in E_{k+1} \cap E_k$. It follows that $\omega \in \limsup (E_{n+1} \cap E_n^c)$. In this case, we still have

$$\omega \in [\limsup (E_{n+1} \cap E_n^c)] \cup \liminf E_n.$$

Therefore, Equation (11.10) holds. By the first Borel-Cantelli lemma (Theorem 2.2) and the assumption, we have

$$P[\limsup (E_{n+1} \cap E_n^c)] = 0.$$

By Fatou's lemma (Theorem 11.1) and the assumption, we have

$$P(\liminf E_n) \leq \liminf P(E_n) = 0.$$

Hence by Equation (11.10), we get

$$P(\limsup E_n) \leq P[\limsup (E_{n+1} \cap E_n^c)] + P(\liminf E_n) = 0.$$

This completes the proof.

11.5 Bibliographic Notes

In this chapter, we introduced some basic concepts in probability theory based on measure theory. In fact, measure theory is necessary for us to study probability theory in a rigorous manner. Rosenthal (2006, Chapter 1) discussed why measure theory is indispensable to a mathematical rigorous theory of probability. Shafer and Vovk (2001) discussed the historical development of probability theory.

Theorems 11.2 and 11.3 are two interesting results that characterize the σ-algebras generated by random variables. For proofs of these two theorems, readers are referred to Çinlar (2011, pp76-78).

For textbooks on measure-theoretic probability, readers are referred to Loeve (1978), Williams (1991), Billingsley (1995), Kallenberg (1997), Chung (2000), Pollard (2001), Jacod and Protter (2004), Roussas (2005), Athreya and Lahiri (2006), Rao and Swift (2006), Koralov and Sinai (2007), Bhattacharya and Waymire (2007), Durrett (2010), and Çinlar (2011). For probability theory without measure theory, readers may consult Hsu (1996), Roussas (1997), Stirzaker (2003), Dekking et al. (2005), Lefebvre (2006), Bartoszynski and Niewiadomska-Bugaj (2007), and Gut (2009). Most probability concepts and results are also introduced in statistics books, such as Roussas (1997) and Shao (2003).

For problem books on probability, readers are referred to Mosteller (1987), Capinski and Zastawniak (2000), Mills (2001), Grimmett and Stirzaker (2001), and Chaumont and Yor (2003).

CHAPTER 12

INDEPENDENCE

Independence is an important concept in probability theory. Intuitively, two events are independent if the occurrence of one event does not make the other happen more or less. In this chapter, we will introduce the concept of independence of events and random variables.

12.1 Basic Concepts and Facts

Definition 12.1 (Independent σ-Algebras). Let $(\Omega, \mathscr{F}, P)$ be a probability space. Sub-σ-algebras $\mathcal{G}_1, \mathcal{G}_2, \ldots, \mathcal{G}_n$ of $\mathscr{F}$ are considered independent if and only if

$$P\left(\bigcap_{i=1}^{n} G_i\right) = \prod_{i=1}^{n} P(G_i)$$

for all sets $G_i \in \mathcal{G}_i$ $(1 \le i \le n)$.

Let $\mathcal{G}_i$, $i \in I$ (an arbitrary index set) be an arbitrary family of sub-σ-algebras of $\mathscr{F}$. The $\mathcal{G}_i$ are deemed independent if and only if $\mathcal{G}_{i_1}, \mathcal{G}_{i_2}, \ldots, \mathcal{G}_{i_n}$ are independent for each finite set $\{i_1, i_2, \ldots, i_n\}$ of distinct indices in I.

Definition 12.2 (Independent Random Variable). Let $(\Omega, \mathscr{F}, P)$ be a probability space and $X_i,\ i \in I$ be a collection of random variables on Ω, where I is an index set. The random variables X_i are deemed independent if and only if the sub-σ-algebras $\sigma(X_i),\ i \in I$ are independent, where $\sigma(X_i)$ is the σ-algebra generated by X_i for $i \in I$ (see Definition 11.9).

Definition 12.3 (Independent Random Variable and σ-Algebra). A random variable X is said to be independent of a σ-algebra $\mathscr{F}$ if and only if $\sigma(X)$ is independent of $\mathscr{F}$.

Definition 12.4 (Independent Events). Let $(\Omega, \mathscr{F}, P)$ be a probability space and $E_i \in \mathscr{F}$ $(i \in I)$ be a collection of events, where I is an index set. The events E_i are considered independent if and only if the σ-algebras

$$\mathcal{E}_i = \{\emptyset, E_i, \Omega \backslash E_i, \Omega\}, \quad i \in I$$

are independent.

Definition 12.5 (Tail σ-Algebra). Let $\{X_n\}_{n \in \mathbf{N}}$ be a sequence of random variables. The tail σ-algebra $\mathcal{T}$ of the sequence is defined as

$$\mathcal{T} = \bigcap_{n=1}^{\infty} \mathcal{T}_n,$$

where

$$\mathcal{T}_n = \sigma(X_{n+1}, X_{n+2}, \ldots).$$

12.2 Problems

12.1. Let X be a random variable on some probability space $(\Omega, \mathscr{F}, P)$. Let $\mathscr{F}_0 = \{\emptyset, \Omega\}$ be the trivial σ-algebra. Show that X is independent of $\mathscr{F}_0$, that is, $\sigma(X)$ is independent of $\mathscr{F}_0$.

12.2. Let $(\Omega, \mathscr{F}, P)$ be a probability space. Let $\mathcal{G}_1, \mathcal{G}_2, \ldots$ be sub-σ-algebras of $\mathscr{F}$. Then $\mathcal{G}_1, \mathcal{G}_2, \ldots$ are independent if and only if

$$P\left(\bigcap_{j=1}^{r} G_{i_j}\right) = \prod_{j=1}^{r} P(G_{i_j})$$

for any $1 \le i_1 < i_2 < \cdots < i_r$ and all subsets $G_{i_j} \in \mathcal{G}_{i_j}$ $(1 \le j \le r)$.

12.3. Let $(\Omega, \mathscr{F}, P)$ be a probability space and $\mathcal{G}$ and $\mathcal{H}$ be sub-σ-algebras of $\mathscr{F}$. Let $\mathcal{I}$ and $\mathcal{J}$ be π-systems (see Definition 2.15) such that

$$\mathcal{G} = \sigma(\mathcal{I}), \quad \mathcal{H} = \sigma(\mathcal{J}).$$

Then $\mathcal{G}$ and $\mathcal{H}$ are independent if and only if $\mathcal{I}$ and $\mathcal{J}$ are independent in that

$$P(I \cap J) = P(I)P(J), \quad \forall I \in \mathcal{I}, \forall J \in \mathcal{J}.$$

12.4. Let $(\Omega, \mathscr{F}, P)$ be a probability space and $\mathcal{G}_i$ $(1 \le i \le n)$ sub-σ-algebras of $\mathscr{F}$. Let $\mathcal{I}_i$ $(1 \le i \le n)$ be π-systems such that

$$\mathcal{G}_i = \sigma(\mathcal{I}_i), \quad i = 1, 2, \ldots, n.$$

Then $\mathcal{G}_1, \mathcal{G}_2, \ldots, \mathcal{G}_n$ are independent if and only if $\mathcal{I}_1, \mathcal{I}_2, \ldots, \mathcal{I}_n$ are independent in that

$$P\left(\bigcap_{i=1}^n I_i\right) = \prod_{i=1}^n P(I_i)$$

for all sets $I_i \in \mathcal{I}_i$.

12.5. Let $X_1, X_2, \ldots, X_n$ be random variables on $(\Omega, \mathscr{F}, P)$. Show that X_1, X_2, ..., X_n are independent if and only if

$$P\{X_1 \in B_1, X_2 \in B_2, \ldots, X_n \in B_n\} = \prod_{i=1}^n P\{X_i \in B_i\}$$

for all sets $B_1, B_2, \ldots, B_n \in \mathcal{B}(\mathbf{R})$.

12.6. Let $X_1, X_2, \ldots, X_n$ be discrete random variables on a probability space $(\Omega, \mathscr{F}, P)$; that is, X_i assumes either a finite number of values or an infinite sequence of values. Show that X_1, X_2, ..., X_n are independent if and only if

$$P\{X_1 = x_1, X_2 = x_2, \ldots, X_n = x_n\} = \prod_{i=1}^n P\{X_i = x_i\}$$

for all $x_i \in \{X_i(\omega) : \omega \in \Omega\}$, $i = 1, 2, \ldots, n$.

12.7. Let $X_1, X_2, \ldots, X_n$ be random variables on a probability space $(\Omega, \mathscr{F}, P)$. Let $f_i : \mathbf{R} \to \mathbf{R}$ be a Borel measurable function for $i = 1, 2, \ldots, n$. Show that if $X_1, X_2, \ldots, X_n$ are independent, then $f_1(X_1), f_2(X_2), \ldots, f_n(X_n)$ are also independent.

12.8. Let $X_1, X_2, \ldots, X_n$ be random variables on $(\Omega, \mathscr{F}, P)$. Let F_i be the distribution function of X_i for $i = 1, 2, \ldots, n$, and let F be the distribution function of $X = (X_1, X_2, \ldots, X_n)$. Show that $X_1, X_2, \ldots, X_n$ are independent if and only if

$$F(x_1, x_2, \ldots, x_n) = \prod_{i=1}^n F(x_i)$$

for all $x_1, x_2, \ldots, x_n \in \mathbf{R}$.

12.9 (Second Borel-Cantelli Lemma). Let $(\Omega, \mathscr{F}, P)$ be a probability space and $\{E_n\}_{n\geq 1}$ be a sequence of independent events on Ω. If

$$\sum_{n=1}^{\infty} P(E_n) = \infty,$$

then

$$P(\limsup E_n) = 1.$$

12.10 (Kolmogorov's 0-1 Law). Let $(\Omega, \mathscr{F}, P)$ be a probability space and $\{X_n\}_{n\geq 1}$ be a sequence of independent random variables on Ω. Let $\mathcal{T}$ be the tail σ-algebra for X_n $(n \in \mathbf{N})$. Then $\mathcal{T}$ is P-trivial in that

(a) If $F \in \mathcal{T}$, then $P(F) = 0$ or $P(F) = 1$.

(b) If Y is a $\mathcal{T}$-measurable random variable, then $P(Y = c) = 1$ for some constant $c \in [-\infty, \infty]$.

12.11. Let $(\Omega, \mathscr{F}, P)$ be a probability space. Let A and B be two events. Show that A and B are independent if $P(A \cap B) = P(A)P(B)$.

12.12. Let $(\Omega, \mathscr{F}, P)$ be a probability space. Let A and B be two independent events. Show that A and B^c are independent, A^c and B^c are independent, and A^c and B^c are independent.

12.13. Give an example showing that a set of events $E_1, E_2, \ldots, E_n$ are pairwise independent but are not independent.

12.14. Let $(\Omega, \mathscr{F}, P)$ be a probability space and $\mathcal{I}_i$ $(i = 1, 2, 3)$ be three π-systems on Ω such that

$$\mathcal{I}_i \subseteq \mathscr{F} \text{ and } \Omega \in \mathcal{I}_i, \quad i = 1, 2, 3.$$

Show that if

$$P(I_1 \cap I_2 \cap I_3) = P(I_1)P(I_2)P(I_3), \quad \forall I_i \in \mathcal{I}_i,\ i = 1, 2, 3, \tag{12.1}$$

then $\sigma(I_1)$, $\sigma(I_2)$, and $\sigma(I_3)$ are independent.

12.15. Let $s > 1$ and $\varsigma(s) = \sum_{n=1}^{\infty} n^{-s}$. Let $(\Omega, \mathscr{F}, P)$ be a probability space, and let X and Y be two independent $\mathbf{N}$-valued random variables on $(\Omega, \mathscr{F}, P)$ with

$$P(X = n) = P(Y = n) = \frac{n^{-s}}{\varsigma(s)}, \quad \forall n \in \mathbf{N}.$$

Show that

(a) The events E_p $(p \in \mathbf{P})$ are independent, where $E_p = \{\omega \in \Omega : p \mid X(\omega)\}$ and $\mathbf{P}$ is the set of all prime numbers in $\mathbf{N}$. Here $a \mid b$ means b is divisible by a.

(b) $1/\varsigma(s) = \prod_{p\in\mathbf{P}}(1 - p^{-s})$.

(c) $P(\{\omega \in \Omega : m^2 \nmid X(\omega), \forall m > 1\}) = 1/\varsigma(2s)$, where $a \nmid b$ means b is not divisible by a.

(d) $P(H = n) = n^{-2s}/\varsigma(2s)$, where H is the highest common factor of X and Y.

12.16. Let $(\Omega, \mathscr{F}, P)$ be a probability space and X_n $(n \in \mathbf{N})$ be a sequence of independent random variables that are exponentially distributed with rate 1:

$$P(X_n > x) = e^{-x}, \quad x \geq 0.$$

Show that

$$P\left(\limsup \frac{X_n}{\log n} \geq 1\right) = 1.$$

12.17. Let $(\Omega, \mathscr{F}, P)$ be a probability space and X_n $(n \in \mathbf{N})$ be random variables on Ω. Let $\mathcal{T}$ be the tail σ-algebra of X_n $(n \in \mathbf{N})$. Show that F_1, F_2, and F_3 are $\mathcal{T}$-measurable, where

$$\begin{aligned}
F_1 &= \left(\lim_{n\to\infty} X_n \text{ exists}\right) = \left\{\omega \in \Omega : \lim_{n\to\infty} X_n(\omega) \text{ exists}\right\}, \\
F_2 &= \left(\lim_{n\to\infty} \sum_{k=1}^{n} X_k \text{ exists}\right), \\
F_3 &= \left(\lim_{n\to\infty} \frac{1}{n} \sum_{k=1}^{n} X_k \text{ exists}\right).
\end{aligned}$$

12.3 Hints

12.1. Follow the definition of independence (Definition 12.1).

12.2. Use the definition of independent σ-algebras (Definition 12.1) and note that Ω is an element of all sub-σ-algebras of $\mathscr{F}$.

12.3. The "only if" part is straightforward. To prove the "if" part, one needs to use the unique extension theorem (Theorem 3.2). To do so, one can define two set functions $\mu_1(H) = P(I \cap H)$ and $\mu_2(H) = P(I)P(H)$ on $\mathcal{H}$ for fixed $I \in \mathcal{I}$ and show that $\mu_1 = \mu_2$ on $\mathcal{J}$.

12.4. Apply the same argument used to prove Problem 12.3.

12.5. Follow the definition of independent random variables (Definition 12.2) and use the results of Problems 12.4 and 11.10.

12.6. Use the result of Problem 12.5

12.7. Use the result of Problem 12.5

12.8. Use the results of Problems 12.5, 11.10, and 12.4 and note that

$$F(x_1, x_2, \ldots, x_n) = P\{X_1 \leq x_1, X_2 \leq x_2, \ldots, X_n \leq x_n\}.$$

12.9. Show that $P(\bigcap_{n\geq m} E_n^c) = 0$ and apply the result of Problem 2.19.

12.10. For part (a), one needs to show that $\mathcal{T}$ and $\mathcal{T}$ are independent. To do this, one can show that $\sigma(X_1, X_2, \cdots, X_n)$ and $\mathcal{T}$ are independent. Then show that $\sigma(X_n : n \in \mathbf{N})$ and $\mathcal{T}$ are independent. Part (b) is just a corollary of part (a).

12.11. Apply the result of Problem 12.3.

12.12. Apply the result of Problem 12.11.

12.13. Find an example such that $P(E_i \cap E_j) = P(E_i)P(E_j)$ for all $i \neq j$ and $P(E_1 \cap E_2 \cap \cdots \cap E_n) \neq P(E_1)P(E_2)\cdots P(E_n)$.

12.14. Apply the result of Problem 12.4 or apply the result of Problem 12.3 to prove directly.

12.15. For part (a), one can show that E_p $(p \in \mathbf{P})$ are independent by establishing $P\left(\bigcap_{j=1}^r E_{p_{i_j}}\right) = \prod_{j=1}^r P(E_{p_{i_j}})$ for all $1 \leq i_1 < i_2 < \cdots < i_r$, where p_{i_j} are prime numbers. Parts (b)-(d) can be derived from part (a).

12.16. Apply the results of Problems 11.12 and 12.9.

12.17. For the first equation, use the fact that $\lim X_n(\omega)$ exists if and only if

$$\limsup X_n(\omega) = \liminf X_n(\omega).$$

Other equations can be proved similarly.

12.4 Solutions

12.1. By Definition 12.1, we only need to show that for every $A \in \sigma(X)$ and every $B \in \mathscr{F}_0$,

$$P(A \cap B) = P(A) \cdot P(B). \tag{12.2}$$

Since $\mathscr{F}_0$ contains only two elements, we only need to verify that Equation (12.2) holds for $B = \emptyset$ and $B = \Omega$.

If $B = \emptyset$, then $P(A \cap \emptyset) = 0 = P(A) \cdot P(\emptyset)$. If $B = \Omega$, then $P(A \cap \Omega) = P(A) = P(A) \cdot P(\Omega)$. This completes the proof.

12.2. The "if" part is straightforward. Suppose that

$$P\left(\bigcap_{j=1}^r G_{i_j}\right) = \prod_{j=1}^r P(G_{i_j})$$

for any $1 \leq i_1 < i_2 < \cdots < i_r$ and all subsets $G_{i_j} \in \mathcal{G}_{i_j}$ $(1 \leq j \leq r)$. Then by choosing $r = n$ and $i_j = j$ for $j = 1, 2, \ldots, n$, we have

$$P\left(\bigcap_{i=1}^{n} G_i\right) = \prod_{i=1}^{n} P(G_i).$$

Hence these sub-σ-algebras are independent, by Definition 12.1.

Now let us prove the "only if" part. Suppose that $\mathcal{G}_1$, $\mathcal{G}_2$, $\ldots$ are independent. Then we have

$$P\left(\bigcap_{i=1}^{n} G_i\right) = \prod_{i=1}^{n} P(G_i)$$

for any $n \geq 1$ and all subsets $G_i \in \mathcal{G}_i$ $(1 \leq i \leq n)$.

Let $1 \leq i_1 < i_2 < \cdots < i_r$ and $G_{i_j} \in \mathcal{G}_{i_j}$ $(1 \leq j \leq r)$. Then we define $H_1, H_2, \ldots, H_{i_r}$ as follows:

$$H_k = \begin{cases} \Omega, & \text{if } k \notin \{i_1, i_2, \ldots, i_r\}; \\ G_{i_j}, & \text{if } k = i_j \text{ for some } j. \end{cases}$$

By assumption and the fact that $P(\Omega) = 1$, we have

$$P\left(\bigcap_{j=1}^{r} G_{i_j}\right) = P\left(\bigcap_{k=1}^{i_r} H_k\right) = \prod_{k=1}^{i_r} P(H_k) = \prod_{j=1}^{r} P(G_{i_j}).$$

Hence the "only if" part is valid. This completes the proof.

12.3. First, suppose that $\mathcal{G}$ and $\mathcal{H}$ are independent; then by definition, we have

$$P(G \cap H) = P(G)P(H), \quad \forall G \in \mathcal{G},\ \forall H \in \mathcal{H}.$$

By hypothesis, we have $\mathcal{I} \subseteq \mathcal{G}$ and $\mathcal{J} \subseteq \mathcal{H}$. Thus

$$P(I \cap J) = P(I)P(J), \quad \forall I \in \mathcal{I},\ \forall J \in \mathcal{J}.$$

Next, suppose that $\mathcal{I}$ and $\mathcal{J}$ are independent. For fixed $I \in \mathcal{I}$, we define two functions $\mu_1 : \mathcal{H} \to [0, 1]$ and $\mu_2 : \mathcal{H} \to [0, 1]$ as

$$\mu_1(H) = P(I \cap H), \quad \mu_2(H) = P(I)P(H).$$

We claim that μ_1 and μ_2 are measures on $(\Omega, \mathcal{H})$. In fact, noting that P is a probability measure on $(\Omega, \mathcal{H})$, we have

$$\mu_1(\emptyset) = P(\emptyset \cap H) = P(\emptyset) = 0, \quad \mu_2(\emptyset) = P(\emptyset)P(H) = 0,$$

and

$$\begin{aligned}
\mu_1\left(\bigcup_{n=1}^{\infty} H_n\right) &= P\left(I \cap \left(\bigcup_{n=1}^{\infty} H_n\right)\right) \\
&= P\left(\bigcup_{n=1}^{\infty} (I \cap H_n)\right) \\
&= \sum_{n=1}^{\infty} \mu_1(I \cap H_n), \\
\mu_2\left(\bigcup_{n=1}^{\infty} H_n\right) &= P(I)P\left(\bigcup_{n=1}^{\infty} H_n\right) \\
&= P(I)\sum_{n=1}^{\infty} P(H_n) \\
&= \sum_{n=1}^{\infty} \mu_2(H_n),
\end{aligned}$$

where H_n $(n \in \mathbf{N})$ are mutually disjoint elements of $\mathcal{H}$. Therefore, μ_1 and μ_2 are measures on $(\Omega, \mathcal{H})$. Since μ_1 and μ_2 are finite and agree on $\mathcal{J}$, by Theorem 3.2 they agree on $\sigma(\mathcal{J}) = \mathcal{H}$. Hence

$$\mu_1(H) = \mu_2(H), \quad \forall I \in \mathcal{I}, \forall H \in \mathcal{H}.$$

For fixed $H \in \mathcal{H}$, we define two functions $\nu_1 : \mathcal{G} \to [0,1]$ and $\nu_2 : \mathcal{G} \to [0,1]$ as

$$\nu_1(G) = P(G \cap H), \quad \nu_2(G) = P(H)P(G).$$

Similarly, we can show that ν_1 and ν_2 are measures on $(\Omega, \mathcal{G})$ and agree on $\mathcal{I}$. By Theorem 3.2, they agree on $\sigma(\mathcal{I}) = \mathcal{G}$. Thus, $\mathcal{G}$ and $\mathcal{H}$ are independent.

12.4. The argument used to prove this problem is similar to that in the proof of Problem 12.3. The "only if" part is obvious. We only prove the "if" part here.

To do that, we fix $I_i \in \mathcal{I}_i$ $(i = 2, 3, \ldots, n)$ and define two functions $\mu_1 : \mathcal{G}_1 \to [0,1]$ and $\nu_1 : \mathcal{G}_1 \to [0,1]$ as

$$\begin{aligned}
\mu_1(G_1) &= P\left(G_1 \bigcap \left(\bigcap_{i=2}^{n} I_i\right)\right), \\
\nu_1(G_1) &= P(G_1)\prod_{i=2}^{n} P(I_i).
\end{aligned}$$

In the same way as in Problem 12.3, we can verify that μ_1 and ν_1 are measures on $(\Omega, \mathcal{G}_1)$. By hypothesis, μ_1 and ν_1 agree on $\mathcal{I}_1$. By Theorem 3.2, they agree on $\mathcal{G}_1 = \sigma(\mathcal{I}_1)$.

Then we fix $I_1 \in \mathcal{G}_1$ and $I_i \in \mathcal{I}_i$ $(i = 3, 4, \ldots, n)$ and define two functions $\mu_2 : \mathcal{G}_2 \to [0, 1]$ and $\nu_2 : \mathcal{G}_2 \to [0, 1]$ as

$$\begin{aligned} \mu_1(G_2) &= P\left(I_1 \bigcap G_2 \bigcap \left(\bigcap_{i=3}^{n} I_i\right)\right), \\ \nu_1(G_2) &= P(I_1)P(G_2)\prod_{i=3}^{n} P(I_i). \end{aligned}$$

We can verify that μ_2 and ν_2 are measures on $(\Omega, \mathcal{G}_2)$ and agree on $\mathcal{I}_2$. By Theorem 3.2, they therefore agree on $\mathcal{G}_2 = \sigma(\mathcal{I}_2)$.

Repeating this process, we can show that

$$P\left(\bigcap_{i=1}^{n} I_i\right) = \prod_{i=1}^{n} P(I_i)$$

for $I_i \in \mathcal{G}_i$. Since the I_i are arbitrary, by definition $\mathcal{G}_n$ $(n \in \mathbf{N})$ are independent. This completes the proof.

12.5. First let us prove the "if" part. Let $\pi(X_i) = \{\{X_i \le c\} : c \in \mathbf{R}\} = \{\{\omega \in \Omega : X_i(\omega) \le c\} : c \in \mathbf{R}\}$ for $i = 1, 2, \ldots, n$. Then $\pi(X_i)$ is a π-system and, by Problem 11.10, $\sigma(X_i) = \sigma(\pi(X_i))$.

Let $I_i \in \pi(X_i)$. Then $I_i = \{X_i \le c_i\}$ for some $c_i \in \mathbf{R}$. Then by assumption we have

$$P\left(\bigcap_{i=1}^{n} I_i\right) = \prod_{i=1}^{n} P(I_i)$$

By Problem 12.4, we know that $X_1, X_2, \ldots, X_n$ are independent.

The "only if" part is obvious. Suppose that $X_1, X_2, \ldots, X_n$ are independent. Let $B_i \in \mathcal{B}$. Then $\{X_i \in B_i\} \in \sigma(X_i)$. By assumption we have

$$P\{X_1 \in B_1, X_2 \in B_2, \ldots, X_n \in B_n\} = \prod_{i=1}^{n} P\{X_i \in B_i\}.$$

This completes the proof.

12.6. The "only if" follows from Problem 12.5 directly. We only need to prove the "if" part. Suppose that

$$P\{X_1 = x_1, X_2 = x_2, \ldots, X_n = x_n\} = \prod_{i=1}^{n} P\{X_i = x_i\}$$

for all $x_i \in \{X_i(\omega) : \omega \in \Omega\}$, $i = 1, 2, \ldots, n$. Let $B_1, B_2, \ldots, B_n$ be arbitrary sets in $\mathcal{B}(\mathbf{R})$, and let $V_i = \{X_i(\omega) : \omega \in \Omega\}$ for $i = 1, 2, \ldots, n$. Then we have

$$\begin{aligned}
& P\{X_1 \in B_1, X_2 \in B_2, \ldots, X_n \in B_n\} \\
= & \sum_{x_1 \in B_1 \cap V_1} P\{X_1 = x_1, X_2 \in B_2, \ldots, X_n \in B_n\} \\
= & \sum_{x_1 \in B_1 \cap V_1} \cdots \sum_{x_n \in B_n \cap V_n} P\{X_1 = x_1, \ldots, X_n = x_n\} \\
= & \sum_{x_1 \in B_1 \cap V_1} \cdots \sum_{x_n \in B_n \cap V_n} \prod_{i=1}^{n} P\{X_i = x_i\} \\
= & P\{X_n \in B_n\} \sum_{x_1 \in B_1 \cap V_1} \cdots \sum_{x_{n-1} \in B_{n-1} \cap V_{n-1}} \prod_{i=1}^{n-1} P\{X_i = x_i\} \\
= & \prod_{i=1}^{n} P\{X_i \in B_i\}.
\end{aligned}$$

By Problem 12.5, X_1, X_2, ..., X_n are independent. This completes the proof.

12.7. By Problem 12.5, we only need to show that

$$P\{f_1(X_1) \in B_1, f_2(X_2) \in B_2, \ldots, f_n(X_n) \in B_n\} = \prod_{k=1}^{n} P\{f_k(X_k) \in B_k\}$$

for all $B_1, B_2, \ldots, B_n \in \mathcal{B}$. Note that

$$\begin{aligned}
\{f_k(X_k) \in B_k\} &= \{\omega \in \Omega : f_k(X_k(\omega)) \in B_k\} \\
&= \{\omega \in \Omega : X_k(\omega) \in f_k^{-1}(B_k)\} = \{X_k \in f_k^{-1}(B_k)\}
\end{aligned}$$

for $k = 1, 2, \ldots, n$. Since $X_1, X_2, \ldots, X_n$ are independent random variables, we have

$$P\{X_1 \in f_1^{-1}(B_1), \ldots, X_n \in f_n^{-1}(B_n)\} = \prod_{k=1}^{n} P\{X_k \in f_k^{-1}(B_k)\},$$

which leads to the results. This completes the proof.

12.8. The "only if" part follows from Problem 12.5. Let us prove the "if" part. Suppose that

$$F(x_1, x_2, \ldots, x_n) = \prod_{i=1}^{n} F(x_i)$$

for all $x_1, x_2, \ldots, x_n \in \mathbf{R}$. Then

$$P\{X_1 \le x_1, X_2 \le x_2, \ldots, X_n \le x_n\} = \prod_{i=1}^{n} P\{X_i \le x_i\}.$$

for all $x_1, x_2, \ldots, x_n \in \mathbf{R}$.

Now let $\pi(X_i) = \{\{X_i \le c\} : c \in \mathbf{R}\} = \{\{\omega \in \Omega : X_i(\omega) \le c\} : c \in \mathbf{R}\}$ for $i = 1, 2, \ldots, n$. Then $\pi(X_i)$ is a π-system and, by Problem 11.10, $\sigma(X_i) = \sigma(\pi(X_i))$. Then by assumption we have

$$P\left(\bigcap_{i=1}^{n} I_i\right) = \prod_{i=1}^{n} P(I_i)$$

for all $I_i \in \pi(X_i), i = 1, 2, \ldots, n$. By Problem 12.4, we know that $X_1, X_2, \ldots, X_n$ are independent. This completes the proof.

12.9. By Definition 11.4, we have

$$(\limsup E_n)^c = \bigcup_{m \ge 1} G_m,$$

where

$$G_m = \bigcap_{n \ge m} E_n^c.$$

Since E_n $(n \in \mathbf{N})$ are independent and $1 - x \le e^{-x}$, $\forall x \ge 0$, we have

$$P(G_m) = \prod_{n \ge m} P(E_n^c) = \prod_{n \ge m} (1 - P(E_n)) \le \exp\left(-\sum_{n \ge m} P(E_n)\right).$$

Noting that $\sum_{n=1}^{\infty} P(E_n) = \infty$, we have $P(G_m) = 0$ $(m \in \mathbf{N})$. Since the union of P-null sets are P-null (see Problem 2.19), we have

$$P(\limsup E_n) = 1 - P((\limsup E_n)^c) = 1 - P\left(\bigcup_{m \ge 1} G_m\right) = 1.$$

This completes the proof.

12.10. Let

$$\begin{aligned} \mathcal{X}_n &= \sigma(X_1, X_2, \ldots, X_n), \\ \mathcal{T}_n &= \sigma(X_{n+1}, X_{n+1}, \ldots), \end{aligned}$$

and

$$\begin{aligned} \mathcal{K}_n &= \{\{\omega : X_i(\omega) \le c_i,\ 1 \le i \le n\} : c_i \in \mathbf{R} \cup \{\infty\}\}, \\ \mathcal{J}_n &= \{\{\omega : X_i(\omega) \le c_i,\ n+1 \le i\} : c_i \in \mathbf{R} \cup \{\infty\}\}. \end{aligned}$$

From Problem 11.11, we know that $\mathcal{K}_n$ and $\mathcal{J}_n$ are π-systems that generate $\mathcal{X}_n$ and $\mathcal{T}_n$, respectively. Let $K \in \mathcal{K}_n$ and $J \in \mathcal{T}_n$; then $K = \{\omega : X_i(\omega) \le c_i,\ 1 \le i \le n\}$

and $J = \{\omega : X_i(\omega) \leq c_i,\ n+1 \leq i\}$ for some $c_i \in \mathbf{R} \cup \{\infty\}$ $(i \in \mathbf{N})$. Noting that X_n $(n \in \mathbf{N})$ are independent random variables, we have

$$\begin{aligned}
P(K \cap J) &= P(\{\omega : X_i(\omega) \leq c_i,\ 1 \leq i \leq n\} \cap \{\omega : X_i(\omega) \leq c_i,\ n+1 \leq i\}) \\
&= P\left(\left(\bigcap_{i=1}^{n}\{\omega : X_i(\omega) \leq c_i\}\right) \bigcap \left(\bigcap_{i=n+1}^{\infty}\{\omega : X_i(\omega) \leq c_i\right)\right) \\
&= P\left(\bigcap_{i=1}^{n}\{\omega : X_i(\omega) \leq c_i\}\right) P\left(\bigcap_{i=n+1}^{\infty}\{\omega : X_i(\omega) \leq c_i\right) \\
&= P(K)P(J),
\end{aligned}$$

which implies that $\mathcal{K}_n$ and $\mathcal{J}_n$ are independent. By Problem 12.3, we know that $\mathcal{X}_n$ and $\mathcal{T}_n$ are independent. Since $\mathcal{T} \subseteq \mathcal{T}_n$, it follows that $\mathcal{X}_n$ and $\mathcal{T}$ are independent.

Let $\mathcal{X}_\infty = \sigma(X_n : n \in \mathbf{N})$ and $\mathcal{K}_\infty = \bigcup_{n=1}^{\infty} \mathcal{X}_n$. We claim that $\mathcal{K}_\infty$ is a π-system and $\sigma(\mathcal{K}_\infty) = \mathcal{X}_\infty$. In fact, let $A, B \in \mathcal{K}_\infty$, then $A \in \mathcal{X}_r$ and $B \in \mathcal{X}_s$ for some $r, s \in \mathbf{N}$. Assuming $r \leq s$ and noting that $\mathcal{X}_r \subseteq \mathcal{X}_s$, we have $A, B \in \mathcal{X}_s$, which implies $A \cap B \in \mathcal{K}_\infty$. Thus, $\mathcal{K}_\infty$ is a π-system. Noting that $\mathcal{X}_n \subseteq \mathcal{X}_\infty$ for all $n \in \mathbf{N}$, we have $\mathcal{K}_\infty \subseteq \mathcal{X}_\infty$ which implies $\sigma(\mathcal{K}_\infty) \subseteq \mathcal{X}_\infty$. Also, by definition, we have

$$\begin{aligned}
\mathcal{X}_\infty &= \sigma(X_n : n \in \mathbf{N}) \\
&= \sigma(\{\omega : X_n(\omega) \in B\} : n \in \mathbf{N}, B \in \mathcal{B}) \\
&= \sigma\left(\bigcup_{n=1}^{\infty} \sigma(X_n)\right),
\end{aligned}$$

we have $\mathcal{X}_\infty \subseteq \sigma(\mathcal{K}_\infty)$, since $\bigcup_{n=1}^{\infty} \sigma(X_n) \subseteq \mathcal{K}_\infty$. Thus $\mathcal{X}_\infty = \sigma(\mathcal{K}_\infty)$; that is, $\mathcal{K}_\infty$ generates $\mathcal{X}_\infty$.

Since $\mathcal{X}_n$ and $\mathcal{T}$ are independent for all $n \in \mathbf{N}$, it follows that $\mathcal{K}_\infty$ and $\mathcal{T}$ are independent. By Problem 12.3, $\mathcal{X}_\infty$ and $\mathcal{T}$ are independent.

Since $\mathcal{T} = \bigcap_{n=1}^{\infty} \mathcal{T}_n \subseteq \mathcal{X}_\infty$, $\mathcal{T}$ is independent of $\mathcal{T}$. Therefore, for $F \in \mathcal{T}$, we have

$$P(F) = P(F \cap F) = P(F)P(F) = P(F)^2,$$

which gives $P(F) = 0$ or $P(F) = 1$.

If Y is a $\mathcal{T}$-measurable random variable, then

$$P(Y \leq y) = 0 \text{ or } 1, \quad \forall y \in \mathbf{R}.$$

Let $c \in \mathbf{R} \cup \{-\infty, \infty\}$ be such that

$$c = \sup\{y : P(Y \leq y) = 0\}.$$

If $c = -\infty$, then $P(Y = -\infty) = 1$. If $c = \infty$, then $P(Y = \infty) = 1$. If $-\infty < c < \infty$, then, by Problems 2.19 and 11.1, we have

$$\begin{aligned} P(Y < c) &= P\left(\bigcup_{n=1}^{\infty}\left\{Y \le c - \frac{1}{n}\right\}\right) = 0, \\ P(Y \le c) &= P\left(\bigcap_{n=1}^{\infty}\left\{Y \le c + \frac{1}{n}\right\}\right) = 1, \end{aligned}$$

which gives

$$P(Y = c) = 1.$$

This completes the proof.

12.11. By definition, events A and B are independent if the σ-algebras $\mathcal{E}_1 = \{\emptyset, A, A^c, \Omega\}$ and $\mathcal{E}_2 = \{\emptyset, B, B^c, \Omega\}$ are independent. However, $\mathcal{E}_1 = \sigma(\{A\})$ and $\mathcal{E}_2 = \sigma(\{B\})$. Thus the result follows from Problem 12.3.

12.12. By Problem 12.11, to show that A and B^c are independent we only need to prove $P(A \cap B^c) = P(A)P(B^c)$. But $P(A \cap B^c) = P(A) - P(A \cap B) = P(A) - P(A)P(B) = P(A)P(B^c)$. Thus A and B^c are independent. Other implications can be proved similarly.

12.13. Let $\Omega = \{1, 2, 3, 4\}$, $\mathscr{F} = 2^{\Omega}$, and $P(i) = 0.25$ for $i = 1, 2, 3, 4$. Let $E_1 = \{1, 2\}$, $E_2 = \{2, 3\}$, and $E_3 = \{1, 3\}$. Then $P(E_i \cap E_j) = 0.25 = P(E_i)P(E_j)$ for $i \neq j$. But $P(E_1 \cap E_2 \cap E_3) = P(\emptyset) = 0 \neq P(E_1)P(E_2)P(E_3)$.

12.14. By hypothesis $\Omega \in \mathcal{I}_i$ $(i = 1, 2, 3)$ and I_i $(i = 1, 2, 3)$ are π-systems on Ω, and from (12.1) we have

$$P(I_1 \cap I_2 \cap \Omega) = P(I_1)P(I_2)P(\Omega), \quad \forall I_i \in \mathcal{I}_i,\ i = 1, 2.$$

Since $P(\Omega) = 1$, the above equation gives

$$P(I_1 \cap I_2) = P(I_1)P(I_2), \quad \forall I_i \in \mathcal{I}_i,\ i = 1, 2.$$

By Problem 12.3, $\sigma(I_1)$ and $\sigma(I_2)$ are independent.

Now we define $\mathcal{C}$ as

$$\mathcal{C} = \{I_1 \cap I_2 : I_1 \in \mathcal{I}_1,\ I_2 \in \mathcal{I}_2\}. \tag{12.3}$$

We can verify that $\mathcal{C}$ is a π-system on Ω. In fact, let $I, J \in \mathcal{C}$, then $I = I_1 \cap I_2$ and $J = J_1 \cap J_2$ for some $I_1, J_1 \in \mathcal{I}_1$ and $I_2, J_2 \in \mathcal{I}_2$, respectively. But, $I \cap J = (I_1 \cap I_2) \cap (J_1 \cap J_2) = (I_1 \cap J_1) \cap (I_2 \cap J_2)$. Noting that $\mathcal{I}_1$ and $\mathcal{I}_2$ are π-systems, we have $I_1 \cap J_1 \in \mathcal{I}_1$ and $I_2 \cap J_2 \in \mathcal{I}_2$, which lead to $I \cap J \in \mathcal{C}$. Thus, $\mathcal{C}$ is a π-system.

We claim that $\sigma(\mathcal{C})$ and $\sigma(\mathcal{I}_3)$ are independent. In fact, let $I \in \sigma(\mathcal{C})$ and $I_3 \in \mathcal{I}_3$, then $I = I_1 \cap I_2$ for some $I_1 \in \mathcal{I}_1$ and $I_2 \in \mathcal{I}_2$ and

$$P(I \cap I_3) = P(I_1 \cap I_2 \cap I_3) = P(I_1)P(I_2)P(I_3). \tag{12.4}$$

Since $\sigma(\mathcal{I}_1)$ and $\sigma(\mathcal{I}_2)$ are independent, we have $P(I) = P(I_1 \cap I_2) = P(I_1)P(I_2)$. Thus from (12.4) we have

$$P(I \cap I_3) = P(I)P(I_3).$$

Since I and I_3 are arbitrary, by Problem 12.3, $\sigma(\mathcal{C})$ and $\sigma(I_3)$ are independent.

Since $\Omega \in \mathcal{I}_i$ $(i = 1, 2)$, from (12.3) we have $\mathcal{I}_i \subseteq \mathcal{C}$ $(i = 1, 2)$. Thus, $\sigma(\mathcal{I}_i) \subseteq \sigma(\mathcal{C})$ $(i = 1, 2)$. Now let $A_i \in \sigma(\mathcal{I}_i)$ $(i = 1, 2, 3)$, then $A_1, A_2 \in \sigma(\mathcal{C})$. Since $\sigma(\mathcal{C})$ and $\sigma(\mathcal{I}_3)$ are independent, we have $P(A_1 \cap A_2 \cap A_3) = P(A_1 \cap A_2)P(A_3)$. Since $\sigma(I_1)$ and $\sigma(I_2)$ are independent, we have $P(A_1 \cap A_2) = P(A_1)P(A_2)$. Therefore

$$P(A_1 \cap A_2 \cap A_3) = P(A_1 \cap A_2)P(A_3) = P(A_1)P(A_2)P(A_3),$$

which implies that $\sigma(I_1)$, $\sigma(I_2)$, and $\sigma(I_3)$ are independent.

12.15.

(a) By definition of independent events, to show that E_p $(p \in \mathbf{P})$ are independent we need to show that

$$\mathcal{E}_p = \{\emptyset, E_p, \Omega \backslash E_p, \Omega\}, \quad p \in \mathbf{P}$$

are independent. Note that E_p $(p \in \mathbf{P})$ are π-systems and $\mathcal{E}_p = \sigma(\{E_p\})$, by Problem 12.4 we only need to show $\{E_p\}$ $(p \in \mathbf{P})$ are independent.

To do this, let $\mathbf{P} = \{p_1, p_2, \ldots\}$ such that $p_1 < p_2 < p_3 < \cdots$, and let $r \in \mathbf{N}$ and $i_j \in \mathbf{N}$ $(j = 1, 2, \ldots, r)$ such that $1 \le i_1 < i_2 < \cdots < i_r$; then $\bigcap_{j=1}^r E_{p_{i_j}}$ is the set of samples ω for which $X(\omega)$ are divisible by $\prod_{j=1}^r p_{i_j}$:

$$\begin{aligned} \bigcap_{j=1}^r E_{p_{i_j}} &= \left\{\omega \in \Omega : \prod_{j=1}^r p_{i_j} \mid X(\omega)\right\} \\ &= \left\{\omega \in \Omega : X(\omega) = n \prod_{j=1}^r p_{i_j},\ n \in \mathbf{N}\right\}. \end{aligned}$$

Thus,

$$P\left(\bigcap_{j=1}^r E_{p_{i_j}}\right) = \sum_{n=1}^{\infty} \frac{1}{\varsigma(s)} \left(n \prod_{j=1}^r p_{i_j}\right)^{-s} = \prod_{j=1}^r p_{i_j}^{-s}. \tag{12.5}$$

Note that, for any $p \in \mathbf{P}$, we have

$$E_p = \{\omega : p \mid X(\omega)\} = \{\omega : X(\omega) = np,\ n \in \mathbf{N}\},$$

which gives

$$P(E_p) = \sum_{n=1}^{\infty} \frac{1}{\varsigma(s)} (np)^{-s} = p^{-s}. \tag{12.6}$$

In particular, we have

$$\prod_{j=1}^{r} P(E_{p_{i_j}}) = \prod_{j=1}^{r} p_{i_j}^{-s}. \tag{12.7}$$

Combining (12.5) and (12.7) gives

$$P\left(\bigcap_{j=1}^{r} E_{p_{i_j}}\right) = \prod_{j=1}^{r} P(E_{p_{i_j}}).$$

Therefore, E_p $(p \in \mathbf{P})$ are independent.

(b) On one hand, we have $P(X = 1) = 1/\varsigma(s)$. On the other hand, $X = 1$ if and only if $p \nmid X$, $\forall p \in \mathbf{P}$; that is,

$$\{\omega : X(\omega) = 1\} = \{\omega : p \nmid X(\omega),\ \forall p \in \mathbf{P}\} = \bigcap_{p \in \mathbf{P}} E_p^c.$$

Since E_p $(p \in \mathbf{P})$ are independent, we have

$$P(X = 1) = P(\{\omega : X(\omega) = 1\}) = P\left(\bigcap_{p\in\mathbf{P}} E_p^c\right) = \prod_{p \in \mathbf{P}} (1 - p^{-s}).$$

Therefore,

$$\frac{1}{\varsigma(s)} = \prod_{p \in \mathbf{P}} (1 - p^{-s}). \tag{12.8}$$

(c) Noting that $m^2 \nmid X\ \forall m > 1$ if and only $p^2 \nmid X\ \forall p \in \mathbf{P}$, we have

$$\begin{aligned} \{\omega \in \Omega : m^2 \nmid X(\omega),\ \forall m > 1\} &= \{\omega \in \Omega : p^2 \nmid X(\omega),\ \forall p \in \mathbf{P}\} \\ &= \bigcap_{p \in \mathbf{P}} G_p^c, \end{aligned}$$

where

$$G_p = \{\omega \in \Omega : p^2 \mid X(\omega)\} = \{\omega \in \Omega : X(\omega) = np^2\ \forall n \in \mathbf{N}\}.$$

Using the same argument as in the first part of this proof, we can show that G_p $(p \in \mathbf{P})$ are independent and

$$P(G_p) = p^{-2s}.$$

Then

$$\begin{aligned} P\left(\{\omega \in \Omega : m^2 \nmid X(\omega),\ \forall m > 1\}\right) &= P\left(\bigcap_{p\in\mathbf{P}} G_p^c\right) \\ &= \prod_{p \in \mathbf{P}} (1 - p^{-2s}). \end{aligned} \tag{12.9}$$

Since (12.8) holds for any $s > 1$, we have

$$\frac{1}{\varsigma(2s)} = \prod_{p\in \mathbf{P}} (1 - p^{-2s}). \tag{12.10}$$

Combining (12.9) and (12.10) gives

$$P\left(\{\omega \in \Omega : m^2 \nmid X(\omega),\ \forall m > 1\}\right) = \frac{1}{\varsigma(2s)}.$$

(d) Note that $\gcd(X, Y) = n$ if and only if there exist $k, l \in \mathbf{N}$ such that $X = kn$, $Y = ln$, and $\gcd(k, l) = 1$, where $\gcd(X, Y)$ denote the highest common factor of X and Y. For $n \in \mathbf{N}$, we define events F_n as

$$F_n = \{H = n\} = \{\omega : X(\omega) = kn,\ Y(\omega) = ln,\ (k, l) \in I\},$$

where $I = \{(k, l) : k, l \in \mathbf{N}, \gcd(k, l) = 1\}$. Since X and Y are independent, we then have

$$\begin{aligned} P(F_n) &= \sum_{(k,l)\in I} P(X = kn, Y = ln) = \sum_{(k,l)\in I} P(X = kn)P(Y = ln) \\ &= \sum_{(k,l)\in I} \frac{(kn)^{-s}}{\varsigma(s)} \cdot \frac{(ln)^{-s}}{\varsigma(s)} \\ &= n^{-2s} \sum_{(k,l)\in I} \frac{k^{-s}}{\varsigma(s)} \cdot \frac{l^{-s}}{\varsigma(s)} \\ &= n^{-2s} P(F_1). \end{aligned} \tag{12.11}$$

Noting that F_n $(n \in \mathbf{N})$ are mutually disjoint and $\bigcup_{n=1}^{\infty} F_n = \Omega$, we have

$$1 = P\left(\bigcup_{n=1}^{\infty} F_n\right) = \sum_{n=1}^{\infty} P(F_n) = \sum_{n=1}^{\infty} n^{-2s} P(F_1),$$

which results in

$$P(F_1) = \frac{1}{\sum_{n=1}^{\infty} n^{-2s}} = \frac{1}{\varsigma(2s)}.$$

Thus, from (12.11) we get

$$P(F_n) = \frac{n^{-2s}}{\varsigma(2s)}.$$

12.16. We define events E_n $(n \in \mathbf{N})$ as

$$E_n = \{X_n \geq \log n\} = \left\{\omega \in \Omega : \frac{X(\omega)}{\log n} \geq 1\right\}.$$

Then

$$\left\{\limsup \frac{X_n}{\log n} \geq 1\right\} = \limsup E_n.$$

Note that

$$P(E_n) = P(X_n \geq \log n) = \frac{1}{n},$$

and

$$\sum_{n=1}^{\infty} P(E_n) = \sum_{n=1}^{\infty} \frac{1}{n} = \infty.$$

By the second Borel-Cantelli theorem (see Problem 12.9), we have

$$P(\limsup E_n) = 1,$$

which gives

$$P\left(\limsup \frac{X_n}{\log n} \geq 1\right) = 1.$$

12.17. Let $\mathcal{T}_n = \sigma(X_{n+1}, X_{n+2}, \cdots)$ $(n \in \mathbf{N})$. By Definition 11.9, we have

$$\mathcal{T}_n = \sigma(\{\omega \in \Omega : X_m(\omega) \in B\},\ m \geq n+1, B \in \mathcal{B}).$$

To prove that F_1, F_2, and F_3 are $\mathcal{T}$-measurable, we only need to show that F_1, F_2, and F_3 are $\mathcal{T}_n$-measurable for all $n \in \mathbf{N}$.

Noting that $\lim_{n\to\infty} a_n$ exists if and only if $\limsup a_n = \liminf a_n$, we have

$$\begin{aligned}
F_1 &= \{\omega \in \Omega : \limsup X_n(\omega) = \liminf X_n(\omega)\} = h_1^{-1}(0), \\
F_2 &= \left\{\omega \in \Omega : \limsup \sum_{k=1}^{n} X_k(\omega) = \liminf X_k(\omega)\right\} = h_2^{-1}(0), \\
F_3 &= \left\{\omega \in \Omega : \limsup \frac{1}{n}\sum_{k=1}^{n} X_k(\omega) = \liminf \frac{1}{n}\sum_{k=1}^{n} X_k(\omega)\right\} = h_3^{-1}(0),
\end{aligned}$$

where

$$\begin{aligned}
h_1(X_n) &= \limsup X_n - \liminf X_n, \\
h_2\left(\sum_{k=1}^{n} X_k\right) &= \limsup \sum_{k=1}^{n} X_k - \liminf \sum_{k=1}^{n} X_k, \\
h_3\left(\frac{1}{n}\sum_{k=1}^{n} X_k\right) &= \limsup \frac{1}{n}\sum_{k=1}^{n} X_k - \liminf \frac{1}{n}\sum_{k=1}^{n} X_k.
\end{aligned}$$

For fixed $N \in \mathbf{N}$, to show that F_1 is $\mathcal{T}_N$-measurable, we only need to show that $\limsup X_n$ and $\liminf X_n$ are $\mathcal{T}_N$-measurable. By Problem 5.7, we only need to show that

$$\{\limsup X_n \leq c\} \in \mathcal{T}_N, \quad \{\liminf X_n \leq c\} \in \mathcal{T}_N$$

for all $c \in \mathbf{R}$. Note that

$$\{\limsup X_n \leq c\} = \bigcap_{j\geq 1}\bigcup_{i\geq j}\{X_i \leq c\} = \bigcap_{j\geq N+1}\bigcup_{i\geq j}\{X_i \leq c\},$$

and

$$\{X_i > c\} \in \mathcal{T}_N, \quad \forall i \geq N+1,$$

We have $\{\limsup X_n \leq c\} \in \mathcal{T}_N$. Similarly, we can show that $\{\liminf X_n \leq c\} \in \mathcal{T}_N$. Hence, F_1 is $\mathcal{T}_N$-measurable.

Regarding F_2, we have

$$\begin{aligned}
\left\{\limsup \sum_{k=1}^{n} X_k \leq c\right\} &= \bigcap_{j\geq 1}\bigcup_{i\geq j}\left\{\sum_{k=1}^{i} X_k \leq c\right\} \\
&= \bigcap_{j\geq N+1}\bigcup_{i\geq j}\left\{\sum_{k=1}^{i} X_k \leq c\right\} \\
&= \bigcap_{j\geq N+1}\bigcup_{i\geq j}\left\{\sum_{k=N+1}^{i} X_k \leq c - \sum_{k=1}^{N} X_k\right\}.
\end{aligned}$$

Since $\left\{\sum_{k=N+1}^{i} X_k \leq c - \sum_{k=1}^{N} X_k\right\}$ are $\mathcal{T}_N$-measurable, we have

$$\left\{\limsup \sum_{k=1}^{n} X_k \leq c\right\} \in \mathcal{T}_N.$$

Similarly, we can show that

$$\left\{\liminf \sum_{k=1}^{n} X_k \leq c\right\} \in \mathcal{T}_N.$$

Thus, F_2 is $\mathcal{T}_N$-measurable.

Similarly, for F_3 we have

$$\begin{aligned}
\left\{\limsup \frac{1}{n}\sum_{k=1}^{n} X_k \leq c\right\} &= \bigcap_{j\geq 1}\bigcup_{i\geq j}\left\{\frac{1}{i}\sum_{k=1}^{i} X_k \leq c\right\} \\
&= \bigcap_{j\geq N+1}\bigcup_{i\geq j}\left\{\frac{1}{i}\sum_{k=1}^{i} X_k \leq c\right\} \\
&= \bigcap_{j\geq N+1}\bigcup_{i\geq j}\left\{\sum_{k=N+1}^{i} X_k \leq ic - \sum_{k=1}^{N} X_k\right\}.
\end{aligned}$$

Since $\left\{\sum_{k=N+1}^{i} X_k \leq ic - \sum_{k=1}^{N} X_k\right\}$ are $\mathcal{T}_N$-measurable, we have

$$\left\{\limsup \frac{1}{n}\sum_{k=1}^{n} X_k \leq c\right\} \in \mathcal{T}_N.$$

Similarly, we can show that

$$\left\{\liminf \frac{1}{n}\sum_{k=1}^{n} X_k \leq c\right\} \in \mathcal{T}_N.$$

Thus, F_3 is $\mathcal{T}_N$-measurable.

12.5 Bibliographic Notes

In this chapter, we introduced the concept of independence in probability theory. Mathematically, definition of the independence of random variables and events is based on the independence of σ-algebras. For more information on independence, readers are referred to Williams (1991), Ash and Doleans-Dade (1999), and Çinlar (2011).

CHAPTER 13

EXPECTATION

In probability theory, the expectation of a random variable is the integral of the random variable with respect to its probability measure. For example, the expectation of a discrete random variable is the weighted average of all possible values that the random variable can take on. In this chapter, we shall introduce the definition of expectation and relevant concepts.

13.1 Basic Concepts and Facts

Definition 13.1 (Expectation)**.** Let $(\Omega, \mathscr{F}, P)$ be a probability space and $X \in L^1(\Omega, \mathscr{F}, P)$, where $L^1(\Omega, \mathscr{F}, P)$ denotes the set of all P-integrable functions on Ω (see Definition 6.4). The expectation $E(X)$ of X is defined as

$$E(X) = \int_\Omega X(\omega)P(\mathrm{d}\omega) = \int_\Omega X\mathrm{d}P.$$

Definition 13.2 (Expectation over Subsets). Let $(\Omega, \mathscr{F}, P)$ be a probability space, $F \in \mathscr{F}$, and X be a random variable on Ω. Then $E(X; F)$ is defined as

$$E(X; F) = \int_F X(\omega) P(\mathrm{d}\omega) = \int_F X \mathrm{d}P = E(X I_F),$$

where

$$I_F(\omega) = \begin{cases} 1, & \text{if } \omega \in F; \\ 0, & \text{if } \omega \notin F. \end{cases}$$

Definition 13.3 (Integration with Respect to Distribution Functions). Let F be a distribution function on $\mathbf{R}^n$ and let μ be the corresponding Lebesgue-Stieltjes measure. Let g be a Borel function from $\mathbf{R}^n$ to $\mathbf{R}$. Then the integration of g with respect to the distribution function F is defined as

$$\int_{\mathbf{R}^n} g(\mathbf{x}) \mathrm{d}F(\mathbf{x}) = \int_{\mathbf{R}^n} g \mathrm{d}\mu.$$

Definition 13.4 (Covariance and Variance). Let $(\Omega, \mathscr{F}, P)$ be a probability space and $X, Y \in L^2(\Omega, \mathscr{F}, P)$. Then, by the monotonicity of norms (see Problem 8.5), $X, Y \in L^1(\Omega, \mathscr{F}, P)$. Let $\tilde{X} = X - \mu_X$ and $\tilde{Y} = Y - \mu_Y$, where $\mu_X = E(X)$ and $\mu_Y = E(Y)$. Then $\tilde{X}, \tilde{Y} \in L^2(\Omega, \mathscr{F}, P)$ and so $\tilde{X}\tilde{Y} \in L^1(\Omega, \mathscr{F}, P)$. The covariance of X and Y is defined as

$$\mathrm{Cov}(X, Y) = E(\tilde{X}\tilde{Y}) = E[(X - \mu_X)(Y - \mu_Y)].$$

The variance of X is defined as

$$\mathrm{Var}(X) = \mathrm{Cov}(X, X).$$

Definition 13.5 (Moment, Central Moment and Absolute Moment). Let X be a random variable on the probability space $(\Omega, \mathscr{F}, P)$. Let $k > 0$. Then the kth moment of X is defined as

$$E(X^k).$$

The kth absolute moment of X is defined as

$$E(|X|^k).$$

If $E(X)$ is finite, then the kth central moment of X is defined as

$$E[(X - E(X))^k],$$

and the kth absolute central moment of X is defined as

$$E(|X - E(x)|^k).$$

The first moment of X is also called the *mean* of X. The second absolute central moment of X is also called the *variance* of X.

Definition 13.6 (Correlation). Let $X, Y \in L^2(\Omega, \mathscr{F}, P)$. Then the correlation of X and Y is defined as

$$\operatorname{Corr}(X, Y) = \frac{\operatorname{Cov}(X, Y)}{\sqrt{\operatorname{Var}(X)\operatorname{Var}(Y)}} = \frac{E[(X - \mu_X)(Y - \mu_Y)]}{||X||_2 \cdot ||Y||_2},$$

where $\mu_X = E(X)$ and $\mu_Y = E(Y)$.

Theorem 13.1. *Let X be a random variable on the probability space $(\Omega, \mathscr{F}, P)$, and let g a Borel function from $\mathbf{R}$ to $\mathbf{R}$. Let $Y = g \circ X$. Then*

$$E(Y) = \int_{\mathbf{R}} g \mathrm{d}P_X = \int_{\mathbf{R}} g(x)\mathrm{d}F(x),$$

where F is the distribution function of X.

If X has a density function f, we have

$$E(Y) = \int_{\mathbf{R}} g(x)f(x)\mathrm{d}x.$$

Theorem 13.2. *Let $(\Omega, \mathscr{F}, P)$ be a probability space, and let X and Y be independent random variables on Ω such that $X, Y \in L^1(\Omega, \mathscr{F}, P)$. Then $XY \in L^1(\Omega, \mathscr{F}, P)$ and*

$$E(XY) = E(X)E(Y).$$

13.2 Problems

13.1. Let $\Omega = (2, \infty)$ and $\mathscr{F} = \mathcal{B}(\Omega)$. Let X be a random variable on $(\Omega, \mathscr{F})$ with the following distribution function:

$$F(x) = \begin{cases} \frac{1}{3} & \text{if } x \in (\infty, 2]; \\ \frac{1}{3} + \frac{2}{3}(x - 2) & \text{if } x \in (2, 3); \\ 1 & \text{if } x \in [3, \infty). \end{cases}$$

Compute $E(X^2)$.

13.2. Let $(\Omega, \mathscr{F}, P)$ be a probability space and $X \in (m\mathscr{F})^+$ such that $E(X) < \infty$. Show that $P(X < \infty) = 1$.

13.3. Let $(\Omega, \mathscr{F}, P)$ be a probability space, and let $\{Z_n\}_{n \geq 1}$ be a sequence of elements of $(m\mathscr{F})^+$. Show that

$$E\left(\sum_{n=1}^{\infty} Z_n\right) = \sum_{n=1}^{\infty} E(Z_n).$$

13.4. Let X_i $(i = 1, 2, \ldots, n)$ be random variables on a probability space $(\Omega, \mathscr{F}, P)$ with finite expectations. Show that

$$E\left(\sum_{i=1}^{n} X_i\right) = \sum_{i=1}^{n} E(X_i).$$

13.5. Let $(\Omega, \mathscr{F}, P)$ be a probability space and $\{Z_n\}_{n\geq 1}$ be a sequence of elements of $(m\mathscr{F})^+$ such that

$$\sum_{n=1}^{\infty} E(Z_n) < \infty.$$

Show that $\sum_{n=1}^{\infty} Z_n < \infty$ a.s. and $Z_n \to 0$ a.s.

13.6. Let $X_i \in L^2(\Omega, \mathscr{F}, P)$ for $i = 1, 2, \ldots, n$. Show that

$$\text{Var}\left(\sum_{i=1}^{n} X_i\right) = \sum_{i=1}^{n} \text{Var}(X_i) + 2\sum_{i=1}^{n}\sum_{j=i+1}^{n} \text{Cov}(X_i, X_j).$$

13.7. Let X, Y be independent elements of $L^2(\Omega, \mathscr{F}, P)$. Show that

$$\text{Cov}(X, Y) = 0$$

and

$$\text{Var}(X + Y) = \text{Var}(X) + \text{Var}(Y).$$

13.8. Let X, Y be elements of $L^2(\Omega, \mathscr{F}, P)$. Show that X and Y are independent if and only if

$$E[g(X)h(Y)] = E[g(X)] \cdot E[h(Y)]$$

for all $g, h : \mathbf{R} \to \mathbf{R}$ such that the expectations exist.

13.9. Let $(\Omega, \mathscr{F}, P)$ be a probability space. Suppose that X is a random variable on $(\Omega, \mathscr{F}, P)$ that assumes only nonnegative integer values. Show that

$$E(X) = \sum_{k=0}^{\infty} P\{X > k\}.$$

13.10. Let $(\Omega, \mathscr{F}, P)$ be a probability space. Suppose that X is a nonnegative random variable on $(\Omega, \mathscr{F}, P)$. Show that

$$E(X) = \int_0^{\infty} P\{X > \lambda\} \mathrm{d}\lambda.$$

13.11. Let $(\Omega, \mathscr{F}, P)$ be a probability space. Let X and Y be random variables on $(\Omega, \mathscr{F}, P)$. Suppose that Y is nonnegative. Show that

$$E(XY) = \int_0^{\infty} E\left[XI_{\{Y>\lambda\}}\right] \mathrm{d}\lambda.$$

13.12. Let X be a nonnegative random variable on some probability space $(\Omega, \mathscr{F}, P)$. Let f be a function on $[0, \infty)$ such that f has a nonnegative continuous first derivative (i.e., $f' \geq 0$). Show that

$$E[f(X)] = f(0) + \int_0^\infty f'(\lambda) P\{X \geq \lambda\} \mathrm{d}\lambda.$$

13.13. Let X be a nonnegative random variable on the probability space $(\Omega, \mathscr{F}, P)$. Suppose that $E(X) < \infty$. Show that

$$\lim_{n \to \infty} nP\{X > n\} = 0.$$

13.14. Let $X_1, X_2, \ldots$ be a sequence of identically distributed random variables on a probability space $(\Omega, \mathscr{F}, P)$. Suppose that

$$E\left(e^{|X_1|}\right) < \infty.$$

Show that

$$P\left\{\limsup \frac{|X_n|}{\ln n} \leq 1\right\} = 1.$$

13.3 Hints

13.1. Use Theorem 13.1.

13.2. Consider the sequence of sets $A_n = \{X < n\}$ for $n = 1, 2, \ldots$ and use the results of Problem 15.2 and Theorem 2.1.

13.3. Consider the partial sums $S_n = Z_1 + Z_2 + \cdots + Z_n$ for $n \geq 1$ and apply Theorem 6.2.

13.4. Use the definition of integration (see Definition 6.4) and the result of Problem 13.3.

13.5. The first conclusion is implied from Problems 13.2 and 13.3. For the second conclusion, one can use the inequality $0 \leq Z_n \leq S - S_{n-1}$, where $S_n = Z_1 + Z_2 + \cdots + Z_n$ and $S = \sum_{n=1}^\infty Z_n$.

13.6. Follow the definition of variances and covariances (Definition 13.4).

13.7. Apply Theorem 13.2 and the result of Problem 13.6.

13.8. The "if" part can be proved by using the result of Problem 12.5 and special cases of g and h. The "only if" part can be proved by using the result of Problem 12.5 and Theorem 13.2.

13.9. Follow the definition of expectation (Definition 13.1) and use the result of Problem 6.21 to show that

$$E[X] = \sum_{k=1}^{\infty} kP\{X = k\}.$$

13.10. Apply Tonelli's theorem (Theorem 10.1).

13.11. Apply Tonelli's theorem (Theorem 10.1). The proof is similar to Problem 13.10.

13.12. Use Tonelli's theorem (Theorem 10.1).

13.13. Note that

$$E(X) = \int_{\{X>n\}} X\mathrm{d}P + \int_{\{X\leq n\}} X\mathrm{d}P$$

and use Theorem 6.2.

13.14. Use the results of Problems 11.12 and 13.9 and Theorem 2.2.

13.4 Solutions

13.1. The distribution function F has the following density function:

$$f(x) = \begin{cases} 0 & \text{if } x \in (\infty, 2]; \\ \frac{2}{3} & \text{if } x \in (2, 3); \\ 0 & \text{if } x \in [3, \infty). \end{cases}$$

Then, by Theorem 13.1, we have

$$E(X^2) = \int_{\mathbf{R}} x^2 f(x)\mathrm{d}x = \int_2^3 \frac{2}{3} x^2 \mathrm{d}x = \frac{38}{9}.$$

13.2. Let $A_n = \{X < n\}$ for $n = 1, 2, \ldots$. Then $A_n \uparrow \{X < \infty\}$. By hypothesis, X is nonnegative and $E(X) < \infty$. Then, by Problem 15.2, we obtain

$$nP(X \geq n) \leq E(X) < \infty,$$

which implies

$$P(A_n) = 1 - P(X \geq n) \geq 1 - \frac{1}{n}E(X).$$

Therefore, $P(A_n) \to 1$. Recalling that $A_n \uparrow \{X < \infty\}$, we have, by Theorem 2.1, $P(A_n) \uparrow P(\{X < \infty\})$. Hence $P(\{X < \infty\}) = 1$.

13.3. By hypothesis, Z_n $(n \geq 1)$ are nonnegative random variables on Ω. Then $S_n = Z_1 + Z_2 + \cdots + Z_n$ $(n \geq 1)$ are also nonnegative random variables on Ω. By Problem 6.12, we have

$$E(S_n) = E\left(\sum_{k=1}^{n} Z_k\right) = \sum_{k=1}^{n} E(Z_k), \quad n \geq 1. \tag{13.1}$$

Noting that $S_n \in (m\mathscr{F})^+$ and $S_n \uparrow \sum_{k=1}^{\infty} Z_k$, by Theorem 6.2, we have

$$E(S_n) \uparrow E\left(\sum_{k=1}^{\infty} Z_k\right). \tag{13.2}$$

Combining Equations (13.1) and (13.2), we get

$$\sum_{k=1}^{\infty} E(Z_k) = E\left(\sum_{k=1}^{\infty} Z_k\right).$$

This completes the proof.

13.4. Let X_i^+ and X_i^- be the positive and negative parts of X_i, respectively. Then $X_i = X_i^+ - X_i^-$. By the assumption that X_i has finite expectation and by Definition 6.4, we have $E(X_i^+) < \infty$, $E(X_i^-) < \infty$, and

$$E(X_i) = E(X_i^+) - E(X_i^-).$$

By Problem 13.3, we have

$$E\left(\sum_{i=1}^{n} X_i^+\right) = \sum_{i=1}^{n} E(X_i^+) \tag{13.3}$$

and

$$E\left(\sum_{i=1}^{n} X_i^-\right) = \sum_{i=1}^{n} E(X_i^-). \tag{13.4}$$

Subtracting Equation (13.4) from Equation (13.3) gives

$$E\left(\sum_{i=1}^{n} X_i\right) = \sum_{i=1}^{n} E(X_i).$$

This completes the proof.

13.5. We first show that $\sum_{n=1}^{\infty} Z_n < \infty$ a.s. Since, by hypothesis, $\sum_{n=1}^{\infty} E(Z_n) < \infty$, we have, by Problem 13.3, $E\left(\sum_{n=1}^{\infty} Z_n\right) < \infty$. Therefore, by Problem 13.2, $\sum_{n=1}^{\infty} Z_n < \infty$ a.s.

Next, we show that $Z_n \to 0$ a.s. Let $S_n = Z_1 + Z_2 + \cdots + Z_n$ and $S = \sum_{n=1}^{\infty} Z_n$. Then $S_n \uparrow S$. Recall that $S < \infty$ a.s., we have $S - S_n \downarrow 0$ a.s. Since $0 \leq Z_n \leq S - S_{n-1}$ for every $n \geq 2$, we have $Z_n \to 0$ a.s. This completes the proof.

13.6. Since, by hypothesis, $X_i \in L^2(\Omega, \mathscr{F}, P)$ for $i = 1, 2, \ldots, n$, we have $X_1 + X_2 + \cdots + X_n \in L^2(\Omega, \mathscr{F}, P)$ and $X_i, X_j, X_iX_j \in L^1(\Omega, \mathscr{F}, P)$. Then

$$\begin{aligned}
\text{Var}\left(\sum_{i=1}^n X_i\right) &= E\left[\left(\sum_{i=1}^n (X_i - \mu_i)\right)^2\right] \\
&= E\left[\sum_{i=1}^n (X_i - \mu_i)^2 + 2\sum_{i=1}^n \sum_{j=i+1}^n (X_i - \mu_i)(X_j - \mu_j)\right] \\
&= \sum_{i=1}^n E[(X_i - \mu_i)^2] + 2\sum_{i=1}^n \sum_{j=i+1}^n E[(X_i - \mu_i)(X_j - \mu_j)] \\
&= \sum_{i=1}^n \text{Var}(X_i) + 2\sum_{i=1}^n \sum_{j=i+1}^n \text{Cov}(X_i, X_j),
\end{aligned}$$

where $\mu_i = E(X_i)$ for $i = 1, 2, \ldots, n$.

13.7. Since X and Y are independent, we have, by Theorem 13.2, $E(XY) = E(X)E(Y)$. Then

$$\text{Cov}(X, Y) = E[(X - E(X))(Y - E(Y))] = E(XY) - E(X)E(Y) = 0,$$

and so, by Problem 13.6, we have

$$\text{Var}(X + Y) = \text{Var}(X) + \text{Var}(Y) + 2\text{Cov}(X, Y) = \text{Var}(X) + \text{Var}(Y).$$

13.8. Let us first prove the "if" part of this problem. To do that, we assume that

$$E[g(X)h(Y)] = E[g(X)] \cdot E[h(Y)]$$

for all $g, h : \mathbf{R} \to \mathbf{R}$ such that the expectations exist. Let $B_1, B_2 \in \mathcal{B}(\mathbf{R})$ and define $f_1, f_2 : \mathbf{R} \to \mathbf{R}$ as

$$f_1(X) = \begin{cases} 1, & \text{if } X \in B_1; \\ 0, & \text{if } X \notin B_1, \end{cases}$$

and

$$f_2(Y) = \begin{cases} 1, & \text{if } Y \in B_2; \\ 0, & \text{if } Y \notin B_2, \end{cases}$$

Then by assumption, we have $E[f_1(X)f_2(Y)] = E[f_1(X)] \cdot E[f_2(Y)]$. But note that

$$E[f_1(X)f_2(Y)] = E[I_{B_1 \times B_2}(X, Y)] = P(X_1 \in B_1, Y \in B_2)$$

and

$$E[f_1(X)] = E[I_{B_1}(X)] = P(X \in B_1), \quad E[f_2(Y)] = E[I_{B_2}(Y)] = P(Y \in B_2),$$

we have $P(X \in B_1, Y \in B_2) = P(X \in B_1) \cdot P(Y \in B_2)$. By Problem 12.5, we know that X and Y are independent.

Now we prove the "only if" part. Suppose that X and Y are independent. We claim that $g(X)$ and $h(Y)$ are also independent for any functions g and h. To prove this claim, let $B_1, B_2 \in \mathcal{B}(\mathbf{R})$. Then we have $g^{-1}(B_1), h^{-1}(B_2) \in \mathcal{B}(\mathbf{R})$. Hence, by Problem 12.5, we have

$$\begin{aligned} P(g(X) \in B_1, h(Y) \in B_2) &= P(X \in g^{-1}(B_1), Y \in h^{-1}(B_2)) \\ &= P(X \in g^{-1}(B_1)) \cdot P(Y \in h^{-1}(B_2)) \\ &= P(g(X) \in B_1) \cdot P(h(Y) \in B_2). \end{aligned}$$

Hence $g(X)$ and $h(Y)$ are independent. The result follows from Theorem 13.2.

This completes the proof.

13.9. By Definition 13.1 and Problem 6.21, we have

$$\begin{aligned} E[X] &= \int_\Omega X \mathrm{d}P \\ &= \int_\Omega \sum_{k=1}^{\infty} k I_{\{X=k\}} \mathrm{d}P \\ &= \sum_{k=1}^{\infty} k \int_\Omega I_{\{X=k\}} \mathrm{d}P \\ &= \sum_{k=1}^{\infty} k P\{X = k\}. \end{aligned}$$

Note that $P\{X = k\} = P\{X > k-1\} - P\{X > k\}$. We have

$$\begin{aligned} \sum_{k=1}^{\infty} k P\{X = k\} &= \sum_{k=1}^{\infty} k (P\{X > k-1\} - P\{X > k\}) \\ &= \sum_{k=1}^{\infty} k P\{X > k-1\} - \sum_{k=1}^{\infty} k P\{X > k\} \\ &= \sum_{k=0}^{\infty} (k+1) P\{X > k\} - \sum_{k=1}^{\infty} k P\{X > k\} \\ &= \sum_{k=0}^{\infty} P\{X > k\}. \end{aligned}$$

This completes the proof.

13.10. Note that for each $\omega \in \Omega$, we have

$$X(\omega) = \int_0^{X(\omega)} \mathrm{d}\lambda = \int_0^{\infty} I_{[0, X(\omega))}(\lambda) \mathrm{d}\lambda.$$

Then, by Theorem 10.1, we have

$$\begin{aligned} E(X) &= \int_\Omega X \mathrm{d}P \\ &= \int_\Omega \int_0^\infty I_{[0,X)}(\lambda) \mathrm{d}\lambda \mathrm{d}P \\ &= \int_0^\infty \int_\Omega I_{[0,X)}(\lambda) \mathrm{d}P \mathrm{d}\lambda \\ &= \int_0^\infty \int_\Omega I_{\{X>\lambda\}} \mathrm{d}P \mathrm{d}\lambda \\ &= \int_0^\infty P\{X > \lambda\} \mathrm{d}\lambda. \end{aligned}$$

This completes the proof.

13.11. Note that for each $\omega \in \Omega$, we have

$$Y(\omega) = \int_0^{Y(\omega)} \mathrm{d}\lambda = \int_0^\infty I_{[0,Y(\omega))}(\lambda) \mathrm{d}\lambda.$$

Then by Theorem 10.1, we have

$$\begin{aligned} E(XY) &= \int_\Omega XY \mathrm{d}P \\ &= \int_\Omega X \int_0^\infty I_{[0,Y)}(\lambda) \mathrm{d}\lambda \mathrm{d}P \\ &= \int_0^\infty \int_\Omega X I_{[0,Y)}(\lambda) \mathrm{d}P \mathrm{d}\lambda \\ &= \int_0^\infty \int_\Omega X I_{\{Y>\lambda\}} \mathrm{d}P \mathrm{d}\lambda \\ &= \int_0^\infty E\left[X I_{\{Y>\lambda\}}\right] \mathrm{d}\lambda. \end{aligned}$$

This completes the proof.

13.12. By the definition of expectations, we have

$$E[f(X)] = \int_\Omega f(X) \mathrm{d}P = \int_0^\infty f(x) P(\mathrm{d}x) = \int_0^\infty \left(f(0) + \int_0^x f'(\lambda) \mathrm{d}\lambda \right) P(\mathrm{d}x).$$

Then by Theorem 10.1, we have

$$\begin{aligned} E[f(X)] &= f(0) + \int_0^\infty f'(\lambda) \int_\lambda^\infty P(\mathrm{d}x) \mathrm{d}\lambda \\ &= f(0) + \int_0^\infty f'(\lambda) P\{X \geq \lambda\} \mathrm{d}\lambda. \end{aligned}$$

This completes the proof.

13.13. For arbitrary integer n, we have

$$E(X) = \int_{\{X>n\}} X\mathrm{d}P + \int_{\{X\leq n\}} X\mathrm{d}P.$$

Since X is nonnegative, the above equation leads to

$$E(X) \geq nP\{X > n\} + \int_{\{X\leq n\}} X\mathrm{d}P.$$

Note that $XI_{\{X\leq n\}} \uparrow X$ as $n \to \infty$. By Theorem 6.2, we have

$$\lim_{n\to\infty} \int_{\{X\leq n\}} X\mathrm{d}P = E(X).$$

The result follows by the assumption that $E(X)$ is finite. This completes the proof.

13.14. To prove this problem, we only need to show that

$$P\left\{\limsup \frac{|X_n|}{\ln n} > 1\right\} = 0.$$

By Problem 11.12, we have

$$\left\{\limsup \frac{|X_n|}{\ln n} > 1\right\} = \limsup \left\{\frac{|X_n|}{\ln n} > 1\right\}.$$

Hence we only need to show that

$$P\left(\limsup \left\{\frac{|X_n|}{\ln n} > 1\right\}\right) = 0. \tag{13.5}$$

Note the assumption that $X_1, X_2, \ldots$ are identically distributed and that

$$E\left[\exp(|X_1|)\right] < \infty.$$

By Problem 13.9, we have

$$\begin{aligned}
\sum_{k=1}^{\infty} P\{|X_k| > \ln k\} &= \sum_{k=1}^{\infty} P\{|X_1| > \ln k\} \\
&= \sum_{k=1}^{\infty} P\left\{e^{|X_1|} > k\right\} \\
&= E\left[e^{|X_1|}\right] - P\left\{e^{|X_1|} > 0\right\} \\
&< \infty.
\end{aligned}$$

By Theorem 2.2, Equation (13.5) holds. This completes the proof.

13.5 Bibliographic Notes

In this chapter, we introduced the concept of expectation, variance, covariance, and moments. We defined expectations on the basis of integration with respect to probability measures. In some books such as Bartoszynski and Niewiadomska-Bugaj (2007), expectations are defined in terms of densities. For more information on expectations, readers are referred to Williams (1991, Chapters 6 and 7), Ash and Doleans-Dade (1999), and Çinlar (2011).

CHAPTER 14

CONDITIONAL EXPECTATION

In probability theory, the conditional expectation is the expected value of a random variable given some information represented by a σ-algebra. The conditional expectation defined on the basis of a σ-algebra is again a random variable satisfying certain properties. In this chapter, we present the definition of conditional expectations based on this approach.

14.1 Basic Concepts and Facts

Definition 14.1 (Conditional Expectation of a Random Variable Given a σ-Algebra)**.** Let $(\Omega, \mathscr{F}, P)$ be a probability space, and X a random variable with $E(|X|) < \infty$. Let $\mathcal{G}$ be a sub-σ-algebra of $\mathscr{F}$. Then a version of the conditional expectation of X given $\mathcal{G}$ is defined to be a random variable Y (see Theorem 14.1) such that

(a) Y is $\mathcal{G}$-measurable.

(b) $E(|Y|) < \infty$.

(c) For every $G \in \mathcal{G}$, we have

$$\int_G Y \mathrm{d}P = \int_G X \mathrm{d}P.$$

The version of the conditional expectation of X given $\mathcal{G}$ is written as $Y = E(X|\mathcal{G})$ a.s.

Definition 14.2 (Conditional Expectation of a Random Variable Given a Random Variable). Let X and Y be random variables on some probability space $(\Omega, \mathscr{F}, P)$ with $E(|X|) < \infty$. The conditional expectation of Y given X, written as $E[Y|X]$, is defined as $E[Y|\sigma(X)]$. Similarly, $E(X|Z_1, Z_2, \ldots)$ stands for $E(X|\sigma(Z_1, Z_2, \ldots))$.

Definition 14.3 (Conditional Expectation of a Random Variable Given an Event). Let Y be a random variable on some probability space $(\Omega, \mathscr{F}, P)$. Let $H \in \mathscr{F}$. Then the conditional expectation of Y given H, written as $E[Y|H]$, is defined as $E[Y|\sigma(I_H)]$, where I_H is the indicator function of H.

Definition 14.4 (Conditional Probability of an Event Given a σ-Algebra). Let $(\Omega, \mathscr{F}, P)$ be a probability space and $\mathcal{G}$ a sub-σ-algebra of $\mathscr{F}$. Let $B \in \mathscr{F}$. Then a version of the conditional probability of B given $\mathcal{G}$ is defined to be a function $Y : (\Omega, \mathcal{G}) \to (\mathbf{R}, \mathcal{B})$ (see Theorem 14.2) such that

$$P(A \cap B) = \int_A Y \mathrm{d}P, \quad A \in \mathcal{G}.$$

The version of the conditional probability of B given $\mathcal{G}$ is written as $P(B|\mathcal{G})$.

Note that $E[I_B|\mathcal{G}]$ is a version of the conditional probability of B given $\mathcal{G}$. We also write

$$P(B|\mathcal{G}) = E[I_B|\mathcal{G}].$$

Definition 14.5 (Conditional Probability of an Event Given a Random Variable). Let $(\Omega, \mathscr{F}, P)$ be a probability space and $\mathcal{G}$ be a sub-σ-algebra of $\mathscr{F}$. Let $B \in \mathscr{F}$, and let X be a random variable on the probability space. Then a version of the conditional probability of B given X is defined as

$$P(B|X) = P(B|\sigma(X)) = E[I_B|\sigma(X)].$$

Then $P(B|X)$ is a $\sigma(X)$-memasurable function. By Theorem 11.2, there exists a Borel function f such that $P(B|X) = f(X)$. The conditional probability $P(B|X = x)$ is defined as

$$P(B|X = x) = f(x).$$

Definition 14.6 (Conditional Probability of an Event Given an Event). Let G and H be two events. The conditional probability of H given G is defined to be any number $P(H|G)$ in $[0, 1]$ satisfying

$$P(G \cap H) = P(G) \cdot P(H|G).$$

This number is unique if $P(G) > 0$.

Theorem 14.1. *Let $(\Omega, \mathscr{F}, P)$ be a probability space, and X a random variable with $E(|X|) < \infty$. Let $\mathcal{G}$ be a sub-σ-algebra of $\mathscr{F}$. Then there exists a random variable Y such that*

(a) Y is $\mathcal{G}$-measurable.

(b) $E(|Y|) < \infty$.

(c) For every $G \in \mathcal{G}$, we have

$$\int_G Y \mathrm{d}P = \int_G X \mathrm{d}P.$$

Moreover, if $\tilde{Y}$ is another random variable satisfying (a)-(c), then $\tilde{Y} = Y$ a.s.

Theorem 14.2. *Let $(\Omega, \mathscr{F}, P)$ be a probability space and $\mathcal{G}$ a sub-σ-algebra of $\mathscr{F}$. Let $B \in \mathscr{F}$. Then there exists a function $Y : (\Omega, \mathcal{G}) \to (\mathbf{R}, \mathcal{B})$ such that*

$$P(A \cap B) = \int_A Y \mathrm{d}P, \quad A \in \mathcal{G}.$$

In addition, if $\tilde{Y}$ is another such function, then $\tilde{Y} = Y$ a.s.

14.2 Problems

14.1. Let $(\Omega, \mathscr{F}, P)$ be a probability space and $\mathcal{G}$ a sub-σ-field of $\mathscr{F}$. Let $X \in L^1(\Omega, \mathscr{F}, P)$. Use the Radon-Nikodym theorem to show that there is a $\mathcal{G}$-measurable random variable Y such that $E(|Y|) < \infty$ and

$$\int_A X \mathrm{d}P = \int_A Y \mathrm{d}P, \quad \forall A \in \mathcal{G}.$$

14.2. Let $(\Omega, \mathscr{F}, P)$ be a probability space, and let X be a random variable with $E(|X|) < \infty$. Let $\mathcal{G}$ be a sub-σ-algebra of $\mathscr{F}$. Suppose that Y is random variable with the following properties:

(a) Y is $\mathcal{G}$-measurable.

(b) $E(|Y|) < \infty$.

(c) For every $G \in \mathcal{I}$, where $\mathcal{I}$ is a π-system that contains Ω and generates $\mathcal{G}$, we have

$$\int_G Y \mathrm{d}P = \int_G X \mathrm{d}P.$$

Show that $Y = E(X|\mathcal{G})$ a.s.

14.3. Let $(\Omega, \mathscr{F}, P)$ be a probability space and X be a random variable on Ω with $E(|X|) < \infty$. Let $\mathcal{G}$ be a sub-σ-algebra of $\mathscr{F}$. Show that if Y is a version of $E(X|\mathcal{G})$, then $E(Y) = E(X)$.

14.4. Let $(\Omega, \mathscr{F}, P)$ be a probability space and X be a random variable on Ω with $E(|X|) < \infty$. Let $\mathcal{G}$ be a sub-σ-algebra of $\mathscr{F}$. Show that if X is $\mathcal{G}$-measurable, then $E(X|\mathcal{G}) = X$ a.s.

14.5 (Linearity)**.** Let $(\Omega, \mathscr{F}, P)$ be a probability space, and let X_1 and X_2 be random variables on Ω with $E(|X_1|) < \infty$ and $E(|X_2|) < \infty$. Let $\mathcal{G}$ be a sub-σ-algebra of $\mathscr{F}$. Show that for $a_1, a_2 \in \mathbf{R}$, $E(a_1X_1 + a_2X_2|\mathcal{G}) = a_1E(X_1|\mathcal{G}) + a_2E(X_2|\mathcal{G})$ a.s.

14.6 (Positivity)**.** Let $(\Omega, \mathscr{F}, P)$ be a probability space and X be a random variable on Ω with $E(|X|) < \infty$. Let $\mathcal{G}$ be a sub-σ-algebra of $\mathscr{F}$. Show that if $X \geq 0$, then $E(X|\mathcal{G}) \geq 0$ a.s.

14.7 (Conditional Monotone Convergence)**.** Let $(\Omega, \mathscr{F}, P)$ be a probability space, and let X be a random variable on Ω with $E(|X|) < \infty$. Let $\mathcal{G}$ be a sub-σ-algebra of $\mathscr{F}$. Show that if $0 \leq X_n \uparrow X$, then $E(X_n|\mathcal{G}) \uparrow E(X|\mathcal{G})$ a.s.

14.8 (Conditional Fatou's Lemma)**.** Let $(\Omega, \mathscr{F}, P)$ be a probability space and let $\{X_n\}_{n\geq 1}$ be a sequence of nonnegative random variables on Ω with $E(|X_n|) < \infty$. Let $\mathcal{G}$ be a sub-σ-algebra of $\mathscr{F}$. Show that $E(\liminf X_n|\mathcal{G}) \leq \liminf E(X_n|\mathcal{G})$ a.s.

14.9 (Conditional Dominated Convergence)**.** Let $(\Omega, \mathscr{F}, P)$ be a probability space. Let $\{X_n\}_{n\geq 1}$ be a sequence of random variables on Ω such that $|X_n| \leq V$ for $n \geq 1$, where V is a random variable with $E(V) < \infty$, and $X_n \to X$ a.s. Let $\mathcal{G}$ be a sub-σ-algebra of $\mathscr{F}$. Show that $E(X_n|\mathcal{G}) \to E(X|\mathcal{G})$ a.s.

14.10 (Tower Property)**.** Let $(\Omega, \mathscr{F}, P)$ be a probability space and X be a random variable on Ω with $E(|X|) < \infty$. Let $\mathcal{G}$ be a sub-σ-algebra of $\mathscr{F}$ and $\mathcal{H}$ be a sub-σ-algebra of $\mathcal{G}$. Show that

$$E(X|\mathcal{G}|\mathcal{H}) = E(X|\mathcal{H}),$$

where $E(X|\mathcal{G}|\mathcal{H}) = E[E(X|\mathcal{G})|\mathcal{H}]$.

14.11. Let X be a random variable on some probability space $(\Omega, \mathscr{F}, P)$. Show that

$$E[E[X]] = E[X] \quad \text{a.s.}$$

14.12. Let $(\Omega, \mathscr{F}, P)$ be a probability space and X be a random variable on Ω with $E(|X|) < \infty$. Let $\mathcal{G}$ be a sub-σ-algebra of $\mathscr{F}$. Show that if Z is $\mathcal{G}$-measurable and bounded, then $E(ZX|\mathcal{G}) = ZE(X|\mathcal{G})$ a.s.

14.13. Let $X : (\Omega, \mathscr{G}) \to (E, \mathscr{E})$ and $Y : (\Omega, \mathscr{H}) \to (F, \mathscr{F})$ be two random variables. Let Φ be a nonnegative Borel function on $(E \times F, \mathscr{E} \otimes \mathscr{F})$. Let $\varphi : (E, \mathscr{E}) \to ([0, \infty), \mathcal{B}[0, \infty))$ be defined by

$$\varphi(x) = E[\Phi(x, Y)], \quad x \in E.$$

Suppose that Y is independent of $\mathscr{G}$. Show that

$$E[\Phi(X,Y)|\mathscr{G}] = \varphi(X) \quad \text{a.s.}$$

14.14. Let $(\Omega, \mathscr{F}, P)$ be a probability space. Let $p > 1$ and $1/p + 1/q = 1$. Suppose that $X \in L^p(\Omega, \mathscr{F}, P)$ and $Z \in L^q(\Omega, \mathcal{G}, P)$, where $\mathcal{G}$ is a sub-σ-algebra of $\mathscr{F}$. Show that $E(ZX|\mathcal{G}) = ZE(X|\mathcal{G})$ a.s.

14.15. Let $(\Omega, \mathscr{F}, P)$ be a probability space. Let $\mathcal{G}$ be a sub-σ-algebra of $\mathscr{F}$. Let X and Z be random variables such that $X \in (m\mathscr{F})^+$, $Z \in (m\mathcal{G})^+$, $E(X) < \infty$, and $E(ZX) < \infty$. Show that $E(ZX|\mathcal{G}) = ZE(X|\mathcal{G})$ a.s.

14.16. Let $(\Omega, \mathscr{F}, P)$ be a probability space and X be a random variable on Ω with $E(|X|) < \infty$. Let $\mathcal{G}$ and $\mathcal{H}$ be sub-σ-algebras of $\mathscr{F}$. Show that

(a) If $\mathcal{H}$ is independent of $\sigma(\sigma(X), \mathcal{G})$, then

$$E(X|\sigma(\mathcal{G}, \mathcal{H})) = E(X|\mathcal{G}) \quad \text{a.s.}$$

(b) If X is independent of $\mathcal{H}$, then $E(X|\mathcal{H}) = E(X)$.

14.17. Let $(\Omega, \mathscr{F}, P)$ be a probability space and $\mathcal{G}$ a sub-σ-algebra of $\mathscr{F}$. Let $B \in \mathscr{F}$. Show that

$$P(B|\mathcal{G}) = E(I_B|\mathcal{G}), \quad a.s.$$

14.18. Let X be a random variable on a probability space $(\Omega, \mathscr{F}, P)$ with $E[|X|] < \infty$. Let $H \in \mathscr{F}$. Show that

(a)

$$E[XI_H] = E[X|H] \cdot E[I_H] = E[X|H]P(H),$$

where I_H is the indicator function of H.

(b)

$$E[X] = E[X|H]P(H) + E[X|H^c]P(H^c),$$

where $H^c = \Omega \backslash H$.

14.19. Let X be a real random variable on a probability space $(\Omega, \mathscr{F}, P)$. Let $\mathscr{G}$ be a sub-σ-algebra of $\mathscr{F}$. Suppose that X is discrete and takes values in $\{x_1, x_2, \ldots, x_n\}$. Show that

$$E[X|\mathscr{G}] = \sum_{k=1}^{n} x_k P(X = x_k|\mathscr{G}).$$

14.20 (Bayes' Formula)**.** Let $(\Omega, \mathscr{F}, P)$ be a probability space. Let Q be a probability measure on $(\Omega, \mathscr{F})$ such that $Q \ll P$. Let $\mathscr{G}$ be a sub-σ-algebra of $\mathscr{F}$ and $X \in L^1(\Omega, \mathscr{F}, Q)$. Let $L = \mathrm{d}Q/\mathrm{d}P$ be the Radon-Nikodym derivative. Show that

(a) $Q\{E_P[L|\mathscr{G}] > 0\} = 1$.

(b)

$$E_Q[X|\mathscr{G}] = \frac{E_P[XL|\mathscr{G}]}{E_P[L|\mathscr{G}]}.$$

14.3 Hints

14.1. Consider the following set function

$$\lambda(A) = \int_A X \mathrm{d}P, \quad A \in \mathcal{G}$$

and apply the Radon-Nikodym theorem (Theorem 7.4).

14.2. By Definition 14.1, one only needs to show that $E(Y; G) = E(X; G)$ for $G \in \mathcal{G}$. This can be achieved by showing that $\{G \in \mathcal{G} : E(Y; G) = E(X; G)\} = \mathcal{G}$. The result of Problem 2.7 is helpful here.

14.3. This problem can be proved by using the definition of conditional expectations (Definition 14.1) and the fact that $\Omega \in \mathcal{G}$.

14.4. Use the definition of conditional expectations (Definition 14.1) to show that X is a version of $E(X|\mathcal{G})$.

14.5. Show that $a_1 Y_1 + a_2 Y_2$ is a version of $E(a_1 X_1 + a_2 X_2|\mathcal{G})$, where $Y_1 = E(X_1|\mathcal{G})$ and $Y_2 = E(X_2|\mathcal{G})$.

14.6. To show that $E(X|\mathcal{G}) \geq 0$ a.s., one only needs to prove that $P(E(X|\mathcal{G}) < 0) = 0$. To do that, one can use the definition of conditional expectations and Theorem 2.1.

14.7. Using Theorem 6.2, one can show that $\limsup Y_n$ is a version of $E(X|\mathcal{G})$, where $Y_n = E(X_n|\mathcal{G})$.

14.8. Consider $Y_m = \inf_{n \geq m} X_n$ for $m \geq 1$ and use the result of Problem 14.7.

14.9. Apply the conditional Fatou's lemma (see Problem 14.8) to the sequences $V + X_n$ and $V - X_n$. The result of Problem 1.8 is also helpful here.

14.10. Use the definition of conditional expectations to show that $E(X|\mathcal{H})$ is a version of $E(Y|\mathcal{H})$.

14.11. Use the Tower property (see Problem 14.10) and the result of Problem 12.1.

14.12. Use the definition of conditional expectations and the standard technique (i.e., from simple functions to general functions) to show that $ZE(X|\mathcal{G})$ is a version of $E(ZX|\mathcal{G})$.

14.13. Use the definition of conditional expectation (Definition 14.1), Tonelli's theorem (Theorem 10.1), and change of variables, specifically,

$$\int_A \varphi(X) \mathrm{d}P = \int_{X(A)} \varphi(x) P_X(\mathrm{d}x),$$

where $X(A) = \{X(\omega) : \omega \in A\}$ and P_X is the probability law of X.

14.14. Use the same hint as Hint 14.12.

14.15. Use the same hint as Hint 14.12.

14.16. Show that $Y = E(X|\mathcal{G})$ is also a version of $E(X|\sigma(\mathcal{G}, \mathcal{H}))$. This can be achieved by using the results of Problem 14.2, Problem 2.16, and Theorem 13.2.

14.17. Use the definition of conditional expectations (Definition 14.1) and the definition of conditional probabilities (Definition 14.4).

14.18. Follow the definition of the expectation of a random variable given an event (Definition 14.3) to and the tower property (see Problem 14.10) prove part (a). Use part (a) to prove part (b).

14.19. Use the definition of conditional expectations (Definition 14.1) and conditional probabilities (Definition 14.4).

14.20. Follow the definition of conditional expectations (Definition 14.1) and Radon-Nikodym derivative (Theorem 7.4) to prove part (a). Use the definition of conditional expectations and the result of Problem 14.12 to prove part (b).

14.4 Solutions

14.1. Let $\lambda : \mathcal{G} \to \mathbf{R}$ be defined as

$$\lambda(A) = \int_A X \mathrm{d}P, \quad A \in \mathcal{G}.$$

We claim that λ is a signed measure on $(\Omega, \mathcal{G})$. By the definition of the integral over subsets (Definition 6.5), we have

$$\lambda(\emptyset) = \int_\emptyset X \mathrm{d}P = \int_\Omega X I_\emptyset \mathrm{d}P = 0.$$

Let A_n, $n = 1, 2, \ldots$, be disjoint sets in $\mathcal{G}$. Then

$$\lambda\left(\bigcup_{n=1}^\infty A_n\right) = \int_\Omega X I_{\bigcup_{n=1}^\infty A_n} \mathrm{d}P = \sum_{n=1}^\infty \int_\Omega X I_{A_n} \mathrm{d}P = \sum_{n=1}^\infty \lambda(A_n).$$

Hence λ is a signed measure on $(\Omega, \mathcal{G})$.

Now if $P(A) = 0$ and $A \in \mathcal{G}$, then by Problem 6.10, we have $\lambda(A) = 0$. Hence λ is absolutely continuous with respect to P. By the Radon-Nikodym theorem, there is a $\mathcal{G}$-measurable random variable Y such that

$$\int_A X \mathrm{d}P = \int_A Y \mathrm{d}P, \quad \forall A \in \mathcal{G}.$$

Note that $X \in L^1(\Omega, \mathscr{F}, P)$. We have $E(|X|) < \infty$. But

$$-E(|X|) \leq E(X) = E(Y) \leq E(|X|),$$

which implies that $E(|Y|) < \infty$. This completes the proof.

14.2. We only need to show that $E(Y; G) = E(X; G)$ for $G \in \mathcal{G}$. To do this, let $\mathcal{C} = \{G \in \mathcal{G} : E(Y; G) = E(X; G)\}$. We claim that $\mathcal{C}$ is a d-system on Ω. By hypothesis, we have $E(Y; \Omega) = E(X; \Omega)$. Hence $\Omega \in \mathcal{C}$. Suppose that $A, B \in \mathcal{C}$ with $A \subseteq B$. Then $E(Y; B \backslash A) = E(Y I_{B \backslash A}) = E(Y I_B - Y I_A) = E(Y; B) - E(Y; A) = E(X; B) - E(X; A) = E(X; B \backslash A)$. Hence $B \backslash A \in \mathcal{C}$. Suppose that $A_n \in \mathcal{C}$ and $A_n \uparrow A$. Then $X I_{A_n} \uparrow X I_A$ and $Y I_{A_n} \uparrow Y I_A$ and so, by Theorem 6.2, $E(X; A_n) = E(X I_{A_n}) \uparrow E(X I_A) = E(X; A)$ and $E(Y; A_n) \uparrow E(Y; A)$. Since $E(Y; A_n) = E(X; A_n)$, we have $E(X; A) = E(Y; A)$. Thus $A \in \mathcal{C}$. Therefore $\mathcal{C}$ is a d-system.

Noting that $\mathcal{I} \subseteq \mathcal{C} \subseteq \mathcal{G}$, we have $d(\mathcal{I}) \subseteq \mathcal{C}$ and so, by Problem 2.7, $\sigma(\mathcal{I}) \subseteq \mathcal{C}$. But with $\sigma(\mathcal{I}) = \mathcal{G}$, we have $\mathcal{C} = \mathcal{G}$. This completes the proof.

14.3. Let Y be a version of $E(X|\mathcal{G})$. Then, by definition of conditional expectations, we have $E(Y; G) = E(X; G)$ for all $G \in \mathcal{G}$. Since $\mathcal{G}$ is a sub-σ-algebra of $\mathscr{F}$, we have $\Omega \in \mathcal{G}$. Thus $E(Y; \Omega) = E(X; \Omega)$, which is $E(Y) = E(X)$.

14.4. By hypothesis, we know that X is $\mathcal{G}$-measurable and $E(|X|) < \infty$. Also, we have

$$\int_G X \mathrm{d}P = \int_G X \mathrm{d}P, \quad G \in \mathcal{G}.$$

Thus X is a version of $E(X|\mathcal{G})$. Hence $X = E(X|\mathcal{G})$, a.s.

14.5. Let $Y_1 = E(X_1|\mathcal{G})$ and $Y_2 = E(X_2|\mathcal{G})$. We only need to show that $a_1 Y_1 + a_2 Y_2$ is a version of $E(a_1 X_1 + a_2 X_2|\mathcal{G})$. Since Y_1 and Y_2 are $\mathcal{G}$-measurable, we have $a_1 Y_1 + a_2 Y_2 \in m\mathcal{G}$. Since $E(|Y_1|) < \infty$ and $E(|Y_2|) < \infty$, we have

$$E(|a_1 Y_1 + a_2 Y_2|) \leq |a_1| E(|Y_1|) + |a_2| E(|Y_2|) < \infty.$$

Moreover, for $G \in \mathcal{G}$, we have

$$\begin{aligned} E(a_1 Y_1 + a_2 Y_2; G) &= a_1 E(Y_1; G) + a_2 E(Y_2; G) \\ &= a_1 E(X_1; G) + a_2 E(X_2; G) \\ &= E(a_1 X_1 + a_2 X_2; G). \end{aligned}$$

Hence $a_1 Y_1 + a_2 Y_2$ is a version of $E(a_1 X_1 + a_2 X_2|\mathcal{G})$.

14.6. Let $Y = E(X|\mathcal{G})$. Since $P(Y \geq 0) = 1 - P(Y < 0)$, we only need to show that $P(Y < 0) = 0$. To do this, let $A_n = \{Y < -1/n\}$ for $n \geq 1$. Then $A_n \uparrow \{Y < 0\}$. Since $X \geq 0$, we have

$$0 \leq E(X; A_n) = E(Y; A_n) \leq -\frac{1}{n} P(A_n) \leq 0, \quad n \geq 1,$$

which implies $P(A_n) = 0$ for $n \geq 1$. Thus by Theorem 2.1, $P(A) = 0$, that is, $P(Y < 0) = 0$.

14.7. Let $Y_n = E(X_n|\mathcal{G})$. Then by Problems 14.6 and 14.5, we have $Y_n \geq 0$ a.s. and $Y_{n+1} - Y_n = E(X_{n+1} - X_n|\mathcal{G}) \geq 0$ a.s. for $n \geq 1$. Let $Y = \limsup Y_n$. Then Y is $\mathcal{G}$-measurable and $Y_n \uparrow Y$ a.s. Since $E(Y_n; G) = E(X_n; G)$ for $G \in \mathcal{G}$ and $n \geq 1$, it follows from Theorem 6.2 that $E(Y; G) = E(X; G)$ for $G \in \mathcal{G}$. In particular, $E(Y; \Omega) = E(X; \Omega)$, which is $E(Y) = E(X)$. Thus $E(Y) < \infty$. Hence Y is a version of $E(X|\mathcal{G})$. Therefore $E(X_n|\mathcal{G}) \uparrow E(X|\mathcal{G})$ a.s.

14.8. Let $Y_m = \inf_{n \geq m} X_n$ for $m \geq 1$. Then $Y_m \uparrow \liminf X_n$ as $m \to \infty$. Since $Y_m \leq X_n$ for $n \geq m$, by Problem 14.6, we have

$$E(Y_m|\mathcal{G}) \leq E(X_n|\mathcal{G}), \quad n \geq m,$$

which gives

$$E(Y_m|\mathcal{G}) \leq \inf_{n \geq m} E(X_n|\mathcal{G}) \leq \sup_{m \geq 1} \inf_{n \geq m} E(X_n|\mathcal{G}) = \liminf E(X_n|\mathcal{G}).$$

Since $Y_m \uparrow \liminf X_n$, by Problem 14.7, we have $E(Y_m|\mathcal{G}) \uparrow E(\liminf X_n|\mathcal{G})$. Therefore

$$E(\liminf X_n|\mathcal{G}) \leq \liminf E(X_n|\mathcal{G}).$$

14.9. Since $|X_n| \leq V$ for $n \geq 1$, we have $V + X_n \geq 0$ and $V - X_n \geq 0$. Then by Problem 14.8, we have

$$E(\liminf(V + X_n)|\mathcal{G}) \leq \liminf E(V + X_n|\mathcal{G}), \quad \text{a.s.} \tag{14.1}$$

and

$$E(\liminf(V - X_n)|\mathcal{G}) \leq \liminf E(V - X_n|\mathcal{G}), \quad \text{a.s.} \tag{14.2}$$

Since $X_n \to X$, we have $\liminf(V + X_n) = V + X$ and $\liminf(V - X_n) = V - X$. In addition, by Problem 1.8, we have $\liminf E(V + X_n|\mathcal{G}) = E(V|\mathcal{G}) + \liminf E(X_n|\mathcal{G})$ and $\liminf E(V - X_n|\mathcal{G}) = E(V|\mathcal{G}) - \limsup E(X_n|\mathcal{G})$. Thus it follows from (14.1) and (14.2) that

$$\limsup E(X_n|\mathcal{G}) \leq E(X|\mathcal{G}) \leq \liminf E(X_n|\mathcal{G}), \quad \text{a.s.},$$

which implies $E(X_n|\mathcal{G}) \to E(X|\mathcal{G})$, a.s.

14.10. Let $Y = E(X|\mathcal{G})$ and $Z = E(X|\mathcal{H})$. We only need to show that Z is a version of $E(Y|\mathcal{H})$. By definition, Z is $\mathcal{H}$-measurable. Let $Z_+ = E(|X||\mathcal{H})$. Then by Problem 14.6, we have $|Z| = |E(X|\mathcal{H})| \leq E(|X||\mathcal{H}) = Z_+$ and

$$E(|Z|) \leq E(Z_+) = E(Z_+; \Omega) = E(|X|; \Omega) = E(|X|) < \infty,$$

because $\Omega \in \mathcal{H}$. In addition, we have

$$\int_G X \mathrm{d}P = \int_G Y \mathrm{d}P, \quad G \in \mathcal{G},$$

and

$$\int_H X \mathrm{d}P = \int_H Z \mathrm{d}P, \quad H \in \mathcal{H}.$$

Since $\mathcal{H}$ is a sub-σ-algebra of $\mathcal{G}$, we have

$$\int_H Y \mathrm{d}P = \int_H Z \mathrm{d}P, \quad H \in \mathcal{H}.$$

Therefore Z is a version of $E(Y|\mathcal{H})$.

14.11. Let $\mathscr{F}_0 = \{\emptyset, \Omega\}$ be the trivial σ-field. Then by Problems 12.1 and 14.10, we have

$$E[E[X]] = E[E[X|\mathscr{F}_0]|\mathscr{F}_0] = E[X|\mathscr{F}_0] = E[X].$$

This finishes the proof.

14.12. Let Y be a version of $E(X|\mathcal{G})$. We show that ZY is a version of $E(ZX|\mathcal{G})$. Since Z is $\mathcal{G}$-measurable and bounded, we have $ZY \in m\mathcal{G}$ (see Theorem 5.1) and

$$E(|ZY|) \leq \left(\sup_{\omega \in \Omega} Z(\omega) \right) E(|Y|) < \infty.$$

Therefore, to show $ZY = E(ZX|\mathcal{G})$, we only need to show that

$$E(ZY; G) = E(ZX; G), \quad G \in \mathcal{G}. \tag{14.3}$$

If we can show that (14.3) holds for $X \geq 0$, then $E(ZY;G) = E(ZY_1;G) - E(ZY_2;G) = E(ZX^+;G) - E(ZX^-;G) = E(ZX;G)$ for $G \in \mathcal{G}$, where $Y_1 = E(X^+|\mathcal{G})$ and $Y_2 = E(X^-|\mathcal{G})$. Therefore, we can assume that $X \geq 0$. By Problem 14.6, $Y \geq 0$, a.s.

First, we consider the case when Z is an indicator function. Since Z is $\mathcal{G}$-measurable, we have $Z = I_A$ for some $A \in \mathcal{G}$ ($A = Z^{-1}(1) \in \mathcal{G}$). Then $E(ZY;G) = E(YI_A;G) = E(Y; A \cap G) = E(X; A \cap G) = E(XI_A;G) = E(XZ;G)$ for $G \in \mathcal{G}$. Hence (14.3) is true.

Next, we consider the case when Z is a simple function. Then Z has finitely many values. Let $a_i \geq 0$, $i = 1, 2, \ldots, n$ be all the distinct values of Z. Then

$$Z = \sum_{i=1}^n a_i I_{A_i},$$

where $A_i = Z^{-1}(a_i) \in \mathcal{G}$. For $G \in \mathcal{G}$, we have

$$\begin{aligned}
E(ZY;G) &= \int_G Y \sum_{i=1}^n a_i I_{A_i} \mathrm{d}P \\
&= \sum_{i=1}^n a_i \int_G Y I_{A_I} \mathrm{d}P \\
&= \sum_{i=1}^n a_i \int_G X I_{A_I} \mathrm{d}P \\
&= E(ZX;G).
\end{aligned}$$

Next, we consider the case when Z is a nonnegative function. Then there exists a sequence $\{Z_n\}_{n\geq 1}$ of $\mathcal{G}$-measurable simple functions such that $Z_n \uparrow Z$ (see Problem 5.11). We have already shown that $E(Z_nY;G) = E(Z_nX;G)$ for $G \in \mathcal{G}$. Noting that $Y \geq 0$, a.s., and $X \geq o$, we have $Z_nY \uparrow ZY$, a.s., and $Z_nX \uparrow ZX$, a.s. By Theorem 6.2, $E(ZY;G) = E(ZX;G)$ for $G \in \mathcal{G}$.

Finally, we consider thc casc when Z is a general function. Let $Z = Z^+ - Z^-$. Then $E(ZY;G) = E(Z^+Y;G) - E(Z^-Y;G) = E(Z^+X;G) - E(Z^-X;G) = E(ZX;G)$ for $G \in \mathcal{G}$. This completes the proof.

14.13. By Theorem 10.1, $\varphi(x)$ is a $\mathscr{E}$-measurable function. In consequence, we know that $\varphi(X)$ is $\mathscr{G}/\mathscr{E}$-measurable. Since Φ is nonnegative, $\varphi(X)$ is also nonnegative. Hence $E[\varphi(X)]$ is defined. Thus, for every $A \in \mathscr{G}$, we only need to show that

$$\int_A \varphi(X)\mathrm{d}P = \int_A \Phi(X,Y)\mathrm{d}P. \tag{14.4}$$

Then by Theorem 10.1 and Problem 14.11, we have

$$\begin{aligned}
\int_A \varphi(X)\mathrm{d}P &= \int_{X(A)} \varphi(x)P_X(\mathrm{d}x) \\
&= \int_{X(A)} \left(\int_\Omega \Phi(x,Y)\mathrm{d}P\right) P_X(\mathrm{d}x) \\
&= \int_{X(A)} \int_F \Phi(x,y)P_Y(\mathrm{d}y)P_X(\mathrm{d}x) \\
&= \int_F \int_{X(A)} \Phi(x,y)P_X(\mathrm{d}x)P_Y(\mathrm{d}y) \\
&= \int_F \left(\int_A \Phi(X,y)\mathrm{d}P\right) P_Y(\mathrm{d}y) \\
&= \int_F E[I_A\Phi(X,y)]P_Y(\mathrm{d}y) \\
&= E[E[I_A\Phi(X,Y)]] \\
&= E[I_A\Phi(X,Y)].
\end{aligned}$$

Equation (14.4) follows immediately from

$$E[I_A\Phi(X,Y)] = \int_A \Phi(X,Y)\mathrm{d}P.$$

This completes the proof.

14.14. Since $p > 1$, $X \in L^p(\Omega, \mathscr{F}, P)$, and $Z \in L^q(\Omega, \mathcal{G}, P) \subseteq L^q(\Omega, \mathscr{F}, P)$, by Theorem 8.1, we have

$$\begin{aligned} E(|ZX|) &= \int_\Omega |ZX| \mathrm{d}P \\ &\leq \left(\int_\Omega |X|^p \mathrm{d}P \right)^{\frac{1}{p}} \left(\int_\Omega |Z|^q \mathrm{d}P \right)^{\frac{1}{q}} \\ &< \infty. \end{aligned}$$

Thus $E(ZX|\mathcal{G})$ exists. By Problem 8.5, we have $X \in L^1(\Omega, \mathscr{F}, P)$, which gives $E(|X|) < \infty$. Hence $E(X|\mathcal{G})$ exists.

Let $Y = E(X|\mathcal{G})$. We show that ZY is a version of $E(ZX|\mathcal{G})$. By definition, Y is $\mathcal{G}$-measurable. Hence ZY is $\mathcal{G}$-measurable. By Problem 15.8, we know that $Y \in L^p(\Omega, \mathscr{F}, P)$ and so, by Theorem 8.1, we have

$$\begin{aligned} E(|ZY|) &= \int_\Omega |ZY| \mathrm{d}P \\ &\leq \left(\int_\Omega |Y|^p \mathrm{d}P \right)^{\frac{1}{p}} \left(\int_\Omega |Z|^q \mathrm{d}P \right)^{\frac{1}{q}} \\ &< \infty. \end{aligned}$$

Using the same argument in Problem 14.12, we have $E(ZY; G) = E(ZX; G)$ for $G \in \mathcal{G}$. Hence ZY is a version of $E(ZX|\mathcal{G})$.

14.15. Since $X \in (m\mathscr{F})^+$ and $E(X) < \infty$, we know that $E(X|\mathcal{G})$ exists and, by Problem 14.6, $Y = E(X|\mathcal{G}) \geq 0$ a.s. Since $Z \in (m\mathcal{G})^+$, we know that $ZY \in m\mathcal{G}$. Using a similar argument in Problem 14.12, we can show that $E(ZX; G) = E(ZY; G)$ for $G \in \mathcal{G}$. Note that $\Omega \in \mathcal{G}$ and $E(ZX) < \infty$, we have $E(|ZY|) = E(ZY) = E(ZY; \Omega) = E(ZX; \Omega) = E(ZX) < \infty$. Hence $ZY = E(ZX|\mathcal{G})$.

14.16. Let Y be a version of $E(X|\mathcal{G})$. Then Y is $\mathcal{G}$-measurable, $E(|Y|) < \infty$, and

$$E(YI_G) = E(XI_G), \quad \forall G \in \mathcal{G}. \tag{14.5}$$

Note that $\mathcal{G} \subseteq \sigma(\mathcal{G}, \mathcal{H})$, we have Y is $\sigma(\mathcal{G}, \mathcal{H})$-measurable. By Problems 14.2 and 2.16, we only need to show that

$$E(YI_G I_H) = E(XI_G I_H), \quad \forall G \in \mathcal{G},\ H \in \mathcal{H}.$$

In fact, by Theorem 13.2, Equation (14.5), and the assumption that $\mathcal{H}$ is independent of $\sigma(\sigma(X), \mathcal{G})$, for all $G \in \mathcal{G},\ H \in \mathcal{H}$, we have

$$E(YI_G I_H) = E(YI_G)E(I_H) = E(XI_G)E(I_H) = E(XI_G I_H).$$

This completes the proof.

14.17. We only need to show that $E(I_B|\mathcal{G})$ is a version of the conditional probability $P(B|\mathcal{G})$. By Definition 14.1, $E(I_B|\mathcal{G})$ is $\mathcal{G}$-measurable and

$$\int_A I_B \mathrm{d}P = \int_A E(I_B|\mathcal{G})\mathrm{d}P, \quad A \in \mathcal{G}.$$

Integrating out the left term of the above equation gives

$$P(A \cap B) = \int_A E(I_B|\mathcal{G})\mathrm{d}P, \quad A \in \mathcal{G}.$$

Hence $E(I_B|\mathcal{G})$ is a version of the conditional probability $P(B|\mathcal{G})$. This completes the proof.

14.18.

(a) By Problems 14.10 and 14.12, we have

$$E[XI_H] = E[E[XI_H|H]] = E[I_H E[X|H]].$$

Hence we only need to show that

$$E[I_H E[X|H]] = E[X|H] \cdot E[I_H]. \tag{14.6}$$

Since $E[X|H] = f(I_H)$ for some Borel function f (see Theorem 11.2), we know that $E[X|H]$ is constant on the event H. Therefore, we have

$$\begin{aligned} E[I_H E[X|H]] &= \int_\Omega I_H E[X|H]\mathrm{d}P \\ &= \int_H E[X|H]\mathrm{d}P \\ &= E[X|H] \int_H \mathrm{d}P \\ &= E[X|H]E(I_H) \\ &= E[X|H]P(H). \end{aligned}$$

Hence Equation (14.6) holds.

(b) By part (a) of this problem, we have

$$E[X] = E[XI_H] + E[XI_{H^c}] = E[X|H]P(H) + E[X|H^c]P(H^c).$$

This completes the proof.

14.19. We only need to show that $\sum_{k=1}^n x_k P(X = x_k|\mathscr{G})$ is a version of $E[X|\mathscr{G}]$. Since $P(X = x_k|\mathscr{G})$ is $\mathscr{G}$-measurable, $\sum_{k=1}^n x_k P(X = x_k|\mathscr{G})$ is $\mathscr{G}$-measurable. In addition, we have

$$E\left[\left|\sum_{k=1}^n x_k P(X = x_k|\mathscr{G})\right|\right] < \sum_{k=1}^n |x_k| < \infty.$$

Now let $G \in \mathscr{G}$. Then by Definition 14.4 we have

$$\begin{aligned}
\int_G \sum_{k=1}^{n} x_k P(X = x_k|\mathscr{G}) \mathrm{d}P &= \sum_{k=1}^{n} x_k \int_G P(X = x_k|\mathscr{G}) \mathrm{d}P \\
&= \sum_{k=1}^{n} x_k P(\{X = x_k\} \cap G) \\
&= \sum_{k=1}^{n} \int_{\{X = x_k\} \cap G} X \mathrm{d}P \\
&= \int_G X \mathrm{d}P.
\end{aligned}$$

Hence $\sum_{k=1}^{n} x_k P(X = x_k|\mathscr{G})$ is a version of $E[X|\mathscr{G}]$. This completes the proof.

14.20.

(a) Since $E_P[L|\mathscr{G}]$ is $\mathscr{G}$-measurable, we have $\{E_P[L|\mathscr{G}] > 0\} \in \mathscr{G}$. Hence, by Definition 14.1, we have

$$\begin{aligned}
Q\{E_P[L|\mathscr{G}] = 0\} &= \int_{\{E_P[L|\mathscr{G}]=0\}} \mathrm{d}Q \\
&= \int_{\{E_P[L|\mathscr{G}]=0\}} L \mathrm{d}P \\
&= \int_{\{E_P[L|\mathscr{G}]=0\}} E_P[L|\mathscr{G}] \mathrm{d}P \\
&= 0,
\end{aligned}$$

which gives $Q\{E_P[L|\mathscr{G}] > 0\} = 1$.

(b) Since $E_P[L|\mathscr{G}]$ and $E_Q[X|\mathscr{G}]$ are $\mathscr{G}$-measurable, it follows from Problem 14.12 that

$$E_P[L|\mathscr{G}] \cdot E_Q[X|\mathscr{G}] = E_P(L E_Q[X|\mathscr{G}]|\mathscr{G}).$$

Hence for every $G \in \mathscr{G}$, we have

$$\begin{aligned}
\int_G E_P[L|\mathscr{G}] \cdot E_Q[X|\mathscr{G}] \mathrm{d}P &= \int_G E_P(L E_Q[X|\mathscr{G}]|\mathscr{G}) \mathrm{d}P \\
&= \int_G L E_Q[X|\mathscr{G}] \mathrm{d}P \\
&= \int_G E_Q[X|\mathscr{G}] \mathrm{d}Q \\
&= \int_G X \mathrm{d}Q \\
&= \int_G X L \mathrm{d}P.
\end{aligned}$$

It follows from Definition 14.1 that

$$E_P[L|\mathscr{G}] \cdot E_Q[X|\mathscr{G}] = E_P[XL].$$

This completes the proof.

14.5 Bibliographic Notes

In this chapter, we introduced conditional expectations. Many properties of conditional expectations are presented as problems. Those properties are very important at a later stage when we study martingales. For more information about conditional expectations, readers are referred to Williams (1991), Ash and Doleans-Dade (1999), Rosenthal (2006), Kuo (2006), and Çinlar (2011). Wise and Hall (1993) and Stoyanov (1997) present several counterexamples related to conditional expectations.

In particular, Çinlar (2011) presents a detailed introduction to conditional expectations and conditional probabilities. As Çinlar (2011, p147) pointed out, for example, the proper interpretation of $P(X|Y = y)$ when $P\{Y = y\} = 0$ is $f(y)$, where $f(\cdot)$ is defined as $f(Y) = P(X|Y)$. When $P\{Y = y\} > 0$, $P(X|Y = y)$ can be interpreted as $P(X|\{Y = y\})$ or $f(y)$ as the two quantities are identical.

CHAPTER 15

INEQUALITIES

We introduced several important inequalities in Chapter 8 when we presented the L^p spaces. In this chapter, we shall introduce Jensen's inequality for convex functions. Jensen's inequality can be used to prove many other inequalities.

15.1 Basic Concepts and Facts

Definition 15.1 (Convex Functions)**.** Let G be an open subinterval of $\mathbf{R}$. A function $\varphi : G \to \mathbf{R}$ is considered convex on G if its graph lies below any of its chords; that is, for any $x, y \in G$ and $0 \leq p \leq 1$, we have

$$\varphi(px + qy) \leq p\varphi(x) + q\varphi(y),$$

where $q = 1 - p$.

Theorem 15.1 (Mean Value Theorem)**.** *Let $f(x)$ be a continuous function on the closed interval $[a, b]$. Suppose that $f(x)$ is differentiable on the open interval (a, b). Then there exists a point $c \in (a, b)$ such that*

$$f'(c) = \frac{f(b) - f(a)}{b - a}.$$

Theorem 15.2 (Jensen's Inequality). *Let $(\Omega, \mathscr{F}, P)$ be a probability space and $\varphi : G \to \mathbf{R}$ be a convex function on G, where G is an open subinterval of $\mathbf{R}$. Suppose that X is a random variable on Ω such that $E(|X|) < \infty$, $P(X \in G) = 1$, and $E(|\varphi(X)|) < \infty$. Then*

$$E(\varphi(X)) \geq \varphi(E(X)).$$

15.2 Problems

15.1 (Chebyshev's Inequality). Let $X \in L^2(\Omega, \mathscr{F}, P)$ and $c > 0$. Then

$$P(|X - \mu| > c) \leq \frac{\operatorname{Var}(X)}{c^2},$$

where $\mu = E(X)$.

15.2 (Markov's Inequality). Let $(\Omega, \mathscr{F}, P)$ be a probability space and $Z \in m\mathscr{F}$. Suppose that $g : \mathbf{R} \to [0, \infty]$ is $\mathcal{B}$-measurable and nondecreasing. Show that

$$Eg(Z) \geq E(g(Z); Z \geq c) \geq g(c)P(Z \geq c), \quad c \in \mathbf{R}.$$

15.3. Let $\varphi : G \to \mathbf{R}$ be a convex function on an open subinterval G of $\mathbf{R}$. Show that

(a) $\Delta_{u,v} \leq \Delta_{u,w} \leq \Delta_{v,w}$ for $u, v, w \in G$ and $u < v < w$, where

$$\Delta_{a,b} = \frac{\varphi(a) - \varphi(b)}{a - b}, \quad a, b \in G,\ a \neq b.$$

(b) φ is continuous.

15.4. Let $\varphi : G \to \mathbf{R}$ be a convex function on an open subinterval G of $\mathbf{R}$. Suppose that the second derivative $\varphi''(x) \geq 0$ for all $x \in G$. Show that φ is a convex function on G.

15.5 (Line of Support Theorem). Let $\varphi : G \to \mathbf{R}$ be a convex function on an open subinterval G of $\mathbf{R}$. Show that there exists a sequence $\{(a_n, b_n)\}_{n\geq 1}$ of points in $\mathbf{R}^2$ such that

$$\varphi(x) = \sup_{n\geq 1}(a_n x + b_n), \quad x \in G.$$

15.6. Let $n \geq 1$ and $a_1, a_2, \ldots, a_n \in \mathbf{R}$. Show that

$$\left(\sum_{i=1}^{n} a_i\right)^2 \leq n\left(\sum_{i=1}^{n} a_i^2\right).$$

15.7 (Conditional Jensen's Inequality). Let $(\Omega, \mathscr{F}, P)$ be a probability space and X be a random variable on Ω with $E(|X|) < \infty$. Let $\mathcal{G}$ be a sub-σ-algebra of $\mathscr{F}$. Show that if $\varphi : \mathbf{R} \to \mathbf{R}$ is convex, then

$$\varphi(E(X|\mathcal{G})) \leq E(\varphi(X)|\mathcal{G}), \quad \text{a.s.}$$

15.8. Let $(\Omega, \mathscr{F}, P)$ be a probability space and X be a random variable on Ω with $E(|X|) < \infty$. Let $\mathcal{G}$ be a sub-σ-algebra of $\mathscr{F}$. Let $p \geq 1$. Show that

$$\|E(X|\mathcal{G})\|_p \leq E(\|X\|_p|\mathcal{G}), \quad \text{a.s.},$$

where

$$\|Y\|_p = \left(\int_\Omega |Y|^p \mathrm{d}P\right)^{\frac{1}{p}}.$$

15.9 (Kolmogorov's Inequality). Let $X_1, X_2, \ldots, X_n$ be independent random variables on the probability space $(\Omega, \mathscr{F}, P)$ with $|E(X_i)| < \infty$, $i = 1, 2, \ldots, n$. Let $S_j = X_1 + X_2 + \cdots + X_j$, $j = 1, 2, \ldots, n$. For any $\epsilon > 0$, show that

$$P\left\{\max_{1\leq j\leq n} |S_j - E(S_j)| \geq \epsilon\right\} \leq \frac{\text{Var}(S_n)}{\epsilon^2}. \tag{15.1}$$

15.10. Let Y be a nonnegative random variable on a probability space $(\Omega, \mathscr{F}, P)$. Show that

$$\sum_{n=1}^{\infty} P\{Y \geq n\} \leq E(Y) \leq 1 + \sum_{n=1}^{\infty} P\{Y \geq n\}. \tag{15.2}$$

15.11 (Lévy's Inequality). Let $X_1, X_2, \ldots, X_N$ be independent random variables that are distributed symmetrically about zero. Let

$$S_j = \sum_{i=1}^{j} X_i, \quad j = 1, 2, \ldots, N.$$

For $y > 0$, show that

$$P\left\{\max_{1\leq j\leq N} S_j \geq y\right\} \leq 2P\{S_N \geq y\} - P\{S_N = y\}.$$

15.3 Hints

15.1. This problem is a special case of Problem 15.1.

15.2. Use the fact that $g(Z) \geq g(Z)I_{Z\geq c} \geq g(c)I_{Z\geq c}$ and properties of integration.

15.3. Follow the definition of convex functions (Definition 15.1) to prove part (a). Apply part (a) to prove part (b).

15.4. Note that the first derivative $\varphi'(x)$ is a nondecreasing function and use the mean value theorem (Theorem 15.1).

15.5. Use the result of Problem 15.3 and the fact that rational numbers are countable and dense.

15.6. Construct a discrete random variable with values in $\{a_1, a_2, \ldots, a_n\}$ and apply Jensen's inequality (Theorem 15.2).

15.7. Use the result of Problem 15.5.

15.8. Show that $\|\cdot\|_p$ is convex, and then use the result of Problem 15.7.

15.9. First break the set $A = \{\max_{1\le j\le n} |S_j - E(S_j)| \ge \epsilon\}$ into a disjoint union of n sets in the following way:

$$A_k = \{|S_k - E(S_k)| \ge \epsilon\} \cap \left(\bigcap_{j=1}^{k-1} \{|S_j - E(S_j)| < \epsilon\} \right), \quad k = 1, 2, \ldots, n.$$

Then consider the variance of S_n and use properties of integration and independence.

15.10. Break the range of Y into the following sequence of intervals: $[0, 1)$, $[1, 2)$, ..., and then use properties of integration.

15.11. Note that

$$\left\{ \max_{1\le j\le N} S_j \ge y \right\} = \{S_N \ge y\} \cup \left(\bigcup_{j=1}^{N-1} A_j \cap \{S_N < y\} \right),$$

where $A_j = \{S_1 < y, S_2 < y, \ldots, S_{j-1} < y, S_j \ge y\}$ for $j = 1, 2, \ldots, N$.

15.4 Solutions

15.1. By definition, we have

$$\begin{aligned} \mathrm{Var}(X) &= E[(X-\mu)^2] \\ &= \int_\Omega |X(\omega) - \mu|^2 P(\mathrm{d}\omega) \\ &\ge \int_{|X-\mu|>c} c^2 \mathrm{d}P = c^2 P(|X-\mu| > c), \end{aligned}$$

from which the result follows.

15.2. By hypothesis, g is nonnegative and nondecreasing. Hence

$$g(Z) \ge g(Z) I_{Z\ge c} \ge g(c) I_{Z\ge c}.$$

Then we have

$$\begin{aligned} Eg(Z) &\ge E(g(Z) I_{Z\ge c}) \\ &\ge E(g(c) I_{Z\ge c}) \\ &= \int_{Z\ge c} g(c) \mathrm{d}P \\ &= g(c) P(Z \ge c). \end{aligned}$$

This completes the proof.

15.3. Let $p = (v-u)/(w-u)$ and $q = 1-p = (w-v)/(w-u)$. Then, $0 < p < 1$ and so, by the definition of convex functions, we obtain

$$\varphi(pw + qu) \leq p\varphi(w) + q\varphi(u).$$

Noting that $pw + qu = v$, we have

$$(w-u)\varphi(v) \leq (v-u)\varphi(w) + (w-v)\varphi(u), \tag{15.3}$$

which is equivalent to

$$(w-u)(\varphi(u) - \varphi(v)) \geq (u-v)(\varphi(w) - \varphi(u)).$$

Dividing both sides of the above equation by $(w-u)(u-v)$, we get $\Delta_{u,v} \leq \Delta_{u,w}$. Note that Equation (15.3) is also equivalent to

$$(u-w)(\varphi(v) - \varphi(w)) \geq (v-w)(\varphi(u) - \varphi(w)).$$

Dividing both sides of this equation by $(u-w)(v-w)$, we get $\Delta_{u,w} \leq \Delta_{v,w}$.

Now we show that φ is continuous. Let $x_0 \in G$. To show that φ is continuous at x_0, we only need to find a δ for every $\epsilon > 0$ such that $|\varphi(x) - \varphi(x_0)| < \epsilon$ whenever $|x - x_0| < \delta$. To do this, let $y_1, y_2 \in G$ such that $y_1 < x_0 < y_2$ (Since G is open, such y_1, y_2 exist.). Then we have $\Delta_{y_1,x_0} \leq \Delta_{x_0,x} \leq \Delta_{x_0,y_2}$ for $x \in (y_1, x_0) \cup (x_0, y_2)$. Consequently, we have

$$|\Delta_{x,x_0}| \leq L, \quad x \in (y_1, x_0) \cup (x_0, y_2),$$

where $L = \max(|\Delta_{y_1,x_0}|, |\Delta_{x_0,y_2}|)$. Let $\delta = \epsilon/L$. Then

$$|\varphi(x) - \varphi(x_0)| \leq L|x - x_0| < \epsilon, \quad x \in (x_0 - \delta, x_0 + \delta).$$

Therefore, $\varphi(x)$ is continuous at x_0. Since this is true for every $x_0 \in G$, $\varphi(x)$ is continuous on G.

15.4. Let $x, y \in G$ such that $x < y$. Let $p \in (0,1)$. Let $z = px + qy$, where $q = 1-p$. Then by Theorem 15.1, we have

$$\varphi'(\xi_1) = \frac{\varphi(z) - \varphi(x)}{z - x} = \frac{\varphi(z) - \varphi(x)}{q(y-x)}$$

and

$$\varphi'(\xi_2) = \frac{\varphi(y) - \varphi(z)}{y - z} = \frac{\varphi(y) - \varphi(z)}{p(y-x)},$$

where for some $\xi_1 \in (x, z)$ and $\xi_2 \in (z, y)$. But by Theorem 15.1, we have

$$\varphi''(c) = \frac{\varphi'(\xi_2) - \varphi'(\xi_1)}{\xi_2 - \xi_2} \geq 0,$$

which gives $\varphi'(\xi_1) \leq \varphi'(\xi_1)$. Thus we have

$$\frac{\varphi(z) - \varphi(x)}{q(y-x)} \leq \frac{\varphi(y) - \varphi(z)}{p(y-x)},$$

which leads to

$$\varphi(z) \leq p\varphi(x) + q\varphi(y).$$

Therefore, φ is convex on G. This completes the proof.

15.5. For $q \in G$, let $\{y_n\}_{n\geq 1}$ be a sequence of points in G such that $y_n \uparrow q$. Then by Problem 15.3, we have $d_n \leq d_{n+1} \leq d_+$ for $n \geq 1$, where $d_n = (\varphi(y_n) - \varphi(q))/(y_n - q)$ and $d_+ = (\varphi(q_+) - \varphi(q))/(q_+ - q)$ for some $q_+ \in G$ and $q_+ > q$. Thus $\lim_{n\to\infty} d_n$ exists and is bounded. Let

$$\Delta_-(q) = \lim_{y\uparrow q} \frac{\varphi(y) - \varphi(q)}{y - q}, \quad q \in G.$$

Let $x \in G$. Then by Problem 15.3, we have

$$\varphi(x) - \varphi(q) \geq \Delta_-(q)(x - q)$$

or

$$\varphi(x) \geq \Delta_-(q)(x - q) + \varphi(q)$$

for all $q \in G$. Thus

$$\varphi(x) = \sup_{q\in G}[\Delta_-(q)(x - q) + \varphi(q)], \quad x \in G.$$

Now we show that

$$\varphi(x) = \sum_{q\in G\cap Q} [\Delta_-(q)(x - q) + \varphi(q)], \quad x \in G, \tag{15.4}$$

where Q is the set of all rational numbers. To do this, we fix $x \in G$. Since G is open, we can choose an $\epsilon > 0$ such that $(x - \epsilon, x + \epsilon) \subseteq G$. Let $\{q_n\}_{n\geq q}$ be a sequence of rational numbers in $(x - \epsilon, x + \epsilon)$ such that $q_n \to x$. Then it follows from Problem 15.3 that

$$\Delta_-(x - \epsilon) \leq \Delta_-(q_n) \leq \Delta_-(x + \epsilon), \quad n \geq 1.$$

Thus $|\Delta_-(q_n)| \leq L = \max(|\Delta_-(x - \epsilon)|, |\Delta_-(x + \epsilon)|)$ for $n \geq 1$ and so

$$|\Delta_-(q_n)(x - q_n) + \varphi(q_n) - \varphi(x)| \leq L|x - q_n| + |\varphi(q_n) - \varphi(x)|. \tag{15.5}$$

By Problem 15.3, φ is continuous on G. Therefore, it follows from Equation (15.5) that $\Delta_-(q_n)(x - q_n) + \varphi(q_n) \to \varphi(x)$. Therefore (15.4) is true. Since $G \cap Q$ is countable, we can label all the points in $G \cap Q$ as a sequence $\{x_n\}_{n\geq 1}$ and let $a_n = \Delta_-(x_n)$ and $b_n = \varphi(q_n) - \Delta_-(q_n)q_n$ for $n \geq 1$. Then

$$\varphi(x) = \sup_{n\geq 1}(a_n x + b_n), \quad x \in G.$$

15.6. Let X be a random variable such that

$$P\{X = a_i\} = \frac{1}{n}, \quad i = 1, 2, \ldots, n.$$

Then we have

$$E[X] = \sum_{i=1}^{n} \frac{a_i}{n}.$$

Let $\varphi(x) = x^2$. Since $\varphi(x)$ is a convex function, it follows from Theorem 15.2 that

$$\varphi(E[X]) \leq E[\varphi(x)],$$

which is

$$\left(\sum_{i=1}^{n} \frac{a_i}{n}\right)^2 \leq \sum_{i=1}^{n} \frac{a_i^2}{n}.$$

The resulting inequality follows by multiplying n^2 in both sides of the above equation. This completes the proof.

15.7. Since φ is convex, by Problem 15.5, we have

$$\varphi(X) = \sup_{n\geq 1}(a_n X + b_n),$$

where $\{(a_n, b_n)\}_{n\geq 1}$ is a sequence in $\mathbf{R}^2$. Since $E(|\varphi(X)|) < \infty$, $E(\varphi(X)|\mathcal{G})$ exists. Let Y be a version of $E(\varphi(X)|\mathcal{G})$ and Z be a version of $E(X|\mathcal{G})$. Since for each fixed n, $\varphi(X) \geq a_n X + b_n$, by Problems 14.6 and 14.5, we have $Y \geq a_n Z + b_n$, a.s. For each $n \geq 1$, let A_n be defined as

$$A_n = \{\omega \in \Omega : Y(\omega) \geq a_n Z(\omega) + b_n\}.$$

Then $P(A_n) = 1$ for $n \geq 1$ and so, by Problem 11.1, we obtain

$$P(A) = P\left(\bigcap_{n\geq 1} A_n\right) = 1,$$

where $A = \bigcap_{n\geq 1} A_n$. Since $Y(\omega) \geq a_n Z(\omega) + b_n$ for all $n \geq 1$ and $\omega \in A$, we have $Y \geq a_n Z + b_n$ simultaneously for all $n \geq 1$, a.s., and so

$$Y \geq \sup_{n\geq 1}(a_n Z + b_n) = \varphi(Z), \quad \text{a.s.},$$

which is

$$E(\varphi(X)|\mathcal{G}) \geq \varphi(E(X|\mathcal{G}), \quad \text{a.s.}$$

This completes the proof.

15.8. We show that $\|\cdot\|_p$ is a convex function for $p \geq 1$. Let $0 \leq a = 1 - b \leq 1$. Then by Theorem 8.2, we have

$$\begin{aligned}
\|aY + bZ\|_p &= \left(\int_\Omega |aY + bZ|^p \mathrm{d}P\right)^{\frac{1}{p}} \\
&\leq \left(\int_\Omega |aY|^p \mathrm{d}P\right)^{\frac{1}{p}} + \left(\int_\Omega |bZ|^p \mathrm{d}P\right)^{\frac{1}{p}} \\
&= a\|Y\|_p + b\|Z\|_p.
\end{aligned}$$

Thus $\|\cdot\|_p$ is convex. The result follows from Problem 15.7.

15.9. Let $A = \{\max_{1\leq j\leq n} |S_j - E(S_j)| \geq \epsilon\}$ and

$$A_k = \{|S_k - E(S_k)| \geq \epsilon\} \cap \left(\bigcap_{j=1}^{k-1} \{|S_j - E(S_j)| < \epsilon\}\right), \quad k = 1, 2, \ldots, n.$$

Then $A_1, A_2, \ldots, A_n$ are mutually disjoint and $A = \bigcup_{k=1}^n A_k$. By definition, we have

$$\begin{aligned}
\mathrm{Var}(S_n) &= E\left[(S_n - E(S_n))^2\right] \\
&= \int_\Omega (S_n - E(S_n))^2 \mathrm{d}P \geq \int_A (S_n - E(S_n))^2 \mathrm{d}P \\
&= \sum_{k=1}^n \int_{A_k} (S_n - E(S_n))^2 \mathrm{d}P. \qquad (15.6)
\end{aligned}$$

Now let $Y_k = S_n - S_k$, $k = 1, 2, \ldots, n$. Then S_k and Y_k are independent. Hence

$$\begin{aligned}
\int_{A_k} (S_n - E(S_n))^2 \mathrm{d}P &= \int_{A_k} [(S_k - E(S_k)) + (Y_k - E(Y_k))]\mathrm{d}P \\
&= \int_{A_k} (S_k - E(S_k))^2 \mathrm{d}P + \int_{A_k} (Y_k - E(Y_k))^2 \mathrm{d}P \\
&\quad +2\int_\Omega I_{A_k}(S_k - E(S_k))(Y_k - E(Y_k))\mathrm{d}P \\
&= \int_{A_k} (S_k - E(S_k))^2 \mathrm{d}P + \int_{A_k} (Y_k - E(Y_k))^2 \mathrm{d}P \\
&\geq \int_{A_k} (S_k - E(S_k))^2 \mathrm{d}P \\
&\geq \epsilon^2 P(A_k), \quad k = 1, 2, \ldots, n. \qquad (15.7)
\end{aligned}$$

Combining Equations (15.6) and (15.7) gives

$$\mathrm{Var}(S_n) \geq \sum_{k=1}^n \epsilon^2 P(A_k) = \epsilon^2 P(A).$$

This completes the proof.

15.10. Let $A_n = Y \geq n$, $n = 0, 1, \ldots$. By the assumption that Y is nonnegative, we have

$$\begin{aligned} E(Y) &= \int_\Omega Y \mathrm{d}P \\ &= \sum_{n=0}^{\infty} \int_{n \leq Y < n+1} Y \mathrm{d}P \\ &\geq \sum_{n=0}^{\infty} nP\{n \leq Y < n+1\} \\ &= \sum_{n=0}^{\infty} n(P(A_n) - P(A_{n+1})) \\ &= \sum_{n=1}^{\infty} nP(A_n) - \sum_{n=2}^{\infty} (n-1)P(A_n) \\ &= \sum_{n=1}^{\infty} P(A_n). \end{aligned}$$

Similarly, we have

$$\begin{aligned} E(Y) &= \int_\Omega Y \mathrm{d}P \\ &= \sum_{n=0}^{\infty} \int_{n \leq Y < n+1} Y \mathrm{d}P \\ &\leq \sum_{n=0}^{\infty} (n+1)P\{n \leq Y < n+1\} \\ &= \sum_{n=0}^{\infty} (n+1)(P(A_n) - P(A_{n+1})) \\ &= \sum_{n=0}^{\infty} (n+1)P(A_n) - \sum_{n=1}^{\infty} nP(A_n) \\ &\leq 1 + \sum_{n=1}^{\infty} P(A_n). \end{aligned}$$

This completes the proof.

15.11. Let $A_j = \{S_1 < y, S_2 < y, \ldots, S_{j-1} < y, S_j \geq y\}$ for $j = 1, 2, \ldots, N$. Then we have

$$\left\{ \max_{1 \leq j \leq N} S_j \geq y \right\} = \{S_N \geq y\} \cup \left(\bigcup_{j=1}^{N-1} A_j \cap \{S_N < y\} \right),$$

which gives

$$P\left\{\max_{1\leq j\leq N} S_j \geq y\right\} = P\{S_N \geq y\} + \sum_{j=1}^{N-1} P\left(A_j \cap \{S_N < y\}\right).$$

Note that $A_j \cap \{S_N < y\} \subseteq A_j \cap \{S_N - S_j < 0\}$ and that A_j and $\{S_N - S_j < 0\}$ are independent. We have

$$\begin{aligned} P\left\{\max_{1\leq j\leq N} S_j \geq y\right\} &= P\{S_N \geq y\} + \sum_{j=1}^{N-1} P(A_j)P\left(\{S_N < y\}\right) \\ &\leq P\{S_N \geq y\} + \sum_{j=1}^{N-1} P(A_j)P\left(\{S_N - S_j < 0\}\right). \end{aligned}$$

By the assumption that $X_1, X_2, \ldots, X_N$ are symmetrically distributed about zero, we have $P\{S_N - S_j > 0\} = P\{S_N - S_j > 0\}$. Hence we have

$$\begin{aligned} P\left\{\max_{1\leq j\leq N} S_j \geq y\right\} &= P\{S_N \geq y\} + \sum_{j=1}^{N-1} P(A_j)P\left(\{S_N - S_j > 0\}\right) \\ &= P\{S_N \geq y\} + \sum_{j=1}^{N-1} P\left(A_j \cap \{S_N - S_j > 0\}\right) \\ &\leq P\{S_N \geq y\} + \sum_{j=1}^{N-1} P\left(A_j \cap \{S_N > y\}\right) \\ &\leq P\{S_N \geq y\} + P\{S_N > y\} \\ &= 2P\{S_N \geq y\} - P\{S_N = y\}. \end{aligned}$$

This completes the proof.

15.5 Bibliographic Notes

In this chapter, we introduced Jensen's inequality. Jensen's inequality is very powerful in that it allows us to prove many other inequalities. A proof of Jensen's inequality can be found in Williams (1991). For more information about probability inequalities, readers are referred to Jacod and Protter (2004), Rosenthal (2006), Ross and Pekoz (2007), Gut (2007), and Lin and Bai (2010). In particular, the book by Lin and Bai (2010) is devoted to probability inequalities.

CHAPTER 16

LAW OF LARGE NUMBERS

The law of large numbers is a theorem in probability theory that describes the average of a large number of random variables. The law states that the average of a large number of random variables converges to the expected value in some sense under certain conditions. In this chapter, we introduce some laws of large numbers.

16.1 Basic Concepts and Facts

Definition 16.1 (Almost Surely Convergence). Let $X, X_1, X_2, \ldots$ be random variables on a probability space $(\Omega, \mathscr{F}, P)$. The sequence $\{X_n\}_{n\geq 1}$ is said to converge to X almost surely if and only if it converges to X almost everywhere; that is, there is a set $A \in \mathscr{F}$ such that $X_n \to X$ on A and $P(A^c) = P(\Omega \backslash A) = 0$.

Definition 16.2 (Convergence in Probability). Let $X_1, X_2, \ldots,$ be a sequence of random variables on $(\Omega, \mathscr{F}, P)$. The sequence X_i is said to be convergent to a random variable X on $(\Omega, \mathscr{F}, P)$ if for every $\epsilon > 0$, we have

$$\lim_{n\to\infty} P\left(|X_n - X| \geq \epsilon\right) = 0.$$

Definition 16.3 (Identically Distributed Random Variables and IID). A sequence of random variables $\{X_n\}_{n\geq 1}$ is are said to be identically distributed if they have the same distribution. These random variables are called *IID random variables* if they are independent and identically distributed.

Theorem 16.1 (Weak Law of Large Numbers). *Let $X_1, X_2, \ldots,$ be a sequence of independent random variables on $(\Omega, \mathscr{F}, P)$. Suppose that $|E(X_i)| < \infty$ and $\text{Var}(X_i) < M$ for all $i = 1, 2, \ldots$, where $M < \infty$. Let $S_n = X_1 + X_2 + \cdots + X_n$. Then $[S_n - E(S_n)]/n$ converges to 0 in probability; that is, for every $\epsilon > 0$, we have*

$$\lim_{n\to\infty} P\left(\left|\frac{S_n - E(S_n)}{n}\right| \geq \epsilon\right) = 0.$$

Theorem 16.2 (Kolmogorov Strong Law of Large Numbers). *Let $X_1, X_2, \ldots,$ be a sequence of independent random variables on $(\Omega, \mathscr{F}, P)$, each with finite mean and variance. Let $\{b_n\}_{n\geq 1}$ be a sequence of positive real numbers such that $b_n \uparrow \infty$. If*

$$\sum_{n=1}^{\infty} \frac{\text{Var}(X_n)}{b_n^2} < \infty,$$

then

$$\lim_{n\to\infty} \frac{S_n - E(S_n)}{b_n} = 0 \quad \textit{a.s.},$$

where $S_n = \sum_{j=1}^n X_j$, $n = 1, 2, \ldots$.

Theorem 16.3 (Strong Law of Large Numbers, IID Case). *Let $\{X_n\}_{n\geq 1}$ be a sequence of IID random variables on the probability space $(\Omega, \mathscr{F}, P)$ with finite expectation m. Let $S_n = \sum_{j=1}^n X_j$, $n = 1, 2, \ldots$. Then*

$$\lim_{n\to\infty} \frac{S_n}{n} = m \quad \textit{a.s.}$$

16.2 Problems

16.1. Prove Theorem 16.1.

16.2. Let $(\Omega, \mathscr{F}, P)$ be a probability space. Suppose that X_n $(n \geq 1)$ are independent random variables on Ω such that $E(X_k) = 0$ and $E(X_k^4) \leq K$ for all $k \geq 1$, where $K \in [0, \infty)$ is a constant. Show that

$$P\left(\frac{S_n}{n} \to 0\right) = 1,$$

where $S_n = X_1 + X_2 + \cdots + X_n$.

16.3. Let $A = (a_{ij})_{i\geq 1, j\geq 1}$ be an infinite matrix of real numbers with the following properties:

(a) For fixed j, we have

$$\lim_{n\to\infty} a_{nj} = 0;$$

(b) For some $c \geq 0$, we have

$$\sum_{j=1}^{\infty} |a_{nj}| \leq c, \quad \forall n \geq 1.$$

Let $x_{nn\geq 1}$ be a sequence of bounded real numbers and define

$$y_n = \sum_{j=1}^{\infty} a_{nj} x_j, \quad n = 1, 2, \ldots.$$

Show that

(a) If $\lim_{n\to\infty} x_n = 0$, then $\lim_{n\to\infty} y_n = 0$.

(b) If $\lim_{n\to\infty} \sum_{j=1}^{\infty} a_{nj} = 1$ and $\lim_{n\to\infty} x_n = x$ for some $x \in \mathbf{R}$, then

$$\lim_{n\to\infty} y_n = x.$$

16.4 (Toeplitz Lemma)**.** Let $\{a_n\}_{n\geq 1}$ be a sequence of nonnegative real numbers. Let b_n be the partial sum defined as

$$b_n = \sum_{j=1}^{n} a_j, \quad n = 1, 2, \ldots.$$

Suppose that $b_n > 0$ for all $n \geq 1$ and $b_n \to \infty$ as $n \to \infty$. Let $\{x_n\}_{n\geq 1}$ be a sequence of real numbers such that $\lim_{n\to\infty} x_n = x$. Show that

$$\lim_{n\to\infty} \frac{1}{b_n} \sum_{j=1}^{n} a_j x_j = x.$$

16.5 (Kronecker Lemma)**.** Let $\{b_n\}_{n\geq 1}$ be a sequence of positive real numbers such that $b_n \uparrow \infty$, and let $\{x_n\}_{n\geq 1}$ be a sequence of real numbers such that $\sum_{n=1}^{\infty} x_n = x$, where $|x| < \infty$. Show that

$$\lim_{n\to\infty} \frac{1}{b_n} \sum_{j=1}^{n} b_j x_j = 0.$$

16.6 (Mean Square Summability Criterion)**.** Let $\{X_n\}_{n\geq 1}$ be a sequence of independent random variables on the probability space $(\Omega, \mathscr{F}, P)$ with $|E(X_n)| < \infty$, $n = 1, 2, \ldots$. Show that if

$$\sum_{n=1}^{\infty} \mathrm{Var}(X_n) < \infty,$$

then $\sum_{n=1}^{\infty}(X_n - E(X_n))$ exists a.s.

16.7. Prove the Kolmogorov strong law of large numbers (Theorem 16.2).

16.8. Let $\{X_n\}_{n\geq 1}$ be a sequence of IID random variables on the probability space $(\Omega, \mathscr{F}, P)$. Assume that

$$\lim_{n\to\infty} \frac{1}{n}\sum_{j=1}^{n} X_j = m \quad \text{a.s.},$$

where $|m| < \infty$. Show that $E(X_1)$ is finite and $E(X_1) = m$ a.s.

16.9. Let $\{X_n\}_{n\geq 1}$ be a sequence of independent random variables. Suppose that for every $n \geq 1$, we have $P(X_n = 1) = p_n$ and $P(X_n = 0) = 1 - p_n$. Show that

(a) $X_n \xrightarrow{P} 0$ if and only if $\lim_{n\to\infty} p_n = 0$.

(b) $X_n \xrightarrow{\text{a.s.}} 0$ if and only if $\sum_{n=1}^{\infty} p_n < \infty$.

16.10. Let $\{X_n : n \geq 1\}$ be a sequence of independent and identically distributed random variables such that $E[|\ln X_1|] < \infty$.

(a) Find the a.s. limiting geometric mean

$$\lim_{n\to\infty}\left(\prod_{i=1}^{n} X_i\right)^{\frac{1}{n}}.$$

(b) Suppose that X_1 is uniformly distributed on $(0, 1)$. Calculate the numerical value of the a.s. limiting geometric mean.

16.11 (Kolmogorov's Three-Series Criterion: Sufficiency Part)**.** Let $\{X_n\}_{n\geq 1}$ be a sequence of independent random variables. Suppose that there exists a number $a > 0$ such that

(a) $\sum_{n=1}^{\infty} P\{|X_n| > a\}$ converges.

(b) $\sum_{n=1}^{\infty} E[X_n I_{\{|X_n|\leq a\}}]$ converges.

(c) $\sum_{n=1}^{\infty} \text{Var}(X_n I_{\{|X_n|\leq a\}})$ converges.

Show that $\sum_{n=1}^{\infty} X_n$ converges with probability 1.

16.12. Let $\{Y_n\}_{n\geq 1}$ be a sequence of independent random variables. Suppose that for all $n \geq 1$, we have $E[Y_n] = 0$ and $|Y_n| \leq c$ a.s. for some $c < \infty$. Show that if $\sum_{n=1}^{\infty} Y_n$ is convergent a.s., then

$$\sum_{n=1}^{\infty} E\left[Y_n^2\right] < \infty.$$

Furthermore, for all $\lambda > 0$, the following inequality holds, where $S_n = Y_1 + Y_2 + \cdots + Y_n$:

$$\sum_{n=1}^{\infty} E\left[Y_n^2\right] \le \frac{(\lambda + c)^2}{P\{\sup_{n\ge 1} |S_n| \le \lambda\}}.$$

16.13. Let $\{Y_n\}_{n\ge1}$ be a sequence of independent random variables. Suppose that for all $n \ge 1$, $|Y_n| \le c$ a.s. for some $c < \infty$. Show that if $\sum_{n=1}^{\infty} Y_n$ is convergent a.s., then $\sum_{n=1}^{\infty} E[Y_n]$ converges.

16.14 (Kolmogorov's Three-Series Criterion: Necessary Part)**.** Let $\{X_n\}_{n\ge1}$ be a sequence of independent random variables. Suppose that $\sum_{n=1}^{\infty} X_n$ converges with probability 1. Let $a > 0$. Show that

(a) $\sum_{n=1}^{\infty} P\{|X_n| > a\}$ converges.

(b) $\sum_{n=1}^{\infty} E[X_n I_{\{|X_n|\le a\}}]$ converges.

(c) $\sum_{n=1}^{\infty} \text{Var}(X_n I_{\{|X_n|\le a\}})$ converges.

16.15. Let $\{X_n\}_{n\ge1}$ be a sequence of independent and identically distributed random variables defined as

$$P\{X_1 = 1\} = P\{X_1 = -1\} = \frac{1}{2}.$$

Let $\{a_n\}_{n\ge1}$ be a sequence of real numbers. Show that $\sum_{n=1}^{\infty} a_n X_n$ converges a.s. if and only if

$$\sum_{n=1}^{\infty} a_n^2 < \infty.$$

16.3 Hints

16.1. Use Chebyshev's inequality (see Problem 15.1) and the result of Problem 13.7.

16.2. First try to establish that

$$E\left[\sum_{n=1}^{\infty} \left(\frac{S_n}{n}\right)^4\right] < \infty,$$

and then use the result of Problem 13.5.

16.3. Part (a) can be proved using the standard technique; that is, for every $\epsilon > 0$, we need to find an integer N_ϵ such that

$$|y_n| < \epsilon, \quad \forall n > N_\epsilon.$$

We can derive part (b) from part (a).

16.4. Use the result of Problem 16.3.

16.5. Use the Toeplitz lemma (see Problem 16.4).

16.6. Use the result of Problem 9.8 and Kolmogorov's inequality (see Problem 15.9).

16.7. Use the results of Problems 16.6 and 16.5.

16.8. Use the second Borel-Cantelli lemma (see Problem 12.9), the result of Problem 15.10, and the strong law of large numbers for the IID case (see Theorem 16.3).

16.9. Part (a) can be proved using the definition of convergence in probability (Definition 16.2). Part (b) can be proved using the first Borel-Cantelli lemma (Theorem 2.2) and the second Borel-Cantelli lemma (see Problem 12.9).

16.10. Consider the random variables $Y_n = \ln X_n$, $n \geq 1$, and apply Theorem 16.3.

16.11. Use the results of Problems 9.8 and 16.6 to show that both the sequences

$$\sum_{n=1}^{\infty} X_n I_{\{|X_n| \leq a\}}$$

and

$$\sum_{n=1}^{\infty} X_n I_{\{|X_n| > a\}}$$

converge a.s.

16.12. For a fixed λ, introduce a random variable $\tau = \inf\{n \geq 1 : |S_n| > \lambda\}$ and note that $\{\tau = j\}$ is independent of $Y_{j+1}, Y_{j+2}, \ldots$.

16.13. Let $(\Omega, \mathscr{F}, P)$ be the probability space on which $\{Y_n\}_{n\geq 1}$ are defined. Consider the random variable Z_n on $(\Omega \times \Omega, \mathscr{F} \otimes \mathscr{F}, P \times P)$

$$Z_n(\omega_1, \omega_2) = Y_n(\omega_1) - Y_n(\omega_2),$$

and use Fubini's theorem (Theorem 10.2) to establish

$$E\left[Z_n^2\right] = 2\text{Var}(Y_n).$$

Then use the result of Problems 16.12 and 16.6.

16.14. The convergence of the first series can be proved by the method of contradiction with Problem 9.8 and the second Borel-Cantelli lemma (see Problem 12.9). The convergence of the second series can be proved by Problem 16.13. The convergence of the third series can be proved by Problem 16.12.

16.15. The "if" part is implied by Problem 16.6. The "only if" part can be proved by the result of Problem 16.14.

16.4 Solutions

16.1. Let $\epsilon > 0$. Then by Chebyshev's inequality (see Problem 15.1), Problem 13.7, and the assumption, we have

$$\begin{aligned} P\left(\left|\frac{S_n - E(S_n)}{n}\right| \geq \epsilon\right) &\leq \frac{1}{\epsilon^2} E\left[\left(\frac{S_n - E(S_n)}{n}\right)^2\right] \\ &= \frac{1}{\epsilon^2 n^2} \text{Var}(S_n) \\ &= \frac{1}{\epsilon^2 n^2} \sum_{i=1}^{n} \text{Var}(X_i) \\ &< \frac{M}{\epsilon^2 n}. \end{aligned}$$

Hence

$$\lim_{n\to\infty} P\left(\left|\frac{S_n - E(S_n)}{n}\right| \geq \epsilon\right) = 0.$$

This completes the proof.

16.2. By the assumption that $E\left[X_k^4\right] \leq K < \infty$ and Problem 8.5, we get $E\left[X_k^2\right] < \infty$ and $E\left[X_k^3\right] < \infty$. Also from the assumption that X_n $(n \geq 1)$ are independent and $E(X_n) = 0 (n \geq 1)$, we have $E[X_i X_j X_k X_l] = 0$, $E[X_i X_j X_k^2] = 0$, and $E[X_i X_j^3] = 0$ for all distinct i, j, k, l. Therefore, we have

$$\begin{aligned} E\left[S_n^4\right] &= E\left[(X_1 + X_2 + \cdots + X_n)^4\right] \\ &= E\left[\sum_{i=1}^{n} X_i^4 + 6 \sum_{i<j} X_i^2 X_j^2\right]. \end{aligned}$$

By Schwarz's inequality (Theorem 8.3), we have

$$E\left[X_i^2 X_j^2\right] \leq \|X_i^2\|_2 \cdot \|X_j^2\|_2 \leq \left(E[X_i^4] E[X_j^4]\right)^{\frac{1}{2}} \leq K.$$

Combining the above two equations, we have

$$E\left[S_n^4\right] \leq nK + 3n(n-1)K \leq 3n^2 K,$$

which implies that

$$E\left[\left(\frac{S_n}{n}\right)^4\right] \leq \frac{3K}{n^2}.$$

By Problem 13.3 and the above inequality, we have

$$E\left[\sum_{n=1}^{\infty} \left(\frac{S_n}{n}\right)^4\right] \leq \sum_{n=1}^{\infty} \frac{3K}{n^2} < \infty.$$

By Problem 13.5, $\left(\frac{S_n}{n}\right)^4 \to 0$ a.s. Hence we have $\frac{S_n}{n} \to 0$ a.s. This completes the proof.

16.3.

(a) Let $\epsilon > 0$. By assumption that $\lim_{n\to\infty} x_n = 0$, there exists an integer $N_{0,\epsilon}$ such that

$$|x_j| < \frac{\epsilon}{2c}, \quad \forall j > N_{0,\epsilon}.$$

Then by the property of A, we have

$$\begin{aligned} |y_n| &= \left|\sum_{j=1}^{\infty} a_{nj} x_j\right| \\ &\leq \sum_{j=1}^{N_{0,\epsilon}} |a_{n,j}| \cdot |x_j| + \sum_{j=N_{0,\epsilon}+1}^{\infty} |a_{n,j}| \cdot |x_j| \\ &< \sum_{j=1}^{N_{0,\epsilon}} |a_{n,j}| \cdot |x_j| + \frac{\epsilon}{2}. \end{aligned} \tag{16.1}$$

By the assumption that $a_{nj} \to 0$ as $n \to \infty$ for fixed j, there exists an integer $N_{j,\epsilon}$ such that

$$|a_{nj}| < \frac{\epsilon}{N_{0,\epsilon}(1 + |x_j|)}, \quad \forall n > N_{j,\epsilon}.$$

Combining the above inequality into Equation (16.1) gives

$$|y_n| < \epsilon, \quad \forall n > \max(N_{1,\epsilon}, N_{2,\epsilon}, \ldots, N_{N_{0,\epsilon},\epsilon}).$$

Since $\epsilon > 0$ is arbitrary, we have $y_n \to 0$ as $n \to \infty$.

(b) We write y_n as

$$y_n - x = \sum_{j=1}^{\infty} a_{nj}(x_j - x) + \left(\sum_{j=1}^{\infty} a_{nj} - 1\right) x, \quad n = 1, 2, \ldots.$$

By the first item, we have $\sum_{j=1}^{\infty} a_{nj}(x_j - x) \to 0$ as $n \to \infty$. By the assumption that $\lim_{n\to\infty} \sum_{j=1}^{\infty} a_{nj} = 1$, we have $y_n - x \to 0$ as $n \to \infty$. Hence $\lim_{n\to\infty} y_n = x$.

This completes the proof.

16.4. Let $C = (c_{ij})_{i\geq 1, j\geq 1}$ be an infinite matrix defined by

$$c_{ij} = \begin{cases} \frac{a_j}{b_i} & \text{if } j \leq i; \\ 0 & \text{if } j > i. \end{cases}$$

By the assumption that $b_n \to \infty$ as $n \to \infty$, we have $c_{nj} \to 0$ as $n \to \infty$ for fixed j. In addition, we have

$$\sum_{j=1}^{\infty} |c_{nj}| = \sum_{j=1}^{n} \frac{a_j}{b_n} = 1.$$

Therefore, by Problem 16.3, we have

$$\lim_{n\to\infty} \sum_{j=1}^{\infty} c_{nj} x_j = x,$$

which is

$$\lim_{n\to\infty} \frac{1}{b_n} \sum_{j=1}^{n} a_j x_j = x.$$

This completes the proof.

16.5. Let s_n be defined as

$$s_{n+1} = \sum_{j=1}^{n} x_j, \quad n \geq 1.$$

Taking $b_0 = 0$ and $s_1 = 0$, we have

$$\begin{aligned} \frac{1}{b_n} \sum_{j=1}^{n} b_j x_j &= \frac{1}{b_n} \sum_{j=1}^{n} b_j (s_{j+1} - s_j) \\ &= s_{n+1} + \frac{1}{b_n} \left(\sum_{j=0}^{n-1} b_j s_{j+1} - \sum_{j=1}^{n} b_j s_j \right) \\ &= s_{n+1} + \frac{1}{b_n} \left(\sum_{j=1}^{n} (b_{j-1} - b_j) s_j \right). \end{aligned} \tag{16.2}$$

Let $a_n = b_n - b_{n-1}$ for $n \geq 1$. Then $a_n \geq 0$ and $\sum_{j=1}^{n} a_j = b_n$. By Problem 16.4, we have

$$\lim_{n\to\infty} \frac{1}{b_n} \left(\sum_{j=1}^{n} (b_{j-1} - b_j) s_j \right) = - \lim_{n\to\infty} \frac{1}{b_n} \left(\sum_{j=1}^{n} a_j s_j \right) = -x.$$

Note that $s_{n+1} \to x$ as $n \to \infty$. From Equation (16.2) we have

$$\lim_{n\to\infty} \frac{1}{b_n} \sum_{j=1}^{n} b_j x_j = 0.$$

This completes the proof.

16.6. Let $S_n = \sum_{j=1}^n (X_j - E(X_j))$, $n = 1, 2, \ldots$. Then by Problem 9.8, to show that $\lim_{n\to\infty} S_n$ exists a.s., we only need to show that for every $\epsilon > 0$,

$$P\left(\bigcup_{j=1}^{\infty}\{|S_{n+j} - S_n| \geq \epsilon\}\right) \to 0 \quad \text{as} \quad n \to \infty. \tag{16.3}$$

By Problem 15.9, we have

$$\begin{aligned} P\left(\bigcup_{j=1}^{\infty}\{|S_{n+j} - S_n| \geq \epsilon\}\right) &= \lim_{m\to\infty} P\left(\bigcup_{j=1}^{m}\{|S_{n+j} - S_n| \geq \epsilon\}\right) \\ &= \lim_{m\to\infty} P\left(\max_{1\leq j\leq m}\{|S_{n+j} - S_n| \geq \epsilon\}\right) \\ &\leq \frac{1}{\epsilon^2}\limsup_{m\to\infty} \text{Var}(S_{n+m} - S_n) \\ &\leq \frac{1}{\epsilon^2}\limsup_{m\to\infty}\sum_{j=1}^{m}\text{Var}(X_{n+j}). \end{aligned}$$

Note that $\sum_{n=1}^{\infty}\text{Var}(X_n) < \infty$, we have

$$\limsup_{m\to\infty}\sum_{j=1}^{m}\text{Var}(X_{n+j}) \to 0 \quad \text{as} \quad n \to \infty.$$

Combining the above two equations leads to Equation (16.3). This completes the proof.

16.7. By assumption, we have

$$\sum_{n=1}^{\infty}\text{Var}\left(\frac{X_n - E(X_n)}{b_n}\right) = \sum_{n=1}^{\infty}\frac{\text{Var}(X_n)}{b_n^2} < \infty.$$

Then by Problem 16.6, we know that

$$\sum_{n=1}^{\infty}\frac{X_n - E(X_n)}{b_n}$$

exists a.s. Note that

$$\frac{S_n - E(S_n)}{b_n} = \frac{1}{b_n}\sum_{k=1}^{n} b_k \frac{X_k - E(X_k)}{b_k}, \qquad n = 1, 2, \ldots.$$

By Problem 16.5, we have

$$\lim_{n\to\infty}\frac{S_n - E(S_n)}{b_n} = 0 \quad \text{a.s.}$$

This completes the proof.

16.8. Let $S_n = \sum_{j=1}^{n} X_j$, $n = 1, 2, \ldots$. Note that

$$\frac{X_n}{n} = \frac{S_n}{n} - \frac{S_{n-1}}{n-1} - \frac{S_{n-1}}{n(n-1)}, \quad n \geq 1.$$

Since $S_n/n \to m$ as $n \to \infty$ a.s. and $|m| < \infty$, we have

$$\lim_{n \to} \frac{X_n}{n} = 0 \quad \text{a.s.}$$

which means that

$$P\left\{\liminf \frac{X_n}{n} = \limsup \frac{X_n}{n} = 0\right\} = 1.$$

Hence

$$P\left(\limsup \{|X_n| \geq n\}\right) = P\left\{\limsup \frac{|X_n|}{n} \geq 1\right\} = 0.$$

By the second Borel-Cantelli lemma (see Problem 12.9), we must have

$$\sum_{n=1}^{\infty} P(|X_n| \geq n) < \infty.$$

By Problem 15.10, $E(|X_1|)$ is finite a.s. By Theorem 16.3, we have $E(|X_1|) = m$ a.s. This completes the proof.

16.9.

(a) By definition, $X_n \xrightarrow{P} 0$ means that for every $\epsilon > 0$, we have

$$\lim_{n \to \infty} P\{|X_n| \geq \epsilon\} = 0.$$

Note that $P\{|X_n| \geq \epsilon\} = P\{X_n = 1\} = p_n$. Hence $X_n \xrightarrow{P} 0$ is equivalent to

$$\lim_{n \to \infty} p_n = 0.$$

(b) First let us prove the "if" part. Suppose that $\sum_{n=1}^{\infty} p_n < \infty$. Then by the first Borel-Cantelli lemma (Theorem 2.2), we have

$$P\{\limsup X_n = 1\} = P\{\limsup\{X_n = 1\}\} = 0.$$

Hence $P\{\limsup X_n = 0\} = 1$ as $\limsup X_n$ can be either 0 or 1. Noting that $\limsup X_n \geq \liminf X_n \geq 0$, we have

$$P\{\limsup X_n = \liminf X_n = 0\} = 1,$$

which implies that $X_n \xrightarrow{\text{a.s.}} 0$.

Next, let us prove the "only if" part. Suppose that $X_n \xrightarrow{\text{a.s.}} 0$. Then we have

$$P\{\limsup X_n = \liminf X_n = 0\} = 1,$$

which implies that $P\{\limsup X_n = 1\} = 0$. By the second Borel-Cantelli lemma (see Problem 12.9), we have

$$\sum_{n=1}^{\infty} p_n < \infty.$$

This completes the proof.

16.10.

(a) Let $Y_n = \ln X_n$ for $n \geq 1$. Then by the assumption, we have $E[|Y_1|] < \infty$. Hence by Theorem 16.3, we have

$$\lim_{n\to\infty} \frac{Y_1 + Y_2 + \cdots + Y_n}{n} = E[Y_1], \quad \text{a.s.}$$

But

$$\frac{Y_1 + Y_2 + \cdots + Y_n}{n} = \ln\left[\left(\prod_{i=1}^{n} X_i\right)^{\frac{1}{n}}\right].$$

Hence we have

$$\lim_{n\to\infty} \left(\prod_{i=1}^{n} X_i\right)^{\frac{1}{n}} = e^{E[Y_1]} = e^{E[\ln X_1]}, \quad \text{a.s.}$$

(b) If X_1 is a uniformly distributed random variable on $(0, 1)$, we have

$$E[\ln X_1] = \int_0^1 \ln x \mathrm{d}x = x \ln x|_0^1 - \int_0^1 1\mathrm{d}x = -1.$$

Hence the numerical value is $e^P - 1$.

This completes the solution.

16.11. By the assumption and Problem 16.6, we know that the sequence

$$\sum_{n=1}^{\infty} X_n I_{\{|X_n| \leq a\}}$$

converges with probability one. Now let

$$S_n = \sum_{i=1}^{n} X_i I_{\{|X_i| > a\}}, \quad n \geq 1.$$

Let $\epsilon > 0$. Then for each $n \geq 1$, we have

$$\begin{aligned}\bigcup_{j=1}^{\infty}\{|S_{n+j} - S_n| > \epsilon\} &= \bigcup_{j=1}^{\infty}\left\{\left|\sum_{i=1}^{j} X_{n+i} I_{\{|X_{n+i}|>a\}}\right| > \epsilon\right\} \\ &\subseteq \bigcup_{j=1}^{\infty}\{|X_{n+j}| > a\}. \end{aligned} \tag{16.4}$$

In fact, let

$$\omega \in \bigcup_{j=1}^{\infty}\left\{\left|\sum_{i=1}^{j} X_{n+i} I_{\{|X_{n+i}|>a\}}\right| > \epsilon\right\}.$$

Then we must have $\omega \in \{|X_{n+k}| > a\}$ for some $k \geq 1$. This gives

$$\omega \in \bigcup_{j=1}^{\infty}\{|X_{n+j}| > a\}.$$

Hence Equation (16.4) holds. By the assumption that $\sum_{n=1}^{\infty} P\{|X_n| > a\}$ converges, we have

$$\lim_{n\to\infty} P\{|X_{n+j}| > a\} = 0.$$

Hence by Equation (16.4), we have

$$\lim_{n\to\infty} P\left(\bigcup_{j=1}^{\infty}\{|S_{n+j} - S_n| > \epsilon\}\right) = 0.$$

It follows from Problem 9.8 that $\lim_{n\to\infty} S_n$ exists a.s. Note that

$$\sum_{n=1}^{\infty} X_n = \sum_{n=1}^{\infty} X_n I_{\{|X_n|\leq a\}} + \sum_{n=1}^{\infty} X_n I_{\{|X_n|>a\}}.$$

The result holds. This completes the proof.

16.12. Let $\lambda > 0$. Let τ be defined as $\tau = \inf\{n \geq 1 : |S_n| > \lambda\}$. Since

$$\{\tau = j\} = \{|S_1| \leq \lambda, |S_2| \leq \lambda, \ldots, |S_{j-1}| \leq \lambda, |S_j| > \lambda\},$$

$\{\tau = j\}$ is independent of $Y_{j+1}, Y_{j+2}, \ldots$. Then for $N \geq 1$, we have

$$\begin{aligned}
E[S_N^2 I_{\{\tau \leq N\}}] &= \sum_{j=1}^{N} E[S_N^2 I_{\{\tau=j\}}] \\
&= \sum_{j=1}^{N} E\left[\left(S_j^2 + 2S_j(S_N - S_j) + (S_N - S_j)^2\right) I_{\{\tau=j\}}\right)\right] \\
&= \sum_{j=1}^{N} E\left[S_j^2 I_{\{\tau=j\}}\right] + \sum_{j=1}^{N} E[2S_j I_{\{\tau=j\}}] E[S_N - S_j] \\
&\quad + \sum_{j=1}^{N} E\left[(S_N - S_j)^2\right] E\left[I_{\{\tau=j\}}\right] \\
&\leq \sum_{j=1}^{N} E\left[(S_{j-1} + Y_j)^2 I_{\{\tau=j\}}\right] + \sum_{j=1}^{N} E\left[S_N^2\right] P\{\tau = j\} \\
&\leq \sum_{j=1}^{N} E\left[(\lambda + c)^2 I_{\{\tau=j\}}\right] + E\left[S_N^2\right] \sum_{j=1}^{N} P\{\tau = j\} \\
&= \left[(\lambda + c)^2 + E\left[S_N^2\right]\right] P\{\tau \leq N\} \\
&\leq (\lambda + c)^2 + E\left[S_N^2\right] P\{\tau \leq N\}.
\end{aligned}$$

Hence we have

$$\begin{aligned}
E\left[S_N^2\right] &= E\left[S_N^2 I_{\{\tau \leq N\}}\right] + E\left[S_N^2 I_{\{\tau > N\}}\right] \\
&\leq (\lambda + c)^2 + E\left[S_N^2\right] P\{\tau \leq N\} + \lambda^2 P\{\tau > N\},
\end{aligned}$$

which gives

$$E\left[S_N^2\right] \leq \frac{(\lambda + c)^2}{1 - P\{\tau \leq N\}} \leq \frac{(\lambda + c)^2}{P\{\tau = \infty\}} = \frac{(\lambda + c)^2}{P\{\sup_{n \geq 1} |S_n| \leq \lambda\}}.$$

By the assumption that S_n converges a.s., we have

$$P\left\{\sup_{n \geq 1} |S_n| < \infty\right\} = 1,$$

which gives

$$\lim_{\lambda \to \infty} P\left\{\sup_{n \geq 1} |S_n| < \lambda\right\} = 1.$$

Hence for sufficiently large λ, we have $P\left\{\sup_{n \geq 1} |S_n| < \lambda\right\} > 0$. Therefore

$$\sum_{j=1}^{\infty} E\left[Y_j^2\right] = \lim_{N \to \infty} E\left[S_N^2\right] \leq \frac{(\lambda + c)^2}{P\{\sup_{n \geq 1} |S_n| \leq \lambda\}} < \infty.$$

This completes the proof.

16.13. Let $(\Omega, \mathscr{F}, P)$ be the probability space on which $\{Y_n\}_{n\geq 1}$ are defined. Let Z_n be random variables on $(\Omega \times \Omega, \mathscr{F} \otimes \mathscr{F}, P \times P)$ that are defined as

$$Z_n(\omega_1, \omega_2) = Y_n(\omega_1) - Y_n(\omega_2), \quad (\omega_1, \omega_2) \in \Omega \times \Omega.$$

Then we have $E[Z_n] = 0$ and by Theorem 10.2, we obtain

$$\begin{aligned} E\left[Z_n^2\right] &= \int_{\Omega\times\Omega} Z_n^2 \mathrm{d}(P \times P) \\ &= \int_\Omega \left(\int_\Omega \left[Y_n^2(\omega_1) - 2Y_n(\omega_1)Y_n(\omega_2) + Y_n^2(\omega_2)\right] P(\mathrm{d}\omega_1) \right) P(\mathrm{d}\omega_2) \\ &= \int_\Omega \left(E\left[Y_n^2\right] - 2E[Y_n]Y_n(\omega_2) + Y_n^2(\omega_2)\right) P(\mathrm{d}\omega_2) \\ &= E\left[Y_n^2\right] - 2E[Y_n]E[Y_n] + E\left[Y_n^2\right] \\ &= 2\text{Var}(Y_n). \end{aligned}$$

Note that $|Z_n| \leq 2c$ for all $n \geq 1$ and $\sum_{n=1}^\infty Z_n$ converges a.s. By Problem 16.12, we have

$$\sum_{n=1}^\infty E\left[Z_n^2\right] < \infty.$$

Hence we have

$$\sum_{n=1}^\infty \text{Var}(Y_n) < \infty.$$

It follows from Problem 16.6 that $\sum_{n=1}^\infty (Y_n - E[Y_n])$ converges a.s. By the assumption that $\sum_{n=1}^\infty Y_n$ converges a.s., $\sum_{n=1}^\infty E[Y_n]$ converges a.s. This completes the proof.

16.14.

(a) Assume that $\sum_{n=1}^\infty P\{|X_n| > a\}$ does not converge. Then we have

$$\sum_{n=1}^\infty P\{|X_n| > a\} = \infty.$$

Then by the second Borel-Cantelli lemma (see Problem 12.9), we have

$$P(\limsup\{|X_n| > a\}) = 1.$$

However, with

$$\limsup\{|X_n| > a\} = \bigcap_{i\geq 1} \bigcup_{j \geq i} \{|X_j| > a\} \subseteq \bigcup_{j=1}^\infty \left\{ \left| \sum_{k=1}^j X_{n+k} \right| \geq a \right\}, \quad n \geq 1,$$

we have

$$P\left(\bigcup_{j=1}^{\infty}\left\{\left|\sum_{k=1}^{j} X_{n+k}\right| \geq a\right\}\right) = 1.$$

By Problem 9.8, $\sum_{n=1}^{\infty} X_n$ does not converge. This contradicts the assumption. Hence $\sum_{n=1}^{\infty} P\{|X_n| > a\}$ converges.

(b) Let

$$S_n = \sum_{i=1}^{n} X_n I_{\{|X_n|>a\}}, \quad n \geq 1.$$

Then, for $\epsilon > 0$, we have

$$\bigcup_{j=1}^{\infty} \{|S_{n+j} - S_n| \geq \epsilon\} \subseteq \bigcup_{j=1}^{\infty} \{|X_{n+j}| > a\},$$

which gives

$$P\left(\bigcup_{j=1}^{\infty} \{|S_{n+j} - S_n| \geq \epsilon\}\right) \leq P\left(\bigcup_{j=1}^{\infty} \{|X_{n+j}| > a\}\right) \leq \sum_{j=1}^{\infty} P\{|X_{n+j}| > a\}.$$

By part (a) of this proof, we have

$$\lim_{n\to\infty} P\left(\bigcup_{j=1}^{\infty} \{|S_{n+j} - S_n| \geq \epsilon\}\right) = 0.$$

It follows from Problem 9.8 that S_n converges a.s. Therefore the sequence

$$\sum_{n=1}^{\infty} X_n I_{\{|X_n|\leq a\}} = \sum_{n=1}^{\infty} X_n - \sum_{n=1}^{\infty} X_n I_{\{|X_n|>a\}}$$

converges a.s. Then by Problem 16.13, $\sum_{n=1}^{\infty} E[X_n I_{\{|X_n|\leq a\}}]$ converges.

(c) Let

$$Y_n = X_n I_{\{|X_n|\leq a\}} - E[X_n I_{\{|X_n|\leq a\}}], \quad n \geq 1.$$

By part (b) of this proof, we know that $\sum_{n=1}^{\infty} Y_n$ converges a.s. Also note that

$$E[Y_n] = 0$$

and

$$|Y_n| \leq |X_n I_{\{|X_n|\leq a\}}| + E[|X_n I_{\{|X_n|\leq a\}}|] \leq 2a.$$

It follows from Problem 16.12 that

$$\sum_{n=1}^{\infty} \mathrm{Var}(X_n I_{\{|X_n|\leq a\}}) = \sum_{n=1}^{\infty} E\left[Y_n^2\right] < \infty.$$

This completes the proof.

16.15. Let us prove the sufficiency first. Suppose that

$$\sum_{n=1}^{\infty} a_n^2 < \infty.$$

Then we have

$$\sum_{n=1}^{\infty} \text{Var}\,(a_n X_n) = \sum_{n=1}^{\infty} a_n^2 < \infty.$$

It follows from Problem 16.6 that $\sum_{n=1}^{\infty} a_n X_n$ converges a.s.

Now let us prove the necessity. Suppose that $\sum_{n=1}^{\infty} a_n X_n$ converges a.s. Let $a > 0$. Then by Problem 16.14, we know that $\sum_{n=1}^{\infty} P\{|a_n X_n| > a\}$ converges. But $|a_n X_n| = |a_n|$. Hence there exists an integer $N > 0$ such that $|a_n| \le a$ for all $n \ge N$. Therefore, we have

$$\sum_{n=N}^{\infty} a_n^2 = \sum_{n=N}^{\infty} \text{Var}(a_n X_n I_{\{|a_n X_n| \le a\}}).$$

Also by Problem 16.14, we have

$$\sum_{n=1}^{\infty} \text{Var}(a_n X_n I_{\{|a_n X_n| \le a\}}) < \infty,$$

which leads to

$$\sum_{n=N}^{\infty} a_n^2 < \infty.$$

This completes the proof.

16.5 Bibliographic Notes

In this chapter, we presented some results on law of large numbers. The weak law of large numbers is about a sequence of random variables converging in probability, while the strong law of large numbers is about a sequence of random variables converging almost surely. A proof of Theorem 16.3 based on Chebyshev's inequality can be found in Bhattacharya and Waymire (2007, p51).

For more information about the law of large numbers, readers can consult textbooks such as Billingsley (1999), Chung (2000), and Bhattacharya and Waymire (2007).

CHAPTER 17

CHARACTERISTIC FUNCTIONS

Characteristic functions provide an important tool to study the distributional properties of random variables and sums of independent random variables. In particular, characteristic functions can be used to characterize the probability distribution of a random variable and to identify the limit distribution of a sequence of random variables. In this chapter, we present characteristic functions and relevant results.

17.1 Basic Concepts and Facts

Definition 17.1 (Moment Generating Function)**.** Let X be a random variable. Then the moment generating function of X is defined as

$$M_X(t) = E\left(e^{tX}\right), \quad t \in \mathbf{R}. \tag{17.1}$$

The moment generating function of a random vector $\mathbf{X} = (X_1, X_2, \ldots, X_n)^T$ is defined as

$$M_{\mathbf{X}}(\mathbf{t}) = E\left(e^{\mathbf{t}^T\mathbf{X}}\right) = E\left[\exp\left(\sum_{i=1}^{n} t_i X_i\right)\right], \quad \mathbf{t} \in \mathbf{R}^n. \tag{17.2}$$

Measure, Probability, and Mathematical Finance.

Definition 17.2 (Cumulant Generating Function). Let X be a random variable. Then the cumulant generating function of X is defined as

$$C_X(h) = \ln(M_X(h)), \tag{17.3}$$

where $M_X(h)$ is the moment generating function of X.

Definition 17.3 (Characteristic Function of Random Variables). Let X be a random variable. Then the characteristic function of X is defined as

$$\phi_X(t) = E\left(e^{itX}\right) = E[\cos(tX)] + iE[\sin(tX)], \quad t \in \mathbf{R}, \tag{17.4}$$

where $i = \sqrt{-1}$ is the imaginary number.

The characteristic function of a random vector $\mathbf{X} = (X_1, X_2, \ldots, X_n)^T$ is defined as

$$\phi_{\mathbf{X}}(\mathbf{t}) = E\left(e^{i\mathbf{t}^T\mathbf{X}}\right) = E\left[\exp\left(i\sum_{i=1}^{n} t_i X_i\right)\right], \quad \mathbf{t} \in \mathbf{R}^n. \tag{17.5}$$

Definition 17.4 (Characteristic Function of Probability Measures). Let μ be a probability measure on $(\mathbf{R}, \mathcal{B})$. Then the characteristic function of μ is defined as

$$\hat{\mu}(t) = \int_{\mathbf{R}} e^{itx}\mu(\mathrm{d}x), \quad t \in \mathbf{R}. \tag{17.6}$$

Definition 17.5 (Characteristic Function of Distribution Functions). Let F be a cumulative distribution function on $(\mathbf{R}, \mathcal{B})$. Then the characteristic function of F is defined as $\hat{\mu}_F(t)$, where μ_F is the Lebesgue-Stieltjes measure corresponding to F.

Theorem 17.1 (Fourier Uniqueness Theorem). *Let X and Y be two random variables. Then $\phi_X(t) = \phi_Y(t)$ for all $t \in \mathbf{R}$ if and only if $\mathcal{L}_X = \mathcal{L}_Y$, that is, if and only if X and Y have the same distribution.*

Theorem 17.2 (Kac's Theorem). *Let X_1, X_2, ..., X_n be $\mathbf{R}^d$-valued random variables. Then X_1, X_2, ..., X_n are independent if and only if for all $\mathbf{u}_1, \mathbf{u}_2, \ldots, \mathbf{u}_n \in \mathbf{R}^d$, we have*

$$E\left[\exp\left(i\sum_{j=1}^{n} \mathbf{u}_j^T X_j\right)\right] = \prod_{j=1}^{n} \phi_{X_j}(\mathbf{u}_j).$$

17.2 Problems

17.1. Let X and Y be independent random variables. Show that

$$M_{X+Y}(t) = M_X(t)M_Y(t), \quad t \in \mathbf{R},$$

where $M_{X+Y}(t)$, $M_X(t)$, and $M_Y(t)$ are the moment generating functions of $X+Y$, X, and Y, respectively.

17.2. Let X and Y be independent random variables. Show that

$$\phi_{X+Y}(t) = \phi_X(t)\phi_Y(t), \quad t \in \mathbf{R},$$

where $\phi_{X+Y}(t)$, $\phi_X(t)$, and $\phi_Y(t)$ are the characteristic functions of $X+Y$, X, and Y, respectively.

17.3. Let $\{F_t\}_{a<t<b}$ be a collection of random variables on the probability space $(\Omega, \mathscr{F}, P)$ with finite expectations. Suppose that for each $\omega \in \Omega$ and each $t \in (a, b)$, the derivative

$$F'_t(\omega) = \frac{\partial}{\partial t} F_t(\omega)$$

exists. Show that

(a) For every $t \in (a, b)$, F'_t is a random variable.

(b) For every $t, t+h \in (a, b)$, we have

$$\left| \frac{F_{t+h} - F_t}{h} \right| \leq Y.$$

(c) If there exists a random variable Y on $(\Omega, \mathscr{F}, P)$ with a finite expectation such that for all $t \in (a, b)$, we have

$$|F'_t| \leq Y,$$

then for all $t \in (a, b)$, the function $f(t) = E(F_t)$ is differentiable and has a finite derivative.

17.4. Let X be a random variable with $M_X(t) < \infty$ for $t \in (-t_0, t_0)$, for some $t_0 > 0$. Show that

(a) $E(|X^n|) < \infty$ for all $n \geq 0$.

(b) For any $t \in (-t_0, t_0)$, we have

$$M_X(t) = \sum_{n=0}^{\infty} \frac{t^n E(X^n)}{n!}.$$

(c) For any $j \geq 0$, the jth derivative at $t = 0$ is

$$M_X^{(j)}(0) = E(X^j).$$

17.5. Let X be a random variable. Suppose that $M_X(t) < \infty$ for all $t \in (r - t_0, r + t_0)$, where $t_0 > 0$, $r \in \mathbf{R}$, and $M_X(t)$ is the moment generating function of X. Show that

(a) $E\left[\left|X^n e^{rX}\right|\right] < \infty$ for all $n \geq 0$;

(b) For any $j \geq 0$, we have

$$M_X^{(j)}(r) = E\left[X^j e^{rX}\right].$$

17.6. Let X be a nondegenerate random variable. Suppose that $M_X(t) < \infty$ for all $t \in \mathbf{R}$, where $M_X(t)$ is the moment generating function of X. Show that

(a) $M_X(t)$ is a convex function on $\mathbf{R}$.

(b) $C_X(t)$ is a convex function, where $C_X(t)$ is the cumulant generating function.

17.7. Let $\{X_n\}_{n\geq 1}$ be a sequence of independent and identically distributed random variables with common mean m and with $M_{X_1}(t) < \infty$ for $t \in (-a, b)$, where $a > 0$ and $b > 0$. Show that for any $\epsilon > 0$, we have

(a)

$$\inf_{0<t<b} \frac{M_{X_1}(t)}{\exp(t(m+\epsilon))} < 1.$$

(b)

$$P\left(\frac{X_1 + X_2 + \cdots + X_n}{n} \geq m + \epsilon\right) \leq \rho^n,$$

where

$$\rho = \inf_{0<t<b} \frac{M_{X_1}(t)}{\exp(t(m+\epsilon))}.$$

17.8. Let X be a random variable on a probability space $(\Omega, \mathscr{F}, P)$. Let $|\phi_X(t)|$ be the norm given by

$$|\phi_X(t)| = \left[(E[\cos tX])^2 + (E[\sin tX])^2\right]^{\frac{1}{2}}.$$

Show that $|\phi_X(t_0)| = 1$ for some $t_0 \neq 0$ if and only if there exist $a \in \mathbf{R}$ and $h \neq 0$ such that

$$P\{X \in \{a + jh : j \in \mathbf{Z}\}\} = 1.$$

17.3 Hints

17.1. Use the result of Problem 12.7 and Theorem 13.2.

17.2. Follow the definition of characteristic functions (Definition 17.3) and use trigonometric identities to expand $\phi_{X+Y}(t)$ as

$$\begin{aligned}
\phi_{X+Y}(t) &= E[\cos(tX + tY)] + iE[\sin(tX + tY)] \\
&= E[\cos(tX)\cos(tY) - \sin(tX)\sin(tY)] \\
&\quad + iE[\sin(tX)\cos(tY) + \cos(tX)\sin(tY)].
\end{aligned}$$

Then use the result of Problem 12.7 and Theorem 13.2.

17.3. To prove part (a), follow the definition of random variables (Definition 11.6) and use the result of Problem 5.9. To prove part (b), use the mean value theorem (Theorem 15.1). Use part (b) and the dominated convergence theorem (Theorem 6.4) to prove part (c).

17.4. Part part (a) can be proved by using the inequality $|x| < e^{|x|}$. To prove part (b), consider the following sequence

$$S_n = \sum_{k=0}^{n} \frac{t^k X^k}{k!}, \quad n \geq 0.$$

Then use the dominated convergence theorem (Theorem 6.4) to show that

$$E\left(\lim_{n\to\infty} S_n\right) = \lim_{n\to\infty} E\left(S_n\right).$$

Part (c) is implied from part (b).

17.5. Consider the random variable X under the probability measure $\tilde{P}$ defined as

$$\tilde{P}(\mathrm{d}x) = \frac{e^{rx}}{M_X(r)} P(\mathrm{d}x),$$

and use the result of Problem 17.4.

17.6. Use the results of Problems 17.5 and 15.4 to prove part (a). Use Schwarz's inequality (Theorem 8.3) to prove part (b).

17.7. Part (a) can be proved by considering the function

$$f(t) = E\left[e^{t(X-m-\epsilon)}\right]$$

and its derivative at 0. Part (b) can be proved by consider the integration of $X_1 + X_2 + \cdots + X_n$ over the set $\{(X_1 + X_2 + \cdots + X_n)/n \geq m + \epsilon\}$.

17.8. To prove the "only if" part, consider the characteristic function of $X - a_0/t_0$, where a_0 is a real number such that $\phi_X(t_0) = e^{ia_0}$.

17.4 Solutions

17.1. Let $t \in \mathbf{R}$. By Theorem 5.2, the functions $f(x) = e^{tx}$ and $g(y) = e^{ty}$ are Borel measurable. Then by Problem 12.7 and the assumption that X and Y are independent, e^{tX} and e^{tY} are independent. By Theorem 13.2, we have

$$\begin{aligned} M_{X+Y}(t) &= E\left[e^{t(X+Y)}\right] \\ &= E\left[e^{tX} e^{tY)}\right] \\ &= E\left[e^{tX}\right] E\left[e^{tY)}\right] \\ &= M_X(t) M_Y(t). \end{aligned}$$

This completes the proof.

17.2. By Definition 17.3 and the trigonometric identities, we have

$$\begin{aligned}\phi_{X+Y}(t) &= E[\cos(tX+tY)] + iE[\sin(tX+tY)] \\ &= E[\cos(tX)\cos(tY) - \sin(tX)\sin(tY)] \\ &\quad + iE[\sin(tX)\cos(tY) + \cos(tX)\sin(tY)].\end{aligned}$$

By the assumption that X and Y are independent, Problem 12.7, and Theorem 13.2, the above equation can be written as

$$\begin{aligned}\phi_{X+Y}(t) &= E[\cos(tX)]E[\cos(tY)] - E[\sin(tX)]E[\sin(tY)] \\ &\quad + iE[\sin(tX)]E[\cos(tY)] + iE[\cos(tX)]E[\sin(tY)] \\ &= (E[\cos(tX)] + iE[\sin(tX)])(E[\cos(tY)] + iE[\sin(tY)]) \\ &= \phi_X(t)\phi_Y(t).\end{aligned}$$

This completes the proof.

17.3.

(a) Let $t \in (a,b)$ and $t_n = t + 1/n$ for $n = 1, 2, \ldots$. By the assumption, we have

$$\lim_{n\to\infty} n(F_{t_n} - F_t) = F_t'.$$

It follows from Problem 5.9 that F_t' is also $\mathscr{F}$-measurable. Hence F_t' is a random variable.

(b) Let $t, t+h \in (a,b)$. Then by Theorem 15.1, we have

$$\frac{F_{t+h} - F_t}{h} = F_c'$$

for some c between t and $t+h$. Thus

$$\left|\frac{F_{t+h} - F_t}{h}\right| = |F_c'| \leq Y.$$

(c) Let $t \in (a,b)$. Let $\{t_n\}_{n\geq 1}$ be an arbitrary sequence of real numbers in (a,b) such that $t_n \to t$ as $n \to \infty$. Then by part (b), for every $n \geq 1$, we have

$$\left|\frac{F_{t_n} - F_t}{t_n - t}\right| \leq Y.$$

It follows from Theorem 6.4 that

$$f_t' = \lim_{n\to\infty} E\left[\frac{F_{t_n} - F_t}{t_n - t}\right] = E\left[\lim_{n\to\infty} \frac{F_{t_n} - F_t}{t_n - t}\right] = E[F_t'].$$

Since $|E[F_t']| \leq E[|F_t'|] \leq E[Y] < \infty$, it follows that f_t' is finite.

This completes the proof.

17.4.

(a) If $n = 0$, the result holds obviously. Let $n > 0$ and $t \in (-t_0, t_0)$. Since $|x| < e^{|x|}$ for all $x \in \mathbf{R}$, we have

$$\frac{|tX|}{n} < \exp\left(\frac{|tX|}{n}\right),$$

which gives

$$|X|^n < \frac{n^n e^{|tX|}}{t^n}.$$

Then we have

$$E\left[|X|^n\right] < E\left[\frac{n^n e^{|tX|}}{t^n}\right] = \frac{n^n (M_X(t) + M_X(-t))}{t^n} < \infty.$$

(b) Let $t \in (-t_0, t_0)$. Define

$$S_n = \sum_{k=0}^{n} \frac{t^k X^k}{k!}, \quad n \geq 0.$$

Then we have

$$|S_n| \leq \sum_{k=0}^{n} \frac{|tX|^k}{k!} < e^{|tX|}.$$

Note that

$$E\left[e^{|tX|}\right] \leq E\left[e^{tX}\right] + E\left[e^{-tX}\right] = M_X(t) + M_X(-t) < \infty$$

and $S_n \to e^{tX}$. By Theorem 6.4, we have

$$\lim_{n\to\infty} E(S_n) = E\left[\lim_{n\to\infty} S_n\right],$$

which gives

$$M_X(t) = \sum_{n=0}^{\infty} \frac{t^n E(X^n)}{n!}.$$

(c) By the second item, $M_X(t)$ is a polynomial of t. Hence the result is obtained by taking derivatives of $M_X(t)$.

This completes the proof.

17.5. Let $\tilde{P}$ be a probability measure defined as

$$\tilde{P}(\mathrm{d}x) = \frac{e^{rx}}{M_X(r)} P(\mathrm{d}x),$$

where P is the underlying probability for X. Let $\tilde{M}_X(t)$ be the moment generating function of X under the probability measure $\tilde{P}$. Then we have

$$\tilde{M}_X(t) = \int_\Omega e^{tX} \mathrm{d}\tilde{P} = \int_\Omega e^{tX} \frac{e^{rX}}{M_X(r)} \mathrm{d}P = \frac{M_X(r+t)}{M_X(r)}.$$

(a) By the assumption that $M_X(t) < \infty$ in the neighborhood of r, we know that $\tilde{M}_X(t) < \infty$ in the neighborhood of r. By Problem 17.4, we have

$$E_{\tilde{P}}[|X^n|] < \infty, \quad n \geq 0,$$

where $E_{\tilde{P}}$ is the expectation under the measure $\tilde{P}$. But

$$E_{\tilde{P}}[|X^n|] = \int_\Omega |X^n| \mathrm{d}\tilde{P} = \int_\Omega |X^n| \frac{e^{rX}}{M_x(r)} \mathrm{d}P = \frac{E[|X^n e^{rX}|]}{M_X(r)},$$

which gives

$$E[|X^n e^{rX}|] < \infty, \quad n \geq 0.$$

(b) By Problem 17.4, for $j \geq 0$, we have

$$\tilde{M}_X^{(j)}(t) = E_{\tilde{P}}[X^j].$$

Hence

$$\frac{M_X^{(j)}(r+t)}{M_X(r)} = \int_\Omega X^j \mathrm{d}\tilde{P} = \frac{E[X^j e^{rX}]}{M_X(r)},$$

which leads to

$$M_X^{(j)}(r+t) = E[X^j e^{rX}].$$

This completes the proof.

17.6.

(a) By Problem 17.5, we have

$$M_X^{(2)}(t) = E[X^2 e^{tX}] > 0, \quad t \in \mathbf{R}.$$

It follows from Problem 15.4 that $M_X(t)$ is a convex function on $\mathbf{R}$.

(b) By Problem 17.5, we have

$$\begin{aligned} C_X^{(2)}(t) &= \frac{M_X^{(2)}(t) M_X(t) - (M_X^{(1)}(t))^2}{M_X^2(t)} \\ &= \frac{E[X^2 e^{tX}] E[e^{tX}] - (E[X e^{tX}])^2}{M_X^2(t)}. \end{aligned}$$

By Theorem 8.3, we have

$$\left(E[Xe^{tX}]\right)^2 = \left(E[Xe^{tX/2}e^{tX/2}]\right)^2 \leq E[X^2e^{tX}]E[e^{tX}].$$

Hence we have $C_X^{(2)}(t) \geq 0$. By Problem 15.4, $C_X(t)$ is a convex function on $\mathbf{R}$.

This completes the proof.

17.7.

Let $f(t)$ be defined as

$$f(t) = E\left[e^{t(X-m-\epsilon)}\right], \quad t \in (-a, b).$$

Then we have $f(0) = 1$ and, by Problem 17.4, we have

$$f'(0) = E(X - m - \epsilon) = -\epsilon < 0.$$

Hence f is decreasing around 0. Therefore we must have $f(t) < 1$ for all positive t that are sufficiently close to 0.

(a) Let $t \in (0, b)$ and $S_n = X_1 + X_2 + \cdots + X_n$. Note that the X_n are independent and identically distributed. By Problem 17.1, we have

$$E\left[e^{tS_n}\right] = M_{X_1+X_2+\cdots+X_n}(t) = M_{X_1}(t)^n$$

But

$$\begin{aligned} E\left[e^{tS_n}\right] &= \int_\Omega e^{tS_n} \mathrm{d}P \\ &\geq \int_{\{S_n \geq n(m+\epsilon)\}} e^{tS_n} \mathrm{d}P \\ &\geq \int_{\{S_n \geq n(m+\epsilon)\}} e^{nt(m+\epsilon)} \mathrm{d}P \\ &= P\left(\frac{X_1 + X_2 + \cdots + X_n}{n} \geq m + \epsilon\right) e^{nt(m+\epsilon)}. \end{aligned}$$

Hence we have

$$P\left(\frac{X_1 + X_2 + \cdots + X_n}{n} \geq m + \epsilon\right) \leq e^{-nt(m+\epsilon)} M_{X_1}(t)^n.$$

Since t is an arbitrary number in $(0, b)$, we have

$$P\left(\frac{X_1 + X_2 + \cdots + X_n}{n} \geq m + \epsilon\right) \leq \inf_{0<t<b} e^{-nt(m+\epsilon)} M_{X_1}(t)^n = \rho^n.$$

This completes the proof.

17.8. Let us first prove the "if" part. Suppose that there exist $a \in \mathbf{R}$ and $h \neq 0$ such that

$$P\{X \in \{a + jh : j \in \mathbf{Z}\}\} = 1.$$

Let $P\{X = a + jh\} = p_j$ for $j \in \mathbf{Z}$. Noting that $\cos(x - y) = \cos x \cos y + \sin x \sin y$, we have

$$\begin{aligned} |\phi_X(t)|^2 &= \left(\sum_{j\in\mathbf{Z}} p_j \cos(t(a + jh))\right)^2 + \left(\sum_{j\in\mathbf{Z}} p_j \sin(t(a + jh))\right)^2 \\ &= \sum_{j\in\mathbf{Z}} p_j^2 + \sum_{i\neq j} p_i p_j \cos(th(j - i)). \end{aligned}$$

Letting $t_0 = 2\pi/h$ gives $|\phi_X(t_0)| = 1$.

Now let us prove the "only if" part. Suppose that there exists $t_0 =\neq 0$ such that $|\phi_X(t_0)| = 1$. Then there exists $a_0 \in \mathbf{R}$ such that

$$\phi_X(t_0) = e^{ia_0}.$$

Let $a = a_0/t_0$. Then the characteristic function of $X - a$ is $e^{-ia}\phi_X(t)$. Hence we have $E[\exp(it_0(X - a)] = 1$. Equating the real parts gives

$$E[\cos(t_0(X - a))] = 1.$$

Since $|\cos\theta| \leq 1$ and $\cos\theta = 1$ if and only if $\theta = 2\pi j$ for some $j \in \mathbf{Z}$, the above equation implies that

$$P\{t_0(X - a) \in \{2\pi j : j \in \mathbf{Z}\}\} = 1.$$

The result holds with $h = 2\pi/t_0$. This completes the proof.

17.5 Bibliographic Notes

Most material in this chapter was taken from Rosenthal (2006, Chapters 9 and 11). For more information about characteristic function, readers are referred to Lukacs (1970), Billingsley (1995), Chung (2000), Athreya and Lahiri (2006), Klenke (2006), Rao and Swift (2006), Koralov and Sinai (2007), and Bhattacharya and Waymire (2007). In particular, Athreya and Lahiri (2006) discussed the criteria for a function to be a characteristic function.

A proof of Theorem 17.1 can be found in Rosenthal (2006, Corollary 11.1.7). Theorem 17.2 was adopted from Applebaum (2009, Theorem 1.1.16). For a proof of Theorem 17.2, readers are referred to Pascucci (2011, Corollary A.91).

CHAPTER 18

DISCRETE DISTRIBUTIONS

Discrete distributions are used to describe random variables that can only take countably many different values. In this chapter, we present several discrete probability distributions and their properties.

18.1 Basic Concepts and Facts

Definition 18.1 (Bernoulli Distribution). Let $\Omega = \{0, 1\}$ and $p \in (0, 1)$. A random variable X on $(\Omega, 2^{\Omega})$ is said to have a Bernoulli distribution with parameter p, written as $X \sim Be(p)$, if and only if

$$P\{X = k\} = p^k (1-p)^{1-k}, \quad k = 0, 1.$$

Definition 18.2 (Binomial Distribution). Let $\Omega = \{0, 1, 2, \ldots, n\}$, where n is a positive integer, and $p \in (0, 1)$. A random variable X on $(\Omega, 2^{\Omega})$ is said to have a binomial distribution with parameters (n, p), written as $X \sim B(n, p)$, if and only if

$$P\{X = k\} = \binom{n}{k} p^k q^{n-k}, \quad k = 0, 1, \ldots, n,$$

where $q = 1 - p$ and

$$\binom{n}{k} = \frac{n!}{k!(n-k)!} = \frac{n(n-1)\cdots(n-k+1)}{k(k-1)\cdots 1}.$$

When $n = 1$, the binomial distribution is the same as the Bernoulli distribution.

Definition 18.3 (Poisson Distribution). Let $\Omega = \{0, 1, 2, \ldots\}$ and $\theta > 0$. A random variable X on $(\Omega, 2^\Omega)$ is said to have a Poisson distribution with parameter θ, written as $X \sim P(\theta)$, if and only if

$$P\{X = k\} = \frac{e^{-\theta}\theta^k}{k!}, \quad k = 0, 1, 2, \ldots.$$

Definition 18.4 (Geometric Distribution). Let $\Omega = \{0, 1, 2, \ldots\}$ and $p \in (0, 1)$. A random variable X on $(\Omega, 2^\Omega)$ is said to have a geometric distribution with parameters p, written as $X \sim G(p)$, if and only if

$$P\{X = k\} = (1-p)p^k, \quad k = 0, 1, 2, \ldots.$$

Definition 18.5 (Negative Binomial Distribution). Let $\Omega = \{0, 1, 2, \ldots\}$, $r > 0$, and $p \in (0, 1)$. A random variable X on $(\Omega, 2^\Omega)$ is said to have a negative binomial distribution with parameters (r, p), written as $X \sim NB(r, p)$, if and only if

$$P\{X = k\} = \binom{k+r-1}{k}(1-p)^r p^k, \quad k = 0, 1, 2, \ldots.$$

When $r = 1$, the negative binomial distribution is the same as the geometric distribution.

Definition 18.6 (Hypergeometric Distribution). Let N, m, and n be integers such that $N \geq 1, 0 \leq m \leq N$, and $1 \leq n \leq N$. Let $\Omega = \{k \in \mathbf{Z} : \max(0, n+m-N) \leq k \leq \min(m, n)\}$. A random variable X on $(\Omega, 2^\Omega)$ is said to have a hypergeometric distribution with parameters (N, m, n), written as $X \sim HG(N, m, n)$, if and only if

$$P\{X = k\} = \frac{\binom{m}{k}\binom{N-m}{n-k}}{\binom{N}{n}}.$$

18.2 Problems

18.1. Let $\Omega = \{0, 1, 2, \ldots, n\}$, where n is a positive integer. Let X be a binomial random variable with parameters (n, p) on $(\Omega, 2^\Omega)$. Show that

$$\phi_X(t) = \left(1 - p + pe^{it}\right)^n$$

and

$$M_X(t) = \left(1 - p + pe^t\right)^n,$$

where $\phi_X(t)$ and $M_X(t)$ are the characteristic function and the moment generating function of X, respectively.

18.2. Let n be a positive integer and $p_1, p_2, \ldots, p_n \in (0, 1)$. Let $X_1, X_2, \ldots, X_n$ be n independent random variables such that

$$X_k \sim Be(p_k), \quad k = 1, 2, \ldots, n.$$

Let M be a random variable defined as

$$M = \max\{X_1, X_2, \ldots, X_n\}.$$

Show that

$$M \sim Be(\lambda),$$

where

$$\lambda = 1 - \prod_{k=1}^{n} (1 - p_k).$$

18.3. Let X be a binomial random variable with parameters (n, p). Show that

$$E(X) = np$$

and

$$\text{Var}(X) = np(1 - p).$$

18.4. Let n_1 and n_2 be positive integers and $p \in (0, 1)$. Let X_1 and X_2 be independent random variables having binomial distributions with parameters (n_1, p) and (n_2, p), respectively. Find

(a) The distribution of $X_1 + X_2$.

(b) The conditional probability distribution of X_1 given $X_1 + X_2 = k$, where k is an integer in $[0, n_1 + n_2]$.

18.5. Let $X \sim B(n, p)$; that is, X is binomially distributed with parameter (n, p). Show that for any integer $m \in [1, n]$, we have

$$\sum_{k=m}^{n} \binom{n}{k} p^k (1 - p)^{n-k} = \frac{n!}{(m - 1)!(n - m)!} \int_0^p x^{m-1} (1 - x)^{n-m} \mathrm{d}x.$$

18.6. Let $\lambda > 0$ be a real number. For each integer $n > \lambda$, let $X_n \sim B(n, \lambda/n)$. Show that for any fixed integer m, we have

$$\lim_{n \to \infty} P\{X_n = m\} = e^{-\lambda} \frac{\lambda^m}{m!}.$$

18.7. Let X be a Poisson random variable with parameter θ, that is, $X \sim P(\theta)$. Show that

$$\phi_X(t) = \exp\left[\theta\left(e^{it} - 1\right)\right],$$

and

$$M_X(t) = \exp\left[\theta\left(e^{t} - 1\right)\right],$$

where $\phi_X(t)$ and $M_X(t)$ are the characteristic function and the moment generating function of X, respectively.

18.8. Let X be a Poisson random variable with parameter θ, that is, $X \sim P(\theta)$. Show that

$$E(X) = \theta$$

and

$$\text{Var}(X) = \theta.$$

18.9. Let $\lambda_1 > 0$ and $\lambda_2 > 0$ be real numbers. Let X_1 and X_2 be independent random variables having Poisson distributions with parameters λ_1 and λ_2, respectively. Find

(a) The distribution of $X_1 + X_2$.

(b) The conditional probability distribution of X_1 given $X_1 + X_2 = k$, where $k \geq 0$ is an integer.

18.10. Let n be a positive integer and $p_1, p_2, \ldots, p_n \in (0, 1)$. Let $X_1, X_2, \ldots, X_n$ be n independent random variables such that

$$X_k \sim G(p_k), \quad k = 1, 2, \ldots, n,$$

where $G(p_k)$ denotes the geometric distribution with parameter p_k. Let Y be a random variable defined as

$$Y = \min\{X_1, X_2, \ldots, X_n\}.$$

Show that $Y \sim G(p)$, where

$$p = \prod_{k=1}^{n} p_k.$$

18.3 Hints

18.1. Note that the characteristic function is defined as $E\left[e^{itX}\right]$, and use the binomial expansion

$$(p+q)^n = \sum_{k=0}^{n} \binom{n}{k} p^k q^{n-k}.$$

18.2. Follow the definition of Bernoulli distribution (Definition 18.1) and use the result of Problem 12.5.

18.3. Use the moment generating function and the result of Problem 17.4.

18.4. To prove part (a), use the results of Problems 18.1 and 17.2, and the Fourier uniqueness theorem (Theorem 17.1). To prove part (b), use the result of part (a) and follow the definition of conditional probabilities (Definition 14.6).

18.5. Use integration by parts and the method of mathematical induction.

18.6. Consider $\ln P\{X_n = m\}$ and use l'Hôpital's rules.

18.7. Use the Taylor series for the exponential function e^x:

$$e^x = \sum_{k=0}^{\infty} \frac{x^k}{k!}.$$

18.8. Use the moment generating function of X and the result of Problem 17.4.

18.9. To prove part (a), use the results of Problems 18.7 and 17.2, and the Fourier uniqueness theorem (Theorem 17.1). To prove part (b), use the result of part (a) and follow the definition of conditional probabilities (Definition 14.6).

18.10. Note that

$$P\{Y = m\} = P\{Y > m - 1\} - P\{Y > m\},$$

and use the definition of geometric distribution (Definition 18.4) and the result of Problem 12.5.

18.4 Solutions

18.1. By definition, we have

$$\begin{aligned}
\phi_X(t) = E\left[e^{itX}\right] &= \sum_{k=0}^{n} e^{itk} P\{X = k\} \\
&= \sum_{k=0}^{n} e^{itk} \binom{n}{k} p^k q^{n-k} \\
&= \sum_{k=0}^{n} \binom{n}{k} \left(pe^{it}\right)^k q^{n-k} \\
&= \left(pe^{it} + q\right)^n.
\end{aligned}$$

Since $q = 1 - p$, the result follows from the above equation. Similarly, we can show that

$$M_X(t) = \left(1 - p + pe^t\right)^n.$$

18.2. Since $X_1, X_2, \ldots, X_n$ are independent, by Problem 12.5 we have

$$\begin{aligned} P\{M = 0\} &= P\{X_1 = 0, X_2 = 0, \ldots, X_n = 0\} \\ &= \prod_{k=1}^{n} P\{X_k = 0\} \\ &= \prod_{k=1}^{n} (1 - p_k) \\ &= 1 - \lambda. \end{aligned}$$

Since $X_1, X_2, \ldots, X_n$ are all Bernoulli random variables, we know that $M = 0$ or 1. Hence

$$P\{M = 1\} = 1 - P\{M = 0\} = \lambda.$$

Therefore, $M \sim Be(\lambda)$. This completes the proof.

18.3. By Problem 17.4, we have

$$E(X) = M_X^{(1)}(0)$$

and

$$\text{Var}(X) = E\left[X^2\right] - (E[X])^2 = M_X^{(2)}(0) - (E[X])^2.$$

By Problem 18.1, we have $M_X(t) = (1 - p + pe^t)^n$. Hence

$$M_X^{(1)}(t) = npe^t \left(1 - p + pe^t\right)^{n-1}$$

and

$$M_X^{(2)}(t) = npe^t \left(1 - p + pe^t\right)^{n-1} + n(n-1)p^2 e^{2t} \left(1 - p + pe^t\right)^{n-2}.$$

Therefore, we have

$$E(x) = np$$

and

$$\text{Var}(X) = np(1 - p).$$

This completes the proof.

18.4.

(a) Since X_1 and X_2 are independent, it follows from Problems 18.1 and 17.2 that

$$\phi_{X+Y}(t) = \left(1 - p + pe^{it}\right)^{n_1 + n_2}.$$

By the Fourier uniqueness theorem (Theorem 17.1), we know that $X_1 + X_2 \sim B(n_1 + n_2, p)$, that is, $X_1 + X_2$ has a binomial distribution with parameter $(n_1 + n_2, p)$.

(b) By Definition 14.6, we have

$$\begin{aligned} P\{X_1 = x | X_1 + X_2 = k\} &= \frac{P(\{X_1 = x\} \cap \{X_1 + X_2 = k\})}{P\{X_1 + X_2 = k\}} \\ &= \frac{P(\{X_1 = x\} \cap \{X_2 = k - x\})}{P\{X_1 + X_2 = k\}} \\ &= \frac{P\{X_1 = x\} P\{X_2 = k - x\}}{P\{X_1 + X_2 = k\}}. \end{aligned}$$

By part (a) and the assumption, we have

$$P\{X_1 = x\} = \binom{n_1}{x} p^x (1-p)^{n_1 - x},$$

$$P\{X_2 = k - x\} = \binom{n_2}{k - x} p^{k-x} (1-p)^{n_2 - k + x},$$

and

$$P\{X_1 + X_2 = k\} = \binom{n_1 + n_2}{k} p^k (1-p)^{n_1 + n_2 - k}.$$

Hence we have

$$\begin{aligned} & P\{X_1 = x | X_1 + X_2 = k\} \\ = & \frac{\binom{n_1}{x} p^x (1-p)^{n_1 - x} \binom{n_2}{k - x} p^{k-x} (1-p)^{n_2 - k + x}}{\binom{n_1 + n_2}{k} p^k (1-p)^{n_1 + n_2 - k}} \\ = & \frac{\binom{n_1}{x} \binom{n_2}{k - x}}{\binom{n_1 + n_2}{k}}, \end{aligned}$$

which is a hypergeometric distribution with parameter $(n_1 + n_2, n_1, k)$.

This completes the solution.

18.5. We use the method of mathematical induction on m. When $m = 1$, we have

$$\sum_{k=1}^{n} \binom{n}{k} p^k (1-p)^{n-k} = 1 - (1-p)^n$$

and

$$\frac{n!}{(n-1)!}\int_0^p (1-x)^{n-1}\mathrm{d}x = -(1-x)^n|_0^p = 1-(1-p)^n.$$

Hence the result holds for $m = 1$.

Suppose that the result holds for $m = m_0$ for some integer $m_0 \in [1, n]$:

$$\sum_{k=m_0}^{n}\binom{n}{k}p^k(1-p)^{n-k} = \frac{n!}{(m_0-1)!(n-m_0)!}\int_0^p x^{m_0-1}(1-x)^{n-m_0}\mathrm{d}x. \tag{18.1}$$

We consider the case when $m = m_0 + 1$. In this case, the integration by parts gives

$$\begin{aligned}
&\frac{n!}{m_0!(n-m_0-1)!}\int_0^p x^{m_0}(1-x)^{n-m_0-1}\mathrm{d}x \\
=\ &\frac{n!}{m_0!(n-m_0-1)!}\left[\left.\frac{-x^{m_0}(1-x)^{n-m_0}}{n-m_0}\right|_0^p\right. \\
&\left.+\frac{m_0}{n-m_0}\int_0^p x^{m_0-1}(1-x)^{n-m_0}\mathrm{d}x\right] \\
=\ &\frac{n!}{(m_0-1)!(n-m_0)!}\int_0^p x^{m_0-1}(1-x)^{n-m_0}\mathrm{d}x - \\
&\binom{n}{m_0}p^{m_0}(1-p)^{n-m_0}.
\end{aligned} \tag{18.2}$$

Combining Equations (18.1) and (18.2) gives

$$\frac{n!}{m_0!(n-m_0-1)!}\int_0^p x^{m_0}(1-x)^{n-m_0-1}\mathrm{d}x = \sum_{k=m_0+1}^{n}\binom{n}{k}p^k(1-p)^{n-k},$$

which shows that the result also holds from $m = m_0 + 1$. Therefore, the result holds for all integers in $[1, n]$. This completes the proof.

18.6. Since $X_n \sim B(n, \lambda/n)$, we have

$$P\{X_n = m\} = \binom{n}{m}\left(\frac{\lambda}{n}\right)^m\left(1-\frac{\lambda}{n}\right)^{n-m},$$

which gives

$$\ln P\{X_n = m\} = \ln\left(\frac{\lambda^m}{m!}\right) + (n-m)\ln\left(1-\frac{\lambda}{n}\right) + \sum_{j=0}^{m-1}\ln\left(1-\frac{j}{n}\right).$$

Since

$$\lim_{n\to\infty}\sum_{j=0}^{m-1}\ln\left(1-\frac{j}{n}\right) = 0$$

and by l'Hôpital's rules,

$$\begin{aligned}\lim_{n\to\infty}(n-m)\ln\left(1-\frac{\lambda}{n}\right) &= \lim_{x\to\infty}\frac{\ln\left(1-\frac{\lambda}{x}\right)}{\frac{1}{(x-m)}}\\ &= -\lambda\lim_{x\to\infty}\frac{(x-m)^2}{x^2-\lambda x}\\ &= -\lambda,\end{aligned}$$

we get

$$\lim_{n\to\infty}\ln P\{X_n=m\}=\ln\left(\frac{\lambda^m}{m!}\right)-\lambda.$$

Hence

$$\lim_{n\to\infty}P\{X_n=m\}=e^{-\lambda}\frac{\lambda^m}{m!}.$$

This completes the proof.

18.7. By definition, we have

$$\begin{aligned}\phi_X(t)=E\left[e^{itX}\right] &= \sum_{k=0}^{\infty}e^{itk}P\{X=k\}\\ &= \sum_{k=0}^{\infty}e^{itk}\frac{e^{-\theta}\theta^k}{k!}\\ &= \sum_{k=0}^{\infty}e^{-\theta}\frac{\left(\theta e^{it}\right)^k}{k!}\\ &= \exp\left[\theta\left(e^{it}-1\right)\right].\end{aligned}$$

Similary, we can show that

$$M_X(t)=\exp\left[\theta\left(e^t-1\right)\right],$$

This completes the proof.

18.8. To calculate the mean and variance of X, we use the moment generating function of the Poisson distribution from Problem 18.7:

$$M_X(t)=\exp\left[\theta\left(e^t-1\right)\right].$$

The first derivative and the second derivative of the moment generating function are

$$M_X^{(1)}(t)=\theta\exp\left[\theta e^t-\theta+t\right]$$

and

$$M_X^{(2)}(t)=\theta\exp\left[\theta e^t-\theta+t\right]\left(\theta e^t+1\right),$$

respectively. Hence we have

$$E(X) = M_X^{(1)}(0) = \theta$$

and

$$\mathrm{Var}(X) = E\left[X^2\right] - (E[X])^2 = M_X^{(2)}(0) - (E[X])^2 = \theta.$$

This completes the proof.

18.9.

(a) Since X_1 and X_2 are independent, it follows from Problems 18.7 and 17.2 that

$$\phi_{X+Y}(t) = \exp\left((\lambda_1 + \lambda_2)(e^{it} - 1)\right).$$

By the Fourier uniqueness theorem (Theorem 17.1), we know that $X_1 + X_2 \sim P(\lambda_1 + \lambda_2)$, that is, $X_1 + X_2$ has a Poisson distribution with parameter $\lambda_1 + \lambda_2$.

(b) By Definition 14.6, we have

$$\begin{aligned} P\{X_1 = x | X_1 + X_2 = k\} &= \frac{P(\{X_1 = x\} \cap \{X_1 + X_2 = k\})}{P\{X_1 + X_2 = k\}} \\ &= \frac{P(\{X_1 = x\} \cap \{X_2 = k - x\})}{P\{X_1 + X_2 = k\}} \\ &= \frac{P\{X_1 = x\} P\{X_2 = k - x\}}{P\{X_1 + X_2 = k\}}. \end{aligned}$$

By part (a) and the assumption, we have

$$P\{X_1 = x\} = e^{-\lambda_1} \frac{\lambda_1^x}{x!},$$

$$P\{X_2 = k - x\} = e^{-\lambda_2} \frac{\lambda_2^{k-x}}{(k - x)!},$$

and

$$P\{X_1 + X_2 = k\} = e^{-\lambda_1 - \lambda_2} \frac{(\lambda_1 + \lambda_2)^k}{k!}.$$

Hence we have

$$\begin{aligned} & P\{X_1 = x | X_1 + X_2 = k\} \\ = & \frac{e^{-\lambda_1} \frac{\lambda_1^x}{x!} e^{-\lambda_2} \frac{\lambda_2^{k-x}}{(k-x)!}}{e^{-\lambda_1 - \lambda_2} \frac{(\lambda_1 + \lambda_2)^k}{k!}} \\ = & \binom{k}{x} \left(\frac{\lambda_1}{\lambda_1 + \lambda_2}\right)^x \left(\frac{\lambda_2}{\lambda_1 + \lambda_2}\right)^{k-x}, \end{aligned}$$

which is a binomial distribution with parameter $(k, \lambda_1/(\lambda_1 + \lambda_2))$.

This completes the solution.

18.10. Let $m \geq 0$ be an integer. Since $X_1, X_2, \ldots, X_n$ are independent, by Problem 12.5 we have

$$\begin{aligned} P\{Y > m\} &= P\{X_1 > m, X_2 > m, \ldots, X_n > m\} \\ &= \prod_{k=1}^{n} P\{X_k > m\} \\ &= \prod_{k=1}^{n} p_k^{m+1} \\ &= p^{m+1}. \end{aligned}$$

Hence we have

$$P\{Y = m\} = P\{Y > m - 1\} - P\{Y > m\} = p^m - p^{m+1} = (1 - p)p^m,$$

which shows that Y is geometrically distributed with parameter p. This completes the proof.

18.5 Bibliographic Notes

In this chapter, we presented several discrete distributions. These discrete distributions are probably discussed in many books. Johnson et al. (2005) discussed many advanced results of discrete distributions regarding their history, properties, estimation, applications, and recent developments. Kinney (2009) introduced many examples of statistical distributions including discrete distributions. Balakrishnan and Nevzorov (2003) is a textbook on statistical distributions, which also discussed these discrete distributions. For more information about discrete distributions, readers are referred to Feller (1968), Wimmer and Altmann (1999), and Evans et al. (2000).

CHAPTER 19

CONTINUOUS DISTRIBUTIONS

Unlike the discrete distributions described in the previous chapter, continuous distributions are used to describe random variables that can take any number of values. In this chapter, we present some common continuous distributions and their properties.

19.1 Basic Concepts and Facts

Definition 19.1 (Univariate Normal Distribution). A random variable X on $(\mathbf{R}, \mathcal{B})$ is said to be normally distributed with mean μ and standard deviation σ, written as $X \sim N(\mu, \sigma^2)$, if and only if its probability density function is given by

$$\varphi(\mu, \sigma, x) = \frac{1}{\sigma\sqrt{2\pi}} \exp\left(-\frac{(x-\mu)^2}{2\sigma^2}\right), \quad x \in \mathbf{R}. \tag{19.1}$$

A normal random variable is called a *standard normal random variable* if it has mean 0 and standard deviation 1. The probability density function of a standard normal random variable is denoted by $\varphi(x)$.

The cumulative density function of a normal random variable with mean μ and standard deviation σ is given by

$$\Phi(\mu, \sigma, x) = \int_{-\infty}^{x} \varphi(\mu, \sigma, t)\mathrm{d}t = \frac{1}{\sigma\sqrt{2\pi}} \int_{-\infty}^{x} \exp\left(-\frac{(t-\mu)^2}{2\sigma^2}\right) \mathrm{d}t. \tag{19.2}$$

The cumulative density function of a standard normal random variable is denoted by $\Phi(x)$ or $N(x)$.

Definition 19.2 (Lognormal Distribution). Let $\mu \in \mathbf{R}$ and $\sigma > 0$. A random variable X on $((0,\infty), \mathcal{B}(0,\infty))$ is said to have a lognormal distribution with parameters (μ, σ^2), written as $X \sim LN(\mu, \sigma^2)$, if and only if X has the following probability density function

$$f(x) = \frac{1}{x\sigma\sqrt{2\pi}} \exp\left[-\frac{(\ln x - \mu)^2}{2\sigma^2}\right], \quad x > 0. \tag{19.3}$$

Definition 19.3 (Student's t-Distribution). Let $\nu > 0$. A random variable X on $(\mathbf{R}, \mathcal{B})$ is said to have a Student's t-distribution, written as $X \sim ST(\nu)$, if and only if X has the following probability density function:

$$f(x) = \frac{\Gamma\left(\frac{\nu+1}{2}\right)}{\sqrt{\nu\pi}\Gamma\left(\frac{\nu+1}{2}\right)} \left(1 + \frac{x^2}{\nu}\right)^{-\frac{\nu+1}{2}}, \quad x \in \mathbf{R}, \tag{19.4}$$

where $\Gamma(\cdot)$ is the gamma function defined as

$$\Gamma(x) = \int_0^\infty t^{x-1}e^{-t}\mathrm{d}t. \tag{19.5}$$

Definition 19.4 (Cauchy Distribution). Let $a, h \in \mathbf{R}$. A random variable X on $(\mathbf{R}, \mathcal{B})$ is said to have a Cauchy distribution, written as $X \sim C(a, h)$, if and only if X has the following probability density function:

$$f(x) = \frac{h}{(h^2 + (x-a)^2)\,\pi}, \quad x \in \mathbf{R}. \tag{19.6}$$

Definition 19.5 (Pareto Distribution). Let $\alpha > 0$ and $h > 0$. A random variable X on $([h,\infty), \mathcal{B}[h,\infty))$ is said to have a Pareto distribution, written as $X \sim Pa(\alpha, h)$, if and only if X has the following probability density function:

$$f(x) = \frac{\alpha h^\alpha}{x^{\alpha+1}}, \quad x \geq h. \tag{19.7}$$

Definition 19.6 (Beta Distribution). Let $p > 0$ and $q > 0$. A random variable X on $((0,1), \mathcal{B}(0,1))$ is said to have a standard beta distribution, written as $X \sim Beta(p,q)$, if and only if X has the following probability density function

$$f(x) = \frac{1}{B(p,q)} x^{p-1}(1-x)^{q-1}, \quad x \in (0,1), \tag{19.8}$$

where $B(\cdot,\cdot)$ is the beta function defined as

$$B(p,q)=\int_0^1 x^{p-1}(1-x)^{q-1}\mathrm{d}x. \tag{19.9}$$

Definition 19.7 (Gamma Distribution). Let $\alpha>0$ and $\theta>0$. A random variable X on $([0,\infty),\mathcal{B}[0,\infty))$ is said to have a gamma distribution with parameters α and θ, written as $X\sim\Gamma(\alpha,\theta)$, if and only if X has the following probability density function, where $\Gamma(\cdot,\cdot)$ is the gamma function defined in Equation (19.5):

$$f(x)=\frac{1}{\Gamma(\alpha)\theta^{\alpha}}x^{\alpha-1}e^{-\frac{x}{\theta}},\quad x\in[0,\infty). \tag{19.10}$$

The parameters α and θ are known as the shape and scale parameters, respectively. If $\theta=1$, then the distribution is called the standard gamma distribution.

Definition 19.8 (Exponential Distribution). Let $\lambda>0$. A random variable X on $((0,\infty),\mathcal{B}(0,\infty))$ is said to have an exponential distribution, written as $X\sim Exp(\lambda)$, if and only if X has the following probability density function

$$f(x)=\lambda e^{-\lambda x},\quad x\in(0,\infty). \tag{19.11}$$

Definition 19.9 (Multivariate Normal Distribution). A random vector $\mathbf{X}=(X_1,X_2,\ldots,X_n)^T$ on $(\mathbf{R}^n,\mathcal{B}(\mathbf{R}^n))$ is said to have a multivariate normal distribution if its probability density function is given by

$$\varphi(\boldsymbol{\mu},\boldsymbol{\Sigma},\mathbf{x})=(2\pi)^{-\frac{n}{2}}|\Sigma|^{-\frac{1}{2}}\exp\left(-\frac{1}{2}(\mathbf{x}-\boldsymbol{\mu})^T\Sigma^{-1}(\mathbf{x}-\boldsymbol{\mu})\right), \tag{19.12}$$

where $\boldsymbol{\mu}=(\mu_1,\mu_2,\ldots,\mu_n)^T$ is the mean of $\mathbf{X}$ and Σ is the covariance matrix of $\mathbf{X}$ given by

$$\Sigma_{ij}=\mathrm{Cov}(X_i,X_j),\quad i,j=1,2,\ldots,n.$$

Theorem 19.1. *Let $X\sim C(a,h)$. Then the characteristic function of X is given by*

$$\phi_X(t)=e^{ita-h|t|}.$$

19.2 Problems

19.1. Let X be a normal random variable with mean μ and standard deviation σ. Show that

$$M_X(t)=\exp\left(\mu t+\frac{1}{2}\sigma^2t^2\right)$$

and

$$\phi_X(t)=\mathrm{cxp}\left(\mu it-\frac{1}{2}\sigma^2t^2\right),$$

where $M_X(t)$ and $\phi_X(t)$ are the moment generating function and the characteristic function of X, respectively.

19.2. Let $\mathbf{X} = (X_1, X_2, \ldots, X_n)^T$ on $(\mathbf{R}^n, \mathcal{B}(\mathbf{R}^n))$ be a multivariate normal random vector with a probability density function given by

$$\varphi(\boldsymbol{\mu}, \boldsymbol{\Sigma}, \mathbf{x}) = (2\pi)^{-\frac{n}{2}} |\Sigma|^{-\frac{1}{2}} \exp\left(-\frac{1}{2}(\mathbf{x} - \boldsymbol{\mu})^T \Sigma^{-1} (\mathbf{x} - \boldsymbol{\mu})\right),$$

where $\boldsymbol{\mu} = (\mu_1, \mu_2, \ldots, \mu_n)^T$ is the mean vector and Σ is the covariance matrix. Show that the characteristic function of $\mathbf{X}$ is

$$\phi_{\mathbf{X}}(\mathbf{t}) = \exp\left(i\mathbf{t}^T \boldsymbol{\mu} - \frac{1}{2}\mathbf{t}^T \Sigma \mathbf{t}\right).$$

19.3. Let X and Y be two independent normal random variables with means μ_1 and μ_2, and variances σ_1^2 and σ_2^2. Show that $aX + bY$ is normally distributed with mean $a\mu_1 + b\mu_2$ and variance $a^2\sigma_1^2 + b^2\sigma_2^2$.

19.4. Let (X, Y) be bivariate normally distributed. Show that the linear combination of X and Y is a normal random variable, that is, $aX + bY$ is normally distributed for $a, b \in \mathbf{R}$.

19.5. Let $\mathbf{X} = (X_1, X_2, \ldots, X_n)^T$ be a random vector on $(\mathbf{R}^n, \mathcal{B}(\mathbf{R}^n))$. Then $\mathbf{X}$ has a multivariate normal distribution if and only if, for every $\mathbf{a} = (a_1, a_2, \ldots, a_n)^T$, the random variable

$$\mathbf{a}^T \mathbf{X} = \sum_{i=1}^{n} a_i X_i$$

is a normal random variable.

19.6. Let $(X_1, X_2, \ldots, X_n)^T$ be multivariate normally distributed. Show that X_1, X_2, …, X_n are independent if and only if

$$E(X_i X_j) = E(X_i)E(X_j), \quad \text{for all } i \neq j.$$

19.7. Let $X \sim N(0, 1)$. Show that

$$E\left(|X|^{2n+1}\right) = 2^n n! \sqrt{\frac{2}{\pi}}. \tag{19.13}$$

19.8. Let $N(d)$ be defined as

$$N(d) = \frac{1}{\sqrt{2\pi}} \int_{-\infty}^{d} e^{-\frac{x^2}{2}} \mathrm{d}x. \tag{19.14}$$

Show that

(a) $N(\infty) = 1$:

$$\frac{1}{\sqrt{2\pi}} \int_{-\infty}^{\infty} e^{-\frac{x^2}{2}} \mathrm{d}x = 1;$$

(b) For $d \in (-\infty, \infty)$, $N(d) + N(-d) = 1$.

19.9. Let $a, b, \beta, \gamma \in \mathbf{R}$ and $T > 0$. Let $N(\cdot)$ as be defined in Problem 19.8. Show that

(a)

$$\begin{aligned} & \frac{1}{\sqrt{2\pi T}} \int_a^b \exp\left(\beta + \gamma x - \frac{1}{2T}x^2\right) \mathrm{d}x \\ = \; & \exp\left(\frac{1}{2}\gamma^2 T + \beta\right) \left[N\left(\frac{-a+\gamma T}{\sqrt{T}}\right) - N\left(\frac{-b+\gamma T}{\sqrt{T}}\right) \right]. \end{aligned}$$

(b)

$$\begin{aligned} & \frac{1}{\sqrt{2\pi T}} \int_{-\infty}^b \exp\left(\beta + \gamma x - \frac{1}{2T}x^2\right) \mathrm{d}x \\ = \; & \exp\left(\frac{1}{2}\gamma^2 T + \beta\right) \left[1 - N\left(\frac{-b+\gamma T}{\sqrt{T}}\right) \right]. \end{aligned}$$

(c)

$$\begin{aligned} & \frac{1}{\sqrt{2\pi T}} \int_a^{\infty} \exp\left(\beta + \gamma x - \frac{1}{2T}x^2\right) \mathrm{d}x \\ = \; & \exp\left(\frac{1}{2}\gamma^2 T + \beta\right) N\left(\frac{-a+\gamma T}{\sqrt{T}}\right). \end{aligned}$$

19.10. Let a, b, c, and d be constants such that $a \neq 0$ and $b < 0$. Show that

$$\begin{aligned} & \int_d^{\infty} e^{ay} N(by + c) \mathrm{d}y \\ = \; & -\frac{1}{a} e^{ad} N(bd + c) + \frac{1}{a} \exp\left(\frac{a^2 - 2abc}{2b^2}\right) N\left(bd + c - \frac{a}{b}\right), \end{aligned}$$

where $N(\cdot)$ is as defined in Problem 19.8.

19.11. Let $X_i \sim C(a_i, h_i)$ for $i = 1, 2$. Suppose that X_1 and X_2 are independent. Show that

$$X_1 + X_2 \sim C(a_1 + a_2, h_1 + h_2).$$

19.12. Let $X \sim C(a, h)$ and $Y = 1/X$. Show that

$$Y \sim C\left(\frac{a}{a^2 + h^2}, \frac{h}{a^2 + h^2}\right).$$

19.13. Let $X_i \sim Pa(\alpha_i, h)$ for $i = 1, 2, \ldots, n$ and suppose that $X_1, X_2, \ldots, X_n$ are independent. Let $Y = \min\{X_1, X_2, \ldots, X_n\}$. Show that

$$Y \sim Pa(\alpha_1 + \alpha_2 + \cdots + \alpha_n, h).$$

19.14. Let $B(\cdot,\cdot)$ and $\Gamma(\cdot)$ be the beta function and the gamma function, respectively. Show that

(a) For any $x \in \mathbf{R}$,

$$\Gamma(x) = (x-1)\Gamma(x-1).$$

(b) If n is a positive integer, then

$$\Gamma(n) = (n-1)!.$$

(c) For any $p > 0$ and $q > 0$,

$$B(p,q) = \frac{\Gamma(p)\Gamma(q)}{\Gamma(p+q)}.$$

19.15. Let m and n be integers such that $m \leq n$. Let $p \in (0,1)$. Let $X \sim Beta(m, n-m+1)$ and $Y \sim B(n,p)$; that is, X has a beta distribution with parameters $(m, n-m+1)$ and Y has a binomial distribution with parameters (n,p). Show that

$$P\{X < p\} = P\{Y \geq m\}.$$

19.16. Let $X \sim N(0,1)$, that is, X is a standard normal random variable. Find the distribution of X^2.

19.17 (Feller's Tail Probability Estimates)**.** Let $Z \sim N(0,1)$ and $x > 1$. Show that

(a)

$$P\{|Z| \geq x\} \leq \frac{1}{x}\sqrt{\frac{2}{\pi}}\exp\left(-\frac{x^2}{2}\right).$$

(b)

$$P\{|Z| \geq x\} \geq \left(\frac{1}{x} - \frac{1}{x^3}\right)\sqrt{\frac{2}{\pi}}\exp\left(-\frac{x^2}{2}\right).$$

(c)

$$\lim_{x\to\infty} \frac{P\{|Z| \geq x\}}{\frac{1}{x}\sqrt{\frac{2}{\pi}}\exp\left(-\frac{x^2}{2}\right)} = 1.$$

19.3 Hints

19.1. Use the fact that $P(\Omega) = 1$, that is, the integration of the probability density function is 1.

19.2. Note that

$$\begin{aligned} & i\mathbf{t}^T\mathbf{x} - \frac{1}{2}(\mathbf{x}-\boldsymbol{\mu})^T\Sigma^{-1}(\mathbf{x}-\boldsymbol{\mu}) \\ = \;& -\frac{1}{2}(\mathbf{x}-\boldsymbol{\mu}+i\Sigma\mathbf{t})^T\Sigma^{-1}(\mathbf{x}-\boldsymbol{\mu}+i\Sigma\mathbf{t}) + i\mathbf{t}^T\boldsymbol{\mu} - \frac{1}{2}\mathbf{t}^T\Sigma\mathbf{t}. \end{aligned}$$

19.3. Use the characteristic function theorem (Theorem 17.1), the characteristic function of normal random variables (see Problem 19.1), and Theorem 13.2.

19.4. Use the characteristic function theorem (Theorem 17.1) and the characteristic function of normal random variables (see Problem 19.1).

19.5. The "if" part can be proved by the results of Problems 19.1 and 19.2, and the Fourier uniqueness theorem (Theorem 17.1). The "only if" part can be proved by the result of Problem 19.1 and the Fourier uniqueness theorem.

19.6. The "only if" part follows from Theorem 13.2. The "if" part can be proved by using the result of Problem 12.8.

19.7. Note that

$$E\left(|X|^{2n+1}\right) = 2\int_0^\infty x^{2n+1}\frac{1}{\sqrt{2\pi}}e^{-\frac{x^2}{2}}\,\mathrm{d}x$$

and use the method of induction.

19.8. To prove part (a), use Tonelli's theorem (Theorem 10.1) and consider

$$N(\infty)^2 = \frac{2}{\pi}\int_0^\infty e^{-\frac{x^2}{2}}\,\mathrm{d}x\int_0^\infty e^{-\frac{y^2}{2}}\,\mathrm{d}y = \frac{2}{\pi}\int_0^\infty\left(\int_0^\infty e^{-\frac{x^2+y^2}{2}}\,\mathrm{d}y\right)\mathrm{d}x.$$

Then for fixed x, let $y = sx$ and $\mathrm{d}y = x\mathrm{d}s$. Use part (a) to prove part (b).

19.9. Use change of variables and the definition of $N(\cdot)$ to prove part (a). Then use part (a) and the monotone convergence theorem (Theorem 6.2) to prove parts (b) and (c).

19.10. Use integration by parts and l'Hôpital's rule.

19.11. Use the characteristic function of the Cauchy distribution (see Theorem 19.1) and Theorem 17.1.

19.12. Note that

$$F_Y(y) = P(Y \le y) = P\left(\frac{1}{X} \le y\right)$$

and use the probability density function of the Cauchy distribution.

19.13. Note that

$$P(Y \le y) = 1 - P(Y > y) = 1 - P(X_1 > y, X_2 > y, \ldots, X_n > y)$$

and use the result of Problem 12.5.

19.14. Part (a) can be proved by integration by parts. Part (b) can be derived from part (a). Part (c) can be proved by using multivariate integration and changes of variables (i.e., Jacobian).

19.15. Note that

$$P\{X < p\} = \frac{1}{B(m, n-m+1)} \int_0^p x^{m-1}(1-x)^{n-m} \mathrm{d}x$$

and apply integration by parts.

19.16. Try to compute the probability density function by taking the derivative of the distribution function of X^2.

19.17. The first inequality can be proved by observing that

$$\int_x^\infty e^{-\frac{s^2}{2}} \mathrm{d}s \leq \int_x^\infty \frac{s}{x} e^{-\frac{s^2}{2}} \mathrm{d}s.$$

The second inequality can be proved by observing that

$$\left(1 - \frac{3}{s^4}\right) e^{-\frac{s^2}{2}} \leq e^{-\frac{s^2}{2}}.$$

19.4 Solutions

19.1. By definition, we have

$$\begin{aligned}
M_X(t) &= E\left[e^{tX}\right] \\
&= \frac{1}{\sigma\sqrt{2\pi}} \int_{-\infty}^\infty \exp(tx) \cdot \exp\left(-\frac{(x-\mu)^2}{2\sigma^2}\right) \mathrm{d}x \\
&= \frac{1}{\sigma\sqrt{2\pi}} \int_{-\infty}^\infty \exp\left(tx - \frac{(x-\mu)^2}{2\sigma^2}\right) \mathrm{d}x \\
&= \frac{1}{\sigma\sqrt{2\pi}} \int_{-\infty}^\infty \exp\left(-\frac{(x-\sigma^2 t-\mu)^2}{2\sigma^2} + \frac{1}{2}\sigma^2 t^2 + \mu t\right) \mathrm{d}x \\
&= \exp\left(\mu t + \frac{1}{2}\sigma^2 t\right) \frac{1}{\sigma\sqrt{2\pi}} \int_{-\infty}^\infty \exp\left(-\frac{(x-\sigma^2 t-\mu)^2}{2\sigma^2}\right) \mathrm{d}x \\
&= \exp\left(\mu t + \frac{1}{2}\sigma^2 t\right).
\end{aligned}$$

Similarly, we can show that

$$\phi_X(t) = \exp\left(\mu i t - \frac{1}{2}\sigma^2 t^2\right).$$

This completes the proof.

19.2. Let $C = (2\pi)^{-\frac{n}{2}}|\Sigma|^{-\frac{1}{2}}$. By definition, we have

$$\begin{aligned}
& \phi_{\mathbf{X}}(\mathbf{t}) \\
= \ & E\left[e^{i\mathbf{t}^T\mathbf{X}}\right] \\
= \ & C\int_{\Omega} \exp\left(i\mathbf{t}^T\mathbf{x} - \frac{1}{2}(\mathbf{x}-\boldsymbol{\mu})^T\Sigma^{-1}(\mathbf{x}-\boldsymbol{\mu})\right) \mathrm{d}P \\
= \ & C\int_{\Omega} \exp\left(-\frac{1}{2}(\mathbf{x}-\boldsymbol{\mu}+i\Sigma\mathbf{t})^T\Sigma^{-1}(\mathbf{x}-\boldsymbol{\mu}+i\Sigma\mathbf{t}) + i\mathbf{t}^T\boldsymbol{\mu} - \frac{1}{2}\mathbf{t}^T\Sigma\mathbf{t}\right) \mathrm{d}P \\
= \ & \exp\left(i\mathbf{t}^T\boldsymbol{\mu} - \frac{1}{2}\mathbf{t}^T\Sigma\mathbf{t}\right).
\end{aligned}$$

This completes the proof.

19.3. By Problem 19.1, we have

$$\phi_X(t) = \exp\left(i\mu_1 t - \frac{1}{2}\sigma_1^2 t^2\right)$$

and

$$\phi_Y(t) = \exp\left(i\mu_2 t - \frac{1}{2}\sigma_2^2 t^2\right).$$

Hence by Theorem 13.2, we have

$$\begin{aligned}
\phi_{aX+bY}(t) &= E\left[\exp(it(aX+bY))\right] \\
&= E\left[e^{itaX}\right]E\left[e^{itaY}\right] \\
&= \exp\left[i(a\mu_1 + b\mu_2)t - \frac{1}{2}\left(a^2\sigma_1^2 + b^2\sigma_2^2\right)t^2\right].
\end{aligned}$$

The conclusion follows from Theorem 17.1. This completes the proof.

19.4. Suppose that (X, Y) has the following joint probability distribution function:

$$\begin{aligned}
& f(x,y) \\
= \ & \frac{1}{2\pi\sigma_1\sigma_2\sqrt{(1-\rho^2)}} \cdot \\
& \exp\left(-\frac{1}{2(1-\rho^2)}\left[\frac{(x-\mu_1)^2}{\sigma_1^2} + \frac{(y-\mu_2)^2}{\sigma_2^2} - \frac{2\rho(x-\mu_1)(y-\mu_2)}{\sigma_1\sigma_2}\right]\right),
\end{aligned}$$

Then we have

$$\begin{aligned}
& E\left(e^{it(aX+bY)}\right) \\
= \ & \int_{-\infty}^{\infty}\int_{-\infty}^{\infty} f(x,y)e^{it(ax+by)}\mathrm{d}x\mathrm{d}y \\
= \ & \exp\left((a\mu_1 + b\mu_2)it - \frac{1}{2}(a^2\sigma_1^2 + b^2\sigma_2^2 - 2\rho ab\sigma_1\sigma_2)t^2\right).
\end{aligned}$$

By Theorem 17.1, the random variable $aX + bY$ is normally distributed with mean $a\mu_1 + b\mu_2$ and variance $(a^2\sigma_1^2 + b^2\sigma_2^2 - 2\rho ab\sigma_1\sigma_2)$.

19.5. Let us prove the sufficiency first. Suppose that for every $\mathbf{a} = (a_1, a_2, \ldots, a_n)^T$, the random variable

$$\mathbf{a}^T\mathbf{X} = \sum_{i=1}^{n} a_i X_i$$

is a normal random variable. Then $\mathbf{t}^T\mathbf{X}$ is a normal random variable. Hence by Problem 19.1, we have

$$\begin{aligned}\phi_{\mathbf{X}}(\mathbf{t}) &= E\left[e^{i\mathbf{t}^T\mathbf{X}}\right] \\ &= \exp\left(i\mu(\mathbf{t}^T\mathbf{X}) - \frac{1}{2}\sigma^2(\mathbf{t}^T\mathbf{X})\right),\end{aligned}$$

where $\mu(\mathbf{t}^T\mathbf{X})$ and $\sigma^2(\mathbf{t}^T\mathbf{X})$ are the mean and variance of $\mathbf{t}^T\mathbf{X}$, respectively. Suppose that $E[\mathbf{X}] = \boldsymbol{\mu}$ and $\mathrm{Cov}(\mathbf{X}) = \Sigma$. Then we have

$$\mu(\mathbf{t}^T\mathbf{X}) = \mathbf{t}^T\boldsymbol{\mu}$$

and

$$\sigma^2(\mathbf{t}^T\mathbf{X}) = \mathbf{t}^T\Sigma\mathbf{t}.$$

Therefore, we get

$$\phi_{\mathbf{X}}(\mathbf{t}) = \exp\left(i\mathbf{t}^T\boldsymbol{\mu} - \frac{1}{2}\mathbf{t}^T\Sigma\mathbf{t}\right).$$

It follows from Theorem 17.1 and Problem 19.2 that $\mathbf{X}$ has a multivariate normal distribution.

Now let us prove the necessity. Suppose that $\mathbf{X}$ has a multivariate normal distribution whose density function is given by

$$\varphi(\boldsymbol{\mu}, \boldsymbol{\Sigma}, \mathbf{x}) = (2\pi)^{-\frac{n}{2}}|\Sigma|^{-\frac{1}{2}}\exp\left(-\frac{1}{2}(\mathbf{x}-\boldsymbol{\mu})^T\Sigma^{-1}(\mathbf{x}-\boldsymbol{\mu})\right).$$

Let $\mathbf{a} = (a_1, a_2, \ldots, a_n)^T$ be a vector. Let $Y = \mathbf{a}^T\mathbf{X}$. Then

$$\begin{aligned}\phi_Y(t) &= E\left[e^{itY}\right] \\ &= E\left[e^{it\mathbf{a}^T\mathbf{X}}\right] \\ &= (2\pi)^{-\frac{n}{2}}|\Sigma|^{-\frac{1}{2}}\int_\Omega \exp\left(it\mathbf{a}^T\mathbf{X} - \frac{1}{2}(\mathbf{X}-\boldsymbol{\mu})^T\Sigma^{-1}(\mathbf{X}-\boldsymbol{\mu})\right)\mathrm{d}P \\ &= \exp\left(it\mathbf{a}^T\boldsymbol{\mu} - \frac{1}{2}t^2\mathbf{a}^T\Sigma\mathbf{a}\right).\end{aligned}$$

By Problem 19.1 and Theorem 17.1, Y is normally distributed with mean $\mathbf{a}^T\boldsymbol{\mu}$ and variance $\mathbf{a}^T\Sigma\mathbf{a}$.

This completes the proof.

19.6. The "only if" part follows from Theorem 13.2. We only need to prove the "if" part. Assume that $E(X_iX_j) = E(X_i)E(X_j)$ for all $i \neq j$. Then $\text{Cov}(X_i, X_j) = E(X_iX_j) - E(X_i)E(X_j) = 0$. Hence $(X_1, X_2, \ldots, X_n)^T$ has the following joint distribution function

$$f(\mathbf{x}) = \frac{1}{(2\pi)^{\frac{n}{2}} \prod_{i=1}^n \sigma_i} \cdot \exp\left(-\frac{1}{2}\sum_{i=1}^n \frac{(x_i - \mu_i)^2}{\sigma_i^2}\right),$$

where μ_i and σ_i are the mean and standard deviation of X_i. Then we have

$$\begin{aligned}
& P(X_1 \leq x_1, X_2 \leq x_2, \ldots, X_n \leq x_n) \\
= & \int_{-\infty}^{x_1} \int_{-\infty}^{x_2} \cdots \int_{-\infty}^{x_n} f(\mathbf{s}) \mathrm{d}s_n \cdots \mathrm{d}s_2 \mathrm{d}s_1 \\
= & \prod_{i=1}^n \left(\frac{1}{\sqrt{2\pi}\sigma_i} \int_{-\infty}^{x_i} \exp\left(-\frac{(s_i - \mu_i)^2}{2\sigma_i^2}\right) \mathrm{d}s_i \right) \\
= & \prod_{i=1}^n P(X_i \leq x_i).
\end{aligned}$$

By Problem 12.8, $X_1, X_2, \ldots, X_n$ are independent. This finishes the proof.

19.7. Note that

$$\begin{aligned}
E\left(|X|^{2n+1}\right) &= \int_{-\infty}^{\infty} |x|^{2n+1} \frac{1}{\sqrt{2\pi}} e^{-\frac{x^2}{2}} \mathrm{d}x \\
&= \int_0^{\infty} x^{2n+1} \frac{1}{\sqrt{2\pi}} e^{-\frac{x^2}{2}} \mathrm{d}x - \int_{-\infty}^0 x^{2n+1} \frac{1}{\sqrt{2\pi}} e^{-\frac{x^2}{2}} \mathrm{d}x \\
&= 2\int_0^{\infty} x^{2n+1} \frac{1}{\sqrt{2\pi}} e^{-\frac{x^2}{2}} \mathrm{d}x.
\end{aligned}$$

Now we show that Equation (19.13) is true by the method of induction. When $n = 0$, integrating by parts gives

$$E\left(|X|\right) = 2\int_0^{\infty} x \frac{1}{\sqrt{2\pi}} e^{-\frac{x^2}{2}} \mathrm{d}x = \int_0^{\infty} \frac{1}{\sqrt{2\pi}} e^{-\frac{x^2}{2}} \mathrm{d}x^2 = \sqrt{\frac{2}{\pi}}.$$

Hence Equation (19.13) holds for $n = 0$. Now suppose that Equation (19.13) holds for $n = k$; it follows from the integration by parts and Equation (19.15) that

$$E\left(|X|^{2k+1}\right) = 2^k k! \sqrt{\frac{2}{\pi}}. \tag{19.15}$$

Then for $n = k+1$, we have

$$\begin{aligned} E\left(|X|^{2(k+1)+1}\right) &= 2\int_0^\infty x^{2k+3}\frac{1}{\sqrt{2\pi}}e^{-\frac{x^2}{2}}\,\mathrm{d}x \\ &= -\sqrt{\frac{2}{\pi}}\int_0^\infty x^{2k+2}\mathrm{d}e^{-\frac{x^2}{2}} \\ &= \sqrt{\frac{2}{\pi}}\int_0^\infty (2k+2)x^{2k+1}e^{-\frac{x^2}{2}}\,\mathrm{d}x \\ &= 2(k+1)E\left(|X|^{2k+1}\right) \\ &= 2^{k+1}(k+1)!\sqrt{\frac{2}{\pi}}. \end{aligned}$$

Hence Equation (19.13) holds for $n = k+1$. This completes the proof.

19.8.

(a) Since

$$N(\infty) = \frac{2}{\sqrt{2\pi}}\int_0^\infty e^{-\frac{x^2}{2}}\,\mathrm{d}x,$$

by Theorem 10.1, we have

$$N(\infty)^2 = \frac{2}{\pi}\int_0^\infty e^{-\frac{x^2}{2}}\,\mathrm{d}x\int_0^\infty e^{-\frac{y^2}{2}}\,\mathrm{d}y = \frac{2}{\pi}\int_0^\infty\left(\int_0^\infty e^{-\frac{x^2+y^2}{2}}\,\mathrm{d}y\right)\mathrm{d}x.$$

Now for fixed x, we let $y = sx$. Then we have

$$\begin{aligned} N(\infty)^2 &= \frac{2}{\pi}\int_0^\infty\left(\int_0^\infty e^{-x^2\frac{1+s^2}{2}}x\mathrm{d}s\right)\mathrm{d}x \\ &= \frac{2}{\pi}\int_0^\infty\left(\int_0^\infty e^{-x^2\frac{1+s^2}{2}}x\mathrm{d}x\right)\mathrm{d}s \\ &= \frac{2}{\pi}\int_0^\infty \frac{1}{1+s^2}\mathrm{d}s \\ &= \frac{2}{\pi}\int_0^{\frac{\pi}{2}} \frac{1}{1+\tan^2(\theta)}\mathrm{d}\tan(\theta) \\ &= 1. \end{aligned}$$

Since $N(\infty) > 0$, we have $N(\infty) = 1$.

(b) By changes of variables and the first item, we have

$$\begin{aligned} N(d) + N(-d) &= \frac{1}{\sqrt{2\pi}}\int_{-\infty}^d e^{-\frac{x^2}{2}}\,\mathrm{d}x + \frac{1}{\sqrt{2\pi}}\int_{-\infty}^{-d} e^{-\frac{x^2}{2}}\,\mathrm{d}x \\ &= \frac{1}{\sqrt{2\pi}}\int_{-\infty}^d e^{-\frac{x^2}{2}}\,\mathrm{d}x + \frac{1}{\sqrt{2\pi}}\int_{\infty}^{d} e^{-\frac{y^2}{2}}\,\mathrm{d}(-y) \\ &= \frac{1}{\sqrt{2\pi}}\int_{-\infty}^d e^{-\frac{x^2}{2}}\,\mathrm{d}x + \frac{1}{\sqrt{2\pi}}\int_{d}^{\infty} e^{-\frac{x^2}{2}}\,\mathrm{d}x \\ &= 1. \end{aligned}$$

This completes the proof.

19.9.

(a) Let

$$y = \frac{x - \gamma T}{\sqrt{T}}.$$

Writing the integral with respect to the variable y gives

$$\begin{aligned}
& \frac{1}{\sqrt{2\pi T}} \int_a^b \exp\left(\beta + \gamma x - \frac{1}{2T}x^2\right) \mathrm{d}x \\
= & \frac{\exp\left(\beta + \frac{1}{2}\gamma^2 T\right)}{\sqrt{2\pi}} \int_{\frac{a-\gamma T}{\sqrt{T}}}^{\frac{b-\gamma T}{\sqrt{T}}} \exp\left(-\frac{1}{2}y^2\right) \mathrm{d}y \\
= & \exp\left(\frac{1}{2}\gamma^2 T + \beta\right)\left[N\left(\frac{b-\gamma T}{\sqrt{T}}\right) - N\left(\frac{a-\gamma T}{\sqrt{T}}\right)\right].
\end{aligned}$$

Since $N(d) = 1 - N(-d)$ (see Problem 19.8) for any $d \in \mathbf{R}$, we have

$$\begin{aligned}
& \frac{1}{\sqrt{2\pi T}} \int_a^b \exp\left(\beta + \gamma x - \frac{1}{2T}x^2\right) \mathrm{d}x \\
= & \exp\left(\frac{1}{2}\gamma^2 T + \beta\right)\left[N\left(\frac{-a+\gamma T}{\sqrt{T}}\right) - N\left(\frac{-b+\gamma T}{\sqrt{T}}\right)\right].
\end{aligned}$$

(b) Let $\{a_n : n = 1, 2, \ldots\}$ be an arbitrary sequence of real numbers such that $a_n \downarrow -\infty$ as $n \to \infty$. By the monotone convergence theorem (Theorem 6.2) and part (a), we have

$$\begin{aligned}
& \frac{1}{\sqrt{2\pi T}} \int_{-\infty}^b \exp\left(\beta + \gamma x - \frac{1}{2T}x^2\right) \mathrm{d}x \\
= & \lim_{n\to\infty} \frac{1}{\sqrt{2\pi T}} \int_{a_n}^b \exp\left(\beta + \gamma x - \frac{1}{2T}x^2\right) \mathrm{d}x \\
= & \lim_{n\to\infty} \exp\left(\frac{1}{2}\gamma^2 T + \beta\right)\left[N\left(\frac{-a_n+\gamma T}{\sqrt{T}}\right) - N\left(\frac{-b+\gamma T}{\sqrt{T}}\right)\right] \\
= & \exp\left(\frac{1}{2}\gamma^2 T + \beta\right)\left[1 - N\left(\frac{-b+\gamma T}{\sqrt{T}}\right)\right].
\end{aligned}$$

(c) Let $\{b_n : n = 1, 2, \ldots\}$ be an arbitrary sequence of real numbers such that $b_n \uparrow \infty$ as $n \to \infty$. By the monotone convergence theorem (Theorem 6.2) and part (a), we have

$$\begin{aligned}
& \frac{1}{\sqrt{2\pi T}} \int_a^\infty \exp\left(\beta + \gamma x - \frac{1}{2T}x^2\right) \mathrm{d}x \\
= & \lim_{n\to\infty} \frac{1}{\sqrt{2\pi T}} \int_a^{b_n} \exp\left(\beta + \gamma x - \frac{1}{2T}x^2\right) \mathrm{d}x \\
= & \lim_{n\to\infty} \exp\left(\frac{1}{2}\gamma^2 T + \beta\right) \left[N\left(\frac{-a+\gamma T}{\sqrt{T}}\right) - N\left(\frac{-b_n+\gamma T}{\sqrt{T}}\right)\right] \\
= & \lim_{n\to\infty} \exp\left(\frac{1}{2}\gamma^2 T + \beta\right) \left[N\left(\frac{b_n-\gamma T}{\sqrt{T}}\right) - N\left(\frac{a-\gamma T}{\sqrt{T}}\right)\right] \\
= & \exp\left(\frac{1}{2}\gamma^2 T + \beta\right) \left[1 - N\left(\frac{a-\gamma T}{\sqrt{T}}\right)\right] \\
= & \exp\left(\frac{1}{2}\gamma^2 T + \beta\right) N\left(\frac{-a+\gamma T}{\sqrt{T}}\right).
\end{aligned}$$

This completes the proof.

19.10. Using integration by parts, we get

$$\begin{aligned}
& \int_d^\infty e^{ay} N(by+c)\mathrm{d}y \\
= & \left.\frac{e^{ay}}{a} N(by+c)\right|_d^\infty - \frac{b}{a\sqrt{2\pi}} \int_d^\infty e^{ay} \exp\left(-\frac{1}{2}(by+c)^2\right) \mathrm{d}y. \quad (19.16)
\end{aligned}$$

By Problem 19.9 and the assumption that $b < 0$, we have

$$\begin{aligned}
& \frac{b}{a\sqrt{2\pi}} \int_d^\infty e^{ay} \exp\left(-\frac{1}{2}(by+c)^2\right) \mathrm{d}y \\
= & -\frac{1}{a} \exp\left(\frac{(a-bc)^2}{2b^2} - \frac{1}{2}c^2\right) N\left(-b\left[-d + \frac{a-bc}{b^2}\right]\right) \\
= & -\frac{1}{a} \exp\left(\frac{a^2-2abc}{2b^2}\right) N\left(bd + c - \frac{a}{b}\right). \quad (19.17)
\end{aligned}$$

By l'Hôpital's rule, we have

$$\begin{aligned}
\lim_{y\to\infty} e^{ay} N(by+c) &= \lim_{y\to\infty} \frac{\frac{b}{\sqrt{2\pi}} \exp\left(-\frac{1}{2}(by+c)^2\right)}{-ae^{-ay}} \\
&= -\frac{b}{a\sqrt{2\pi}} \lim_{y\to\infty} \exp\left(ay - \frac{1}{2}(by+c)^2\right) \\
&= 0. \quad (19.18)
\end{aligned}$$

The result follows immediately by combining Equations (19.18), (19.17), and (19.16). This completes the proof.

19.11. By the assumption that X_1 and X_2 are independent and Theorem 19.1, we have

$$\phi_{X_1+X_2}(t) = E\left[e^{itX_1}e^{itX_2}\right] = E\left[e^{itX_1}\right]E\left[e^{itX_2}\right] = e^{i(a_1+a_2)t-(h_1+h_2)|t|}.$$

The result follows from Theorem 17.1.

19.12. By definition, the distribution function of Y is

$$F_Y(y) = P(Y \le y) = P\left(\frac{1}{X} \le y\right) = \begin{cases} P\left(\frac{1}{y} \le X < 0\right) & \text{if } y < 0; \\ P\left(\frac{1}{y} \le X \text{ or } X < 0\right) & \text{if } y > 0. \end{cases}$$

Note that for $y < 0$, we have

$$P\left(\frac{1}{y} \le X < 0\right) = \int_{\frac{1}{y}}^{0} \frac{h}{(h^2 + (x-a)^2)\pi},$$

which gives

$$\begin{aligned} f_Y(y) &= \frac{\mathrm{d}F_Y(y)}{\mathrm{d}y} \\ &= \frac{h}{\left(h^2 + \left(\frac{1}{y} - a\right)^2\right)\pi} \cdot \frac{1}{y2} \\ &= \frac{\frac{h}{a^2+h^2}}{\left(y - \frac{a}{a^2+h^2}\right)^2 + \left(\frac{h}{a^2+h^2}\right)^2}. \end{aligned}$$

When $y > 0$, the probability density function of Y is the same as above. Hence Y follows a Cauchy distribution with parameters $\left(\frac{a}{a^2+h^2}, \frac{h}{a^2+h^2}\right)$. This completes the proof.

19.13. By the assumption that $X_1, X_2, \ldots, X_n$ are independent and Problem 12.5, we have

$$\begin{aligned} F_Y(y) &= P(Y \le y) = 1 - P(Y > y) \\ &= 1 - P(X_1 > y, X_2 > y, \ldots, X_n > y) \\ &= 1 - \prod_{i=1}^{n} P(X_i > y) \\ &= 1 - \prod_{i=1}^{n} \int_y^{\infty} \frac{\alpha_i h^{\alpha_i}}{x^{\alpha_i+1}} \mathrm{d}x \\ &= 1 - \frac{h^{\alpha}}{y^{\alpha}}, \end{aligned}$$

where $\alpha = \alpha_1 + \alpha_2 + \cdots + \alpha_n$. Taking the derivative of the above equation gives

$$f_Y(y) = \frac{\alpha h^\alpha}{y^{\alpha+1}}.$$

Hence Y is Pareto distributed with parameters $\alpha_1 + \alpha + 2 + \cdots + \alpha_n$ and h. This completes the proof.

19.14.

(a) By integration by parts, we have

$$\begin{aligned}
\Gamma(x) &= \int_0^\infty t^{x-1}e^{-t}\mathrm{d}t \\
&= -\int_0^\infty t^{x-1}\mathrm{d}e^{-t} \\
&= \int_0^\infty (x-1)t^{x-2}e^{-t}\mathrm{d}t \\
&= (x-1)\Gamma(x-1).
\end{aligned}$$

(b) Applying the result of part (a) gives

$$\Gamma(n) = (n-1)!\Gamma(1).$$

Since

$$\Gamma(1) = \int_0^\infty e^{-t}\mathrm{d}t = -e^{-t}|_0^\infty = 1,$$

we get $\Gamma(n) = (n-1)!$.

(c) By the definition of Γ function, we have

$$\begin{aligned}
\Gamma(p)\Gamma(q) &= \int_0^\infty r^{p-1}e^{-r}\mathrm{d}r \int_0^\infty s^{q-1}e^{-s}\mathrm{d}s \\
&= \int_0^\infty \int_0^\infty r^{p-1}s^{q-1}e^{-r-s}\mathrm{d}r\mathrm{d}s.
\end{aligned}$$

Let $r = xt$ and $s = t(1-x)$, where $x \in [0,1]$ and $t \in [0,\infty)$. Then we have

$$\mathrm{d}r\mathrm{d}s = \begin{vmatrix} \frac{\partial r}{\partial x} & \frac{\partial r}{\partial t} \\ \frac{\partial s}{\partial x} & \frac{\partial s}{\partial t} \end{vmatrix} \mathrm{d}x\mathrm{d}t = \begin{vmatrix} t & x \\ -t & 1-x \end{vmatrix} \mathrm{d}x\mathrm{d}t = t\mathrm{d}x\mathrm{d}t.$$

Hence we get

$$\begin{aligned}
\Gamma(p)\Gamma(q) &= \int_0^\infty \int_0^\infty (xt)^{p-1}(t(1-x))^{q-1}e^{-t}t\mathrm{d}x\mathrm{d}t \\
&= \int_0^\infty \int_0^\infty t^{p+1-1}e^{-t}x^{p-1}(1-x)^{q-1}\mathrm{d}x\mathrm{d}t \\
&= \Gamma(p+q)B(p,q).
\end{aligned}$$

This completes the proof.

19.15. Note that $\Gamma(l+1) = l!$ for all integers $l \geq 0$. Then by integration by parts, we have

$$\begin{aligned} P\{X < p\} &= \frac{1}{B(m, n-m+1)} \int_0^p x^{m-1}(1-x)^{n-m} \mathrm{d}x \\ &= \frac{n!}{(m-1)!(n-m)!} \int_0^p \frac{1}{m}(1-x)^{n-m} \mathrm{d}x^m \\ &= \binom{n}{j} p^m (1-p)^{n-m} - \binom{n}{j} \int_0^p x^m (n-m)(1-x)^{n-m-1} \mathrm{d}x \\ &= \cdots \\ &= \sum_{j=m}^{n} \binom{n}{j} p^m (1-p)^{n-m} \\ &= P\{Y \geq m\}. \end{aligned}$$

This completes the proof.

19.16. Let $Y = X^2$. Then for $y \geq 0$, we have

$$\begin{aligned} F_Y(y) &= P\{Y \leq y\} \\ &= P\{X^2 \leq y\} \\ &= P\{-\sqrt{y} \leq X \leq \sqrt{y}\} \\ &= 2\int_0^{\sqrt{y}} \frac{1}{\sqrt{2\pi}} e^{-\frac{x^2}{2}} \mathrm{d}x. \end{aligned}$$

Taking the derivative of $F_Y(y)$ with respective to y gives

$$f_Y(y) = y^{-\frac{1}{2}} \frac{1}{\sqrt{2\pi}} e^{-\frac{y}{2}}.$$

Note that $\Gamma(\frac{1}{2}) = \sqrt{\pi}$. The distribution of X^2 is a gamma distribution with parameters $\frac{1}{2}$ and 2:

$$X^2 \sim \Gamma(\frac{1}{2}, 2).$$

This completes the solution.

19.17.

(a) By definition, we have

$$\begin{aligned} P\{|Z| \geq x\} &= \int_x^\infty e^{-\frac{s^2}{2}} \mathrm{d}s \\ &\leq \int_x^\infty \frac{s}{x} e^{-\frac{s^2}{2}} \mathrm{d}s \\ &= \frac{1}{x}\sqrt{\frac{2}{\pi}} \exp\left(-\frac{x^2}{2}\right). \end{aligned}$$

(b) Noting that for $s > 1$

$$-\frac{\mathrm{d}}{\mathrm{d}s}\left[\left(\frac{1}{s}-\frac{1}{s^3}\right)e^{-\frac{s^2}{2}}\right] = \left(1-\frac{3}{s^4}\right)e^{-\frac{s^2}{2}} \le e^{-\frac{s^2}{2}},$$

we have

$$\begin{aligned} P\{|Z| \ge x\} &= \int_x^\infty e^{-\frac{s^2}{2}}\,\mathrm{d}s \\ &\ge \int_x^\infty \left(1-\frac{3}{s^4}\right)e^{-\frac{s^2}{2}}\,\mathrm{d}s \\ &= \left(\frac{1}{x}-\frac{1}{x^3}\right)\sqrt{\frac{2}{\pi}}\exp\left(-\frac{x^2}{2}\right). \end{aligned}$$

(c) By parts (a) and (b) of this proof, we have

$$\frac{\frac{1}{x}-\frac{1}{x^3}}{\frac{1}{x}} \le \frac{P\{|Z| \ge x\}}{\frac{1}{x}\sqrt{\frac{2}{\pi}}\exp\left(-\frac{x^2}{2}\right)} \le 1.$$

Letting $x \to \infty$ gives the result.

This finishes the proof.

19.5 Bibliographic Notes

In this chapter, we presented several continuous probability distributions that can be used in finance. In general, probability distributions of asset returns should have the following properties (Watsham and Parramore, 1997):

(a) *Stationarity* – that is, the parameters of the distribution are unchanging through time;

(b) *Stability* – that is, linear combinations of the distributions have the same type of distribution.

(c) *Finite variance*.

Johnson et al. (1994), Johnson et al. (1995), Evans et al. (2000), and Forbes et al. (2011) presented a comprehensive discussion of continuous univariate distributions. Haas and Pigorsch (2011) studied several fat-tailed distributions. Balakrishnan and Nevzorov (2003) is a textbook on statistical distributions that includes several continuous probability distributions. Some common continuous distributions are discussed in Borokov (1999), Meucci (2005), Krishnan (2006), and DasGupta (2010).

CHAPTER 20

CENTRAL LIMIT THEOREMS

The central limit theorem is a very important result in probability theory. It states that the mean of a sufficiently large number of independent random variables will be approximately normally distributed under certain conditions. The central limit theorem has several versions. In this chapter, we present some central limit theorems.

20.1 Basic Concepts and Facts

Definition 20.1 (Degenerate Distribution)**.** A distribution function is said to be degenerate if it is equal to Δ_{x_0} for some x_0, where Δ_{x_0} is defined as

$$\Delta_{x_0}(x) = \begin{cases} 0, & \text{if } x < x_0; \\ 1, & \text{if } x \geq x_0. \end{cases} \tag{20.1}$$

Definition 20.2 (Weak Convergence)**.** A sequence $\{F_n\}_{n\geq 1}$ of distribution functions is said to converge weakly to a distribution function F, written as $F_n \Rightarrow F$ or $F_n(x) \Rightarrow F(x)$, if

$$\lim_{n\to\infty} F_n(x) = F(x), \quad \text{for all } x \in C(F),$$

where $C(F) = \{x \in \mathbf{R} : F \text{ is continuous at } x\}$.

Definition 20.3 (Convergence in Distribution). Let $\{X_n\}_{n\geq 1}$ be a sequence of random variables. The sequence is said to converge in distribution (or in law) to a random variable X_0, written as $X_n \xrightarrow{d} X_0$, if $F_n \Rightarrow F_0$, where F_n is the distribution function of X_n for $n \geq 0$.

Theorem 20.1 (Lindeberg's Central Limit Theorem). *For each n, let $X_{n,1}$, $X_{n,2}$, ..., X_{n,k_n} be k_n independent random variables satisfying the following conditions:*

(a) $E(X_{n,j}) = 0$ *for all* $j = 1, 2, \ldots, k_n$.

(b)

$$\sum_{j=1}^{k_n} \sigma_{n,j}^2 = 1,$$

where $\sigma_{n,j} = \sqrt{E[X_{n,j}^2]}$.

(c) (Lindeberg condition) For any $\epsilon > 0$*, we have*

$$\lim_{n\to\infty} \sum_{j=1}^{k_n} E\left(X_{n,j}^2 I_{\{|X_{n,j}|>\epsilon\}}\right) = 0.$$

Then $\sum_{j=1}^{k_n} X_{n,j}$ *converges in distribution to the standard normal random variable, that is,*

$$\sum_{j=1}^{k_n} X_{n,j} \xrightarrow{d} N(0, 1).$$

20.2 Problems

20.1. Let $\{X_n\}_{n\geq 1}$ be a sequence of random variables on some probability space $(\Omega, \mathscr{F}, P)$. Let X_0 be a random variable on $(\Omega, \mathscr{F}, P)$. Show that if $X_n \xrightarrow{P} X_0$, then $X_n \xrightarrow{d} X_0$; that is, if the X_n converges to X_0 in probability, then X_n converges to X_0 in distribution.

20.2. Let $a > 0$ and $\{a_n\}_{n\geq 1}$ a sequence of positive real numbers. Let $\{b_n\}_{n\geq 1}$ be a sequence of real numbers and b a real number. Let F be a distribution function and $\{F_n\}_{n\geq 1}$ a sequence of distribution function.

(a) Suppose that $F_n \Rightarrow F$, $a_n \to a$, and $b_n \to b$. Show that

$$G_n \Rightarrow G,$$

where $G_n(x) = F_n(a_n x + b_n)$ and $G(x) = F(ax + b)$.

(b) Suppose that $F_n \Rightarrow F$ and $a_n \to \infty$. Show that

$$G_n \Rightarrow \Delta_0,$$

where $G_n(x) = F_n(a_n x)$ and Δ_0 is as defined in Definition 20.1.

(c) Suppose that $F_n \Rightarrow F$ and $\{b_n\}_{n\geq 1}$ is unbounded. Let $G_n(x) = F_n(x + b_n)$. Show that G_n cannot converge weakly.

(d) Let $G_n(x) = F_n(a_n x + b_n)$. Suppose that $F_n \Rightarrow F$, $G_n \Rightarrow G$, and F and G are nondegenerate. Show that

$$0 < \inf_{n\geq 1} a_n \leq \sup_{n\geq 1} a_n < \infty$$

and

$$\sup_{n\geq 1} |b_n| < \infty.$$

(e) Suppose that $F(x) = F(ax + b)$ for all $x \in (-\infty, \infty)$ and F is nondegenerate. Show that $a = 1$ and $b = 0$.

20.3. Let $\{a_n\}_{n\geq 1}$ and $\{u_n\}_{n\geq 1}$ be two sequences of positive real numbers, that is, $a_n > 0$ and $u_n > 0$ for all $n \geq 1$. Let $\{F_n\}_{n\geq 1}$ be a sequence of distribution functions such that $F_n(a_n x + b_n) \Rightarrow G(x)$ and $F_n(u_n x + v_n) \Rightarrow F(x)$, where F and G are nondegenerate distribution functions. Then there exist a and b, $a > 0$, such that

$$\lim_{n\to\infty} \frac{a_n}{u_n} = a,$$

$$\lim_{n\to\infty} \frac{b_n - v_n}{u_n} = b,$$

and

$$F(ax + b) = G(x).$$

20.4 **(The Classical Central Limit Theorem).** Let $\{X_n\}_{n\geq 1}$ be a sequence of independent and identically distributed random variables with $E(X_1) = \mu$ and $\sigma = \sqrt{\text{Var}(X_1)} < \infty$. Show that

$$\sum_{j=1}^{n} \frac{X_j - \mu}{\sigma\sqrt{n}} \xrightarrow{d} N(0, 1).$$

20.5 **(Lyapounov's Central Limit Theorem).** For each n, let $X_{n,1}, X_{n,2}, \ldots, X_{n,k_n}$ be k_n independent random variables satisfying the following conditions:

(a)

$$\sum_{j=1}^{k_n} E(X_{n,j}) = \mu.$$

(b)

$$\sum_{j=1}^{k_n} \text{Var}(X_{n,j}) = \sigma^2 > 0.$$

(c) (Lyapounov condition) For some $\delta > 0$,

$$\lim_{n\to\infty} \sum_{j=1}^{k_n} E\left(|X_{n,j} - E(X_{n,j})|^{2+\delta}\right) = 0.$$

Show that $\sum_{j=1}^{k_n} X_{n,j}$ converges in distribution to the normal distribution with parameters (μ, σ^2), that is,

$$\sum_{j=1}^{k_n} X_{n,j} \xrightarrow{d} N(\mu, \sigma^2).$$

20.6. Let $\{X_n : n \geq 1\}$ be a sequence of independent random variables distributed as

$$P\{X_n = 1\} = P\{X_n = -1\} = p_n, \quad P\{X_n = 0\} = 1 - 2p_n, \quad n \geq 1,$$

where $p_n \in [0, \frac{1}{2}]$ and

$$\sum_{n=1}^{\infty} p_n = \infty.$$

Find a suitable choice of scaling constants c_n such that the rescaled sum

$$\frac{1}{c_n} \sum_{i=1}^{n} X_i$$

converges in distribution to the normal distribution with mean 0 and variance 1 as $n \to \infty$.

20.3 Hints

20.1. Use the definition of convergence in probability (Definition 9.1) and the definition of convergence in distribution (Definition 20.3).

20.2. To prove part (a), follow the definition of weak convergence (Definition 20.2) and note that a distribution function has only countably many discontinuities (see Problem 11.13). To prove part (b), consider the cases $x > 0$ and $x < 0$ and follow Definition 20.2. To prove part (c), use the method of contradiction. To prove part (d), use the method of contradiction and parts (a), (b), and (c). Use part (d) to prove part (e).

20.3. Apply the results of Problem 20.2.

20.4. Use Lindeberg's central limit theorem (Theorem 20.1) and the dominated convergence theorem (Theorem 6.4).

20.5. Let

$$Y_{n,j} = \frac{X_{n,j} - E(X_{n,j})}{\sigma}$$

and apply Lindeberg's central limit theorem (Theorem 20.1).

20.6. Find c_n to satisfy the conditions in Lindeberg's central limit theorem (Theorem 20.1).

20.4 Solutions

20.1. Let F_n be the distribution function of X_n for $n \geq 0$. Let $x \in C(F_0)$, where $C(F_0)$ is the set of points at which F_0 is continuous. Let $\delta > 0$. Then by the continuity of F_0 at x, there exists an $\epsilon > 0$ such that

$$|F_0(x+\epsilon) - F_0(x)| < \frac{\delta}{2} \tag{20.2}$$

and

$$|F_0(x-\epsilon) - F_0(x)| < \frac{\delta}{2}. \tag{20.3}$$

By the assumption that $X_n \xrightarrow{P} X$, there exists an integer N such that

$$P\{|X_n - X| \geq \epsilon\} < \frac{\delta}{2}, \quad \text{for all } n \geq N.$$

For $n \geq N$, since

$$\begin{aligned}
& P\{X_n \leq x\} \\
= \; & P(\{X_n \leq x\} \cap \{|X_n - X| < \epsilon\}) + P(\{X_n \leq x\} \cap \{|X_n - X| \geq \epsilon\}),
\end{aligned}$$

$$P(\{X_n \leq x\} \cap \{|X_n - X| < \epsilon\}) \leq P\{X \leq x + \epsilon\},$$

and

$$P(\{X_n \leq x\} \cap \{|X_n - X| \geq \epsilon\}) \leq P\{|X_n - X| \geq \epsilon\} < \frac{\delta}{2},$$

we have

$$P\{X_n \leq x\} \leq \frac{\delta}{2} + P\{X < x + \epsilon\}. \tag{20.4}$$

Similarly, we have

$$\begin{aligned}
& P\{X \leq x - \epsilon\} \\
= \; & P(\{X \leq x - \epsilon\} \cap \{|X_n - X| < \epsilon\}) + \\
& P(\{X \leq x - \epsilon\} \cap \{|X_n - X| \geq \epsilon\}),
\end{aligned}$$

which leads to

$$P\{X \leq x - \epsilon\} \leq \frac{\delta}{2} + P\{X_n \leq x\}. \tag{20.5}$$

Combining Equations (20.2) – (20.5) gives

$$F_0(x) - \delta \leq F_n(x) \leq F_0(x) + \delta.$$

Since δ is arbitrary, $F_n(x) \to F_0(x)$. Hence the conclusion is true. This completes the proof.

20.2.

(a) Let x be a continuity point of $G(x)$ and let $\epsilon > 0$. Then $F(y)$ is continuous at $ax + b$. By Problem 11.13, F has only countably many discontinuities. Then we can choose two continuity points u and v of F such that $u < ax + b < v$ and $F(v) - F(u) < \epsilon$. By the assumption that $F_n \Rightarrow F$, $a_n \to a$, and $b_n \to b$, there exists an integer N_ϵ such that for $n \geq N_\epsilon$, we have

$$|F_n(u) - F(u)| < \epsilon,$$

$$|F_n(v) - F(v)| < \epsilon,$$

and

$$u < a_n x + b_n < v.$$

Therefore

$$\begin{aligned}
& |G_n(x) - G(x)| \\
\leq\ & |F_n(a_n x + b_n) - F(ax + b)| \\
\leq\ & |F_n(a_n x + b_n) - F_n(u)| + |F_n(u) - F(u)| + |F(u) - F(ax + b)| \\
\leq\ & F_n(v) - F_n(u) + |F_n(u) - F(u)| + F(v) - F(u) \\
\leq\ & 5\epsilon.
\end{aligned}$$

Since ϵ is arbitrary, we have $G_n(x) \to G(x)$ as $n \to \infty$.

(b) Let $\epsilon > 0$. Let $x > 0$. Then $\Delta_0(x) = 1$. Let u be a continuity point of F such that $F(u) > 1 - \epsilon/2$. By the assumption that $F_n \Rightarrow F$, there exists an integer N_1 such that

$$|F_n(u) - F(u)| < \frac{\epsilon}{2}, \quad \forall n \geq N_1.$$

Since $a_n \to \infty$ and $x > 0$, there exists an integer N_2 such that $a_n x > u$ for all $n \geq N_2$. Hence for every $n \geq \max(N_1, N_2)$, we have

$$\begin{aligned}
1 - F_n(a_n x) &\leq 1 - F_n(u) \\
&\leq 1 - F(u) + |F_n(u) - F(u)| \\
&\leq \epsilon.
\end{aligned}$$

Since ϵ is arbitrary, we have $F_n(a_n x) \to 1$. Similarly, we can show that $F_n(a_n x) \to 0$ when $x < 0$. Therefore, we have $G_n \Rightarrow \Delta_0$.

(c) Assume that $G_n \Rightarrow G$ for some distribution function G. Since $\{b_n\}_{n\geq 1}$ is unbounded, we suppose that $\{b_n\}_{n\geq 1}$ is unbounded above. Let u be a continuity point of G such that $G(u) < \frac{1}{4}$. By the hypothesis that $G_n \Rightarrow G$, there exists an integer N_1 such that for all $n > N_1$, we have

$$|G_n(u) - G(u)| = |F_n(u + b_n) - G(u)| < \frac{1}{4},$$

which implies that

$$G(u) > F_n(u + b_n) - \frac{1}{4} \tag{20.6}$$

Note the assumption that $F_n \Rightarrow F$. Let v be a continuity point of F such that $F(v) > \frac{3}{4}$. Then there exists an integer N_2 such that for all $n > N_2$, we have

$$|F_n(v) - F(v)| < \frac{1}{4}. \tag{20.7}$$

By the hypothesis that $\{b_n\}_{n\geq 1}$ is unbounded above, there exists an integer $N_3 > \max(N_1, N_2)$ such that $u + b_{N_3} > v$. Hence from Equations (20.6) and (20.7), we have

$$G(u) > F_n(v) - \frac{1}{4} \geq F(v) - |F_n(v) - F(v)| - \frac{1}{4} > \frac{1}{4}.$$

But $G(u) < \frac{1}{4}$. This is a contradiction. Similarly, we can show that G_n cannot converge weakly when $\{b_n\}_{n\geq 1}$ is unbounded below.

(d) Suppose that the sequence $\{a_n\}_{n\geq 1}$ is not bounded above. Then there exists a subsequence $\{a_{k_n}\}_{n\geq 1}$ of $\{a_n\}_{n\geq 1}$ such that $a_{k_n} \to \infty$. By part (b), we have

$$F_{k_n}(a_{k_n} x) \to \Delta_0(x), \quad x \neq 0. \tag{20.8}$$

Since G is nondegenerate and

$$F_{k_n}\left[a_{k_n}\left(x + \frac{b_{k_n}}{a_{k_n}}\right)\right] = F_{k_n}(a_{k_n} x + b_{k_n}) \to G(x),$$

it follows from part (c) that $\{b_{k_n}/a_{k_n}\}_{n\geq 1}$ is bounded. Hence there exists a subsequence $\{b_{i_n}/a_{i_n}\}_{n\geq 1}$ of $\{b_{k_n}/a_{k_n}\}_{n\geq 1}$ such that

$$\left\{\frac{b_{i_n}}{a_{i_n}}\right\}_{n\geq 1} \to c$$

for some constant c. Then by part (a) and Equation (20.8), we have

$$F_{k_n}\left[a_{k_n}\left(x + \frac{b_{k_n}}{a_{k_n}}\right)\right] \to \Delta_0(x + c).$$

This contradicts the assumption that G is nondegenerate. Thus $\{a_n\}_{n\geq 1}$ is bounded above.

Since $G_n \Rightarrow G$ and $G_n(a_n^{-1}x - a_n^{-1}b_n) = F_n(x) \to F(x)$, the same argument shows that $\{a_n^{-1}\}_{n\geq 1}$ is bounded above. Hence we have

$$0 < \inf_{n\geq 1} a_n \leq \sup_{n\geq 1} a_n < \infty.$$

Now suppose that $\{b_n\}_{n\geq 1}$ is unbounded. Then $\{b_n/a_n\}_{n\geq 1}$ is unbounded. There exists a subsequence $\{b_{j_n}/a_{j_n}\}_{n\geq 1}$ such that $b_{j_n}/a_{j_n} \to \pm\infty$ and $a_{j_n} \to a$ for some $a > 0$. Then by part (a), we have

$$F_{j_n}(a_{j_n}x) \to F(ax).$$

Since G is nondegenerate, we have

$$F_{j_n}\left[a_{j_n}\left(x + \frac{b_{j_n}}{a_{j_n}}\right)\right] = F_{j_n}(a_{j_n}x + b_{j_n}) \to G(x),$$

it follows from part (c) that $\{b_{j_n}/a_{j_n}\}_{n\geq 1}$ is bounded. This is a contradiction. Hence $\{b_n\}_{n\geq 1}$ is bounded.

(e) Since $F(x) = F(ax + b)$ for all x, we have the following for every $n \geq 1$:

$$\begin{aligned} F(x) &= F(ax+b) \\ &= F(a(ax+b)+b) \\ &= \cdots \\ &= F(a^n x + (1 + a + \cdots + a^n)b). \end{aligned}$$

It follows from Item (d) that $\{a^n\}_{n\geq 1}$ is bounded away from 0 and ∞. Hence $a = 1$. Then also by part (d), $\{nb\}_{n\geq 1}$ is bounded. Hence $b = 0$.

This completes the proof.

20.3. Let $H_n(x) = F_n(u_n x + v_n)$ for $n \geq 1$. Then we have $H_n(x) \Rightarrow F(x)$ and

$$H_n\left(\frac{a_n}{u_n}x + \frac{b_n - v_n}{u_n}\right) = F_n(a_n x + b_n) \Rightarrow G(x). \tag{20.9}$$

It follows from part (d) of Problem 20.2 that

$$0 < \inf_{n\geq 1} \frac{a_n}{u_n} \leq \sup_{n\geq 1} \frac{a_n}{u_n} < \infty$$

and

$$\sup_{n\geq 1}\left|\frac{b_n - v_n}{u_n}\right| < \infty.$$

Then there exists a subsequence $\{k_n\}_{n\geq 1}$ of $\{1, 2, \ldots\}$ such that

$$\frac{a_{k_n}}{u_{k_n}} \to a, \qquad \frac{b_{k_n} - v_{k_n}}{u_{k_n}} \to b$$

for some $a > 0$ and b. By part (a) of Problem 20.2, we have

$$H_{k_n}\left(\frac{a_{k_n}}{u_{k_n}}x + \frac{b_{k_n} - v_{k_n}}{u_{k_n}}\right) \Rightarrow F(ax + b). \tag{20.10}$$

From Equations (20.9) and (20.10), we get $F(ax + b) = G(x)$.

Suppose that there exists another subsequence $\{j_n\}_{n\geq 1}$ of $\{1, 2, \ldots\}$ such that

$$\frac{a_{j_n}}{u_{j_n}} \to a', \qquad \frac{b_{j_n} - v_{j_n}}{u_{j_n}} \to b'$$

for some $a' > 0$ and b'. Then we have

$$F(a'x + b') = G(x) = F(ax + b).$$

It follows from part (e) of Problem 20.2 that $a' = a$ and $b' = b$. Therefore, we have

$$\frac{a_n}{u_n} \to a, \qquad \frac{b_n - v_n}{u_n} \to b.$$

This completes the proof.

20.4. For each n, let $Y_{n,1}$, $Y_{n,2}$, …, $Y_{n,n}$ be defined as

$$Y_{n,j} = \frac{X_j - \mu}{\sigma\sqrt{n}}, \quad j = 1, 2, \ldots, n.$$

Then it is obvious that $E(Y_{n,j}) = 0$ for all $j = 1, 2, \ldots, n$ and

$$\sum_{j=1}^{n} E\left(\frac{(X_j - \mu)^2}{\sigma^2 n}\right) = 1.$$

Now we show that the $Y_{n,j}$ satisfies the Lindeberg condition. Since $X_1, X_2, \ldots$ are identically distributed, we have

$$\begin{aligned}\sum_{j=1}^{n} E\left(Y_{n,j}^2 I_{\{|Y_{n,j}|>\epsilon\}}\right) &= \sum_{j=1}^{n} E\left(\frac{(X_j - \mu)^2}{\sigma^2 n} I_{\{|X_j-\mu|>\epsilon\sigma\sqrt{n}\}}\right)\\ &= E\left(\frac{(X_1 - \mu)^2}{\sigma^2} I_{\{|X_1-\mu|>\epsilon\sigma\sqrt{n}\}}\right).\end{aligned}$$

Note that

$$E\left(\frac{(X_1 - \mu)^2}{\sigma^2} I_{\{|X_1-\mu|>\epsilon\sigma\sqrt{n}\}}\right) \leq E\left(\frac{(X_1 - \mu)^2}{\sigma^2}\right) = 1$$

and

$$\lim_{n\to\infty} \frac{(X_1 - \mu)^2}{\sigma^2} I_{\{|X_1 - \mu| > \epsilon\sigma\sqrt{n}\}} = 0.$$

By Theorem 6.4, the $Y_{n,j}$ satisfies the Lindeberg condition. Hence the conclusion follows from Theorem 20.1. This completes the proof.

20.5. Let

$$Y_{n,j} = \frac{X_{n,j} - E(X_{n,j})}{\sigma}.$$

Then the $Y_{n,j}$ satisfies the first two conditions given in Theorem 20.1. Let $\epsilon > 0$. Since

$$I_{\{|Y_{n,j}|>\epsilon\}} \le \frac{1}{\epsilon^\delta} |Y_{n,j}|^\delta,$$

we have

$$\sum_{j=1}^{k_n} E\left(|Y_{n,j}|^2 I_{\{|Y_{n,j}|>\epsilon\}}\right) \le \sum_{j=1}^{k_n} \frac{1}{\epsilon^\delta} E\left(|Y_{n,j}|^{2+\delta}\right).$$

Hence by the Liyapounov condition, we have

$$\lim_{n\to\infty} \sum_{j=1}^{k_n} E\left(|Y_{n,j}|^2 I_{\{|Y_{n,j}|>\epsilon\}}\right) = 0.$$

It follows from Theorem 20.1 that $\sum_{j=1}^{k_n} Y_{n,j}$ converges in distribution to $N(0,1)$. Therefore $\sum_{j=1}^{k_n} X_{n,j}$ converges in distribution to $N(\mu, \sigma^2)$. This completes the proof.

20.6. Let c_n be defined as

$$c_n = \sqrt{2\sum_{j=1}^{n} p_j}, \quad n \ge 1.$$

Let $X_{n,j}$ be defined as

$$X_{n,j} = \frac{X_j}{c_n}, \quad n \ge 1, j \ge 1.$$

Then we have $E[X_{n,j}] = 0$ for all $j = 1, 2, \ldots, n$. Let $\sigma_{n,j}^2 = E[X_{n,j}^2]$. Then we have

$$\sum_{j=1}^{n} \sigma_{n,j}^2 = \sum_{j=1}^{n} \frac{2p_j}{c_n^2} = 1.$$

Let $\epsilon > 0$. Then for $n > 1/\epsilon$, we have

$$|X_{n,j}| = \frac{|X_j|}{c_n} \le \frac{1}{c_n} < \epsilon, \quad j = 1, 2, \ldots, n,$$

which gives

$$\sum_{j=1}^{n} E[X_{n,j}^2 I_{\{|X_{n,j}|>\epsilon\}}] = 0.$$

Hence

$$\lim_{n\to\infty} \sum_{j=1}^{n} E[X_{n,j}^2 I_{\{|X_{n,j}|>\epsilon\}}] = 0.$$

Therefore, it follows from Theorem 20.1 that

$$\sum_{j=1}^{n} X_{n,j} \xrightarrow{d} N(0,1).$$

This completes the solution.

20.5 Bibliographic Notes

In this chapter, we introduced the concepts of weak convergence and convergence in distribution, which are classic topics in probability theory and discussed in many texts such as Billingsley (1995). Unlike other notions of convergence, convergence in distribution does not require the random variables to be defined on a common probability space.

Lindeberg's central limit theorem is a very general central limit theorem, which can be used to prove other central limit theorems. Lindeberg's central limit theorem can be proved in several ways. For example, Bhattacharya and Waymire (2007) presented a proof without relying on the characteristic function. Athreya and Lahiri (2006) presented a proof using the characteristic function.

PART III

STOCHASTIC PROCESSES

CHAPTER 21

STOCHASTIC PROCESSES

A stochastic process is a collection of random variables. Stochastic processes play an important role in mathematical finance as they are used to model the stock prices. In this chapter, we shall present some general concepts and results about stochastic processes.

21.1 Basic Concepts and Facts

Definition 21.1 (Stochastic Process). Let I be a set and $(E, \mathscr{E})$ a measurable space. A stochastic process indexed by I is a family of measurable functions X_t, $t \in I$, from a probability space $(\Omega, \mathscr{F}, P)$ into $(E, \mathscr{E})$. The measurable space $(E, \mathscr{E})$ is referred to as the *state space*.

For each $\omega \in \Omega$, the mapping $t \to X_t(\omega)$ is called a *trajectory* or a *sample path* of X.

Definition 21.2 (Probability Law of Stochastic Process). Let $\{X_t : t \in I\}$ be a stochastic process on some probability space $(\Omega, \mathscr{F}, P)$ with some state space $(E, \mathscr{E})$. Let $X = (X_t)_{t \in I}$ be the corresponding random variable from $(\Omega, \mathscr{F})$ into the product space $(F, \mathscr{G}) = (E^I, \mathscr{E}^I)$, where

$$E^I = \mathop{\times}_{t \in I} E$$

and

$$\mathscr{E}^I = \bigotimes_{t \in I} \mathscr{E}.$$

The probability law or distribution of the stochastic process $\{X_t : t \in I\}$ is defined as $P \circ X^{-1}$, namely, the probability law of the random variable X.

Definition 21.3 (Continuous and Right-Continuous Stochastic Process)**.** A stochastic process $\{X_t : t \in I\}$ is said to be continuous if almost all sample paths of the X_t are continuous. A stochastic process $\{X_t : t \in I\}$ is said to be right-continuous if almost all sample paths of the X_t are right-continuous.

Definition 21.4 (Càdlàg Process)**.** A stochastic process $\{X_t : t \in I\}$ is called a *càdlàg process* if almost all its sample paths are right-continuous with left-hand limits.

Definition 21.5 (Independent Increments)**.** A stochastic process $\{X_t : t \in I\}$ is said to have independent increments if $X_t - X_s$ is independent of $\sigma(X_u : u \leq s)$ for all $s \leq t$.

Definition 21.6 (Stationary Increments)**.** A stochastic process $\{X_t : t \in I\}$ is said to have stationary increments if for any $s, t, u, v \in I$, $s < t$, $u < v$, $t - s = v - u$; then the distribution $X_t - X_s$ is the same as that of $X_v - X_u$.

Definition 21.7 (Finite-Dimensional Distribution)**.** Let $\{X_t : t \in I\}$ be a stochastic process with values on $(E, \mathscr{E})$. Then the finite-dimensional distributions of a stochastic process $\{X_t : t \in I\}$ are probability measures defined as

$$\mu_{t_1, t_2, \ldots, t_k}(H) = P\{(X_{t_1}, X_{t_2}, \ldots, X_{t_k}) \in H\}, \quad H \in \mathscr{E}^k,$$

where $(t_1, t_2, \ldots, t_k)$ is k-tuple of distinct elements of I.

Definition 21.8 (Consistency Conditions)**.** Let I be a set and $(E, \mathscr{E})$ a measurable space. For each n and each n-tuple $(t_1, t_2, \ldots, t_n)$ of distinct elements of I, let $\mu_{t_1, t_2, \ldots, t_n}$ be a probability measure on $(E^n, \mathscr{E}^n)$. Then the collection

$$\{\mu_{t_1, t_2, \ldots, t_n} : n \geq 1,\ t_1, t_2, \ldots, t_n \text{ are distinct elements of } I\}$$

of probability measures is said to be consistent if it satisfies the following two consistency conditions:

(a) For all $t_1, t_2, \ldots, t_k \in I$, each permutation π of $\{1, 2, \ldots, k\}$, and all H_1, H_2, $\ldots$, $H_k \in \mathscr{E}$, we have

$$\mu_{t_1, \ldots, t_k}(H_1 \times \cdots \times H_k) = \mu_{t_{\pi(1)}, \ldots, t_{\pi(k)}}(H_{\pi(1)} \times \cdots \times H_{\pi(k)}).$$

(b) For all $t_1, t_2, \ldots, t_{k+1} \in I$ and $H_1, H_2, \ldots, H_k \in \mathscr{E}$, we have

$$\mu_{t_1,\ldots,t_{k+1}}(H_1 \times \cdots \times H_k \times E) = \mu_{t_1,\ldots,t_k}(H_1 \times \cdots \times H_k).$$

Definition 21.9 (Version)**.** Let I be a set and $(E, \mathscr{E})$ a measurable space. Let $\{X_t : t \in I\}$ and $\{Y_t : t \in I\}$ be two stochastic processes with the same state space $(E, \mathscr{E})$, defined on the probability spaces $(\Omega, \mathscr{F}, P)$ and $(\Omega', \mathscr{F}', P')$, respectively. Then the two stochastic processes are called *versions* of each other if for all $t_1, t_2, \ldots, t_k \in I$ and $H_1, H_2, \ldots, H_k \in \mathscr{E}$, we have

$$P(X_{t_1} \in H_1, \ldots, X_{t_k} \in H_k) = P'(Y_{t_1} \in H_1, \ldots, Y_{t_k} \in H_k).$$

Two stochastic processes are also said to be equivalent if they are versions of each other.

Definition 21.10 (Modification)**.** Let I be a set and $(E, \mathscr{E})$ a measurable space. Let $\{X_t : t \in I\}$ and $\{Y_t : t \in I\}$ be two stochastic processes with the same state space $(E, \mathscr{E})$, defined on the same probability space $(\Omega, \mathscr{F}, P)$. The two stochastic processes are said to be modifications of each other if for any $t \in I$,

$$P\{X_t = Y_t\} = 1.$$

Definition 21.11 (Indistinguishable Process)**.** Let I be a set and $(E, \mathscr{E})$ a measurable space. Let $\{X_t : t \in I\}$ and $\{Y_t : t \in I\}$ be two stochastic processes with the same state space $(E, \mathscr{E})$, defined on the same probability space $(\Omega, \mathscr{F}, P)$. The two stochastic processes are said to be indistinguishable if almost all their sample paths agree:

$$P\{X_t = Y_t : t \in I\} = 1.$$

Definition 21.12 (Dominance)**.** Let $\{X_t : t \in I\}$ and $\{Y_t : t \in I\}$ be two stochastic processes defined on the same probability space $(\Omega, \mathscr{F}, P)$, where I is an index set. The process $\{X_t : t \in I\}$ is said to dominate $\{Y_t : t \in I\}$ if

$$P\{X_t \ge Y_t : t \in I\} = 1.$$

Definition 21.13 (Independent Stochastic Processes)**.** Let $\{X_t^{(1)} : t \in I\}$, $\{X_t^{(2)} : t \in I\}$, ..., $\{X_t^{(n)} : t \in I\}$ be n stochastic processes defined on the same probability space. The n stochastic processes are said to be independent if the σ-algebras $\mathscr{F}^{X^{(1)}}$, $\mathscr{F}^{X^{(2)}}$, ..., $\mathscr{F}^{X^{(n)}}$ are independent, where

$$\mathscr{F}^{X^{(k)}} = \sigma(X_t^{(k)} : t \in I), \quad 1 \le k \le n.$$

Definition 21.14 (Processes with Same Finite-Dimensional Distribution)**.** Let I be an index set. Let $\{X_t : t \in I\}$ and $\{Y_t : t \in I\}$ be two stochastic processes defined on probability spaces $(\Omega_1, \mathscr{F}_1, P_1)$ and $(\Omega_2, \mathscr{F}_2, P_2)$, respectively, and having the same state space $(E, \mathscr{E})$. Then X and Y are said to have the same finite-dimensional distribution if, for any integer $n \geq 1$, $t_1, t_2, \ldots, t_n \in I$, and $A \in \mathscr{E} \otimes \cdots \otimes \mathscr{E}$, we have

$$P_1((X_{t_1}, X_{t_2}, \ldots, X_{t_n}) \in A) = P_2((Y_{t_1}, Y_{t_2}, \ldots, Y_{t_n}) \in A).$$

Definition 21.15 (Measurable Process)**.** A stochastic process $\{X_t : t \geq 0\}$, defined on a probability space $(\Omega, \mathscr{F}, P)$ and having a state space $(E, \mathscr{E})$, is said to be measurable if, for every $A \in \mathscr{E}$, we have

$$\{(t, \omega) : X_t(\omega) \in A\} \in \mathcal{B}[0, \infty) \otimes \mathscr{F}.$$

Definition 21.16 (Gaussian Process)**.** A stochastic process $\{X_t : t \geq 0\}$ with the state space $(\mathbf{R}^d, \mathcal{B}(\mathbf{R}^d))$ is called a *Gaussian process* if its finite-dimensional distributions are (multidimensional) normal distributions (see Definition 19.9).

Definition 21.17 (Random Time)**.** A random variable T on a measurable space $(\Omega, \mathscr{F})$ with values in $[0, \infty]$ is called a *random time*.

Definition 21.18 (Filtration)**.** Let I be a set and $(\Omega, \mathscr{F})$ a measurable space. A filtration on $(\Omega, \mathscr{F})$ is a family $\{\mathscr{F}_t : t \in I\}$ of sub-σ-algebras of $\mathscr{F}$ such that

$$\mathscr{F}_s \subseteq \mathscr{F}_t$$

for all $s \leq t$.

If the set I is a discrete set (e.g., $I = \{1, 2, \ldots\}$), the filtration is called a *discrete filtration*. If the set I is a continuous set (e.g., $I = \{t \in \mathbf{R} : t \geq 0\}$), then the filtration is called a *continuous filtration*.

Definition 21.19 (Right Limit and Left Limit of Filtration)**.** Let $\{\mathscr{F}_t : a \leq t \leq b\}$ be a continuous filtration. Let $t \in [a, b]$. Then the right limit of the filtration at t is defined as

$$\mathscr{F}_{t+} = \bigcap_{n=1}^{\infty} \mathscr{F}_{t+\frac{1}{n}},$$

where $\mathscr{F}_t = \mathscr{F}_b$ for $t > b$.

The left limit of the filtration at t is defined as

$$\mathscr{F}_{t-} = \bigcap_{n=1}^{\infty} \mathscr{F}_{t-\frac{1}{n}},$$

where $\mathscr{F}_t = \mathscr{F}_a$ for $t < a$.

Definition 21.20 (Right-Continuous and Left-Continuous Filtration)**.** A continuous filtration $\{\mathscr{F}_t : a \leq t \leq b\}$ is said to be a right-continuous filtration if

$$\mathscr{F}_t = \mathscr{F}_{t+} \quad \text{for all } t \in [a, b].$$

A continuous filtration $\{\mathscr{F}_t : a \leq t \leq b\}$ is said to be a left-continuous filtration if

$$\mathscr{F}_t = \mathscr{F}_{t-} \quad \text{for all } t \in [a, b].$$

Definition 21.21 (Natural Filtration). Let $X = (X_t : t \in I)$ be a stochastic process. The natural filtration $\{\mathscr{F}_t^X : t \in I\}$ of X is defined as

$$\mathscr{F}_t^X = \sigma(X_s : s \leq t), \quad t \in I.$$

The natural filtration of X is also referred to as the filtration generated by X.

Definition 21.22 (Adapted Process). Let $X = (X_t : t \in I)$ be a stochastic process and $\{\mathscr{F}_t : t \in I\}$ a filtration, where I is an index set. The stochastic process is considered adapted to the filtration if X_t is $\mathscr{F}_t$-measurable for all $t \in I$.

Definition 21.23 (Progressively Measurable Process). Let $I = [0, \infty)$ or $I = [0, t_0]$ for some $t_0 < \infty$. A stochastic process $\{X_t : t \in I\}$ with values in a measurable space $(E, \mathscr{E})$ is said to be progressively measurable with respect to a filtration $\{\mathscr{F}_t : t \in I\}$ if for each $t \in I$, the map $(s, \omega) \to X_s(\omega)$ from $[0, t] \times \Omega$ to E is $\mathcal{B}[0, t] \times \mathscr{F}_t/\mathscr{E}$-measurable (i.e., measurable with respect to σ-fields $\mathcal{B}[0, t] \times \mathscr{F}_t$ and $\mathscr{E}$).

Definition 21.24 (Stochastic Continuity). A stochastic process $\{X_t : t \geq 0\}$ is said to be stochastically continuous if for all $\delta > 0$ and for all $t \geq 0$, we have

$$\lim_{h \to 0} P\{|X_{t+h} - X_t| > \delta\} = 0.$$

Theorem 21.1 (Kolmogorov's Existence Theorem). *Let I be a set. For each n and each n-tuple $(t_1, t_2, \ldots, t_n)$ of distinct elements of I, let $\mu_{t_1, t_2, \ldots, t_n}$ be a probability measure on $(\mathbf{R}^n, \mathcal{B}(\mathbf{R})^n)$. If the collection*

$$\{\mu_{t_1, t_2, \ldots, t_n} : n \geq 1,\ t_1, t_2, \ldots, t_n \textit{ are distinct elements of } T\}$$

satisfies the two consistency conditions given in Definition 21.8, then on some probability space $(\Omega, \mathscr{F}, P)$ there exists a stochastic process $\{X_t : t \in I\}$ having the collection of probability measures as its finite-dimensional distributions.

Theorem 21.2 (Kolmogorov's Continuity Criterion). *Let $\{X_t : 0 \leq t \leq T\}$ be a stochastic process with the state space $(\mathbf{R}^d, \mathcal{B}(\mathbf{R}^d))$. Suppose that there exist constants α, β, and K such that*

$$E\left(\|X_t - X_s\|^\alpha\right) \leq K|t - s|^{1+\beta}$$

for all $s, t \in [0, T]$. Then there exists a continuous modification of $\{X_t : 0 \leq t \leq T\}$.

21.2 Problems

21.1. Let $\{X_t : t \in I\}$ be a stochastic process with values on $(E, \mathscr{E})$. Let $\mu_{t_1, t_2, \ldots, t_n}$, $n \geq 1, t_1, t_2, \ldots, t_n$ distinct elements of I, be the finite-dimensional distributions of $\{X_t : t \in I\}$. Show that the collection of finite-dimensional distributions satisfies the two consistency conditions given in Definition 21.8.

21.2. Let $\{X_t : t \in I\}$ and $\{Y_t : t \in I\}$ be two stochastic processes defined on the same probability space. Show that if the two processes are modifications of each other, then they are versions of each other.

21.3. Let $\Omega = [0, \infty)$ and $\mathscr{F} = \mathcal{B}[0, \infty))$. Let P be a probability measure on $(\Omega, \mathscr{F})$ such that $P\{x\} = 0$ for all $x \in [0, \infty)$. Let $\{X_t : t \geq 0\}$ be a stochastic process on the probability space $(\Omega, \mathscr{F}, P)$ defined as

$$X_t(\omega) = \begin{cases} 1, & \text{if } t = \omega; \\ 0, & \text{otherwise.} \end{cases}$$

Let $\{Y_t : t \geq 0\}$ be a stochastic process on the probability space $(\Omega, \mathscr{F}, P)$ defined as

$$Y_t(\omega) = 0, \quad \text{for all } (t, \omega) \in [0, \infty) \times [0, \infty).$$

Show that

(a) The process $\{X_t : t \geq 0\}$ is a modification of the $\{Y_t : t \geq 0\}$.

(b) The two processes $\{X_t : t \geq 0\}$ and $\{Y_t : t \geq 0\}$ have the same finite-dimensional distributions.

21.4. Let $X = \{X_t : t \geq 0\}$ and $Y = \{Y_t : t \geq 0\}$ be two stochastic processes. Suppose that X and Y are modifications of each other and that almost all sample paths of X and Y are right-continuous. Show that X and Y are indistinguishable.

21.5. Let $X = \{X_t : t \geq 0\}$ and $Y = \{Y_t : t \geq 0\}$ be two stochastic processes defined on the same probability space $(\Omega, \mathscr{F}, P)$. Suppose that almost all sample paths of X and Y are right-continuous and that for every $t \geq 0$,

$$P\{X_t \geq Y_t\} = 1.$$

Show that X dominates Y:

$$P\{X_t \geq Y_t : t \geq 0\} = 1.$$

21.6. Let $X = \{X_t : t \geq 0\}$ be a stochastic process on some measurable space $(\Omega, \mathscr{F})$. Assume that every sample path of X is right-continuous on $[0, \infty)$ with finite left-hand limits on $(0, \infty)$. Let A be the event defined as

$$A = \{\omega \in \Omega : X_t(\omega) \text{ is continuous on } [0, t_0)\},$$

where t_0 is some positive real number. Show that

$$A \in \mathscr{F}^X_{t_0},$$

where $\mathscr{F}^X_{t_0}$ is as defined in Definition 21.21

21.7. Let $I = [0, \infty)$ or $I = [0, t_0]$ for some $t_0 < \infty$. Let $\{X_t : t \in I\}$ be a stochastic process with values in a measurable space $(E, \mathscr{E})$, where E is a metric space and $\mathscr{E}$ is its Borel σ-field. Suppose that X is adapted to a filtration $\{\mathscr{F}_t : t \in I\}$ and that every sample path of X is right-continuous. Show that X is progressively measurable with respect to the filtration.

21.8. Let $(\Omega, \mathscr{F})$ be a measurable space, and let I be an index set, countable or uncountable. For each $t \in I$, let X_t be a random variable from $(\Omega, \mathscr{F})$ into some measurable space $(E_t, \mathscr{E}_t)$. Let $X = (X_t)_{t\in I}$ be the random variable from $(\Omega, \mathscr{F})$ into the product space

$$(F, \mathscr{G}) = \left(\mathop{\times}_{t\in I} E_t, \bigotimes_{t\in I} \mathscr{E}_t \right).$$

For each $t \in I$, let $\mathcal{D}_t$ be a π-system that generates $\mathscr{E}_t$. Let $\mathscr{G}_0$ be the collection of all rectangles G in F having the form

$$G = \mathop{\times}_{t\in I} A_t,$$

where $A_t \in \mathcal{D}_t$ and A_t differs from E_t for only a finite number of t. Let $\mathcal{C}$ be given by

$$\mathcal{C} = \{\{X \in B\} : B \in \mathscr{G}_0\}.$$

Show that

(a) The set $\mathcal{C}$ is a π-system.

(b) $\mathscr{G} = \sigma(\mathscr{G}_0)$.

(c) $\sigma(\mathcal{C}) = \sigma(X)$.

21.9. Let $(\Omega, \mathscr{F})$ be a measurable space, and let I be an index set, countable or uncountable. For each $t \in I$, let X_t be a random variable from $(\Omega, \mathscr{F})$ into some measurable space $(E_t, \mathscr{E}_t)$. Let $X = (X_t)_{t\in I}$ be the random variable from $(\Omega, \mathscr{F})$ into the product space

$$(F, \mathscr{G}) = \left(\mathop{\times}_{t\in I} E_t, \bigotimes_{t\in I} \mathscr{E}_t \right).$$

Show that

$$\sigma(X) = \sigma(X_t : t \in I),$$

where

$$\sigma(X) = \{\{\omega \in \Omega : X(\omega) \in B\} : B \in \mathscr{G}\}$$

and

$$\sigma(X_t : t \in I) = \sigma\{\{\omega \in \Omega : X_t(\omega) \in B\} : B \in \mathscr{E}_t, t \in I\}.$$

21.10. Let $\{X_t : t \geq 0\}$ be a real-valued stochastic process on some probability space $(\Omega, \mathscr{F}, P)$, and let $\{\mathscr{F}_t : t \geq 0\}$ be its natural filtration:

$$\mathscr{F}_t = \sigma(X_s : 0 \leq s \leq t), \quad t \geq 0.$$

Let Y be a random variable defined on the same probability space $(\Omega, \mathscr{F}, P)$. Show that Y is independent of $\mathscr{F}_t$ (see Definition 12.3) if and only if for any integer m, any $0 \leq s_1 < s_2 < \ldots < s_m \leq t$, and any $c_1, c_2, \ldots, c_m, c \in \mathbf{R}$,

$$\begin{aligned} & P\{X_{s_1} \leq c_1, X_{s_2} \leq c_2, \ldots, X_{s_m} \leq c_m, Y \leq c\} \\ = \; & P\{X_{s_1} \leq c_1, X_{s_2} \leq c_2, \ldots, X_{s_m} \leq c_m\} P\{Y \leq c\}. \end{aligned}$$

21.11. Let $(\Omega, \mathscr{F})$ be a measurable space and let I be an index set, countable or uncountable. For each $t \in I$, let X_t be a random variable from $(\Omega, \mathscr{F})$ into some measurable space $(E_t, \mathscr{E}_t)$. Let $X = (X_t)_{t \in I}$ be the random variable from $(\Omega, \mathscr{F})$ into the product space

$$(F, \mathscr{G}) = \left(\underset{t \in I}{\times} E_t, \bigotimes_{t \in I} \mathscr{E}_t \right).$$

For each $t \in I$, let $\mathcal{D}_t$ be a π-system that generates $\mathscr{E}_t$. Let $\mathcal{M}$ be a monotone class (see Definition 5.4) of mappings from Ω to $[-\infty, \infty]$. Suppose that $\mathcal{M}$ includes every mapping $V : \Omega \to [0, 1]$ having the form

$$V = \prod_{t \in I} I_{A_t}(X_t),$$

where $A_t \in \mathcal{D}_t$ and A_t differs from E_t for only a finite number of t. Show that $\mathcal{M}$ includes every positive $\sigma(X)$-measurable function.

21.12. Let $\{X_t : t \in I\}$ be a stochastic process on some probability space $(\Omega, \mathscr{F}, P)$ with state space $(\mathbf{R}, \mathcal{B})$. Let $\mathcal{M}$ be a monotone class of mappings from Ω to $[-\infty, \infty]$. Suppose that $\mathcal{M}$ includes every function V of the form

$$V = \prod_{i=1}^{n} f_i(X_{t_i}),$$

where $n \geq 1, t_1, \ldots, t_n \in I$, and $f_1, \ldots, f_n$ are bounded continuous functions from $\mathbf{R}$ to $\mathbf{R}$. Show that $\mathcal{M}$ contains all positive $\sigma(X_t : t \in I)$-measurable functions.

21.13. Let $\{X_t : t \geq 0\}$ be a stochastically continuous process. Show that for every $u \in \mathbf{R}$, the map

$$t \to \phi_{X_t}(u)$$

is continuous, where $\phi_{X_t}(\cdot)$ is the characteristic function of X_t.

21.3 Hints

21.1. Follow the definition of the consistency conditions.

21.2. Try to establish that

$$\left(\bigcap_{i=1}^{k} \{X_{t_i} \in H_i\} \right) \cap \left(\bigcap_{i=1}^{k} \{Y_{t_i} \in H_i\} \right)^c \subseteq \bigcup_{i=1}^{k} \{X_{t_i} \neq Y_{t_i}\}$$

and

$$\left(\bigcap_{i=1}^{k}\{Y_{t_i} \in H_i\}\right) \cap \left(\bigcap_{i=1}^{k}\{X_{t_i} \in H_i\}\right)^c \subseteq \bigcup_{i=1}^{k}\{X_{t_i} \neq Y_{t_i}\}.$$

21.3. part (a) can be proved by verifying the definition directly. Part (b) can be proved by part (a) and the result of Problem 21.2.

21.4. Use the density of rational numbers to show that

$$\{X_t = Y_t : t \geq 0\} = \bigcap_{r \in [0,\infty)\cap \mathbf{Q}} \{X_r = Y_r\},$$

where $\mathbf{Q}$ is the set of all rational numbers. Then use the result of Problem 11.1.

21.5. The proof is similar to that of Problem 21.4.

21.6. Try to prove that

$$A = \left(\bigcup_{n=1}^{\infty} A_n\right)^c,$$

where

$$A_n = \bigcap_{m=1}^{\infty} \bigcup_{r_1, r_2 \in \mathbf{Q}\cap[0,t_0), |r_1 - r_2| < \frac{1}{m}} \left\{|X_{r_1} - X_{r_2}| > \frac{1}{n}\right\}.$$

Here $\mathbf{Q}$ is the set of all rational numbers.

21.7. For a fixed $t \in I$, construct a sequence $\{X_s^{(n)} : s \in [0,t]\}_{n\geq 1}$ of stochastic processes as

$$X_s^{(n)}(\omega) = X_{kt/2^n}(\omega) \text{ for } \frac{(k-1)t}{2^n} \leq s < \frac{kt}{2^n}, \quad X_t^{(n)} = X_t,$$

where $k = 1, 2, \ldots, 2^n$. Noting that for $B \in \mathscr{E}$, we have

$$\begin{aligned}&\{(s,\omega) \in [0,t] \times \Omega : X_s^{(n)} \in B\}\\ =\ &\bigcup_{k=1}^{2^n}\left(\left[\frac{(k-1)t}{2^n}, \frac{kt}{2^n}\right) \times \{\omega : X_{kt/2^n}(\omega) \in B\}\right)\\ &\cup(\{t\} \times \{\omega : X_t(\omega) \in B\}).\end{aligned}$$

21.8. Use the definition of measurable rectangles (Definition 10.4) to prove part (a). To prove part (b), consider the collection $\mathscr{G}_1$ of all measurable rectangles in F and show that $\mathscr{G}_1 \subseteq \sigma(\mathscr{G}_0)$. Part (c) can be proved by using the result of Problem 5.5 and part (b).

21.9. Use the result of Problem 21.8 to get

$$\sigma(X) = \sigma\left\{\{\omega \in \Omega : X(\omega) \in B\} : B \in \mathscr{G}_0\right\},$$

where $\mathscr{G}_0$ is the collection of measurable rectangles generating $\mathscr{G}$. Then show that

$$\sigma\left\{\{\omega \in \Omega : X(\omega) \in B\} : B \in \mathscr{G}_0\right\} \subseteq \sigma(X_t : t \in I)$$

and

$$\sigma(X_t : t \in I) \subseteq \sigma\left\{\{\omega \in \Omega : X(\omega) \in B\} : B \in \mathscr{G}_0\right\}.$$

21.10. Use the results of Problem 21.8, 21.9, 12.3, and 2.4.

21.11. Use the results of Problems 5.12 and 21.8.

21.12. Use the results of Problems 2.4, 21.11, and 21.9. Let $A = (-\infty, c]$ and show that I_A is the limit of an increasing sequence of bounded continuous functions.

21.13. Use the definition of stochastic continuity (Definition 21.24) and the $\epsilon - \delta$ technique; that is, to show that for any $\epsilon > 0$, there exists a $\delta > 0$ such that for any $h \in (-\delta, \delta)$, we have

$$|\phi_{X_{t+h}}(u) - \phi_{X_t}(u)| < \epsilon.$$

21.4 Solutions

21.1. Let $t_1, t_2, \ldots, t_k, t_{k+1} \in I$ and $H_1, H_2, \ldots, H_k \in \mathscr{E}$. Let π be a permutation of $\{1, 2, \ldots, k\}$. Since

$$\{(X_{t_1}, X_{t_2}, \ldots, X_{t_k}) \in H_1 \times H_2 \times \cdots \times H_k\}$$

and

$$\{(X_{t_{\pi(1)}}, X_{t_{\pi(2)}}, \ldots, X_{t_{\pi(k)}}) \in H_{\pi(1)} \times H_{\pi(2)} \times \cdots \times H_{\pi(k)}\}$$

are the same event, we have

$$\mu_{t_1,\ldots,t_k}(H_1 \times \cdots \times H_k) = \mu_{t_{\pi(1)},\ldots,t_{\pi(k)}}(H_{\pi(1)} \times \cdots \times H_{\pi(k)}).$$

Now since

$$\{(X_{t_1}, X_{t_2}, \ldots, X_{t_k}) \in H_1 \times H_2 \times \cdots \times H_k\}$$

and

$$\{(X_{t_1}, X_{t_2}, \ldots, X_{t_k}, X_{t_{k+1}}) \in H_1 \times H_2 \times \cdots \times H_k \times E\}$$

are the same event, we have

$$\mu_{t_1,\ldots,t_{k+1}}(H_1 \times \cdots \times H_k \times E) = \mu_{t_1,\ldots,t_k}(H_1 \times \cdots \times H_k).$$

Hence the collection of finite-dimensional distributions satisfies the two consistency conditions. This completes the proof.

21.2. Let $(E, \mathscr{E})$ be the state space of the two stochastic processes. Let $k \geq 1$ and $t_1, t_2, \ldots, t_k$ be distinct elements of I. Let $H_1, H_2, \ldots, H_k \in \mathscr{E}$. Let A and B be given by

$$A = \bigcap_{i=1}^{k} \{X_{t_i} \in H_i\}$$

and

$$B = \bigcap_{i=1}^{k} \{Y_{t_i} \in H_i\}.$$

Now we claim that

$$A \cap B^c \subseteq \bigcup_{i=1}^{k} \{X_{t_i} \neq Y_{t_i}\}. \tag{21.1}$$

To show this, let $\omega \in A \cap B^c$. Then $\omega \in A$ and $\omega \notin B$; that is, $X_{t_i}(\omega) \in H_i$ for all $i = 1, 2, \ldots, k$, and $Y_{t_j}(\omega) \notin H_j$ for some $1 \leq j \leq k$. Hence we have $X_{t_j}(\omega) \neq Y_{t_j}(\omega)$, that is, $\omega \in \{X_{t_j} \neq Y_{t_j}\}$. This gives

$$\omega \in \bigcup_{i=1}^{k} \{X_{t_i} \neq Y_{t_i}\}.$$

Since ω is arbitrary, Equation (21.1) is true. From Equation (21.1), we have

$$P(A) - P(B) \leq P(A \cap B^c) \leq P\left(\bigcup_{i=1}^{k} \{X_{t_i} \neq Y_{t_i}\}\right).$$

By the assumption that the two processes are modifications of each other, we get $P(A) - P(B) \leq 0$. Similarly we can show that $P(B) - P(A) \leq 0$. Hence we have $P(A) = P(B)$. Therefore, the two processes are versions of each other. This finishes the proof.

21.3.

(a) Let $t \geq 0$. Then we have

$$\{X_t = Y_t\} = \{X_t = 0\} = [0, t) \cup (t, \infty).$$

By the assumption that $P\{x\} = 0$ for all $x \geq 0$, we have

$$P\{X_t = Y_t\} = 1 - P\{t\} = 1.$$

Hence the two processes are modifications of each other.

(b) By part (a) and Problem 21.2, the two processes $\{X_t : t \geq 0\}$ and $\{Y_t : t \geq 0\}$ are versions of each other. By the definition of versions, the two processes have the same finite-dimensional distributions.

This completes the proof.

21.4. Let $\mathbf{Q}$ be the set of all rational numbers. We first show that

$$\{X_t = Y_t : t \geq 0\} = \bigcap_{r \in [0,\infty) \cap \mathbf{Q}} \{X_r = Y_r\}. \tag{21.2}$$

In fact, it is clear that

$$\{X_t = Y_t : t \geq 0\} \subseteq \bigcap_{r \in [0,\infty) \cap \mathbf{Q}} \{X_r = Y_r\}.$$

To show that Equation (21.2) holds, we only need to verify that

$$\bigcap_{r \in [0,\infty) \cap \mathbf{Q}} \{X_r = Y_r\} \subseteq \{X_t = Y_t : t \geq 0\}. \tag{21.3}$$

To do that, let $\omega \in \bigcap_{r \in [0,\infty) \cap \mathbf{Q}} \{X_r = Y_r\}$, then $X_r(\omega) = Y_r(\omega)$ for all $r \in [0, \infty) \cap \mathbf{Q}$. Let $t \geq 0$. Then there exists a sequence of rational numbers $r_1, r_2, \ldots$ such that $r_n \geq t$ and $r_n \to t$ as $n \to \infty$. By the assumption that X and Y are right-continuous, we have $X_{r_n}(\omega) \to X_t(\omega)$ and $Y_{r_n}(\omega) \to Y_t(\omega)$ as $n \to \infty$. Noting that $X_{r_n}(\omega) = Y_{r_n}(\omega)$ for all $n \geq 1$, we have $X_t(\omega) = Y_t(\omega)$. Since t is arbitrary, we have $X_t(\omega) = Y_t(\omega)$ for all $t \geq 0$. Hence Equation (21.3) is true.

By the assumption that X and Y are modifications of each other, for all $n \geq 1$, we have

$$P\{X_{r_n} = Y_{r_n}\} = 1.$$

Therefore, by Problem 11.1 and Equation (21.2), we have

$$P\{X_t = Y_t : t \geq 0\} = P\left(\bigcap_{r \in [0,\infty) \cap \mathbf{Q}} \{X_r = Y_r\} \right) = 1,$$

which shows that X and Y are indistinguishable. This completes the proof.

21.5. Let $\mathbf{Q}$ be the set of all rational numbers. We claim that

$$\{X_t \geq Y_t : t \geq 0\} = \bigcap_{r \in [0,\infty) \cap \mathbf{Q}} \{X_r \geq Y_r\}. \tag{21.4}$$

In fact, it is clear that

$$\{X_t \geq Y_t : t \geq 0\} \subseteq \bigcap_{r \in [0,\infty) \cap \mathbf{Q}} \{X_r \geq Y_r\}.$$

To show that Equation (21.4) holds, we only need to show that

$$\bigcap_{r \in [0,\infty) \cap \mathbf{Q}} \{X_r \geq Y_r\} \subseteq \{X_t \geq Y_t : t \geq 0\}. \tag{21.5}$$

To do that, let $\omega \in \bigcap_{r\in[0,\infty)\cap\mathbf{Q}}\{X_r \geq Y_r\}$, then $X_r(\omega) \geq Y_r(\omega)$ for all $r \in [0,\infty)\cap\mathbf{Q}$. Let $t \geq 0$. Then there exists a sequence of rational numbers $r_1, r_2, \ldots$ such that $r_n \geq t$ and $r_n \to t$ as $n \to \infty$. By the assumption that X and Y are right-continuous, we have $X_{r_n}(\omega) \to X_t(\omega)$ and $Y_{r_n}(\omega) \to Y_t(\omega)$ as $n \to \infty$. Since $X_{r_n}(\omega) \geq Y_{r_n}(\omega)$ for all $n \geq 1$, we have $X_t(\omega) \geq Y_t(\omega)$. Since t is arbitrary, we have $X_t(\omega) \geq Y_t(\omega)$ for all $t \geq 0$. Hence Equation (21.5) is true.

By assumption, we have $P\{X_{r_n} = Y_{r_n}\} = 1$ for all $n \geq 1$. It follows from Problem 11.1 and Equation (21.4) that

$$P\{X_t \geq Y_t : t \geq 0\} = P\left(\bigcap_{r\in[0,\infty)\cap\mathbf{Q}}\{X_r \geq Y_r\}\right) = 1,$$

which shows that X dominates Y. This completes the proof.

21.6. Let $A_1, A_2, \ldots$ be events defined as

$$A_n = \bigcap_{m=1}^{\infty} \bigcup_{r_1,r_2\in\mathbf{Q}\cap[0,t_0),|r_1-r_2|<\frac{1}{m}} \left\{|X_{r_1} - X_{r_2}| > \frac{1}{n}\right\},$$

where $\mathbf{Q}$ is the set of all rational numbers. Since A_n is an intersection and union of countably many $\mathscr{F}^X_{t_0}$-measurable sets, A_n is $\mathscr{F}^X_{t_0}$-measurable.

We claim that

$$A = \left(\bigcup_{n=1}^{\infty} A_n\right)^c. \tag{21.6}$$

To do that, let $\omega \in A^c$. Then $X_t(\omega)$ is not continuous on $[0, t_0)$. Hence there exists some $s_0 \in (0, t_0)$ such that $X_t(\omega)$ is not continuous at s_0. In other words, there exists an $\epsilon > 0$ such that for each $m \geq 1$, there exists some s_m with $|s_m - s_0| < 1/m$ and $|X_{s_m}(\omega) - X_{s_0}(\omega)| > 3\epsilon$. By the assumption that X is right-continuous, we can find two positive numbers δ_1 and δ_2 such that

$$|X_s(\omega) - X_{s_m}(\omega)| < \epsilon \quad \text{for all } |s - s_m| < \delta_1,$$

and

$$|X_s(\omega) - X_{s_0}(\omega)| < \epsilon \quad \text{for all } |s - s_0| < \delta_2.$$

Hence we can find two rational numbers $r_{m,1}$ and $r_{m,2}$ such that

$$|r_{m,1} - s_m| < \min\left(\delta_1, \frac{1}{3m}\right),$$

$$|r_{m,2} - s_0| < \min\left(\delta_2, \frac{1}{3m}\right),$$

$$|X_{r_{m,1}}(\omega) - X_{s_m}(\omega)| < \epsilon,$$

and

$$|X_{r_{m,2}}(\omega) - X_{s_0}(\omega)| < \epsilon.$$

From the above inequalities we get

$$|r_{m,1} - r_{m,2}| \le |r_{m,1} - s_m| + |s_m - s_0| + |s_0 - r_{m,2}| < \frac{1}{m}$$

and

$$|X_{r_{m,2}}(\omega) - X_{r_{m,1}}(\omega)| > \epsilon.$$

Therefore, we have

$$\omega \in A_N \subseteq \bigcup_{n=1}^{\infty} A_n,$$

where N is some integer greater than $\frac{1}{\epsilon}$. Since ω is arbitrary, we get

$$A^c \subseteq \bigcup_{n=1}^{\infty} A_n. \tag{21.7}$$

It is clear that

$$\bigcup_{n=1}^{\infty} A_n \subseteq A^c. \tag{21.8}$$

Combining Equations (21.7) and (21.8) gives Equation (21.6). Hence A is $\mathscr{F}_{t_0}^X$-measurable. This completes the proof.

21.7. Let $t \in I$. For $n \ge 1$, let $\left\{X_s^{(n)} : s \in [0,t]\right\}$ be a stochastic process defined as

$$X_s^{(n)}(\omega) = X_{kt/2^n}(\omega) \text{ for } \frac{(k-1)t}{2^n} \le s < \frac{kt}{2^n}, \quad X_t^{(n)} = X_t,$$

where $k = 1, 2, \ldots, 2^n$. Let $B \in \mathscr{E}$. Then we have

$$\begin{aligned} & \{(s,\omega) \in [0,t] \times \Omega : X_s^{(n)} \in B\} \\ = & \bigcup_{k=1}^{2^n} \left(\left[\frac{(k-1)t}{2^n}, \frac{kt}{2^n} \right) \times \{\omega : X_{kt/2^n}(\omega) \in B\} \right) \\ & \cup (\{t\} \times \{\omega : X_t(\omega) \in B\}) \\ \in & \mathcal{B}[0,t] \otimes \mathscr{F}_t. \end{aligned}$$

By the assumption that X is right-continuous, we have

$$\lim_{n\to\infty} \{(s,\omega) \in [0,t] \times \Omega : X_s^{(n)} \in B\} = \{(s,\omega) \in [0,t] \times \Omega : X_s \in B\}.$$

Hence we have

$$\{(s,\omega) \in [0,t] \times \Omega : X_s \in B\} \in \mathcal{B}[0,t] \otimes \mathscr{F}_t.$$

Therefore, X is progressively measurable with respect to the filtration.

21.8.

(a) Let $C_1 \in \mathcal{C}$ and $C_2 \in \mathcal{C}$. Then there exist B_1 and B_2 in $\mathscr{G}_0$ such that $C_i = \{X \in B_i\}$, $i = 1, 2$. Since B_1 and B_2 are rectangles, by Definition 10.4, $B_1 \cap B_2$ is also a rectangle. Note that $\mathcal{D}_t$, $t \in I$, are π-systems. We have $B_1 \cap B_2 \in \mathscr{G}_0$. Thus

$$C_1 \cap C_2 = \{X \in B_1\} \cap \{X \in B_2\} = \{X \in B_1 \cap B_2\} \in \mathcal{C}.$$

Therefore, $\mathcal{C}$ is a π-system.

(b) Let $\mathscr{G}_1$ be the collection of all measurable rectangles in F. Then by Definition 10.4, $\mathscr{G} = \sigma(\mathscr{G}_1)$. Hence we only need to show that

$$\sigma(\mathscr{G}_1) = \sigma(\mathscr{G}_0). \tag{21.9}$$

By the assumption that $\mathscr{E}_t = \sigma(\mathcal{D}_t)$, $t \in I$, we have $\mathscr{G}_0 \subseteq \mathscr{G}_1$, which implies that $\sigma(\mathscr{G}_0) \subseteq \sigma(\mathscr{G}_1)$. To show that Equation (21.9) holds, we need to show that $\sigma(\mathscr{G}_1) \subseteq \sigma(\mathscr{G}_0)$. To do this, it is sufficient to show that

$$\mathscr{G}_1 \subseteq \sigma(\mathscr{G}_0). \tag{21.10}$$

Let $G \in \mathscr{G}_1$. Then

$$G = \underset{t\in I}{\times} A_t,$$

where $A_t \in \mathscr{E}_t$ and A_t differs from E_t for only a finite number of t. Let $S = \{t \in I : A_t \neq E_t\}$. For each $s \in S$, let

$$G_s = \underset{t\in I}{\times} B_t,$$

where $B_t = E_t$ for $t \neq s$ and $B_s = A_s$. Then we have

$$G = \bigcap_{s\in S} G_s.$$

But $G_s \in \sigma(\mathscr{G}_0)$. It follows that Equation (21.10) holds.

(c) By part (b) of this proof, $\mathscr{G} = \sigma(\mathscr{G}_0)$. It follows from Problem 5.5 that $\sigma(X) = \sigma(\mathcal{C})$.

This finishes the proof.

21.9. Let $\mathscr{G}_0$ be the collection of all measurable rectangles in F. Then by Definition 10.4, we have $\mathscr{G} = \sigma(\mathscr{G}_0)$. By Problem 5.5, we have

$$\sigma(X) = \sigma\left\{\{\omega \in \Omega : X(\omega) \in B\} : B \in \mathscr{G}_0\right\}.$$

Hence we only need to show that

$$\sigma\left\{\{\omega \in \Omega : X(\omega) \in B\} : B \in \mathscr{G}_0\right\} = \sigma(X_t : t \in I). \tag{21.11}$$

To do that, let us first show that

$$\sigma\left\{\{\omega \in \Omega : X(\omega) \in B\} : B \in \mathscr{G}_0\right\} \subseteq \sigma(X_t : t \in I). \tag{21.12}$$

Let $A \in \{\{\omega \in \Omega : X(\omega) \in B\} : B \in \mathscr{G}_0\}$. Then there exists a set $B \in \mathscr{G}_0$ such that $A = \{X \in B\}$. Since $\mathscr{G}_0$ is the collection of all measurable rectangles on F, we have

$$B = \times_{t \in I} B_t, \quad B_t \in \mathscr{E}_t,$$

where B_t differs from E_t for only a finite number of t. Then we have

$$A = \bigcap_{t \in I} \{X_t \in B_t\} \in \{X_t \in B_t\} \in \sigma(X_t : t \in I).$$

Since A is arbitrary, it follows that Equation (21.12) holds.

Next we show that

$$\sigma(X_t : t \in I) \subseteq \sigma\left\{\{\omega \in \Omega : X(\omega) \in B\} : B \in \mathscr{G}_0\right\}. \tag{21.13}$$

Let $A \in \{\{\omega \in \Omega : X_t(\omega) \in B\} : B \in \mathscr{E}_t, t \in I\}$. Then there exists a $s \in I$ and a set $B_s \in \mathscr{E}_s$ such that $A = \{X_s \in B_s\}$. Let C be a rectangle on F given by

$$C = \times_{t \in I} C_t,$$

where $C_t = E_t$ for $t \neq s$ and $C_s = B_s$. Then we have

$$A = \{X_s \in B_s\} = \bigcap_{t \in I} \{X_t \in B_t\} = \{X \in C\} \in \left\{\{X \in B\} : B \in \mathscr{G}_0\right\},$$

which shows that Equation (21.13) holds.

Combining Equations (21.12) and (21.13) gives Equation (21.11). This completes the proof.

21.10. Let $t \geq 0$ be fixed. By Problem 21.9, we have

$$\mathscr{F}_t = \sigma(X_s : 0 \leq s \leq t) = \sigma(Z),$$

where $Z = (X_s)_{0 \leq s \leq t}$ is a random variable from $(\Omega, \mathscr{F})$ to the product space

$$\left(\times_{0 \leq s \leq t} \mathbf{R}, \bigotimes_{0 \leq s \leq t} \mathcal{B}(\mathbf{R}) \right).$$

By Problem 2.4, we have

$$\mathcal{B}(\mathbf{R}) = \sigma(\mathcal{D}),$$

where $\mathcal{D} = \{(-\infty, x] : x \in \mathbf{R}\}$. Let $\mathscr{G}_0$ be the collection of all rectangles G in $\times_{0 \leq s \leq t} \mathbf{R}$ having the form

$$G = \times_{0 \leq s \leq t} A_s,$$

where $A_s \in \mathcal{D}$ and A_s differs from $\mathbf{R}$ for only a finite number of s. Let $\mathcal{C}_1$ and $\mathcal{C}_2$ be given by

$$\mathcal{C}_1 = \{\{Z \in B\} : B \in \mathscr{G}_0\}, \quad \mathcal{C}_2 = \{\{Y \in B\} : B \in \mathcal{D}\}.$$

Then by Problem 21.8, we find that $\mathcal{C}_1$ and $\mathcal{C}_2$ are π-systems and that $\sigma(\mathcal{C}_1) = \sigma(Z)$ and $\sigma(\mathcal{C}_2) = \sigma(Y)$.

By Problem 12.3, $\sigma(Y)$ and $\mathscr{F}_t$ are independent if and only if

$$P(I \cap J) = P(I)P(J)$$

for all $I \in \mathcal{C}_1$ and $J \in \mathcal{C}_2$. This completes the proof.

21.11. Let $\mathscr{G}_0$ be the collection of all rectangles G in F having the form

$$G = \times_{t \in I} A_t,$$

where $A_t \in \mathcal{D}_t$ and A_t differs from E_t for only a finite number of t. Let $\mathcal{C}$ be given by

$$\mathcal{C} = \{\{X \in B\} : B \in \mathscr{G}_0\}.$$

Then by Problem 21.8, we know that $\mathcal{C}$ is a π-system and $\sigma(X) = \sigma(\mathcal{C})$. Hence by Problem 5.12, it is sufficient to show that for every $A \in \mathcal{C}$, we have

$$I_A \in \mathcal{M}. \tag{21.14}$$

To show this, we let $A \in \mathcal{C}$. Then there exists a set $B \in \mathscr{G}_0$ such that $A = \{X \in B\}$. Since B is a rectangle, we have

$$B = \times_{t \in I} B_t,$$

where $B_t \in \mathcal{D}_t$ and B_t differs from E_t for only a finite number of t. Then we have

$$I_A = I_B(X) = \prod_{t \in I} I_{B_t}(X_t).$$

By the assumption, Equation (21.14) holds. This completes the proof.

21.12. Let $\mathcal{D} = \{(-\infty, c] : c \in \mathbf{R}\}$. Then $\mathcal{D}$ is a π-system and by Problem 2.4, $\mathcal{B} = \sigma(\mathcal{D})$. By Problem 21.9, we have $\sigma(X_t : t \in I) = \sigma(X)$. By Problem 21.11, it is sufficient to show that $\mathcal{M}$ includes every mapping $V : \Omega \to [0, 1]$ having the form

$$V = \prod_{t \in I} I_{A_t}(X_t),$$

where $A_t \in \mathcal{D}$ and A_t differs from $\mathbf{R}$ for only a finite number of t.

To do this, let $S = \{s \in I : A_s \neq \mathbf{R}\}$. For each $s \in S$, let $A_s = (-\infty, c_s]$ for some $c_s < \infty$. Let $f_{s,n} : \mathbf{R} \to [0, 1]$ be defined as

$$f_{s,n}(x) = \begin{cases} 1, & x \in (-\infty, c_s - 1/n]; \\ n(c_s - x), & x \in (c_s - 1/n, c_s); \\ 0, & x \in [c_s, \infty), \end{cases} \qquad n \geq 1.$$

Then $f_{s,n}$, $n \geq 1$, are bounded continuous functions and $f_{s,n} \uparrow I_{A_s}$ as $n \to \infty$. Hence

$$\prod_{s \in S} f_{s,n}(X_s) \uparrow \prod_{s \in S} I_{A_s}(X_s).$$

The result follows from the assumption. This completes the proof.

21.13. Let $u \in \mathbf{R}$ be fixed. We show that the map is continuous at an arbitrary point t. To do that, let $\epsilon > 0$. Since $y \to e^{iuy}$ is continuous at the origin, we can find $\delta_1 > 0$ such that

$$|e^{iuy} - 1| < \frac{\epsilon}{2}$$

for all $y \in (-\delta_1, \delta_1)$. Also by the assumption that $\{X_t : t \geq 0\}$ is stochastically continuous, we can find $\delta_2 > 0$ such that

$$P\{|X_{t+h} - X_t| > \delta_1\} < \frac{\epsilon}{4}$$

for all $h \in (-\delta_2, \delta_2)$.

Then for all $h \in (-\delta_2, \delta_2)$, we have

$$\begin{aligned}
& |\phi_{X_{t+h}}(u) - \phi_{X_t}(u)| \\
= & \left| \int_\Omega e^{iuX_{t+h}(\omega)} - e^{iuX_t(\omega)} P(\mathrm{d}\omega) \right| \\
\leq & \int_\Omega \left| e^{iu[X_{t+h}(\omega) - X_t(\omega)]} - 1 \right| P(\mathrm{d}\omega) \\
= & \int_{-\infty}^{\infty} \left| e^{iuy} - 1 \right| F_{X_{t+h} - X_t}(\mathrm{d}y) \\
= & \int_{-\delta_1}^{\delta_1} \left| e^{iuy} - 1 \right| F_{X_{t+h} - X_t}(\mathrm{d}y) + \int_{y \notin (-\delta_1, \delta_1)} \left| e^{iuy} - 1 \right| F_{X_{t+h} - X_t}(\mathrm{d}y) \\
\leq & \frac{\epsilon}{2} + 2 \int_{y \notin (-\delta_1, \delta_1)} F_{X_{t+h} - X_t}(\mathrm{d}y) \\
< & \epsilon.
\end{aligned}$$

Since ϵ is arbitrary, the map is continuous. This completes the proof.

21.5 Bibliographic Notes

In this chapter, we introduced stochastic processes. The general definition of stochastic process is adopted from Kallenberg (1997) and Revuz and Yor (1999). Karatzas and Shreve (1988) studied stochastic processes with $(\mathbf{R}^d, \mathcal{B}(\mathbf{R}^d))$ as the state space. For a proof of Kolmogorov's existence theorem, readers are referred to Billingsley (1995). For a proof of Kolmogorov's continuity theorem, readers are referred to Stroock and Varadhan (1979).

Brzezniak and Zastawniak (1999) is a textbook that introduces stochastic processes base on a measure theoretic framework. Other books that introduce stochastic processes on a measure theoretic framework include Çinlar (1974), Liptser and Shiryayev (1977), Kannan (1979), Doob (1990), Billingsley (1995), Koralov and Sinai (2007), and Çinlar (2011). For introduction to stochastic processes without measure theory, readers are referred to Karlin and Taylor (1975), Karlin and Taylor (1981), Resnick (1992), Ross (1995), Stirzaker (2005), and Ross (2010).

There are also several problem books on stochastic processes. For example, Brzezniak and Zastawniak (1999) is a problem book that is suitable for final year undergraduate students. Takacs (1960), Baldi et al. (2002), and Gusak et al. (2010) are also problem books on stochastic processes.

CHAPTER 22

MARTINGALES

A martingale is a special stochastic process whose increments over an interval in future have zero expectation given information of the past history of the process. Martingales play a central role in the theory of stochastic processes and stochastic calculus. In this chapter, we present the definition of martingales.

22.1 Basic Concepts and Facts

Definition 22.1 (Martingale). Let I be a set. A stochastic process $X = (X_t : t \in I)$ is called a *martingale* relative to a filtration $(\{\mathscr{F}_t : t \in I\}, P)$ if the process satisfies the following three conditions:

(a) X is adapted to the filtration.

(b) $E(|X_t|) < \infty$ for all $t \in I$.

(c) $E[X_t|\mathscr{F}_s] = X_s$ a.s. for all $s \leq t$.

Definition 22.2 (Supermartingale). Let I be a set. A stochastic process $X = (X_t : t \in I)$ is called a *supermartingale* relative to a filtration $(\{\mathscr{F}_t : t \in I\}, P)$ if the process satisfies the following three conditions:

(a) X is adapted to the filtration;

(b) $E(|X_t|) < \infty$ for all $t \in I$;

(c) $E[X_t|\mathscr{F}_s] \leq X_s$ a.s. for all $s \leq t$.

Definition 22.3 (Submartingale)**.** Let I be a set. A stochastic process $X = (X_t : t \in I)$ is called a *submartingale* relative to a filtration $(\{\mathscr{F}_t : t \in I\}, P)$ if the process satisfies the following three conditions:

(a) X is adapted to the filtration.

(b) $E(|X_t|) < \infty$ for all $t \in I$.

(c) $E[X_t|\mathscr{F}_s] \geq X_s$ a.s. for all $s \leq t$.

22.2 Problems

22.1. Let $X = (X_n : n \geq 0)$ be a martingale with respect to a filtration $(\{\mathscr{F}_n : n \geq 0\}, P)$. Let $m < n$. Show that

$$E[X_n|\mathscr{F}_m] = X_m.$$

22.2. Let $X_1, X_2, \ldots$ be a sequence of independent random variable on a measure space $(\Omega, \mathscr{F}, P)$ with $E(|X_n|) < \infty$ and $E(X_n) = 0$ for all $n \geq 1$. Let $S_0 = 0$, $\mathscr{F}_0 = \{\emptyset, \Omega\}$,

$$S_n = X_1 + X_2 + \cdots + X_n, \quad n \geq 1,$$

and

$$\mathscr{F}_n = \sigma(X_1, X_2, \cdots, X_n), \quad n \geq 1.$$

Show that $S = (S_n : n \geq 1)$ is a martingale with respect to $\{\mathscr{F}_n : n \geq 0\}$.

22.3. Let $\{X_n\}_{n\geq 0}$ be a sequence of integrable random variables on the probability space $(\Omega, \mathscr{F}, P)$ and $\{\mathscr{F}_n : n \geq 0\}$ a filtration of $\mathscr{F}$. Assume that $\{X_n\}_{n\geq 0}$ is adapted to $\{\mathscr{F}_n : n \geq 0\}$. Show that

(a) $\{X_n\}_{n\geq 0}$ is a martingale if and only if

$$\int_A X_n \mathrm{d}P = \int_A X_{n+1} \mathrm{d}P, \quad \forall A \in \mathscr{F}_n,\ n = 1, 2, \ldots.$$

(b) $\{X_n\}_{n\geq 0}$ is a submartingale if and only if

$$\int_A X_n \mathrm{d}P \leq \int_A X_{n+1} \mathrm{d}P, \quad \forall A \in \mathscr{F}_n,\ n = 1, 2, \ldots.$$

22.4. Let $(X_n : n \geq 0)$ and $(Y_n : n \geq 0)$ be submartingales with respect to the filtration $\{\mathscr{F}_n : n \geq 0\}$. Show that $(\max(X_n, Y_n) : n \geq 0)$ is also a submartingale with respect to $\{\mathscr{F}_n : n \geq 0\}$.

22.5. Let $\{Y_n\}_{n\geq 1}$ be a sequence of independent random variables with $E(Y_n) = a_n \neq 0$. Let

$$X_n = \prod_{j=1}^{n} \frac{Y_j}{a_j}, \quad n \geq 1$$

and

$$\mathscr{F}_n = \sigma(Y_1, Y_2, \ldots, Y_n), \quad n \geq 1.$$

Show that $\{X_n\}_{n\geq 1}$ is a martingale with respect to the filtration $\{\mathscr{F}_n : n \geq 1\}$.

22.6 (Doob-Lévy Martingale)**.** Let Y be a random variable on the probability space $(\Omega, \mathscr{F}, P)$ with $E(|Y|) < \infty$. Let $\{\mathscr{F}_n\}_{n\geq 0}$ be an increasing sequence of sub-σ-fields of $\mathscr{F}$. Let $X_n = E(Y|\mathscr{F}_n)$, $n = 0, 1, \ldots$. Show that $\{X_n\}_{n\geq 0}$ is a martingale with respect to the filtration $\{\mathscr{F}_n : n \geq 0\}$.

22.7. Let $\{X_n\}_{n\geq 1}$ be a submartingale with respect to the filtration $\{\mathscr{F}_n : n \geq 1\}$. Let g be a convex and increasing function from $\mathbf{R}$ to $\mathbf{R}$. Assume that $g(X_n)$ is integrable for all $n \geq 1$. Show that $\{g(X_n))\}_{n\geq 1}$ is a submartingale with respect to the filtration $\{\mathscr{F}_n : n \geq 1\}$.

22.8. Let $\{X_n\}_{n\geq 1}$ be a martingale with respect to the filtration $\{\mathscr{F}_n : n \geq 1\}$. Let g be a convex function from $\mathbf{R}$ to $\mathbf{R}$. Assume that $g(X_n)$ is integrable for all $n \geq 1$. Show that $\{g(X_n)\}_{n\geq 1}$ is a submartingale with respect to the filtration $\{\mathscr{F}_n : n \geq 1\}$.

22.9 (Optional Skipping Theorem)**.** Let $\{X_n\}_{n\geq 1}$ be a submartingale with respect to the filtration $\{\mathscr{F}_n : n \geq 1\}$. Let M_k $(k \geq 1)$ be Borel measurable functions of $X_1, X_2, \ldots$ defined as

$$M_k(X_1, X_2, \ldots, X_k) = \begin{cases} 1 & \text{if } (X_1, X_2, \ldots, X_k) \in B_k; \\ 0 & \text{if } (X_1, X_2, \ldots, X_k) \notin B_k, \end{cases}$$

where $B_1, B_2, \ldots$ are arbitrary sets in $\mathcal{B}(\mathbf{R}^k)$. Let Y_n $(n \geq 1)$ be defined as

$$Y_n = X_1 + \sum_{j=1}^{n-1} M_j(X_{j+1} - X_j).$$

Show that

(a) $\{Y_n\}_{n\geq 1}$ is a submartingale with respect to the filtration $\{\mathscr{F}_n : n \geq 1\}$.

(b) $E(Y_n) \leq E(X_n)$ for all $n \geq 1$.

(c) If $\{X_n\}_{n\geq 1}$ is a martingale with respect to the filtration $\{\mathscr{F}_n : n \geq 1\}$, then $\{Y_n\}_{n\geq 1}$ is also a martingale with respect to the filtration $\{\mathscr{F}_n : n \geq 1\}$ and $E(Y_n) = E(X_n)$ for all $n \geq 1$.

22.10. Let $\{X_n\}_{n\geq 0}$ be a submartingale with respect to the filtration $\{\mathscr{F}_n : n \geq 0\}$. Let $\{A_n\}_{n\geq 0}$ be a sequence of random variable defined as

$$A_n = \begin{cases} 0, & \text{if } n = 0; \\ \sum_{i=1}^{n} (E[X_i|\mathscr{F}_{i-1}] - X_{i-1}), & \text{if } n \geq 1. \end{cases}$$

Show that $M_n = X_n - A_n$ is a martingale.

22.11 (Doob Decomposition). Let $\{X_n\}_{n\geq 0}$ be a submartingale with respect to the filtration $\{\mathscr{F}_n : n \geq 0\}$. Show that X_n can be written as

$$X_n = M_n + A_n, \quad n \geq 0,$$

where $\{M_n\}_{n\geq 0}$ is a martingale, A_n is $\mathscr{F}_{n-1}$-measurable for all $n \geq 0$, $A_n \geq A_{n-1}$ for all $n \geq 1$, and $A_0 = 0$.

22.12. Let $\{X_t : 0 \leq t \leq T\}$ be a supermartingale. Show that $\{X_t : 0 \leq t \leq T\}$ is a martingale if and only if

$$E[X_T] = E[X_0].$$

22.3 Hints

22.1. Use the tower property of the conditional expectations (see Problem 14.10).

22.2. Follow the definition of martingales (see Definition 22.1) and use Theorem 11.2 and the results of Problem 13.3, 14.4, 14.5, and 14.16.

22.3. Use the definition of martingales (Definition 22.1) and the definition of conditional expectations (Definition 14.1).

22.4. Use the results of Problems 14.6 and 5.10.

22.5. Use the definition of martingales (Definition 22.1) and Theorem 13.2.

22.6. Use the tower property of the conditional expectations (see Problem 14.10) and the result of Problem 15.7.

22.7. Use the conditional Jensen's inequality (see Problem 15.7).

22.8. Use the conditional Jensen's inequality (see Problem 15.7).

22.9. Part (a) can be proved by using the definition of submartingales (Definition 22.3). Part (b) can be proved by using the tower property of conditional expectations (see Problem 14.10) and using the relation $Y_{n+1}-X_{n+1} = Y_n-X_n+(1-M_n)(X_n-X_{n+1})$ recursively. Part (c) can be proved similarly.

22.10. Follow the definition of discrete martingales (Definition 22.1).

22.11. Use the result of Problem 22.10.

22.12. To prove the sufficiency ("only if") part, use the definition of martingales and the tower property of conditional expectations (see Problem 14.10). Use the method of contradiction to prove the necessity part.

22.4 Solutions

22.1. Note that $\{\mathscr{F}_n : n \geq 0\}$ is a filtration, which is an increasing family of σ-fields. By Problem 14.10, we have

$$E[X_n|\mathscr{F}_m] = E[E[X_n|\mathscr{F}_{n-1}]|\mathscr{F}_m]. \tag{22.1}$$

Since X is a martingale, we have $E[X_n|\mathscr{F}_{n-1}] = X_{n-1}$. Substituting X_{n-1} into Equation (22.1) gives

$$E[X_n|\mathscr{F}_m] = E[X_{n-1}|\mathscr{F}_m].$$

Repeating the above process gives $E[X_n|\mathscr{F}_m] = E[X_m|\mathscr{F}_m]$. By Problem 14.4, we have $E[X_n|\mathscr{F}_m] = X_m$. This completes the proof.

22.2. To show that S is a martingale, we only need to verify that S satisfies the three conditions. First, by Theorem 11.2 we know that S_n is a $\mathscr{F}_n$-measurable function for each $n \geq 0$. Hence S_n is adapted to the filtration. Then, by Problem 13.3 and the assumption, we have

$$\begin{aligned} E(|S_n|) &= E\left(\left|\sum_{i=1}^n X_i\right|\right) \\ &\leq E\left(\sum_{i=1}^n |X_i|\right) \\ &= \sum_{i=1}^n E(|X_i|) \\ &< \infty \end{aligned}$$

for all $n \geq 0$. Finally, by Problem 14.5 and the assumption, we have

$$E(S_n|\mathscr{F}_{n-1}) = E\left(S_{n-1} + X_n|\mathscr{F}_{n-1}\right) = E\left(S_{n-1}|\mathscr{F}_{n-1}\right) + E(X_n|\mathscr{F}_{n-1}).$$

Noting that S_{n-1} is $\mathscr{F}_{n-1}$-measurable, by Problem 14.4 we have $E(S_{n-1}|\mathscr{F}_{n-1}) = S_{n-1}$. Also noting that X_n and $\mathscr{F}_{n-1}$ are independent, by Problem 14.16 we have

$E(X_n|\mathscr{F}_{n-1}) = E(X_n) = 0$. Hence $E(S_n|\mathscr{F}_{n-1}) = S_{n-1}$. This completes the proof.

22.3.

(a) By the assumption that $\{X_n\}_{n\geq 0}$ is integrable, we have $E(|X_n|) < \infty$ for $n \geq 0$. Note that $\{X_n\}_{n\geq 0}$ is adapted to the filtration. We only need to verify that $E(X_{n+1}|\mathscr{F}_n) = X_n$ is equivalent to

$$\int_A X_n \mathrm{d}P = \int_A X_{n+1} \mathrm{d}P, \quad \forall A \in \mathscr{F}_n.$$

But this equation is implied by Definition 14.1 and the assumption. This completes the proof.

(b) The proof is similar to that of part (a).

22.4. By Problem 5.10, we have $\max(X_n, Y_n)$, which is adapted to the filtration $\{\mathscr{F}_n : n \geq 0\}$. In addition,

$$E(|\max(X_n, Y_n)|) \leq E(|X_n| + |Y_n|) = E(|X_n|) + E(|Y_n|) < \infty.$$

Hence we only need to verify that

$$E[\max(X_n, Y_n)|\mathscr{F}_{n-1}] \geq \max(X_{n-1}, Y_{n-1}), \quad n \geq 1. \tag{22.2}$$

By the assumption that $(X_n : n \geq 0)$ and $(Y_n : n \geq 0)$ are submartingales and Problem 14.6, we have

$$E[\max(X_n, Y_n)|\mathscr{F}_{n-1}] \geq E[X_n|\mathscr{F}_{n-1}] \geq X_{n-1}$$

and

$$E[\max(X_n, Y_n)|\mathscr{F}_{n-1}] \geq E[Y_n|\mathscr{F}_{n-1}] \geq Y_{n-1}.$$

The above two inequalities imply that Inequality (22.2) is true. This completes the proof.

22.5. By definition of X_n, X_n is $\mathscr{F}_n$-measurable. By Theorem 13.2, we have

$$E(|X_n|) = E\left(\left|\prod_{j=1}^{n} \frac{Y_j}{a_j}\right|\right) = \prod_{j=1}^{n} E\left(\left|\frac{Y_j}{a_j}\right|\right).$$

Since $E(Y_n), n = 1, 2, \ldots$ exist, we have $E(|X_n|) < \infty$. By Theorem 13.2, we have

$$\begin{aligned} E(X_{n+1}|\mathscr{F}_n) &= E\left(X_n \frac{Y_{n+1}}{a_{n+1}} \middle| \mathscr{F}_n\right) \\ &= X_n E\left(\frac{Y_{n+1}}{a_{n+1}}\right) \\ &= X_n. \end{aligned}$$

Hence $\{X_n\}_{n\geq 1}$ is a martingale with respect to the filtration $\{\mathscr{F}_n : n \geq 1\}$.

22.6. By definition, $X_n = E(Y|\mathscr{F}_n)$ is $\mathscr{F}_n$-measurable. In addition, by Problems 15.7 and 14.10, we have

$$E(|X_n|) = E(|E(Y|\mathscr{F}_n)|) \leq E(E(|Y||\mathscr{F}_n)) = E(|Y|) < \infty.$$

By Problem 14.10, we have

$$\begin{aligned} E(X_n|\mathscr{F}_{n-1}) &= E[E(Y|\mathscr{F}_n)|\mathscr{F}_{n-1}] \\ &= E(Y|\mathscr{F}_{n-1}) \\ &= X_{n-1}, \quad n \geq 1. \end{aligned}$$

Hence $\{X_n\}_{n\geq 0}$ is a martingale. This completes the proof.

22.7. We only need to verify the last condition of submartingales:

$$E(g(X_{n+1})|\mathscr{F}_n) \geq g(X_n), \quad n \geq 1. \tag{22.3}$$

But by Problem 15.7 and the assumption that g is convex, we have

$$E(g(X_{n+1})|\mathscr{F}_n) \geq g(E(X_{n+1}|\mathscr{F}_n), \quad n \geq 1.$$

By the assumption that $\{X_n\}_{n\geq 1}$ is a submartingale, we have

$$E(X_{n+1}|\mathscr{F}_n) \geq X_n, \quad n \geq 1.$$

Note that g is also an increasing function. The inequality (22.3) is true. This completes the proof.

22.8. We only need to verify the last condition of submartingales, i.e,

$$E(g(X_{n+1})|\mathscr{F}_n) \geq g(X_n), \quad n \geq 1. \tag{22.4}$$

But by Problem 15.7 and the assumption that g is convex, we have

$$E(g(X_{n+1})|\mathscr{F}_n) \geq g(E(X_{n+1}|\mathscr{F}_n), \quad n \geq 1.$$

By the assumption that $\{X_n\}_{n\geq 1}$ is a martingale, we have

$$E(X_{n+1}|\mathscr{F}_n) = X_n, \quad n \geq 1.$$

Hence the inequality (22.4) is true. This completes the proof.

22.9.

(a) It is obvious that Y_n is $\mathscr{F}_n$-measurable and $E(|Y_n|) < \infty$ for all $n \geq 1$. Noting that M_n is a Borel measurable function of $X_1, X_2, \ldots, X_n$ and that $\{X_n\}_{n\geq 1}$ is a submartingale, we have

$$\begin{aligned} E(Y_{n+1}|\mathscr{F}_n) &= E(Y_n + M_n(X_{n+1} - X_n)|\mathscr{F}_n) \\ &= Y_n + M_n E(X_{n+1} - X_n|\mathscr{F}_n) \\ &\geq Y_n. \end{aligned}$$

Hence $\{Y_n\}_{n\geq 1}$ is also a submartingale with respect to the filtration $\{\mathscr{F}_n : n \geq 1\}$.

(b) By definition, we have

$$Y_{n+1} = Y_n + M_n(X_{n+1} - X_n).$$

Subtracting X_{n+1} from both sides of the above equation gives

$$Y_{n+1} - X_{n+1} = Y_n + (M_n - 1)X_{n+1} - M_n X_n,$$

which is equivalent to

$$Y_{n+1} - Y_n = X_{n+1} - X_n + (1 - M_n)(X_n - X_{n+1}).$$

Taking expectation of both sides of the above equation gives

$$E(Y_{n+1} - X_{n+1}) = E(Y_n - X_n) + E[(1 - M_n)(X_n - X_{n+1})]. \quad (22.5)$$

By Problem 14.10, we have

$$\begin{aligned} & E[(1 - M_n)(X_n - X_{n+1})] \\ = \; & E(E[(1 - M_n)(X_n - X_{n+1})|\mathscr{F}_n]) \\ = \; & E((1 - M_n)(X_n - E[X_{n+1}|\mathscr{F}_n]). \end{aligned} \quad (22.6)$$

Noting that $\{X_n\}_{n\geq 1}$ is a submartingale, we have

$$E(X_{n+1}|\mathscr{F}_n) \geq X_n.$$

Hence

$$(1 - M_n)(X_n - E(X_{n+1}|\mathscr{F}_n) \leq 0. \quad (22.7)$$

Combining inequalities (22.5), (22.6), and (22.7) gives

$$E(Y_{n+1} - X_{n+1}) \leq E(Y_n - X_n), \quad n \geq 1.$$

Note that $E(Y_1 - X_1) = 0$. We have $E(Y_n - X_n) \leq 0$ for all $n \geq 1$.

(c) The proof is similar to those of parts (a) and (b).

22.10. For each $n \geq 1$, A_n is $\mathcal{F}_{n-1}$-measurable. Hence $X_n - A_n$ is $\mathscr{F}_n$-measurable. Thus, M_n is adapted to the $\mathscr{F}_n$. Since X_n is a submartingale, we have

$$\begin{aligned} E(|M_n|) & \leq E(|X_n|) + E(|A_n|) \\ & \leq E(|X_n|) + \sum_{i=1}^{n} \{E(E(|X_i||\mathscr{F}_{i-1})) + E(|X_{i-1}|)\} \\ & = E(|X_n|) + \sum_{i=1}^{n} \{E(|X_i|) + E(|X_{i-1}|)\} \\ & < \infty. \end{aligned}$$

Now for $n \geq 1$, we have

$$\begin{aligned} & E(M_n|\mathscr{F}_{n-1}) \\ = \ & E(X_n|\mathscr{F}_{n-1}) - E(A_n|\mathscr{F}_{n-1}) \\ = \ & E(X_n|\mathscr{F}_{n-1}) - A_n \\ = \ & E(X_n|\mathscr{F}_{n-1}) - \sum_{i=1}^{n} E(X_i|\mathscr{F}_{i-1}) + \sum_{i=1}^{n} X_{i-1} \\ = \ & M_{n-1}. \end{aligned}$$

Therefore, M_n is a martingale.

22.11. Let $A_0 = 0$ and

$$A_n = \sum_{i=1}^{n} (E[X_i|\mathscr{F}_{i-1}] - X_{i-1}), \quad n \geq 1.$$

Then for each $n \geq 1$, A_n is $\mathscr{F}_{n-1}$-measurable. Since the X_n is a submartingale, we have

$$A_n = E[X_n|\mathscr{F}_{n-1}] - X_{n-1}) + A_{n-1} \geq A_{n-1}, \quad n \geq 1.$$

Now let $M_n = X_n - A_n$. By Problem 22.10, the M_n is a martingale. This completes the proof.

22.12. First, let us prove the sufficiency. Suppose that $\{X_t : 0 \leq t \leq T\}$ is a martingale. Then by Definition 22.1 we have

$$E[X_T|\mathscr{F}_0] = X_0.$$

By Problem 14.10, we get

$$E[X_T] = E(E[X_T|\mathscr{F}_0]) = E[X_0].$$

Now let us prove the necessity. Suppose that $E[X_T] = E[X_0]$. Since $\{X_t : 0 \leq t \leq T\}$ be a supermartingale, we have

$$E[X_t|\mathscr{F}_s] \leq X_s \quad \text{a.s.}$$

for all $s \leq t$. Hence we have

$$E[X_T] \leq E[X_t] \leq E[X_s] \leq E[X_0]$$

for all $0 \leq s \leq t \leq T$. By the assumption, we get $E[X_s] = E[X_t]$ for all $s \leq t$. Let $A_n = \{X_s - E[X_t|\mathscr{F}_s] > 1/n\}$ for $n \geq 1$. Then for $s < t$, we have

$$0 = E[X_s] - E[X_t] = E(X_s - E[X_t|\mathscr{F}_s]) \geq \frac{1}{n} P(A_n),$$

which implies $P(A_n) = 0$. By Theorem 2.2, we have $P(\limsup A_n) = 0$. But

$$\{X_s - E[X_t|\mathscr{F}_s] > 0\} \subseteq \limsup A_n.$$

Hence we have $P\{X_s - E[X_t|\mathscr{F}_s] > 0\} = 0$; that is, $X_s - E[X_t|\mathscr{F}_s] = 0$ a.s. Therefore, $\{X_t : 0 \leq t \leq T\}$ is a martingale. This completes the proof.

22.5 Bibliographic Notes

In this chapter, we defined the martingale. Since the martingale is the most basic concept in stochastic processes and stochastic calculus, it has been studied in many books such as Karlin and Taylor (1975), Williams (1991), Resnick (1999), Shorack (2000), Chung (2000), Klebaner (2005), Gut (2007), and Pascucci (2011). Baldi et al. (2002) is a problem book with martingales. For some interesting applications of martingales, readers are referred to Ross (1995).

CHAPTER 23

STOPPING TIMES

Stopping times are important tools that allow us to analyze stochastic processes by viewing them at random times. In this chapter, we present a definition and relevant results of stopping times.

23.1 Basic Concepts and Facts

Definition 23.1 (Discrete Stopping Time)**.** Let $(\Omega, \mathscr{F}, P)$ be a probability space and $\{\mathscr{F}_n, n = 0, 1, 2, \ldots\}$ an increasing sequence of sub-σ-fields of $\mathscr{F}$. A stopping time for $\{\mathscr{F}_n : n \geq 0\}$ is a function

$$T : \Omega \to \{0, 1, \ldots, \infty\}$$

such that $\{T \leq n\} \in \mathscr{F}_n$ for each $n \geq 0$.

A stopping time for a sequence of random variables, $\{X_n\}_{n \geq 0}$, is a stopping time relative to the σ-fields $\mathscr{F}_n = \sigma(X_0, X_1, \ldots, X_n)$.

In the above definition, the index starts from 0. The definition can be modified in the obvious way if the index starts from 1.

Definition 23.2 (Continuous Stopping Time)**.** Let $(\Omega, \mathscr{F}, P)$ be a probability space and $\{\mathscr{F}_t : a \leq t \leq b\}$ be a filtration. A stopping time with respect to the filtration is a function

$$\tau : \Omega \to [a, b]$$

such that $\{\tau \leq t\} \in \mathscr{F}_t$ for all $t \in [a, b]$.

Definition 23.3 (Events Prior to T)**.** Let $\{\mathscr{F}_n, n = 0, 1, 2, \ldots\}$ be an increasing sequence of sub-σ-fields of $\mathscr{F}$. Let T be a stopping time for $\{\mathscr{F}_n : n \geq 0\}$. An event $A \in \mathscr{F}$ is said to be prior to T if and only if

$$A \cap \{T \leq n\} \in \mathscr{F}_n, \quad \forall n \geq 0.$$

The collection of all sets prior to T is denoted by $\mathscr{F}_T$:

$$\mathscr{F}_T = \{A \in \mathscr{F} : A \cap \{T \leq n\} \in \mathscr{F}_n, n = 0, 1, \ldots\}.$$

For a continuous stopping time τ with respect to a filtration $\{\mathscr{F}_t : a \leq t \leq b\}$, the quantity $\mathscr{F}_\tau$ is defined as

$$\mathscr{F}_\tau = \{A \in \mathscr{F} : A \cap \{\tau \leq t\} \in \mathscr{F}_t \text{ for all } t \in [a, b]\}.$$

Definition 23.4 (Hitting Time and Exit Time)**.** Let $\{X_t : t \geq 0\}$ be a stochastic process adapted to some filtration $\{\mathscr{F}_t : t \geq 0\}$. Let A be a measurable subset of $\mathbf{R}$. The hitting time of A is defined as

$$T_A = \begin{cases} \inf\{t \geq 0 : X_t \in A\}, & \text{if } \{t \geq 0 : X_t \in A\} \neq \emptyset; \\ \infty, & \text{if otherwise.} \end{cases}$$

The exit time from the set A is defined as

$$\tau_A = T_{\mathbf{R} \setminus A} = \begin{cases} \inf\{t \geq 0 : X_t \notin A\}, & \text{if } \{t \geq 0 : X_t \notin A\} \neq \emptyset; \\ \infty, & \text{if otherwise.} \end{cases}$$

Theorem 23.1 (Optimal Stopping Theorem)**.** *Let $I = [0, \infty)$ or $I = \mathbf{N}$. Let $\{X_t : t \in I\}$ be a right-continuous supermartingale on a probability space $(\Omega, \mathscr{F}, P)$ with respect to a filtration $\{\mathscr{F}_t : t \in I\}$. Suppose that τ_1 and τ_2 are bounded stopping times such that $\tau_1 \leq \tau_2$. Then*

$$E[X_{\tau_2} | \mathscr{F}_{\tau_1}] \leq X_{\tau_1},$$

where

$$\mathscr{F}_{\tau_1} = \{A \in \mathscr{F} : A \cap \{\tau_1 \leq t\} \in \mathscr{F}_t, t \in I\}.$$

In addition, if the X_t is a right-continuous martingale, then

$$E[X_{\tau_2} | \mathscr{F}_{\tau_1}] = X_{\tau_1}.$$

If the X_t is a right-continuous submartingale, then

$$E[X_{\tau_2}|\mathscr{F}_{\tau_1}] \geq X_{\tau_1}.$$

Theorem 23.2 (Optimal Sampling Theorem). *Let $I = [0, \infty)$ or $I = \mathbf{N}$. Let $\{X_t : t \in I\}$ be a right-continuous martingale on a probability space $(\Omega, \mathscr{F}, P)$ with respect to a filtration $\{\mathscr{F}_t : t \in I\}$. If τ is a stopping time (not necessarily finite), then $\{X_{\tau\wedge t} : t \in I\}$ is a martingale with respect to the filtration $\{\mathscr{F}_{\tau\wedge t} : t \in I\}$.*

23.2 Problems

23.1. Let $\{\mathscr{F}_n : n \geq 0\}$ be an increasing sequence of σ-fields. Let S and T be stopping times for $\{\mathscr{F}_n : n \geq 0\}$. Show that $S \vee T$ and $S \wedge T$ are also stopping times for $\{\mathscr{F}_n : n \geq 0\}$, where

$$S \vee T = \max(S, T), \quad S \wedge T = \min(S, T).$$

23.2 (Hitting Time). Let $\{X_n\}_{n\geq 0}$ be a sequence of random variables on the probability space $(\Omega, \mathscr{F}, P)$ and $B \in \mathcal{B}$. Let $T : \Omega \to \{0, 1, \ldots, \infty\}$ be defined as

$$T(\omega) = \begin{cases} \min\{n : X_n(\omega) \in B\} & \text{if } X_n(\omega) \in B \text{ for some } n; \\ \infty & \text{if } X_n(\omega) \text{ is never in } B. \end{cases}$$

Show that T is a stopping time for $\{\mathscr{F}_n : n \geq 0\}$, where

$$\mathscr{F}_n = \sigma(X_0, X_1, \ldots, X_n), \quad n = 0, 1, \ldots.$$

23.3. Let $(\Omega, \mathscr{F}, P)$ be a probability space. Let $\{\mathscr{F}_n, n = 0, 1, 2, \ldots\}$ be an increasing sequence of sub-σ-fields of $\mathscr{F}$. Let T be a stopping time for $\{\mathscr{F}_n : n \geq 0\}$. Show that $\mathscr{F}_T$ is a sub-σ-field of $\mathscr{F}$.

23.4. Let S and T be stopping times for $\{\mathscr{F}_n, n = 0, 1, 2, \ldots\}$. Show that if $S \leq T$, then $\mathscr{F}_S \subseteq \mathscr{F}_T$, where $\mathscr{F}_S$ and $\mathscr{F}_T$ are the collections of all sets prior to S and T, respectively.

23.5. Let $(\Omega, \mathscr{F}, P)$ be a probability space. Let $\{\mathscr{F}_n, n = 0, 1, 2, \ldots\}$ be an increasing sequence of sub-σ-fields of $\mathscr{F}$. Let T be a stopping time for $\{\mathscr{F}_n : n \geq 0\}$. Let $\{X_n\}_{n\geq 0}$ be a sequence of random variables adapted to $\{\mathscr{F}_n : n = 0, 1, \ldots\}$. Let $X_T : \Omega \to \mathbf{R}$ be defined as

$$X_T(\omega) = X_n(\omega) \quad \text{if } T(\omega) = n, \omega \in \Omega.$$

Show that X_T is $\mathscr{F}_T$-measurable and is a random variable.

23.6. Let $\{X_n : n = 1, 2, \ldots, m\}$ be a submartingale with respect to the filtration $\{\mathscr{F}_n : n = 1, 2, \ldots, m\}$. Let T be a stopping time for the $\mathscr{F}_n$. Show that

$$E(|X_T|) \leq 2E(X_m^+) - E(X_1).$$

23.7. Let $\{X_n : n = 1, 2, \ldots, m\}$ be a martingale with respect to the filtration $\{\mathscr{F}_n : n = 1, 2, \ldots, m\}$. Let $\{T_n\}_{n\geq 1}$ be a nondecreasing sequence of finite stopping times for $\{\mathscr{F}_n : n = 1, 2, \ldots, m\}$. Then, $T_1 \leq T_2 \leq \ldots \leq m$. Show that

(a) $\{\mathscr{F}_{T_i} : i \geq 1\}$ is a filtration.

(b) $\{X_{T_i} : i \geq 1\}$ is a martingale.

(c) $\{X_{T_i} : i \geq 1\}$ is a submartingale with respect to the filtration $\{\mathscr{F}_{T_i} : i \geq 1\}$ if $\{X_n : n = 1, 2, \ldots, m\}$ is a submartingale.

23.8. Let τ_1 and τ_2 be stopping times for a filtration $\{\mathscr{F}_t : t \geq 0\}$. Suppose that $\tau_1 \leq \tau_2$. Show that $\mathscr{F}_{\tau_1} \subseteq \mathscr{F}_{\tau_2}$.

23.9. Let τ_1 and τ_2 be stopping times with respect to a filtration $\{\mathscr{F}_t : t \geq 0\}$. Show that

(a) $\mathscr{F}_{\tau_1 \wedge \tau_2} = \mathscr{F}_{\tau_1} \cap \mathscr{F}_{\tau_2}$.

(b) The events $\{\tau_1 < \tau_2\}$, $\{\tau_1 \leq \tau_2\}$, $\{\tau_1 > \tau_2\}$, $\{\tau_1 \geq \tau_2\}$, and $\{\tau_1 = \tau_2\}$ belong to $\mathscr{F}_{\tau_1 \wedge \tau_2}$.

23.10. Let $\{\mathscr{F}_t : a \leq t \leq b\}$ be a right-continuous filtration. Show that a random variable $\tau : \Omega \to [a, b]$ is a stopping time with respect to $\{\mathscr{F}_t : a \leq t \leq b\}$ if and only if

$$\{\tau < t\} \in \mathscr{F}_t \quad \text{for all } t \in [a, b].$$

23.11. Let $I = [0, \infty)$ or $I = [0, t_0]$ for some $t_0 < \infty$. Let $\{X_t : t \in I\}$ be a stochastic process with values in a measurable space $(E, \mathscr{E})$. Suppose that $\{X_t : t \in I\}$ is progressively measurable (see Definition 21.23) with respect to a filtration $\{\mathscr{F}_t : t \in I\}$. Let τ be a stopping time for the filtration. Show that X_τ is $\mathscr{F}_\tau$-measurable, that is, $\{X_\tau \in B\} \cap \{\tau \leq t\} \in \mathscr{F}_t$ for each $B \in \mathscr{E}$ and $t \in I$.

23.12. Let $(\Omega, \mathscr{F}, P)$ be a probability space and $\{\mathscr{F}_t : t \geq 0\}$ be a filtration. Let τ be a bounded stopping time with respect to the filtration. Let n be a positive integer. Let $\tau^{(n)} : \Omega \to [0, \infty)$ be defined as

$$\tau^{(n)}(\omega) = \begin{cases} \frac{k+1}{2^n} t_0, & \text{if } \tau(\omega) \in \left[\frac{k}{2^n} t_0, \frac{k+1}{2^n} t_0\right), \, k = 0, 1, \ldots, 2^n - 1; \\ t_0, & \text{if } \tau(\omega) = t_0, \end{cases}$$

where t_0 is the bound of τ, that is, $\tau \leq t_0$. Show that $\tau^{(n)}$ is a stopping time with respect to the filtration.

23.13. Let $\{X_t : t \geq 0\}$ be a right-continuous stochastic process and adapted to a right-continuous filtration $\{\mathscr{F}_t : t \geq 0\}$. Show that

(a) If A is an open set, then T_A is a stopping time.

(b) If D is a closed set, then τ_D is a stopping time.

Here τ_D and T_A are the exit time and the hitting time defined in Definition 23.4, respectively.

23.14. Let $\{X_t : t \geq 0\}$ be a right-continuous stochastic process and adapted to some filtration $\{\mathscr{F}_t : t \geq 0\}$. Let A be a closed subset of $\mathbf{R}$. Show that the hitting time T_A is a stopping time.

23.15. Let $I = [0, \infty)$ or $I = \mathbf{N}$. Let $\{X_t : t \in I\}$ be a right-continuous martingale on a probability space $(\Omega, \mathscr{F}, P)$ with respect to a filtration $\{\mathscr{F}_t : t \in I\}$. Let τ be a stopping time such that

(a) $P\{\tau < \infty\} = 1$.

(b) $X_{\tau \wedge t} (t \in I)$ is uniformly integrable.

Show that

$$E[X_\tau] = E[X_0].$$

23.16. Let $\{\mathscr{F}_n : n \geq 0\}$ be a filtration and Z an integrable random variable. Let $X_n = E[Z|\mathscr{F}_n]$ for all $n \geq 0$. Let τ be a stopping time with respect to the filtration. Suppose that Z is $\mathscr{F}_\infty$-measurable, where $\mathscr{F}_\infty = \sigma\left(\bigcup_{n=0}^{\infty} \mathscr{F}_n\right)$. Show that $\{X_\tau, Z\}$ is a martingale with respect to the filtration $\{\mathscr{F}_\tau, \mathscr{F}_\infty\}$:

$$X_\tau = E[Z|\mathscr{F}_\tau].$$

23.17. Let $\{X_n : n \geq 0\}$ be a martingale adapted to a filtration $\{\mathscr{F}_n : n \geq 0\}$. Let τ be a bounded stopping time with respect to the filtration, that is, $P\{\tau \leq M\} = 1$ for some constant M. Show that $\{X_\tau, X_M\}$ is a martingale.

23.18. Let $\{X_n : n \geq 0\}$ be a submartingale adapted to a filtration $\{\mathscr{F}_n : n \geq 0\}$. Let τ be a bounded stopping time with respect to the filtration, that is, $P\{\tau \leq M\} = 1$ for some constant M. Show that $\{X_\tau, X_M\}$ is a submartingale, i.e.,

$$X_\tau \leq E[X_M|\mathscr{F}_\tau] \quad \text{a.s.}$$

23.19. Let $\{X_t : t \geq 0\}$ be a right-continuous martingale with $X_0 = 0$. Let $A, B > 0$ and

$$\tau = \inf\{t : X_t = A \text{ or } X_t = -B\}.$$

Suppose that $P\{\tau < \infty\} = 1$. Show that $E[X_\tau] = 0$ and

$$P\{X_\tau = A\} = \frac{B}{A + B}.$$

23.3 Hints

23.1. Use the definition of stopping times (Definition 23.1).

23.2. Note that

$$\{T \le n\} = \bigcup_{k=0}^{n} \{X_k \in B\},$$

and then use the definition of stopping times (Definition 23.1).

23.3. Use the definition of σ-field (Definition 2.2).

23.4. Follow the definition of events prior to T (Definition 23.3) and use the standard technique, that is, show that $A \in \mathscr{F}_S$ implies $A \in \mathscr{F}_T$.

23.5. Follow the definition of measurable functions (Definition 5.1) and the definition of random variables (Definition 11.6).

23.6. Note that

$$\begin{aligned} E(|X_T|) &= \int_\Omega |X_T| \mathrm{d}P \\ &= \sum_{i=1}^{m} \int_{\{T=i\}} |X_i| \mathrm{d}P \\ &= 2\sum_{i=1}^{m} \int_{\{T=i\}} X_i^+ \mathrm{d}P - \sum_{i=1}^{m} \int_{\{T=i\}} X_i \mathrm{d}P \end{aligned}$$

and that $\{X_n^+\}_{n=1,2,\ldots}$ is a submartingle (see Problem 22.7). Then use the result of Problem 22.3.

23.7. Part (a) is implied by Problem 23.4. Parts (b) and (c) can be proved by using the result of Problem 22.3 and the technique used to prove Problem 23.6.

23.8. For every $A \in \mathscr{F}_{\tau_1}$, consider the decomposition

$$A \cap \{\tau_2 \le t\} = [A \cap \{\tau_2 \le t\} \cap \{\tau_1 \le t\}] \cup [A \cap \{\tau_2 \le t\} \cap \{\tau_1 > t\}].$$

23.9. To prove part (a), use the result of Problem 23.8 and Definition 23.3. To prove part (b), consider the decompositions

$$\{\tau_1 \le \tau_2\} \cap \{\tau_2 \le t\} = \{\tau_1 \wedge t \le \tau_2 \wedge t\} \cap \{\tau_1 \le t\} \cap \{\tau_2 \le t\}$$

and

$$\{\tau_1 < \tau_2\} \cap \{\tau_2 \le t\} = \{\tau_1 \wedge t < \tau_2 \wedge t\} \cap \{\tau_1 \le t\} \cap \{\tau_2 \le t\}.$$

23.10. Use the definition of right-continuous filtrations (Definition 21.20) and note that

$$\{\tau \le t\} = \bigcap_{n=1}^{\infty} \left\{\tau < t + \frac{1}{n}\right\}$$

and

$$\{\tau < t\} = \bigcup_{n=1}^{\infty} \left\{\tau \le t - \frac{1}{n}\right\}.$$

23.11. For a fixed $t \in I$, note that X_τ is the composition of the map from $\{\tau \le t\}$ to $[0, t] \times \{\tau \le t\}$ and the map from $[0, t] \times \{\tau \le t\}$ to E.

23.12. Try to construct $\{\tau^{(n)} \le t\}$ by a union of events of the form $\{\tau \le t'\}$ and then follow the definition of stopping times.

23.13. Use the result of Problem 23.10 to prove part (a). Part (b) is equivalent to part (a).

23.14. Consider the sequence of sets

$$A_n = \left\{x \in \mathbf{R} : \inf\{|y - x| : y \in A\} < \frac{1}{n}\right\}, \quad n \ge 1,$$

and try to establish

$$\{T_A \le t\} = \{X_t \in A\} \cup \left(\bigcap_{n=1}^{\infty} \bigcup_{q \in \mathbf{Q} \cap [0,t)} \{X_q \in A_n\} \right).$$

23.15. Use the result of Problem 9.10 to show that $E[X_{\tau \wedge t}] \to E[X_\tau]$ as $t \to \infty$. Then use the optimal sampling theorem (Theorem 23.2).

23.16. Follow the definition of conditional expectations (Definition 14.1) and note that for $A \in \mathscr{F}_\tau$, we have

$$A = \bigcup_{n \in \{0,1,\ldots,\infty\}} A \cap \{\tau = n\}.$$

23.17. Use the result of Problem 23.16.

23.18. Use the definition of conditional expectations (Definition 14.1) and consider

$$A = \{X_\tau > E[X_M | \mathscr{F}_\tau]\}.$$

23.19. Use the optimal stopping theorem (Theorem 23.1) and the result of Problem 14.18.

23.4 Solutions

23.1. By definition, we only need to show that $\{S \vee T \le n\}$ and $\{S \wedge T \le n\}$ are elements of $\mathscr{F}_n$ for $n \ge 0$. But

$$\{S \vee T \le n\} = \{S \le n\} \cap \{T \le n\}$$

and

$$\{S \wedge T \leq n\} = \{S \leq n\} \cup \{T \leq n\}.$$

By the assumption that S and T are stopping times, we have $\{S \vee T \leq n\} \in \mathscr{F}_n$ and $\{S \wedge T \leq n\} \in \mathscr{F}_n$. This completes the proof.

23.2. By definition of stopping times, we only need to show that $\{T \leq n\} \in \mathscr{F}_n$, $n \geq 0$. Note that

$$\{T \leq n\} = \bigcup_{k=0}^{n} \{X_k \in B\}, \quad n \geq 0.$$

We have $\{T \leq n\} \in \mathscr{F}_n$. Hence T is a stopping time.

23.3. By the definition of stopping times, we have $\Omega \cap \{T \leq n\} = \{T \leq n\} \in \mathscr{F}_n$, $n = 0, 1, \ldots$. Hence $\Omega \in \mathscr{F}_T$.

Now let $A \in \mathscr{F}_T$. Then $A \cap \{T \leq n\} \in \mathscr{F}_n$, $n = 0, 1, \ldots$. Note that

$$A^c \cap \{T \leq n\} = (A \cap \{T \leq n\})^c \cap \{T \leq n\}$$

and $\mathscr{F}_n$ is a σ-field. We have $A^c \cap \{T \leq n\} \in \mathscr{F}_n$, $n = 0, 1, \ldots$. Hence $A^c \in \mathscr{F}_T$.

Finally, let $A_i \in \mathscr{F}_T$, $i = 0, 1, \ldots$. Then $A_i \cap \{T \leq n\} \in \mathscr{F}_n$, $n = 0, 1, \ldots$. Note that

$$\left(\bigcup_{i=0}^{\infty} A_i\right) \cap \{T \leq n\} = \bigcup_{i=0}^{\infty} [A_i \cap \{T \leq n\}].$$

We have $\bigcup_{i=0}^{\infty} A_i \in \mathscr{F}_T$. Therefore, $\mathscr{F}_T$ is a σ-field.

23.4. Let $A \in \mathscr{F}_S$. Then

$$A \cap \{S \leq n\} \in \mathscr{F}_n, \quad \forall n \geq 0.$$

By the assumption that $S \leq T$, we have

$$\{T \leq n\} \subseteq \bigcup_{k=0}^{n} \{S = k\}.$$

Hence

$$\begin{aligned} A \cap \{T \leq n\} &= A \cap \left(\bigcup_{k=0}^{n} \{S = k\}\right) \cap \{T \leq n\} \\ &= \left(\bigcup_{k=0}^{n} [A \cap \{S = k\}]\right) \cap \{T \leq n\}. \end{aligned}$$

Since S is a stopping time for $\{\mathscr{F}_n : n \geq 0\}$, we have $A \cap \{S = k\} \in \mathscr{F}_k \subseteq \mathscr{F}_n$, $k = 0, 1, \ldots, n$. Therefore, the above equation implies that $A \cap \{T \leq n\} \in \mathscr{F}_n$. Hence $A \in \mathscr{F}_T$. Since A is arbitrary, we have $\mathscr{F}_S \subseteq \mathscr{F}_T$. This completes the proof.

23.5. Let $B \in \mathcal{B}$. Note that $X_T^{-1}(B) = \{\omega \in \Omega : X_T(\omega) \in B\} = \{X_T \in B\}$. To show that X_T is $\mathscr{F}_T$-measurable, we only need show that $\{X_T \in B\} \in \mathscr{F}_T$. Note that

$$\{X_T \in B\} \cap \{T \leq n\} = \bigcup_{k=0}^{n} [\{X_k \in B\} \cap \{T = k\}]$$

and $\{X_n\}_{n \geq 0}$ is adapted to $\{\mathscr{F}_n, n = 0, 1, 2, \ldots\}$. We have

$$\{X_T \in B\} \cap \{T \leq n\} \in \mathscr{F}_n.$$

Hence $\{X_T \in B\} \in \mathscr{F}_T$. Since B is arbitrary, X_T is $\mathscr{F}_T$-measurable.

Since $\mathscr{F}_T \subseteq \mathscr{F}$, X_T is also $\mathscr{F}$-measurable, that is, X_T is a random variable on $(\Omega, \mathscr{F}, P)$. This completes the proof.

23.6. Since

$$\begin{aligned}
E(|X_T|) &= \int_{\Omega} |X_T| \mathrm{d}P \\
&= \sum_{i=1}^{m} \int_{\{T=i\}} |X_i| \mathrm{d}P \\
&= 2 \sum_{i=1}^{m} \int_{\{T=i\}} X_i^+ \mathrm{d}P - \sum_{i=1}^{m} \int_{\{T=i\}} X_i \mathrm{d}P,
\end{aligned}$$

it suffices to show that

$$\sum_{i=1}^{m} \int_{\{T=i\}} X_i^+ \mathrm{d}P \leq E(X_m^+) \tag{23.1}$$

and

$$\sum_{i=1}^{m} \int_{\{T=i\}} X_i \mathrm{d}P \geq E(X_1). \tag{23.2}$$

First, let us show that inequality (23.1) is true. By Problem 22.7, $\{X_n^+\}_{n=1,2,\ldots,m}$ is a submartingale with respect to the $\mathscr{F}_n$. Since $\{T = 1\} \in \mathscr{F}_1$, by Problem 22.3, we have

$$\int_{\{T=1\}} X_1^+ \mathrm{d}P \leq \int_{\{T=1\}} X_2^+ \mathrm{d}P.$$

Hence we have

$$\begin{aligned}
\sum_{i=1}^{m} \int_{\{T=i\}} X_i^+ \mathrm{d}P &= \sum_{i=2}^{m} \int_{\{T=i\}} X_i^+ \mathrm{d}P + \int_{\{T=1\}} X_1^+ \mathrm{d}P \\
&\leq \sum_{i=2}^{m} \int_{\{T=i\}} X_i^+ \mathrm{d}P + \int_{\{T=1\}} X_2^+ \mathrm{d}P \\
&= \sum_{i=3}^{m} \int_{\{T=i\}} X_i^+ \mathrm{d}P + \int_{\{T\leq 2\}} X_2^+ \mathrm{d}P.
\end{aligned}$$

Repeating the above process leads to

$$\sum_{i=1}^{m}\int_{\{T=i\}} X_i^+ \mathrm{d}P \le \int_{\{T\le m\}} X_m^+ \mathrm{d}P.$$

But $\{T \le m\} = \Omega$. It follows that inequality (23.1) is true.

Next, let us prove inequality (23.2). Since $\{X_n\}_{n=1,2,\ldots,m}$ is a submartingale and $\{T = m\} = \{T \ge m\} = \{T \le m-1\}^c \in \mathscr{F}_{m-1}$, by Problem 22.3, we have

$$\int_{\{T=m\}} X_m \mathrm{d}P \ge \int_{\{T=m\}} X_{m-1} \mathrm{d}P.$$

Hence

$$\begin{aligned}
\sum_{i=1}^{m}\int_{\{T=i\}} X_i \mathrm{d}P &= \sum_{i=1}^{m-1}\int_{\{T=i\}} X_i \mathrm{d}P + \int_{\{T=m\}} X_m \mathrm{d}P \\
&\ge \sum_{i=1}^{m-1}\int_{\{T=i\}} X_i \mathrm{d}P + \int_{\{T=m\}} X_{m-1} \mathrm{d}P \\
&= \sum_{i=1}^{m-2}\int_{\{T=i\}} X_i \mathrm{d}P + \int_{\{T\ge m-1\}} X_{m-1} \mathrm{d}P.
\end{aligned}$$

Repeating the above process leads to

$$\sum_{i=1}^{m}\int_{\{T=i\}} X_i \mathrm{d}P \ge \int_{\{T\ge 1\}} X_1 \mathrm{d}P.$$

But $\{T \ge 1\} = \Omega$. It follows that inequality (23.2) is true. This completes the proof.

23.7.

(a) By Problems 23.4 and 23.3, and the assumption that $\{T_n : n \ge 1\}$ is increasing, $\{\mathscr{F}_{T_i} : i \ge 1\}$ is a filtration.

(b) Let $Y_i = X_{T_i}$, $i = 1, 2, \ldots$. Since

$$\int_{\Omega} |Y_i| \mathrm{d}P = \sum_{k=1}^{m}\int_{\{T_i=k\}} |X_k| \mathrm{d}P \le \sum_{k=1}^{m} E(|X_k|) < \infty,$$

the Y_i are integrable. By Problem 22.3, to show that $\{Y_i : i \ge 1\}$ is a martingale with respect to $\{T_i : i \ge 1\}$, we only need to show that

$$\int_A Y_{i+1} \mathrm{d}P = \int_A Y_i \mathrm{d}P, \quad \forall A \in \mathscr{F}_{T_i},\ i \ge 1. \tag{23.3}$$

Let $i \ge 1$ and $A \in \mathscr{F}_{T_i}$. Since T_i is finite, we have

$$A = \bigcup_{k=1}^{m} [A \cap \{T_i = k\}].$$

Then, to show that Equation (23.3) holds, it is sufficient to show that

$$\int_{D_k} Y_{i+1}\mathrm{d}P = \int_{D_k} Y_i\mathrm{d}P = \int_{D_k} X_k\mathrm{d}P, \quad k = 1, 2, \ldots, m, \tag{23.4}$$

where $D_k = A \cap \{T_i = k\}$.

Since T_{i+1} is finite and $T_{i+1} \geq T_i$, we have

$$\begin{aligned}\int_{D_k} Y_{i+1}\mathrm{d}P &= \sum_{j=k}^{m} \int_{D_k \cap \{T_{i+1}=j\}} Y_{i+1}\mathrm{d}P \\ &= \sum_{j=k}^{m} \int_{D_k \cap \{T_{i+1}=j\}} X_j\mathrm{d}P. \end{aligned} \tag{23.5}$$

Since $\{T_{i+1} = m\} = \{T_{i+1} \geq m\} = \{T_{i+1} \leq m-1\}^c \in \mathscr{F}_{m-1}$, we have $D_k \cap \{T_{i+1} = m\} \in \mathscr{F}_{m-1}$. By the assumption that $\{X_n\}_{n\geq 1}$ is a martingale and Problem 22.3, We have

$$\int_{D_k \cap \{T_{i+1}=m\}} X_m\mathrm{d}P = \int_{D_k \cap \{T_{i+1}=m\}} X_{m-1}\mathrm{d}P.$$

Hence, Equation (23.5) becomes

$$\begin{aligned}&\int_{D_k} Y_{i+1}\mathrm{d}P \\ =\;& \sum_{j=k}^{m-1} \int_{D_k \cap \{T_{i+1}=j\}} X_j\mathrm{d}P + \int_{D_k \cap \{T_{i+1}=m\}} X_{m-1}\mathrm{d}P \\ =\;& \sum_{j=k}^{m-2} \int_{D_k \cap \{T_{i+1}=j\}} X_j\mathrm{d}P + \int_{D_k \cap \{T_{i+1}\geq m-1\}} X_{m-1}\mathrm{d}P. \end{aligned} \tag{23.6}$$

Since $\{T_{i+1} \geq m-1\} = \{T_{i+1} \leq m-2\}^c \in \mathscr{F}_{m-2}$, we have $D_k \cap \{T_{i+1} \geq m-1\} \in \mathscr{F}_{m-2}$. Again by Problem 22.3, we have

$$\int_{D_k \cap \{T_{i+1}\geq m-1\}} X_{m-1}\mathrm{d}P = \int_{D_k \cap \{T_{i+1}\geq m-1\}} X_{m-2}\mathrm{d}P.$$

Hence, Equation (23.6) becomes

$$\begin{aligned}&\int_{D_k} Y_{i+1}\mathrm{d}P \\ =\;& \sum_{j=k}^{m-2} \int_{D_k \cap \{T_{i+1}=j\}} X_j\mathrm{d}P + \int_{D_k \cap \{T_{i+1}\geq m-1\}} X_{m-2}\mathrm{d}P \\ =\;& \sum_{j=k}^{m-3} \int_{D_k \cap \{T_{i+1}=j\}} X_j\mathrm{d}P + \int_{D_k \cap \{T_{i+1}\geq m-2\}} X_{m-2}\mathrm{d}P. \end{aligned} \tag{23.7}$$

Repeating the above process, we get

$$\int_{D_k} Y_{i+1} \mathrm{d}P = \int_{D_k \cap \{T_{i+1} \geq k\}} X_k \mathrm{d}P.$$

Since $D_k \subseteq \{T_{i+1} \geq k\}$, we have $D_k \cap \{T_{i+1} \geq k\} = D_k$. Hence the above equation becomes

$$\int_{D_k} Y_{i+1} \mathrm{d}P = \int_{D_k} X_k \mathrm{d}P,$$

which shows that Equation (23.4) is true. Hence $\{X_{T_i} : i \geq 1\}$ is a martingale with respect to the $\mathscr{F}_{T_i}$.

(c) The proof is similar to that of part (b).

23.8. Let $A \in \mathscr{F}_{\tau_1}$ and $t \geq 0$. Then we have

$$A \cap \{\tau_2 \leq t\} = [A \cap \{\tau_2 \leq t\} \cap \{\tau_1 \leq t\}] \cup [A \cap \{\tau_2 \leq t\} \cap \{\tau_1 > t\}].$$

By the assumption that $\tau_1 \leq \tau_2$, we have $\{\tau_2 \leq t\} \cap \{\tau_1 > t\} = \emptyset$. Hence

$$A \cap \{\tau_2 \leq t\} = A \cap \{\tau_2 \leq t\} \cap \{\tau_1 \leq t\}.$$

Since $A \cap \{\tau_1 \leq t\} \in \mathscr{F}_t$ and $\{\tau_2 \leq t\} \in \mathscr{F}_t$, by the above equation we get

$$A \cap \{\tau_2 \leq t\} \in \mathscr{F}_t.$$

Since t is arbitrary, we have $A \in \mathscr{F}_{\tau_2}$. Hence $\mathscr{F}_{\tau_1} \subseteq \mathscr{F}_{\tau_2}$. This completes the proof.

23.9.

(a) By Problem 23.8, we have $\mathscr{F}_{\tau_1 \wedge \tau_2} \subseteq \mathscr{F}_{\tau_1} \cap \mathscr{F}_{\tau_2}$. Hence we only need to prove

$$\mathscr{F}_{\tau_1} \cap \mathscr{F}_{\tau_2} \subseteq \mathscr{F}_{\tau_1 \wedge \tau_2}. \tag{23.8}$$

To do that, let $A \in \mathscr{F}_{\tau_1} \cap \mathscr{F}_{\tau_2}$ and $t \geq 0$. Then we have $A \cap \{\tau_1 \leq t\} \in \mathscr{F}_t$ and $A \cap \{\tau_2 \leq t\} \in \mathscr{F}_t$ for all $t \geq 0$. Note that

$$A \cap \{\tau_1 \wedge \tau_2 \leq t\} = [A \cap \{\tau_1 \leq t\}] \cup [A \cap \{\tau_2 \leq t\}].$$

Hence we have $A \cap \{\tau_1 \wedge \tau_2 \leq t\} \in \mathscr{F}_t$. Since t is arbitrary, we have $A \in \mathscr{F}_{\tau_1 \wedge \tau_2}$. Since A is arbitrary, Equation (23.8) holds.

(b) Let $t \geq 0$. Since

$$\{\tau_1 \leq \tau_2\} \cap \{\tau_2 \leq t\} = \{\tau_1 \wedge t \leq \tau_2 \wedge t\} \cap \{\tau_1 \leq t\} \cap \{\tau_2 \leq t\}$$

and $\{\tau_1 \wedge t \leq \tau_2 \wedge t\}$, $\{\tau_1 \leq t\}$, and $\{\tau_2 \leq t\}$ are events in $\mathscr{F}_t$, we have $\{\tau_1 \leq \tau_2\} \in \mathscr{F}_{\tau_2}$. Similarly, we have $\{\tau_1 < \tau_2\} \in \mathscr{F}_{\tau_2}$. Since $\mathscr{F}_{\tau_2}$ is σ-algebra, we have $\{\tau_1 = \tau_2\} \in \mathscr{F}_{\tau_2}$, $\{\tau_1 > \tau_2\} \in \mathscr{F}_{\tau_2}$, and $\{\tau_1 \geq \tau_2\} \in \mathscr{F}_{\tau_2}$.

If we interchange the roles of τ_1 and τ_2, we see that these events also belong to $\mathscr{F}_{\tau_1}$. Hence by the first item, these events belong to $\mathscr{F}_{\tau_1 \wedge \tau_2}$.

This completes the proof.

23.10. First let us prove the "if" part. Suppose that

$$\{\tau < t\} \in \mathscr{F}_t \quad \text{for all } t \in [a, b].$$

Then we have

$$\left\{\tau < t + \frac{1}{n}\right\} \in \mathscr{F}_{t+\frac{1}{n}} \quad \text{for all } n \geq 1.$$

By the assumption that the $\mathscr{F}_t$ is right-continuous, we have

$$\mathscr{F}_t = \bigcap_{n=1}^{\infty} \mathscr{F}_{t+\frac{1}{n}}.$$

Therefore, we have

$$\{\tau \leq t\} = \bigcap_{n=1}^{\infty} \left\{\tau < t + \frac{1}{n}\right\} \in \mathscr{F}_t.$$

Since t is arbitrary, τ is the stopping time by definition.

Now let us prove the "only if" part. Suppose that τ is the stopping time with respect to the filtration. Then we have $\{\tau \leq t\} \in \mathscr{F}_t$ for all $t \in [a, b]$. Note that

$$\{\tau < t\} = \bigcup_{n=1}^{\infty} \left\{\tau \leq t - \frac{1}{n}\right\}$$

and

$$\left\{\tau \leq t - \frac{1}{n}\right\} \in \mathscr{F}_{t-\frac{1}{n}} \subseteq \mathscr{F}_t.$$

We have $\{\tau < t\} \in \mathscr{F}_t$. This completes the proof.

23.11. Let $t \in I$ be fixed. Let $\Omega_t = \{\tau \leq t\}$ and $\tilde{\mathscr{F}}_t = \{A \cap \Omega_t : A \in \mathscr{F}_t\}$. Let $f : \Omega_t \to [0, t] \times \Omega_t$ be defined as

$$f(\omega) = (\tau(\omega), \omega).$$

Let $g : [0, t] \times \Omega_t \to E$ be defined as

$$g(s, \omega) = X_s(\omega).$$

Then $X_\tau = g \circ f$, that is, X_τ is the composition of maps f and g. By the assumption that X is progressively measurable, g on $[0, t] \times \Omega$ is $B[0, t] \times \mathscr{F}_t/\mathscr{E}$-measurable. Hence the restriction of g on $[0, t] \times \Omega_t$ is measurable with respect to the σ-fields $\{A \cap ([0, t] \times \Omega_t) : A \in \mathcal{B}[0, t] \times \mathscr{F}_t\}$ and $\mathscr{E}$.

Hence we only need to show that f is a measurable map. To do that, let $C \in \{A \cap ([0,t] \times \Omega_t) : A \in \mathcal{B}[0,t] \times \mathscr{F}_t\}$. Then

$$f^{-1}(C) = \{\omega \in \Omega_t : f(\omega) \in C\} = \{\omega \in \Omega_t : \tau(\omega) \leq t\} \in \tilde{\mathscr{F}}_t.$$

Hence f is a measurable map. This completes the proof.

23.12. Let $t \in [0, \infty)$. Let K_t be the largest integer in $\{0, 1, \ldots, 2^n - 1\}$ such that $\frac{K_t+1}{2^n} t_0 \leq t$. Then we have

$$\{\tau^{(n)} \leq t\} = \bigcup_{k=0}^{K_t} \left\{\tau \leq \frac{k+1}{2^n} t_0\right\}.$$

Since τ is a stopping time with respect to $\{\mathscr{F}_t : t \geq 0\}$, we have

$$\left\{\tau \leq \frac{k+1}{2^n} t_0\right\} \in \mathscr{F}_{\frac{k+1}{2^n} t_0} \subseteq \mathscr{F}_t, \quad k = 0, 1, \ldots, K_t.$$

It follows that $\{\tau^{(n)} \leq t\} \in \mathscr{F}_t$. Hence $\tau^{(n)}$ is a stopping time. This completes the proof.

23.13.

By Problem 23.10, it is sufficient to show that

$$\{T_A < t\} \in \mathscr{F}_t, \quad t \geq 0.$$

To do that, we show that

$$\{T_A < t\} = \bigcup_{q \in \mathbf{Q} \cap (0,t)} \{X_q \in A\}, \tag{23.9}$$

where $\mathbf{Q}$ is the set of rational numbers. Let

$$\omega \in \bigcup_{q \in \mathbf{Q} \cap (0,t)} \{X_q \in A\}.$$

Then $\omega \in \{X_q \in A\}$ for some rational number in $(0, t)$. Hence $T_A(\omega) \leq q < t$, which implies that $\omega \in \{T_A < t\}$. Therefore, we have

$$\bigcup_{q \in \mathbf{Q} \cap (0,t)} \{X_q \in A\} \subseteq \{T_A < t\}. \tag{23.10}$$

Now let $\omega \in \{T_A < t\}$. Then we have $X_{T_A(\omega)}(\omega) \in A$. Since A is an open set, there exists an $\epsilon > 0$ such that

$$(X_{T_A(\omega)}(\omega) - \epsilon, X_{T_A(\omega)}(\omega) + \epsilon) \subseteq A.$$

Since X_t is right-continuous, there exists a $\delta \in (0, t - T_A(\omega))$ such that for every $r \in (T_A(\omega), T_A(\omega) + \delta)$,

$$|X_r(\omega) - X_{T_A(\omega)}(\omega)| < \epsilon.$$

Let q be a rational number in $(T_A(\omega), T_A(\omega) + \delta)$. Then we have

$$|X_q(\omega) - X_{T_A(\omega)}(\omega)| < \epsilon.$$

Hence $\omega \in \{X_q \in A\}$. Since ω is arbitray, we have

$$\{T_A < t\} \subseteq \bigcup_{q \in \mathbf{Q} \cap (0,t)} \{X_q \in A\}. \tag{23.11}$$

Equation (23.9) follows from Equations (23.10) and (23.11). Since $\{T_A < t\}$ is a countable union of elements in $\mathscr{F}_t$, we have $\{T_A < t\} \in \mathscr{F}_t$.

(a) This follows from part (a) because $\tau_D = T_{\mathbf{R} \setminus D}$.

This completes the proof.

23.14. For every $n \geq 1$, let

$$A_n = \left\{ x \in \mathbf{R} : \inf\{|y - x| : y \in A\} < \frac{1}{n} \right\}.$$

Then A_n is an open set for all $n \geq 1$. We claim that

$$\{T_A \leq t\} = \{X_t \in A\} \cup \left(\bigcap_{n=1}^{\infty} \bigcup_{q \in \mathbf{Q} \cap [0,t)} \{X_q \in A_n\} \right). \tag{23.12}$$

To show this, let $\omega \in \{T_A \leq t\}$. If $T_A(\omega) = t$, then we have $\omega \in \{X_t \in A\}$. Hence we have

$$\{T_A \leq t\} \subseteq \{X_t \in A\} \cup \left(\bigcap_{n=1}^{\infty} \bigcup_{q \in \mathbf{Q} \cap [0,t)} \{X_q \in A_n\} \right). \tag{23.13}$$

Suppose that $T_A(\omega)) < t$. Note that for every $n \geq 1$, there exists an $\epsilon_n > 0$ such that

$$(X_{T_A(\omega)}(\omega) - \epsilon_n, X_{T_A(\omega)}(\omega) + \epsilon_n) \subseteq A_n.$$

Since the X_t is right-continuous, there exists a $\delta_n \in (0, t - T_A(\omega))$ such that for every $r \in (T_A(\omega), T_A(\omega) + \delta_n)$, we have

$$|X_{T_A(\omega)}(\omega) - X_r(\omega)| < \epsilon_n$$

Let q_n be a rational number in $(T_A(\omega), T_A(\omega)+\delta_n)$. Then we have $X_{q_n}(\omega) \in A_n$, which implies $\omega \in \{X_{q_n} \in A_n\}$. Hence for every $n \geq 1$, we have

$$\omega \in \bigcup_{q \in \mathbf{Q} \cap [0,t)} \{X_q \in A_n\}.$$

Thus Equation (23.13) still holds in this case.

Now let

$$\omega \in \{X_t \in A\} \cup \left(\bigcap_{n=1}^{\infty} \bigcup_{q \in \mathbf{Q} \cap [0,t)} \{X_q \in A_n\} \right).$$

If $\omega \in \{X_t \in A\}$, then $T_A(\omega) \leq t$, which implies $\omega \in \{T_A \leq t\}$. Suppose that

$$\omega \in \left(\bigcap_{n=1}^{\infty} \bigcup_{q \in \mathbf{Q} \cap [0,t)} \{X_q \in A_n\} \right).$$

Then for every $n \geq 1$, we have $\omega \in \{X_{q_n} \in A_n\}$ for some rational number q_n in $[0,t)$. Let $r_n = \inf_{i \geq n} q_n$ and let $r = \lim_{n \to \infty} r_n$. Then we have $r \leq t$ and $X_r(\omega) \in A$. Hence we have $T_A(\omega) \leq r \leq t$, which implies that $\omega \in \{T_A \leq t\}$. Therefore, we have

$$\{X_t \in A\} \cup \left(\bigcap_{n=1}^{\infty} \bigcup_{q \in \mathbf{Q} \cap [0,t)} \{X_q \in A_n\} \right) \subseteq \{T_A \leq t\}. \tag{23.14}$$

It follows from Equations (23.13) and (23.14) that Equation (23.12) holds. Since the right-hand side of Equation (23.12) is a countable union of $\mathscr{F}_t$-measurable sets, we have $\{T_A \leq t\} \in \mathscr{F}_t$. This completes the proof.

23.15. By the assumption that $P(\tau < \infty) = 1$, we know that $\tau \wedge t = \tau$ for sufficiently large t outside a P-null set. Hence $X_{\tau \wedge t} = X_\tau$ a.s. as $t \to \infty$. Now, by the assumption that $X_{\tau \wedge t}$ is uniformly integrable and Problem 9.10, we have

$$\lim_{t \to \infty} E[X_{\tau \wedge t}] = E[X_\tau].$$

By Theorem 23.2, we have

$$E[X_{\tau \wedge t}] = E(E[X_{\tau \wedge t} | \mathscr{F}_0]) = E(X_{\tau \wedge 0}) = E[X_0].$$

Hence $E[X_{\tau \wedge t}] = E[X_0]$. This completes the proof.

23.16. By Problem 23.5, X_τ is $\mathscr{F}_\tau$-measurable. Let $\bar{\mathbf{N}} = \{0, 1, \ldots, \infty\}$. Then by Definition 14.1 and the definition of X_n, we have

$$\begin{aligned}
E[|X_\tau|] &= \int_\Omega |X_\tau| \mathrm{d}P \\
&= \sum_{n \in \bar{\mathbf{N}}} \int_{\{\tau=n\}} |X_\tau| \mathrm{d}P \\
&= \sum_{n \in \bar{\mathbf{N}}} \int_{\{\tau=n\}} |X_n| \mathrm{d}P \\
&\leq \sum_{n \in \bar{\mathbf{N}}} \int_{\{\tau=n\}} E[|Z||\mathscr{F}_n] \mathrm{d}P \\
&= \sum_{n \in \bar{\mathbf{N}}} \int_{\{\tau=n\}} E[|Z|] \mathrm{d}P \\
&= E[|Z|].
\end{aligned}$$

Since Z is integrable, X_τ is also integrable.

Now let $A \in \mathscr{F}_\tau$. Since

$$A = \bigcup_{n \in \bar{\mathbf{N}}} A \cap \{\tau = n\},$$

we have

$$\begin{aligned}
\int_A X_\tau \mathrm{d}P &= \sum_{n \in \bar{\mathbf{N}}} \int_{A \cap \{\tau=n\}} X_\tau \mathrm{d}P \\
&= \sum_{n \in \bar{\mathbf{N}}} \int_{A \cap \{\tau=n\}} X_n \mathrm{d}P \\
&= \sum_{n \in \bar{\mathbf{N}}} \int_{A \cap \{\tau=n\}} E[Z|\mathscr{F}_n] \mathrm{d}P \\
&= \sum_{n \in \bar{\mathbf{N}}} \int_{A \cap \{\tau=n\}} Z \mathrm{d}P \\
&= \int_A Z \mathrm{d}P.
\end{aligned}$$

It follows from Definition 14.1 that $X_\tau = E[Z|\mathscr{F}_\tau]$. This completes the proof.

23.17. Let $Y_n = E[X_M|\mathscr{F}_n]$ for all $n \geq 0$. By Problem 23.16, $\{Y_\tau, X_M\}$ is a martingale. By the assumption that $\{X_n : n \geq 0\}$ is a martingale, we have $Y_n = X_n$ for $0 \leq n \leq M$. Since $\tau \leq M$ a.s., $\{X_\tau, X_M\}$ is a martingale. This completes the proof.

23.18. Let

$$A = \{X_\tau > E[X_M|\mathscr{F}_\tau]\}.$$

Then $A = \bigcup_{k=1}^{\infty} A_k$, where

$$A_k = \left\{ X_\tau - E[X_M|\mathscr{F}_\tau] > \frac{1}{k} \right\}.$$

By Definition 14.1 and the assumption that $\{X_n : n \geq 0\}$ is a submartingale, we have

$$\begin{aligned}
\int_{A_k} X_\tau \mathrm{d}P &= \sum_{n=0}^{M} \int_{A_k \cap \{\tau=n\}} X_\tau \mathrm{d}P \\
&= \sum_{n=0}^{M} \int_{A_k \cap \{\tau=n\}} X_n \mathrm{d}P \\
&\leq \sum_{n=0}^{M} \int_{A_k \cap \{\tau=n\}} E[X_M|\mathscr{F}_n] \mathrm{d}P \\
&= \sum_{n=0}^{M} \int_{A_k \cap \{\tau=n\}} X_M \mathrm{d}P \\
&= \int_{A_k} X_M \mathrm{d}P,
\end{aligned}$$

which implies that $P(A_k) = 0$. Since $A_k \uparrow A$ as $k \to \infty$, it follows from Problem 11.6 that $P(A) = 0$. Hence

$$X_\tau \leq E[X_M|\mathscr{F}_\tau] \quad \text{a.s.}$$

This completes the proof.

23.19. Let

$$\tau_1 = \inf\{t : X_t = A\}, \quad \tau_2 = \inf\{t : X_t = -B\}.$$

Then $\tau = \tau_1 \wedge \tau_2$ and $\{X_\tau = A\} = \{\tau_1 < \tau_2\}$. By Theorem 23.1, we have

$$E[X_\tau] = E(E[X_\tau|\mathscr{F}_0]) = E[X_0] = 0.$$

But by Problem 14.18, we have

$$\begin{aligned}
E[X_\tau] &= E[X_\tau|\tau_1 < \tau_2]P\{\tau_1 < \tau_2\} + E[X_\tau|\tau_1 < \tau_2]P\{\tau_1 > \tau_2\} \\
&= AP\{X_\tau = A\} - B(1 - P\{X_\tau = A\}).
\end{aligned}$$

Combining the above two equations gives

$$P\{X_\tau = A\} = \frac{B}{A+B}.$$

This completes the proof.

23.5 Bibliographic Notes

In this chapter, we introduced the concept of stopping times and relevant results. The proofs of Theorems 23.1 and 23.2 can be found in Bhattacharya and Waymire (2007) and Pascucci (2011). Jeanblanc et al. (2009) studied hitting times for Brownian motions. For more properties of stopping times, readers are referred to Williams (1991) and Gut (2007).

CHAPTER 24

MARTINGALE INEQUALITIES

Martingale inequalities are important in the study of stochastic differential equations. For example, martingale inequalities are used to establish bounds of solutions to stochastic differential equations. In this chapter, we present several important martingale inequalities.

24.1 Basic Concepts and Facts

Theorem 24.1 (Doob's Submartingale Inequality). *Let $\{Y_t : a \leq t \leq b\}$ be a right-continuous submartingale. Then for any $\epsilon > 0$, we have*

$$P\left\{\sup_{a\leq t\leq b} Y_t \geq \epsilon\right\} \leq \frac{1}{\epsilon}E\left[Y_b^+\right], \tag{24.1}$$

where $Y_b^+ = \max(Y_b, 0)$. In particular, if the Y_t is a right-continuous martingale, then for any $\epsilon > 0$, we have

$$P\left\{\sup_{a\leq t\leq b} |Y_t| \geq \epsilon\right\} \leq \frac{1}{\epsilon}E\left[|Y_b|\right]. \tag{24.2}$$

24.2 Problems

24.1. Let $\{X_i : i = 0, 1, \ldots, n\}$ be a submartingale adapted to a filtration $\{\mathscr{F}_i : i = 0, 1, \ldots, n\}$. Show that for every $\lambda > 0$, we have

$$P\{M_n > \lambda\} \leq \frac{1}{\lambda} \int_{\{M_n > \lambda\}} X_n \mathrm{d}P \leq \frac{1}{\lambda} E[X_n^+] \leq \frac{1}{\lambda} E[|X_n|],$$

where $M_n = \max\{X_i : i = 0, 1, \ldots, n\}$.

24.2 (**Doob's Maximal Inequality**). Let $\{X_i : i = 1, 2, \ldots, n\}$ be a martingale or nonnegative submartingale adapted to the filtration $\{\mathscr{F}_i : i = 1, 2, \ldots, n\}$. Suppose that $E[|X_n|^p] < \infty$ for some $p \geq 1$. For all $\lambda > 0$, show that

$$P\{M_n \geq \lambda\} \leq \frac{1}{\lambda^p} \int_{\{M_n \geq \lambda\}} |X_n|^p \mathrm{d}P \leq \frac{1}{\lambda^p} E[|X_n|^p],$$

where

$$M_n = \max\{|X_1|, |X_2|, \ldots, |X_n|\}.$$

24.3. Let $\{X_i : i = 1, 2, \ldots, n\}$ be a martingale adapted to the filtration $\{\mathscr{F}_i : i = 1, 2, \ldots, n\}$. Suppose that $E\left[X_n^2\right] < \infty$. Show that

$$E\left[M_n^2\right] \leq 4E\left[X_n^2\right],$$

where

$$M_n = \max\{|X_1|, |X_2|, \ldots, |X_n|\}.$$

24.4. Let $\{X_t : 0 \leq t \leq b\}$ be a nonnegative right-continuous submartingale adapted to the filtration $\{\mathscr{F}_t : 0 \leq t \leq b\}$. Suppose that $E[X_b^p] < \infty$ for some $p \geq 1$. For all $\lambda > 0$, show that

$$P\{M_b > \lambda\} \leq \frac{1}{\lambda^p} \int_{\{M_b > \lambda\}} X_b^p \mathrm{d}P \leq \frac{1}{\lambda^p} E[X_b^p],$$

where

$$M_b = \sup\{X_t : 0 \leq t \leq b\}.$$

24.5. Let $\{X_t : 0 \leq t \leq b\}$ be a nonnegative right-continuous submartingale adapted to the filtration $\{\mathscr{F}_t : 0 \leq t \leq b\}$. Suppose that $E[X_b^p] < \infty$ for some $p \geq 1$. Show that

$$E\left[M_b^2\right] \leq 4E\left[X_b^2\right],$$

where

$$M_b = \sup\{X_t : 0 \leq t \leq b\}.$$

24.6. Let $\{X_t : 0 \leq t \leq b\}$ be a right-continuous martingale or a nonnegative right-continuous submartingale adapted to the filtration $\{\mathscr{F}_t : 0 \leq t \leq b\}$. For every $\alpha > 1$, show that

$$E\left[\sup_{0 \leq t \leq b} |X_t|^\alpha\right] \leq \left(\frac{\alpha}{\alpha - 1}\right)^\alpha E[|X_b|^\alpha].$$

24.7. Let $\{X_1, X_2, \ldots, X_m\}$ be a submartingale on some probability space $(\Omega, \mathscr{F}, P)$, and let $\lambda \geq 0$. Let A be a set defined as

$$A = \left\{ \max_{1 \leq i \leq m} X_i \geq \lambda \right\}.$$

Show that

$$\lambda P(A) \leq \int_A X_m \mathrm{d}P \leq E(X_m^+).$$

24.8. Let $\{X_n : n \geq 1\}$ be a submartingale on the probability space $(\Omega, \mathscr{F}, P)$ and $\lambda \geq 0$. Show that

$$\lambda P \left\{ \sup_{n \geq 1} X_n > \lambda \right\} \leq \sup_{n \geq 1} E(X_n^+).$$

24.9. Let $\{X_n : n \geq 1\}$ be a supermartingale on the probability space $(\Omega, \mathscr{F}, P)$ and $\lambda \geq 0$. Show that

$$\lambda P \left\{ \sup_{n \geq 1} X_n > \lambda \right\} \leq E(X_1) + \sup_{n \geq 1} E(X_n^-).$$

24.10 (Ward's Theorem)**.** Let $\{Y_n : n \geq 1\}$ be a sequence of independent and identically distributed random variables on the probability space $(\Omega, \mathscr{F}, P)$, with finite mean m. Let $X_n = \sum_{i=1}^n Y_i$, $n = 1, 2, \ldots$ and T be a stopping time for $\{X_n : n \geq 1\}$. Show that

(a) If $Y_n \geq 0$, $n = 1, 2, \ldots$, then $E(X_T) = mE(T)$;

(b) If $E(T) < \infty$, then $E(|X_T|) < \infty$ and $E(X_T) = mE(T)$.

24.11. Let $\{X_t : 0 \leq t \leq b\}$ be a right-continuous supermartingale adapted to the filtration $\{\mathscr{F}_t : 0 \leq t \leq b\}$. Let $\lambda > 0$. Show that

$$P \left\{ \sup_{0 \leq t \leq b} X_t \geq \lambda \right\} \leq \frac{E[X_0] + E[X_b^-]}{\lambda},$$

where $X_b^- = \max\{-X_b, 0\}$.

24.3 Hints

24.1. Consider $\tau = \min\{k : X_k > \lambda\}$, and use the result of Problem 23.18.

24.2. To prove this inequality, one can construct a partition of the event $\{M_n \geq \lambda\}$. For example, let $A_1 = \{|X_1| \geq \lambda\}$, $A_2 = \{|X_1| < \lambda, |X_2| \geq \lambda\}$, and so forth.

24.3. Use the results of Problems 13.10, 13.11, and 24.2.

24.4. Consider the discrete submartingales

$$\left\{X_i : i = 0, \frac{b}{2^n}, \frac{2b}{2^n}, \ldots, b\right\}$$

as $n \to \infty$ and use Doob's maximal inequality (see Problem 24.2).

24.5. The proof is similar to that of Problem 24.3, applying the results of Problems 13.10, 13.11, and 24.4.

24.6. The proof is similar to that of Problem 24.5, but using the results of Problems 13.12 and 13.11, and Hölder's inequality (Theorem 8.1).

24.7. Consider the stopping time τ defined as

$$\tau = \begin{cases} \inf\{k : \sup_{1\le i\le k} X_i \ge \lambda\}, & \text{if } \{k : \sup_{1\le i\le k} X_i \ge \lambda\} \ne \emptyset; \\ m, & \text{if otherwise,} \end{cases}$$

and apply the optimal stopping theorem (Theorem 23.1).

24.8. Use the results of Problem 24.7 and Theorem 2.1.

24.9. First use Theorem 23.1 to prove that for every integer $m \ge 1$, we have

$$\lambda P\left\{\max_{1\le k\le m} X_k > \lambda\right\} \le E(X_1) + E(X_m^-).$$

Then apply Theorem 2.1.

24.10. To prove part (a), consider the stopping time $T \wedge n$ and apply Theorems 23.1 and 6.2. Use part (a) to prove part (b).

24.11. Consider the stopping time

$$\tau = \begin{cases} \inf\{t : \sup_{0\le s\le t} X_s \ge \lambda\}, & \text{if } \{t : \sup_{0\le s\le t} X_s \ge \lambda\} \ne \emptyset; \\ b, & \text{otherwise,} \end{cases}$$

and use the optimal stopping theorem (Theorem 23.1).

24.4 Solutions

24.1. Let

$$\tau = \begin{cases} \min\{k : X_k > \lambda\}, & \text{if } \{k : X_k > \lambda\} \ne \emptyset; \\ n, & \text{otherwise.} \end{cases}$$

Then τ is a bounded stopping time. It follows from Problem 23.18 that $X_\tau \leq E[X_n|\mathscr{F}_\tau]$ a.s. Since $\{M_n > \lambda\} = \{X_\tau > \lambda\}$, we have

$$\begin{aligned}
\lambda P\{M_n > \lambda\} &= \lambda P\{X_\tau > \lambda\} \\
&\leq \int_{\{X_\tau > \lambda\}} X_\tau \mathrm{d}P \\
&\leq \int_{\{X_\tau > \lambda\}} E[X_n|\mathscr{F}_\tau] \mathrm{d}P \\
&= \sum_{k=0}^{n} \int_{\{X_\tau > \lambda\} \cap \{\tau = k\}} E[X_n|\mathscr{F}_\tau] \mathrm{d}P \\
&= \sum_{k=0}^{n} \int_{\{X_\tau > \lambda\} \cap \{\tau = k\}} E[X_n|\mathscr{F}_k] \mathrm{d}P \\
&= \sum_{k=0}^{n} \int_{\{X_\tau > \lambda\} \cap \{\tau = k\}} X_n \mathrm{d}P \\
&= \int_{\{X_\tau > \lambda\}} X_n \mathrm{d}P.
\end{aligned}$$

In addition, we have

$$\int_{\{X_\tau > \lambda\}} X_n \mathrm{d}P \leq \int_{\{X_\tau > \lambda\}} X_n^+ \mathrm{d}P \leq E[X_n^+] \leq E[|X_n|].$$

This completes the proof.

24.2. Let $A_1 = \{|X_1| \geq \lambda\}$ and $A_k = \{|X_1| < \lambda, \ldots, |X_{k-1}| < \lambda, |X_k| \geq \lambda\}$ for $k = 2, 3, \ldots, n$. Then $A_k \in \mathscr{F}_k$ for $k = 1, 2, \ldots, n$ and $\{A_1, A_2, \ldots, A_n\}$ is a partition of $\{M_n \geq \lambda\}$. Hence we have

$$P\{M_n \geq \lambda\} = \sum_{k=1}^{n} P(A_k). \tag{24.3}$$

Note that

$$P(A_k) = \int_{A_k} 1 \mathrm{d}P \leq \int_{A_k} \frac{|X_k|^p}{\lambda^p} \mathrm{d}P, \quad k = 1, 2, \ldots, n. \tag{24.4}$$

Note that $|x|^p$ is a convex function for $p \geq 1$. By Problem 15.7 and the assumption that $\{X_k : k = 1, 2, \ldots, k\}$ is a martingale or nonnegative submartingale, we have

$$E[|X_n|^p|\mathscr{F}_k] \geq |E[X_n|\mathscr{F}_k]|^p \geq |X_k|^p. \tag{24.5}$$

By the definition of conditional expectation (Definition 14.1), we have

$$\int_{A_k} |X_n|^p \mathrm{d}P = \int_{A_k} E[|X_n|^p|\mathscr{F}_k] \mathrm{d}P, \quad k = 1, 2, \ldots, n. \tag{24.6}$$

Combining Equations (24.3), (24.4), (24.5), and (24.6) proves the inequality. This completes the proof.

24.3. By Problem 13.10, we have

$$E\left[M_n^2\right] = \int_0^\infty P\{M_n^2 > s\}\mathrm{d}s = 2\int_0^\infty \lambda P\{M_n > \lambda\}\mathrm{d}\lambda.$$

Then by Doob's maximal inequality (see Problem 24.2) and Problem 13.11, we get

$$\begin{aligned} E\left[M_n^2\right] &\leq 2\int_0^\infty \int_{\{M_n \geq \lambda\}} |X_n|\mathrm{d}P\mathrm{d}\lambda \\ &= 2\int_0^\infty E\left[|X_n| I_{\{M_n \geq \lambda\}}\right]\mathrm{d}\lambda \\ &= 2E[|X_n|M_n]. \end{aligned}$$

Now by Schwarz's inequality (Theorem 8.3), we get

$$E[|X_n|M_n] \leq \sqrt{E(X_n^2) \cdot E(M_n^2)}.$$

Combining the above equations gives

$$E(M_n^2) \leq 4E(X_n^2).$$

This completes the proof.

24.4. For each $n \geq 1$, let

$$M_n = \max\left\{X_i : i = 0, \frac{1}{2^n}b, \frac{2}{2^n}b, \ldots, b\right\}.$$

Let $\lambda > 0$. We claim that $\{M_n > \lambda\} \uparrow \{M_b > \lambda\}$ as $n \to \infty$. In fact, since $\{M_n > \lambda\} \subseteq \{M_{n+1} > \lambda\}$, we have

$$\lim_{n\to\infty} \{M_n > \lambda\} = \bigcup_{n\to\infty} \{M_n > \lambda\} \subseteq \{M_b > \lambda\}.$$

To prove the claim, we only need to show that

$$\{M_b > \lambda\} \subseteq \bigcup_{n\to\infty} \{M_n > \lambda\}.$$

To do that, let $\omega \in \{M_b > \lambda\}$. Then there exists a $s \in [0, b]$ such that $X_s(\omega) > \lambda$. Since X is right-continuous, there exist K and N such that

$$X_{\frac{K}{2^N}b} > \lambda,$$

which gives that

$$\omega \in \{M_N > \lambda\} \subseteq \bigcup_{n\to\infty} \{M_n > \lambda\}.$$

Since ω is arbitrary, the claim is valid.

Now note that for each n, $\left\{X_i : i = 0, \frac{1}{2^n}b, \frac{2}{2^n}b, \ldots, b\right\}$ is a nonnegative submartingale. By Problem 24.2, we have

$$P\{M_n > \lambda\} \le \frac{1}{\lambda^p} \int_{\{M_n > \lambda\}} X_b^p \mathrm{d}P \le \frac{1}{\lambda^p} E\left[X_b^p\right], \quad n \ge 1.$$

The result follows by letting $n \to \infty$. This completes the proof.

24.5. By Problem 13.10, we have

$$E\left[M_b^2\right] = \int_0^\infty P\{M_b^2 > s\}\mathrm{d}s = 2\int_0^\infty \lambda P\{M_b > \lambda\}\mathrm{d}\lambda.$$

By Problem 24.4 and Problem 13.11, we have

$$\begin{aligned} E\left[M_b^2\right] &\le 2\int_0^\infty \int_{\{M_b \ge \lambda\}} X_b \mathrm{d}P\mathrm{d}\lambda \\ &= 2\int_0^\infty E\left[X_b I_{\{M_b \ge \lambda\}}\right] \mathrm{d}\lambda \\ &= 2E[X_b M_b]. \end{aligned}$$

Now by Schwarz's inequality (Theorem 8.3), we get

$$E[X_b M_b] \le \sqrt{E(X_b^2) \cdot E(M_b^2)}.$$

Combining the above equations gives

$$E(M_b^2) \le 4E(X_b^2).$$

This completes the proof.

24.6. Let $M_b = \sup_{0 \le t \le b} |X_t|$. By Problem 13.12, we have

$$\begin{aligned} E\left[M_b^\alpha\right] &= E\left[\left(M_b^{\alpha-1}\right)^{\frac{\alpha}{\alpha-1}}\right] \\ &= \int_0^\infty \frac{\alpha}{\alpha-1}\lambda^{\frac{1}{\alpha-1}} P\left\{M_b^{\alpha-1} \ge \lambda\right\} \mathrm{d}\lambda. \end{aligned}$$

Since $|X_t|$ is a nonnegative submartingale, it follows from Problem 24.4 that for $\lambda > 0$, we have

$$\begin{aligned} P\left\{M_b^{\alpha-1} \ge \lambda\right\} &= P\left\{M_b \ge \lambda^{\frac{1}{\alpha-1}}\right\} \\ &\le \lambda^{-\frac{1}{\alpha-1}} \int_{\{M_b^{\alpha-1} \ge \lambda\}} |X_b| \mathrm{d}P. \end{aligned}$$

By Problem 13.11 and Theorem 8.1, we have

$$\begin{aligned} E\left[M_b^\alpha\right] &\leq \frac{\alpha}{\alpha-1}\int_0^\infty \int_{\{M_b^{\alpha-1}\geq\lambda\}} |X_b|\mathrm{d}P\mathrm{d}\lambda \\ &\leq \frac{\alpha}{\alpha-1}E[|X_b|M_b^{\alpha-1}] \\ &\leq \frac{\alpha}{\alpha-1}E\left[|X_b|^\alpha\right]^{\frac{1}{\alpha}} E\left[M_b^\alpha\right]^{\frac{\alpha-1}{\alpha}}, \end{aligned}$$

which gives

$$E\left[M_b^\alpha\right]^{\frac{1}{\alpha}} \leq \frac{\alpha}{\alpha-1}E\left[|X_b|^\alpha\right]^{\frac{1}{\alpha}}.$$

The result follows immediately by taking powers of α on both sides of the above inequality. This completes the proof.

24.7. Let τ be defined as

$$\tau = \begin{cases} \inf\{k : \sup_{1\leq i\leq k} X_i \geq \lambda\}, & \text{if } \{k : \sup_{1\leq i\leq k} X_i \geq \lambda\} \neq \emptyset; \\ m, & \text{if otherwise.} \end{cases}$$

Then τ is a stopping time with respect to the underlying filtration $\{\mathscr{F}_i : 1 \leq i \leq m\}$. By Theorem 23.1, we have

$$E[X_m] = E[E[X_m|\mathscr{F}_\tau] \geq E[X_\tau]$$

Since $\tau = m$ on the event A^c, we have

$$\begin{aligned} E[X_m] &\geq \int_A X_\tau \mathrm{d}P + \int_{A^c} X_\tau \mathrm{d}P \\ &\geq \int_A \lambda \mathrm{d}P + \int_{A^c} X_m \mathrm{d}P, \end{aligned}$$

which gives

$$\lambda P(A) \leq \int_A X_m \mathrm{d}P \leq \int_A X_m^+ \mathrm{d}P \leq E[X_m^+].$$

This completes the proof.

24.8. By Problem 24.7, for every $m \geq 1$, we have

$$\lambda P\left\{\max_{1\leq k\leq m} X_k \geq \lambda\right\} \leq E[X_m^+] \leq \sup_{n\geq 1} E[X_n^+].$$

Since

$$\left\{\max_{1\leq k\leq m} X_k \geq \lambda\right\} \uparrow \left\{\sup_{n\geq 1} X_n \geq \lambda\right\} \qquad \text{as } m \to \infty,$$

it follows from Theorem 2.1 that

$$P\left\{\max_{1\leq k\leq m} X_k \geq \lambda\right\} \uparrow P\left\{\sup_{n\geq 1} X_n \geq \lambda\right\} \qquad \text{as } m \to \infty.$$

Hence we have

$$\lambda P\left\{\sup_{n\geq 1} X_n > \lambda\right\} \leq \lambda P\left\{\sup_{n\geq 1} X_n \geq \lambda\right\} \leq \sup_{n\geq 1} E(X_n^+).$$

This finishes the proof.

24.9. Let m be a positive integer. We claim that

$$\lambda P\left\{\max_{1\leq k\leq m} X_k > \lambda\right\} \leq E(X_1) + E(X_m^-). \tag{24.7}$$

To prove this claim, let τ be defined as

$$\tau = \begin{cases} \inf\{k : \sup_{1\leq i\leq k} X_i \geq \lambda\}, & \text{if } \{k : \sup_{1\leq i\leq k} X_i \geq \lambda\} \neq \emptyset; \\ m, & \text{if otherwise.} \end{cases}$$

Then τ is a stopping time with respect to the underlying filtration $\{\mathscr{F}_i : 1 \leq i \leq m\}$. By Theorem 23.1, we have

$$E[X_\tau] = E[E[X_\tau|\mathscr{F}_1] \leq E[X_1].$$

Let $A = \{\max_{1\leq k\leq m} X_k \geq \lambda\}$. Then $\tau = m$ on the event A^c. Hence we have

$$E[X_1] \geq \int_A X_\tau \mathrm{d}P + \int_{A^c} X_\tau \mathrm{d}P \geq \int_A \lambda \mathrm{d}P + \int_{A^c} X_m \mathrm{d}P,$$

which gives

$$E[X_1] \geq \lambda P(A) - \int_{A^c} X_m^- \mathrm{d}P \geq \lambda P\left\{\max_{1\leq k\leq m} X_k > \lambda\right\} - E[X_m^-].$$

Hence Equation (24.7) holds. Thus we have

$$\lambda P\left\{\max_{1\leq k\leq m} X_k > \lambda\right\} \leq E(X_1) + E(X_m^-) \leq E(X_1) + \sup_{n\geq 1} E[X_n^-].$$

The result follows immediately from Theorem 2.1 by letting $m \to \infty$. This completes the proof.

24.10.

(a) Let $Z_n = X_n - nm$. Then $\{Z_n : n \geq 1\}$ is a martingale. By Problem 23.1, $T \wedge n$ is a stopping time and is bounded. Then by Theorem 23.1, we have

$$E[Z_{T\wedge n}] = E[Z_1] = 0,$$

which gives

$$E[X_{T\wedge n}] = mE[T \wedge n].$$

Since $Y_j \geq 0$ for all $j \geq 1$ and T is finite, we have $X_{T\wedge n} \uparrow X_T$ and $T \wedge n \uparrow T$ as $n \to \infty$. Hence it follows from Theorem 6.2 that

$$E[X_T] = mE[T].$$

(b) Let

$$X_n^{(1)} = \sum_{j=1}^{n} Y_j^+, \quad X_n^{(2)} = \sum_{j=1}^{n} Y_j^-, \quad n \geq 1.$$

Let $Z_n^{(1)} = X_n^{(1)} - nm_1$ and $Z_n^{(2)} = X_n^{(2)} - nm_2$, where $m_1 = E[Y_1^+]$ and $m_2 = E[Y_1^-]$. Then by part (a) of this problem, we have

$$E[X_T^{(1)}] = m_1 E[T] \text{ and } E[X_T^{(2)}] = m_1 E[T].$$

By the assumption that $E[T] < \infty$, we have

$$E[|X_T|] = E[X_T^{(1)}] + E[X_T^{(2)}] < \infty$$

and

$$E[X_T] = E[X_T^{(1)}] - E[X_T^{(2)}] = (m_1 - m_2)E[T] = mE[T].$$

This completes the proof.

24.11. Let τ be defined as

$$\tau = \begin{cases} \inf\{t : \sup_{0\leq s\leq t} X_s \geq \lambda\}, & \text{if } \{t : \sup_{0\leq s\leq t} X_s \geq \lambda\} \neq \emptyset; \\ b, & \text{otherwise,} \end{cases}$$

Then τ is a stopping time with respect to the filtration $\{\mathscr{F}_t : 0 \leq t \leq b\}$. Since $\{X_t : 0 \leq t \leq b\}$ is a right-continuous supermartingale, by Theorem 23.1 we have

$$\begin{aligned} E[X_0] &\geq E[E[X_\tau|\mathscr{F}_0]] \\ &= E[X_\tau] \\ &= \int_A X_\tau \mathrm{d}P + \int_{A^c} X_\tau \mathrm{d}P, \end{aligned} \tag{24.8}$$

where

$$A = \left\{ \sup_{0\leq t\leq b} X_t \geq \lambda \right\}.$$

Note that $X_\tau \geq \lambda$ on event A and that $\tau = b$ on event A^c. Hence from Equation (24.8), we have

$$\begin{aligned} E[X_0] &\geq \int_A \lambda \mathrm{d}P + \int_{A^c} X_b \mathrm{d}P \\ &\geq \lambda P(A) - \int_{A^c} X_b^- \mathrm{d}P \\ &\geq \lambda P(A) - E[X_b^-], \end{aligned}$$

which gives

$$P(A) \leq \frac{E[X_0] + E[X_b^-]}{\lambda}.$$

This completes the proof.

24.5 Bibliographic Notes

In this chapter, we introduced some inequalities for martingales. Doob's submartingale inequality has two versions: a discrete version (see Problem 24.2) and a continuous version (see Theorem 24.1). The continuous version can be proved by the discrete version. For such a proof, readers are referred to Bhattacharya and Waymire (2007, p41). More information about martingale inequalities can be found in Brzezniak and Zastawniak (1999), Chung (2000), Chung and Lu (2006), and Gut (2007).

CHAPTER 25

MARTINGALE CONVERGENCE THEOREMS

Martingale convergence theorems state that under certain conditions, a martingale, submartingale, or supermartingale converges to a limiting random variable. In this chapter, we shall present several martingale convergence theorems.

25.1 Basic Concepts and Facts

Definition 25.1 (Crossing of Real Numbers). Let $\{x_n : n \geq 0\}$ be a sequence of real numbers. Let $a < b$ be real numbers. The sequence $\{\tau_n(a,b) : n \geq 0\}$ of crossings with respect to the sequence $\{x_n : n \geq 0\}$ is defined by

$$\tau_0(a,b) = 0, \tag{25.1a}$$

and for every $n \geq 1$, we have

$$\tau_{2n-1}(a,b) = \begin{cases} \min A_{2n-1}, & \text{if } A_{2n-1} \neq \emptyset; \\ \infty, & \text{otherwise}, \end{cases} \tag{25.1b}$$

$$\tau_{2n}(a,b) = \begin{cases} \min A_{2n}, & \text{if } A_{2n} \neq \emptyset; \\ \infty, & \text{otherwise}, \end{cases} \tag{25.1c}$$

where $A_{2n-1} = \{k \geq \tau_{2n-2} : x_k \leq a\}$ and $A_{2n} = \{k \geq \tau_{2n-1} : x_k \geq b\}$.

Definition 25.2 (Crossing of Random Variables)**.** Let $\{X_n : n \geq 0\}$ be a sequence of real random variables. Let $a < b$ be real numbers. The sequence $\{\tau_n(a, b) : n \geq 0\}$ of crossings with respect to the sequence $\{X_n : n \geq 0\}$ is defined as follows. For every $\omega \in \Omega$, $\{\tau_n(a, b)(\omega) : n \geq 0\}$ is the sequence of crossings with respect to the sequence $\{X_n(\omega) : n \geq 0\}$.

Definition 25.3 ($\mathscr{F}_\infty$)**.** Let $\{\mathscr{F}_n : n \geq 0\}$ be a filtration. Then $\mathscr{F}_\infty$ is the σ-algebra generated by $\bigcup_{n=0}^{\infty} \mathscr{F}_n$:

$$\mathscr{F}_\infty = \sigma\left(\bigcup_{n=0}^{\infty} \mathscr{F}_n\right).$$

Theorem 25.1 (Submartingale Convergence Theorem)**.** *Let $\{X_n\}_{n\geq 1}$ be a submartingale with respect to the filtration $\{\mathscr{F}_n : n \geq 1\}$. If*

$$\sup_n E(X_n^+) < \infty,$$

then there exists an integrable random variable X_∞ such that

$$X_n \xrightarrow{a.s.} X_\infty.$$

25.2 Problems

25.1. Let $\{x_n : n \geq 0\}$ be a sequence of real numbers. Show that $\{x_n : n \geq 0\}$ converges if and only if

$$\nu(a, b) = \max\{k : \tau_{2k}(a, b) < \infty\} < \infty$$

for all $a, b \in \mathbf{Q}$, $a < b$, where $\tau_n(a, b)$ is as defined in Definition 25.1.

25.2. Let $\{X_n : n \geq 0\}$ be a sequence of real random variables. Show that $\{X_n : n \geq 0\}$ converges almost surely if and only if

$$\nu(a, b) = \max\{k : \tau_{2k}(a, b) < \infty\} < \infty, \quad \text{a.s.}$$

for all $a, b \in \mathbf{Q}$, $a < b$.

25.3 (Upcrossing Theorem)**.** Let $\{X_1, X_2, \ldots, X_n\}$ be a martingale with respect to the filtration $\{\mathscr{F}_1, \mathscr{F}_2, \ldots, \mathscr{F}_n\}$. Let $a, b \in \mathbf{R}$ such that $a < b$. Let U_{ab} be the number of upcrossings by $X_1, X_2, \ldots, X_n$, defined as follows.

Let $T_1 = T_1(\omega)$ be the first integer in $\{1, 2, \ldots, n\}$ such that $X_{T_1} \leq a$. Let T_2 be the first integer greater than T_1 such that $X_{T_2} \geq b$. Let T_3 be the first integer greater than T_2 such that $X_{T_2} \leq a$, and so on. If no such integer exists for T_k, then $T_k = \infty$:

$$T_k = \begin{cases} \min\{i : i \in A_k\} & \text{if } k \text{ is odd and } |A_k| > 0; \\ \min\{i : i \in B_k\} & \text{if } k \text{ is even and } |A_k| > 0; \quad k = 1, 2, \ldots, \\ \infty & \text{if } A_k \text{ is empty,} \end{cases}$$

where $A_k = \{i : T_{k-1} < i \leq n \text{ and } X_i \leq a\}$, $B_k = \{i : T_{k-1} < i \leq n \text{ and } X_i \geq b\}$, and $T_0 = 0$. Then define

$$U_{ab} = \begin{cases} \frac{N}{2} & \text{if } N \text{ is even;} \\ \frac{N-1}{2} & \text{if } N \text{ is odd,} \end{cases}$$

where N is the number of finite T_k. Show that

$$E(U_{ab}) \leq \frac{1}{b-a} E[(X_n - a)^+].$$

25.4. Let Y be an integrable random variable on some probability space $(\Omega, \mathscr{F}, P)$. Let $\mathscr{F}_i$, $i \in I$, be arbitrary sub-σ-fields of $\mathscr{F}$. Show that the random variables $X_i = E(Y|\mathscr{F}_i)$, $i \in I$, are uniformly integrable:

$$\lim_{c\to\infty} \sup_{i\in I} \int_{\{|X_i|\geq c\}} |X_i| \mathrm{d}P = 0.$$

25.5. Let Y be an integrable random variable on some probability space $(\Omega, \mathscr{F}, P)$. Let $\{\mathscr{F}_n\}_{n\geq 1}$ be an increasing sequence of sub-σ-fields of $\mathscr{F}$. Let $X_n = E(Y|\mathscr{F}_n)$, $n = 1, 2, \ldots$. Show that

$$X_n \xrightarrow{\text{a.e.}} E(Y|\mathscr{F}_\infty)$$

and

$$X_n \xrightarrow{L^1} E(Y|\mathscr{F}_\infty),$$

where $\mathscr{F}_\infty$ is as defined in Definition 25.3.

25.6. Let $\{X_n : n = 1, 2, \ldots, \infty\}$ be a nonnegative submartingale with respect to the filtration $\{\mathscr{F}_n : n = 1, 2, \ldots, \infty\}$, where X_∞ is a last element. Show that the X_n are uniformly integrable.

25.7. Let $\{X_n : n \geq 0\}$ be a martingale adapted to a filtration $\{\mathscr{F}_n : n \geq 0\}$. Let $\mathscr{F}_\infty = \bigcup_{n=0}^{\infty} \mathscr{F}_n$. Show that the following are equivalent:

(a) $\{X_n : n \geq 0\}$ is uniformly integrable.

(b) X_n converges in L^1.

(c) $X_n \xrightarrow{\text{a.s.}} X_\infty$ as $n \to \infty$ and $\{X_n : n = 0, 1, \ldots, \infty\}$ is a martingale adapted to $\{\mathscr{F}_n : n = 0, 1, \ldots, \infty\}$, where X_∞ is some integrable random variable.

(d) There exists an integrable random variable Y such that $X_n = E[Y|\mathscr{F}_n]$ for all $n \geq 0$.

25.8. Let $\{X_n : n \geq 0\}$ be a submartingale adapted to a filtration $\{\mathscr{F}_n : n \geq 0\}$. Let $\mathscr{F}_\infty = \bigcup_{n=0}^{\infty} \mathscr{F}_n$. Show that the following are equivalent:

(a) $\{X_n : n \geq 0\}$ is uniformly integrable.

(b) X_n converges in L^1.

(c) $X_n \xrightarrow{\text{a.s.}} X_\infty$ as $n \to \infty$ and $\{X_n : n = 0, 1, \ldots, \infty\}$ is a submartingale adapted to $\{\mathscr{F}_n : n = 0, 1, \ldots, \infty\}$, where X_∞ is some integrable random variable.

25.3 Hints

25.1. Use the definition of convergences (Definition 1.10).

25.2. Use the result of Problems 25.1 and 11.1.

25.3. Use the optional skipping theorem (see Problem 22.9) and the result of Problem 22.7. First consider the case when $a = 0$ and $X_k \geq 0$. Then consider the case $(X_k - a)^+$.

25.4. Use Chebyshev's inequality (see Problem 6.16) and the tower property of conditional expectations (see Problem 14.10).

25.5. Use the results of Problems 22.6 and 25.4, and the submartingale convergence theorem (Theorem 25.1).

25.6. Use the result of Problem 9.11.

25.7. To prove (a) $\Longrightarrow$ (b), use Theorem 25.1 and the results of Problems 9.11, 9.4, 9.2, and 9.12. To prove (b) $\Longrightarrow$ (c), use the result of Problem 9.14. To prove (c) $\Longrightarrow$ (d), consider $Y = X_\infty$. To prove (d) $\Longrightarrow$ (a), use the result of Problem 25.4.

25.8. The proof of (a) $\Longrightarrow$ (b) $\Longrightarrow$ (c) is the same as that of Problem 25.7. The proof of (c) $\Longrightarrow$ (a) is similar to that of Problem 25.4.

25.4 Solutions

25.1. Let use first prove the necessity ("only if") part. Suppose that

$$\nu(a, b) = \max\{k : \tau_{2k}(a, b) < \infty\} < \infty$$

for all $a, b \in \mathbf{Q}$, $a < b$. We claim that

$$\liminf x_n = \limsup x_n. \tag{25.2}$$

If $\liminf x_n \neq \limsup x_n$, there exist rational numbers a and b such that

$$\liminf x_n < a < b < \limsup x_n,$$

which implies that $\nu(a, b) = \infty$. This contradicts our hypothesis. Hence Equation (25.2) must hold. By Definition 1.10, $\{x_n : n \geq 0\}$ converges.

Now let us prove the sufficiency. Suppose that $\{x_n : n \geq 0\}$ converges. Then by Definition 1.10, Equation (25.2) holds. We claim that

$$\nu(a, b) = \max\{k : \tau_{2k}(a, b) < \infty\} < \infty \tag{25.3}$$

for all $a, b \in \mathbf{Q}$, $a < b$. If $\nu(a, b) = \infty$ for some $a, b \in \mathbf{Q}$, $a < b$, then we must have infinitely many x's below a and infinitely many x's above b. Hence we have

$$\liminf x_n \leq a < b \leq \limsup x_n,$$

which implies that $\{x_n : n \geq 0\}$ does not converge. This contradicts our hypothesis. Hence Equation (25.3) must hold. This completes the proof.

25.2. For a given $\omega \in \Omega$, it follows from Problem 25.1 that $\{X_n(\omega) : n \geq 0\}$ if and only if

$$\nu(a, b)(\omega) = \max\{k : \tau_{2k}(a, b)(\omega) < \infty\} < \infty$$

for all $a, b \in \mathbf{Q}$, $a < b$. Hence we have

$$\{\omega \in \Omega : \{X_n(\omega) : n \geq 0\} \text{ converges}\} = \bigcap_{a,b\in\mathbf{Q},a<b} \{\nu(a, b) < \infty\}.$$

If $P\{\omega \in \Omega : \{X_n(\omega) : n \geq 0\} = 1$, then we have

$$P\left(\bigcap_{a,b\in\mathbf{Q},a<b} \{\nu(a, b) < \infty\}\right) = 1,$$

which implies that $P\{\nu(a, b) < \infty\} = 1$ for all $a, b \in \mathbf{Q}$, $a < b$.

If $P\{\nu(a, b) < \infty\} = 1$ for all $a, b \in \mathbf{Q}$, $a < b$, then it follows from Problem 11.1 that

$$P\left(\bigcap_{a,b\in\mathbf{Q},a<b} \{\nu(a, b) < \infty\}\right) = 1,$$

which gives

$$P\{\omega \in \Omega : \{X_n(\omega) : n \geq 0\} = 1.$$

This completes the proof.

25.3. First, assume that $a = 0$ and $X_k \geq 0$ for $k \geq 1$. Then the resulting inequality becomes

$$E(U_{0b}) \leq \frac{1}{b} E[X_n]. \tag{25.4}$$

Let M_k be defined as

$$M_k = \begin{cases} 0 & \text{if } T_{2j-2} \leq k < T_{2j-1} \text{ for some } j \geq 1; \\ 1 & \text{if } T_{2j-1} \leq k < T_{2j} \text{ for some } j \geq 1, \end{cases} \qquad k = 1, 2, \ldots, n-1.$$

Then let Y_n be defined as

$$Y_k = X_1 + \sum_{j=1}^{k-1} M_j(X_{j+1} - X_j), \quad k = 1, 2, \ldots, n.$$

By Problem 22.9, we have

$$E(Y_n) \le E(X_n). \tag{25.5}$$

But

$$Y_n = \begin{cases} X_1 + \sum_{i=1}^{N/2}(X_{T_{2i}} - X_{2i-1}) & \text{if } N \text{ is even;} \\ X_1 + \sum_{i=1}^{(N-1)/2}(X_{T_{2i}} - X_{2i-1}) + (X_{T_N+1} - X_{T_N}) & \text{if } N \text{ is odd.} \end{cases}$$

Noting that $X_1 \ge 0$, we have $Y_n \ge bU_{0b}$, which gives

$$E(Y_n) \ge bE(U_{0b}). \tag{25.6}$$

Combining inequalities (25.5) and (25.6) gives (25.4).

Now we consider the general case. By Problem 22.7, $(X_k - a)^+$ is a submartingale. Since X_k upcrossing (a, b) is equivalent to $(X_k - a)^+$ upcrossing $(0, b - a)$, applying the above result proves the problem. This completes the proof.

25.4. Let $\epsilon > 0$. Note that $|Y|I_{\{|Y|<n\}} \uparrow |Y|$ as $n \to \infty$. By the monotone convergence theorem (Theorem 6.2), we have

$$\int_\Omega |Y|I_{\{|Y|<n\}}\mathrm{d}P = \int_{\{|Y|<n\}} |Y|\mathrm{d}P \uparrow \int_\Omega |Y|\mathrm{d}P \quad \text{as} \quad n \to \infty.$$

Therefore there exists an integer N such that

$$\int_{\{|Y|\ge N\}} |Y|\mathrm{d}P = \int_\Omega |Y|\mathrm{d}P - \int_{\{|Y|<N\}} |Y|\mathrm{d}P < \frac{\epsilon}{2}.$$

By the conditional Jensen's inequality (see Problem 15.7), we have

$$|X_i| = |E(Y|\mathscr{F}_i)| \le E(|Y||\mathscr{F}_i).$$

Hence we have

$$\begin{aligned} \int_{\{|X_i|\ge c\}} |X_i|\mathrm{d}P &\le \int_{\{|X_i|\ge c\}} E(|Y||\mathscr{F}_i)\mathrm{d}P \\ &= \int_{\{|X_i|\ge c\}} |Y|\mathrm{d}P \\ &= \int_{\{|X_i|\ge c\}\cap\{|Y|\ge N\}} |Y|\mathrm{d}P + \int_{\{|X_i|\ge c\}\cap\{|Y|<N\}} |Y|\mathrm{d}P \\ &\le \frac{\epsilon}{2} + NP\{|X_i| \ge c\}. \end{aligned}$$

By Problem 6.16, we have

$$P\{|X_i| \geq c\} \leq \frac{E(|X_i|)}{c} \leq \frac{E[E(|Y||\mathscr{F}_i)]}{c} = \frac{E(|Y|)}{c}$$

If we choose $c > 2NE(|Y|)/\epsilon$, we have

$$\sup_{i \in I} \int_{\{|X_i| \geq c\}} |X_i| \mathrm{d}P < \epsilon.$$

Since ϵ is arbitrary, the conclusion follows. This completes the proof.

25.5. By Problem 22.6, $\{X_n\}$ is a martingale. By Problem 25.4, the X_n are uniformly integrable. By the assumption that Y is integrable and Problem 14.10, we have

$$E(|X_n|) = E(|E(Y|\mathscr{F}_n)|) \leq E(E(|Y||\mathscr{F}_n)) = E(|Y|) < \infty, \quad n \geq 1.$$

Hence, by Theorem 25.1, there exists an integrable random variable X_∞ such that

$$X_n \xrightarrow{\text{a.e.}} X_\infty.$$

Note that the probability measure P is finite. By Egoroff's theorem (see Problem 9.4), X_n converges to X_∞ almost uniformly. By Problem 9.2, $X_n \xrightarrow{\mu} X_\infty$. By Problem 9.12, $X_n \xrightarrow{L^1} X_\infty$.

Now we show that X_∞ is a version of $E(Y|\mathscr{F}_\infty)$. Since X_∞ is integrable, we have $E(|X_\infty|) < \infty$. Since X_n is $\mathscr{F}_\infty$-measurable, X_∞ is also $\mathscr{F}_\infty$-measurable. Let $A \in \mathscr{F}_\infty$. Then

$$\begin{aligned} \left| \int_A Y \mathrm{d}P - \int_A X_\infty \mathrm{d}P \right| &= \left| \int_A E(Y|\mathscr{F}_n) \mathrm{d}P - \int_A X_\infty \mathrm{d}P \right| \\ &\leq \int_A |X_n - X_\infty| \mathrm{d}P. \end{aligned}$$

Since $X_n \xrightarrow{L^1} X_\infty$, we have

$$\int_A Y \mathrm{d}P = \int_A X_\infty \mathrm{d}P.$$

By definition of conditional expectation, we have $X_\infty = E(Y|\mathscr{F}_\infty)$ a.e. This completes the proof.

25.6. By the definition of conditional expectation and the hypothesis, we have

$$\int_{\{X_n \geq c\}} X_n \mathrm{d}P \leq \int_{\{X_n \geq c\}} E(X_\infty|\mathscr{F}_\infty) \mathrm{d}P = \int_{\{X_n \geq c\}} X_\infty \mathrm{d}P.$$

By Chebyshev's inequality (see Problem 6.16), we have

$$P\{X_n \geq c\} \leq \frac{E(X_n)}{c} \leq \frac{E(X_\infty)}{c}.$$

Hence $P\{X_n \geq c\} \to 0$ as $c \to \infty$. Therefore, the integral of X_n is uniformly bounded and uniformly continuous. By Problem 9.11, the X_n are uniformly integrable.

25.7. First, let us prove (a) $\Longrightarrow$ (b). Suppose that $\{X_n : n \geq 0\}$ is uniformly integrable. Then it follows from Problem 9.11 that the $E[|X_n|]$ are uniformly bounded, specifically,

$$\sup_n E[|X_n|] \leq M$$

for some constant $M < \infty$. Hence we have

$$\sup_n E[X_n^+] < \infty.$$

By Theorem 25.1, there exists an integrable random variable X_∞ such that $X_n \xrightarrow{\text{a.s.}} X_\infty$ as $n \to \infty$. Then it follows from Problems 9.4, 9.2, and 9.12 that $X_n \xrightarrow{L^1} X_\infty$ as $n \to \infty$.

Next, let us prove (b) $\Longrightarrow$ (c). Suppose that $X_n \xrightarrow{L^1} X_\infty$ as $n \to \infty$. Then it follows from Problem 9.14 that $\{X_n : n \geq 0\}$ is uniformly integrable. Then from the previous proof, there exists an integrable random variable X_∞ such that $X_n \xrightarrow{a.s.} X_\infty$ as $n \to \infty$. Let $m \geq 0$ and $A \in \mathscr{F}_m$. Since $\{X_n : n \geq 0\}$ is a martingale, we have for very $n \geq m$,

$$\int_A X_m \mathrm{d}P = \int_A X_n \mathrm{d}P.$$

By the hypothesis, we have

$$\int_A X_m \mathrm{d}P = \lim_{n\to\infty} \int_A X_n \mathrm{d}P = \int_A X_\infty \mathrm{d}P.$$

Since A is arbitrary, we get

$$E[X_\infty|\mathscr{F}_m] = X_m.$$

Hence $\{X_n : n = 0, 1, \ldots, \infty\}$ is a martingale.

Next, let us prove (c) $\Longrightarrow$ (d). In fact, this follows immediately if we let $Y = X_\infty$.

Finally, let us prove (d) $\Longrightarrow$ (a). In fact, this follows immediately from Problem 25.4. This completes the proof.

25.8. First, let us prove (a) $\Longrightarrow$ (b). Suppose that $\{X_n : n \geq 0\}$ is uniformly integrable. Then it follows from Problem 9.11 that the $E[|X_n|]$ are uniformly bounded, specifically,

$$\sup_n E[|X_n|] \leq M$$

for some constant $M < \infty$. Hence we have

$$\sup_n E[X_n^+] < \infty.$$

By Theorem 25.1, there exists an integrable random variable X_∞ such that $X_n \xrightarrow{\text{a.s.}} X_\infty$ as $n \to \infty$. Then it follows from Problems 9.4, 9.2, and 9.12 that $X_n \xrightarrow{L^1} X_\infty$ as $n \to \infty$.

Next, let us prove (b) $\Longrightarrow$ (c). Suppose that $X_n \xrightarrow{L^1} X_\infty$ as $n \to \infty$. Then it follows from Problem 9.14 that $\{X_n : n \geq 0\}$ is uniformly integrable. Then from the previous proof, there exists an integrable random variable X_∞ such that $X_n \xrightarrow{a.s.} X_\infty$ as $n \to \infty$. Let $m \geq 0$ and $A \in \mathscr{F}_m$. Since $\{X_n : n \geq 0\}$ is a submartingale, we have for very $n \geq m$ (see Problem 22.3),

$$\int_A X_m \mathrm{d}P \leq \int_A X_n \mathrm{d}P.$$

By the hypothesis that X_n converges in L^1, we have

$$\int_A X_m \mathrm{d}P \leq \lim_{n\to\infty} \int_A X_n \mathrm{d}P = \int_A X_\infty \mathrm{d}P.$$

Since A is arbitrary, we get

$$E[X_\infty | \mathscr{F}_m] \geq X_m.$$

Hence $\{X_n : n = 0, 1, \ldots, \infty\}$ is a submartingale.

Finally, let us prove (c) $\Longrightarrow$ (a). To do that, let $\epsilon > 0$. Since $|X_\infty| I_{\{|X_\infty| < n\}} \uparrow |X_\infty|$ as $n \to \infty$, it follows from the monotone convergence theorem (Theorem 6.2) that

$$\int_\Omega |X_\infty| I_{\{|X_\infty|<n\}} \mathrm{d}P = \int_{\{|X_\infty|<n\}} |X_\infty| \mathrm{d}P \uparrow \int_\Omega |X_\infty| \mathrm{d}P \quad \text{as} \quad n \to \infty.$$

Therefore there exists an integer N such that

$$\int_{\{|X_\infty| \geq N\}} |X_\infty| \mathrm{d}P = \int_\Omega |X_\infty| \mathrm{d}P - \int_{\{|X_\infty|<N\}} |X_\infty| \mathrm{d}P < \frac{\epsilon}{2}.$$

By the hypothesis and the conditional Jensen's inequality (see Problem 15.7), we have $|X_i| \leq |E(X_\infty | \mathscr{F}_i)| \leq E(|X_\infty| | \mathscr{F}_i)$. Hence we have

$$\begin{aligned}
& \int_{\{|X_i| \geq c\}} |X_i| \mathrm{d}P \\
\leq\ & \int_{\{|X_i| \geq c\}} E(|X_\infty| | \mathscr{F}_i) \mathrm{d}P \\
=\ & \int_{\{|X_i| \geq c\}} |X_\infty| \mathrm{d}P \\
=\ & \int_{\{|X_i| \geq c\} \cap \{|X_\infty| \geq N\}} |X_\infty| \mathrm{d}P + \int_{\{|X_i| \geq c\} \cap \{|X_\infty| < N\}} |X_\infty| \mathrm{d}P \\
\leq\ & \frac{\epsilon}{2} + NP\{|X_i| \geq c\}.
\end{aligned}$$

By Problem 6.16, we have

$$P\{|X_i| \geq c\} \leq \frac{E(|X_i|)}{c} \leq \frac{E[E(|X_\infty||\mathscr{F}_i)]}{c} = \frac{E(|X_\infty|)}{c}$$

If we choose $c > 2NE(|X_\infty|)/\epsilon$, we have

$$\sup_{i \in I} \int_{\{|X_i| \geq c\}} |X_i| \mathrm{d}P < \epsilon.$$

Since ϵ is arbitrary, $\{X_n : n \geq 0\}$ is uniformly integrable. This completes the proof.

25.5 Bibliographic Notes

In this chapter, we presented several convergence theorems for submartingales. The proof of Theorem 25.1 can be found in Ash and Doleans-Dade (1999) and Gut (2007). Lawler (2006) presented many examples of applying the martingale convergence theorems. Hall and Heyde (1980) is a book devoted to martingale limit theory.

CHAPTER 26

RANDOM WALKS

A random walk is a simple discrete stochastic process formed by successive summation of independent and identically distributed random variables. Random walks are one of the most widely studied topics in probability theory and have important applications in finance. In this chapter, we present random walks and relevant results.

26.1 Basic Concepts and Facts

Definition 26.1 (Random Walk). A random walk $\{S_n : n \geq 0\}$ starting at x is a stochastic process defined as

$$S_n = x + \sum_{i=1}^{n} Z_i, \quad S_0 = x,$$

where $x \in \mathbf{R}$ and Z_i's are independent and identically distributed random variables.

Definition 26.2 (Simple Random Walk). Let $\{S_n : n \geq 0\}$ be a random walk as defined in Definition 26.1. It is called a *simple random walk* if for every $n \geq 1$,

$$P\{Z_n = 1\} = p, \quad P\{Z_n = -1\} = q = 1 - p,$$

where $p \in (0, 1)$.

26.2 Problems

26.1. Let $\{S_n : n \geq 0\}$ be a random walk as defined in Definition 26.1. Let $\mu = E[Z_1]$ and $\sigma = \sqrt{E[(Z_1 - \mu)^2]}$. Suppose that $|\mu| < \infty$ and $\sigma < \infty$. Show that

(a) $\{S_n - \mu n : n \geq 0\}$ is a martingale.

(b) $\{(S_n - \mu n)^2 - \sigma^2 n : n \geq 0\}$ is a martingale.

(c) For any $u \in \mathbf{R}$, $\{\exp(uS_n - n \ln E[e^{uZ_1}]) : n \geq 0\}$ is a martingale.

(d) If $P(Z_1 = 1) = p$ and $P(Z_1 = -1) = 1 - p$ for some $p \in (0, 1)$, then

$$\left\{ \left(\frac{1-p}{p} \right)^{S_n} : n \geq 0 \right\}$$

is a martingale.

26.2 (Scaled Random Walk). Let $\{Z_n\}_{n \geq 1}$ be a sequence of independent and identically distributed symmetric Bernoulli random variables such that

$$P(Z_n = 1) = P(Z_n = -1) = \frac{1}{2}, \quad n \geq 1.$$

Let $M_0 = 0$ and $M_n = Z_1 + Z_2 + \cdots + Z_n$ for $n \geq 1$. For every $n \geq 1$, let

$$W_n(t) = \frac{nt - [nt]}{\sqrt{n}} M_{[nt]+1} + \frac{[nt] + 1 - nt}{\sqrt{n}} M_{[nt]}, \quad t \geq 0,$$

where $[x]$ denotes the largest integer that is less than or equal to x. Show that for every $t \geq 0$, the distribution of $W_n(t)$ converges weakly to the normal distribution with mean zero and variance t.

26.3. Let $\{Z_n\}_{n \geq 1}$ be a sequence of independent and identically distributed symmetric Bernoulli random variables such that

$$P(Z_n = 1) = P(Z_n = -1) = \frac{1}{2}, \quad n \geq 1.$$

Let $S_0^x = x$ and

$$S_n^x = x + Z_1 + Z_2 + \cdots + Z_n, \quad n \geq 1,$$

where x is an integer. Let N_1 and N_2 be integers such that $N_1 \leq x \leq N_2$. Let τ be defined as

$$\tau = \tau_{N_1} \wedge \tau_{N_2},$$

where

$$\tau_{N_1} = \inf\{n \geq 0 : S_n^x = N_1\}, \quad \tau_{N_2} = \inf\{n \geq 0 : S_n^x = N_2\}.$$

Show that

(a) $\{S_n^x : n \geq 0\}$ is a martingale.

(b) τ_{N_1} and τ_{N_2} are a.s. finite stopping times.

(c) τ is a stopping time and $P\{\tau < \infty\} = 1$.

(d) $S_{\tau\wedge n}^x (n \geq 1)$ is uniformly integrable.

(e)

$$P\{\tau_{N_1} < \tau_{N_2}\} = \frac{N_2 - x}{N_2 - N_1}.$$

26.4. Let $\{S_n^x : n \geq 0\}$ be a simple random walk starting at x (see Definition 26.2), where x is an integer. Let N_1 and N_2 be two integers such that $N_1 \leq x \leq N_2$. Let

$$\tau_{N_1}^x = \inf\{n \geq 0 : S_n^x = N_1\}, \quad \tau_{N_2}^x = \inf\{n \geq 0 : S_n^x = N_2\}.$$

For every integer x, let $\varphi(x)$ be defined as

$$\varphi(x) = P\{\tau_{N_2}^x < \tau_{N_1}^x\}.$$

Show that

(a) For every integer x in $[N_1 + 1, N_2 - 1]$,

$$\varphi(x) = p\varphi(x+1) + q\varphi(x-1);$$

(b) If $p \neq q$, then

$$\varphi(x) = \frac{1 - (q/p)^{x-N_1}}{1 - (q/p)^{N_2 - N_1}};$$

(c) If $p = q$, then

$$\varphi(x) = \frac{x - N_1}{N_2 - N_1}.$$

26.5. Let $\{Z_n\}_{n\geq 1}$ be a sequence of identically and independently distributed random variables such that

$$P(Z_n = 1) = p, \quad P(Z_n = -1) = 1 - p, \quad n \geq 1,$$

where $p \in (0, \frac{1}{2})$. Let $S_0 = 0$ and

$$S_n = Z_1 + Z_2 + \cdots + Z_n, \quad n \geq 1.$$

Show that

(a) For $y \geq 0$, we have

$$P\left\{\sup_{n\geq 0} S_n > y\right\} \leq \left(\frac{p}{1-p}\right)^y.$$

(b)

$$E\left(\sup_{n\geq 0} S_n\right) \leq \frac{1-p}{1-2p}.$$

26.6. Let $Z_1, Z_2, \ldots, Z_N$ be independent and identically distributed symmetric random variables such that

$$P(Z_j = 1) = P(Z_j = -1) = \frac{1}{2}, \quad 1 \leq j \leq N.$$

Let $S_j = Z_1 + Z_2 + \cdots + Z_j$ for $j = 1, 2, \ldots, N$. Let m be a positive integer. Show that

$$P\left\{\max_{1\leq j\leq N} S_j \geq m\right\} = 2P\{S_N \geq m\} - P\{S_N = m\}.$$

26.7. Let $\{Z_n\}_{n\geq 1}$ be a sequence of independent and identically distributed symmetric Bernoulli random variables such that

$$P(Z_n = 1) = P(Z_n = -1) = \frac{1}{2}, \quad n \geq 1.$$

Let $M_0 = 0$ and $M_n = Z_1 + Z_2 + \cdots + Z_n$ for $n \geq 1$. For every $n \geq 1$, let

$$S_n(t) = \left(1 + \frac{\sigma}{\sqrt{n}}\right)^{\frac{1}{2}([nt]+M_{[nt]})} \left(1 - \frac{\sigma}{\sqrt{n}}\right)^{\frac{1}{2}([nt]-M_{[nt]})}, \quad t \geq 0,$$

where $[x]$ denotes the largest integer that is less than or equal to x. Show that for every $t \geq 0$, the distribution of $S_n(t)$ converges weakly to the distribution of

$$S(t) = \exp\left(\sigma B(t) - \frac{1}{2}\sigma^2 t\right),$$

where $B(t)$ is a normal random variable with mean zero and variance t.

26.3 Hints

26.1. Consider the filtration given by

$$\mathscr{F}_0 = \{\emptyset, \Omega\}, \quad \mathscr{F}_n = \sigma(Z_k : k = 1, 2, \ldots, n), \quad n \geq 1$$

and follow the definition of martingales (Definition 22.1).

26.2. Apply the classical central limit theorem (see Problem 20.4) and use the result of Problem 20.2.

26.3. Part (a) can be proved by the definition of martingales. To prove part (b), we follow the definition of stopping times. To prove that the stopping times are a.s. finite, we can establish the following inequality, where $n_0 = N_2 - N_1$:

$$P\{\tau_{N_1} > kn_0\} \leq \left(1 - \frac{1}{2^{n_0}}\right)^k, \quad k \geq 1,$$

Part (c) is implied by part (b) (see Problem 23.1). To prove (d), we follow the definition of uniform integrability (Definition 9.6) and part (c). Part (e) can be proved by the result of Problem 23.15.

26.4. To prove part (a), consider different values of Z_1 and use the result of Problem 14.18. To prove parts (b) and (c), consider $\varphi(N_1)$ and $\varphi(N_2)$ and use part (a).

26.5. Apply Doob's maximal inequality to $\{(q/p)^{S_n} : n \geq 0\}$ to prove part (a). Use the result of Problem 13.10 to prove part (b).

26.6. Note that

$$\left\{\max_{1\leq j\leq N} S_j \geq m\right\} = \{S_N \geq m\} \cup \left(\bigcup_{j=1}^{N-1} A_j \cap \{S_N < m\}\right),$$

where $A_j = \{S_1 < m, S_2 < m, \ldots, S_{j-1} < m, S_j = m\}$ for $j = 1, 2, \ldots, N$.

26.7. Consider $\ln S_n(t)$ and use the classical central limit theorem (see Problem 20.4) and the result of Problem 20.2.

26.4 Solutions

26.1. Let $\mathscr{F}_0 = \{\emptyset, \Omega\}$ and $\mathscr{F}_n = \sigma(Z_k : k = 1, 2, \ldots, n)$ for $n \geq 1$.

(a) By Schwarz's inequality (Theorem 8.3), we have

$$E[|S_n - \mu n|] \leq |x| + n|\mu| + nE[|Z_1 - \mu|] \leq |x| + n\sigma < \infty.$$

Hence $S_n - n\mu$ is integrable. By Problem 14.4, we have

$$\begin{aligned} E[S_n - \mu n|\mathscr{F}_{n-1}] &= S_{n-1} + E[Z_n - \mu|\mathscr{F}_{n-1}] \\ &= S_{n-1} + E[Z_n - \mu] \\ &= S_{n-1}. \end{aligned}$$

Hence $\{S_n - \mu n : n \geq 0\}$ is a martingale.

(b) By Problem 15.6, we have

$$E\left[|(S_n - \mu n)^2 - \sigma^2 n|\right] \leq n\sigma^2 + n^2 E\left[(Z_1 - \mu)^2\right] = (n + n^2)\sigma^2 < \infty,$$

which shows that $(S_n - \mu n)^2 - \sigma^2 n$ is integrable. In addition, by Problem 14.4 we have

$$\begin{aligned} & E[(S_n - \mu n)^2 - \sigma^2 n|\mathscr{F}_{n-1}] \\ = {} & (S_{n-1} - (n-1)\mu)^2 - n\sigma^2 + E[2(Z_n - \mu)(S_{n-1} - (n-1)\mu) \\ & +(Z_n - \mu)^2|\mathscr{F}_{n-1}] \\ = {} & (S_{n-1} - (n-1)\mu)^2 - n\sigma^2 + 2(S_{n-1} - (n-1)\mu)E[(Z_n - \mu)] \\ & +E[(Z_n - \mu)^2] \\ = {} & (S_{n-1} - (n-1)\mu)^2 - (n-1)\sigma^2. \end{aligned}$$

Hence $\{(S_n - \mu n)^2 - \sigma^2 n : n \geq 0\}$ is a martingale.

(c) If $E[e^{uZ_1}] = \infty$, then $\exp(uS_n - n \ln E[e^{uZ_1}]) = 0$. It is a martingale. We assume that $E[e^{uZ_1}] < \infty$. Since

$$\exp(uS_n - n \ln E[e^{uZ_1}]) = (E[e^{uZ_1}])^{-n} \exp(uS_n)$$

and Z_n's are independent, by Theorem 13.2 we have,

$$\begin{aligned} E[\exp(uS_n - n \ln E[e^{uZ_1}])] &= (E[e^{uZ_1}])^{-n} E[\exp(uS_n)] \\ &= (E[e^{uZ_1}])^{-n} e^{ux} \prod_{i=1}^{n} E[e^{uZ_i}] \\ &= e^{ux} < \infty. \end{aligned}$$

Also we have

$$\begin{aligned} E[(E[e^{uZ_1}])^{-n} \exp(uS_n)|\mathscr{F}_{n-1}] &= (E[e^{uZ_1}])^{-n} E[\exp(uS_n)|\mathscr{F}_{n-1}] \\ &= (E[e^{uZ_1}])^{-n} e^{uS_{n-1}} E[\exp(uZ_n)] \\ &= (E[e^{uZ_1}])^{-n+1} e^{uS_{n-1}}. \end{aligned}$$

Hence $\{\exp(uS_n - n \ln E[e^{uZ_1}]) : n \geq 0\}$ is a martingale.

(d) Let u be such that $e^u = (1-p)/p$ in part (c). Then we have

$$E[e^{uZ_1}] = e^u p + e^{-u}(1-p) = 1$$

and

$$\exp(uS_n - n \ln E[e^{uZ_1}]) = \left(\frac{1-p}{p}\right)^{S_n}.$$

By part (c), we obtain that

$$\left\{ \left(\frac{1-p}{p}\right)^{S_n} : n \geq 0 \right\}$$

is a martingale.

This completes the proof.

26.2. If $t = 0$, the conclusion is trivial. Let $t > 0$ be fixed. Let F_n be the distribution function of $M_{[nt]}/\sqrt{[nt]}$:

$$F_n(x) = P\left\{ \frac{M_{[nt]}}{\sqrt{[nt]}} \leq x \right\}.$$

Since the Z_n are independent and identically distributed and $\mathrm{Var}(Z_n) = 1$, it follows from Problem 20.4 that

$$F_n(x) \Rightarrow \Phi(0, 1, x),$$

where $\Phi(0, 1, x)$ is as defined in Equation (19.1). Now let

$$a_n = \frac{\sqrt{n}}{\sqrt{[nt]}}, \quad b_n = -\frac{nt - [nt]}{\sqrt{[nt]}} Z_{[nt]+1}$$

for all n such that $[nt] \geq 1$. Then we have $a_n \to 1/\sqrt{t}$ and $b_n \to 0$ as $n \to \infty$. It follows from part (a) of Problem 20.2 that

$$F_n(a_n x + b_n) \Rightarrow \Phi\left(0, 1, \frac{x}{\sqrt{t}}\right) = \Phi(0, t, x).$$

The result follows immediately by noting that

$$F_n(a_n x + b_n) = P\left\{\frac{M_{[nt]}}{\sqrt{[nt]}} \leq a_n x + b_n\right\} = P\left\{W_n(t) \leq x\right\}.$$

This completes the proof.

26.3. Let $\mathscr{F}_n = \sigma(Z_1, Z_2, \ldots, Z_n), n \geq 1$.

(a) By the assumption that $Z_1, Z_2, \ldots$ are independent, we have

$$E[S_{n+1}^x|\mathscr{F}_n] = S_n^x + E[Z_{n+1}|\mathscr{F}_n] = S_n^x + E[Z_{n+1}] = S_n^x, \quad n \geq 0.$$

Hence $\{S_n^x : n \geq 0\}$ is a martingale.

(b) To show that τ_{N_1} is a stopping time, we only need to show that

$$\{\tau_{N_1} \leq n\} \in \mathscr{F}_n, \quad n \geq 1.$$

Note that

$$\{\tau_{N_1} \leq n\} = \bigcup_{k=0}^{n} \{S_0^x \neq N_1, \ldots, S_{k-1}^x \neq N_1, S_k^x = N_1\}.$$

But $\{S_0^x \neq N_1, \ldots, S_{k-1}^x \neq N_1, S_k^x = N_1\} \in \mathscr{F}_n$ for all $k \leq n$. Hence we have $\{\tau_{N_1} \leq n\} \in \mathscr{F}_n$. Thus τ_{N_1} is a martingale.

Now we show that $P\{\tau_{N_1} < \infty\} = 1$. Let $n_0 = N_2 - N_1$. Then

$$P\{\tau_{N_1} \leq n_0\} \geq P\{Z_1 = -1, Z_2 = -1, \ldots, Z_{n_0} = -1\} = \frac{1}{2^{n_0}},$$

which gives

$$P\{\tau_{N_1} > n_0\} \leq 1 - \frac{1}{2^{n_0}}.$$

For $k \geq 1$, we have

$$\begin{aligned} P\{\tau_{N_1} > kn_0\} &= P\{\tau_{N_1} > (k-1)n_0, \tau_{N_1} > kn_0\} \\ &= E[I_{\{\tau_{N_1} > (k-1)n_0\}} I_{\{\tau_{N_1} > kn_0\}}] \\ &= E[I_{\{\tau_{N_1} > (k-1)n_0\}} E(I_{\{\tau_{N_1} > kn_0\}} | \mathscr{F}_{(k-1)n_0})]. \end{aligned}$$

But by Definition 14.4, we have

$$\begin{aligned}
& E(I_{\{\tau_{N_1}>kn_0\}}|\mathscr{F}_{(k-1)n_0}) \\
= \; & 1 - E(I_{\{\tau_{N_1}\le kn_0\}}|\mathscr{F}_{(k-1)n_0}) \\
\le \; & 1 - E[I_{\{\tau_{N_1}>(k-1)n_0\}}I_{\{\tau_{N_1}\le kn_0\}}|\mathscr{F}_{(k-1)n_0}] \\
= \; & 1 - P(\{\tau_{N_1} > (k-1)n_0\} \cap \{\tau_{N_1} \le kn_0\}|\mathscr{F}_{(k-1)n_0}) \\
\le \; & 1 - P(\{Z_{(k-1)n_0+1} = -1, \ldots, Z_{kn_0} = -1\}|\mathscr{F}_{(k-1)n_0}) \\
= \; & 1 - P(\{Z_{(k-1)n_0+1} = -1, \ldots, Z_{kn_0} = -1\}) \\
= \; & 1 - \frac{1}{2^{n_0}}.
\end{aligned}$$

Hence by induction, we get

$$P\{\tau_{N_1} > kn_0\} \le \left(1 - \frac{1}{2^{n_0}}\right)^k.$$

Letting $k \to \infty$, we get $P\{\tau_{N_1} = \infty\} = 0$.

Similarly, we can show that τ_{N_2} is also an a.s. finite stopping time.

(c) By Problem 23.1, τ is a stopping time. Since

$$\{\tau < \infty\} = \{\tau_{N_1} < \infty\} \cup \{\tau_{N_2} < \infty\},$$

we have $1 \ge P\{\tau < \infty\} \ge P\{\tau_{N_1} < \infty\} = 1$ by part (b). Hence τ is a.s. finite.

(d) By the definition of τ, we have $S^x_{\tau\wedge n} \in [N_1, N_2]$. Hence if $c > \max\{|N_1|, |N_2|\}$, we have

$$\sup_{n\ge 0} \int_{\{|S^x_{\tau\wedge n}|\ge c\}} |S^x_{\tau\wedge n}| \mathrm{d}P = 0,$$

which shows that $S^x_{\tau\wedge n}$ is uniformly integrable.

(e) By parts (a) – (d) and Problem 23.15, we have

$$x = E[S^x_0] = E[S^x_\tau] = N_1 P(\tau_{N_1} < \tau_{N_2}) + N_2 P(\tau_{N_1} > \tau_{N_2}).$$

Solving the above equation gives

$$P(\tau_{N_1} < \tau_{N_2}) = \frac{N_2 - x}{N_2 - N_1}.$$

This completes the proof.

26.4.

(a) Let $H = \{Z_1 = 1\}$. Then we have $H^c = \{Z_1 = -1\}$. It follows from Problem 14.18 that

$$\begin{aligned}
\varphi(x) &= E\left[I_{\{\tau^x_{N_2} < \tau^x_{N_1}\}}\right] \\
&= E\left[I_{\{\tau^x_{N_2} < \tau^x_{N_1}\}} | H\right] P(H) + E\left[I_{\{\tau^x_{N_2} < \tau^x_{N_1}\}} | H^c\right] P(H^c) \\
&= pE\left[I_{\{\tau^x_{N_2} < \tau^x_{N_1}\}} | H\right] + qE\left[I_{\{\tau^x_{N_2} < \tau^x_{N_1}\}} | H^c\right]. \qquad (26.1)
\end{aligned}$$

Let $\tilde{S}^{x+1}_0 = 1 + x$ and $\tilde{S}^{x+1}_n = 1 + x + Z_2 + \cdots + Z_{n+1}$ for $n \geq 1$. Then $\tilde{S}^{x+1}_n$ and S^{x+1}_n have the same distribution. Hence, on the event H, we have

$$\begin{aligned}
\tau^x_{N_1} &= \inf\{n \geq 0 : S^x_n = N_1\} \\
&= \inf\{n \geq 1 : x + 1 + Z_2 + \cdots + Z_n = N_1\} \\
&= \inf\{n \geq 1 : \tilde{S}^{x+1}_{n-1} = N_1\} \\
&= \inf\{n \geq 0 : \tilde{S}^{x+1}_n = N_1\} \\
&= \inf\{n \geq 0 : S^{x+1}_n = N_1\} \\
&= \tau^{x+1}_{N_1} \qquad (26.2)
\end{aligned}$$

and

$$\tau^x_{N_2} = \tau^{x+1}_{N_2}. \qquad (26.3)$$

Similarly, on the event H^c, we have

$$\tau^x_{N_1} = \tau^{x-1}_{N_1}, \quad \tau^x_{N_2} = \tau^{x-1}_{N_2}.. \qquad (26.4)$$

The result follows by combining Equations (26.1), (26.2), (26.3), and (26.4).

(b) From part (a), we have

$$\varphi(x+1) - \varphi(x) = \frac{q}{p}[\varphi(x) - \varphi(x-1)], \quad x \in [N_1 + 1, N_2 - 1],$$

which implies

$$\varphi(x+1) - \varphi(x) = \left(\frac{q}{p}\right)^{x - N_1} [\varphi(N_1 + 1) - \varphi(N_1)], \quad x \in [N_1 + 1, N_2 - 1],$$

By the assumption that $p \neq q$, for every $x \in [N_1 + 1, N_2 - 1]$, we have

$$\begin{aligned}
&\varphi(x+1) - \varphi(N_1 + 1) \\
&= \sum_{i=N_1+1}^{x} [\varphi(i+1) - \varphi(i)] \\
&= [\varphi(N_1 + 1) - \varphi(N_1)] \sum_{i=N_1+1}^{x} \left(\frac{q}{p}\right)^{i - N_1} \\
&= \frac{q/p - (q/p)^{x - N_1 + 1}}{1 - q/p} [\varphi(N_1 + 1) - \varphi(N_1)]. \qquad (26.5)
\end{aligned}$$

Since $\tau_{N_1}^{N_1} = 0$ and $\tau_{N_2}^{N_2} = 0$, we have $\varphi(N_1) = P\{\tau_{N_2}^{N_1} < 0\} = 0$ and $\varphi(N_2) = P\{0 < \tau_{N_1}^{N_1}\} = 1$. Letting $x = N_2 - 1$ in Equation (26.5) gives

$$\varphi(N_2) - \varphi(N_1 + 1) = \frac{q/p - (q/p)^{N_2 - N_1}}{1 - q/p}[\varphi(N_1 + 1) - \varphi(N_1)].$$

Hence we have

$$\varphi(N_1 + 1) = \frac{1 - q/p}{1 - (q/p)^{N_2 - N_1}}. \tag{26.6}$$

Combining Equations (26.5) and (26.6) gives

$$\varphi(x + 1) = \frac{1 - (q/p)^{x - N_1 + 1}}{1 - (q/p)^{N_2 - N_1}}, \quad x \in [N_1 + 1, N_2 - 1]$$

or

$$\varphi(x) = \frac{1 - (q/p)^{x - N_1}}{1 - (q/p)^{N_2 - N_1}}, \quad x \in [N_1 + 2, N_2].$$

Note that $\varphi(N_1) = 0$ and Equation (26.6). We have

$$\varphi(x) = \frac{1 - (q/p)^{x - N_1}}{1 - (q/p)^{N_2 - N_1}}, \quad x \in [N_1, N_2].$$

(c) The proof is similar to that of part (c). When $p = q$, Equation (26.5) becomes

$$\varphi(x+1) - \varphi(N_1+1) = (x - N_1)[\varphi(N_1+1) - \varphi(N_1)], \quad x \in [N_1+1, N_2-1].$$

Since $\varphi(N_1) = 0$ and $\varphi(N_2) = 1$, letting $x = N_2 - 1$ in the above equation gives

$$\varphi(N_1 + 1) = \frac{1}{N_2 - N_1}.$$

Then combining the above equations gives

$$\varphi(x) = \frac{x - N_1}{N_2 - N_1}, \quad x \in [N_1, N_2].$$

This completes the proof.

26.5. Let $q = q - p$.

(a) Let $X_n = \left(\frac{q}{p}\right)^{S_n}$, $n \geq 0$. Let $\mathscr{F}_n = \sigma(Z_1, Z_2, \ldots, Z_n)$. Then

$$E[X_{n+1}|\mathscr{F}_n] = X_n E\left[\left(\frac{q}{p}\right)^{Z_{n+1}} \middle| \mathscr{F}_n\right] = X_n E\left[\left(\frac{q}{p}\right)^{Z_{n+1}}\right] = X_n.$$

Hence $\{X_n : n \geq 0\}$ is a martingale. Let $N > 1$. By Doob's maximal inequality (see Problem 24.2), we have

$$\begin{aligned} P\left\{\sup_{0\leq n\leq N} S_n > y\right\} &= P\left\{\sup_{0\leq n\leq N} X_n > \left(\frac{q}{p}\right)^y\right\} \\ &\leq \left(\frac{p}{q}\right)^y E[X_N] \\ &= \left(\frac{p}{q}\right)^y. \end{aligned}$$

The result follows by letting $N \to \infty$.

(b) Note that $p \in (0, \frac{1}{2})$. We have $p/q < 1$. By part (a) and Problem 13.10, we have

$$\begin{aligned} E\left[\sup_{0\leq n\leq N} S_n\right] &= \int_0^\infty P\left\{\sup_{0\leq n\leq N} S_n > y\right\} dy \\ &\leq \int_0^\infty \left(\frac{p}{q}\right)^y dy \\ &= \sum_{k=0}^\infty \int_k^{k+1} \left(\frac{p}{q}\right)^y dy \\ &\leq \sum_{k=0}^\infty \int_k^{k+1} \left(\frac{p}{q}\right)^k dy \\ &= \sum_{k=0}^\infty \left(\frac{p}{q}\right)^k \\ &= \frac{q}{q-p}. \end{aligned}$$

This completes the proof.

26.6. Let $A_j = \{S_1 < m, S_2 < m, \ldots, S_{j-1} < m, S_j = m\}$ for $j = 1, 2, \ldots, N$. Then we have

$$P\left\{\max_{1\leq j\leq N} S_j \geq m\right\} = P\{S_N \geq m\} + \sum_{j=1}^{N-1} P\left(A_j \cap \{S_N < m\}\right).$$

Note that $A_j \cap \{S_N < m\} = A_j \cap \{S_N < S_j\}$ and that A_j and $\{S_N < S_j\}$ are independent. We have

$$P\left\{\max_{1\leq j\leq N} S_j \geq m\right\} = P\{S_N \geq m\} + \sum_{j=1}^{N-1} P(A_j) P\left(\{S_N < S_j\}\right).$$

By the assumption that $X_1, X_2, \ldots, X_N$ are symmetrically distributed about zero, we have $P\{S_N > S_j\} = P\{S_N > S_j\}$. Hence we have

$$\begin{aligned} P\left\{\max_{1\le j\le N} S_j \ge m\right\} &= P\{S_N \ge m\} + \sum_{j=1}^{N-1} P(A_j)P(\{S_N > S_j\}) \\ &= P\{S_N \ge m\} + \sum_{j=1}^{N-1} P(A_j \cap \{S_N > S_j\}) \\ &= P\{S_N \ge m\} + \sum_{j=1}^{N-1} P(A_j \cap \{S_N > m\}) \\ &= P\{S_N \ge y\} + P\{S_N > m\} \\ &= 2P\{S_N \ge y\} - P\{S_N = m\}. \end{aligned}$$

This completes the proof.

26.7. The case when $t = 0$ is trivial. Suppose that $t > 0$. Then we have

$$\begin{aligned} \ln S_n(t) &= \frac{[nt]}{2} \ln\left(1 - \frac{\sigma^2}{n}\right) + \frac{M_{[nt]}}{2} \ln\left(\frac{1 + \frac{\sigma}{\sqrt{n}}}{1 - \frac{\sigma}{\sqrt{n}}}\right) \\ &= a_n \frac{M_{[nt]}}{\sqrt{[nt]}} + b_n, \end{aligned} \tag{26.7}$$

where

$$a_n = \frac{\sqrt{[nt]}}{2} \ln\left(\frac{1 + \frac{\sigma}{\sqrt{n}}}{1 - \frac{\sigma}{\sqrt{n}}}\right), \quad b_n = \frac{[nt]}{2} \ln\left(1 - \frac{\sigma^2}{n}\right).$$

Applying l'Hospital's rule to a_n and b_n gives

$$a_n \to \sigma\sqrt{t}, \quad b_n \to -\frac{1}{2}\sigma^2 t.$$

Let F_n be the distribution function of $M_{[nt]}/\sqrt{[nt]}$, i.e.,

$$F_n(x) = P\left\{\frac{M_{[nt]}}{\sqrt{[nt]}} \le x\right\}.$$

It follows from Problem 20.4 that $F_n \Rightarrow N(0, 1)$ or

$$F_n(x) \Rightarrow \Phi(0, 1, x),$$

where Φ is as defined in Equation (19.2). By part (a) of Problem 20.2, we have

$$F_n(\frac{x}{a_n} - \frac{b_n}{a_n}) \Rightarrow \Phi\left(0, 1, \frac{x}{\sigma\sqrt{t}} + \frac{1}{2}\sigma\sqrt{t}\right) = \Phi\left(-\frac{1}{2}\sigma^2 t, \sigma^2 t, x\right).$$

Since

$$F_n\left(\frac{x}{a_n} - \frac{b_n}{a_n}\right) = P\left\{\frac{M_{[nt]}}{\sqrt{[nt]}} \leq \frac{x}{a_n} - \frac{b_n}{a_n}\right\} = P\{\ln S_n(t) \leq x\}$$

and the distribution of $\ln S(t)$ is $\Phi\left(-\frac{1}{2}\sigma^2 t, \sigma^2 t, x\right)$, the result follows. This completes the proof.

26.5 Bibliographic Notes

In this chapter, we presented random walks. As one of the most studied topics in probability theory, random walks have been discussed in many books, for example, Bailey (1964), Feller (1968), Loeve (1977), Kannan (1979), Shiryaev and Boas (1995), Grinstead and Snell (1997), Kallenberg (1997), Paul and Baschnagel (1999), Steele (2003), James (2003), Klebaner (2005), Stroock (2005), Lin (2006), Buchanan (2006), Bhattacharya and Waymire (2009), and Durrett (2010). There are several books devoted to random walks, including Spitzer (2001), Rudnick and Gaspari (2004), and Lawler and Limic (2010).

CHAPTER 27

POISSON PROCESSES

A Poisson process is a continuous-time stochastic process that counts the number of events in a given time interval. In mathematical finance, Poisson processes are used to build jump processes for modeling asset prices. In this chapter, we present the mathematical definition of the Poisson process and relevant results.

27.1 Basic Concepts and Facts

Definition 27.1 (Poisson Process). Let $\lambda > 0$. A stochastic process $\{N_t : t \geq 0\}$ is said to be Poisson process with parameter λ if it satisfies the following conditions:

(a) $P\{N_0 = 0\} = P\{\omega : N_0(\omega) = 0\} = 1$.

(b) For any $0 \leq s < t$, the random variable $N_t - N_s$ is a Poisson random variable with parameter $\lambda(t-s)$:

$$P\{N_t - N_s = k\} = e^{-\lambda(t-s)} \frac{(\lambda(t-s))^k}{k!}, \quad k = 0, 1, 2, \ldots.$$

(c) The N_t has independent increments, that is, for any $0 \leq t_1 < t_2 < \ldots < t_n$, the random variables $N_{t_1}, N_{t_2} - N_{t_1}, \ldots, N_{t_n} - N_{t_{n-1}}$ are independent.

(d) Almost all sample paths of $\{N_t : t \geq 0\}$ are right-continuous functions with left-hand limits.

Definition 27.2 (Poisson Process with Respect to Filtrations). Let $\{\mathscr{F}_t : t \geq 0\}$ be a filtration, and let $\lambda > 0$. A stochastic process $\{N_t : t \geq 0\}$ is said to be Poisson process with respect to the filtration $\{\mathscr{F}_t : t \geq 0\}$ and with parameter λ if it satisfies the following conditions:

(a) $P\{N_0 = 0\} = P\{\omega : N_0(\omega) = 0\} = 0$.

(b) For any $0 \leq s < t$, the random variance $N_t - N_s$ is a Poisson random variable with parameter $\lambda(t-s)$:

$$P\{N_t - N_s = k\} = e^{-\lambda(t-s)} \frac{(\lambda(t-s))^k}{k!}, \quad k = 0, 1, 2, \ldots.$$

(c) For any $s \leq t$, the random variable $N_t - N_s$ is independent of the σ-field $\mathscr{F}_s$.

(d) Almost all sample paths of $\{N_t : t \geq 0\}$ are right continuous functions with left-hand limits.

Definition 27.3 (Compensated Poisson Process). Let $\{N_t : t \geq 0\}$ be a Poisson process with parameter $\lambda > 0$. Then the corresponding compensated Poisson process $\tilde{N}_t$ is defined by

$$\tilde{N}_t = N_t - \lambda t, \quad t \geq 0.$$

Definition 27.4 (Compound Poisson Process). A compound Poisson process $\{Q_t : t \geq 0\}$ has the form

$$Q_t = \sum_{i=1}^{N_t} Y_i, \quad t \geq 0,$$

where $\{N_t : t \geq 0\}$ is a Poisson process with some parameter $\lambda > 0$ defined on a probability space $(\Omega, \mathscr{F}, P)$, $\{Y_n : n = 0, 1, \ldots\}$ is a sequence of independent and identically distributed random variables defined on the same probability space $(\Omega, \mathscr{F}, P)$, and the sequence $\{Y_n : n = 0, 1, \ldots\}$ is independent of the Poisson process $\{N_t : t \geq 0\}$.

Definition 27.5 (Counting Process). Let $\{T_n : n \geq 1\}$ be an increasing sequence of positive random variables. Let $T_0 = 0$ a.s. Then the counting process $\{N_t : t \geq 0\}$ associated to the sequence $\{T_n : n \geq 0\}$ is defined as

$$N_t = \sum_{n \geq 1} I_{T_n \leq t}, \quad t \geq 0.$$

Theorem 27.1. *Let $\{T_n : n \geq 0\}$ be an increasing sequence of positive random variables with $P\{T_0 = 0\} = 1$ and $P\{T_n \to \infty\} = 1$. Let $\{N_t : t \geq 0\}$ be a counting process associated with the T_n. If $\{N_t : t \geq 0\}$ has stationary independent increments, then $\{N_t : t \geq 0\}$ is a Poisson process.*

27.2 Problems

27.1. Let $\{N_t : t \geq 0\}$ be a Poisson process with parameter $\lambda > 0$. Show that the corresponding compensated Poisson process $\tilde{N}_t$ is a martingale.

27.2. Let $\{N_t : t \geq 0\}$ be a Poisson process with parameter λ, where $\lambda > 0$. The sample paths of the Poisson process are step functions with jumps. Let T_1 be the time of the first jump. For $n \geq 2$, let T_n be the elapsed time between the $(n-1)$th the jump and the nth jump. Show that $\{T_n : n \geq 1\}$ is a sequence of independent random variables having the same exponential distribution with parameter λ:

$$P\{T_n \leq t\} = \begin{cases} 1 - e^{-\lambda t}, & \text{if } t \geq 0; \\ 0, & \text{if } t < 0. \end{cases}$$

27.3. Let $\{\tau_n\}_{n\geq 1}$ be a sequence of independent and identically distributed exponential random variables with parameter λ. Let $\{T_n\}_{n\geq 1}$ be the sequence of partial sums of the τ_n:

$$T_n = \sum_{i=1}^{n} \tau_i, \quad n \geq 1.$$

Let $\{N_t : t \geq 0\}$ be a stochastic process defined as

$$N_t = \sum_{n\geq 1} I_{t\geq T_n}.$$

Show that $\{N_t : t \geq 0\}$ is a Poisson process with parameter λ.

27.4. Let $\{N_t^{(1)} : t \geq 0\}$ and $\{N_t^{(2)} : t \geq 0\}$ be two independent Poisson processes with parameters λ_1 and λ_2, respectively. Show that $\{N_t^{(1)} + N_t^{(2)} : t \geq 0\}$ is a Poisson process with parameter $\lambda_1 + \lambda_2$.

27.5. Let $\{\tilde{N}_t : t \geq 0\}$ be the compensated Poisson process corresponding to a Poisson process with parameter $\lambda > 0$. Let $0 \leq s < t$. Show that the moment generating function of $\tilde{N}_t - \tilde{N}_s$ is

$$E\left[\exp(x(\tilde{N}_t - \tilde{N}_s))\right] = \exp\left[\lambda(t-s)(e^x - x - 1)\right].$$

In particular, the first four moments are given by

$$\begin{aligned} E\left[\tilde{N}_t - \tilde{N}_s\right] &= 0, \\ E\left[(\tilde{N}_t - \tilde{N}_s)^2\right] &= \lambda(t-s), \\ E\left[(\tilde{N}_t - \tilde{N}_s)^3\right] &= \lambda(t-s), \\ E\left[(\tilde{N}_t - \tilde{N}_s)^4\right] &= \lambda(t-s) + 3\lambda^2(t-s)^2. \end{aligned}$$

27.6. Let $\{\tilde{N}_t : t \geq 0\}$ be the compensated Poisson process corresponding to a Poisson process with parameter $\lambda > 0$. Let $0 \leq a < b < \infty$. Show that

$$\sum_{i=1}^{n} (\tilde{N}_{t_i} - \tilde{N}_{t_{i-1}})^2 \xrightarrow{L^2(\Omega)} \lambda(b-a) + \tilde{N}_b - \tilde{N}_a \quad \text{as} \quad \|\Delta_n\| \to 0,$$

where $\Delta_n = \{t_0, t_1, \ldots, t_n\}$ is a partition of $[a, b]$ and

$$\|\Delta_n\| = \max_{1 \leq i \leq n} (t_i - t_{i-1}).$$

27.7. Let $\{\tilde{N}_t : t \geq 0\}$ be the compensated Poisson process corresponding to a Poisson process with parameter $\lambda > 0$. Show that

$$M_t = \tilde{N}_t^2 - \lambda t$$

is a martingale.

27.8 (Geometric Poisson Process). Let $\{N_t : t \geq 0\}$ be a Poisson process with parameter $\lambda > 0$. Let $\{X_t : t \geq 0\}$ be defined as

$$X_t = \exp\left(N_t \ln(1+\sigma) - \lambda\sigma t\right) = (1+\sigma)^{N_t} e^{-\lambda\sigma t}, \quad t > 0,$$

where $\sigma > -1$ and $X_0 > 0$ are constants. Show that $\{X_t : t \geq 0\}$ is a martingale.

27.9. Let $\{Q_t : t \geq 0\}$ be a compound Poisson process as defined in Definition 27.4. Let $\{\mathscr{F}_t : t \geq 0\}$ be the natural filtration of $\{Q_t : t \geq 0\}$. Show that

(a) For every integer $m \geq 0$, every $t \geq 0$, and every $h \geq 0$, the random variables

$$\sum_{n=1}^{N_{t+h}-N_t} Y_{n+m}$$

and

$$\sum_{n=1}^{N_{t+h}-N_t} Y_{n+N_t}$$

have the same distribution:

$$P\left\{\sum_{n=1}^{N_{t+h}-N_t} Y_{n+m} \leq c\right\} = P\left\{\sum_{n=1}^{N_{t+h}-N_t} Y_{n+N_t} \leq c\right\}, \quad \forall c \in \mathbf{R}.$$

(b) For every $t \geq 0$ and every $h \geq 0$, $Q_{t+h} - Q_t$ has the same distribution as Q_h.

(c) For every $t \geq 0$ and every $h \geq 0$, $Q_{t+h} - Q_t$ is independent of $\mathscr{F}_t$.

27.10. Let $\{Q_t : t \geq 0\}$ be a compound Poisson process as defined in Definition 27.4. Suppose that $E(Y_1) = \beta$. Show that

(a) For every $t \geq 0$, $E[Q_t] = \beta\lambda t$.

(b) The process $\{Q_t - \beta\lambda t : t \geq 0\}$ is a martingale.

27.3 Hints

27.1. Use the definition of compensated Poisson process (Definition 27.3).

27.2. Note that the event $\{T_1 > t\}$ coincides with the event $\{N_t = 0\}$ and use the tower property of conditional expectations (see Problem 14.10).

27.3. Verify the four conditions in Definition 27.1. One can use direct computation to verify the first two conditions. The third condition can be verified by using the result of Problem 12.6. The fourth condition can be proved by using the result of Problem 11.1.

27.4. Verify the four conditions of Poisson processes (Definition 27.1).

27.5. Follow the definition of compensated Poisson processes.

27.6. Consider the random variables $X_1, X_2, \ldots, X_n$ given by

$$X_i = (\tilde{N}_{t_i} - \tilde{N}_{t_{i-1}})^2 - (\tilde{N}_{t_i} - \tilde{N}_{t_{i-1}}) - \lambda(t_i - t_{i-1}).$$

27.7. Follow the definition of martingales (Definition 22.1) and note that $\tilde{N}_t = N_t - \lambda t$.

27.8. Use the definition of Poisson processes (Definition 27.2) and the definition of martingales (Definition 22.1). The results of Problems 14.12 and 14.16 can be used to verify $E[X_t|\mathscr{F}_s] = X_s$.

27.9. To prove parts (a) and (b), note that

$$\left\{ \sum_{n=1}^{N_{t+h}-N_t} Y_{n+N_t} \le c \right\}$$
$$= \bigcup_{i,j\ge 0} \left(\{N_{t+h} - N_t = j\} \cap \{N_t = i\} \cap \left\{ \sum_{n=1}^{j} Y_{n+i} \le c \right\} \right).$$

Use part (a) and the result of Problem 21.10 to prove part (c).

27.10. To prove part (a), follow the definition of compound Poisson process (Definition 27.4) and use the result of Problem 14.18. To prove (b), note that $Q_t - Q_s$ has the same distribution as Q_{t-s} and use part (a).

27.4 Solutions

27.1. By definition, $\tilde{N}(t)$ is $\mathscr{F}_t$-measurable. In addition, we have

$$E(|\tilde{N}_t|) = E(|N_t - \lambda t|) \le E(N_t) + \lambda t = 2\lambda t < \infty.$$

Let $s \le t$. Then we have

$$\begin{aligned} E(\tilde{N}_t|\mathscr{F}_s) &= E(N_t - \lambda t|\mathscr{F}_s) \\ &= E(N_t - N_s + N_s|\mathscr{F}_s) - t\lambda \\ &= E(N_t - N_s) + N_s - \lambda t \\ &= N_s - \lambda s. \end{aligned}$$

Hence the $\tilde{N}_t$ is a martingale. This completes the proof.

27.2. Since $\{T_1 > t\} = \{N_t = 0\}$, we have

$$P\{T_1 > t\} = P\{N_t = 0\} = e^{-\lambda t}, \quad t \ge 0.$$

Since $\{T_1 = s\} \in \mathscr{F}_s$ and $N_{t+s} - N_s$ is independent of $\mathscr{F}_s$, we have

$$\begin{aligned} P(T_2 > t|T_1 = s) &= P\{N_{t+s} - N_s = 0|T_1 = s\} \\ &= P\{N_{t+s} - N_s = 0\} \\ &= e^{-\lambda t}, \end{aligned}$$

which shows that T_2 is independent of T_1. By Definition 14.5, we have

$$P(T_2 > t|T_1) = e^{-\lambda t}.$$

Hence by Problem 14.10, we have

$$\begin{aligned} P\{T_2 > t\} &= E[P\{T_2 > t|T_1\}] \\ &= E\left[e^{-\lambda t}\right] \\ &= e^{-\lambda t}. \end{aligned}$$

Similarly, we can show that T_n is independent of $T_1, T_2, \ldots, T_{n-1}$ and

$$P\{T_n > t\} = e^{-\lambda t}, \quad t \ge 0.$$

This completes the proof.

27.3. To show that $\{N_t : t \ge 0\}$ is a Poisson process, we only need to verify that $\{N_t : t \ge 0\}$ satisfies the four conditions in Definition 27.1. First, let us show that $P\{N_0 = 0\} = 1$. Note that

$$\{\omega : N_0(\omega) = 0\} = \{\omega : T_n(\omega) > 0,\ n = 1, 2, \ldots\} = \{\omega : \tau_1(\omega) > 0\},$$

which gives $P\{N_0 = 0\} = P\{\tau_1 > 0\} = 1$.

Now let us verify the second condition. Let $0 \le s < t$. Then we have

$$N_t - N_s = \sum_{n\ge 1} I_{t\ge T_n} - \sum_{n\ge 1} I_{s\ge T_n} = \sum_{n\ge 1} I_{s<T_n\le t}.$$

Noting that the T_n is an non-decreasing sequence, we have

$$\{N_t - N_s = k\} = \bigcup_{i=0}^{\infty}\{T_i \le s < T_{i+1}, T_{i+k} \le t < T_{i+k+1}\}, \quad k \ge 0,$$

where $T_0 = 0$. Since the sets $\{T_i \le s < T_{i+1}, T_{i+k} \le t < T_{i+k+1}\}$ are mutually disjoint, we have

$$P\{N_t - N_s = k\} = \sum_{i=0}^{\infty} P\{T_i \le s < T_{i+1}, T_{i+k} \le t < T_{i+k+1}\}, \quad k \ge 0. \tag{27.1}$$

But

$$\begin{aligned}
& P\{T_i \le s < T_{i+1}, T_{i+k} \le t < T_{i+k+1}\} \\
= & \int_{T_i \le s < T_{i+1}, T_{i+k} \le t < T_{i+k+1}} \mathrm{d}P \\
= & \int_0^s \mathrm{d}x_1 \cdots \int_0^{s-x_1-\cdots-x_{i-2}} \mathrm{d}x_{i-1} \int_0^{s-x_1-\cdots-x_{i-1}} \mathrm{d}x_i \\
& \int_{s-x_1-\cdots-x_i}^{t-x_1-\cdots-x_i} \mathrm{d}x_{i+1} \int_0^{t-x_1-\cdots-x_i} \mathrm{d}x_{i+2} \cdots \int_0^{t-x_1-\cdots-x_{i+k-1}} \mathrm{d}x_{i+k} \\
& \int_{t-x_1-\cdots-x_{i+k}}^{\infty} \mathrm{d}x_{i+k+1} \lambda^{i+k+1} e^{-\lambda(x_1+\cdots+x_{i+k+1})} \\
= & \int_0^s \mathrm{d}x_1 \cdots \int_0^{s-x_1-\cdots-x_{i-2}} \mathrm{d}x_{i-1} \int_0^{s-x_1-\cdots-x_{i-1}} \mathrm{d}x_i \int_{s-x_1-\cdots-x_i}^{t-x_1-\cdots-x_i} \mathrm{d}x_{i+1} \\
& \int_0^{t-x_1-\cdots-x_i} \mathrm{d}x_{i+2} \cdots \int_0^{t-x_1-\cdots-x_{i+k-1}} \mathrm{d}x_{i+k} \lambda^{i+k} e^{-\lambda t} \\
= & \int_0^s \mathrm{d}x_1 \cdots \int_0^{s-x_1-\cdots-x_{i-2}} \mathrm{d}x_{i-1} \int_0^{s-x_1-\cdots-x_{i-1}} \mathrm{d}x_i \frac{(t-s)^k \lambda^{i+k}}{k!} e^{-\lambda t} \\
= & \int_0^s \mathrm{d}x_1 \cdots \int_0^{s-x_1-\cdots-x_{i-2}} \mathrm{d}x_{i-1} (s - x_1 - \cdots - x_{i-1})(t-s)^k \lambda^{i+k} e^{-\lambda t} \\
= & \frac{s^i}{i!k!}(t-s)^k \lambda^{i+k} e^{-\lambda t}.
\end{aligned} \tag{27.2}$$

Combining Equations (27.1) and (27.2) gives

$$P\{N_t - N_s = k\} = \sum_{i=0}^{\infty} \frac{s^i}{i!k!}(t-s)^k \lambda^{i+k} e^{-\lambda t} = \frac{\lambda^k (t-s)^k}{k!} e^{-\lambda(t-s)}, \quad k \ge 0.$$

Hence $N_t - N_s$ is Poisson distributed with parameter $\lambda(t-s)$.

To verify the third condition, we use the result of Problem 12.6. Let $0 \le t_1 < t_2 < \cdots < t_n$. Then $N_{t_1}, N_{t_2} - N_{t_1}, \ldots, N_{t_n} - N_{t_{n-1}}$ are Poisson distributed. Let

$k_1, k_2, \ldots, k_n \geq 0$ and $j_1 = k_1$, $j_2 = k_1 + k_2$, …, $j_n = k_1 + \cdots + k_n$. Then we have

$$\begin{aligned} & P\{N_{t_1} = k_1, N_{t_2} - N_{t_1} = k_2, \ldots, N_{t_n} - N_{t_{n-1}} = k_n\} \\ = \; & P\{T_{j_1} \leq t_1 < T_{j_1+1}, \ldots, T_{j_n} \leq t_n < T_{j_n+1}\}. \end{aligned}$$

Using the same approach in Equation (27.2), we can get

$$\begin{aligned} & P\{T_{j_1} \leq t_1 < T_{j_1+1}, \ldots, T_{j_n} \leq t_n < T_{j_n+1}\} \\ = \; & e^{-\lambda t_n} \lambda^{j_n} \prod_{i=1}^{n} \frac{(t_i - t_{i-1})_i^k}{k_i!}, \end{aligned}$$

where $t_0 = 0$. Since N_{t_1}, $N_{t_2} - N_{t_1}$, …, $N_{t_n} - N_{t_{n-1}}$ are Poisson distributed, we have

$$\begin{aligned} \prod_{i=1}^{n} P\{N_{t_i} - N_{t_{i-1}} = k_i\} & = \prod_{i=1}^{n} \frac{(t_i - t_{i-1})^{k_i} \lambda^{k_i}}{k_i!} e^{-\lambda(t_i - t_{i-1})} \\ & = e^{-\lambda t_n} \lambda^{j_n} \prod_{i=1}^{n} \frac{(t_i - t_{i-1})^{k_i}}{k_i!}. \end{aligned}$$

Therefore, we have

$$P\{N_{t_1} = k_1, N_{t_2} - N_{t_1} = k_2, \ldots, N_{t_n} - N_{t_{n-1}} = k_n\} = \prod_{i=1}^{n} P\{N_{t_i} - N_{t_{i-1}} = k_i\}.$$

It follows from Problem 12.6 that N_{t_1}, $N_{t_2} - N_{t_1}$, …, $N_{t_n} - N_{t_{n-1}}$ are independent.

Finally, let us verify the fourth condition. By the definition of N_t, for almost every $\omega \in \Omega$, the sample path $t \to N_t(\omega)$ is constant at each interval $(T_{n-1}(\omega), T_n(\omega))$, $n = 1, 2, \ldots$. Hence

$$\begin{aligned} & \left\{\omega \in \Omega : \lim_{s \downarrow t} N_s(\omega) = N_t(\omega) \text{ and } \lim_{s \uparrow t} N_s(\omega) \text{ exists}, \forall t \geq 0\right\} \\ = \; & \bigcap_{n=1}^{\infty} \left\{\omega \in \Omega : \lim_{s \downarrow T_n(\omega)} N_s(\omega) = N_{T_n(\omega)}(\omega) \text{ and } \lim_{s \uparrow T_n(\omega)} N_s(\omega) \text{ exists}\right\}. \end{aligned}$$

But

$$\begin{aligned} & \{\omega \in \Omega : T_{n-1}(\omega) < T_n(\omega) < T_{n+1}(\omega)\} \\ \subseteq \; & \left\{\omega \in \Omega : \lim_{s \downarrow T_n(\omega)} N_s(\omega) = N_{T_n(\omega)}(\omega) \text{ and } \lim_{s \uparrow T_n(\omega)} N_s(\omega) \text{ exists}\right\} \end{aligned}$$

and

$$P\left\{\omega \in \Omega : T_{n-1}(\omega) < T_n(\omega) < T_{n+1}(\omega)\right\} = P\{\tau_n > 0, \tau_{n+1} > 0\} = 1.$$

By Problem 11.1, we have

$$P\left\{\omega \in \Omega : \lim_{s\downarrow t} N_s(\omega) = N_t(\omega) \text{ and } \lim_{s\uparrow t} N_s(\omega) \text{ exists}, \forall t \geq 0\right\} = 1.$$

Therefore, almost all sample paths of $\{N_t : t \geq 0\}$ are right-continuous with left-hand limits. This completes the proof.

27.4. To show that $\{N_t^{(1)} + N_t^{(2)} : t \geq 0\}$ is a Poisson process, we only need to verify that it satisfies the four conditions given in Definition 27.1. Since $\{N_t^{(1)} : t \geq 0\}$ and $\{N_t^{(2)} : t \geq 0\}$ are independent Poisson processes, we have

$$\begin{aligned} P\left\{N_0^{(1)} + N_0^{(2)} = 0\right\} &= P\left\{N_0^{(1)} = 0, N_0^{(2)} = 0\right\} \\ &= P\left\{N_0^{(1)} = 0\right\} P\left\{N_0^{(2)} = 0\right\} \\ &= 1. \end{aligned}$$

Hence the process satisfies the first condition.

Now let $0 \leq s < t$ and $k \geq 0$. Then we have

$$\begin{aligned} &P\left\{N_t^{(1)} + N_t^{(2)} - N_s^{(1)} - N_s^{(2)} = k\right\} \\ =\ & \sum_{i=0}^{k} P\left\{N_t^{(1)} - N_s^{(1)} = i, N_t^{(2)} - N_s^{(2)} = k - i\right\} \\ =\ & \sum_{i=0}^{k} P\left\{N_t^{(1)} - N_s^{(1)} = i\right\} P\left\{N_t^{(2)} - N_s^{(2)} = k - i\right\} \\ =\ & \sum_{i=0}^{k} \frac{\lambda_1^i (t-s)^i}{i!} e^{-\lambda_1(t-s)} \cdot \frac{\lambda_2^{k-i}(t-s)^{k-i}}{(k-i)!} e^{-\lambda_2(t-s)} \\ =\ & \frac{(\lambda_1 + \lambda_2)^k (t-s)^k}{k!} e^{-(\lambda_1+\lambda_2)(t-s)}, \end{aligned}$$

which shows that the process satisfies the second condition.

The third condition and the fourth condition are implied from the properties of Poisson processes and the assumption that $\{N_t^{(1)} : t \geq 0\}$ and $\{N_t^{(2)} : t \geq 0\}$ are independent. This completes the proof.

27.5. Since $\tilde{N}_t - \tilde{N}_s = N_t - N_s - \lambda(t-s)$, we have

$$
\begin{aligned}
E\left[\exp(x(\tilde{N}_t - \tilde{N}_s))\right] &= E\left[\exp(x(N_t - N_s - \lambda(t-s)))\right] \\
&= \sum_{k=0}^{\infty} \exp[x(k - \lambda(t-s))] P\{N_t - N_s = k\} \\
&= \sum_{k=0}^{\infty} \exp[x(k - \lambda(t-s))] e^{-\lambda(t-s)} \frac{(\lambda(t-s))^k}{k!} \\
&= \sum_{k=0}^{\infty} \exp[(\lambda(t-s))(-x-1)] \frac{(e^x \lambda(t-s))^k}{k!} \\
&= \exp\left[\lambda(t-s)(e^x - x - 1)\right].
\end{aligned}
$$

The moments follow from Problem 17.4. This completes the proof.

27.6. Let

$$X_i = (\tilde{N}_{t_i} - \tilde{N}_{t_{i-1}})^2 - (\tilde{N}_{t_i} - \tilde{N}_{t_{i-1}}) - \lambda(t_i - t_{i-1}), \quad i = 1, 2, \ldots, n.$$

Then

$$\sum_{i=1}^{n} (\tilde{N}_{t_i} - \tilde{N}_{t_{i-1}})^2 - (\lambda(b-a) + \tilde{N}_b - \tilde{N}_a) = \sum_{i=1}^{n} X_i.$$

To show that

$$\sum_{i=1}^{n} (\tilde{N}_{t_i} - \tilde{N}_{t_{i-1}})^2 \xrightarrow{L^2(\Omega)} \lambda(b-a) + \tilde{N}_b - \tilde{N}_a \quad \text{as} \quad \|\Delta_n\| \to 0,$$

we only need to prove

$$E\left(\left|\sum_{i=1}^{n} X_i\right|^2\right) \xrightarrow{L^2(\Omega)} 0 \quad \text{as} \quad \|\Delta_n\| \to 0. \tag{27.3}$$

Note that

$$E\left(\left|\sum_{i=1}^{n} X_i\right|^2\right) = \sum_{i=1}^{n} E\left[X_i^2\right] + 2\sum_{i<j} E[X_i X_j].$$

For $i < j$, since X_i and X_j are independent, we have $E[X_i X_j] = E[X_i]E[X_j] = 0$. By Problem 27.5, we have $E\left[X_i^2\right] = 2\lambda^2 (t_i - t_{i-1})^2$. Therefore, we have

$$E\left(\left|\sum_{i=1}^{n} X_i\right|^2\right) = 2\lambda^2 \sum_{i=1}^{n} (t_i - t_{i-1})^2 \leq 2\lambda^2 (b-a)\|\Delta_n\|,$$

which implies that Equation (27.3) is true. This completes the proof.

27.7. Since Poisson processes are square integrable, M_t is integrable. Also the M_t is adapted to the filtration $\mathscr{F}_t$. We are left to verify that $E(M_t|\mathscr{F}_s) = M_s$ for all $0 \leq s < t$.

Let $0 \leq s < t$. Then we have

$$\begin{aligned} & E(M_t|\mathscr{F}_s) \\ = \ & E[(\tilde{N}_t - \tilde{N}_s + \tilde{N}_s)^2 - \lambda t|\mathscr{F}_s] \\ = \ & E[(\tilde{N}_t - \tilde{N}_s)^2|\mathscr{F}_s] + 2E[(\tilde{N}_t - \tilde{N}_s)\tilde{N}_s|\mathscr{F}_s] + E[\tilde{N}_s^2 - \lambda t|\mathscr{F}_s] \\ = \ & E[(\tilde{N}_t - \tilde{N}_s)^2] + \tilde{N}_s^2 - \lambda t \\ = \ & \tilde{N}_s^2 - \lambda s \\ = \ & M_s. \end{aligned}$$

Hence the M_t is a martingale. This completes the proof.

27.8. Suppose that $\{N_t : t \geq 0\}$ is adapted to a filtration $\{\mathscr{F}_t : t \geq 0\}$. Then $\{X_t : t \geq 0\}$ is also adapted to this filtration. Now we show that

$$E(|X_t|) < \infty, \quad \forall t \geq 0. \tag{27.4}$$

In fact, if $t = 0$, then $E(|X_0|) = 0 < \infty$. If $t > 0$, it follows from Definition 27.2 that

$$\begin{aligned} E(|X_t|) &= \sum_{k=0}^{\infty}(1+\sigma)^k e^{-\lambda\sigma t} e^{-\lambda t}\frac{(\lambda t)^k}{k!} \\ &= e^{-\lambda(1+\sigma)t}\sum_{k=0}^{\infty}\frac{[(1+\sigma)\lambda t]^k}{k!} \\ &= e^{-\lambda(1+\sigma)t}\cdot e^{\lambda(1+\sigma)t} \\ &= 1, \end{aligned}$$

which shows that Equation (27.4) holds.

It remains to check that

$$E[X_t|\mathscr{F}_s] = X_s, \quad \text{a.s.} \quad \forall s \leq t. \tag{27.5}$$

To do that, let $0 \leq s < t$ be fixed. By the definition of Poisson processes (Definition 27.2), we know that $N_t - N_s$ is independent of $\mathscr{F}_s$ and N_s is $\mathscr{F}_s$-measurable. Hence it follows from Problems 14.12 and 14.16 that

$$\begin{aligned} E[X_t|\mathscr{F}_s] &= E\left[(1+\sigma)^{N_t}e^{-\lambda\sigma t}|\mathscr{F}_s\right] \\ &= e^{-\lambda\sigma t}(1+\sigma)^{N_s}E\left[(1+\sigma)^{N_t-N_s}|\mathscr{F}_s\right] \\ &= e^{-\lambda\sigma t}(1+\sigma)^{N_s}\sum_{k=0}^{\infty}(1+\sigma)^k e^{-\lambda(t-s)}\frac{[\lambda(t-s)]^k}{k!} \\ &= e^{-\lambda\sigma s}(1+\sigma)^{N_s} \\ &= X_s. \end{aligned}$$

Therefore, Equation (27.5) also holds. This completes the proof.

27.9.

(a) Let $m \geq 0$, $t \geq 0$, and $h \geq 0$ be fixed. Since

$$\begin{aligned}&\left\{\sum_{n=1}^{N_{t+h}-N_t} Y_{n+N_t} \leq c\right\} \\ =\ &\bigcup_{i,j\geq 0}\left(\{N_{t+h}-N_t=j\}\cap\{N_t=i\}\cap\left\{\sum_{n=1}^{j} Y_{n+i}\leq c\right\}\right)\end{aligned}$$

and the sets $\{N_{t+h}-N_t=j\}\cap\{N_t=i\}\left\{\sum_{n=1}^{j} Y_{n+i}\leq c\right\}$ are disjoint for different pairs (i,j), we have

$$\begin{aligned}&P\left\{\sum_{n=1}^{N_{t+h}-N_t} Y_{n+N_t} \leq c\right\} \\ =\ &\sum_{i,j\geq 0} P\left(\{N_{t+h}-N_t=j\}\cap\{N_t=i\}\cap\left\{\sum_{n=1}^{j} Y_{n+i}\leq c\right\}\right).\end{aligned}$$

By the assumption that $\{N_t : t \geq 0\}$ and $\{Y_1, Y_2, \ldots\}$ are independent, we have

$$\begin{aligned}&P\left\{\sum_{n=1}^{N_{t+h}-N_t} Y_{n+N_t} \leq c\right\} \\ =\ &\sum_{i,j\geq 0} P\left(\{N_{t+h}-N_t=j\}\cap\{N_t=i\}\right) P\left\{\sum_{n=1}^{j} Y_{n+i}\leq c\right\}.\end{aligned}$$

Note that Y_{n+i} and Y_{n+m} have the same distribution. We get

$$\begin{aligned}&P\left\{\sum_{n=1}^{N_{t+h}-N_t} Y_{n+N_t} \leq c\right\} \\ =\ &\sum_{i,j\geq 0} P\left(\{N_{t+h}-N_t=j\}\cap\{N_t=i\}\right) P\left\{\sum_{n=1}^{j} Y_{n+m}\leq c\right\} \\ =\ &P\left\{\sum_{n=1}^{N_{t+h}-N_t} Y_{n+m} \leq c\right\}.\end{aligned}$$

(b) Let $t \geq 0$ and $h \geq 0$ be fixed. By the proof of the first item, we have

$$
\begin{aligned}
& P\left\{ \sum_{n=1}^{N_{t+h}-N_t} Y_{n+N_t} \leq c \right\} \\
= & \sum_{i,j\geq 0} P\left(\{N_{t+h} - N_t = j\} \cap \{N_t = i\}\right) P\left\{ \sum_{n=1}^{j} Y_n \leq c \right\}.
\end{aligned}
$$

Since $N_{t+h} - N_t$ has the same distribution as N_h and $N_{t+h} - N_t$ is independent of N_t, the above equation gives

$$
\begin{aligned}
& P\left\{ \sum_{n=1}^{N_{t+h}-N_t} Y_{n+N_t} \leq c \right\} \\
= & \sum_{i,j\geq 0} P\{N_{t+h} - N_t = j\} P\{N_t = i\} P\left\{ \sum_{n=1}^{j} Y_n \leq c \right\} \\
= & \sum_{i,j\geq 0} P\{N_h = j\} P\{N_t = i\} P\left\{ \sum_{n=1}^{j} Y_n \leq c \right\} \\
= & P\left\{ \sum_{n=1}^{N_h} Y_n \leq c \right\}.
\end{aligned}
$$

Hence $Q_{t+h} - Q_t$ has the same distribution as Q_h.

(c) Let $t \geq 0$ and $h \geq 0$ be fixed. By Problem 21.10, we only need to show that for any integer $m \geq 0$, any $0 \leq s_1 < s_2 < \cdots < s_m \leq t$, and any $c_1, c_2, \ldots, c_m, c \in \mathbf{R}$, we have

$$
\begin{aligned}
& P\{Q_{s_1} \leq c_1, \ldots, Q_{s_m} \leq c_m, Q_{t+h} - Q_t \leq c\} \\
= & P\{Q_{s_1} \leq c_1, \ldots, Q_{s_m} \leq c_m\} \cdot P\{Q_{t+h} - Q_t \leq c\}. \quad (27.6)
\end{aligned}
$$

In fact, we have

$$
\begin{aligned}
& P\{Q_{s_1} \le c_1, \ldots, Q_{s_m} \le c_m, Q_{t+h} - Q_t \le c\} \\
= & \sum_{k_1,k_2,\ldots,k_m,i,j\ge 0} P(\{N_{s_1} = k_1, \ldots, N_{s_m} = k_m, N_t = i\} \cap \\
& \{N_{t+h} - N_t = j\} \cap \\
& \left\{\sum_{n=1}^{k_1} Z_n \le c_1, \ldots, \sum_{n=1}^{k_m} Z_n \le c_m, \sum_{n=1}^{j} Z_{n+i} \le c\right\}) \\
= & \sum_{k_1,k_2,\ldots,k_m,i,j\ge 0} P(\{N_{s_1} = k_1, \ldots, N_{s_m} = k_m, N_t = i\}) \cdot \\
& P\{N_{t+h} - N_t = j\} \cdot P\left\{\sum_{n=1}^{k_1} Z_n \le c_1, \ldots, \sum_{n=1}^{k_m} Z_n \le c_m\right\} \cdot \\
& P\left\{\sum_{n=1}^{j} Z_{n+i} \le c\right\}).
\end{aligned}
$$

Since Z_{n+i} has the same distribution as Z_n, we get

$$
\begin{aligned}
& P\{Q_{s_1} \le c_1, \ldots, Q_{s_m} \le c_m, Q_{t+h} - Q_t \le c\} \\
= & \sum_{k_1,k_2,\ldots,k_m,i,j\ge 0} P(\{N_{s_1} = k_1, \ldots, N_{s_m} = k_m, N_t = i\}) \cdot \\
& P\{N_{t+h} - N_t = j\} \cdot P\left\{\sum_{n=1}^{k_1} Z_n \le c_1, \ldots, \sum_{n=1}^{k_m} Z_n \le c_m\right\} \cdot \\
& P\left\{\sum_{n=1}^{j} Z_n \le c\right\}) \\
= & P\{Q_{s_1} \le c_1, \ldots, Q_{s_m} \le c_m\} P\left\{\sum_{n=1}^{N_{t+h}-N_t} Z_n \le c\right\}.
\end{aligned}
$$

Equation (27.6) follows from the above equation and the first item.

This completes the proof.

27.10.

(a) By Problem 14.18 and the assumption that $\{N_t : t \geq 0\}$ and $\{Y_n : n = 0, 1, \ldots\}$ are independent, we have

$$\begin{aligned} E[Q_t] &= E\left[\sum_{i=1}^{N_t} Y_i\right] \\ &= \sum_{k=0}^{\infty} E\left[\sum_{i=1}^{k} Y_i \middle| N_t = k\right] P\{N_t = k\} \\ &= \sum_{k=0}^{\infty} E\left[\sum_{i=1}^{k} Y_i\right] P\{N_t = k\} \\ &= \sum_{k=0}^{\infty} k\beta \frac{(\lambda t)^k}{k!} e^{-\lambda t} \\ &= \beta\lambda t. \end{aligned}$$

(b) Let $\{\mathscr{F}_t : t \geq 0\}$ be the filtration generated by $\{Q_t : t \geq 0\}$. Let $0 \leq s < t$. Then $Q_t - Q_s$ is independent of F_s (see Problem 27.9). In addition, $Q_t - Q_s$ has the same distribution as Q_{t-s} because $N_t - N_s$ has the same distribution as N_{t-s}. Hence

$$\begin{aligned} E[Q_t - \beta\lambda t | \mathscr{F}_s] &= E[Q_t - Q_s | \mathscr{F}_s] + Q_s - \beta\lambda t \\ &= E[Q_{t-s}] + Q_s - \beta\lambda t \\ &= \beta\lambda(t-s) + Q_s - \beta\lambda t \\ &= Q_s - \beta\lambda s, \end{aligned}$$

which shows that $\{Q_t - \beta\lambda t : t \geq 0\}$ is a martingale.

This completes the proof.

27.5 Bibliographic Notes

In this chapter, we introduced Poisson processes. The definition of Poisson process is similar to that of Brownian motion. This definition was adopted from Kuo (2006). In some books, the Poisson process is defined as a counting process [see Cont and Tankov (2003, Section 2.5) and Protter (2003, Section I.3)]. Theorem 27.1 says that if a counting process has stationary independent increments, then it is a Poisson process. For a proof of this theorem, readers are referred to Cont and Tankov (2003, Lemma 2.1). For more information about Poisson process, readers are referred to Ross (1995) and Chung and Aitsahlia (2007).

CHAPTER 28

BROWNIAN MOTION

Brownian motion is a fundamentally important stochastic process in that it is a central notion throughout the theoretical development of stochastic processes. In this chapter, we present the definition and some properties of Brownian motion.

28.1 Basic Concepts and Facts

Definition 28.1 (Standard Brownian Motion). Let $(\Omega, \mathscr{F}, P)$ be a probability space. A stochastic process $\{B_t : t \geq 0\}$ on $(\Omega, \mathscr{F}, P)$ is called a *standard Brownian motion* if it satisfies the following conditions:

(a) $P\{B_0 = 0\} = P\{\omega : B_0(\omega) = 0\} = 1$.

(b) For any $0 \leq s < t$, the increment $B_t - B_s$ is a random variable normally distributed with mean 0 and variance $t - s$, that is, for any $a < b$, we have

$$P\{a \leq B_t - B_s \leq b\} = \frac{1}{\sqrt{2\pi(t-s)}} \int_a^b \exp\left(\frac{-x^2}{2(t-s)}\right) dx.$$

(c) B_t has independent increments, that is, for any $0 \leq t_1 < t_2 < \ldots < t_n$, the random variables $B_{t_1}, B_{t_2} - B_{t_1}, \ldots, B_{t_n} - B_{t_{n-1}}$ are independent.

(d) Almost all sample paths of B_t are continuous functions:

$$P\{\omega : B(t, \omega) = B_t(\omega) \text{ is continuous}\} = 1.$$

Definition 28.2 (Brownian Motion with Respect to Filtrations). Let $\{\mathscr{F}_t : t \geq 0\}$ be a filtration. A stochastic process $\{B_t : t \geq 0\}$ is said to be a *Brownian motion* with respect to the filtration $\{\mathscr{F}_t : t \geq 0\}$ if it satisfies the following conditions:

(a) $P\{B_0 = 0\} = P\{\omega : B_0(\omega) = 0\} = 1$.

(b) For any $0 \leq s < t$, the increment $B_t - B_s$ is a random variable normally distributed with mean 0 and variance $t - s$, that is, for any $a < b$, we have

$$P\{a \leq B_t - B_s \leq b\} = \frac{1}{\sqrt{2\pi(t-s)}} \int_a^b \exp\left(\frac{-x^2}{2(t-s)}\right) dx.$$

(c) For any $s \leq t$, the random variable $B_t - B_s$ is independent of the σ-field $\mathscr{F}_s$.

(d) Almost all sample paths of B_t are continuous functions:

$$P\{\omega : B(t, \omega) = B_t(\omega) \text{ is continuous}\} = 1.$$

Definition 28.3 (m-Dimensional Brownian Motion). Let m be a positive integer and μ a probability measure on $(\mathbf{R}^m, \mathcal{B}(\mathbf{R}^m))$. Let B_t be a continuous process with values in $\mathbf{R}^m$, defined on a probability space $(\Omega, \mathscr{F}, P)$, and adapted to a filtration $\{\mathscr{F}_t : t \geq 0\}$. This process is called an *m-dimensional Brownian motion* with initial distribution μ if

(a) $P\{B_0 \in A\} = \mu(A)$ for all $A \in \mathcal{B}(\mathbf{R}^m)$.

(b) For $0 \leq s < t$, the increment $B_t - B_s$ is independent of $\mathscr{F}_s$ and is normally distributed with mean zero and covariance matrix $(t - s)I_m$, where I_m is the $m \times m$ identity matrix.

Theorem 28.1 (Law of the Iterated Logarithm). *Let $\{B_t : t \geq 0\}$ be a Brownian motion. Then*

$$P\left\{\limsup \frac{B_t}{\sqrt{2t \log \log t}} = 1\right\} = 1$$

and

$$P\left\{\liminf \frac{B_t}{\sqrt{2t \log \log t}} = -1\right\} = 1.$$

Theorem 28.2 (Strong Markov Property). *Let $\{B_t : t \geq 0\}$ be a Brownian motion. Let τ be a stopping time with respective to the filtration generated by the Brownian motion. Then the process*

$$\{B_{\tau+t} - B_\tau : t \geq 0\}$$

is a standard Brownian motion independent of $\mathscr{F}_\tau^+$, *where*

$$\mathscr{F}_\tau^+ = \bigcap_{t>\tau} \sigma(B_s : 0 \le s \le t).$$

28.2 Problems

28.1. Let $\{B_t\}_{t\ge 0}$ be a Brownian motion and $\{\mathscr{F}_t : t \ge 0\}$ be a filtration defined as

$$\mathscr{F}_t = \sigma\{B_s : s \le t\}, \quad t \ge 0.$$

Show that $\{B_t\}_{t\ge 0}$ is a martingale with respect to $\{\mathscr{F}_t : t \ge 0\}$.

28.2. Let $\{B_t\}_{t\ge 0}$ be a Brownian motion. Show that

(a) For any $t > 0$, B_t is normally distributed with mean 0 and variance t.

(b) For any $s, t \ge 0$, $E[B_s B_t] = \min(s, t)$.

28.3 (Translation Invariance). Let $\{B_t\}_{t\ge 0}$ be a Brownian motion. Show that for any fixed $t_0 \ge 0$, the stochastic process $\tilde{B}_t = B_{t+t_0} - B_{t_0}$ is also a Brownian motion.

28.4 (Scaling Invariance). Let $\{B_t\}_{t\ge 0}$ be a Brownian motion. Show that for any $\lambda > 0$, the stochastic process

$$\tilde{B}_t = \frac{B_{\lambda t}}{\sqrt{\lambda}}$$

is also a Brownian motion.

28.5 (Time-Inversion Invariance). Let $\{B_t\}_{t\ge 0}$ be a Brownian motion. Let $W_0 = 0$ and

$$W_t = t B_{\frac{1}{t}}, \quad t > 0.$$

Show that $\{W_t\}_{t\ge 0}$ is a Brownian motion. In particular, show that

$$\lim_{t \downarrow 0} W_t = 0, \quad \text{a.s.}$$

28.6. Let $\{B_t\}_{t\ge 0}$ be a Brownian motion. Show that

$$E\left[(B_t - B_s)^4\right] = 3(t-s)^2.$$

28.7. Let $\{B_t\}_{t\ge 0}$ be a Brownian motion. Let $s, t \in \mathbf{R}$ and $s \le t$. Show that the random variable

$$aB_s + bB_t$$

is normally distributed with mean 0 and variance $(a^2 + 2ab)s + b^2 t$.

28.8. Let $\{B_t\}_{t\ge 0}$ be a Brownian motion, and let $0 < s < t \le u < v$. Show that

(a) The random variables $\frac{1}{t}B_t - \frac{1}{s}B_s$ and $aB_u + bB_v$ are independent for any $a, b \in \mathbf{R}$.

(b) The random variables $aB_s + bB_t$ and $\frac{1}{v}B_v - \frac{1}{u}B_u$ are independent for any $a, b \in \mathbf{R}$ satisfying the condition $as + bt = 0$.

28.9. Let $\{B_t\}_{t\geq 0}$ be a Brownian motion. Let $\{L_t : t \geq 0\}$ be a stochastic process defined as

$$L_t = \frac{1}{2}\left(B_t^2 - t\right), \quad t \geq 0.$$

Show that the L_t is a martingale with respect to the filtration $\mathscr{F}_t = \sigma\{B_s : s \leq t\}$.

28.10. Let $\{W_t : t \geq 0\}$ be a stochastic process adapted to the filtration $\{\mathscr{F}_t : t \geq 0\}$. Suppose that for any $s \leq t$, the random variable $W_t - W_s$ is independent of the σ-field $\mathscr{F}_s$. Show that the process W_t has independent increments.

28.11. Let $\{B_t : t \geq 0\}$ be a Brownian motion. Let $y > 0$. For any $t > 0$, show that

$$P\left\{\max_{0\leq s\leq t} B_s \geq y\right\} \leq 2P\{B_t \geq y\}.$$

28.12. Let $\{B_t : t \geq 0\}$ be a Brownian motion. Show that

$$P\left\{\limsup_{t\to\infty} B_t = \infty\right\} = 1,$$

and

$$P\left\{\liminf_{t\to\infty} B_t = -\infty\right\} = 1.$$

28.13 (Reflection Principle). Let $\{B_t : t \geq 0\}$ be a Brownian motion and τ be a stopping time with respect to the filtration generated by the Brownian motion. Let $\{\tilde{B}_t : t \geq 0\}$ be a process defined by

$$\tilde{B}_t = \begin{cases} B_t, & \text{if } t < \tau; \\ 2B_\tau - B_t & \text{if } t \geq \tau. \end{cases}$$

Show that $\{\tilde{B}_t : t \geq 0\}$ is a standard Brownian motion.

28.14 (Reflection Equality). Let $\{B_t : t \geq 0\}$ be a Brownian motion and $\alpha \in \mathbf{R}$. Let τ_α be a stopping time defined by

$$\tau = \min\{t : B_t = \alpha\}.$$

Suppose that $\alpha > 0$. For every $\beta \leq \alpha$, show that

$$P\{\tau_\alpha \leq t, B_t \leq \beta\} = P\{B_t \geq 2\alpha - \beta\}.$$

28.15. Let $\{B_t : t \geq 0\}$ be a Brownian motion and $\alpha > 0$. Let τ be a stopping time defined by

$$\tau = \min\{t : B_t = \alpha\}.$$

Let $\{M_t : t \geq 0\}$ be defined as

$$M_t = \max_{0 \leq s \leq t} B_s, \quad t \geq 0.$$

Show that

(a) $P\{M_t > \alpha\} = 2P\{B_t > \alpha\}$.

(b) The probability density function of τ is given by

$$f(x) = x^{-\frac{3}{2}} \alpha \varphi\left(\frac{\alpha}{\sqrt{x}}\right), \quad x > 0,$$

where

$$\varphi(t) = \frac{1}{\sqrt{2\pi}} e^{-\frac{t^2}{2}}.$$

28.16. Let $\{B_t : t \geq 0\}$ be a Brownian motion. Let $\{M_t : t \geq 0\}$ be defined as

$$M_t = \max_{0 \leq s \leq t} B_s, \quad t \geq 0.$$

Show that for every $t > 0$, the joint density function of (M_t, B_t) is

$$f_{M_t, B_t}(x, y) = \frac{2(2x - y)}{t\sqrt{2\pi t}} \exp\left(-\frac{(2x - y)^2}{2t}\right), \quad y \leq x, \ x > 0$$

and is zero for other values of x and y.

28.17. Let $B_t = (B_t^{(1)}, B_t^{(2)}, \ldots, B_t^{(m)})$ be an m-dimensional Brownian motion. Show that for $0 \leq s < t$, the increments $B_t^{(1)} - B_s^{(1)}$, $B_t^{(2)} - B_s^{(2)}$, ..., $B_t^{(m)} - B_s^{(m)}$ are independent.

28.3 Hints

28.1. Use the fact that for any $s \leq t$, $B_t - B_s$ is independent of $\mathscr{F}_s$.

28.2. Part (a) is implied by the definition of Brownian motions (Definition 28.1). Part (b) can be proved by the independence property of the increments.

28.3. Use the definition of Brownian motions (Definition 28.1).

28.4. Use the definition of Brownian motions (Definition 28.1).

28.5. Use the definition of Brownian motions (Definition 28.1). The result of Problem 19.3 can be used to prove the second condition. Use the results of Problems 19.5 and 19.6 to prove the third condition. To prove the fourth condition, consider

$$W_{t_n} = \frac{1}{s_n} B_{s_n} = \frac{N_n}{s_n} \frac{1}{N_n} \sum_{i=1}^{N_n} (B_i - B_{i-1}) + \frac{1}{s_n}(B_{s_n} - B_{N_n}),$$

where $s_n = 1/t_n$ and N_n is the largest integer such that $N_n \leq s_n$, and use the strong law of large numbers (Theorem 16.2) and the result of Problem 19.17.

28.6. Use the moment generating function of the normal distribution (see Problem 19.1) and the result of Problem 17.4.

28.7. Use the result of Problem 19.3 and the properties of Brownian motion.

28.8. Use the results of Problems 28.7 and 19.6.

28.9. Use the results of Problems 14.4 and 14.12.

28.10. Follow the definition of independence (Definitions 12.1 and 12.2).

28.11. Use Lévy's inequality (see Problem 15.11).

28.12. Apply the law of the iterated logarithm (Theorem 28.1).

28.13. Apply the strong Markov property (Theorem 28.2).

28.14. Apply the reflection principle (see Problem 28.13).

28.15. To prove part (a), consider the Brownian motion reflected at τ and use the reflection principle (see Problem 28.13). Part (b) can be proved by part (a).

28.16. Use the result of Problem 28.14.

28.17. Use the definition of multidimensional Brownian motions (Definition 28.3) and the result of Problem 19.6.

28.4 Solutions

28.1. Let $s \leq t$. Then $B_t - B_s$ is independent of $\mathscr{F}_s$. Hence we have

$$E[B_t|\mathscr{F}_s] = E[B_t - B_s + B_s|\mathscr{F}_s] = E[B_t - B_s] + E[B_s|\mathscr{F}_s] = B_s.$$

Note that $E(B_t^2) = t < \infty$. By Problem 8.5, we have $E(|B_t|) < \infty$. Hence the B_t is a martingale with respect to the $\mathscr{F}_t$. This completes the proof.

28.2.

(a) By Definition 28.1, $B_t - B_0$ is normally distributed with mean 0 and variance t. But B_0 is equal to 0 a.s. Hence B_t is normally distributed with mean 0 and variance t.

(b) Without loss of generality, we assume that $0 \leq s < t$. Then by definition, $B_s - B_0$ and $B_t - B_s$ are independent. Hence we have

$$E[B_s B_t] = E[B_s(B_t - B_s) + B_s^2] = E[B_s(B_t - B_s)] + E[B_s^2] = 0 + s = s.$$

This completes the proof.

28.3. To show that $\{\tilde{B}_t\}_{t\geq 0}$ is a Brownian motion, we only need to verify that it satisfies the four conditions in Definition 28.1. Since $\tilde{B}_0 = B_{t_0} - B_{t_0} = 0$, the $\tilde{B}_t$ satisfies the first condition. Let $0 \leq s < t$. Then

$$\tilde{B}_t - \tilde{B}_s = B_{t+t_0} - B_{s+t_0},$$

which implies that $\tilde{B}_t - \tilde{B}_s$ is normally distributed with mean 0 and variance $t + t_0 - (s + t_0) = t - s$. Hence the $\tilde{B}_t$ satisfies the second condition.

Now let $0 \leq t_1 < t_2 < \ldots < t_n$. Then we have $0 \leq t_1 + t_0 < t_2 + t_0 < \cdots < t_n + t_0$. The increments $\tilde{B}_{t_1} = B_{t_1+t_0}$, $\tilde{B}_{t_2} - \tilde{B}_{t_1} = B_{t_2+t_0} - B_{t_1+t_0}$, $\ldots$, $\tilde{B}_{t_n} - \tilde{B}_{t_{n-1}} = B_{t_n+t_0} - B_{t_{n-1}+t_0}$ are independent. Hence $\tilde{B}_t$ satisfies the third condition. Since almost all sample paths of the B_t are continuous, so are the sample paths of the $\tilde{(B)}_t = B_{t+t_0} - B_{t_0}$. Hence the $\tilde{B}_t$ satisfies the fourth condition.

28.4. It is obvious that the $\tilde{B}_t$ satisfies the first, third, and fourth conditions in Definition 28.1. Let $0 \leq s < t$. Then

$$\tilde{B}_t - \tilde{B}_s = \frac{B_{\lambda t} - B_{\lambda s}}{\sqrt{\lambda}}.$$

Note that $B_{\lambda t} - B_{\lambda s}$ is normally distributed with mean 0 and variance $\lambda(t - s)$. We know that $\tilde{B}_t - \tilde{B}_s$ is normally distributed with mean 0 and variance $t - s$. Hence the $\tilde{B}_t$ satisfies the second condition. Therefore the $\tilde{B}_t$ is a Brownian motion.

28.5. To show that $\{W_t\}_{t\geq 0}$ is a Brownian motion, we only need to verify that W_t satisfies the four conditions in Definition 28.1. By the definition of W_0, the first condition is satisfied. Let $0 < s < t$. Then

$$W_t - W_s = tB_{\frac{1}{t}} - sB_{\frac{1}{s}} = (t - s)B_{\frac{1}{t}} - s\left(B_{\frac{1}{s}} - B_{\frac{1}{t}}\right).$$

Note that $B_{\frac{1}{t}}$ and $B_{\frac{1}{s}} - B_{\frac{1}{t}}$ are independent normal random variables. By Problem 19.3, we know that $W_t - W_s$ is normally distributed with mean 0 and variance

$$(t - s)^2 \frac{1}{t} + s^2 \left(\frac{1}{s} - \frac{1}{t}\right) = t - s.$$

Similarly, $W_t - W_0 = tB_{\frac{1}{t}}$ is also normally distributed with mean 0 and variance t. Hence W_t satisfies the second condition.

Let $0 < t_1 < t_2 < \ldots < t_n$. Let $(a_1, a_2, \ldots, a_n)^T$ be a vector. Then we have

$$\begin{aligned}
& a_1 W_1 + \sum_{i=2}^{n} a_i (W_{t_i} - W_{t_{i-1}}) \\
= & \sum_{i=1}^{n-1} (a_i - a_{i+1}) t_i B_{\frac{1}{t_i}} + a_n t_n B_{\frac{1}{t_n}} \\
= & \sum_{i=1}^{n-1} \left[\left(\sum_{j=1}^{i} (a_j - a_{j+1}) t_j \right) \left(B_{\frac{1}{t_i}} - B_{\frac{1}{t_{i+1}}} \right) \right] \\
& + \left(\sum_{j=1}^{n-1} (a_j - a_{j+1}) t_j + a_n t_n \right) B_{\frac{1}{t_n}},
\end{aligned}$$

which shows the the linear combination of $W_{t_1}, W_{t_2} - W_{t_1}, \ldots, W_{t_n} - W_{t_{n-1}}$ is a normal random variable. It follows from Problem 19.5 that the vector $(W_{t_1}, W_{t_2} - W_{t_1}, \ldots, W_{t_n} - W_{t_{n-1}})^T$ is a multivariate normal random variable. Now by Problem 28.2, for $i > 1$, we have

$$E[W_{t_1}(W_{t_i} - W_{t_{i-1}})] = t_1 t_i E\left[B_{\frac{1}{t_1}} B_{\frac{1}{t_i}}\right] - t_1 t_{i-1} E\left[B_{\frac{1}{t_1}} B_{\frac{1}{t_{i-1}}}\right] = 0,$$

and for $i > j > 1$, we have

$$\begin{aligned}
& E[(W_{t_i} - W_{t_{i-1}})(W_{t_j} - W_{t_{j-1}})] \\
= & t_i t_j E\left[B_{\frac{1}{t_i}} B_{\frac{1}{t_j}}\right] - t_{i-1} t_j E\left[B_{\frac{1}{t_{i-1}}} B_{\frac{1}{t_j}}\right] - t_i t_{j-1} E\left[B_{\frac{1}{t_i}} B_{\frac{1}{t_{j-1}}}\right] \\
& + t_{i-1} t_{j-1} E\left[B_{\frac{1}{t_{i-1}}} B_{\frac{1}{t_{j-1}}}\right] \\
= & t_j - t_j - t_{j-1} + t_{j-1} \\
= & 0.
\end{aligned}$$

Therefore, by Problem 19.6, the increments are independent. Hence the third condition is satisfied.

To show that the W_t satisfies the fourth condition. We only need to show that

$$\lim_{t \downarrow 0} W_t = 0, \quad \text{a.s.}$$

To do that, let $t_n \downarrow 0$ and $s_n = 1/t_n$. Let N_n be the largest integer less than or equal to s_n. Then we have

$$W_{t_n} = \frac{1}{s_n} B_{s_n} = \frac{N_n}{s_n} \frac{1}{N_n} \sum_{i=1}^{N_n} (B_i - B_{i-1}) + \frac{1}{s_n} (B_{s_n} - B_{N_n}).$$

But by Theorem 16.2, we have

$$\lim_{n \to \infty} \frac{1}{N_n} \sum_{i=1}^{N_n} (B_i - B_{i-1}) = 0, \quad \text{a.s.}$$

Let $\delta > 0$. By Problem 19.17, for sufficiently large n, we have

$$\begin{aligned}
P\left(\bigcup_{k=n}^{\infty}\left\{\left|\frac{B_{s_k}-B_{N_k}}{s_k}\right|\geq\delta\right\}\right) &\leq \sum_{k=n}^{\infty} P\left\{\left|\frac{B_{s_k}-B_{N_k}}{s_k}\right|\geq\delta\right\} \\
&\leq \sum_{k=n}^{\infty}\frac{s_k-N_k}{s_k\delta}\sqrt{\frac{2}{\pi}}\exp\left(-\frac{s_k^2\delta^2}{2(s_k-N_k)^2}\right) \\
&\leq 2\sum_{k=n}^{\infty}\frac{1}{s_k\delta}\exp\left(-\frac{s_k^2\delta^2}{2}\right) \\
&\leq 2\sum_{k=n}^{\infty}\frac{1}{s_k^2\delta^2} \\
&\leq \frac{2}{\delta^2}\left(\frac{1}{s_n^2}+\int_{s_n}^{\infty}\frac{1}{x^2}\mathrm{d}x\right) \\
&= \frac{2}{\delta^2}\left(\frac{1}{s_n^2}+\frac{1}{s_n}\right).
\end{aligned}$$

Hence

$$\lim_{n\to\infty} P\left(\bigcup_{k=n}^{\infty}\left\{\left|\frac{B_{s_k}-B_{N_k}}{s_k}\right|\geq\delta\right\}\right)=0.$$

By Problem 9.3, we have

$$\lim_{n\to\infty}\frac{1}{s_n}(B_{s_n}-B_{N_n})=0, \quad \text{a.s.}$$

Therefore, $\lim_{t_n\downarrow 0} W_{t_n} = 0$. This completes the proof.

28.6. Let $s < t$. Then by definition, $X = B_t - B_s$ is normally distributed with mean 0 and variance $t - s$. By Problem 17.4, we have

$$E\left[(B_t-B_s)^4\right] = M_X^{(4)}(0).$$

By Problem 19.1, we have

$$M_X(t) = e^{\frac{1}{2}\sigma^2 t^2}.$$

The fourth derivative of $M_X(t)$ is

$$M_X^{(4)}(t) = \left(3\sigma^4 + 6\sigma^6 t^2 + \sigma^8 t^4\right) e^{\frac{1}{2}\sigma^2 t^2},$$

which gives

$$M_X^{(4)}(0) = 3\sigma^4 = 3(t-s)^2.$$

This completes the proof.

28.7. The case when $s = t$ is trivial. Suppose that $s < t$. Then by the definition of Brownian motion, the random variables B_s and $B_t - B_s$ are independent normal random variables. Hence by Problem 19.3, $aB_s + bB_t = (a+b)B_s + b(B_t - B_s)$ is normally distributed. In addition, the mean of $aB_s + bB_t$ is 0 and the variance is $(a+b)^2 s + b^2(t-s) = (a^2+2ab)s + b^2 t$. This completes the proof.

28.8.

(a) By Problem 28.7, the random variables $\frac{1}{t}B_t - \frac{1}{s}B_s$ and $aB_u + bB_v$ are normal random variables. By Problem 28.7, $\frac{1}{t}B_t - \frac{1}{s}B_s$ and $aB_u + bB_v$ are independent if and only if

$$E\left[\left(\frac{1}{t}B_t - \frac{1}{s}B_s\right)(aB_u + bB_v)\right] = E\left[\frac{1}{t}B_t - \frac{1}{s}B_s\right] E[aB_u + bB_v]. \tag{28.1}$$

But $E\left[\frac{1}{t}B_t - \frac{1}{s}B_s\right] = 0$ and $E[aB_u + bB_v] = 0$. By Problem 28.2, we have

$$\begin{aligned} & E\left[\left(\frac{1}{t}B_t - \frac{1}{s}B_s\right)(aB_u + bB_v)\right] \\ = & \frac{a}{t}E[B_tB_u] - \frac{a}{s}E[B_sB_u] + \frac{b}{t}E[B_tB_v] - \frac{b}{s}E[B_sB_v] \\ = & a - a + b - b \\ = & 0. \end{aligned}$$

Hence Equation (28.1) holds for all $a, b \in \mathbf{R}$. Therefore, $\frac{1}{t}B_t - \frac{1}{s}B_s$ and $aB_u + bB_v$ are independent.

(b) Similar to part (a), $aB_s + bB_t$ and $\frac{1}{v}B_v - \frac{1}{u}B_u$ are independent if and only if

$$E\left[(aB_s + bB_t)\left(\frac{1}{v}B_v - \frac{1}{u}B_u\right)\right] = 0. \tag{28.2}$$

Expanding the left-hand side of Equation (28.2) gives

$$\frac{a}{v}s + \frac{b}{v}t - \frac{a}{u}s - \frac{b}{u}t = 0,$$

which is equivalent to

$$(as + bt)\left(\frac{1}{v} - \frac{1}{u}\right) = 0.$$

Note that $u < v$. The above equation gives $as + bt = 0$.

This completes the proof.

28.9. Let $s \le t$. Since $B_t - B_s$ and B_u are independent for all $u \le s$, it follows that $B_t - B_s$ and $\mathscr{F}_s$ are independent. By Problems 14.4 and 14.12, we have

$$\begin{aligned} E[L_t|\mathscr{F}_s] &= \frac{1}{2}E\left[B_t^2|\mathscr{F}_s\right] - \frac{1}{2}t \\ &= \frac{1}{2}E\left[(B_t - B_s + B_s)^2|\mathscr{F}_s\right] - \frac{1}{2}t \\ &= \frac{1}{2}E\left[(B_t - B_s)^2 + 2(B_t - B_s)B_s + B_s^2|\mathscr{F}_s\right] - \frac{1}{2}t \\ &= \frac{1}{2}\left(E[(B_t - B_s)^2] + 2B_sE[B_t - B_s] + E[B_s^2|\mathscr{F}_s]\right) - \frac{1}{2}t \\ &= \frac{1}{2}(t-s) + \frac{1}{2}B_s^2 - \frac{1}{2}t \\ &= L_s. \end{aligned}$$

28.10. Let $0 \le t_1 < t_2 < \cdots < t_n$. Let $G_i \in \sigma(W_{t_i} - W_{t_{i-1}})$ for $i = 1, 2, \ldots, n$, where $W_{t_0} = 0$. By Definitions 12.2 and 12.1, we only need to show that

$$P\left(\bigcap_{i=1}^{n} G_i\right) = \prod_{i=1}^{n} P(G_i). \tag{28.3}$$

By the assumption that $W_{t_n} - W_{t_{n-1}}$ is independent of $\mathscr{F}_{t_{n-1}}$ and the fact that $G_1 \cap G_2 \cap \cdots \cap G_{n-1} \in \mathscr{F}_{t_{n-1}}$, we have

$$P\left(\bigcap_{i=1}^{n} G_i\right) = P\left(\bigcap_{i=1}^{n-1} G_i\right) P(G_n).$$

Repeating the above calculation gives Equation (28.3). This completes the proof.

28.11. Let N be an integer. Let $t_j = \frac{j}{N}t$ for $1 \le j \le N$. Let $X_1 = B_{t_1}$ and $X_j = B_{t_{j+1}} - B_{t_j}$ for $j = 2, 3, \ldots, N$. Then $X_1, X_2, \ldots, X_N$ are independent and symmetrically distributed normal random variables. Let

$$S_j = \sum_{i=1}^{j} X_i, \quad j = 1, 2, \ldots, N.$$

Then by Lévy's inequality (see Problem 15.11), we have

$$P\left\{\max_{1\le j\le N} S_j \ge y\right\} \le 2P\{S_n \ge y\} = 2P\{B_t \ge y\}.$$

Now we show that

$$\lim_{N\to\infty} P\left\{\max_{1\le j\le N} S_j \ge y\right\} = P\left\{\max_{0\le s\le t} B_s \ge y\right\}. \tag{28.4}$$

To do that, let $A = \{\omega : B_t(\omega) \text{ is continuous}\}$. Then $P(A) = 1$. Since

$$\lim_{N\to\infty} \left\{ \max_{1\le j\le N} S_j \ge y \right\} \cap A = \left\{ \max_{0\le s\le t} B_s \ge y \right\} \cap A,$$

we have

$$P\left(\lim_{N\to\infty} \left\{ \max_{1\le j\le N} S_j \ge y \right\} \cap A \right) = P\left(\left\{ \max_{0\le s\le t} B_s \ge y \right\} \cap A \right).$$

Since $P(A) = 1$, Equation (28.4) is true. This completes the proof.

28.12. Let E_n be events defined as

$$E_n = \left\{ \limsup_{t\to\infty} B_t > n \right\}, \quad n = 1, 2, \ldots.$$

Since

$$\left\{ \limsup_{t\to\infty} \frac{B_t}{\sqrt{2\log\log t}} = 1 \right\} \subseteq E_n$$

for all $n = 1, 2, \ldots$, by Theorem 28.1, we have

$$P(E_n) = 1, \quad n \ge 1.$$

But by Problem 11.1, we have

$$P\left\{ \limsup_{t\to\infty} B_t = \infty \right\} = P\left(\bigcap_{n=1}^{\infty} E_n \right) = 1.$$

Similarly, we can show that

$$P\left\{ \liminf_{t\to\infty} B_t = -\infty \right\} = 1.$$

This completes the proof.

28.13. By Theorem 28.2, the two processes

$$B_{t+\tau} - B_\tau, \quad -(B_{t+\tau} - B_\tau)$$

are standard Brownian motions independent of $\mathscr{F}_\tau^+$. Hence the concatenations

$$(\{B_t : 0 \le t \le \tau\}, \{B_\tau + (B_{t+\tau} - B_\tau) : t \ge 0\})$$

and

$$(\{B_t : 0 \le t \le \tau\}, \{B_\tau - (B_{t+\tau} - B_\tau) : t \ge 0\})$$

have the same distribution. But note that the first concatenation is just $\{B_t : t \ge 0\}$ and the second one is just $\{\tilde{B}_t : t \ge 0\}$. Hence $\{\tilde{B}_t : t \ge 0\}$ is a standard Brownian motion. This completes the proof.

28.14. Let $\{\tilde{B}_t : t \geq 0\}$ be a process defined by

$$\tilde{B}_t = \begin{cases} B_t, & \text{if } t < \tau_\alpha; \\ 2B_{\tau_\alpha} - B_t & \text{if } t \geq \tau_\alpha. \end{cases}$$

Then it follows from Problem 28.13 that $\{\tilde{B}_t : t \geq 0\}$ is a standard Brownian motion. Hence for every $t \geq \tau_\alpha$, we have $2B_{\tau_\alpha} - B_t = \tilde{B}_t$ and B_t have the same distribution, which gives

$$P\{\tau_\alpha \leq t, B_t \leq \beta\} = P\{\tau_\alpha \leq t, 2B_{\tau_\alpha} - B_t \leq \beta\}.$$

Since $B_{\tau_\alpha} = \alpha$, the above equation becomes

$$P\{\tau_\alpha \leq t, B_t \leq \beta\} = P\{\tau_\alpha \leq t, 2\alpha - \beta \leq B_t\}.$$

By the assumption that $\beta \leq \alpha$, we have $2\alpha - \beta \geq \alpha$. Hence $B_t \geq 2\alpha - \beta$ implies that $t \geq \tau_\alpha$. Therefore, we get

$$P\{\tau_\alpha \leq t, B_t \leq \beta\} = P\{2\alpha - \beta \leq B_t\}.$$

This completes the proof.

28.15.

(a) Let $\{\tilde{B}_t : t \geq 0\}$ be the Brownian motion reflected at τ (see Problem 28.13). Then we have

$$\{M_t > \alpha, B_t \leq \alpha\} = \{\tilde{B}_t \geq \alpha\}.$$

Note that

$$\{M_t > \alpha\} = \{B_t > \alpha\} \cup \{M_t > \alpha, B_t \leq \alpha\}.$$

We have

$$\{M_t > \alpha\} = \{B_t > \alpha\} \cup \{\tilde{B}_t \geq \alpha\},$$

which gives

$$P\{M_t > \alpha\} = P\{B_t > \alpha\} + P\{\tilde{B}_t \geq \alpha\} = 2P\{B_t > \alpha\}.$$

(b) Note that

$$\{\tau > x\} = \left\{ \max_{0 \leq t \leq x} B_t < \alpha \right\}.$$

By part (a), we have

$$P\{\tau > x\} = P\left\{ \max_{0 \leq t \leq x} B_t < \alpha \right\} = 1 - 2P\{B_x > \alpha\}.$$

Therefore, the cumulative density function is given by

$$F(x) = P\{\tau \leq x\} = 2P\{B_x > \alpha\} - 2\int_\alpha^\infty \frac{1}{\sqrt{2\pi x}} \exp\left(-\frac{s^2}{2x}\right) ds.$$

Taking the derivative of $F(x)$ gives

$$f(x) = x^{-\frac{3}{2}} \alpha \varphi\left(\frac{\alpha}{\sqrt{x}}\right).$$

This completes the proof.

28.16. Let $t > 0$, $x > 0$ and $y \le x$. Then

$$P\{M_t \ge x, B_t \le y\} = \int_x^\infty \int_{-\infty}^y f_{M_t, B_t}(r, s) \mathrm{d}s\mathrm{d}r. \tag{28.5}$$

Let $\tau_x = \min\{s \ge 0 : B_s = x\}$. Then $\{M_t \ge x\} = \{\tau_x \le t\}$, which gives

$$P\{M_t \ge x, B_t \le y\} = P\{\tau_x \le t, B_t \le y\}. \tag{28.6}$$

From Problem 28.14, we have

$$P\{\tau_x \le t, B_t \le y\} = P\{B_t \ge 2x - y\}. \tag{28.7}$$

Hence combining Equations (28.5), (28.6), and (28.7) gives

$$\int_x^\infty \int_{-\infty}^y f_{M_t, B_t}(r, s) \mathrm{d}s\mathrm{d}r = \int_{2x-y}^\infty \frac{1}{\sqrt{2\pi t}} \exp\left(-\frac{s^2}{2t}\right) \mathrm{d}s.$$

Differentiating with respect to x in the above equation gives

$$-\int_{-\infty}^y f_{M_t, B_t}(x, s) \mathrm{d}s = -\frac{2}{\sqrt{2\pi t}} \exp\left(-\frac{(2x-y)^2}{2t}\right).$$

Differentiating with respect to y in the above equation gives

$$-f_{M_t, B_t}(x, y) = -\frac{2(2x-y)}{t\sqrt{2\pi t}} \exp\left(-\frac{(2x-y)^2}{2t}\right).$$

This completes the proof.

28.17. Let $X_i = B_t^{(i)} - B_s^{(i)}$ for $i = 1, 2, \ldots, m$. Then by Definition 28.3, we have

$$\mathrm{Cov}(X_i, X_j) = (t - s)\delta_{ij}, \quad 1 \le i, j \le m,$$

where δ_{ij} is the Kronecker delta. Hence we have

$$E(X_i X_j) - E(X_i)E(X_j) = 0, \quad \forall i \ne j.$$

The independence of $X_1, X_2, \ldots, X_m$ follows immediately from Problem 19.6. This completes the proof.

28.5 Bibliographic Notes

In this chapter, we presented Brownian motion and related problems. A Brownian motion is defined as a stochastic process satisfying certain properties. In fact, such stochastic processes exist and can be constructed mathematically in several ways. For example, Kuo (2006, Chapter 3) introduced three approaches of constructing Brownian motion: Wiener's method, Kolmogorov's method, and Lévy's interpolation method. The history of Brownian motion is discussed in Bhattacharya and Waymire (2007, Chapter XIII).

Theorem 28.1 is a very interesting result of Brownian motions. Readers can find a proof of this theorem in Bhattacharya and Waymire (2007, p144). For a proof of Theorem 28.2, readers are referred to Hunt (1956).

The definition of m-dimensional Brownian motion (see Definition 28.3) was adopted from Karatzas and Shreve (1991, Section 2.5). This definition implies that the increments of the components are independent (see Problem 28.17). For more information about Brownian motion, readers are referred to Borodin and Salminen (2002), Shreve (2004), Wiersema (2008), and Mörters and Peres (2010).

CHAPTER 29

MARKOV PROCESSES

A Markov process is a stochastic process that has the Markov property. In other words, a stochastic process is called a Markov process if at every time t, the conditional probability law of the process given the past depends only on the present state. Intuitively, a Markov process is a process that does not have memory. In this chapter, we present the mathematical definition of Markov processes and relevant results.

29.1 Basic Concepts and Facts

Definition 29.1 (Conditional Independence). Let $(\Omega, \mathscr{F}, P)$ be a probability space. Let $\mathscr{H}$, $\mathscr{F}_1$, ..., $\mathscr{F}_n$ be sub-σ-algebras of $\mathscr{F}$. Then $\mathscr{F}_1$, ..., $\mathscr{F}_n$ are said to be conditionally independent given $\mathscr{H}$ if

$$E[V_1 V_2 \cdots V_n | \mathscr{H}] = \prod_{i=1}^{n} E[V_i | \mathscr{H}],$$

where V_i is an arbitrary positive random variable in $m\mathscr{F}_i$, $i = 1, 2, \ldots, n$. Here $m\mathscr{F}_i$ is the set of all $\mathscr{F}_i$-measurable functions.

Definition 29.2 (Markov Process). Let I be an index set. Let $(\Omega, \mathscr{F}, P)$ be a probability space and $\{\mathscr{F}_t : t \in I\}$ be a filtration over the probability space. Let $\{X_t : t \in I\}$ be a stochastic process adapted to the filtration. The process $\{X_t : t \in I\}$ is called a *Markov process* relative to the filtration if for every $t \in I$, the past $\mathscr{F}_t$ and the future $\mathscr{G}_\infty^t = \sigma(X_u : u \geq t,\ u \in I)$ are conditionally independent given the present state X_t.

Definition 29.3 (Transition Kernel). Let $(E, \mathscr{E})$ and $(F, \mathscr{F})$ be two measurable spaces. A transition kernel from $(E, \mathscr{E})$ to $(F, \mathscr{F})$ is a mapping $K : (E, \mathscr{F}) \to [0, \infty]$ with the following properties:

(a) For every set $B \in \mathscr{F}$, the mapping $x \to K(x, B)$ is $\mathscr{E}$-measurable.

(b) For every $x \in E$, the mapping $B \to K(x, B)$ is a measure on $(F, \mathscr{F})$.

A transition kernel from $(E, \mathscr{E})$ to $(E, \mathscr{E})$ is called a *transition kernel* on $(E, \mathscr{E})$.

Definition 29.4 (Markov Kernel). Let $(E, \mathscr{E})$ be a measurable space. A Markov kernel on $(E, \mathscr{E})$ is a transition kernel K on $(E, \mathscr{E})$ such that for every $x \in E$, we have

$$K(x, E) = 1.$$

Definition 29.5 (Identity Kernel). Let $(E, \mathscr{E})$ be a measurable space. An identity kernel on $(E, \mathscr{E})$ is a transition kernel I on $(E, \mathscr{E})$ such that for every $x \in E$,

$$I(x, A) = I_A(x), \quad A \in \mathscr{E},$$

where $I_A(\cdot)$ is the indicator function.

Definition 29.6 (Product of Kernels). Let K be a transition kernel from $(E, \mathscr{E})$ into $(F, \mathscr{F})$, and let L be a transition kernel from $(F, \mathscr{F})$ into $(G, \mathscr{G})$. Then their product is a transition kernel KL from $(E, \mathscr{E})$ into $(G, \mathscr{G})$ defined as

$$KL(x, B) = \int_F K(x, \mathrm{d}y) L(y, B), \quad x \in E,\ B \in \mathscr{G}.$$

Definition 29.7 (Markovian Transition Function). Let $(E, \mathscr{E})$ be a measurable space. Let $\{P_{t,u} : 0 \leq t \leq u\}$ be a family of Markov kernels on $(E, \mathscr{E})$ indexed by pairs of times t and u. It is called a *Markovian transition function* on $(E, \mathscr{E})$ if it satisfies the Chapman-Kolmogorov equation

$$P_{s,t} P_{t,u} = P_{s,u}, \quad 0 \leq s < t \leq u. \tag{29.1}$$

Definition 29.8 (Markov Process with a Transition Function). Let $\{X_t : t \in I\}$ be a Markov process with some state space $(E, \mathscr{E})$. Let $\{P_{t,u} : t, u \in I, t \leq u\}$ be a transition function on $(E, \mathscr{E})$. The Markov process is said to admit the transition function if for every positive $\mathscr{E}$-measurable function, we have

$$E[f(X_u) | X_t] = (P_{t,u} f)(X_t), \quad t < u,$$

where $P_{t,u}f$ is the product of $P_{t,u}$ and f (see Theorem 29.1).

Definition 29.9 (Time-Homogeneous Markov Process)**.** Let $\{X_t : t \in I\}$ be a Markov process that admits $\{P_{t,u} : 0 \le t \le u\}$ as a transition function. The Markov process is said to be time-homogeneous if for every time t and time $u > t$, the dependence of $P_{t,u}$ on the pair (t, u) is through $u - t$ only, that is,

$$P_{t,u} = P_{u-t}$$

for some Markov kernel P_{u-t}.

Definition 29.10 (Markov Chain)**.** A stochastic process $\{X_t : t \in I\}$ is said to be a Markov chain if the process is a time-homogeneous Markov process and $I = \mathbf{N}$.

Theorem 29.1 (Measure, Kernel, and Function)**.** *Let K be a transition kernel from $(E, \mathscr{E})$ into $(F, \mathscr{F})$. Let f be a positive $\mathscr{F}$-measurable function and μ a measure on $(E, \mathscr{E})$. Then*

(a) The function Kf defined as

$$Kf(x) = \int_F K(x, \mathrm{d}y) f(y), \quad x \in E,$$

is a positive $\mathscr{E}$-measurable function.

(b) The function μK defined as

$$\mu K(B) = \int_E \mu(\mathrm{d}x) K(x, B), \quad B \in \mathscr{F},$$

is a measure on $(F, \mathscr{F})$.

(c)

$$(\mu K) f = \mu(Kf) = \int_E \mu(\mathrm{d}x) \int_F K(x, \mathrm{d}y) f(y).$$

29.2 Problems

29.1. Let $(\Omega, \mathscr{F}, P)$ be a probability space. Let $\mathscr{H}$, $\mathscr{F}_1$, and $\mathscr{F}_2$ be sub-σ-algebras of $\mathscr{F}$. Show that the following are equivalent:

(a) $\mathscr{F}_1$ and $\mathscr{F}_2$ are conditionally independent given $\mathscr{H}$.

(b) For every positive random variable $V_2 \in m\mathscr{F}_2$, we have

$$E[V_2 | \mathscr{H} \vee \mathscr{F}_1] = E[V_2 | \mathscr{H}],$$

where $\mathscr{H} \vee \mathscr{F}_1$ is the σ-algebra generated by $\mathscr{H} \cup \mathscr{F}_1$.

(c) For every positive random variable $V_2 \in m\mathscr{F}_2$, we have

$$E[V_2|\mathscr{H} \vee \mathscr{F}_1] \in m\mathscr{H}.$$

29.2. Let $\{X_t : t \in I\}$ be a stochastic process with some state space $(E, \mathscr{E})$ and adapted to a filtration $\{\mathscr{F}_t : t \in I\}$. Let $\mathscr{G}_\infty^t = \sigma(X_u : u \geq t,\ u \in I)$. Show that the following are equivalent:

(a) The process $\{X_t : t \in I\}$ is a Markov process with respect to the filtration $\{\mathscr{F}_t : t \in I\}$.

(b) For every $t \in I$, $u > t$ $(u \in I)$, and positive function $f \in m\mathscr{E}$, we have

$$E[f(X_u)|\mathscr{F}_t] = E[f(X_u)|X_t]. \tag{29.2}$$

(c) For every $t \in I$ and positive random variable $V \in m\mathscr{G}_\infty^t$, we have

$$E[V|\mathscr{F}_t] = E[V|X_t]. \tag{29.3}$$

(d) For every $t \in I$ and every V of the form

$$V = \prod_{i=1}^{n} f_i(X_{u_i}),$$

where $n \geq 1$, $t \leq u_1 \leq u_2 \leq \cdots \leq u_n$, and $f_1, \ldots, f_n$ are bounded continuous functions in $m\mathscr{E}$, the following equation holds:

$$E[V|\mathscr{F}_t] = E[V|X_t]. \tag{29.4}$$

(e) For every $t \in I$ and positive random variable $V \in m\mathscr{G}_\infty^t$, we have

$$E[V|\mathscr{F}_t] \in \sigma(X_t), \tag{29.5}$$

that is, $E[V|\mathscr{F}_t]$ is $\sigma(X_t)$-measurable.

29.3. Let K be a transition kernel from $(E, \mathscr{E})$ into $(F, \mathscr{F})$, and let L be a transition kernel from $(F, \mathscr{F})$ into $(G, \mathscr{G})$. Show that

(a) For every $x \in E$ and $B \in \mathscr{F}$, we have

$$K(x, B) = KI_B(x),$$

where $I_B(\cdot)$ is the indicator function.

(b) For every positive $\mathscr{G}$-measurable function f, we have

$$(KL)f = K(Lf),$$

where KL is the product of K and L.

29.4. Let $\{X_t : t \in I\}$ be a Markov process with some state space $(E, \mathscr{E})$ that admits $\{P_{t,u} : t, u \in I, t \leq u\}$ as a transition function. Let $x \in E$. Show that

$$P_{t,u}(x, A) = P\{X_u \in A | X_t = x\}, \quad A \in \mathscr{E}.$$

29.5. Let $\{X_t : t \geq 0\}$ be a Markov process with state space $(E, \mathscr{E})$ and transition function $\{P_{t,u} : 0 \leq t \leq u\}$. Let $\{Y_t : t \geq 0\}$ be a stochastic process defined as

$$Y_t = (t, X_t), \quad t \in [0, \infty).$$

Show that

(a) The process $\{Y_t : t \geq 0\}$ is a Markov process with state space $(F, \mathscr{F}) = ([0, \infty) \times E, \mathcal{B}[0, \infty) \otimes \mathscr{E})$.

(b) The process $\{Y_t : t \geq 0\}$ is time-homogeneous.

(c) The transition function for $\{Y_t : t \geq 0\}$ is given by

$$Q_t f(y) = \int_E P_{s,s+t}(x, \mathrm{d}z) f(s + t, z), \quad y = (s, x) \in F,$$

where f is an arbitrary positive $\mathscr{F}$-measurable function.

29.6. Let $\{P_{t,u} : 0 \leq t \leq u\}$ be the transition function of a Markov process $\{X_t : t \geq 0\}$, which is adapted to some filtration $\{\mathscr{F}_t : t \geq 0\}$. Show that the transition function is Markovian:

$$P_{s,t} P_{t,u} = P_{s,u}, \quad 0 \leq s < t \leq u.$$

29.7. Let $\{B_t : t \geq 0\}$ be a Brownian motion adapted to some filtration $\{\mathscr{F}_t : t \geq 0\}$. Show that

(a) The Brownian motion $\{B_t : t \geq 0\}$ is a Markov process relative to the filtration.

(b) The transition function for the Brownian motion is given by

$$P_{t,u}(x, A) = \int_A \frac{1}{\sqrt{2\pi(u - t)}} \exp\left(-\frac{(y - x)^2}{2(u - t)}\right) \mathrm{d}y, \quad 0 \leq t < u.$$

29.3 Hints

29.1. Use the results of Problems 14.2, 14.15, and 2.16 to show that part (a) is equivalent to part (b). Then show that parts (b) and (c) are equivalent.

29.2. Use the definition of Markov processes (Definition 29.2) and the results of Problems 21.12 and 29.1. First show that part (a) is equivalent to part (c) and that part (d) is equivalent to part (e). Then show that part (c) $\Longrightarrow$ part (b) $\Longrightarrow$ part (d). Finally, show that part (e) $\Longrightarrow$ part (c).

29.3. Part (a) can be proved by using the definition of kernel and function composition (see Theorem 29.1). To prove part (b), use the third result of Theorem 29.1 to show that $(KL)f(x) = K(Lf)(x)$ for $x \in E$.

29.4. Use the definition of Markov processes with a transition function (Definition 29.8) to establish

$$P(X_u \in A|X_t) = P_{t,u}(X_t, A).$$

Then note that how $P(X_u \in A|X_t = x)$ is interpreted (see Section 14.5).

29.5. Part (a) follows from the definition of Markov processes directly. To prove part (b), one can show that

$$E[f(Y_{s_1+t})|Y_{s_1}] = E[f(Y_{s_2+t})|Y_{s_2}], \quad s_1, s_2, t \geq 0.$$

To prove part (c), one can use the following equation

$$Q_t f(y) = E[f(Y_{t+s})|Y_s = y] = E[f(Y_{t+s})|Y_s]|_{Y_s=y}, \quad y \in [0, \infty) \times E.$$

29.6. Use the definition of Markov processes with a transition function (Definitionn 29.8), the result of part (b) of Problem 29.2, and the tower property of conditional expectations (see Problem 14.10).

29.7. Use the results of Problems 29.2 and 14.13 to prove part (a). Use Definition 29.8, part (a) of this problem, and the result of Problem 14.13 to prove part (b).

29.4 Solutions

29.1. We first show that part (a) is equivalent to part (b). Note that part (a) is equivalent to

$$E[V_1V_2|\mathscr{H}] = E[V_1|\mathscr{H}] \cdot E[V_2|\mathscr{H}], \tag{29.6}$$

where V_1 and V_2 are arbitrary positive random variables in $m\mathscr{F}_1$ and $m\mathscr{F}_2$, respectively. By Problem 14.15, Equation (29.6) is equivalent to

$$E[V_1V_2|\mathscr{H}] = E[V_1 \cdot E(V_2|\mathscr{H})|\mathscr{H}]. \tag{29.7}$$

By the definition of conditional expectation (Definition 14.1), Equation (29.7) is equivalent to

$$\int_A V_1V_2 \mathrm{d}P = \int_A V_1 \cdot E(V_2|\mathscr{H}) \mathrm{d}P, \quad \forall A \in \mathscr{H}. \tag{29.8}$$

By the monotone convergence theorem (Theorem 6.2), we can show that Equation (29.8) is equivalent to

$$\int_{A\cap B} V_2 \mathrm{d}P = \int_{A\cap B} E(V_2|\mathscr{H}) \mathrm{d}P, \quad \forall A \in \mathscr{H},\ B \in \mathscr{F}_1. \tag{29.9}$$

But by Problem 14.2, Equation (29.9) is equivalent to

$$E[V_2|\sigma(\mathscr{H}, \mathscr{F}_1)] = E[V_2|\mathscr{H}],$$

which is part (b).

Now we show that part (b) is equivalent to part (c). It is clear that part (b) implies part (c). Conversely, assume that part (c) holds. Then by the definition of conditional expectation, we have

$$\int_A E[V_2|\mathscr{H} \vee \mathscr{F}_1]\mathrm{d}P = \int_A V_2\mathrm{d}P, \quad \forall A \in \mathscr{H} \vee \mathscr{F}_1, \tag{29.10}$$

which implies that

$$\int_A E[V_2|\mathscr{H} \vee \mathscr{F}_1]\mathrm{d}P = \int_A V_2\mathrm{d}P, \quad \forall A \in \mathscr{H}. \tag{29.11}$$

Note that $E[V_2|\mathscr{H} \vee \mathscr{F}_1] \in m\mathscr{H}$. We have

$$E[V_2|\mathscr{H} \vee \mathscr{F}_1] = E[V_2|\mathscr{H}].$$

Hence part (b) is true. This completes the proof.

29.2. Let V be a positive random variable that is measurable with respect to $\mathscr{G}^t_\infty$. Then by Problem 29.1, part (a) is equivalent to

$$E[V|\mathscr{F}_t \vee \sigma(X_t)] = E[V|\sigma(X_t)].$$

Since $\{X_t : t \in I\}$ is adapted to $\{\mathscr{F}_t : t \in I\}$, we have $\sigma(X_t) \subseteq \mathscr{F}_t$. Hence part (a) is equivalent to

$$E[V|\mathscr{F}_t] = E[V|\sigma(X_t)] = E[V|X_t].$$

Therefore, part (a) is equivalent to part (c). It follows from Problem 21.12 that part (d) is equivalent to part (e).

It is obvious that part (c) implies part (b). Now we show that part (b) $\implies$ part (d). To do that, we use the induction method. For $k = 1$, we have

$$\begin{aligned} E[V_1|\mathscr{F}_t] &= E[f_1(X_{u_1})|\mathscr{F}_t] \\ &= E[f_1^+(X_{u_1})|\mathscr{F}_t] - E[f_1^-(X_{u_1})|\mathscr{F}_t] \\ &= E[f_1^+(X_{u_1})|X_t] - E[f_1^-(X_{u_1})|X_t] \\ &= E[f_1(X_{u_1})|X_t] \in m\sigma(X_t). \end{aligned}$$

Hence part (d) holds. Suppose that part (d) holds for k. Then for $n = k + 1$,

$$\begin{aligned} E[V_{k+1}|\mathscr{F}_t] &= E(E[V_{k+1}|\mathscr{F}_{u_k}]|\mathscr{F}_t) \\ &= E(V_k E[f_{k+1}(X_{u_{k+1}})|\mathscr{F}_{u_k}]|\mathscr{F}_t) \\ &= E(V_k g(X_{u_k})|\mathscr{F}_t) \\ &\in m\sigma(X_t). \end{aligned}$$

Here $E[f_{k+1}(X_{u_{k+1}})|\mathscr{F}_{u_k}] \in \sigma(X_{u_k})$. By Theorem 11.2, $E[f_{k+1}(X_{u_{k+1}})|\mathscr{F}_{u_k}]$ is a function of X_{u_k}. Hence part (d) holds for $k+1$.

It follows from Problem 29.1 that part (e) implies part (c). This completes the proof.

29.3.

(a) By Theorem 29.1, we have

$$KI_B(x) = \int_F K(x, \mathrm{d}y) I_B(y) = \int_B K(x, \mathrm{d}y). \tag{29.12}$$

Note that by the definition of transition kernels (Definition 29.3), $K(x, \cdot)$ is a measure for every x. Hence from Equation (29.12), we have

$$KI_B(x) = K(x, B).$$

(b) Let $x \in E$ be fixed. Let $\mu_x(B) = K(x, B)$ for $B \in \mathscr{F}$. Then by Theorem 29.1, we have

$$\mu_x L(C) = \int_F \mu_x(\mathrm{d}y) L(y, C) = \int_F K(x, \mathrm{d}y) L(y, C), \quad C \in \mathscr{G}. \tag{29.13}$$

Again by Theorem 29.1 and Equation (29.13), we have

$$\begin{aligned}
(KL)f(x) &= \int_G KL(x, \mathrm{d}z) f(z) \\
&= \int_G f(z) \int_F K(x, \mathrm{d}y) L(y, \mathrm{d}z) \\
&= \int_G f(z) \mu_x L(\mathrm{d}z) \\
&= (\mu_x L) f \\
&= \mu_x (Lf) \\
&= \int_F \mu_x(\mathrm{d}y) \int_G L(y, \mathrm{d}z) f(z) \\
&= K(Lf)(x).
\end{aligned}$$

Since x is arbitrary, it follows that $(KL)f = K(Lf)$.

This completes the proof.

29.4. Let $x \in E$ and $A \in \mathscr{E}$ be fixed. Letting $f(\cdot) = I_A(\cdot)$ in Definition 29.8 gives

$$E[I_A(X_u)|X_t] = (P_{t,u} I_A)(X_t) = \int_E P_{t,u}(X_t, \mathrm{d}y) I_A(y). \tag{29.14}$$

Note that $P(X_u \in A|X_t) = E[I_A(X_u)|X_t]$ and $P_{t,u}(X_t, \cdot)$ is a measure. From Equation (29.14), we have

$$P(X_u \in A|X_t) = P_{t,u}(X_t, A). \tag{29.15}$$

Hence $P_{t,u}(X_t, A)$ is the conditional probability of $\{X_u \in A\}$ given $\sigma(X_t)$. Therefore, we have

$$P_{t,u}(x, A) = P(X_u \in A | X_t = x).$$

This completes the proof.

29.5.

(a) Let $\{\mathscr{H}_t : t \geq 0\}$ be the filtration to which the process $\{X_t : t \geq 0\}$ is adapted. By Problem 29.2, to prove that $\{Y_t : t \geq 0\}$ is a Markov process, we only need to show that

$$E[f(Y_u)|\mathscr{H}_t] = E[f(Y_u)|Y_t], \quad 0 \leq t \leq u, \tag{29.16}$$

where f is a positive $\mathscr{F}$-measurable function. But since $f(t, \cdot)$ is a nonnegative $\mathscr{E}$-measurable function, we have

$$E[f(Y_u)|\mathscr{H}_t] = E[f(u, X_u)|\mathscr{H}_t] = E[f(u, X_u)|X_t]. \tag{29.17}$$

Now by the definition of conditional expectation (Definition 14.1), for any $C \in \sigma(X_t)$, we have

$$\begin{aligned} \int_{\{u\}\times C} E[f(Y_u)|Y_t]\mathrm{d}P &= \int_{\{u\}\times C} f(Y_u)\mathrm{d}P \\ &= \int_C f(u, X_u)\mathrm{d}P \\ &= \int_C E[f(u, X_u)|X_t]\mathrm{d}P. \end{aligned} \tag{29.18}$$

Since Equation (29.18) holds for any $C \in \sigma(X_t)$, we have $E[f(Y_u)|Y_t] = E[f(Y_u)|X_t]$ a.s. Hence Equation (29.16) holds.

(b) To show that $\{Y_t : t \geq 0\}$ is time-homogeneous, we only need to show that

$$E[f(Y_{s_1+t})|Y_{s_1}] = E[f(Y_{s_2+t})|Y_{s_2}], \quad s_1, s_2, t \geq 0. \tag{29.19}$$

But

$$E[f(Y_{s_1+t})|Y_{s_1} = (s, x)] = E[f(Y_{s+t})|X_s = x]$$

and

$$E[f(Y_{s_2+t})|Y_{s_2} = (s, x)] = E[f(Y_{s+t})|X_s = x],$$

it follows that Equation (29.19) holds.

(c) Since $\{Y_t : t \geq 0\}$ is time-homogeneous, for $y = (s, x)$, we have

$$\begin{aligned} Q_t f(y) &= E[f(Y_{r+t})|Y_r = y] \\ &= E[f(Y_{s+t})|X_s = x] \\ &= E[f(s+t, X_{s+t})|X_s = x] \\ &= P_{s,s+t} f(s+t, x) \\ &= \int_E P_{s,s+t}(x, \mathrm{d}z) f(s+t, z). \end{aligned}$$

This finishes the proof.

29.6. Let f be a positive Borel function. Let $0 \le s < t \le u$. Then by Problems 29.2 and 14.10, and Definition 29.8, we have

$$\begin{aligned} P_{s,u}f(X_s) &= E[f(X_u)|X_s] \\ &= E[f(X_u)|\mathscr{F}_s] \\ &= E(E[f(X_u)|\mathscr{F}_t]|\mathscr{F}_s) \\ &= E(P_{t,u}f(X_t)|\mathscr{F}_s) \\ &= P_{s,t}(P_{t,u}f)(X_s) \end{aligned}$$

But by Theorem 29.1, we have

$$P_{s,t}(P_{t,u}f)(X_s) = (P_{s,t}P_{t,u})f(X_s).$$

Hence $P_{s,u} = P_{s,t}P_{t,u}$. This finishes the proof.

29.7. Let $0 \le t < u$ be fixed.

(a) By Problem 29.2, we only need to prove that

$$E[f(B_u)|\mathscr{F}_t] = E[f(B_u)|B_t]. \tag{29.20}$$

To do this, let $\varphi(x) = E[f(B_u - B_t + x)]$. Since $B_u - B_t$ is independent of $\mathscr{F}_t$ and B_t is $\mathscr{F}_t$-measurable, it follows from Problem 14.13 that

$$E[f(B_u)|\mathscr{F}_t] = E[f(B_u - B_t + B_t)|\mathscr{F}_t] = \varphi(B_t). \tag{29.21}$$

Similarly, since $B_u - B_t$ is independent of $\sigma(B_t)$ and B_t is $\sigma(B_t)$-measurable, it follows from Problem 14.13 that

$$E[f(B_u)|B_t] = E[f(B_u - B_t + B_t)|B_t] = \varphi(B_t). \tag{29.22}$$

Hence Equation (29.20) follows by combining Equations (29.21) and (29.22).

(b) By Theorem 29.1, we have

$$P_{t,u}(x, A) = P_{t,u}I_A(x).$$

By Definition 29.8 and the first item, we have

$$\begin{aligned} P_{t,u}I_A(B_t) &= E[I_A(B_u)|X_t] \\ &= E[I_A(B_u - B_t + B_t)|X_t] \\ &= \varphi(B_t), \end{aligned}$$

where

$$\varphi(x) = E[I_A(B_u - B_t + x))] = P\{B_u - B_t + x \in A\}.$$

Note that $B_u - B_t + x \sim N(x, u-t)$, that is, $B_u - B_t + x$ is a normal random variable with mean x and variance $u - t$. Then we have

$$\varphi(x) = \int_A \frac{1}{\sqrt{2\pi(u-t)}} \exp\left(-\frac{(y-x)^2}{2(u-t)}\right) dy.$$

Combining the above equations gives

$$P_{t,u}(x, A) = \int_A \frac{1}{\sqrt{2\pi(u-t)}} \exp\left(-\frac{(y-x)^2}{2(u-t)}\right) dy.$$

This completes the proof.

29.5 Bibliographic Notes

In this chapter, we defined Markov processes and relevant concepts. The definition of Markov processes (see Definition 29.2) is adopted from Çinlar (2011). For more information about Markov processes, readers are referred to Friedman (1975), Ethier and Kurtz (1986), Kannan and Lakshmikantham (2001), Klebaner (2005), Stroock (2005), and Çinlar (2011). For information about Markov chains, readers are referred to Hoel et al. (1972), Ross (1995), Kao (1997), Tijms (2003), Stirzaker (2005), Lawler (2006), Beichelt (2006), and Bhattacharya and Waymire (2009).

CHAPTER 30

LÉVY PROCESSES

Lévy processes are stochastic processes that have independent and stationary increments. The basic theory of Lévy processes was established in the 1930s. Recently, Lévy processes have been used as asset price models in mathematical finance. In this chapter, we present Lévy processes and their main properties.

30.1 Basic Concepts and Facts

Definition 30.1 (Lévy Process). Let $(\Omega, \mathscr{F}, P)$ be a probability space and $\{\mathscr{F}_t : t \geq 0\}$ a filtration on the probability space. A stochastic process $X = \{X_t : t \geq 0\}$ is called a *Lévy process* in $\mathbf{R}^d$ with respect to $\{\mathscr{F}_t : t \geq 0\}$ if it is adapted to the filtration and

(a) For almost every ω, the path $t \to X_t(\omega)$ is right-continuous with left limit and $X_0(\omega) = 0$.

(b) For every t and h in $[0, \infty)$, the increment $X_{t+h} - X_t$ is independent of $\mathscr{F}_t$ and has the same distribution as X_h.

A process $\{Y_t : t \geq 0\}$ is considered a Lévy process, without mentioning a filtration, if it is a Lévy process with respect to the filtration generated by itself.

Definition 30.2 (Jump). Let $\{X_t : t \geq 0\}$ be a Lévy process. For every $\omega \in \Omega$ and every $t \in [0, \infty)$, let

$$X_{t-}(\omega) = \lim_{x \uparrow t} X_x(\omega), \quad X_{t+}(\omega) = \lim_{s \downarrow t} X_s(\omega) = X_t(\omega).$$

Here $X_{t-}(\omega) = 0$ for $t = 0$. If $X_{t-}(\omega) \neq X_t(\omega)$, we say that the path $X_t(\omega)$ jumps from its left limit $X_{t-}(\omega)$ to its right-hand value $X_t(\omega)$. The size of the jump is given by

$$\Delta X_t(\omega) = X_t(\omega) - X_{t-}(\omega).$$

The jump magnitude is $|\Delta X_t(\omega)|$. The discontinuity set for the path $X_t(\omega)$ is denoted by

$$D_\omega = \{t > 0 : \Delta X_t(\omega) \neq 0\}.$$

Definition 30.3 (Pure-Jump Process). A Lévy process is called a *pure-jump process* if for every $t \geq 0$, X_t is equal to the sum of the sizes of its jumps during $[0, t]$. That is, for almost every ω, we have

$$X_t(\omega) = \sum_{s \in D_\omega \cap [0,t]} \Delta X_s(\omega), \quad t \in [0, \infty),$$

where the sum on the right-hand side converges absolutely:

$$V_t(\omega) = \sum_{s \in D_\omega \cap [0,t]} |\Delta X_s(\omega)| < \infty.$$

Here $V_t(\omega)$ denotes the total variation of the path $X_t(\omega)$ over $[0, t]$.

Definition 30.4 (Jump-Diffusion Process). A stochastic process $\{X_t : t \geq 0\}$ is called a *jump-diffusion process* if it has the following form

$$X_t = at + \sigma B_t + \sum_{n=1}^{N_t} Z_n,$$

where $a \in \mathbf{R}$, $\sigma \geq 0$, $\{B_t : t \geq 0\}$ is a standard Brownian motion, $\{N_t : t \geq 0\}$ is a Poisson process, and $\{Z_n : n = 1, 2, \ldots\}$ is a sequence of independent and identically distributed random variables that is independent of $\{N_t : t \geq 0\}$. The Poisson process $\{N_t : t \geq 0\}$, the sequence $\{Z_n : n = 1, 2, \ldots\}$, and the Brownian motion $\{B_t : t \geq 0\}$ are assumed to be independent.

Definition 30.5 (Infinitely Divisibility). A probability distribution F on $(\mathbf{R}, \mathcal{B})$ is said to be infinitely divisible if for any integer $n \geq 2$, there exists n independent and identically random variables $Y_1, Y_2, \ldots, Y_n$ such that $Y_1 + Y_2 + \cdots + Y_n$ has distribution F.

A random variable X is said to be infinitely divisible if and only if for each $n \geq 1$, X has the same distribution as the sum of n independent and identically distributed random variables.

Definition 30.6 (Random Measure). Let $(\Omega, \mathscr{F}, P)$ be a probability space and $(E, \mathscr{E})$ a measurable space. A random measure on $(E, \mathscr{E})$ is a transition kernel from $(\Omega, \mathscr{F})$ into $(E, \mathscr{E})$; that is, a mapping $M : \Omega \times \mathscr{E} \to [0, \infty]$ if

(a) For every $A \in \mathscr{E}$, $M(A) : \omega \to M(\omega, A)$ is a random variable.

(b) For every $\omega \in \Omega$, $M_\omega : A \to M(\omega, A)$ is a measure on $(E, \mathscr{E})$.

Definition 30.7 (Lévy Measure). A Lévy measure ν is a Borel measure defined on $\mathbf{R}\backslash\{0\} = \{x \in \mathbf{R} : x \neq 0\}$ such that

$$\int_{\mathbf{R}\backslash\{0\}} (y^2 \wedge 1)\nu(\mathrm{d}y) < \infty.$$

A Lévy measure ν can be extended to the whole real line $\mathbf{R}$ by assigning

$$\nu(\{0\}) = 0.$$

Definition 30.8 (Poisson Random Measure). Let $(E, \mathscr{E})$ be a measurable space. Let ν be a measure on $(E, \mathscr{E})$. A random measure N on $(E, \mathscr{E})$ is called a *Poisson random measure* with mean ν if it satisfies the following conditions:

(a) For every $A \in \mathscr{E}$, the random variable $N(A)$ has a Poisson distribution with mean $\nu(A)$.

(b) For every integer $n \geq 2$, if $A_1, A_2, \ldots, A_n \in \mathscr{E}$ are disjoint, then the random variables $N(A_1)$, $N(A_2)$, ..., $N(A_n)$ are independent.

Theorem 30.1 (Lévy-Khintchine Formula). *A probability distribution F of a real-valued random variable is infinitely divisible if and only if there exists a triplet (a, σ, ν), where $a \in \mathbf{R}$, $\sigma \geq 0$, and ν is a Lévy measure, such that for all $\theta \in \mathbf{R}$,*

$$\begin{aligned} & \int_{\mathbf{R}} e^{i\theta x} \mu_F(\mathrm{d}x) \\ = & \int_{\mathbf{R}} e^{i\theta x} \mathrm{d}F(x) \\ = & \exp\left(ia\theta - \frac{1}{2}\sigma^2\theta - \int_{\mathbf{R}} \left(1 - e^{i\theta x} + i\theta x I_{(-1,1)}(x)\right) \nu(\mathrm{d}x) \right), \end{aligned}$$

where μ_F is the Lebesgue-Stieltjes measure corresponding to F and $I_{(-1,1)}(\cdot)$ is an indicator function.

Theorem 30.2 (Lévy Processes and Infinitely Divisible Distributions). *If $\{X_t : t \geq 0\}$ is a Lévy process, then for every t, X_t has an infinitely divisible distribution.*

Conversely, if F is an infinitely divisible distribution, then there exists a Lévy process $\{X_t : t \geq 0\}$ such that the distribution of X_1 is F.

Theorem 30.3 (Characteristic Function of a Lévy Process). *Let $\{X_t : t \geq 0\}$ be a $\mathbf{R}^d$-valued Lévy process. Then for each $t \geq 0$, the characteristic function of X_t is given by*

$$\phi_{X_t}(u) = e^{t\eta(u)}, \quad \forall u \in \mathbf{R}^d,$$

where $\eta(u)$ is a continuous function, which is said to be the characteristic exponent of the process.

Theorem 30.4 (Existence of Poisson Random Measure). *For any given σ-finite measure space (S, Σ, λ), there exists a Poisson random measure M on some probability space $(\Omega, \mathscr{F}, P)$ such that $\lambda(A) = E(M(A))$ for all $A \in \Sigma$.*

Theorem 30.5 (Lévy-Itô Decomposition). *Let $\{X_t : t \geq 0\}$ be a $\mathbf{R}$-valued Lévy process. Then*

$$X_t = bt + \sigma B_t + \int_{|x|<1} x\tilde{N}_t(\mathrm{d}x) + \int_{|x|\geq 1} xN_t(\mathrm{d}x),$$

where

$$b = E\left[X_1 - \int_{|x|\geq 1} xN_1(\mathrm{d}x)\right],$$

$\sigma \geq 0$, $\{B_t : t \geq 0\}$ is a standard Brownian motion, N_t is a Poisson random measure on $(\mathbf{R}\backslash\{0\}, \mathcal{B}(\mathbf{R}\backslash\{0\}))$ defined as

$$N_t(A)(\omega) = |\{0 \leq s \leq t : X_s(\omega) - X_{s-}(\omega) \in A\}|, \quad t \geq 0, A \in \mathcal{B}(\mathbf{R}\backslash\{0\}),$$

and

$$\tilde{N}_t(A) = N_t(A) - tE[N_1(A)].$$

Here $|\cdot|$ denotes the number of elements in a set.

30.2 Problems

30.1. Let $\{X_t : t \geq 0\}$ be a Lévy process. For every $t \geq 0$, show that X_t is an infinitely divisible random variable.

30.2. Let F be a Poisson distribution with parameter $\lambda > 0$. Show that F is infinitely divisible.

30.3. Let N be a Poisson random variable with parameter $\lambda > 0$ and $\{Y_i : i = 1, 2, \ldots\}$ a sequence of independent and identically distributed random variables having distribution F. Suppose that N is independent of $\{Y_i : i = 1, 2, \ldots\}$ and that $\mu_F(\{0\}) = 0$, where μ_F is the Lebesgue-Stieltjes measure corresponding to F. Let X be a compound Poisson random variable defined as

$$X = \sum_{i=1}^{N} Y_i.$$

Show that X has an infinitely divisible distribution with the Lévy-Khintchine triplet (a, σ, ν) given by

$$a = \lambda \int_{-1}^{1} x\mu_F(\mathrm{d}x), \quad \sigma = 0, \quad \nu = \lambda\mu_F.$$

30.4. Let X be a normal random variable with a distribution F given by

$$\mathrm{d}F(x) = \frac{1}{s\sqrt{2\pi}} \exp\left(-\frac{(x-\gamma)^2}{2s^2}\right) \mathrm{d}x,$$

where $s > 0$ and $\gamma \in \mathbf{R}$ are constants. Show that F is an infinitely divisible distribution with the Lévy-Khintchine triplet (a, σ, ν) given by

$$a = \gamma, \quad \sigma = s, \quad \nu = 0.$$

30.5. Let $\{Y_t : t \geq 0\}$ be a compound Poisson process defined as

$$Y_t = \sum_{n=1}^{N_t} Z_n, \quad t \geq 0,$$

where $\{N_t : t \geq 0\}$ is a Poisson process and $\{Z_n : n = 1, 2, \ldots\}$ is a sequence of independent and identically distributed random variables that is independent of $\{N_t : t \geq 0\}$. Show that $\{Y_t : t \geq 0\}$ is a Lévy process.

30.6. Let $\{X_t : t \geq 0\}$ and $\{Y_t : t \geq 0\}$ be two independent Lévy processes. Show that $\{X_t + Y_t : t \geq 0\}$ is also a Lévy process.

30.7. Let $\{X_t : t \geq 0\}$ be a Lévy process with characteristic exponent $\eta(\cdot)$. For any $\theta \in \mathbf{R}$, show that

$$\{\exp(i\theta X_t - t\eta(\theta)) : t \geq 0\}$$

is a martingale.

30.8 (Markov Property)**.** Let $\{X_t : t \geq 0\}$ be a Lévy process adapted to some filtration $\{\mathscr{F}_t : t \geq 0\}$. Show that $\{X_t : t \geq 0\}$ is a Markov process with respect to the filtration.

30.9. Let $\{X_t : t \geq 0\}$ be a jump-diffusion process as defined in Definition 30.4. For every $t \geq 0$ and $A \in \mathcal{B}(\mathbf{R})$, define

$$M_t(A) = \sum_{n=0}^{N_t} I_A(Z_n),$$

where $I_A(\cdot)$ is an indicator function. Let ν be a set function defined as

$$\nu(A) = E[M_1(A)], \quad A \in \mathcal{B}(\mathbf{R}).$$

Suppose that N_1 has a Poisson distribution with parameter λ and Z_1 has a distribution function F. Show that

(a) The process $\{X_t : t \geq 0\}$ is a Lévy process.

(b) For every $t \geq 0$, M_t is a random measure.

(c) For every $t \geq 0$, M_t is a Poisson random measure with mean $t\nu$.

(d) The set function ν is a finite measure on $(\mathbf{R}, \mathcal{B}(\mathbf{R}))$ and

$$\nu(A) = \lambda \mu_F(A), \quad A \in \mathcal{B}(\mathbf{R}),$$

where μ_F is the Lebesgue-Stieltjes measure corresponding to F.

(e) The characteristic exponent of $\{X_t : t \geq 0\}$ is given by

$$\eta(u) = iau - \frac{1}{2}\sigma^2 u + \int_{\mathbf{R}} \left(e^{iux} - 1\right) \nu(\mathrm{d}x).$$

30.10. Let $\{X_t : t \geq 0\}$ be a jump-diffusion process as defined in Definition 30.4. For every $t \geq 0$, let M_t be a Poisson random measure (see Problem 30.9) defined as

$$M_t(A) = \sum_{n=0}^{N_t} I_A(Z_n), \quad A \in \mathcal{B}(\mathbf{R}).$$

Let ν be a measure (see Problem 30.9) defined as

$$\nu(A) = E[M_1(A)], \quad A \in \mathcal{B}(\mathbf{R}).$$

Let $\{Y_t : t \geq 0\}$ be a process defined by

$$Y_t = \int_{\mathbf{R}} x\tilde{M}_t(\mathrm{d}x) = \int_{\mathbf{R}} xM_t(\mathrm{d}x) - \int_{\mathbf{R}} xt\nu(\mathrm{d}x),$$

where $\tilde{M}_t = M_t - t\nu$. Show that

(a) For every $t > 0$, we have

$$\sum_{0<s\leq t} \Delta X_s = \sum_{n=1}^{N_t} Z_n, \quad \text{a.s.},$$

where $\Delta X_s = X_s - X_{s-}$.

(b) For every $t > 0$, we have

$$\sum_{n=1}^{N_t} Z_n = \int_{\mathbf{R}} x M_t(\mathrm{d}x), \quad \text{a.s.}$$

(c) If $\left|\int_{\mathbf{R}} x\nu(\mathrm{d}x)\right| < \infty$, then $\{Y_t : t \geq 0\}$ is a martingale and $E[Y_t] = 0$:

$$E\left[\int_{\mathbf{R}} x M_t(\mathrm{d}x)\right] = t \int_{\mathbf{R}} x\nu(\mathrm{d}x).$$

(d) If $\left|\int_{\mathbf{R}} x^2\nu(\mathrm{d}x)\right| < \infty$, then

$$\mathrm{Var}(Y_t) = t \int_{\mathbf{R}} x^2\nu(\mathrm{d}x).$$

30.3 Hints

30.1. Follow the definition of Lévy processes (Definition 30.1) and the definition of infinitely divisible random variables (Definition 30.5).

30.2. Apply the Lévy-Khintchine formula (Theorem 30.1).

30.3. Apply the Lévy-Khintchine formula (Theorem 30.1).

30.4. Use the result of Problem 19.1 and apply the Lévy-Khintchine formula (Theorem 30.1).

30.5. Follow the definition of Lévy processes (Definition 30.1) and the definition of Poisson processes (Definition 27.1).

30.6. Follow the definition of Lévy processes (Definition 30.1).

30.7. Follow the definition of characteristic exponent (see Theorem 30.3) and the definition of martingales (Definition 22.1).

30.8. Use the result of Problem 14.13 to show that for all positive Borel function f,

$$E[f(X_t)|\mathscr{F}_s] = E[f(X_t)|X_s], \quad 0 \leq s < t.$$

Then apply the result of Problem 29.2.

30.9. To prove part (a), use the results of Problems 30.6 and 27.9, and the definitions of Brownian motion (Definition 28.1). To prove part (b), follow the definition of random measures (Definition 30.6). To prove part (c), follow the definition of Poisson random measures (Definition 30.8) and use Kac's theorem (Theorem 17.2). To prove part (d), first show that $\nu(A) = \lambda\mu_F(A)$. To prove part (e), use the results of Problems 30.3 and 30.4.

30.10. To prove part (a), note that almost all paths of $\{N_s : 0 \leq s \leq t\}$ have finite number of jumps. To prove part (b), use the result of Problem 5.11 and the monotone convergence theorem (Theorem 6.2). Use the result of part (b) to prove part (c). Use the results of parts (b) and (c) to prove part (d).

30.4 Solutions

30.1. Let $n \geq 2$ be an integer. For each t, we can express X_t as

$$X_t = \sum_{k=1}^{n} Y_t^{(k)},$$

where

$$Y_t^{(k)} = X_{\frac{kt}{n}} - X_{\frac{(k-1)t}{n}}.$$

By the definition of Lévy processes (Definition 30.1), $Y_t^{(1)}, Y_t^{(2)}, \ldots, Y_t^{(n)}$ are independent and identically distributed random variables. Hence X_t is an infinitely divisible random variable. This completes the proof.

30.2. Let μ_F be the Lebesgue-Stieltjes measure corresponding to F. Then we have

$$\mu_F(\{k\}) = e^{-\lambda}\frac{\lambda^k}{k!}, \quad k \geq 0$$

and

$$\int_{\mathbf{R}} e^{itx}\mu_F(\mathrm{d}x) = \sum_{k=0}^{\infty} e^{itk}e^{-\lambda}\frac{\lambda^k}{k!} = \exp\left(-\lambda + \lambda e^{it}\right).$$

Let $a = \sigma = 0$ and $\nu = \lambda\delta_1$, where δ_1 is the Dirac measure supported on $\{1\}$:

$$\delta_1(A) = \begin{cases} 1, & \text{if } 1 \in A; \\ 0, & \text{if } 1 \notin A, \end{cases} \qquad \forall A \in \mathcal{B}(\mathbf{R}).$$

Then we have

$$iat - \frac{1}{2}\sigma^2 t - \int_{\mathbf{R}} \left(1 - e^{itx} + itxI_{(-1,1)}(x)\right)\nu(\mathrm{d}x) = -\lambda + \lambda e^{it}.$$

It follows from Theorem 30.1 that F is infinitely divisible. This completes the proof.

30.3. Let G be the distribution of X. Then we have

$$\begin{aligned}
\int_{\mathbf{R}} e^{i\theta x}\mu_G(\mathrm{d}x) &= E\left[e^{i\theta X}\right] \\
&= \sum_{n=0}^{\infty} E\left[\exp\left(i\theta\sum_{j=1}^{n} Y_j\right)\right] P\{N = n\} \\
&= \sum_{n=0}^{\infty}\left(\int_{\mathbf{R}} e^{i\theta}\mu_F(\mathrm{d}x)\right)^n e^{-\lambda}\frac{\lambda^n}{n!} \\
&= \exp\left(-\lambda\int_{\mathbf{R}}(1 - e^{i\theta x})\mu_F(\mathrm{d}x)\right).
\end{aligned}$$

Comparing the above equation with the Lévy-Khintchine formula gives

$$a = \lambda \int_{-1}^{1} x \mu_F(\mathrm{d}x), \quad \sigma = 0, \quad \nu = \lambda \mu_F.$$

Hence it follows from Theorem 30.1 that G is infinitely divisible. This completes the proof.

30.4. By Problem 19.1, we have

$$\int_{\mathbf{R}} e^{i\theta x} \mu_F(\mathrm{d}x) = E\left[e^{i\theta X}\right] = \exp\left(i\gamma\theta - \frac{1}{2}s^2\theta^2\right).$$

Comparing the above equation with the Lévy-Khintchine formula gives

$$a = \gamma, \quad \sigma = s, \quad \nu = 0.$$

By Theorem 30.1, F is infinitely divisible. This finishes the proof.

30.5. By Definition 27.1, $\{N_t : t \geq 0\}$ is right-continuous with left limit and $P\{N_0 = 0\} = 1$. Hence $\{Y_t : t \geq 0\}$ is right-continuous with left limit and $P\{Y_0 = 0\} = P\{N_0 = 0\} = 1$.

Let $h \geq 0$. Then

$$Y_{t+h} - Y_t = \sum_{n=N_t+1}^{N_{t+h}} Z_n.$$

30.6. Let $Z_t = X_t + Y_t$ for all $t \geq 0$. To show that $\{Z_t : t \geq 0\}$ is a Lévy process, we only need to verify that it satisfies the two conditions in Definition 30.1.

Since $\{X_t : t \geq 0\}$ and $\{Y_t : t \geq 0\}$ are Lévy processes, we have $P(\{X_0 = 0\} \cap \{Y_0 = 0\}) = 1$ (see Problem 11.1). Hence $Z_0 = X_0 + Y_0 = 0$ a.s. In addition, by Problem 1.10, for every $s \geq 0$, we have

$$\lim_{t \downarrow s} Z_t = \lim_{t \downarrow s} X_t + \lim_{t \downarrow s} Y_t = X_s + Y_s,$$

and for every $s > 0$, we have

$$\lim_{t \uparrow s} Z_t = \lim_{t \uparrow s} X_t + \lim_{t \uparrow s} Y_t = X_{s-} + Y_{s-}.$$

Hence $\{Z_t : t \geq 0\}$ satisfies the first condition.

Let $\mathscr{F}_t = \sigma(Z_s : 0 \leq s \leq t)$ and $\mathscr{H}_t = \sigma(X_s, Y_s : s \leq t)$ for all $t \geq 0$. Then we have

$$\mathscr{F}_t \subseteq \mathscr{H}_t, \quad t \geq 0.$$

Since $\{X_t : t \geq 0\}$ and $\{Y_t : t \geq 0\}$ are independent Lévy processes, $Z_t - Z_s = (X_t - X_s) + (Y_t - Y_s)$ is independent of $\mathscr{H}_s$ and thus $Z_t - Z_s$ is independent of $\mathscr{F}_s$. In addition, $Z_t - Z_s$ has the same distribution as $X_{t-s} + Y_{t-s} = Z_{t-s}$. Hence $\{Z_t : t \geq 0\}$ satisfies the first condition.

This completes the proof.

30.7. Let $\{\mathscr{F}_t : t \geq 0\}$ be the filtration generated by $\{X_t : t \geq 0\}$. Then the process $\{\exp(i\theta X_t - t\eta(\theta)) : t \geq 0\}$ is adapted to the filtration. In addition, we have

$$E[|\exp(i\theta X_t - t\eta(\theta))|] = 1 < \infty.$$

For $0 \leq s < t$, we have

$$\begin{aligned}
& E[\exp(i\theta X_t - t\eta(\theta)) | \mathscr{F}_s] \\
= \ & E[\exp(i\theta(X_t - X_s)) \exp(i\theta X_s - t\eta(\theta)) | \mathscr{F}_s] \\
= \ & E[\exp(i\theta(X_t - X_s))] \exp(i\theta X_s - t\eta(\theta))) \\
= \ & \exp(i\theta X_s - t\eta(\theta)) e^{i(t-s)\eta(\theta)} \\
= \ & \exp(i\theta X_s - s\eta(\theta)).
\end{aligned}$$

This completes the proof.

30.8. Let $0 \leq s < t$ be fixed. Let f be a positive Borel function. Let $Y = X_t - X_s$. Since $\{X_t : t \geq 0\}$ is a Lévy process, Y is independent of $\mathscr{F}_s$. Then by Problem 14.13, we have

$$E[f(X_t)|\mathscr{F}_s] = E[f(Y + X_s)|\mathscr{F}_s] = g(X_s),$$

where $g(x) = E[f(Y + x)]$. Since Y is also independent of $\sigma(X_s)$, by Problem 14.13 we have

$$E[f(X_t)|X_s] = E[f(Y + X_s)|\sigma(X_s)] = g(X_s).$$

Hence

$$E[f(X_t)|\mathscr{F}_s] = E[f(X_t)|X_s].$$

It follows from Problem 29.2 that $\{X_t : t \geq 0\}$ is a Markov process. This completes the proof.

30.9.

(a) By Definition 28.1, the Brownian motion $\{\sigma B_t : t \geq 0\}$ is a Lévy process. By Problem 30.6, we only need to show that the compound Poisson process $\{\sum_{n=1}^{N_t} Z_n : t \geq 0\}$ is a Lévy process. To do that, let

$$Y_t = \sum_{n=1}^{N_t} Z_n, \quad t \geq 0.$$

Then $P\{Y_0 = 0\} = P\{N_0 = 0\} = 1$. By Problem 27.9, $Y_{t+h} - Y_t$ is independent of $\sigma(Y_s : 0 \leq s \leq t)$ and $Y_{t+h} - Y_t$ has the same distribution as Y_h. Hence $\{Y_t : t \geq 0\}$ is a Lévy process.

(b) Suppose that $\{N_t : t \geq 0\}$ and $\{Z_1, Z_2, \ldots\}$ are defined on a probability space $(\Omega, \mathscr{F}, P)$.

Let $A \in \mathcal{B}(\mathbf{R})$ and $t \geq 0$ be fixed. We show that $M_t(A)$ is a random variable. To do that, let $H \in \mathcal{B}(\mathbf{R})$. Then

$$\{M_t(A) \in H\} = \bigcup_{k \in H \cap \bar{\mathbf{N}}} \left[\bigcup_{i \geq 0} \left(\{N_t = i\} \cap \left\{ \sum_{n=0}^{i} I_A(Z_n) = k \right\} \right) \right] \in \mathscr{F},$$

where $\bar{\mathbf{N}} = \{0, 1, 2, \ldots\}$. Since H is arbitrary, $M_t(A)$ is $\mathscr{F}$-measurable. Hence $M_t(A)$ is a random variable.

Now let $\omega \in \Omega$ and $t \geq 0$ be fixed. We show that $M_{t,\omega}$ is a measure on $(\mathbf{R}, \mathcal{B}(\mathbf{R}))$. In fact, by the definition of M_t, we have $M_{t,\omega}(\emptyset) = 0$ and $M_{t,\omega}(A) \geq 0$ for all $A \in \mathcal{B}(\mathbf{R})$. Let $G_1, G_2, \ldots \in \mathcal{B}(\mathbf{R})$ be a sequence of mutually disjoint sets. Then

$$\begin{aligned} M_{t,\omega}\left(\bigcup_{n=1}^{\infty} G_n \right) &= \sum_{i=1}^{N_t} I_{\bigcup_{n=1}^{\infty} G_n}(Z_i) \\ &= \sum_{i=1}^{N_t} \sum_{n=1}^{\infty} I_{G_n}(Z_i) \\ &= \sum_{n=1}^{\infty} M_{t,\omega}(G_n), \end{aligned}$$

which shows that $M_{t,\omega}$ is countably additive. Therefore, $M_{t,\omega}$ is a measure.

(c) By part (b), M_t is a random measure. To show that M_t is a Poisson random measure with mean ν, we first show that for every $A \in \mathcal{B}(\mathbf{R})$, $M_t(A)$ has a Poisson distribution with mean $\nu(A)$. Let $k \geq 0$ be an integer. Then by Problem 14.18, we have

$$\begin{aligned} P\{M_t(A) = k\} &= P\left\{ \sum_{n=1}^{N_t} I_A(Z_n) = k \right\} \\ &= \sum_{i=0}^{\infty} P\{N_t = i\} P\left\{ \sum_{n=1}^{i} I_A(Z_n) = k \right\} \\ &= \sum_{i=k}^{\infty} e^{-\lambda t} \frac{(\lambda t)^i}{i!} \binom{i}{k} P\{Z_1 \in A\}^k P\{Z_1 \notin A\}^{i-k} \\ &= e^{-\lambda t P\{Z_1 \in A\}} \frac{(\lambda t P\{Z_1 \in A\})^k}{k!}, \end{aligned}$$

which shows that $M_t(A)$ has a Poisson distribution with mean $\lambda t P\{Z_1 \in A\}$. Hence we have

$$\nu(A) = E[M_1(A)] = \lambda P\{Z_1 \in A\} = \lambda \mu_F(A).$$

Now let $n \geq 2$, and let $A_1, \ldots, A_n \in \mathcal{B}(\mathbf{R})$ be disjoint. To show that $M_t(A_1)$, $M_t(A_2)$, ..., $M_t(A_n)$ are independent, we use Kac's theorem (Theorem 17.2) to show that

$$E\left[\exp\left(i\sum_{k=1}^{n}\theta_k M_t(A_k)\right)\right] = \prod_{k=1}^{n} E\left[\exp\left(i\theta_k M_t(A_k)\right)\right]. \tag{30.1}$$

In fact, by Problems 14.18 and 12.7 we have

$$\begin{aligned}
& E\left[\exp\left(i\sum_{k=1}^{n}\theta_k M_t(A_k)\right)\right] \\
= \; & E\left[\exp\left(i\sum_{k=1}^{n}\theta_k \sum_{j=1}^{N_t} I_{A_k}(Z_j)\right)\right] \\
= \; & \sum_{m=0}^{\infty} E\left[\exp\left(i\sum_{j=1}^{m}\sum_{k=1}^{n}\theta_k I_{A_k}(Z_j)\right)\right] P\{N_t = m\} \\
= \; & \sum_{m=0}^{\infty}\prod_{j=1}^{m} E\left[\exp\left(i\sum_{k=1}^{n}\theta_k I_{A_k}(Z_j)\right)\right] e^{-\lambda t}\frac{(\lambda t)^m}{m!}
\end{aligned}$$

Since $Z_1, Z_2, \ldots$ have the same distribution and $A_1, A_2, \ldots, A_n$ are disjoint, the above equation gives

$$\begin{aligned}
& E\left[\exp\left(i\sum_{k=1}^{n}\theta_k M_t(A_k)\right)\right] \\
= \; & \exp\left(\lambda t E\left[\exp\left(i\sum_{k=1}^{n}\theta_k I_{A_k}(Z_1)\right)\right] - \lambda t\right) \\
= \; & \exp\left(\lambda t \int_{\mathbf{R}} \exp\left(i\sum_{k=1}^{n}\theta_k I_{A_k}(x)\right)\mu_F(\mathrm{d}x) - \lambda t\right) \\
= \; & \exp\left(\lambda t\left[\sum_{k=1}^{n}(1+e^{i\theta_k})\mu_F(A_k)\right]\right).
\end{aligned} \tag{30.2}$$

Since $Z_1, Z_2, \ldots$ are independent, by Problem 14.18 and Theorem 17.2 we have

$$\begin{aligned}
& \prod_{k=1}^{n} E\left[\exp\left(i\theta_k M_t(A_k)\right)\right] \\
= & \prod_{k=1}^{n} \sum_{m=0}^{\infty} E\left[\exp\left(i\theta_k \sum_{j=1}^{m} I_{A_k}(Z_j)\right)\right] P\{N_t = m\} \\
= & \prod_{k=1}^{n} \sum_{m=0}^{\infty} \left(E\left[\exp\left(i\theta_k I_{A_k}(Z_1)\right)\right]\right)^m e^{-\lambda t} \frac{(\lambda t)^m}{m!} \\
= & \prod_{k=1}^{n} \exp\left(\lambda t E\left[\exp\left(i\theta_k I_{A_k}(Z_1)\right)\right] - \lambda t\right) \\
= & \prod_{k=1}^{n} \exp\left(\lambda t[1 + e^{i\theta_k}]\mu_F(A_k)\right). \qquad (30.3)
\end{aligned}$$

Hence Equation (30.1) follows from Equations (30.2) and (30.3). Thus, M_t is a Poisson random measure.

(d) By Problem 14.18, we have

$$\begin{aligned}
\nu(A) &= E\left[\sum_{n=1}^{N_1} I_A(Z_n)\right] \\
&= \sum_{m=0}^{\infty} P\{N_t = m\} E\left[\sum_{n=1}^{m} I_A(Z_n)\right] \\
&= \sum_{m=0}^{\infty} e^{-\lambda} \frac{\lambda^m}{m!} m \mu_F(A) \\
&= \lambda \mu_F(A).
\end{aligned}$$

Since μ_F is a finite measure, ν is also a finite measure.

(e) By the assumption that $\{N_t : t \geq 0\}$, $\{Z_n : n = 1, 2, \ldots\}$, and $\{B_t : t \geq 0\}$ are independent, we have

$$E\left[e^{iuX_t}\right] = E\left[e^{iu(at+\sigma B_t)}\right] E\left[e^{iu\sum_{n=1}^{N_t} Z_n}\right].$$

By Problems 30.3 and 30.4, we have

$$E\left[e^{iu(at+\sigma B_t)}\right] = \exp\left(iuat - \frac{1}{2}\sigma^2 tu\right)$$

and

$$
\begin{aligned}
& E\left[e^{iu\sum_{n=1}^{N_t} Z_n}\right] \\
= \ & \exp\left(iut\int_{-1}^{1} x\nu(\mathrm{d}x) - t\int_{\mathbf{R}} \left[1 - e^{iux} + iuxI_{(-1,1)}(x)\right]\nu(\mathrm{d}x)\right) \\
= \ & \exp\left(-t\int_{\mathbf{R}} \left[1 - e^{iux}\right]\nu(\mathrm{d}x)\right).
\end{aligned}
$$

Combining the above three equations gives

$$
E\left[e^{iuX_t}\right] = \exp\left(iuat - \frac{1}{2}\sigma^2 tu - t\int_{\mathbf{R}} \left[1 - e^{iux}\right]\nu(\mathrm{d}x)\right).
$$

This completes the proof.

30.10.

(a) Let Ω_0 be the set such that for any $\omega \in \Omega_0$, the path $\{N_s(\omega) : 0 \le s \le t\}$ is right-continuous with left limit and the path $\{as + \sigma B_s(\omega) : 0 \le s \le t\}$ is continuous. Then we have $P(\Omega_0) = 1$. Let $\omega \in \Omega_0$. Since $N_t(\omega) < \infty$, the path $\{N_s(\omega) : 0 \le s \le t\}$ has at most a finite number of jumps. Let $s_1, s_2, \ldots, s_m$ be the jump points of the path. Then we have

$$
\sum_{0<s\le t} \Delta X_s = \sum_{i=1}^{m} \sum_{n=N_{s_{i-}}+1}^{N_{s_i}} Z_n = \sum_{n=1}^{N_t} Z_n.
$$

(b) For every $n \ge 1$, let $A_{n,0} = \{0\}$,

$$
A_{n,i} = \left\{x \in \mathbf{R} : (i-1)2^{-n} < x \le i2^{-n}\right\}, \quad i = 1, 2, \ldots, n2^n,
$$

and $A_{n,n2^n+1} = \{x \in \mathbf{R} : x > n\}$. For every $n \ge 1$, let

$$
\alpha_n(x) = \begin{cases} 0, & \text{if } x = 0; \\ (i-1)2^{-n}, & \text{if } x \in A_{n,i},\ i = 1, 2, \ldots, n2^n; \\ n, & \text{if } n < x. \end{cases}
$$

Then by Problem 5.11, we have for any $x \in \mathbf{R}$, $\alpha_n(x^+) \uparrow x^+$ as $n \to \infty$, where $x^+ = \max(x, 0)$. By Theorem 6.2, we have

$$
\int_{\mathbf{R}} \alpha_n(x^+) M_t(\mathrm{d}x) \uparrow \int_{\mathbf{R}} x^+ M_t(\mathrm{d}x) \quad \text{as } n \to \infty
$$

and

$$
\int_{\mathbf{R}} \alpha_n(x^-) M_t(\mathrm{d}x) \uparrow \int_{\mathbf{R}} x^- M_t(\mathrm{d}x) \quad \text{as } n \to \infty,
$$

where $x^- = \max(-x, 0)$. But

$$
\begin{aligned}
\int_{\mathbf{R}} \alpha_n(x^+) M_t(\mathrm{d}x) &= \sum_{i=0}^{n2^n+1} \int_{A_{n,i}} \alpha_n(x^+) M_t(\mathrm{d}x) \\
&= \sum_{i=1}^{n2^n+1} \frac{i-1}{2^n} M_t(A_{n,i}) \\
&= \sum_{i=1}^{n2^n+1} \frac{i-1}{2^n} \sum_{j=1}^{N_t} I_{A_{n,i}}(Z_j),
\end{aligned}
$$

which gives

$$
\begin{aligned}
&\left| \int_{\mathbf{R}} \alpha_n(x^+) M_t(\mathrm{d}x) - \sum_{j=1}^{N_t} Z_j^+ \right| \\
= &\left| \sum_{i=1}^{n2^n+1} \frac{i-1}{2^n} \sum_{j=1}^{N_t} I_{A_{n,i}}(Z_j) - \sum_{j=1}^{N_t} Z_j \sum_{i=1}^{n2^n+1} I_{A_{n,i}}(Z_j) \right| \\
\leq &\sum_{i=1}^{n2^n+1} \sum_{j=1}^{N_t} \left| \frac{i-1}{2^n} - Z_j \right| I_{A_{n,i}}(Z_j) \\
\leq &\sum_{i=1}^{n2^n+1} \sum_{j=1}^{N_t} \frac{1}{2^n} I_{A_{n,i}}(Z_j) \\
= &\sum_{j=1}^{N_t} \left(\frac{1}{2^n} I_{(0,n]}(Z_j) + I_{(n,\infty)}(Z_j) \right).
\end{aligned}
$$

Hence we have

$$
\lim_{n\to\infty} \int_{\mathbf{R}} \alpha_n(x^+) M_t(\mathrm{d}x) = \sum_{j=1}^{N_t} Z_j^+.
$$

Similarly, we have

$$
\lim_{n\to\infty} \int_{\mathbf{R}} \alpha_n(x^-) M_t(\mathrm{d}x) = \sum_{j=1}^{N_t} Z_j^-.
$$

Therefore, we have

$$
\int_{\mathbf{R}} x M_t(\mathrm{d}x) = \sum_{j=1}^{N_t} Z_j.
$$

(c) By part (b), we only need to show that

$$
E\left[\sum_{n=1}^{N_t} Z_n \right] = t \int_{\mathbf{R}} x \nu(\mathrm{d}x). \tag{30.4}
$$

But by Problem 27.10, we have

$$E\left[\sum_{n=1}^{N_t} Z_n\right] = tE[N_1]E[Z_1]. \tag{30.5}$$

By Problem 30.9, we have

$$t\int_{\mathbf{R}} x\nu(\mathrm{d}x) = t\int_{\mathbf{R}} xE[N_1]\mu_F(\mathrm{d}x) = tE[N_1]E[Z_1], \tag{30.6}$$

where F is the distribution function of Z_1 and μ_F is the Lebesgue-Stieltjes measure with respect to F. Hence Equation (30.4) follows from Equations (30.5) and (30.6). It follows from Problem 27.10 that $\{Y_t : t \geq 0\}$ is a martingale.

(d) By parts (c) and (b), we have

$$\begin{aligned} \text{Var}(Y_t) &= E\left[Y_t^2\right] = E\left[\left(\sum_{j=1}^{N_t} Z_j - t\int_{\mathbf{R}} x\nu(\mathrm{d}x)\right)^2\right] \\ &= E\left[\left(\sum_{j=1}^{N_t} Z_j\right)^2\right] - \left(t\int_{\mathbf{R}} x\nu(\mathrm{d}x)\right)^2. \end{aligned}$$

But

$$\begin{aligned} E\left[\left(\sum_{j=1}^{N_t} Z_j\right)^2\right] &= \sum_{m=0}^{\infty} P\{N_t = m\} E\left[\left(\sum_{j=1}^{m} Z_j\right)^2\right] \\ &= \sum_{m=1}^{\infty} e^{-\lambda t}\frac{(\lambda t)^m}{m!}\left[mE[Z_1^2] + m(m-1)E[Z_1]E[Z_1]\right] \\ &= \lambda t E[Z_1^2] + (\lambda t E[Z_1])^2 \\ &= t\int_{\mathbf{R}} x^2\lambda\mu_F(\mathrm{d}x) + \left(t\int_{\mathbf{R}} x\lambda\mu_F(\mathrm{d}x)\right)^2 \\ &= t\int_{\mathbf{R}} x^2\nu(\mathrm{d}x) + \left(t\int_{\mathbf{R}} x\nu(\mathrm{d}x)\right)^2, \end{aligned}$$

where $\lambda = E[N_1]$, F is the distribution function of Z_1 and μ_F is the corresponding Lebesgue-Stieltjes measure of F. The result follows by combining the above two equations.

This finishes the proof.

30.5 Bibliographic Notes

The Lévy process introduced in this chapter forms a rich class of stochastic processes. For example, the Poisson process and Brownian motion are special cases of the Lévy process. Our definition of the Lévy process is adopted from Çinlar (2011). For an alternative definition, readers are referred to Protter (2003) and Capasso and Bakstein (2005). For stochastic calculus with Lévy processes, readers are referred to Applebaum (2009) and Pascucci (2011). For more information about Lévy processes, readers are referred to Bertoin (1996), Sato (1999), Applebaum (2004), Kyprianou (2006), Pascucci (2011), and Stroock (2011).

Lévy processes with jumps have been used to model stock prices for the purpose of risk management and derivative pricing. For more information about financial modeling with Lévy processes, readers are referred to Cont and Tankov (2003), Schoutens (2003), and Schoutens and Cariboni (2009).

The Lévy-Khintchine formula (see Theorem 30.1) describes the structure of the characteristic exponent of an infinitely divisible distribution. A proof of this theorem can be found in Lukacs (1970) and Sato (1999). Theorem 30.2 states the relationship between Lévy processes and infinitely divisible distributions. For a proof of the converse part, readers are referred to Sato (1999, Corollary 11.6). A proof of Theorem 30.3 can be found in Applebaum (2009, Theorem 1.3.3).

Theorem 30.4 says that Poisson random measures exist. A proof of this theorem can be found in Sato (1999, Proposition 19.4). For more information about Poisson random measures, readers are referred to Çinlar (2011, Chapter VI).

The Lévy-Itô decomposition (see Theorem 30.5) is an important result, which describes the path properties of Lévy processes. For a proof of this theorem, readers are referred to Applebaum (2009, Section 2.4).

PART IV

STOCHASTIC CALCULUS

CHAPTER 31

THE WIENER INTEGRAL

The Wiener integral is a simple stochastic integral where the integrand is a deterministic function. In this chapter, we present a definition and some results of the Wiener integral.

31.1 Basic Concepts and Facts

Definition 31.1 (Function Space $L^2[a,b]$). The function space

$$L^2[a,b] = L^2([a,b], \mathcal{B}([a,b]), \mu)$$

denotes the space of all real-valued square Lebesgue integrable functions on $[a,b]$, where μ is the Lebesgue measure.

Definition 31.2 (Function Space $L^2(\Omega, \mathscr{F}, P)$). The function space

$$L^2(\Omega, \mathscr{F}, P)$$

denotes the space of all square integrable real-valued random variables with inner product $\langle X, Y \rangle = E(XY)$.

Definition 31.3 (Wiener Integral of Step Functions). Let f be a step function in $L^2[a, b]$ given by

$$f(x) = \sum_{i=1}^{n} a_i I_{[t_{i-1}, t_i)}(x), \quad x \in [a, b],$$

where $a = t_0 < t_1 < \cdots < t_n = b$. The Wiener integral of f is defined as

$$I(f) = \sum_{i=1}^{n} a_i (B_{t_i} - B_{t_{i-1}}), \tag{31.1}$$

where $\{B_t\}_{t \geq 0}$ is the standard Brownian motion.

Definition 31.4 (Wiener Integral of General Functions). Let $f \in L^2[a, b]$, and let $\{f_n\}_{n \geq 1}$ be a sequence of step functions such that

$$f_n \xrightarrow{L^2[a,b]} f.$$

Then the sequence $\{I(f_n)\}_{n \geq 1}$ is Cauchy in $L^2(\Omega, \mathscr{F}, P)$ and thus converges in $L^2(\Omega, \mathscr{F}, P)$. The Wiener integral of f is defined as

$$I(f) = \lim_{n \to \infty} I(f_n) \quad in \quad L^2(\Omega, \mathscr{F}, P). \tag{31.2}$$

The Wiener integral $I(f)$ is also denoted by $\int_a^b f(t) \mathrm{d}B_t$ or $\int_a^b f(t) \mathrm{d}B_t(\omega)$.

Definition 31.5 (Riemann-Stieltjes Integral). Let f and g be bounded functions on a finite closed interval $[a, b]$. The function f is said to be Riemann-Stieltjes integrable with respect to g if the following limit exists:

$$\int_a^b f(t) \mathrm{d}g(t) = \lim_{\|\Delta_n\| \to 0} \sum_{i=1}^{n} f(\tau_i)(g(t_i) - g(t_{i-1})),$$

where $\Delta_n = \{t_0, t_1, \ldots, t_n\}$ is a partition of $[a, b]$, that is, $a = t_0 < t_1 < \cdots < t_n = b$, $\Delta_n = \max_{1 \leq i \leq n}(t_i - t_{i-1})$, and $\tau_i \in [t_{i-1}, t_i]$.

Theorem 31.1. *For each $f \in L^2[a, b]$, the Wiener integral $\int_a^b f(t) \mathrm{d}B_t$ is a random variable normally distributed with mean 0 and variance $\|f\|^2 = \int_a^b f(t)^2 \mathrm{d}t$.*

Theorem 31.2. *Let f be a continuous function in $L^2[a, b]$. Then for almost all $\omega \in \Omega$, we have*

$$\left(\int_a^b f(t) \mathrm{d}B_t \right)(\omega) = (RS) \int_a^b f(t) \mathrm{d}B_t(\omega),$$

where the right-hand side is the Riemann-Stieltjes integral and the left-hand side is the Wiener integral.

31.2 Problems

31.1. Let f be a step function in $L^2[a,b]$ given by

$$f(x) = \sum_{i=1}^{n} a_i I_{[t_{i-1},t_i)}(x), \quad x \in [a,b],$$

where $a = t_0 < t_1 < \cdots < t_n = b$. Let $I(f)$ be the Wiener integral defined in Equation (31.1). Show that

$$E\left[I(f)^2\right] = \int_a^b f(t)^2 \mathrm{d}t.$$

31.2. Show that the Wiener integral $I(f)$ in Equation (31.1) is well defined a.s. In other words, if f has two representations

$$f(x) = \sum_{i=1}^{n} a_i I_{[t_{i-1},t_i)}(x) = \sum_{j=1}^{m} b_j I_{[s_{j-1},s_j)}(x), \quad x \in [a,b],$$

where $a = t_0 < t_1 < \cdots < t_n = b$ and $a = s_0 < s_1 < \cdots < s_m = b$, then

$$\sum_{i=1}^{n} a_i (B_{t_i} - B_{t_{i-1}}) = \sum_{j=1}^{m} b_j (B_{s_i} - B_{s_{i-1}}), \quad \text{a.s.}$$

31.3. Let f and g be step functions in $L^2[a,b]$. Let $\alpha, \beta \in \mathbf{R}$. Show that

$$I(\alpha f + \beta g) = \alpha I(f) + \beta I(g),$$

where $I(\cdot)$ is the Wiener integral defined in Equation (31.1).

31.4. Show that the Wiener integral in Equation (31.2) is well-defined a.s. In other words, if $\{f_n\}_{n\geq 1}$ and $\{g_n\}_{n\geq 1}$ are two sequences of step functions such that

$$f_n \xrightarrow{L^2[a,b]} f, \quad g_n \xrightarrow{L^2[a,b]} f,$$

then we have

$$\lim_{n\to\infty} I(f_n) = \lim_{n\to\infty} I(g_n), \quad \text{a.s.}$$

31.5. Let $f, g \in L^2[a,b]$. Show that

(a)

$$E[I(f)I(g)] = \int_a^b f(t)g(t)\mathrm{d}t.$$

(b) If f and g are orthogonal (i.e., $\int_a^b f(t)g(t)\mathrm{d}t = 0$), then the normal random variables $I(f)$ and $I(g)$ are independent.

31.6. Let $\{B_t\}_{t\geq 0}$ be a Brownian motion and

$$X = \int_0^1 B_t \mathrm{d}t.$$

Find the distribution of X.

31.7. Let $\{B_t\}_{t\geq 0}$ be a Brownian motion and let $f \in L^2[a,b]$. Show that the stochastic process

$$M_t = \int_a^t f(s)\mathrm{d}B_s, \quad t \in [a,b],$$

is a martingale with respect to the filtration $\mathscr{F}_t = \sigma\{B_s : s \leq t\}$.

31.8 (Integration by Parts)**.** Let $\{B_t\}_{t\geq 0}$ be a Brownian motion and let $f \in L^2[a,b]$. Show that

$$\int_a^b f(t)\mathrm{d}B_t = f(b)B_b - \int_a^b B_t \mathrm{d}f(t).$$

31.3 Hints

31.1. Use the properties of Brownian motion.

31.2. Consider the partition produced by points $t_0, t_1, \ldots, t_n$ and s_0, s_1, ..., s_m and use the results of Problems 31.1 and 6.8.

31.3. Suppose that f and g are given by

$$f(x) = \sum_{i=1}^n a_i I_{[t_{i-1},t_i)}(x), \quad x \in [a,b)$$

and

$$g(x) = \sum_{i=1}^m b_i I_{[s_{j-1},s_j)}(x), \quad x \in [a,b),$$

where $a = t_0 < t_1 < \cdots < t_n = b$ and $a = s_0 < s_1 < \cdots < s_m$. Then consider the partition produced by points $t_0, t_1, \ldots, t_n, s_0, s_1, \ldots, s_m$ and use the result of Problem 31.2.

31.4. Use the results of Problems 31.1 and 31.3.

31.5. Part (a) can be proved by considering the expectation $E[(I(f) + I(g))^2]$ and the linearity of I (see Problem 31.3). Part (b) is implied from part (a) and Problem 19.6.

31.6. Use the integration by parts and Theorem 31.2.

31.7. Use the idea of how the Wiener integral is defined and the fact that the B_t is a martingale with respect to the $\mathscr{F}_t$ (see Problem 28.1); that is, first consider step functions and then general functions.

31.8. Apply Theorem 31.2.

31.4 Solutions

31.1. By Definition 31.3, we have

$$I(f)^2 = \left(\sum_{i=1}^{n} a_i(B_{t_i} - B_{t_{i-1}})\right)^2 = \sum_{i=1}^{n}\sum_{j=1}^{n} a_i a_j (B_{t_i} - B_{t_{i-1}})(B_{t_j} - B_{t_{j-1}}).$$

Since the B_t is a Brownian motion, $E[(B_{t_i} - B_{t_{i-1}})(B_{t_j} - B_{t_{j-1}})] = 0$ if $i \neq j$, and $E[(B_{t_i} - B_{t_{i-1}})(B_{t_j} - B_{t_{j-1}})] = t_i - t_{i-1}$ if $i = j$. Therefore we have

$$E\left[I(f)^2\right] = \sum_{i=1}^{n} a_i^2(t_i - t_{i-1}) = \int_a^b f(t)^2 \mathrm{d}t.$$

This completes the proof.

31.2. Let $a = r_0 < r_1 < \cdots < r_l = b$ be distinct points of $t_0, t_1, \ldots, t_n$ and $s_0, s_1, \ldots, s_m$. Then we have

$$\bigcup_{i=1}^{n}[t_{i-1}, t_i) = \bigcup_{j=1}^{m}[s_{j-1}, s_j) = \bigcup_{k=1}^{l}[r_{k-1}, r_k) = [a, b).$$

Let c_k be the value of f on $[r_{k-1}, r_k)$, $k = 1, 2, \ldots, l$. Then $c_k = a_i = b_j$ if $[r_{k-1}, r_k) \subseteq [t_{i-1}, t_i) \cap [s_{j-1}, s_j)$.

Now let

$$I_1(f) = \sum_{i=1}^{n} a_i(B_{t_i} - B_{t_{i-1}})$$

and

$$I_2(f) = \sum_{j=1}^{m} b_j(B_{s_i} - B_{s_{i-1}}).$$

Then, by Problem 31.1, we have

$$\begin{aligned}
& E\left[(I_1(f) - I_2(f))^2\right] \\
= \; & E\left[(I_1(f))^2 + (I_2(f))^2 - 2I_1(f)I_2(f)\right] \\
= \; & 2\int_a^b f(x)^2 \mathrm{d}x - 2\sum_{i=1}^{n}\sum_{j=1}^{m} a_i b_j E[(B_{t_i} - B_{t_{i-1}})(B_{s_j} - B_{s_{j-1}})] \\
= \; & 2\int_a^b f(x)^2 \mathrm{d}x - 2\sum_{k=1}^{l} c_k^2(r_k - r_{k-1}) \\
= \; & 0.
\end{aligned}$$

By Problem 6.8, we have $|I_1(f) - I_2(f)| = 0$ a.s. Hence we have $I_1(f) = I_2(f)$ a.s. This completes the proof.

31.3. Let f and g be given by

$$f(x) = \sum_{i=1}^{n} a_i I_{[t_{i-1}, t_i)}(x), \quad x \in [a, b)$$

and

$$g(x) = \sum_{i=1}^{m} b_i I_{[s_{j-1}, s_j)}(x), \quad x \in [a, b),$$

where $a = t_0 < t_1 < \cdots < t_n = b$ and $a = s_0 < s_1 < \cdots < s_m$. Then we have

$$\alpha I(f) + \beta I(g) = \sum_{i=1}^{n} \alpha a_i (B_{t_i} - B_{t_{i-1}}) + \sum_{j=1}^{m} \beta b_j (B_{s_j} - B_{s_{j-1}}). \tag{31.3}$$

Let $0 = r_0 < r_1 < \cdots < r_l = b$ be the distinct points of $t_0, t_1, \ldots, t_n, s_0, s_1, \ldots, s_m$. Then we have

$$\alpha f + \beta g = \sum_{k=1}^{l} c_k I_{[r_{k-1}, r_k)},$$

where $c_k = \alpha a_i + \beta b_j$ if $[r_{k-1}, r_k) \subseteq [a_{i-1}, a_i) \cap [s_{j-1}, s_j)$. Hence we have

$$I(\alpha f + \beta g) = \sum_{k=1}^{l} c_k (B_{r_k} - B_{r_{k-1}}). \tag{31.4}$$

Note that

$$\bigcup_{i=1}^{n} [t_{i-1}, t_i) = \bigcup_{j=1}^{m} [s_{j-1}, s_j) = \bigcup_{k=1}^{l} [r_{k-1}, r_k) = [a, b).$$

It follows that

$$\sum_{i=1}^{n} \alpha a_i (B_{t_i} - B_{t_{i-1}}) + \sum_{j=1}^{m} \beta b_j (B_{s_j} - B_{s_{j-1}}) = \sum_{k=1}^{l} c_k (B_{r_k} - B_{r_{k-1}}).$$

By Equations (31.3) and (31.4), we have $I(af + bg) = aI(f) + bI(g)$. This completes the proof.

31.4. By Problems 31.3 and 31.1, we have

$$E\left[(I(f_n) - I(g_m))^2\right] = E\left[(I(f_n - g_m))^2\right] = \int_a^b (f_n(x) - g_m(x))^2 \mathrm{d}x.$$

But

$$\begin{aligned}
& \int_a^b (f_n(x) - g_m(x))^2 \mathrm{d}x \\
= \; & \int_a^b [(f_n(x) - f(x)) - (g_m(x) - f(t)))^2 \mathrm{d}x \\
\leq \; & 2 \int_a^b (f_n(x) - f(x))^2 \mathrm{d}x + 2 \int_a^b (g_m(x) - f(t)))^2 \mathrm{d}x.
\end{aligned}$$

By the assumption that

$$f_n \xrightarrow{L^2[a,b]} f, \quad g_n \xrightarrow{L^2[a,b]} f,$$

we have

$$E\left[(I(f_n) - I(g_m))^2\right] \to 0 \quad \text{as} \quad n, m \to \infty.$$

It follows that

$$\lim_{n\to\infty} I(f_n) = \lim_{n\to\infty} I(g_n), \quad \text{a.s.}$$

This completes the proof.

31.5.

(a) By Problems 31.3 and 31.1, we have

$$\begin{aligned} & E[(I(f) + I(g))^2] \\ = \; & E[(I(f+g))^2] \\ = \; & \int_a^b (f(t) + g(t))^2 \mathrm{d}t \\ = \; & \int_a^b f(t)^2 \mathrm{d}t + 2\int_a^b f(t)g(t)\mathrm{d}t + \int_a^b g(t)^2 \mathrm{d}t. \end{aligned} \tag{31.5}$$

But

$$\begin{aligned} & E[(I(f) + I(g))^2] \\ = \; & E[I(f)^2] + 2E[I(f)I(g)] + E[I(g)^2] \\ = \; & \int_a^b f(t)^2 \mathrm{d}t + 2E[I(f)I(g)] + \int_a^b g(t)^2 \mathrm{d}t. \end{aligned} \tag{31.6}$$

The result follows by combining Equations (31.5) and (31.6).

(b) The independence follows from part (a) and Problem 19.6.

31.6. Integrating by parts, we get

$$X(\omega) = \int_0^1 B_t(\omega)\mathrm{d}t = B_t(\omega)(t-1)|_0^1 - \int_0^1 (t-1)\mathrm{d}B_t(\omega) = \int_0^1 (1-t)\mathrm{d}B_t(\omega).$$

By Theorem 31.2, we have

$$X = \int_0^1 (1-t)\mathrm{d}B_t.$$

By Theorem 31.1, X is normally distributed with mean 0 and variance

$$\int_0^1 (1-t)^2 \mathrm{d}t = \frac{1}{3}.$$

31.7. By the definition of the $\mathscr{F}_t$, M_t is adapted to the $\mathscr{F}_t$. Note that

$$E[|M_t|^2] = \int_a^t f(r)^2 \mathrm{d}r < \infty$$

and $E(|M_t|) \leq (E[|M_t|^2])^{\frac{1}{2}}$. We have $E(|M_t|) < \infty$.

It is left to show that

$$E(M_t|\mathscr{F}_s) = M_s, \quad s \leq t.$$

But

$$M_t = M_s + \int_s^t f(r)\mathrm{d}B_r$$

and M_s is $\mathscr{F}_s$-measurable, we only need to show that

$$E\left(\int_s^t f(r)\mathrm{d}B_r \middle| \mathscr{F}_s\right) = 0, \quad \text{a.s.} \tag{31.7}$$

To to that, we first consider the case when f is a step function. Suppose that f is given by

$$f(r) = \sum_{i=1}^n a_i I_{[t_{i-1}, t_i)}(r), \quad r \in [a, b),$$

where $s = t_0 < t_1 < \cdots < t_n = t$. Note that the B_t is a martingale. We have $E(B_{t_i} - B_{t_{i-1}}|\mathscr{F}_s) = 0$ since $t_i \geq s$ for all $i = 0, 1, \ldots, n$. Then

$$\begin{aligned} E\left(\int_s^t f(r)\mathrm{d}B_r \middle| \mathscr{F}_s\right) &= E\left(\sum_{i=1}^n a_i(B_{t_i} - B_{t_{i-1}}) \middle| \mathscr{F}_s\right) \\ &= \sum_{i=1}^n a_i E(B_{t_i} - B_{t_{i-1}}|\mathscr{F}_s) \\ &= 0. \end{aligned}$$

Hence Equation (31.7) holds for step functions.

Now consider $f \in L^2[a, b]$. Let $\{f_n\}_{n\geq 1}$ be a sequence of step functions converging to f in $L^2[a, b]$. Let the X_n be random variables defined as

$$X_n = \int_s^t f_n(r)\mathrm{d}B_r, \quad n \geq 1.$$

Then $E(X_n|\mathscr{F}_s) = 0$. By the conditional Jensen's inequality (see Problem 15.7) with $\varphi(x) = x^2$, we have

$$\left(E\left[X_n - \int_s^t f(r)\mathrm{d}B_r \middle| \mathscr{F}_s\right]\right)^2 \leq E\left[\left(X_n - \int_s^t f(r)\mathrm{d}B_r\right)^2 \middle| \mathscr{F}_s\right],$$

which gives

$$E\left[\left(E\left[X_n - \int_s^t f(r)\mathrm{d}B_r \middle| \mathscr{F}_s\right]\right)^2\right] \leq \left(X_n - \int_s^t f(r)\mathrm{d}B_r\right)^2$$
$$\leq \int_s^t (f_n(r) - f(r))^2 \,\mathrm{d}B_r.$$

Hence

$$E(X_n|\mathscr{F}_s) \xrightarrow{L^2} E\left(\int_s^t f(r)\mathrm{d}B_r|\mathscr{F}_s\right) \text{ as } n \to \infty.$$

By Problems 9.1 and 9.6, there is a subsequence $\{X_{n_i}\}_{i\geq 1}$ such that

$$E(X_{n_i}|\mathscr{F}_s) \xrightarrow{\text{a.s.}} E\left(\int_s^t f(r)\mathrm{d}B_r|\mathscr{F}_s\right) \text{ as } i \to \infty.$$

Therefore Equation (31.7) holds for general functions in $L^2[a, b]$. This completes the proof.

31.8. By Theorem 31.2 and integration by parts from calculus, we have

$$\left(\int_a^b f(t)\mathrm{d}B_t\right)(\omega) = (RS)\int_a^b f(t)\mathrm{d}B_t(\omega)$$
$$= f(b)B_b(\omega) - \int_a^b B_t(\omega)\mathrm{d}f(t).$$

This completes the proof.

31.5 Bibliographic Notes

The Wiener integral is a simple type of stochastic integral, which integrates a deterministic function with respect to a standard Brownian motion. Most material in this chapter is taken from Kuo (2006), where readers can find proofs of Theorems 31.1 and 31.2.

CHAPTER 32

THE ITÔ INTEGRAL

The Itô integral defines an integral of a stochastic process with respect to a standard Brownian motion. The Itô integral is more general than the Wiener integral introduced in the previous chapter. In this chapter, we shall introduce the concept of Itô integrals.

32.1 Basic Concepts and Facts

Definition 32.1 (Stochastic Process Space $L^2_{ad}([a,b]\times\Omega)$). Let $\{\mathscr{F}_t : a \le t \le b\}$ be a filtration under consideration. The space $L^2_{ad}([a,b]\times\Omega)$ is defined to be the space of all stochastic processes $f(t,\omega)$, $t\in[a,b]$, $\omega\in\Omega$, satisfying the following conditions:

(a) $f(t)$ is adapted to the $\mathscr{F}_t$.

(b) $\int_a^b E(|f(t)|^2)\mathrm{d}t < \infty$.

Definition 32.2 (Itô Integral of Step Stochastic Processes). Let $\{\mathscr{F}_t : a \le t \le b\}$ be a filtration. Let $\{B_t : a \le t \le b\}$ be a Brownian motion satisfying the following conditions:

(a) For each t, B_t is $\mathscr{F}_t$-measurable.

(b) For any $s \leq t$, the random variable $B_t - B_s$ is independent of $\mathscr{F}_s$.

Let f be a step stochastic process in $L^2_{ad}([a,b] \times \Omega)$ given by

$$f(t,\omega) = \sum_{i=1}^{n} \xi_{i-1}(\omega) I_{[t_{i-1},t_i)}(t),$$

where $a = t_0 < t_1 < \ldots < t_n = b$ and ξ_{i-1} is $\mathscr{F}_{t_{i-1}}$-measurable with $E(\xi_{i-1}^2) < \infty$ for $i = 1, 2, \ldots, n$. Then the stochastic integral of f is defined as

$$I(f) = \sum_{i=1}^{n} \xi_{i-1}(B_{t_i} - B_{t_{i-1}}). \tag{32.1}$$

Definition 32.3 (Itô Integral of General Stochastic Processes). Let $\{\mathscr{F}_t : a \leq t \leq b\}$ be a filtration. Let $\{B_t : a \leq t \leq b\}$ be a Brownian motion satisfying the following conditions:

(a) For each t, B_t is $\mathscr{F}_t$-measurable.

(b) For any $s \leq t$, the random variable $B_t - B_s$ is independent of $\mathscr{F}_s$.

Let f be a stochastic process in $L^2_{ad}([a,b] \times \Omega)$. Then there exists a sequence $\{f_n : n \geq 1\}$ of step stochastic processes such that

$$\lim_{n\to\infty} \int_a^b E\left(|f(t) - f_n(t)|^2\right) \mathrm{d}t = 0.$$

The stochastic integral of f is defined as

$$I(f) = \lim_{n\to\infty} I(f_n), \quad \text{in} \quad L^2(\Omega),$$

where $I(f_n)$ is the stochastic integral of the step stochastic process f_n. The stochastic integral $I(f)$ is said to be the Itô integral of f and is written as $\int_a^b f(t)\mathrm{d}B_t$.

Theorem 32.1. *Let $f \in L^2_{ad}([a,b] \times \Omega)$. Then the Itô integral*

$$I(f) = \int_a^b f(t)\mathrm{d}B_t$$

is a random variable with mean $E[I(f)] = 0$ and variance

$$E\left(|I(f)|^2\right) = \int_a^b E\left(|f(t)|^2\right) \mathrm{d}t.$$

Theorem 32.2 (Martingale Property). *Let $f \in L^2_{ad}([a,b] \times \Omega)$. Then the stochastic process*

$$X_t = \int_a^t f(u)\mathrm{d}B_u, \quad a \leq t \leq b,$$

is a martingale with respect to the filtration $\{\mathscr{F}_t : a \le t \le b\}$.

Theorem 32.3 (Continuity Property). *Let* $f \in L^2_{ad}([a,b] \times \Omega)$. *Then the stochastic process*

$$X_t = \int_a^t f(u)\mathrm{d}B_u, \quad a \le t \le b,$$

is continuous; that is, almost all of its sample paths are continuous functions on $[a,b]$.

Theorem 32.4. *Let* $f \in L^2_{ad}([a,b] \times \Omega)$. *Suppose that* $E[f(t)f(s)]$ *is a continuous function of* $(t,s) \in [a,b] \times [a,b]$. *Then*

$$\int_a^b f(t)\mathrm{d}B_t = \lim_{\|\Delta_n\| \to \infty} \sum_{i=1}^n f(t_{i-1})(B_{t_i} - B_{t_{i-1}}), \quad in \quad L^2(\Omega),$$

where $\Delta_n = \{t_0, t_1, \ldots, t_n\}$ *are partitions of* $[a,b]$ *and*

$$\|\Delta_n\| = \max_{1 \le i \le n} (t_i - t_{i-1}).$$

32.2 Problems

32.1. Let f be a step stochastic process in $L^2_{ad}([a,b] \times \Omega)$ given by

$$f(t,\omega) = \sum_{i=1}^n \xi_{i-1}(\omega) I_{[t_{i-1}, t_i)}(t),$$

where $a = t_0 < t_1 < \cdots < t_n = b$ and $\xi_{t_{i-1}}$ is $\mathscr{F}_{t_{i-1}}$-measurable with $E(\xi_{i-1}^2) < \infty$ for $i = 1, 2, \ldots, n$. Show that

$$E[I(f)] = 0$$

and

$$E\left(|I(f)|^2\right) = \int_a^b E\left(|f|^2\right)\mathrm{d}t,$$

where $I(f)$ is the stochastic integral defined in Equation (32.1).

32.2. Show that the stochastic integral $I(f)$ defined in Equation (32.1) is well defined. In other words, if f has two representations given by

$$f(t,\omega) = \sum_{i=1}^n \xi_{i-1}(\omega) I_{[t_{i-1}, t_i)}(t) = \sum_{j=1}^m \eta_{j-1}(\omega) I_{[s_{j-1}, s_i)}(t),$$

where $a = t_0 < t_1 < \cdots < t_n = b$, $a = s_0 < s_1 < \cdots < s_m = b$, ξ_{i-1} is $\mathscr{F}_{t_{i-1}}$-measurable for $i = 1, 2, \ldots, n$, and η_{j-1} is $\mathscr{F}_{s_{j-1}}$-measurable for $j = 1, 2, \ldots, m$, then

$$\sum_{i=1}^n \xi_{i-1}(B_{t_i} - B_{t_{i-1}}) = \sum_{j=1}^m \eta_{j-1}(B_{s_j} - B_{s_{j-1}}), \quad \text{a.s.}$$

32.3. Let f and g be two step stochastic processes in $L^2_{ad}([a,b]\times\Omega)$. Show that

$$I(\alpha f+\beta g)=\alpha I(f)+\beta I(g),\quad \alpha,\beta\in\mathbf{R},$$

where $I(\cdot)$ is as defined in Equation (32.1).

32.4. Let $f(t)$ be a step stochastic process in $L^2_{ad}([a,b]\times\Omega)$. Let $C>0$, and let $f_C(t)$ be a stochastic process defined as

$$f_C(t,\omega)=\begin{cases} f(t,\omega) & \text{if } \int_a^t |f(u,\omega)|^2\mathrm{d}u\le C;\\ 0 & \text{otherwise.}\end{cases}$$

Show that

$$\left\{\int_a^b f(t)\mathrm{d}B_t \ne \int_a^b f_C(t)\mathrm{d}B_t\right\}\subseteq\left\{\int_a^b |f(t)|^2\mathrm{d}t>C\right\}.$$

32.5. Let $f(t)$ be a step stochastic process in $L^2_{ad}([a,b]\times\Omega)$. Show that for any $\epsilon>0$ and $C>0$, we have

$$P\left\{\left|\int_a^b f(t)\mathrm{d}B_t\right|>\epsilon\right\}\le\frac{C}{\epsilon^2}+P\left\{\int_a^b |f(t)|^2\mathrm{d}t>C\right\}.$$

32.6. Let f be a function in $L^2_{ad}([a,b]\times\Omega)$ such that $E[f(t)f(s)]$ is a continuous function of $(t,s)\in[a,b]\times[a,b]$. Show that there exists a sequence $\{f_n\}_{n\ge1}$ of step stochastic processes in $L^2_{ad}([a,b]\times\Omega)$ such that

$$\lim_{n\to\infty}\int_a^b E\left(|f(t)-f_n(t)|^2\right)\mathrm{d}t=0.$$

32.7. Let f be a bounded function in $L^2_{ad}([a,b]\times\Omega)$. Then $f(t)$ is adapted to some filtration $\{\mathscr{F}_t : a\le t\le b\}$. Let $\{g_n\}_{n\ge1}$ be a sequence of stochastic processes defined by

$$g_n(t,\omega)=\int_0^{n(t-a)} e^{-\tau} f\left(t-\frac{\tau}{n},\omega\right)\mathrm{d}\tau.$$

Show that

(a) For each $n\ge1$, $g_n(t)$ is adapted to the filtration $\{\mathscr{F}_t : a\le t\le b\}$.

(b) For each $n\ge1$,

$$\int_a^b E\left(|g_n(t)|^2\right)\mathrm{d}t<\infty.$$

(c) For each $n\ge1$, $E[g_n(t)g_n(s)]$ is a continuous function of $(t,s)\in[a,b]\times[a,b]$.

(d)

$$\lim_{n\to\infty}\int_a^b E\left(|f(t)-g_n(t)|^2\right)\mathrm{d}t = 0.$$

32.8. Let $\{\mathscr{F}_t : 0 \le t \le b\}$ be a filtration and $\{B_t : 0 \le t \le b\}$ be a Brownian motion adapted to the filtration. Let $f(t)$ be a continuous stochastic process adapted to the filtration. Let

$$M_t = \left(E\left[\sup_{0\le s\le t}|f(s)|^2\right]\right)^{\frac{1}{2}}, \quad t\in[0,b].$$

Suppose that $M_b < \infty$. Show that

(a) For every $t \in [0,b]$, we have

$$E\left[\sup_{0\le s\le t}\left|\int_0^s f(u)\mathrm{d}u\right|^2\right] \le t\int_a^t M_s^2\mathrm{d}s.$$

(b) For every $t \in [0,b]$, we have

$$E\left[\sup_{0\le s\le t}\left|\int_0^s f(u)\mathrm{d}B_u\right|^2\right] \le 4\int_0^t M_s^2\mathrm{d}s.$$

32.9. Let f and g be two stochastic processes in $L^2_{ad}([a,b]\times\Omega)$. Show that

$$E\left[\int_a^b f(t)\mathrm{d}B_t\int_a^b g(t)\mathrm{d}B_t\right] = \int_a^b E[f(t)g(t)]\mathrm{d}t.$$

32.10. Let $\{B_t\}_{t\ge0}$ be a Brownian motion and $[a,b]\subset\mathbf{R}$. Let $\Delta_n = \{t_{n,0}, t_{n,1},\ldots,t_{n,n}\}$ be a sequence of partitions of $[a,b]$. Show that

$$\sum_{i=1}^n (B_{t_{n,i}} - B_{t_{n,i-1}})^2 \xrightarrow{L^2(\Omega)} b-a \quad \text{as} \quad \|\Delta_n\|\to 0,$$

where $\|\Delta_n\| = \max_{1\le i\le n}(t_{n,i}-t_{n,i-1})$.

32.11. Let $\{B_t\}_{t\ge0}$ be a Brownian motion and $[a,b]\subset\mathbf{R}$. Let $\Delta_n = \{t_{n,0}, t_{n,1},\ldots,t_{n,n}\}$ be a sequence of partitions of $[a,b]$. Show that

$$\sum_{i=1}^n B_{t_{n,i}}(B_{t_{n,i}} - B_{t_{n,i-1}}) \xrightarrow{L^2(\Omega)} \frac{1}{2}\left(B_b^2 - B_a^2 + b - a\right) \quad \text{as} \quad \|\Delta_n\|\to 0,$$

and

$$\sum_{i=1}^n B_{t_{n,i-1}}(B_{t_{n,i}} - B_{t_{n,i-1}}) \xrightarrow{L^2(\Omega)} \frac{1}{2}\left(B_b^2 - B_a^2 - (b - a)\right) \quad \text{as} \quad \|\Delta_n\|\to 0,$$

where $\|\Delta_n\| = \max_{1\le i\le n}(t_{n,i} - t_{n,i-1})$.

32.12. Let $\{B_t\}_{t\ge 0}$ be a Brownian motion and $[a,b] \subset \mathbf{R}$. Show that

$$\int_a^b B_t \mathrm{d}B_t = \frac{1}{2}\left[B_b^2 - B_a^2 - (b-a)\right].$$

32.13. Let $\{B_t\}_{t\ge 0}$ be a Brownian motion and $[a,b] \subset \mathbf{R}$. Show that

$$\int_a^b B_t^2 \mathrm{d}B_t = \frac{1}{3}\left[B_b^3 - B_a^3\right] - \int_a^b B_t \mathrm{d}t.$$

32.14. Let $\{B_t\}_{t\ge 0}$ be a Brownian motion. Define

$$X_t = \int_0^t B_u^2 \mathrm{d}B_u, \quad t \ge 0.$$

Show that the X_t is a martingale.

32.15 (Exponential Inequality). Let $f \in L^2_{ad}([0,b]\times\Omega)$ such that

$$\int_0^b f(s)^2 \mathrm{d}s \le K \quad \text{a.s.}$$

for some constant K. Let

$$X_t = \int_0^t f(s)\mathrm{d}B_s, \quad t \in [0,b].$$

Show that

(a) For every $\alpha \in \mathbf{R}$, the process

$$Y_t = \exp\left(\alpha X_t - \frac{\alpha^2}{2}\int_0^t f(s)^2 \mathrm{d}s\right)$$

is a martingale.

(b) For every $\lambda > 0$,

$$P\left\{\sup_{0\le t\le b} |X_t| \ge \lambda\right\} \le 2\exp\left(-\frac{\lambda^2}{2K}\right).$$

32.16. Let $\{B_t : 0 \le t \le b\}$ be a Brownian motion adapted to some filtration $\{\mathscr{F}_t : 0 \le t \le b\}$. Let $f(t)$ be a process adapted to $\{\mathscr{F}_t : 0 \le t \le b\}$. Suppose that

$$|f(t)| \le k \quad a.s., \quad \forall t \in [0,b],$$

where k is some constant. For every $t \in [0,b]$, show that

$$E\left[\exp\left(\pm\int_0^t f(s)\mathrm{d}B_s\right)\right] \le e^{k^2 t}.$$

32.3 Hints

32.1. Use the tower property of conditional expectations (see Problem 14.10) and note that for $i < j$, $\xi_{i-1}\xi_{j-1}(B_{t_i} - B_{t_{i-1}})$ is $\mathscr{F}_{t_{j-1}}$-measurable.

32.2. Consider the partition produced by $t_0, t_1, \ldots, t_n, s_0, s_1, \ldots, s_m$ and use the result of Problem 32.1.

32.3. Suppose that f and g are given by

$$f(t,\omega) = \sum_{i=1}^{n} \xi_{i-1}(\omega) I_{[t_{i-1},t_i)}(t)$$

and

$$g(t,\omega) = \sum_{i=1}^{m} \eta_{j-1}(\omega) I_{[s_{j-1},s_j)}(t),$$

where $a = t_0 < t_1 < \cdots < t_n = b$, $a = s_0 < s_1 < \cdots < s_m$, ξ_{i-1} is $\mathscr{F}_{t_{i-1}}$-measurable for $i = 1, 2, \ldots, n$, and η_{j-1} is $\mathscr{F}_{s_{j-1}}$-measurable. Then consider the partition produced by the points $t_0, t_1, \ldots, t_n, s_0, s_1, \ldots, s_m$.

32.4. Note that the result is equivalent to

$$\left\{ \int_a^b |f(t)|^2 \mathrm{d}t \le C \right\} \subseteq \left\{ \int_a^b f(t)\mathrm{d}B_t = \int_a^b f_C(t)\mathrm{d}B_t \right\}.$$

32.5. Consider the stochastic process f_C defined in Problem 32.4 and note that

$$\begin{aligned} &\left\{ \left| \int_a^b f(t)\mathrm{d}B_t \right| > \epsilon \right\} \\ \subseteq\ &\left\{ \left| \int_a^b f_C(t)\mathrm{d}B_t \right| > \epsilon \right\} \cup \left\{ \int_a^b f(t)\mathrm{d}B_t \neq \int_a^b f_C(t)\mathrm{d}B_t \right\}. \end{aligned}$$

32.6. Let $\Delta_n = \{t_0, t_1, \ldots, t_n\}$ be a partition of $[a, b]$. Consider the step stochastic processes $f_n(t,\omega) = f(t_{i-1}, \omega)$ if $t \in (t_{i-1}, t_i]$.

32.7. Part (a) can be proved by the result of Problem 5.9. Part (b) follows from the fact that f is bounded. To prove (c), try to establish that $E[g_n(t)g_n(s)]$ is continuous if

$$\lim_{s\to s_0} E\left[(g_n(s) - g_n(s_0))^2\right] = 0, \quad s \in [a, b].$$

Part (d) can be proved by the dominated convergence theorem (Theorem 6.4).

32.8. Use Hölder's inequality (Theorem 8.1) to prove part (a). Use Theorem 32.2 and the result of Problem 24.5 to prove part (b).

32.9. Consider the expectation of $[I(f) + I(g)]^2$ and use Theorem 32.1.

32.10. Note that the $(B_{t_{n,i}} - B_{t_{n,i-1}})^2$ are independent and that

$$E(B_{t_{n,i}} - B_{t_{n,i-1}})^2 = t_{n,i} - t_{n,i-1}.$$

32.11. Use the result of Problem 32.10.

32.12. Follow the definition of the Itô integral and use the results of Problems 32.11 and 32.6.

32.13. Construct a sequence of step stochastic processes

$$f_n(t,\omega) = B_{t_{i-1}}(\omega)^2$$

and calculate the limit of $I(f_n)$.

32.14. Use the result of Problem 32.13.

32.15. To prove part (a), try to show that

$$E\left[\exp\left(\alpha(X_t - X_s) - \frac{\alpha^2}{2}\int_s^t f(u)^2 \mathrm{d}u\right)\middle| \mathscr{F}_s\right] = 1.$$

Use part (a) and the result of Problem 24.11 to prove part (b).

32.16. Use the result of Problem 32.15 and Hölder's inequality (Theorem 8.1).

32.4 Solutions

32.1. First, we show that $E[I(f)] = 0$. By definition, we have

$$I(f) = \sum_{i=1}^{n} \xi_{i-1}(B_{t_i} - B_{t_{i-1}}).$$

By Problem 14.10, we have

$$\begin{aligned} E[\xi_{i-1}(B_{t_i} - B_{t_{i-1}})] &= E\{E[\xi_{i-1}(B_{t_i} - B_{t_{i-1}})|\mathscr{F}_{t_{i-1}}]\} \\ &= E\{\xi_{i-1}E[B_{t_i} - B_{t_{i-1}}|\mathscr{F}_{t_{i-1}}]\} \\ &= E\{\xi_{i-1}E[B_{t_i} - B_{t_{i-1}}]\} \\ &= 0. \end{aligned}$$

Hence $E[I(f)] = 0$.

Second, let us prove

$$E\left(|I(f)|^2\right) = \int_a^b E\left(|f|^2\right) \mathrm{d}t. \tag{32.2}$$

By definition we have

$$|I(f)|^2 = \sum_{i=1}^{n} \xi_{i-1}^2 (B_{t_i} - B_{t_{i-1}})^2 + 2\sum_{i<j} \xi_{i-1}\xi_{j-1}(B_{t_i} - B_{t_{i-1}})(B_{t_j} - B_{t_{j-1}}).$$

Note that ξ_{i-1} is $\mathscr{F}_{i-1}$-measurable. We have

$$\begin{aligned} E[\xi_{i-1}^2(B_{t_i} - B_{t_{i-1}})^2] &= E\{E[\xi_{i-1}^2(B_{t_i} - B_{t_{i-1}})^2|\mathscr{F}_{t_{i-1}}]\} \\ &= E\{\xi_{i-1}^2 E[(B_{t_i} - B_{t_{i-1}})^2|\mathscr{F}_{t_{i-1}}]\} \\ &= E\{\xi_{i-1}^2 E[(B_{t_i} - B_{t_{i-1}})^2]\} \\ &= (t_i - t_{i-1})E\left(\xi_{i-1}^2\right). \end{aligned}$$

Since $\xi_{i-1}\xi_{j-1}(B_{t_i} - B_{t_{i-1}})$ is $\mathscr{F}_{t_{j-1}}$-measurable for $i < j$, we have

$$\begin{aligned} & E[\xi_{i-1}\xi_{j-1}(B_{t_i} - B_{t_{i-1}})(B_{t_j} - B_{t_{j-1}})] \\ = \;& E\{E[\xi_{i-1}\xi_{j-1}(B_{t_i} - B_{t_{i-1}})(B_{t_j} - B_{t_{j-1}})|\mathscr{F}_{t_{j-1}}]\} \\ = \;& E\{\xi_{i-1}\xi_{j-1}(B_{t_i} - B_{t_{i-1}})E[(B_{t_j} - B_{t_{j-1}})|\mathscr{F}_{t_{j-1}}]\} \\ = \;& E\{\xi_{i-1}\xi_{j-1}(B_{t_i} - B_{t_{i-1}})E[(B_{t_j} - B_{t_{j-1}})]\} \\ = \;& 0. \end{aligned}$$

Hence we have

$$E\left(|I(f)|^2\right) = \sum_{i=1}^{n}(t_i - t_{i-1})E\left(\xi_{i-1}^2\right). \tag{32.3}$$

On the other hand, we have

$$\begin{aligned} & \int_a^b E\left(|f|^2\right) \mathrm{d}t \\ = \;& \int_a^b E\left(\sum_{i=1}^{n} \xi_{i-1}^2 I_{[t_{i-1},t_i)}(t) + 2\sum_{i<j} \xi_{i-1}\xi_{j-1} I_{[t_{i-1},t_i)}(t) I_{[t_{j-1},t_j)}(t)\right) \mathrm{d}t \\ = \;& \sum_{i=1}^{n} E\left(\xi_{i-1}^2\right) \int_a^b I_{[t_{i-1},t_i)}(t)\mathrm{d}t \\ = \;& \sum_{i=1}^{n}(t_i - t_{i-1})E\left(\xi_{i-1}^2\right). \end{aligned} \tag{32.4}$$

Equation (32.2) follows from Equation (32.3) and Equation (32.4). This completes the proof.

32.2. Let $a = r_0 < r_1 < \cdots < r_l = b$ be the distinct points from points $t_0, t_1, \ldots, t_n, s_0, s_1, \ldots, s_m$. Then $t_i \in \{r_0, r_1, \ldots, r_l\}$ for $i = 0, 1, \ldots, n$ and $s_j \in \{r_0, r_1, \ldots, r_l\}$ for $j = 0, 1, \ldots, m$. Let

$$\theta_{k-1} = \xi_{i-1} - \eta_{j-1} \quad \text{if} \quad r_{k-1} \in [t_{i-1}, t_i) \cap [s_{j-1}, s_j)$$

for $k = 1, 2, \ldots, l$. Then we have

$$\begin{aligned} 0 &= \sum_{i=1}^{n} \xi_{i-1} I_{[t_{i-1}, t_i)}(t) - \sum_{j=1}^{m} \eta_{j-1} I_{[s_{j-1}, s_i)}(t) \\ &= \sum_{k=1}^{l} \theta_{k-1} I_{[r_{k-1}, r_k)}. \end{aligned}$$

By the definition of the stochastic integral of step functions, we have

$$I(0) = \sum_{k=1}^{l} \theta_{k-1}(B_{r_k} - B_{r_{k-1}}).$$

But

$$\sum_{k=1}^{l} \theta_{k-1}(B_{r_k} - B_{r_{k-1}}) = I_1(f) - I_2(f),$$

where

$$I_1(f) = \sum_{i=1}^{n} \xi_{i-1}(B_{t_i} - B_{t_{i-1}}), \quad I_2(f) = \sum_{j=1}^{m} \eta_{j-1}(B_{s_j} - B_{s_{j-1}}).$$

Therefore, by Problem 32.1, we have

$$E\left[(I_1(f) - I_2(f))^2\right] = E\left[(I(0))^2\right] = \int_a^b E(0)\mathrm{d}t = 0,$$

which gives $I_1(f) = I_2(f)$ a.s. This completes the proof.

32.3. Let f and g be given by

$$f(t, \omega) = \sum_{i=1}^{n} \xi_{i-1}(\omega) I_{[t_{i-1}, t_i)}(t)$$

and

$$g(t, \omega) = \sum_{i=1}^{m} \eta_{j-1}(\omega) I_{[s_{j-1}, s_j)}(t),$$

where $a = t_0 < t_1 < \cdots < t_n = b$, $a = s_0 < s_1 < \cdots < s_m$, ξ_{i-1} is $\mathscr{F}_{t_{i-1}}$-measurable for $i = 1, 2, \ldots, n$, and η_{j-1} is $\mathscr{F}_{s_{j-1}}$-measurable. Then by definition, we have

$$I(f) = \sum_{i=1}^{n} \xi_{i-1}(B_{t_i} - B_{t_{i-1}}), \quad I(g) = \sum_{j=1}^{m} \eta_{j-1}(B_{s_j} - B_{s_{j-1}}).$$

Let $a = r_0 < r_1 < \cdots < r_l = b$ be the distinct points from points $t_0, t_1, \ldots,$ $t_n, s_0, s_1, \ldots, s_m$, and let

$$\theta_{k-1} = \alpha\xi_{i-1} + \beta\eta_{j-1} \quad \text{if} \quad r_{k-1} \in [t_{i-1}, t_i) \cap [s_{j-1}, s_j)$$

for $k = 1, 2, \ldots, l$. Then we have

$$\alpha f + \beta g = \sum_{k=1}^{l} \theta_{k-1}(\omega) I_{[r_{k-1}, r_k)}(t),$$

which gives

$$I(\alpha f + \beta g) = \sum_{k=1}^{l} \theta_{k-1}(B_{r_k} - B_{r_{k-1}}).$$

But

$$\begin{aligned}\alpha I(f) + \beta I(g) &= \sum_{i=1}^{n} \xi_{i-1}(B_{t_i} - B_{t_{i-1}}) + \sum_{j=1}^{m} \eta_{j-1}(B_{s_j} - B_{s_{j-1}}) \\ &= \sum_{k=1}^{l} \theta_{k-1}(B_{r_k} - B_{r_{k-1}}).\end{aligned}$$

Therefore we have $I(\alpha f + \beta g) = \alpha I(f) + \beta I(g)$. This completes the proof.

32.4. To prove this problem, we only need to show that

$$\left\{ \int_a^b |f(t)|^2 \mathrm{d}t \le C \right\} \subseteq \left\{ \int_a^b f(t) \mathrm{d}B_t = \int_a^b f_C(t) \mathrm{d}B_t \right\}.$$

To do that, let

$$\omega \in \left\{ \int_a^b |f(t)|^2 \mathrm{d}t \le C \right\}.$$

Then we have $\int_a^t |f(u,\omega)|^2 \mathrm{d}u \le \int_a^b |f(u,\omega)|^2 \mathrm{d}u \le C$ for all $t \in [a,b]$. Hence we have $f_C(t,\omega) = f(t,\omega)$ for all $t \in [a,b]$. It follows that $\int_a^b f(t,\omega) \mathrm{d}B_t = \int_a^b f_C(t,\omega) \mathrm{d}B_t$. Therefore

$$\omega \in \left\{ \int_a^b f(t) \mathrm{d}B_t = \int_a^b f_C(t) \mathrm{d}B_t \right\}.$$

Since ω is arbitrary, the result holds. This completes the proof.

32.5. Let $f_C(t)$ be a stochastic process defined as

$$f_C(t,\omega) = \begin{cases} f(t,\omega) & \text{if } \int_a^t |f(u,\omega)|^2 \mathrm{d}u \le C; \\ 0 & \text{otherwise.} \end{cases}$$

Then we have

$$\begin{aligned} & \left\{\left|\int_a^b f(t)\mathrm{d}B_t\right| > \epsilon\right\} \\ \subseteq & \left\{\left|\int_a^b f_C(t)\mathrm{d}B_t\right| > \epsilon\right\} \cup \left\{\int_a^b f(t)\mathrm{d}B_t \neq \int_a^b f_C(t)\mathrm{d}B_t\right\}, \end{aligned}$$

which gives

$$\begin{aligned} & P\left\{\left|\int_a^b f(t)\mathrm{d}B_t\right| > \epsilon\right\} \\ \leq & P\left\{\left|\int_a^b f_C(t)\mathrm{d}B_t\right| > \epsilon\right\} + P\left\{\int_a^b f(t)\mathrm{d}B_t \neq \int_a^b f_C(t)\mathrm{d}B_t\right\}. \end{aligned}$$

By Problem 32.4, we have

$$P\left\{\int_a^b f(t)\mathrm{d}B_t \neq \int_a^b f_C(t)\mathrm{d}B_t\right\} \leq P\left\{\int_a^b |f(t)|^2 \mathrm{d}t > C\right\}.$$

Also by Chebyshev inequality (see Problem 6.16), we have

$$\begin{aligned} & P\left\{\left|\int_a^b f_C(t)\mathrm{d}B_t\right| > \epsilon\right\} \\ \leq & \frac{1}{\epsilon^2} E\left[\left|\int_a^b f_C(t)\mathrm{d}B_t\right|^2\right] \leq \frac{1}{\epsilon^2}\int_a^b E(|f_C(t)|^2)\mathrm{d}t \leq \frac{C}{\epsilon^2}. \end{aligned}$$

The result follows by combining the above inequalities.

32.6. Let $\Delta_n = \{t_0, t_1, \ldots, t_n\}$ be a partition of $[a, b]$. Let $f_n(t, \omega)$ be a step stochastic process defined by

$$f_n(t, \omega) = f(t_{i-1}, \omega), \quad t \in (t_{i-1}, t_i].$$

Since f is an adapted stochastic process to some filtration $\{\mathscr{F}_t : a \leq t \leq b\}$, it follows that f_n is also adapted to the filtration. By the assumption that $E[f(t)f(s)]$ is a continuous function on $[a, b] \times [a, b]$, we have

$$\lim_{n\to\infty} E\left(|f(t) - f_n(t)|^2\right) = 0.$$

Note that

$$E\left(|f(t) - f_n(t)|^2\right) \leq 2E\left(|f(t)|^2 + |f_n(t)|^2\right) \leq 4 \sup_{a\leq s\leq b} E\left(|f(s)|^2\right).$$

By Theorem 6.4, we have

$$\lim_{n\to\infty}\int_a^b E\left(|f(t)-f_n(t)|^2\right)\mathrm{d}t = \int_a^b \lim_{n\to\infty} E\left(|f(t)-f_n(t)|^2\right)\mathrm{d}t = 0.$$

This completes the proof.

32.7. By changing of variable, we can write g_n as

$$g_n(t,\omega) = \int_a^t ne^{-n(t-x)} f(x,\omega)\,\mathrm{d}x.$$

(a) Let $\Delta_m = \{t_0, t_1, \ldots, t_m\}$ be a sequence of partitions of $[a,t]$ such that $\|\Delta_m\| \to 0$ as $m\to\infty$, where $\|\Delta_m\| = \max_{1\le k\le m}(t_k - t_{k-1})$. Define

$$X_m(\omega) = \sum_{k=1}^m ne^{-n(t-t_{k-1})} f(t_{k-1},\omega), \quad \omega\in\Omega.$$

Then X_m are random variables and are $\mathscr{F}_t$-measurable. By the assumption that $f\in L^2_{ad}([a,b]\times\Omega)$, $f(t,\omega)$ is an integrable function of t for each ω. Hence $X_m \to g_n(t,\cdot)$ as $m\to\infty$. By Problem 5.9, $g_n(t,\cdot)$ is $\mathscr{F}_t$-measurable. Therefore, g_n is adapted to the $\mathscr{F}_t$.

(b) Since f is bounded, we have

$$|g_n(t)| \le \int_a^t ne^{-n(t-x)}|f(x,\omega)|\mathrm{d}x \le M\int_a^t ne^{-n(t-x)}\mathrm{d}x < \infty,$$

where $M = \sup_{(x,\omega)\in[a,b]\times\Omega}|f(x,\omega)|$. Hence we have

$$\int_a^b E\left(|g_n(t)|^2\right)\mathrm{d}t < \infty.$$

(c) Let $(t_0,s_0)\in[a,b]\times[a,b]$. We need to show that

$$\lim_{(t,s)\to(t_0,s_0)} E[g_n(t)g_n(s)] = E[g_n(t_0)g_n(s_0)].$$

By Jensen's inequality with $\phi(x) = x^2$ (see Theorem 15.2) and the inequality $(x+y)^2 \le 2(x^2+y^2)$, we have

$$\begin{aligned}
& \left(E[g_n(t)g_n(s)] - E[g_n(t_0)g_n(s_0)]\right)^2 \\
\le\ & E\left[(g_n(t)g_n(s) - g_n(t_0)g_n(s_0))^2\right] \\
=\ & E\left[\{g_n(t)(g_n(s)-g_n(s_0)) - g_n(s_0)(g_n(t)-g_n(t_0))\}^2\right] \\
\le\ & 2E\left[g_n(t)^2(g_n(s)-g_n(s_0))^2\right] + 2E\left[g_n(s_0)^2(g_n(t)-g_n(t_0))^2\right] \\
\le\ & 2M^2(E\left[(g_n(s)-g_n(s_0))^2\right] + E\left[(g_n(t)-g_n(t_0))^2\right]), \qquad (32.5)
\end{aligned}$$

where $M = \sup_{(x,\omega)\in[a,b]\times\Omega} |g_n(x,\omega)|$. By inequality (32.5), we only need to show that

$$\lim_{s\to s_0} E\left[(g_n(s) - g_n(s_0))^2\right] = 0, \quad s \in [a,b]. \tag{32.6}$$

Note that

$$\begin{aligned} & g_n(s) - g_n(s_0) \\ = & \int_{s_0}^{s} ne^{-n(s-x)} f(s)\mathrm{d}x + \left[e^{-ns} - e^{-ns_0}\right] \int_a^{s_0} ne^{nx} f(x)\mathrm{d}x. \end{aligned}$$

We have $g_n(s) \to g_n(s_0)$ as $s \to s_0$. Inequality (32.6) follows from the dominated convergence theorem 6.4. Therefore $E[g_n(t)g_n(s)]$ is a continuous function of (t,s).

(d) Suppose that $f(t) = 0$ for $t < a$. Then we have

$$f(t) - g_n(t) = \int_0^\infty e^{-\tau}\left[f(t) - f\left(t - \frac{\tau}{n}\right)\right] \mathrm{d}\tau.$$

By Schwarz's inequality (Theorem 8.3), we have

$$\begin{aligned} |f(t) - g_n(t)|^2 &= \left|\int_0^\infty e^{-\frac{\tau}{2}} e^{-\frac{\tau}{2}} \left[f(t) - f\left(t - \frac{\tau}{n}\right)\right] \mathrm{d}\tau\right|^2 \\ &\leq \int_0^\infty e^{-\tau}\left[f(t) - f\left(t - \frac{\tau}{n}\right)\right]^2 \mathrm{d}\tau \end{aligned}$$

Hence we have

$$\begin{aligned} & \int_a^b E\left[|f(t) - g_n(t)|^2\right] \mathrm{d}t \\ \leq & \int_a^b E\left[\int_0^\infty e^{-\tau}\left[f(t) - f\left(t - \frac{\tau}{n}\right)\right]^2 \mathrm{d}\tau\right] \mathrm{d}t \\ = & \int_0^\infty e^{-\tau} E\left(\int_a^b \left[f(t) - f\left(t - \frac{\tau}{n}\right)\right]^2 \mathrm{d}t\right) \mathrm{d}\tau. \end{aligned}$$

By the assumption that f is bounded, we have

$$\int_a^b \left[f(t) - f\left(t - \frac{\tau}{n}\right)\right]^2 \mathrm{d}t \to 0 \quad \text{a.s. as} \quad n \to \infty.$$

It follows that

$$\lim_{n\to\infty} \int_a^b E\left[|f(t) - g_n(t)|^2\right] \mathrm{d}t = 0.$$

32.8.

(a) By Theorem 8.1, for every $s \in [0, b]$, we have

$$\begin{aligned}\left|\int_0^s f(u)\mathrm{d}u\right|^2 &\leq \left(\int_0^s |f(u)|\mathrm{d}u\right)^2 \\ &\leq s\int_0^s |f(u)|^2\mathrm{d}u \\ &\leq t\int_0^t |f(u)|^2\mathrm{d}u \\ &\leq t\int_0^t \sup_{0\leq r\leq u} |f(r)|^2\mathrm{d}u.\end{aligned}$$

Since the above inequality holds for all $s \in [0, b]$, we have

$$\sup_{0\leq s\leq t}\left|\int_0^s f(u)\mathrm{d}u\right|^2 \leq t\int_0^t \sup_{0\leq r\leq u} |f(r)|^2\mathrm{d}u.$$

Hence the result follows by taking expectation in both sides of the above inequality and Tonelli's theorem (Theorem 10.1).

(b) For every $s \in [0, b]$, let

$$X_s = \int_0^s f(u)\mathrm{d}B_u.$$

Then by Theorem 32.2, $\{X_s : 0 \leq s \leq b\}$ is a martingale adapted to the filtration $\{\mathscr{F}_s : 0 \leq s \leq b\}$. By Problem 22.8, $\{|X_s| : 0 \leq s \leq b\}$ is a submartingale. Hence it follows from Problem 24.5 that

$$E\left[\sup_{0\leq s\leq t} |X_s|^2\right] \leq 4E\left[|X_t|^2\right], \quad t \in [0, b]. \tag{32.7}$$

By Theorem 32.1, we have

$$E\left[|X_t|^2\right] = \int_0^t E\left[|f(u)|^2\right]\mathrm{d}u \leq \int_0^t M_u^2\mathrm{d}u. \tag{32.8}$$

We get the result by combining Equations (32.7) and (32.8).

This finishes the proof.

32.9. On one hand, by Theorem 32.1, we have

$$\begin{aligned}&E\left[|I(f)+I(g)|^2\right] \\ =\ & E\left[|I(f+g)|^2\right] \\ =\ & \int_a^b E\left[|f(t)+g(t)|^2\right]\mathrm{d}t \\ =\ & \int_a^b E\left[|f(t)|^2\right]\mathrm{d}t + 2\int_a^b E\left[f(t)g(t)\right]\mathrm{d}t + \int_a^b E\left[|g(t)|^2\right]\mathrm{d}t.\end{aligned}$$

On the other hand, we have

$$\begin{aligned}
& E\left[|I(f)+I(g)|^2\right] \\
= \quad & E\left[|I(f)|^2 + 2I(f)I(g) + |I(g)|^2\right] \\
= \quad & \int_a^b E\left[|f(t)|^2\right] \mathrm{d}t + 2E[I(f)I(g)] + \int_a^b E\left[|g(t)|^2\right] \mathrm{d}t.
\end{aligned}$$

Combining the above equations gives

$$E[I(f)I(g)] = \int_a^b E\left[f(t)g(t)\right] \mathrm{d}t.$$

This completes the proof.

32.10. By the definition of L^2 convergence, we need to show that

$$\lim_{\|\Delta_n\|\to 0} E\left[\left(\sum_{i=1}^n (B_{t_{n,i}} - B_{t_{n,i-1}})^2 - (b-a)\right)^2\right] = 0.$$

To do that, we let $X_{n,i} = (B_{t_{n,i}} - B_{t_{n,i-1}})^2 - (t_{n,i} - t_{n,i-1})$ for $i = 1, 2, \ldots, n$ and $n \geq 1$. Then $X_{n,1}, X_{n,2}, \ldots, X_{n,n}$ are independent and

$$\begin{aligned}
E(X_{n,i}^2) \quad = \quad & E\left[\left((B_{t_{n,i}} - B_{t_{n,i-1}})^2 - (t_{n,i} - t_{n,i-1})\right)^2\right] \\
= \quad & E\left[(B_{t_{n,i}} - B_{t_{n,i-1}})^2\right] - 2E\left[(B_{t_{n,i}} - B_{t_{n,i-1}})^2 (t_{n,i} - t_{n,i-1})\right] \\
& + E\left[(t_{n,i} - t_{n,i-1})^2\right] \\
= \quad & 2(t_{n,i} - t_{n,i-1})^2.
\end{aligned}$$

Hence we have

$$\begin{aligned}
& E\left[\left(\sum_{i=1}^n (B_{t_{n,i}} - B_{t_{n,i-1}})^2 - (b-a)\right)^2\right] \\
= \quad & E\left[\left(\sum_{i=1}^n X_{n,i}\right)^2\right] \\
= \quad & \sum_{i=1}^n E(X_{n,i}^2) + 2\sum_{i<j} E(X_{n,i}X_{n,j}) \\
= \quad & \sum_{i=1}^n 2(t_{n,i} - t_{n,i-1})^2 \\
\leq \quad & 2\Delta_n \sum_{i=1}^n (t_{n,i} - t_{n,i-1}) \\
= \quad & 2\Delta_n (b-a).
\end{aligned}$$

Therefore the result is true. This completes the proof.

32.11. Let R_n and L_n be defined as

$$R_n = \sum_{i=1}^{n} B_{t_{n,i}}(B_{t_{n,i}} - B_{t_{n,i-1}}), \quad n \geq 1,$$

$$L_n = \sum_{i=1}^{n} B_{t_{n,i-1}}(B_{t_{n,i}} - B_{t_{n,i-1}}), \quad n \geq 1.$$

Then we have

$$R_n - L_n = \sum_{i=1}^{n} (B_{t_{n,i}} - B_{t_{n,i-1}})^2$$

and

$$R_n + L_n = \sum_{i=1}^{n} (B^2_{t_{n,i}} - B^2_{t_{n,i-1}}) = B_b^2 - B_a^2.$$

From the above two equations we can solve R_n and L_n as

$$R_n = \frac{1}{2}\left(B_b^2 - B_a^2 + \sum_{i=1}^{n} (B_{t_{n,i}} - B_{t_{n,i-1}})^2 \right),$$

$$L_n = \frac{1}{2}\left(B_b^2 - B_a^2 - \sum_{i=1}^{n} (B_{t_{n,i}} - B_{t_{n,i-1}})^2 \right).$$

The results follow from Problem 32.10. This completes the proof.

32.12. By Problem 28.2, $E[B_t B_s] = \min(s,t)$, which is a continuous function. Let $\Delta_n = \{t_{n,0}, t_{n,1}, \ldots, t_{n,n}\}$ be a sequence of partitions of $[a,b]$ such that $\|\Delta_n\| \to 0$ as $n \to \infty$. Let $\{f_n\}_{n \geq 1}$ be a sequence of step stochastic processes defined by

$$f_n(t,\omega) = \sum_{i=1}^{n} B_{t_{n,i-1}} I_{[t_{n,i-1}, t_{n,i})}(t).$$

By Problem 32.6, we have

$$\int_a^b B_t \mathrm{d}B_t = \lim_{n\to\infty} I(f_n), \quad \text{in} \quad L^2(\Omega).$$

But by definition, we have

$$I(f_n) = \sum_{i=1}^{n} B_{t_{n,i-1}}(B_{t_{n,i}} - B_{t_{n,i-1}}).$$

Thus it follows from Problem 32.11 that

$$\int_a^b B_t \mathrm{d}B_t = \frac{1}{2}\left[B_b^2 - B_a^2 - (b-a) \right].$$

This completes the proof.

32.13. Note that $E[B_t^2 B_s^2] = ts + 2\min(s,t)^2$, which is a continuous function of (t,s). Let $\Delta_n = \{t_{n,0}, t_{n,1}, \ldots, t_{n,n}\}$ be a sequence of partitions of $[a,b]$ such that $\|\Delta_n\| \to 0$ as $n \to \infty$. Let $\{f_n\}_{n\geq 1}$ be a sequence of step stochastic processes defined by

$$f_n(t,\omega) = \sum_{i=1}^n B_{t_{n,i-1}}^2 I_{[t_{n,i-1},t_{n,i})}(t).$$

By Problem 32.6, we have

$$\int_a^b B_t^2 \mathrm{d}B_t = \lim_{n\to\infty} I(f_n), \quad \text{in} \quad L^2(\Omega).$$

By definition, we have

$$\begin{aligned} I(f_n) &= \sum_{i=1}^n B_{t_{n,i-1}}^2 (B_{t_{n,i}} - B_{t_{n,i-1}}) \\ &= \frac{1}{3}\left(B_b^3 - B_a^3\right) - \frac{1}{3}\sum_{i=1}^n (B_{t_{n,i}} - B_{t_{n,i-1}})^3 \\ &\quad - \sum_{i=1}^n B_{t_{n,i-1}} (B_{t_{n,i}} - B_{t_{n,i-1}})^2. \end{aligned} \tag{32.9}$$

Since $E[|B_t - B_s|^6] = 15|t-s|^3$, we have

$$\begin{aligned} E\left[\left(\sum_{i=1}^n (B_{t_{n,i}} - B_{t_{n,i-1}})^3\right)^2\right] &= 15\sum_{i=1}^n (t_{n,i} - t_{n,i-1})^3 \\ &\leq 15\|\Delta_n\|^2 (b-a), \end{aligned}$$

which implies that

$$\sum_{i=1}^n (B_{t_{n,i}} - B_{t_{n,i-1}})^3 \xrightarrow{L^2(\Omega)} 0 \quad \text{as} \quad n \to \infty. \tag{32.10}$$

Let $X_i = (B_{t_{n,i}} - B_{t_{n,i-1}})^2 - (t_{n,i} - t_{n,i-1})$ for $i = 1, 2, \ldots, n$. Since $E[(B_t - B_s)^2] = |t-s|$, $E(X_i) = 0$ for $i = 1, 2, \ldots, n$. By the tower property of conditional expectations, for $i < j$, we have

$$\begin{aligned} E\left[B_{t_{n,i-1}} X_i B_{t_{n,j-1}} X_j\right] &= E\{E\left[B_{t_{n,i-1}} X_i B_{t_{n,j-1}} X_j | \mathscr{F}_{t_{n,j-1}}\right]\} \\ &= E\{B_{t_{n,i-1}} X_i B_{t_{n,j-1}} E\left[X_j | \mathscr{F}_{t_{n,j-1}}\right]\} \\ &= E\{B_{t_{n,i-1}} X_i B_{t_{n,j-1}} E\left[X_j\right]\} \\ &= 0, \end{aligned}$$

since $B_{t_{n,i-1}}$, X_i, and $B_{t_{n,j-1}}$ are $\mathscr{F}_{t_{n,j-1}}$-measurable, and X_j is independent of $\mathscr{F}_{t_{n,j-1}}$.

Since $B_{t_{n,i-1}}$ and X_i are independent, we have

$$E\left[B^2_{t_{n,i-1}} X_i^2\right] = E\left[B^2_{t_{n,i-1}}\right] E\left[X_i^2\right] = 2t_{n,i-1}(t_{n,i} - t_{n,i-1})^2.$$

Therefore

$$\begin{aligned} & E\left[\left(\sum_{i=1}^n B_{t_{n,i-1}}(B_{t_{n,i}} - B_{t_{n,i-1}})^2 - \sum_{i=1}^n B_{t_{n,i}}(t_{n,i} - t_{n,i-1})\right)^2\right] \\ = & \sum_{i=1}^n 2t_{n,i-1}(t_{n,i} - t_{n,i-1})^2 \\ \leq & 2b\Delta_n(b-a), \end{aligned}$$

which implies that

$$\sum_{i=1}^n B_{t_{n,i-1}}(B_{t_{n,i}} - B_{t_{n,i-1}})^2 \xrightarrow{L^2(\Omega)} \int_a^b B_t \mathrm{d}t \quad \text{as} \quad n \to \infty. \tag{32.11}$$

The result follows from Equations (32.9), (32.10), and (32.11). This completes the proof.

32.14. Let $s \leq t$. Then by Problem 32.13, we have

$$\begin{aligned} E[X_t|\mathscr{F}_s] &= E\left[\frac{1}{3}B_t^3 - \int_0^t B_u \mathrm{d}u \middle| \mathscr{F}_s\right] \\ &= \frac{1}{3}E\left[B_t^3|\mathscr{F}_s\right] - \int_0^s B_u \mathrm{d}u - (t-s)B_s. \end{aligned}$$

Note that

$$\begin{aligned} E\left[B_t^3|\mathscr{F}_s\right] &= E\left[(B_t - B_s)^3 + 3(B_t - B_s)^2 B_s + 3(B_t - B_s)B_s^2 + B_s^3|\mathscr{F}_s\right] \\ &= 3(t-s)B_s + B_s^3. \end{aligned}$$

Combining the above two equations gives

$$E[X_t|\mathscr{F}_s] = \frac{1}{3}B_s^3 - \int_0^s B_u \mathrm{d}u = X_s,$$

which shows that X_t is a martingale. This completes the proof.

32.15.

(a) Since $f \in L^2_{ad}([0,b] \times \Omega)$, it follows from Theorem 32.2 that $\{X_t : 0 \le t \le b\}$ is a martingale. Now for $\alpha > 0$, $0 \le s \le t$, we have

$$\begin{aligned} E[Y_t|\mathscr{F}_s] &= E\left[\exp\left(\alpha X_t - \frac{\alpha^2}{2}\int_0^t f(u)^2 \mathrm{d}u\right)\middle| \mathscr{F}_s\right] \\ &= \exp\left(\alpha X_s - \frac{\alpha^2}{2}\int_0^s f(u)^2 \mathrm{d}u\right) \\ &\quad \times E\left[\exp\left(\alpha(X_t - X_s) - \frac{\alpha^2}{2}\int_s^t f(u)^2 \mathrm{d}u\right)\middle| \mathscr{F}_s\right] \\ &= Y_s \cdot E\left[\exp\left(\alpha(X_t - X_s) - \frac{\alpha^2}{2}\int_s^t f(u)^2 \mathrm{d}u\right)\middle| \mathscr{F}_s\right]. \end{aligned}$$

Note that $\exp\left(\alpha(X_t - X_s) - \frac{\alpha^2}{2}\int_s^t f(u)^2 \mathrm{d}u\right)$ is independent of $\mathscr{F}_s$ and $X_t - X_s \sim N(0, \sigma^2)$, where

$$\sigma^2 = \int_s^t f(u)^2 \mathrm{d}u.$$

Hence we have

$$\begin{aligned} & E\left[\exp\left(\alpha(X_t - X_s) - \frac{\alpha^2}{2}\int_s^t f(u)^2 \mathrm{d}u\right)\middle| \mathscr{F}_s\right] \\ =\ & E\left[\exp\left(\alpha(X_t - X_s) - \frac{\alpha^2\sigma^2}{2}\right)\right] \\ =\ & \frac{1}{\sqrt{2\pi}\sigma}\int_{-\infty}^{\infty} \exp\left(\alpha y - \frac{\alpha^2\sigma^2}{2}\right) \cdot \exp\left(-\frac{y^2}{2\sigma^2}\right) dy \\ =\ & \frac{1}{\sqrt{2\pi}\sigma}\int_{-\infty}^{\infty} \exp\left(-\frac{(y - \alpha\sigma^2)^2}{2\sigma^2}\right) dy \\ =\ & 1. \end{aligned}$$

Combining the above two equations shows that $\{Y_t : 0 \le t \le b\}$ is a martingale.

(b) Let $\{Y_t : 0 \le t \le b\}$ be the process given in the first item of this problem. Then we have

$$\left\{\sup_{0 \le t \le b} X_t \ge \lambda\right\} \subseteq \left\{\sup_{0 \le t \le b} Y_t \ge \exp\left(\alpha\lambda - \frac{\alpha^2}{2}K\right)\right\}. \tag{32.12}$$

Note that $Y_0 = 1$ and $Y_b^- = 0$. By Problem 24.11, we have

$$P\left\{\sup_{0 \le t \le b} Y_t \ge \exp\left(\alpha\lambda - \frac{\alpha^2}{2}K\right)\right\} \le \exp\left(-\alpha\lambda + \frac{\alpha^2}{2}K\right). \tag{32.13}$$

If we let $\alpha = \frac{\lambda}{K}$, then Equations (32.12) and (32.13) give

$$P\left\{\sup_{0 \le t \le b} X_t \ge \lambda\right\} \le \exp\left(-\frac{\lambda^2}{2K}\right).$$

Similarly, we have

$$P\left\{\sup_{0\le t\le b}(-X_t)\ge\lambda\right\}\le\exp\left(-\frac{\lambda^2}{2K}\right).$$

Hence the result follows from the above two inequalities.

This completes the proof.

32.16. By Problem 32.15, the process

$$Y_t=\exp\left(2\int_0^t f(s)\mathrm{d}B_s-2\int_0^t f^2(s)\mathrm{d}s\right)$$

is a martingale. Hence we have

$$E\left[\exp\left(2\int_0^t f(s)\mathrm{d}B_s-2\int_0^t f^2(s)\mathrm{d}s\right)\right]=E[Y_0]=1.$$

By Theorem 8.1, we have

$$\begin{aligned}
& E\left[\exp\left(\int_0^t f(s)\mathrm{d}B_s\right)\right]\\
= & E\left[\exp\left(\int_0^t f(s)\mathrm{d}B_s-\int_0^t f^2(s)\mathrm{d}s\right)\cdot\exp\left(\int_0^t f^2(s)\mathrm{d}s\right)\right]\\
\le & \left(E\left[\exp\left(2\int_0^t f(s)\mathrm{d}B_s-2\int_0^t f^2(s)\mathrm{d}s\right)\right]\right)^{\frac{1}{2}}\\
& \cdot\left(E\left[\exp\left(2\int_0^t f^2(s)\mathrm{d}s\right)\right]\right)^{\frac{1}{2}}\\
= & \left(E\left[\exp\left(2\int_0^t f^2(s)\mathrm{d}s\right)\right]\right)^{\frac{1}{2}},
\end{aligned}$$

which gives

$$E\left[\exp\left(\int_0^t f(s)\mathrm{d}B_s\right)\right]\le\left(E\left[\exp\left(2k^2t\right)\right]\right)^{\frac{1}{2}}=e^{k^2t}.$$

Similarly, we can show that

$$E\left[\exp\left(-\int_0^t f(s)\mathrm{d}B_s\right)\right]\le e^{k^2t}.$$

This completes the proof.

32.5 Bibliographic Notes

The stochastic integral was first introduced by K. Itô in his 1944 paper (Itô, 1944). In this chapter, we defined the stochastic integral for stochastic processes $f(t)$ that are adapted to a filtration $\{\mathscr{F}_t : a \le t \le b\}$ and satisfy the condition

$$\int_a^b E(|f(t)|^2)\mathrm{d}t < \infty.$$

The stochastic integral

$$\int_a^b f(t)\mathrm{d}B_t$$

is defined in such a way that the stochastic process

$$M_t = \int_a^t f(s)\mathrm{d}B_s$$

is a martingale.

The proofs of Theorems 32.2 and 32.3 can be found in Kuo (2006, Chapter 4). The two theorems can be proved by first considering step stochastic processes and then general stochastic processes.

There are several texts on stochastic calculus. Karatzas and Shreve (1991) introduces stochastic calculus for Brownian motions. Baxter and Rennie (1996) gives an intuitive introduction to stochastic calculus. Durrett (1996) gives a concise introduction to stochastic calculus. Mikosch (1999) is a short book on stochastic calculus at an elementary level. Steele (2003) introduces stochastic calculus with a focus on finance applications. Klebaner (2005) introduced stochastic calculate with some applications in finance. Sondermann (2007) is a short course on stochastic calculus for finance. Roberts (2009) and Kopp (2011) are short books on Itó calculus.

Malliavin and Thalmaier (2006) is a book on Malliavin calculus (i.e., stochastic calculus of variations), which provides mechanics to calculate derivatives of random variables.

CHAPTER 33

EXTENSION OF THE ITÔ INTEGRAL

In this chapter, we define the Itô integral for stochastic processes in a larger space.

33.1 Basic Concepts and Facts

Definition 33.1 (Stochastic Process Space $\mathcal{L}_{ad}(\Omega, L^2[a,b])$). Let $\{\mathscr{F}_t : a \leq t \leq b\}$ be a filtration under consideration. The space $\mathcal{L}_{ad}(\Omega, L^2[a,b])$ is defined to be the space of all stochastic processes $f(t,\omega)$, $t \in [a,b]$, $\omega \in \Omega$, satisfying the following conditions:

(a) $f(t)$ is adapted to the $\mathscr{F}_t$.

(b) $\int_a^b |f(t)|^2 \mathrm{d}t < \infty$ a.s.

Definition 33.2 (Extension of the Itô Integral). Let $\{\mathscr{F}_t : a \leq t \leq b\}$ be a filtration. Let $\{B_t : a \leq t \leq b\}$ be a Brownian motion satisfying the following conditions:

(a) For each t, B_t is $\mathscr{F}_t$-measurable.

(b) For any $s \leq t$, the random variable $B_t - B_s$ is independent of $\mathscr{F}_s$.

Let $f \in \mathcal{L}_{ad}(\Omega, L^2[a,b])$. Then there exists a sequence $\{f_n(t)\}_{n\geq 1}$ of step stochastic processes in $L^2_{ad}([a,b]\times\Omega)$ such that

$$\int_a^b |f_n(t) - f(t)|^2 \mathrm{d}t \xrightarrow{P} 0 \quad \text{as} \quad n \to \infty.$$

The Itô integral of f is defined as

$$\int_a^b f(t)\mathrm{d}B_t = \lim_{n\to\infty} I(f_n) \quad \text{in probability},$$

where $I(\cdot)$ is as defined in Equation (32.1).

Definition 33.3 (Local Martingale). Let $\{X_t : a \leq t \leq b\}$ be a stochastic process adapted to the filtration $\{\mathscr{F}_t : a \leq t \leq b\}$. The stochastic process X_t is said to be a local martingale with respect to the $\mathscr{F}_t$ if there exists a nondecreasing sequence $\{\tau_n : n \geq 1\}$ of stopping times of $\{\mathscr{F}_t : a \leq t \leq b\}$ such that

(a) $\tau_n \to b$ a.s. as $n \to \infty$.

(b) For each $n \geq 1$, $X_{t\wedge\tau_n}$ is a martingale with respect to the filtration $\{\mathscr{F}_t : a \leq t \leq b\}$.

The stopping times $\{\tau_n : n \geq 1\}$ are said to reduce $\{X_t : a \leq t \leq b\}$. Local submartingales and local boundedness are defined in the same way.

Theorem 33.1. *Let f be a continuous stochastic process adapted to the filtration $\{\mathscr{F}_t : a \leq t \leq b\}$. Then*

(a) $f \in \mathcal{L}_{ad}(\Omega, L^2[a,b])$*;*

(b)

$$\int_a^b f(t)\mathrm{d}B_t = \lim_{\|\Delta_n\|\to 0} \sum_{i=1}^n f(t_{i-1})(B_{t_i} - B_{t_{i-1}}), \quad \textit{in probability},$$

where $\Delta_n = \{t_0, t_1, \ldots, t_n\}$ *is a partition of the finite interval* $[a,b]$ *and* $\|\Delta_n\| = \max_{1\leq i\leq n}(t_i - t_{i-1})$.

Theorem 33.2. *Let* $f \in \mathcal{L}_{ad}(\Omega, L^2[a,b])$. *Let* $\{X_t : a \leq t \leq b\}$ *be a stochastic process defined as*

$$X_t = \int_a^t f(u)\mathrm{d}B_u, \quad t \in [a,b].$$

Then $\{X_t : a \leq t \leq b\}$ *is a local martingale with respect to the underlying filtration* $\{\mathscr{F}_t : a \leq t \leq b\}$.

Theorem 33.3 (Continuous Realization). *Let* $f \in \mathcal{L}_{ad}(\Omega, L^2[a,b])$. *Then the process*

$$X_t = \int_a^t f(u)\mathrm{d}B_u, \quad t \in [a,b]$$

has a continuous realization; that is, there exists a continuous stochastic process $\{Y_t : a \leq t \leq b\}$ *such that for each* $t \in [a,b]$, $Y_t = X_t$ *a.s.*

33.2 Problems

33.1. Let $L^2_{ad}([a,b]\times\Omega)$ be the space of stochastic processes defined in Definition 32.1. Let $\mathcal{L}_{ad}(\Omega, L^2[a,b])$ be the space of stochastic processes defined in Definition 33.1. Show that

$$L^2_{ad}([a,b]\times\Omega) \subseteq \mathcal{L}_{ad}(\Omega, L^2[a,b]).$$

33.2. Let $f \in \mathcal{L}_{ad}(\Omega, L^2[a,b])$. Show that there exists a sequence $\{f_n\}_{n\geq 1}$ of functions in $L^2_{ad}([a,b]\times\Omega)$ such that

(a)

$$\int_a^b |f_n(t) - f(t)|^2 \mathrm{d}t \xrightarrow{\text{a.s.}} 0,$$

where $\xrightarrow{\text{a.s.}}$ denotes convergence almost surely.

(b)

$$\int_a^b |f_n(t) - f(t)|^2 \mathrm{d}t \xrightarrow{P} 0,$$

where $\xrightarrow{P}$ denotes convergence in probability.

33.3. Let $f \in \mathcal{L}_{ad}(\Omega, L^2[a,b])$. For each fixed $n \geq 1$, let τ_n be defined as

$$\tau_n(\omega) = \begin{cases} \inf A_n(\omega) & \text{if } A_n(\omega) \neq \emptyset; \\ b & \text{if } A_n(\omega) = \emptyset, \end{cases} \tag{33.1}$$

where

$$A_n(\omega) = \left\{ t \in [a,b] : \int_a^t |f(u,\omega)|^2 \mathrm{d}u > n \right\}.$$

Show that the τ_n is a stopping time.

33.4. Let τ be a stopping time with respect to a filtration $\{\mathscr{F}_t : t \geq 0\}$. Let $\{X_t : t \geq 0\}$ be a process adapted to $\{\mathscr{F}_t : t \geq 0\}$. Show that $\{X_{t\wedge\tau} : t \geq 0\}$ is a martingale with respect to the filtration $\{\mathscr{F}_{t\wedge\tau} : t \geq 0\}$ if and only if it is a martingale with respect to the filtration $\{\mathscr{F}_t : t \geq 0\}$.

33.5. Let $f \in \mathcal{L}_{ad}(\Omega, L^2[a,b])$. Let $\{X_t : a \leq t \leq b\}$ be a stochastic process defined as

$$X_t = \int_a^t f(u)\mathrm{d}B_u, \quad t \in [a,b].$$

Let τ_n be as defined in Equation (33.1). Show that for each n, the stochastic process $X_{t\wedge\tau_n}$ is a martingale, where $t\wedge\tau_n = \min(t, \tau_n)$.

33.6. Let f and g stochastic processes in $L^2_{ad}([a,b]\times\Omega)$. Let A be an event defined as

$$A = \{\omega \in \Omega : f(t,\omega) = g(t,\omega) \text{ for all } t \in [a,b]\}.$$

Show that

$$\int_a^b f(t)\mathrm{d}B_t = \int_a^b g(t)\mathrm{d}B_t$$

almost surely on A:

$$P\left\{\omega \in A : \int_a^b f(t,\omega)\mathrm{d}B_t(\omega) \neq \int_a^b g(t,\omega)\mathrm{d}B_t(\omega)\right\} = 0.$$

33.7. Let $f \in \mathcal{L}_{ad}(\Omega, L^2[a,b])$. For every $t \in [a,b]$, let M_t be defined as

$$M_t = \int_a^t f(s)\mathrm{d}B_s.$$

Suppose that $\sup_{a\leq t\leq b} M_t$ is square-integrable. Show that $f \in L^2_{ad}([a,b]\times\Omega)$.

33.3 Hints

33.1. Use the Fibini theorem (Theorem 10.2).

33.2. Part (a) can be proved by considering the sequence

$$f_n(t,\omega) = \begin{cases} f(t,\omega) & \text{if } \int_a^t |f(u,\omega)|^2\mathrm{d}u \leq n; \\ 0 & \text{otherwise.} \end{cases}$$

Part (b) is implied by part (a).

33.3. Use the result of Problem 23.10 and note that

$$\{\tau_n < t\} = \bigcup_{a<r<t, r\in\mathbf{Q}} \left\{\int_a^r |f(u)|^2\mathrm{d}u > n\right\}.$$

33.4. The "if" part can be proved by noting that $\mathscr{F}_{s\wedge\tau} \subseteq \mathscr{F}_s$. To prove the "only if" part, use the definition of conditional expectations (Definition 14.1) and consider

$$C = (C \cap \{\tau \leq s\}) \cup (C \cap \{\tau > s\})$$

to prove, for every $C \in \mathscr{F}_s$, that

$$\int_C X_{s\wedge\tau}\mathrm{d}P = \int_C X_{t\wedge\tau}\mathrm{d}P.$$

33.5. Consider the stochastic process

$$f_n(t,\omega) = \begin{cases} f(t,\omega) & \text{if } \int_a^t |f(u,\omega)|^2\mathrm{d}u \leq n; \\ 0 & \text{otherwise,} \end{cases}$$

and use the martingale property of the Itô integral (Theorem 32.2).

33.6. Consider the random variable $X = \int_a^\tau [f(u) - g(u)]\mathrm{d}B_u$, where τ is a stopping time defined as

$$\tau(\omega) = \begin{cases} \inf\{t : f(t,\omega) \neq g(t,\omega)\}, & \text{if } \{t : f(t,\omega) \neq g(t,\omega)\} \neq \emptyset; \\ b, & \text{if } \{t : f(t,\omega) \neq g(t,\omega)\} = \emptyset. \end{cases}$$

33.7. Use the results of Problems 33.3 and 33.5 to show that

$$E\left[M^2_{t\wedge\tau_n}\right] = E\left[\int_0^{t\wedge\tau_n} f(s)^2 \mathrm{d}s\right],$$

where τ_n is as given in Problem 33.3.

33.4 Solutions

33.1. Suppose that $f \in L^2_{ad}([a,b] \times \Omega)$. By Theorem 10.2, we have

$$E\left[\int_a^b |f(t)|^2 \mathrm{d}t\right] = \int_a^b E(|f(t)|^2)\mathrm{d}t < \infty. \tag{33.2}$$

Let A_n be sets defined as

$$A_n = \left\{\omega : \int_a^b |f(t,\omega)|^2 \mathrm{d}t < n\right\}, \quad n \geq 1.$$

Then

$$\left\{\int_a^b |f(t)|^2 \mathrm{d}t < \infty\right\} = \bigcup_{n=1}^{\infty} A_n.$$

We claim that

$$P\left(\bigcap_{n=1}^{\infty} A_n^c\right) = 0. \tag{33.3}$$

In fact, if this is not the case, then there exists an $\epsilon > 0$ such that

$$P\left(\bigcap_{n=1}^{\infty} A_n^c\right) > \epsilon,$$

which implies that $P(A_n^c) > \epsilon$ for all $n \geq 1$. Then

$$E\left[\int_a^b |f(t)|^2 \mathrm{d}t\right] \geq \int_{A_n^c} \int_a^b |f(t)|^2 \mathrm{d}t\mathrm{d}P \geq nP(A_n^c) > n\epsilon$$

for all $n \geq 1$. This contradicts Inequality (33.2). Hence Equation (33.3) is true. Therefore $f \in \mathcal{L}_{ad}(\Omega, L^2[a,b])$. This completes the proof.

33.2.

(a) Let $\{f_n\}_{n\geq 1}$ be a sequence of functions defined as

$$f_n(t,\omega) = \begin{cases} f(t,\omega) & \text{if } \int_a^t |f(u,\omega)|^2 \mathrm{d}u \leq n; \\ 0 & \text{otherwise.} \end{cases}, \quad n \geq 1.$$

By the assumption that $f \in \mathcal{L}_{ad}(\Omega, L^2[a,b])$, we have

$$\int_a^b |f(t)|^2 \mathrm{d}t < \infty, \quad \text{a.s.}$$

Then we have

$$\int_a^b |f_n(t)|^2 \mathrm{d}t = \int_a^{\tau_n} |f(t)|^2 \mathrm{d}t \leq \int_a^b |f(t)|^2 \mathrm{d}t < \infty, \quad \text{a.s.},$$

where $\tau_n = \sup\{t : \int_a^t |f(u)|^2 \mathrm{d}u \leq n\}$. Hence we have

$$\int_a^b E\left(|f_n(t)|^2\right) \mathrm{d}t = E\left(\int_a^b |f_n(t)|^2 \mathrm{d}t\right) \leq n,$$

which implies that $f_n \in L^2_{ad}([a,b] \times \Omega)$.

Now we show that

$$\int_a^b |f_n(t) - f(t)|^2 \mathrm{d}t \xrightarrow{\text{a.s.}} 0.$$

To do that, let

$$A = \left\{\omega : \int_a^b |f(t,\omega)|^2 < \infty\right\}.$$

Then $P(A) = 1$ by assumption. Note that

$$A = \bigcup_{n=1}^{\infty} \left\{\omega : \int_a^b |f(t,\omega)|^2 \leq n\right\}.$$

Let $\omega \in A$. Then

$$\omega \in \left\{\omega : \int_a^b |f(t,\omega)|^2 \leq N_\omega\right\}$$

for some $N_\omega \geq 1$. Hence for any $n \geq N_\omega$, we have $f_n(t,\omega) = f(t,\omega)$ for all $t \in [a,b]$. Therefore

$$\int_a^b |f_n(t) - f(t)|^2 \mathrm{d}t = \int_a^b 0 \mathrm{d}t = 0 \quad \text{for all} \quad n \geq N_\omega,$$

which implies that

$$\int_a^b |f_n(t,\omega) - f(t,\omega)|^2 \mathrm{d}t \to 0.$$

Since $P(A) = 1$, the almost surely convergence is true.

(b) By Problem 9.3, $\int_a^b |f_n(t) - f(t)|^2 \mathrm{d}t$ converges to 0 almost uniformly. By Problem 9.2, the sequence also converges in probability.

This completes the proof.

33.3. Since $\{\tau_n < a\} = \emptyset$, we have $\{\tau_n < a\} \in \mathscr{F}_a$. Now let $t \in (a, b]$. We claim that

$$\{\tau_n < t\} = \bigcup_{a<r<t, r\in\mathbf{Q}} \left\{ \int_a^r |f(u)|^2 \mathrm{d}u > n \right\}, \tag{33.4}$$

where $\mathbf{Q}$ denotes the set of all rational numbers. Let $\omega \in \{\tau_n < t\}$. Since

$$\{\tau_n < t\} = \bigcup_{m=1}^{\infty} \left\{ \tau_n \le t - \frac{1}{m} \right\},$$

we have

$$\omega \in \left\{ \tau_n \le t - \frac{1}{m} \right\}$$

for some $m > 1$. Hencc wc have

$$\int_a^{t-\frac{1}{m}} |f(u,\omega)|^2 \mathrm{d}u > n.$$

Thus there exists a rational number $r \in (t - \frac{1}{m}, t)$ such that

$$\int_a^r |f(u,\omega)|^2 \mathrm{d}u \ge \int_a^{t-\frac{1}{m}} |f(u,\omega)|^2 \mathrm{d}u > n,$$

which implies that

$$\omega \in \left\{ \int_a^r |f(u)|^2 \mathrm{d}u > n \right\}.$$

Hence we have

$$\{\tau_n < t\} \subseteq \bigcup_{a<r<t, r\in\mathbf{Q}} \left\{ \int_a^r |f(u)|^2 \mathrm{d}u > n \right\}.$$

On the other hand, it is obvious that

$$\bigcup_{a<r<t, r\in\mathbf{Q}} \left\{ \int_a^r |f(u)|^2 \mathrm{d}u > n \right\} \subseteq \{\tau_n < t\}.$$

Therefore Equation (33.4) is true.

Since Equation (33.4) holds, we have $\{\tau_n < t\} \in \mathscr{F}_t$. Hence by Problem 23.10, the τ_n is a stopping time.

33.4. Let us first prove the "if" part. Suppose that $\{X_{t\wedge\tau} : t \geq 0\}$ is a martingale with respect to the filtration $\{\mathscr{F}_t : t \geq 0\}$. Then for all $0 \leq s < t$, we have

$$E[X_{t\wedge\tau}|\mathscr{F}_s] = X_{s\wedge\tau}.$$

Since $\mathscr{F}_{s\wedge\tau} \subseteq \mathscr{F}_s$ (see Problem 23.4) and $X_{s\wedge\tau}$ is $\mathscr{F}_{s\wedge\tau}$-measurable (see Problem 23.11), it follows from Problems 14.10 and 14.4 that

$$E[X_{t\wedge\tau}|\mathscr{F}_{s\wedge\tau}] = E(E[X_{t\wedge\tau}|\mathscr{F}_s]|\mathscr{F}_{s\wedge\tau}) = E[X_{s\wedge\tau}|\mathscr{F}_{s\wedge\tau}] = X_{s\wedge\tau}.$$

Hence $\{X_{t\wedge\tau} : t \geq 0\}$ is a martingale with respect to the filtration $\{\mathscr{F}_{t\wedge\tau} : t \geq 0\}$.

Now let us prove the "only if" part. Suppose that $\{X_{t\wedge\tau} : t \geq 0\}$ is a martingale with respect to the filtration $\{\mathscr{F}_{t\wedge\tau} : t \geq 0\}$. Let $0 \leq s < t$. Then, for all $A \in \mathscr{F}_{s\wedge\tau}$ (see Problem 22.3), we have

$$\int_A X_{s\wedge\tau}\mathrm{d}P = \int_A X_{t\wedge\tau}\mathrm{d}P.$$

Let $C \in \mathscr{F}_s$. Then

$$C = (C \cap \{\tau \leq s\}) \cup (C \cap \{\tau > s\}).$$

Since $C \cap \{\tau > s\} \cap \{\tau \leq r\} = C \cap \{s < \tau \leq r\} \in \mathscr{F}_r$ for all $r \geq 0$, it follows from Definition 23.3 that $C \cap \{\tau > s\} \in \mathscr{F}_{s\wedge\tau}$. Hence we have

$$\int_{C\cap\{\tau>s\}} X_{s\wedge\tau}\mathrm{d}P = \int_{C\cap\{\tau>s\}} X_{t\wedge\tau}\mathrm{d}P. \tag{33.5}$$

Also we have

$$\int_{C\cap\{\tau\leq s\}} X_{s\wedge\tau}\mathrm{d}P = \int_{C\cap\{\tau\leq s\}} X_{\tau}\mathrm{d}P = \int_{C\cap\{\tau\leq s\}} X_{t\wedge\tau}\mathrm{d}P. \tag{33.6}$$

Combining Equations (33.5) and (33.6) gives

$$\int_C X_{s\wedge\tau}\mathrm{d}P = \int_C X_{t\wedge\tau}\mathrm{d}P.$$

Therefore, we have

$$E[X_{t\wedge\tau}|\mathscr{F}_s] = X_{s\wedge\tau}.$$

This completes the proof.

33.5. Let n be fixed and let f_n be a stochastic process defined as

$$f_n(t,\omega) = \begin{cases} f(t,\omega) & \text{if } \int_a^t |f(u,\omega)|^2\mathrm{d}u \leq n; \\ 0 & \text{otherwise.} \end{cases}$$

Then we have

$$X_{t\wedge\tau_n} = \int_a^{t\wedge\tau_n} f(u)\mathrm{d}B_u = \int_a^t f_n(u)\mathrm{d}B_u, \quad t \in [a,b].$$

Note that $\int_a^b |f_n(t)|^2\mathrm{d}t \leq n$ a.s. We have $\int_a^b E(|f_n(t)|^2)\mathrm{d}t \leq n$, which implies that $f_n \in L^2_{ad}([a,b]\times\Omega)$. By Theorem 32.2, $X_{t\wedge\tau_n}$ is the martingale. This completes the proof.

33.6. Let $\tau : \Omega \to [a,b]$ be a random variable defined as

$$\tau(\omega) = \begin{cases} \inf\{t : f(t,\omega) \neq g(t,\omega)\}, & \text{if } \{t : f(t,\omega) \neq g(t,\omega)\} \neq \emptyset; \\ b, & \text{if } \{t : f(t,\omega) \neq g(t,\omega)\} = \emptyset. \end{cases}$$

Then τ is the stopping time. Let $X(\tau)$ be a random variable defined as

$$X(\tau) = \int_a^\tau (f(u) - g(u))\mathrm{d}B_u = \int_a^b I_{[a,\tau]}(u)(f(u) - g(u))\mathrm{d}B_u.$$

Since $I_{[a,\tau]}(u)(f(u) - g(u)) \in L^2_{ad}([a,b]\times\Omega)$, the $X(\tau)$ is defined. Then, by Theorem 32.1, we have

$$E(|X(\tau)|^2) = E\left[\int_a^b I_{[a,\tau]}(u)|f(u) - g(u)|^2\mathrm{d}u\right].$$

But

$$\int_a^b I_{[a,\tau]}(u)|f(u) - g(u)|^2\mathrm{d}u = \int_a^\tau |f(u) - g(u)|^2\mathrm{d}u = 0, \quad \text{a.s.}$$

Therefore we have $E(|X(\tau)|^2) = 0$, which implies that $X(\tau) = 0$ a.s., that is, $P(X(\tau) = 0) = 1$. Now let $\omega \in A$. Then $\tau(\omega) = b$ by the definition of A. Hence we have $X(\tau(\omega)) = \int_a^b (f(u) - g(u))\mathrm{d}u$. Hence we have

$$\left\{\omega \in A : \int_a^b f(t,\omega)\mathrm{d}B_t(\omega) \neq \int_a^b g(t,\omega)\mathrm{d}B_t(\omega)\right\} \subseteq \{X(\tau) \neq 0\},$$

which implies that

$$P\left\{\omega \in A : \int_a^b f(t,\omega)\mathrm{d}B_t(\omega) \neq \int_a^b g(t,\omega)\mathrm{d}B_t(\omega)\right\} = 0.$$

This completes the proof.

33.7. Let $\{\tau_n : n \geq 1\}$ be the sequence of stopping times defined in Problem 33.3. For each $n \geq 1$, let $f_n(t) = f(t)I_{\{t\leq\tau_n\}}$. Then we have

$$M_{t\wedge\tau_n} = \int_a^{t\wedge\tau_n} f(s)\mathrm{d}B_s = \int_a^b f_n(s)\mathrm{d}B_s.$$

Since

$$\begin{aligned}\int_a^b E\left[f_n(s)^2\right] \mathrm{d}s &= E\left[\int_a^b f_n(s)^2 \mathrm{d}s\right] \\ &= E\left[\int_a^{b\wedge\tau_n} f(s)^2 \mathrm{d}s\right] \\ &\leq n,\end{aligned}$$

we have $f_n \in L^2_{ad}([a,b]\times\Omega)$. Then it follows from Theorem 32.1 that

$$E\left[M^2_{t\wedge\tau_n}\right] = E\left[\int_a^{b\wedge\tau_n} f(s)^2 \mathrm{d}s\right]. \tag{33.7}$$

By the assumption that $\sup_{a\leq t\leq b} M_t$ is square-integrable, we have

$$E\left[M^2_{t\wedge\tau_n}\right] \leq E\left[\sup_{a\leq t\leq b} M_t^2\right] = L < \infty, \tag{33.8}$$

where L is some positive constant. From Equations (33.7) and (33.8), for every $n \geq 1$, we have

$$E\left[\int_a^{b\wedge\tau_n} f(s)^2 \mathrm{d}s\right] \leq L.$$

Since $\tau_n \to b$ as $n \to \infty$, letting $n \to \infty$ in the above equation gives

$$E\left[\int_a^b f(s)^2 \mathrm{d}s\right] \leq L < \infty.$$

Hence $f \in L^2_{ad}([a,b]\times\Omega)$. This completes the proof.

33.5 Bibliographic Notes

In this chapter, we introduced the Itô integral for stochastic processes $f(t)$ satisfying the condition

$$\int_a^b |f(t)|^2 \mathrm{d}t < \infty, \quad \text{a.s.}$$

The resulting Itô integral is a random variable, which in general is not integrable. The associated stochastic process

$$X_t = \int_a^t f(u)\mathrm{d}B_u, \quad t \in [a,b],$$

may not be a martingale but is a local martingale. For more information about the extension of Itô integrals, readers are referred to Karatzas and Shreve (1988), Lamberton and Lapeyre (1996), Steele (2003), and Kuo (2006, Chapter 5). For properties of local martingales, readers are referred to Durrett (1996), Meyer (2000), Steele (2003), and Medvegyev (2007).

CHAPTER 34

MARTINGALE STOCHASTIC INTEGRALS

In the previous two chapters, we introduced stochastic integrals with respect to standard Brownian motions. In this chapter, we expand the definition of stochastic integral such that the integrators are right-continuous and square-integrable martingales.

34.1 Basic Concepts and Facts

Definition 34.1 (Stochastic Process Space $\mathbb{L}([a,b] \times \Omega)$). Let $\{\mathscr{F}_t : a \le t \le b\}$ be a right-continuous filtration. The stochastic process space $\mathbb{L}([a,b] \times \Omega)$ is the space of all stochastic processes $X(t,\omega)$ satisfying the following conditions:

(a) $X(t)$ is adapted to the filtration $\{\mathscr{F}_t : a \le t \le b\}$.

(b) Almost all sample paths of X_t are left-continuous.

Definition 34.2 (Predictable Stochastic Process). Let $\mathcal{P}$ be the smallest σ-field of subsets of $[a,b] \times \Omega$ with respect to which all stochastic processes in $\mathbb{L}([a,b] \times \Omega)$ are measurable. A stochastic process $\{X_t : a \le t \le b\}$ is said to be predictable if the function

$$(t,\omega) \to X_t(\omega), \quad (t,\omega) \in [a,b] \times \Omega,$$

is $\mathcal{P}$-measurable.

Definition 34.3 (Increasing Process). A stochastic process $\{X_t : a \leq t \leq b\}$ on a probability space $(\Omega, \mathscr{F}, P)$ is said to be increasing if there exists a set $\Omega_0 \subseteq \Omega$ with $P(\Omega_0) = 1$ such that for every $\omega \in \Omega_0$,

(a) $X_0(\omega) = 0$.

(b) The function $t \to X_t(\omega)$ is a nondecreasing, right-continuous function,

and $E(X_t) < \infty$ holds for every $t \in [a, b]$.

Definition 34.4 (Natural Process). An increasing process $\{A_t : a \leq t \leq b\}$ is considered natural if for every bounded, right-continuous martingale $\{M_t : a \leq t \leq b\}$ with left-hand limits, we have

$$E\left(\int_{(0,t]} M_s \mathrm{d}A_s\right) == E\left(\int_{(0,t]} M_{s-} \mathrm{d}A_s\right),$$

where M_{s-} is the left-hand limit of M_t at s.

Definition 34.5 (Square-Integrable Stochastic Process). A stochastic process $\{X_t : a \leq t \leq b\}$ is said to be square-integrable if and only if for each $t \in [a, b]$, we have

$$E\left(X_t^2\right) < \infty.$$

Definition 34.6 (p-th Variation Process). Let $X = \{X_t : a \leq t \leq b\}$ be a stochastic process. Let $p > 0$. Then the p-th variation process of X is defined as

$$[X]_t^{(p)} = \lim_{\|\Delta\| \to 0} \sum_{k=1}^{n} |X_{t_k} - X_{t_{k-1}}|^p, \quad \text{(in probability)}$$

where $\Delta = \{t_0, t_1, \ldots, t_n\}$ is a partition of the interval $[a, t]$ and

$$\|\Delta\| = \max_{1 \leq k \leq n} |t_k - t_{k-1}|.$$

In particular, if $p = 1$, the process is referred to as the *total variation process*. If $p = 2$, the process is called the *quadratic variation process* and is written as $[X]_t$.

Definition 34.7 (Stochastic Process Sapce $L^2_{pred}([a, b]_{\langle M \rangle} \times \Omega)$). The space

$$L^2_{pred}([a, b]_{\langle M \rangle} \times \Omega)$$

denotes the space of all predictable stochastic process $f : [a, b] \times \Omega \to \mathbf{R}$ satisfying the following condition:

$$E\left[\int_a^b |f(t)|^2 \mathrm{d}\langle M \rangle_t\right] < \infty.$$

Definition 34.8 (Martingale Stochastic Integral for Step Stochastic Processes in Stochastic Process Space $L^2_{pred}([a,b]_{\langle M\rangle}\times\Omega)$). Let $\{M_t : a \le t \le b\}$ be a right-continuous, square-integrable martingale with left-hand limits. Let f be a step stochastic process in $L^2_{pred}([a,b]_{\langle M\rangle} \times \Omega)$ given by

$$f(t,\omega) = \sum_{i=1}^{n} \xi_{i-1}(\omega) I_{(t_{i-1},t_i]}(t),$$

where ξ_{i-1} is $\mathscr{F}_{t_{i-1}}$-measurable for $i = 1, 2, \ldots, n$. Then the stochastic integral of f with respect to the martingale M_t is defined as

$$I(f) = \sum_{i=1}^{n} \xi_{i-1}(M_{t_i} - M_{t_{i-1}}). \tag{34.1}$$

Definition 34.9 (Martingale Stochastic Integral for General Stochastic Processes in Stochastic Process Space $L^2_{pred}([a,b]_{\langle M\rangle} \times \Omega)$). Let $\{M_t : a \le t \le b\}$ be a right-continuous, square-integrable martingale with left-hand limits. Let f be a stochastic process in $L^2_{pred}([a,b]_{\langle M\rangle} \times \Omega)$. Then there exists a sequence $\{f_n\}_{n\ge1}$ of step stochastic processes in $L^2_{pred}([a,b]_{\langle M\rangle} \times \Omega)$ such that

$$\lim_{n\to\infty} E\left[\int_a^b |f(t) - f_n(t)|^2 \mathrm{d}\langle M\rangle_t\right] = 0.$$

Then the stochastic integral of f with respect to the martingale M_t, denoted by $\int_a^b f(t)\mathrm{d}M_t$, is defined as

$$\int_a^b f(t)\mathrm{d}M_t = \lim_{n\to\infty} I(f_n), \quad \text{in} \quad L^2(\Omega), \tag{34.2}$$

where $I(\cdot)$ is as defined in Equation (34.1).

Definition 34.10 (Stochastic Process Sapce $\mathcal{L}_{pred}(\Omega, L^2[a,b]_{\langle M\rangle})$). The space

$$\mathcal{L}_{pred}(\Omega, L^2[a,b]_{\langle M\rangle})$$

denotes the space of all predictable stochastic process $f : [a,b] \times \Omega \to \mathbf{R}$ satisfying the following condition:

$$\int_a^b |f(t)|^2 \mathrm{d}\langle M\rangle_t < \infty, \quad \text{a.s.}$$

Definition 34.11 (Martingale Stochastic Integral for Stochastic Processes in the Process Space $\mathcal{L}_{pred}(\Omega, L^2[a,b]_{\langle M\rangle})$). Let $\{M_t : a \le t \le b\}$ be a right-continuous, square-integrable martingale with left-hand limits. Let f be a stochastic process in the space $\mathcal{L}_{pred}(\Omega, L^2[a,b]_{\langle M \rangle})$. Then there exists a sequence $\{f_n\}_{n\geq 1}$ of stochastic processes in $L^2_{pred}([a,b]_{\langle M \rangle} \times \Omega)$ such that

$$\lim_{n\to\infty} \int_a^b |f(t) - f_n(t)|^2 \mathrm{d}\langle M \rangle_t = 0, \quad \text{a.s.}$$

Then the stochastic integral of f with respect to the martingale M_t, denoted by $\int_a^b f(t)\mathrm{d}M_t$, is defined as

$$\int_a^b f(t)\mathrm{d}M_t = \lim_{n\to\infty} \int_a^b f_n(t)\mathrm{d}M_t, \quad \text{in probability.} \tag{34.3}$$

Theorem 34.1 (Doob-Meyer Decomposition for M_t^2). *Let $\{M_t : a \leq t \leq b\}$ be a right-continuous, square-integrable martingale with left-hand limits. Then the process M_t^2 can be uniquely (up to indistinguishability) decomposed as*

$$M_t^2 = L_t + A_t, \quad a \leq t \leq b,$$

where the L_t is a right-continuous martingale with left-hands limits and the A_t is a natural process.

Process A_t is called the compensator of M_t^2 and is denoted by $\langle M \rangle_t$.

Theorem 34.2. *Let $\{M_t : a \leq t \leq b\}$ be a continuous and square integrable martingale. Then*

$$[M]_t = \langle M \rangle_t, \quad a \leq t \leq b,$$

where $[M]_t$ is the quadratic variation process (see Definition 34.6) and $\langle M \rangle_t$ is the compensator of M_t^2.

Theorem 34.3. *Let $\{M_t : a \leq t \leq b\}$ be a right-continuous, square-integrable martingale with left-hand limits. Let f be a stochastic process in $L^2_{pred}([a,b]_{\langle M \rangle} \times \Omega)$. Then the stochastic integral $\int_a^b f(t)\mathrm{d}M_t$ defined in Equation (34.2) is well defined. In addition, for any $\alpha, \beta \in \mathbf{R}$ and $f, g \in L^2_{pred}([a,b]_{\langle M \rangle} \times \Omega)$, we have*

$$\int_a^b (\alpha f(t) + \beta g(t))\mathrm{d}M_t = \alpha \int_a^b f(t)\mathrm{d}M_t + \beta \int_a^b g(t)\mathrm{d}M_t.$$

Theorem 34.4. *Let $f \in L^2_{pred}([a,b]_{\langle M \rangle} \times \Omega)$. Let $\{X_t : a \leq t \leq b\}$ be a stochastic process defined as*

$$X_t = \int_a^t f(s)\mathrm{d}M_s, \quad a \leq t \leq b.$$

Then

(a) The X_t is a martingale.

(b)

$$E\left(|X_t|^2\right) = E\left[\int_a^t |f(s)|^2 \mathrm{d}\langle M \rangle_s\right].$$

(c) The X_t is right-continuous with left-hand limits, that is, almost all of its sample paths are right-continuous functions with left-hand limits on $[a, b]$;

(d) The compensator of X_t^2 is given by

$$\langle X\rangle_t = \int_a^t |f(s)|^2 \mathrm{d}\langle M\rangle_s.$$

Theorem 34.5. *Let $f \in \mathcal{L}_{pred}(\Omega, L^2[a, b]_{\langle M\rangle})$. Let $\{X_t : a \leq t \leq b\}$ be a stochastic process defined as*

$$X_t = \int_a^t f(s)\mathrm{d}M_s, \quad a \leq t \leq b.$$

Then

(a) The X_t is a local martingale.

(b) The X_t has a right-continuous realization with left-hand limits.

(c) If the M_t is continuous, then the X_t has a continuous realization.

Theorem 34.6. *Let $\{M_t : a \leq t \leq b\}$ be a continuous and square-integrable martingale. Let f be a continuous stochastic process in $\mathcal{L}_{pred}(\Omega, L^2[a, b]_{\langle M\rangle})$. Then*

$$\int_a^b f(t)\mathrm{d}M_t = \lim_{\|\Delta_n\|\to 0} \sum_{i=1}^n f(t_{i-1})(M_{t_i} - M_{t_{i-1}}), \quad \text{in probability},$$

where $\Delta_n = \{t_0, t_1, \ldots, t_n\}$ is a partition of the interval $[a, b]$ and

$$\|\Delta_n\| = \max_{1\leq i\leq n} (t_i - t_{i-1}).$$

Theorem 34.7. *Let $\{M_t : a \leq t \leq b\}$ be a continuous and square-integrable martingale. Let f be a bounded, continuous, and adapted stochastic process. Then*

$$\int_a^b f(t)\mathrm{d}\langle M\rangle_t = \lim_{\|\Delta_n\|\to 0} \sum_{i=1}^n f(t_{i-1})(M_{t_i} - M_{t_{i-1}})^2, \quad \text{in probability},$$

where $\Delta_n = \{t_0, t_1, \ldots, t_n\}$ is a partition of the interval $[a, b]$ and

$$\|\Delta_n\| = \max_{1\leq i\leq n} (t_i - t_{i-1}).$$

The left-hand side is the Riemann-Stieltjes integral $\int_a^b f(t, \omega)\mathrm{d}\langle M\rangle_t(\omega)$ defined for each ω in some Ω_0 with $P(\Omega_0) = 1$.

34.2 Problems

34.1. Let $\{\mathscr{F}_t : a \le t \le b\}$ be a right-continuous filtration. Let $a \le s < t \le b$ and $A \in \mathscr{F}_s$. Let $\{\xi_u, a \le u \le b\}$ be a stochastic process defined as

$$\xi(u,\omega) = I_A(\omega) I_{(s,t]}(u), \quad u \in [a,b],\ \omega \in \Omega.$$

Show that $\xi_u \in \mathbb{L}([a,b] \times \Omega)$.

34.2. Let $\{\mathscr{F}_t : a \le t \le b\}$ be a right-continuous filtration, which is complete (i.e., all σ-fields $\mathscr{F}_t$ are complete). Assume that for each $t \in [a,b]$, the σ-field $\mathscr{F}_t$ is complete. Let $\mathcal{C}$ be a collection of sets defined as

$$\mathcal{C} = \{(s,t] \times A : a \le s < t \le b,\ A \in \mathscr{F}_s\} \cup \{\{a\} \times B : B \in \mathscr{F}_a\}.$$

Show that

$$\mathcal{P} = \sigma(\mathcal{C}),$$

where $\mathcal{P}$ is as defined in Definition 34.2, namely, it is the smallest σ-field of subsets of $[a,b] \times \Omega$ with respect to which all stochastic processes in $\mathbb{L}([a,b] \times \Omega)$ are measurable.

34.3. Let $\{M_t : a \le t \le b\}$ be a square-integrable martingale. Show that $\{M_t^2 : a \le t \le b\}$ is a submartingale.

34.4. Let $\{B_t : t \ge 0\}$ be a standard Brownian motion. Find the Doob-Meyer decomposition of B_t^2.

34.5. Let $\{B_t : t \ge 0\}$ be a standard Brownian motion, and let $M_t = B_t^2 - t$. Then the M_t is a martingale. Find the Doob-Meyer decomposition of M_t^2.

34.6. Let $\{B_t : a \le t \le b\}$ be a standard Brownian motion and $\{\tilde{N}_t : a \le t \le b\}$ be a compensated Poisson process with parameter λ. Let $M_t = B_t + \tilde{N}_t$. Assume that the B_t and the $\tilde{N}_t$ are independent. Calculate $\langle M \rangle_t$ and $[M]_t$.

34.7. Let $\{M_t : a \le t \le b\}$ be a right-continuous, square-integrable martingale with left-hand limits. For any $a \le s < t \le b$, show that

$$E\left[(M_t - M_s)^2 | \mathscr{F}_s\right] = E\left[\langle M \rangle_t - \langle M \rangle_s | \mathscr{F}_s\right].$$

34.8. Let $\{M_t : t \ge 0\}$ be a continuous and square-integrable martingale. Show that

$$\int_0^t M_s \mathrm{d}M_s = \frac{1}{2}\left(M_t^2 - M_0^2 - \langle M \rangle_t\right).$$

34.9. Let f be a step stochastic process and M_t a martingale given in Definition 34.8. Show that

$$E\left(|I(f)|^2\right) = E\left[\int_a^b |f(t)|^2 \mathrm{d}\langle M \rangle_t\right],$$

where $I(f)$ is as defined in Equation (34.1).

34.10. Let $\lambda \in \mathbf{R}$ and $\{B_t : 0 \leq t\}$ be a standard Brownian motion. Let

$$M_t = \exp\left(\lambda B_t - \frac{\lambda^2}{2}t\right), \quad t \geq 0.$$

Show that

(a) The M_t is a martingale.

(b) The compensator of M_t^2 is given by

$$\langle M\rangle_t = \lambda^2 \int_0^t \exp\left(2\lambda B_s - \lambda^2 s\right) \mathrm{d}s.$$

34.3 Hints

34.1. Follow the definition of $\mathbb{L}([a, b] \times \Omega)$ (Definition 34.1).

34.2. Use the standard technique, that is, show that $\mathcal{P} \subseteq \sigma(\mathcal{C})$ and $\sigma(\mathcal{C}) \subseteq \mathcal{P}$. To prove $\sigma(\mathcal{C}) \subseteq \mathcal{P}$, one can choose special functions from $\mathbb{L}([a, b] \times \Omega)$. To prove $\mathcal{P} \subseteq \sigma(\mathcal{C})$, one can show that a process $X \in \mathbb{L}([a, b] \times \Omega)$ is $\sigma(\mathcal{C})$-measurable by constructing a sequence of processes

$$X_n(t, \omega) = \begin{cases} X(a, \omega), & \text{if } t = a; \\ X(r_k, \omega), & \text{if } r_k < t \leq r_{k+1}, \end{cases}$$

where $r_k = \frac{k}{2^n}(b - a) + a$, $k = 0, 1, \ldots, 2^n$.

34.3. Use the conditional Jensen's inequality (see Problem 15.7).

34.4. Note that the $B_t^2 - t$ is a martingale.

34.5. Let $0 \leq s < t$. Suppose that the Doob-Meyer decomposition is $M_t^2 = L_t + A_t$. Then consider $E(M_t^2|\mathscr{F}_s = L_s + E(A_t|\mathscr{F}_s)$.

34.6. Use the independence of the B_t and the $\tilde{N}_t$ to establish that the $B_t\tilde{N}_t$ is a martingale. Then use the results of Problems 32.11, 27.6, and 27.7.

34.7. Use the Doob-Meyer decomposition of M_t^2.

34.8. Use Theorems 34.6 and 34.2.

34.9. Follow the definition of $I(f)$ (Definition 34.8) and use the tower property of conditional expectations (see Problem 14.10) and the result of Problem 34.7.

34.10. Part (a) can be proved by the result of Problem 19.1. To prove part (b), let $M_t^2 = L_t + A_t$ and consider the conditional expectation $E(M_t^2|\mathscr{F}_s)$.

34.4 Solutions

34.1. By the definition of ξ_u, we have

$$\xi_u = \begin{cases} 0, & \text{if } u \le s; \\ I_A, & \text{if } s < u \le t; \\ 0, & \text{if } t < u. \end{cases}$$

Hence for $u < s$, we have

$$\xi_u^{-1}(B) = \begin{cases} \Omega, & \text{if } 0 \in B; \\ \emptyset, & \text{if } 0 \notin B. \end{cases} \quad B \in \mathcal{B},$$

which implies that $\xi_u \in \mathscr{F}_u$ for $u < s$. Similarly, we can show that $\xi_u \in \mathscr{F}_u$ for $u \in (s, t]$ and for $u \in (t, b]$. Hence ξ_u is adapted to the filtration $\{\mathscr{F}_t : a \le t \le b\}$.

It is left to show that ξ_u is left-continuous. Let $\omega \in \Omega$ be fixed. Then

$$\xi(u, \omega) = \begin{cases} I_{(s,t]}(u), & \text{if } \omega \in A; \\ 0, & \text{if } \omega \notin A. \end{cases}, \quad u \in [a, b],$$

which shows that the sample path $\xi(\cdot, \omega)$ is left-continuous. Since ω is arbitrary, ξ_u is a left-continuous process. Therefore, $\xi_u \in \mathbb{L}([a, b] \times \Omega)$. This completes the proof.

34.2. First let us show that $\sigma(\mathcal{C}) \subseteq \mathcal{P}$. Let $a \le s < t \le b$ and $A \in \mathscr{F}_s$. Consider the random variable $\xi(u, \omega)$ defined as

$$\xi(u, \omega) = I_A(\omega) I_{(s,t]}(u), \quad u \in [a, b],\ \omega \in \Omega.$$

By Problem 34.1, we have $\xi_u \in \mathbb{L}([a, b] \times \Omega)$. Then by the definition of $\mathcal{P}$, ξ_u is $\mathcal{P}$-measurable. But

$$\xi_u^{-1}(1) = (s, t] \times A,$$

we have $(s, t] \times A \in \mathcal{P}$. Similarly, we can show that $\{a\} \times B \in \mathcal{P}$ for all $B \in \mathscr{F}_a$. Since $\mathcal{P}$ is the σ-field, we have $\sigma(\mathcal{C}) \subseteq \mathcal{P}$.

Now we show that $\mathcal{P} \subseteq \sigma(\mathcal{C})$. To do this, let $X \in \mathbb{L}([a, b] \times \Omega)$. Then X is adapted to $\{\mathscr{F}_t : a \le t \le b\}$ and almost all sample paths of X are left-continuous. We show that X is $\sigma(\mathcal{C})$-measurable. Let $D \in \mathcal{B}$. Let $\{X_n\}_{n \ge 1}$ be a sequence of processes defined as

$$X_n(t, \omega) = \begin{cases} X(a, \omega), & \text{if } t = a; \\ X(r_k, \omega), & \text{if } r_k < t \le r_{k+1}, \end{cases}$$

where $r_k = \frac{k}{2^n}(b - a) + a$, $k = 0, 1, \ldots, 2^n$. Then we have

$$\begin{aligned} & \{(t, \omega) \in [a, b] \times \Omega : X_n(t, \omega) \in D\} \\ = & \{\{a\} \times \{\omega \in \Omega : X(a, \omega) \in D\}\} \cup \\ & \bigcup_{k=0}^{2^n - 1} \{(r_k, r_{k+1}] \times \{\omega \in \Omega : X(r_k, \omega) \in D\}\}. \end{aligned}$$

Note that X is adapted to the filtration $\mathscr{F}_t$. We have

$$\{(t,\omega) \in [a,b] \times \Omega : X_n(t,\omega) \in D\} \in \mathcal{C}.$$

Hence X_n is $\sigma(\mathcal{C})$-measurable. Since X_n converges to X a.s. and the filtration is complete, X is also $\sigma(\mathcal{C})$-measurable. Therefore, we have $\mathcal{P} \subseteq \sigma(\mathcal{C})$ because $\mathcal{P}$ is the smallest σ-field to which processes in $\mathbb{L}([a,b] \times \Omega)$ are measurable.

This completes the proof.

34.3. Since $\{M_t : a \le t \le b\}$ is a martingale, the M_t^2 is an adapted process. By the assumption that the M_t is square-integrable, we have $E(|M_t|^2) < \infty$. Now by the conditional Jensen's inequality (see Problem 15.7), for $a \le s < t \le b$, we have

$$E\left(M_t^2|\mathscr{F}_s\right) \ge [E(M_t|\mathscr{F}_s)]^2 = M_s^2.$$

Therefore, the M_t^2 is a submartingale.

34.4. Let $0 \le s < t$. Then

$$\begin{aligned} E\left(B_t^2|\mathscr{F}_s\right) &= E\left((B_t - B_s + B_s)^2|\mathscr{F}_s\right) \\ &= B_s^2 + t - s, \end{aligned}$$

which implies that $B_t^2 - t$ is a martingale. Note that

$$B_t^2 = (B_t^2 - t) + t.$$

The Doob-Meyer decomposition is $L_t = B_t^2 - t$ and $\langle B \rangle_t = t$.

34.5. Let $0 \le s < t$. Then

$$E\left(B_t^2|\mathscr{F}_s\right) = B_s^2 - s + t$$

and

$$E\left(B_t^4|\mathscr{F}_s\right) = B_s^4 + 6(t-s)B_s^2 + 3(t-s)^2.$$

Suppose that the Doob-Meyer decomposition of M_t^2 is $L_t + A_t$. Then we have

$$\begin{aligned} E\left(M_t^2|\mathscr{F}_s\right) &= E\left(B_t^4|\mathscr{F}_s\right) - 2tE\left(B_t^2|\mathscr{F}_s\right) + t^2 \\ &= B_s^4 + (4t - 6s)B_s^2 + 3s^2 - 4ts + t^2 \\ &= L_s + E(A_t|\mathscr{F}_s) \\ &= M_s^2 - A_s + E(A_t|\mathscr{F}_s), \end{aligned}$$

which gives

$$E(A_t - A_s|\mathscr{F}_s) = 4(t-s)B_s^2 + 2(t-s)^2.$$

Note that $E\left(B_t^2|\mathscr{F}_s\right) = B_s^2 - s + t$. We have

$$\begin{aligned} E\left(\int_s^t B_u^2 \mathrm{d}u|\mathscr{F}_s\right) &= \int_s^t E(B_u^2|\mathscr{F}_s)\mathrm{d}u \\ &= \int_s^t (B_s^2 + u - s)\mathrm{d}u \\ &- (t-s)B_s^2 + \frac{1}{2}(l-s)^2. \end{aligned}$$

Hence we have

$$A_t = \int_0^t B_u^2 du$$

and

$$L_t = (B_t^2 - t)^2 - \int_0^t B_u^2 du.$$

34.6. Let $0 \le s < t$. Then we have

$$\begin{aligned} E(B_t \tilde{N}_t | \mathscr{F}_s) &= E[(B_t - B_s + B_s)(\tilde{N}_t - \tilde{N}_s + \tilde{N}_s) | \mathscr{F}_s] \\ &= B_s \tilde{N}_s, \end{aligned}$$

which implies that $B_t \tilde{N}_t$ is a martingale. By Problem 27.7, $\tilde{N}_t^2 - \lambda t$ is a martingale. Therefore, the following process

$$M_t^2 - (1+\lambda)t = (B_t^2 - t) + 2B_t \tilde{N}_t + (\tilde{N}_t^2 - \lambda t)$$

is a martingale. Hence we have the following Doob-Meyer decomposition for M_t^2:

$$M_t^2 = (M_t^2 - (1+\lambda)t) + (1+\lambda)t,$$

which gives $\langle M \rangle_t = (1+\lambda)t$.

Now let us calculate $[M]_t$. Let $\Delta_n = \{t_0, t_1, \ldots, t_n\}$ be a partition of the interval $[0, t]$. Then we have

$$\begin{aligned} & \sum_{i=1}^n (M_{t_i} - M_{t_{i-1}})^2 \\ = & \sum_{i=1}^n (B_{t_i} - B_{t_{i-1}} + \tilde{N}_{t_i} - \tilde{N}_{t_{i-1}})^2 \\ = & \sum_{i=1}^n (B_{t_i} - B_{t_{i-1}})^2 + \sum_{i=1}^n 2(B_{t_i} - B_{t_{i-1}})(\tilde{N}_{t_i} - \tilde{N}_{t_{i-1}}) \\ & + \sum_{i=1}^n (\tilde{N}_{t_i} - \tilde{N}_{t_{i-1}})^2. \end{aligned}$$

Hence by Problems 32.11 and 27.6, we have

$$[M]_t = (1+\lambda)t + \tilde{N}_t.$$

34.7. By Theorem 34.1, M_t^2 can be uniquely decomposed as

$$M_t^2 = L_t + \langle M \rangle_t, \quad a \le t \le b,$$

where the L_t is a martingale. Then we have

$$\begin{aligned} E\left[(M_t - M_s)^2 | \mathscr{F}_s\right] &= E\left[M_t^2 - 2M_t M_s + M_s^2 | \mathscr{F}_s\right] \\ &= E[L_t + \langle M \rangle_t | \mathscr{F}_s] - M_s^2 \\ &= E[L_t + \langle M \rangle_t | \mathscr{F}_s] - L_s - \langle M \rangle_s \\ &= E[\langle M \rangle_t - \langle M \rangle_s | \mathscr{F}_s]. \end{aligned}$$

This completes the proof.

34.8. Let R_n and L_n be defined as

$$R_n = \sum_{i=1}^{n} M_{t_i}(M_{t_i} - M_{t_{i-1}}),$$

and

$$L_n = \sum_{i=1}^{n} M_{t_{i-1}}(M_{t_i} - M_{t_{i-1}}),$$

respectively, where $\Delta_n = \{t_0, t_1, \ldots, t_n\}$ is a partition of the interval $[0, t]$. Then we have

$$R_n + L_n = \sum_{i=1}^{n}(M_{t_i} + M_{t_{i-1}})(M_{t_i} - M_{t_{i-1}}) = M_t^2 - M_0^2$$

and

$$R_n - L_n = \sum_{i=1}^{n}(M_{t_i} - M_{t_{i-1}})^2.$$

Solving the above equations gives

$$L_n = \frac{1}{2}\left(M_t^2 - M_0^2 - \sum_{i=1}^{n}(M_{t_i} - M_{t_{i-1}})^2\right).$$

The result follows by letting $\|\Delta_n\| \to 0$ and Theorem 34.2. This completes the proof.

34.9. By the definition of $I(f)$ in Equation (34.1), we have

$$\begin{aligned} & |I(f)|^2 \\ = & \sum_{i=1}^{n} \xi_{i-1}^2 (M_{t_i} - M_{t_{i-1}})^2 + 2\sum_{i<j} \xi_{i-1}\xi_{j-1}(M_{t_i} - M_{t_{i-1}})(M_{t_j} - M_{t_{j-1}}). \end{aligned}$$

By Problems 14.10 and 34.7, we have

$$\begin{aligned} E\left[\xi_{i-1}^2(M_{t_i} - M_{t_{i-1}})^2\right] &= E\left[E\left\{\xi_{i-1}^2(M_{t_i} - M_{t_{i-1}})^2 | \mathscr{F}_{t_{i-1}}\right\}\right] \\ &= E\left[\xi_{i-1}^2 E\left\{(M_{t_i} - M_{t_{i-1}})^2 | \mathscr{F}_{t_{i-1}}\right\}\right] \\ &= E\left[\xi_{i-1}^2 E\left\{(\langle M\rangle_{t_i} - \langle M\rangle_{t_{i-1}}) | \mathscr{F}_{t_{i-1}}\right\}\right] \\ &= E\left[\xi_{i-1}^2(\langle M\rangle_{t_i} - \langle M\rangle_{t_{i-1}})\right]. \end{aligned}$$

Again by Problem 14.10, we have for $i < j$,

$$\begin{aligned} & E\left[\xi_{i-1}\xi_{j-1}(M_{t_i} - M_{t_{i-1}})(M_{t_j} - M_{t_{j-1}})\right] \\ = & E\left[E\left\{\xi_{i-1}\xi_{j-1}(M_{t_i} - M_{t_{i-1}})(M_{t_j} - M_{t_{j-1}}) | \mathscr{F}_{t_{j-1}}\right\}\right] \\ = & E\left[\xi_{i-1}\xi_{j-1}(M_{t_i} - M_{t_{i-1}}) E\left\{M_{t_j} - M_{t_{j-1}} | \mathscr{F}_{t_{j-1}}\right\}\right] \\ = & 0. \end{aligned}$$

Therefore, we have

$$\begin{aligned}
E\left[|I(f)|^2\right] &= E\left[\sum_{i=1}^{n} \xi_{i-1}^2 (M_{t_i} - M_{t_{i-1}})^2\right] \\
&= \sum_{i=1}^{n} E\left[\xi_{i-1}^2 (M_{t_i} - M_{t_{i-1}})^2\right] \\
&= \sum_{i=1}^{n} E\left[\xi_{i-1}^2 (\langle M\rangle_{t_i} - \langle M\rangle_{t_{i-1}})\right] \\
&= E\left[\sum_{i=1}^{n} \xi_{i-1}^2 (\langle M\rangle_{t_i} - \langle M\rangle_{t_{i-1}})\right] \\
&= E\left[\int_a^b |f(t)|^2 \mathrm{d}\langle M\rangle_t\right].
\end{aligned}$$

This completes the proof.

34.10.

(a) It is obvious that the M_t is an adapted process and is integrable. Since $B_t - B_s$ is normally distributed with mean 0 and variance $t - s$, by Problem 19.1 we have

$$\begin{aligned}
E[M_t|\mathscr{F}_s] &= \exp\left(\lambda B_s - \frac{\lambda^2}{2} t\right) E\left[e^{\lambda(B_t - B_s)} \middle| \mathscr{F}_s\right] \\
&= \exp\left(\lambda B_s - \frac{\lambda^2}{2} t\right) \exp\left(\frac{1}{2}\lambda^2 (t-s)\right) \\
&= \exp\left(\lambda B_s - \frac{\lambda^2}{2} s\right) \\
&= M_s.
\end{aligned}$$

Hence the M_t is a martingale.

(b) Let $M_t^2 = L_t + A_t$, where L_t is a martingale. Then we have

$$E\left(M_t^2|\mathscr{F}_s\right) = E(L_t + A_t|\mathscr{F}_s) = L_s + E\left(A_t|\mathscr{F}_s\right) = M_s^2 + E\left(A_t - A_s|\mathscr{F}_s\right)$$

and

$$\begin{aligned}
E\left(M_t^2|\mathscr{F}_s\right) &= E\left(e^{2\lambda B_t - \lambda^2 t}|\mathscr{F}_s\right) \\
&= e^{2\lambda B_s - \lambda^2 t} E\left(e^{2\lambda(B_t - B_s)}|\mathscr{F}_s\right) \\
&= e^{2\lambda B_s - \lambda^2 t} e^{2\lambda^2 (t-s)} \\
&= e^{2\lambda B_s + \lambda^2 t - 2\lambda^2 s}.
\end{aligned}$$

Combining the above two equations gives

$$E(A_t - A_s|\mathscr{F}_s) = e^{2\lambda B_s + \lambda^2 t - 2\lambda^2 s} - e^{2\lambda B_s - \lambda^2 s}. \tag{34.4}$$

On the other hand, we have

$$\begin{aligned} & E\left(\int_s^t \lambda^2 e^{2\lambda B_u - \lambda^2 u} \mathrm{d}u \middle| \mathscr{F}_s\right) && (34.5) \\ = & \int_s^t \lambda^2 e^{2\lambda B_s - \lambda^2 u} E\left[e^{2\lambda(B_u - B_s)}|\mathscr{F}_s\right] \mathrm{d}u \\ = & \int_s^t \lambda^2 e^{2\lambda B_s + \lambda^2(u-2s)} du \\ = & e^{2\lambda B_s + \lambda^2(t-2s)} - e^{2\lambda B_s - \lambda^2 s}. && (34.6) \end{aligned}$$

Comparing Equations (34.4) and (34.6) gives

$$A_t = \lambda^2 \int_0^t \exp\left(2\lambda B_s - \lambda^2 s\right) \mathrm{d}s.$$

This completes proof.

34.5 Bibliographic Notes

For more information on stochastic integration, readers are referred to Karatzas and Shreve (1988), Chung and Williams (1990), Durrett (1996), Oksendal (1998), Meyer (2000), and Protter (2003). Meyer (2000) introduced stochastic integration with many applications in finance. Durrett (1996) introduced stochastic integration with respect to local martingales. Bichteler (2002) presented stochastic integration with jumps.

Theorem 34.1 is a special case of the Doob-Meyer decomposition theorem where the square of a martingale is considered. For a general Doob-Meyer decomposition and its proof, readers are referred to Karatzas and Shreve (1991, Theorem 4.10). Theorem 34.2 states that if a square integrable martingale M_t is continuous, then the compensator of M_t^2 is just the quadratic variation process of M_t. For a proof of Theorem 34.2, readers are referred to Karatzas and Shreve (1991, Theorem 5.8).

CHAPTER 35

THE ITÔ FORMULA

The Itô formula in stochastic calculus is similar to the chain rule in Leibniz-Newton calculus. In this chapter, we introduce the simple Itô formula and its variations.

35.1 Basic Concepts and Facts

Definition 35.1 (C^2-Function). A function f on $[a, b]$ is called a C^2-*function* if and only if f is twice differentiable and the second derivative f'' is continuous on $[a, b]$.

Definition 35.2 (Stochastic Process Space $\mathcal{L}_{ad}(\Omega, L^1[a, b])$). Let $\{\mathscr{F}_t : a \leq t \leq b\}$ be a filtration under consideration. The space $\mathcal{L}_{ad}(\Omega, L^1[a, b])$ is defined to be the space of all stochastic processes $f(t, \omega)$, $t \in [a, b]$, $\omega \in \Omega$, satisfying the following conditions:

(a) $f(t)$ is adapted to the $\mathscr{F}_t$.

(b) $\int_a^b |f(t)| \mathrm{d}t < \infty$ a.s.

Definition 35.3 (Itô Process). A stochastic process $\{X_t : a \leq t \leq b\}$ is called an *Itô process* if it has the form

$$X_t = X_a + \int_a^t f(s)\mathrm{d}B_s + \int_a^t g(s)\mathrm{d}s, \quad a \leq t \leq b, \tag{35.1}$$

where X_a is $\mathscr{F}_a$-measurable, $f \in \mathcal{L}_{ad}(\Omega, L^2[a,b])$, and $g \in \mathcal{L}_{ad}(\Omega, L^1[a,b])$.

For convenience, Equation (35.1) is also written in the following stochastic differential form

$$\mathrm{d}X_t = f(t)\mathrm{d}B_t + g(t)\mathrm{d}t. \tag{35.2}$$

Definition 35.4 (Continuous Part of Quadratic Variation Processes). Let $\{M_t : a \leq t \leq b\}$ be a right-continuous, square-integrable martingale with left-hand limits. Then the continuous part of $[M]_t$ is defined as

$$[M]_t^c = [M]_t - [M]_a - \sum_{a<s\leq t} \Delta[M]_s, \quad a \leq t \leq b,$$

where $\Delta[M]_s$ is the jump of $[M]_s$ at s:

$$\Delta[M]_s = [M]_s - [M]_{s-}.$$

Definition 35.5 (Cross-Variation Process). Let X and Y be right-continuous and square-integrable martingales. Then the cross-variation process $\langle X, Y\rangle_t$ is defined by

$$\langle X, Y\rangle_t = \frac{1}{4}\left(\langle X+Y\rangle_t - \langle X-Y\rangle_t\right), \quad 0 \leq t < \infty.$$

Theorem 35.1. *Let $f \in C^2[a,b]$, that is, let f be a C^2-function on $[a,b]$. Then*

$$f(B_t) - f(B_a) = \int_a^t f'(B_s)\mathrm{d}B_s + \frac{1}{2}\int_a^t f''(B_s)\mathrm{d}s,$$

where the B_t is a standard Brownian motion.

Theorem 35.2 (Itô's Formula). *Let $f(t,x)$ be a continuous function with continuous partial derivatives $\frac{\partial f}{\partial t}$, $\frac{\partial f}{\partial x}$, and $\frac{\partial^2 f}{\partial x^2}$. Then*

$$\begin{aligned} & f(t, B_t) - f(a, B_a) \\ = \; & \int_a^t \frac{\partial f}{\partial x}(s, B_s)\mathrm{d}B_s + \int_a^t \left(\frac{\partial f}{\partial t}(s, B_s) + \frac{1}{2}\frac{\partial^2 f}{\partial x^2}(s, B_s)\right)\mathrm{d}s. \end{aligned}$$

Theorem 35.3 (General Form of Itô's Formula). *Let $\{X_t : a \leq t \leq b\}$ be an Itô process given by*

$$X_t = X_a + \int_a^t f(s)\mathrm{d}B_s + \int_a^t g(s)\mathrm{d}s, \quad a \leq t \leq b,$$

where X_a is $\mathscr{F}_a$-measurable, $f \in \mathcal{L}_{ad}(\Omega, L^2[a,b])$, and $g \in \mathcal{L}_{ad}(\Omega, L^1[a,b])$. Let $\theta(t,x)$ be a continuous function with continuous partial derivatives $\frac{\partial\theta}{\partial t}$, $\frac{\partial\theta}{\partial x}$, and $\frac{\partial^2\theta}{\partial x^2}$. (If $\frac{\partial^2\theta}{\partial x^2}$ is undefined at finitely many points, the formula still holds if $\left|\frac{\partial^2\theta}{\partial x^2}\right|$ is bounded and $\frac{\partial\theta}{\partial x}$ is defined for every $x \in \mathbf{R}$ and is continuous (Shreve, 2004, p205)). Then $\theta(t, X_t)$ is also an Itô process and

$$
\begin{aligned}
& \theta(t, X_t) - \theta(a, X_a) \\
= & \int_a^t \frac{\partial\theta}{\partial x}(s, X_s) f(s) \mathrm{d}B_s \\
& + \int_a^t \left[\frac{\partial\theta}{\partial t}(s, X_s) + \frac{\partial\theta}{\partial x}(s, X_s) g(s) + \frac{1}{2}\frac{\partial^2\theta}{\partial x^2}(s, X_s) f(s)^2 \right] \mathrm{d}s.
\end{aligned}
$$

The general form of Itô's formula can be written in the following stochastic differential form

$$
\mathrm{d}\theta(t, X_t) = \frac{\partial\theta}{\partial t}(t, X_t)\mathrm{d}t + \frac{\partial\theta}{\partial x}(t, X_t)\mathrm{d}X_t + \frac{1}{2}\frac{\partial^2\theta}{\partial x^2}(t, X_t)(\mathrm{d}X_t)^2,
$$

where $(\mathrm{d}X_t)^2 = f(t)^2\mathrm{d}t$, which is derived symbolically using the Itô table given in Table 35.1.

Table 35.1 Itô table.

$\times$	$\mathrm{d}B_t$	$\mathrm{d}t$
$\mathrm{d}B_t$	$\mathrm{d}t$	0
$\mathrm{d}t$	0	0

Theorem 35.4 (Multidimensional Itô's Formula). *Let $B_t^{(1)}, B_t^{(2)}, \ldots, B_t^{(m)}$ be m independent standard Brownian motions. Let $X_t^{(1)}, X_t^{(2)}, \ldots, X_t^{(n)}$ be n Itô processes given by*

$$
X_t^{(i)} = X_a^{(i)} + \sum_{j=1}^{m} \int_a^t f_{ij}(s)\mathrm{d}B_s^{(j)} + \int_a^t g_i(s)\mathrm{d}s, \quad i = 1, 2 \ldots, n,
$$

where $X_a^{(i)}$ is $\mathscr{F}_a$-measurable, $f_{ij} \in \mathcal{L}_{ad}(\Omega, L^2[a,b])$, and $g_i \in \mathcal{L}_{ad}(\Omega, L^1[a,b])$ for all $1 \le i \le n$ and $1 \le j \le m$. Let $\theta(t, x_1, x_2, \ldots, x_n)$ be a continuous function on $[a,b] \times \mathbf{R}^n$ with continuous partial derivatives $\frac{\partial\theta}{\partial t}$, $\frac{\partial\theta}{\partial x_i}$, and $\frac{\partial^2\theta}{\partial x_i \partial x_k}$ for $1 \le i, k \le n$.

Then $\theta(t, X_t^{(1)}, X_t^{(2)}, \ldots, X_t^{(n)})$ *has the following stochastic differential*

$$\begin{aligned}
& \mathrm{d}\theta(t, X_t^{(1)}, X_t^{(2)}, \ldots, X_t^{(n)}) \\
= \; & \frac{\partial \theta}{\partial t}(t, X_t^{(1)}, X_t^{(2)}, \ldots, X_t^{(n)})\mathrm{d}t + \sum_{i=1}^{n} \frac{\partial \theta}{\partial x_i}(t, X_t^{(1)}, X_t^{(2)}, \ldots, X_t^{(n)})\mathrm{d}X_t^{(i)} \\
& + \frac{1}{2}\sum_{i=1}^{n}\sum_{k=1}^{n} \frac{\partial^2 \theta}{\partial x_i \partial x_k}(t, X_t^{(1)}, X_t^{(2)}, \ldots, X_t^{(n)})\mathrm{d}X_t^{(i)}\mathrm{d}X_t^{(k)},
\end{aligned}$$

where $\mathrm{d}X_t^{(i)}\mathrm{d}X_t^{(k)}$ *is computed using the Itô table given in Table 35.2.*

Table 35.2 Multidimensional Itô table. In the table, δ_{ik} is Kronecker delta, i.e., $\delta_{ik} = 1$ for $i = k$ and $\delta_{ik} = 0$ for $i \neq k$.

$\times$	$\mathrm{d}B_t^{(k)}$	$\mathrm{d}t$
$\mathrm{d}B_t^{(i)}$	$\delta_{ik}\mathrm{d}t$	0
$\mathrm{d}t$	0	0

Theorem 35.5 (Itô's Formula for Martingales)**.** *Let* F *be a* C^2*-function and* M_t *a right-continuous, square-integrable martingale with left-hand limits. Then*

$$\begin{aligned}
& F(M_t) - F(M_a) \\
= \; & \int_a^t F'(M_{s-})\mathrm{d}M_s + \frac{1}{2}\int_a^t F''(M_s)\mathrm{d}[M]_s^c \\
& + \sum_{a<s\leq t} \left[F(M_s) - F(M_{s-}) - F'(M_{s-})\Delta M_s\right],
\end{aligned}$$

where M_{s-} *is the left-hand limit of* M_s*,* $[M]_s^c$ *is the continuous part of* $[M]_s$ *(see Definition 35.4), and* $\Delta M_s = M_s - M_{s-}$.

Theorem 35.6 (Itô's Formula for Continuous and Square-Integrable Martingales)**.** *Let* $F(t, x)$ *be a continuous function with continuous partial derivatives* $\frac{\partial F}{\partial t}$, $\frac{\partial F}{\partial x}$, *and* $\frac{\partial^2 F}{\partial x^2}$. *Let* M_t *be a continuous and square-integrable martingale. Then*

$$\begin{aligned}
& F(t, M_t) - F(a, M_a) \\
= \; & \int_a^t \frac{\partial F}{\partial t}(s, M_s)\mathrm{d}s + \int_a^t \frac{\partial F}{\partial x}(s, M_s)\mathrm{d}M_s + \frac{1}{2}\int_a^t \frac{\partial^2 F}{\partial x^2}(s, M_s)\mathrm{d}\langle M\rangle_s.
\end{aligned}$$

Theorem 35.7 (Multidimensional Itô Formula for Continuous and Square-Integrable Martingales)**.** *Let* $M_t^{(1)}, M_t^{(2)}, \ldots, M_t^{(n)}$ *be* n *continuous and square-integrable martingales. Let* $F(t, x_1, x_2, \ldots, x_n)$ *be a continuous function with continuous partial derivatives* $\frac{\partial F}{\partial t}$, $\frac{\partial F}{\partial x_i}$, *and* $\frac{\partial^2 F}{\partial x_i \partial x_k}$ *for all* $1 \leq i, k \leq n$. *Then*

$$\mathrm{d}F = \frac{\partial F}{\partial t}\mathrm{d}t + \sum_{i=1}^{n} \frac{\partial F}{\partial x_i}\mathrm{d}M_t^{(i)} + \frac{1}{2}\sum_{i=1}^{n}\sum_{k=1}^{n} \frac{\partial^2 F}{\partial x_i \partial x_k}\mathrm{d}\langle M^{(i)}, M^{(k)}\rangle_t,$$

where $\langle \cdot, \cdot \rangle$ is the cross variation process.

Theorem 35.8 (Uniqueness of Cross-Variation Process). *Let X and Y be two right-continuous and square-integrable martingales. Then the cross variation process $\langle X, Y\rangle_t$ is (up to indistinguishability) the only process of the form $A = A^{(1)} - A^{(2)}$ with $A^{(1)}$ and $A^{(2)}$ adapted and natural, such that $XY - A$ is a martingale.*

35.2 Problems

35.1. Let $g(x)$ be a continuous function on $\mathbf{R}$. Let X_n be defined as

$$X_n = \sum_{i=1}^{n} \left[g(B_{t_{i-1}} + \lambda_i(B_{t_i} - B_{t_{i-1}})) - g(B_{t_{i-1}})\right](B_{t_i} - B_{t_{i-1}})^2, \quad n \geq 1,$$

where $\Delta_n = \{t_0, t_1, \ldots, t_n\}$ is a partition of the interval $[a, t]$ and $\lambda_i \in (0, 1)$ for $i = 1, 2, \ldots, n$. Show that there exists a subsequence of $\{X_n\}_{n\geq 1}$ converging to 0 a.s. as $\|\Delta_n\| \to 0$, where $\|\Delta_n\| = \max_{1\leq i\leq n}(t_i - t_{i-1})$.

35.2. Let $g(x)$ be a continuous function on $\mathbf{R}$. Let X_n be defined as

$$X_n = \sum_{i=1}^{n} g(B_{t_{i-1}})\left[(B_{t_i} - B_{t_{i-1}})^2 - (t_i - t_{i-1})\right], \quad n \geq 1,$$

where $\Delta_n = \{t_0, t_1, \ldots, t_n\}$ is a partition of the interval $[a, t]$. Show that

$$X_n \xrightarrow{P} 0 \quad \text{as} \quad \|\Delta_n\| \to 0.$$

35.3. Let $\{B_t : t \geq 0\}$ be a standard Brownian motion. Show that

$$tB_t^2 = 2\int_0^t sB_s \mathrm{d}B_s + \int_0^t B_s^2 \mathrm{d}s + \frac{1}{2}t^2.$$

35.4. Let $\{M_t : 0 \leq t\}$ be a process defined by

$$M_t = \exp\left(cB_t - \frac{1}{2}c^2 t\right), \quad t \geq 0,$$

where c is a constant and the B_t is a standard Brownian motion. Show that

(a) The M_t is martingale.

(b) The compensator of M_t^2 is

$$\langle M\rangle_t = c^2 \int_0^t \exp\left(2cB_s - c^2 s\right) \mathrm{d}s.$$

35.5 (Langevin Equation). Let $\alpha \in \mathbf{R}$ and $\beta > 0$. Let $\{X_t : 0 \leq t\}$ be a process satisfying the following equation

$$X_t = x_0 + \alpha B_t - \beta \int_0^t X_s \mathrm{d}s,$$

where $x_0 \in \mathbf{R}$. Show that

$$X_t = x_0 e^{-\beta t} + \alpha \int_0^t e^{-\beta(t-s)} \mathrm{d}B_s.$$

35.6 (Itô's Product Formula). Let X_t and Y_t be two Itô processes given by

$$\mathrm{d}X_t = f(t)\mathrm{d}B_t + \xi(t)\mathrm{d}t$$

and

$$\mathrm{d}Y_t = g(t)\mathrm{d}B_t + \eta(t)\mathrm{d}t,$$

respectively. Show that

$$\begin{aligned} &X_t Y_t - X_a Y_a \\ = &\int_a^t [f(s)Y_s + g(s)X_s]\, \mathrm{d}B_s + \int_a^t [\xi(s)Y_s + \eta(s)X_s + f(s)g(s)]\, \mathrm{d}s. \end{aligned}$$

35.7. Let X, Y, and Z be right-continuous and square-integrable martingales. Let α and β be real numbers. Show that

(a) $\langle X, Y\rangle_t = \langle Y, X\rangle_t$.

(b) $\langle \alpha X + \beta Y, Z\rangle_t = \alpha\langle X, Z\rangle_t + \beta\langle Y, Z\rangle_t$.

(c) $\langle \alpha X\rangle_t = \alpha^2\langle X\rangle_t$.

(d) $\langle X\rangle_t = \langle X, X\rangle_t$.

(e) $|\langle X, Y\rangle_t|^2 \leq \langle X\rangle_t\langle Y\rangle_t$.

35.8. Let $B_t^{(1)}$ and $B_t^{(2)}$ be two independent standard Brownian motions. Calculate $\langle B^{(1)} + B^{(2)}\rangle_t$, $\langle B^{(1)} - B^{(2)}\rangle_t$, and $\langle B^{(1)}, B^{(2)}\rangle_t$.

35.9. Let X_t and Y_t be two stochastic processes defined by

$$X_t = \int_a^t f(s)\mathrm{d}B_s$$

and

$$Y_t = \int_a^t g(s)\mathrm{d}B_s,$$

respectively, where $f, g \in L^2_{ad}([a,b] \times \Omega)$. Show that

$$\langle X, Y \rangle_t = \int_a^t f(s)g(s)\mathrm{d}s.$$

35.10. Let $\{B_t^{(1)} : a \le t \le b\}$ and $\{B_t^{(2)} : a \le t \le b\}$ be two independent Brownian motions. Let X_t and Y_t be two stochastic processes defined by

$$X_t = \int_a^t f(s)\mathrm{d}B_s^{(1)}$$

and

$$Y_t = \int_a^t g(s)\mathrm{d}B_s^{(2)},$$

respectively, where $f, g \in L^2_{ad}([a,b] \times \Omega)$. Show that

$$\langle X, Y \rangle_t = 0.$$

35.11. Let $c, a_1, a_2, \ldots, a_n$ be constants. Let $B_t^{(1)}$, $B_t^{(2)}$, ..., $B_t^{(m)}$ be m independent standard Brownian motions. Let

$$X_t = \exp\left(ct + \sum_{j=1}^m a_j B_t^{(j)} \right).$$

Show that

$$\mathrm{d}X_t = \left(c + \frac{1}{2}\sum_{j=1}^m a_j^2 \right) X_t \mathrm{d}t + X_t \left(\sum_{j=1}^m \mathrm{d}B_t^{(j)} \right).$$

35.12. Let n be a positive integer. Let f be a stochastic process such that $f^n \in L^2_{ad}([a,b] \times \Omega)$. Show that for every $t \in [a,b]$, we have

$$E\left[\left| \int_a^t f(s)\mathrm{d}B_s \right|^{2n} \right] \le C_n (t-a)^{n-1} \int_a^t E\left[f^{2n}(s) \right] \mathrm{d}s,$$

where C_n is a constant depending only on n.

35.3 Hints

35.1. Consider the sequence $\{\xi_n\}_{n \ge 1}$ defined by

$$\xi_n = \max_{1 \le i \le n, 0 < \lambda < 1} \left| g(B_{t_{i-1}} + \lambda(B_{t_i} - B_{t_{i-1}})) - g(B_{t_{i-1}}) \right|, \quad n \ge 1$$

and first prove that $\xi_n \to 0$ a.s. as $\|\Delta_n\| \to 0$ by using the continuity of g and B_t.

35.2. Let $L > 0$ be a large number. Consider the sequence $S_{n,L}$ given by

$$S_{n,L} = \sum_{i=1}^{n} g(B_{t_{i-1}}) \left[(B_{t_i} - B_{t_{i-1}})^2 - (t_i - t_{i-1}) \right] I_{A_{i-1,L}},$$

where $A_{i-1,L} = \left\{ |B_{t_j}| \le L \text{ for all } j \le i-1 \right\}$ for $i = 1, 2, \ldots, n$. First show that $S_{n,L}$ converges to 0 in $L^2(\Omega)$ and in probability.

35.3. Use Itô's formula with $f(t,x) = tx^2$.

35.4. Part (a) can be proved by using Itô's formula with

$$f(t,x) = \exp\left(cx - \frac{1}{2}c^2 t \right)$$

and Theorem 32.2. Part (b) can be proved similarly.

35.5. Use the general form of Itô's formula with $\theta(t,x) = e^{\beta t}x$.

35.6. Apply the multidimensional Itô formula (Theorem 35.4).

35.7. Part (a) can be proved by the definition of cross-variation processes (Definition 35.5) and the fact that $(X - Y)^2 = (Y - X)^2$. Part (b) can be proved by using Theorem 35.8. Part (c) is implied by the definition. Part (d) can be derived from parts (b) and (c). Part (e), one can first establish that

$$4|\langle X, Y\rangle_t|^2 \le (\langle X\rangle_t - \langle Y\rangle_t)^2 + 4\langle X\rangle_t \langle Y\rangle_t$$

and then use the method of contradiction.

35.8. First calculate $\langle B^{(1)}, B^{(2)}\rangle_t$ by using the fact that $B^{(1)}B^{(2)} - \langle B^{(1)}, B^{(2)}\rangle_t$ is a martingale. Other two compensators can be derived by using the results of Problem 35.7.

35.9. Use the definition of $\langle X, Y\rangle_t$ (Definition 35.5), the result of Problem 34.8, and Theorem 35.3.

35.10. The proof is similar to that of Problem 35.9, namely, using the definition of $\langle X, Y\rangle_t$ (Definition 35.5), the result of Problem 34.8, and Theorem 35.4.

35.11. Consider

$$Y_t = \sum_{j=1}^{m} a_j B_t^{(j)}$$

and $\theta(t,x) = \exp(ct + x)$. Then apply the multidimensional Itô formula (Theorem 35.4).

35.12. First consider the case when f is bounded. Use Itô's lemma (Theorem 35.3), the result of Problem 32.16, and Hölder's inequality (Theorem 8.1) to prove this case. Then consider the general case by constructing a sequence of bounded processes.

35.4 Solutions

35.1. Let ξ_n be defined as

$$\xi_n = \max_{1\le i\le n, 0<\lambda<1} \left|g(B_{t_{i-1}} + \lambda(B_{t_i} - B_{t_{i-1}})) - g(B_{t_{i-1}})\right|, \quad n \ge 1.$$

By the definition of the Brownian motion, almost all sample paths of the B_t are continuous. Suppose that $B_t(\omega)$ is continuous for all $\omega \in A$ for some $A \subseteq \Omega$ with $P(A) = 1$. Then for each $\omega \in A$, we know that $\xi_n(\omega)$ converges to 0 as $\|\Delta_n\| \to 0$ by the continuity of $g(x)$ and $B_t(\omega)$.

By Problem 32.10, $\sum_{i=1}^n (B_{t_i} - B_{t_{i-1}})^2$ converges to $t-a$ in $L^2(\Omega)$. By Problems 9.1 and 9.6, there exists a subsequence of $\sum_{i=1}^n (B_{t_i} - B_{t_{i-1}})^2$ converging to $t - a$ a.s. Since

$$|X_n| \le \xi_n \sum_{i=1}^n (B_{t_i} - B_{t_{i-1}})^2,$$

there exists a subsequence of X_n converging to 0 a.s. This completes the proof.

35.2. Let $L > 0$. Let $A_{i-1,L}$ be events defined as

$$A_{i-1,L} = \{|B_{t_j}| \le L \text{ for all } j \le i-1\}, \quad i = 1, 2, \ldots, n.$$

Let $S_{n,L}$ be a sequence defined as

$$S_{n,L} = \sum_{i=1}^n g(B_{t_{i-1}}) \left[(B_{t_i} - B_{t_{i-1}})^2 - (t_i - t_{i-1})\right] I_{A_{i-1,L}}, \quad n \ge 1.$$

Let $Y_i = \left[(B_{t_i} - B_{t_{i-1}})^2 - (t_i - t_{i-1})\right]$ and $Z_{i,L} = g(B_{t_{i-1}}) Y_i I_{A_{i-1,L}}$ for $i = 1, 2, \ldots, n$. Let $\mathscr{F}_t = \sigma(B_s : s \le t)$. Then for $i < j$, we have

$$\begin{aligned} E(Z_{i,L} Z_{j,L}) &= E[E(Z_{i,L} Z_{j,L} | \mathscr{F}_{t_{j-1}})] \\ &= E[g(B_{t_{i-1}}) Y_i g(B_{t_{j-1}}) I_{A_{i-1,L}} I_{A_{j-1,L}} E(Y_j | \mathscr{F}_{t_{j-1}})] \\ &= 0. \end{aligned}$$

Since $Z_{i,L}^2 \le Y_i^2 \max_{|x|\le L} |g(x)|^2$, we have

$$E\left[Z_{i,L}^2\right] \le E\left[Y_i^2\right] \max_{|x|\le L} |g(x)|^2 = 2(t_i - t_{i-1})^2 \max_{|x|\le L} |g(x)|^2.$$

Hence we have

$$\begin{aligned} E\left[S_{n,L}^2\right] &= \sum_{i=1}^n E\left[Z_{i,L}^2\right] + 2\sum_{i<j} E(Z_{i,L} Z_{j,L}) \\ &\le 2 \max_{|x|\le L} |g(x)|^2 \sum_{i=1}^n (t_i - t_{i-1})^2 \\ &\le 2 \max_{|x|\le L} |g(x)|^2 \|\Delta_n\| (t - a), \end{aligned}$$

which implies that $S_{n,L} \xrightarrow{L^2(\Omega)} 0$ as $\|\Delta_n\| \to 0$. By Problem 9.1, $S_{n,L}$ converges to 0 in probability.

Now let $\epsilon > 0$. Then

$$P\{|S_n| \geq \epsilon\} \leq P\{|S_{n,L}| \geq \epsilon\} + P\{S_n \neq S_{n,L}\}. \tag{35.3}$$

Since

$$\{S_n \neq S_{n,L}\} \subseteq A^c_{n-1,L} \subseteq \left\{ \max_{a \leq s \leq t} |B_s| > L \right\},$$

we have by Doob submartingale inequality (see Theorem 24.1),

$$P\{S_n \neq S_{n,L}\} \leq P\left\{ \max_{a \leq s \leq t} |B_s| > L \right\} \leq \frac{1}{L} E(|B_t|) = \frac{1}{L}\sqrt{\frac{2t}{\pi}}.$$

Let $\delta > 0$. Let $L_\delta = \frac{2}{\delta}\sqrt{\frac{2t}{\pi}}$. Then

$$P\{S_n \neq S_{n,L_\delta}\} \leq \frac{\delta}{2}.$$

Since S_{n,L_δ} converges to 0 in probability, there exist an $N_{\epsilon,\delta}$ such that

$$P(|S_{n,L_\delta}| \geq \epsilon) < \delta/2$$

for all $n \geq N_{\epsilon,\delta}$. Thus for $n \geq N_{\epsilon,\delta}$, we have by Equation (35.3),

$$P\{|S_n| \geq \epsilon\} < \delta.$$

Since ϵ and δ are arbitrary, S_n converges to 0 in probability. This completes the proof.

35.3. Let $f(t, x) = tx^2$. Then we have $\frac{\partial f}{\partial t} = x^2$, $\frac{\partial f}{\partial x} = 2tx$, and $\frac{\partial^2 f}{\partial x^2} = 2t$. By Theorem 35.2, we have

$$\begin{aligned} tB_t^2 &= \int_0^t \frac{\partial f}{\partial x}(s, B_s)\mathrm{d}B_s + \int_0^t \left(\frac{\partial f}{\partial t}(s, B_s) + \frac{1}{2}\frac{\partial^2 f}{\partial x^2}(s, B_s) \right) \mathrm{d}s \\ &= \int_0^t 2sB_s\mathrm{d}B_s + \int_0^t \left(B_s^2 + s\right) \mathrm{d}s \\ &= 2\int_0^t sB_s\mathrm{d}B_s + \int_0^t B_s^2\mathrm{d}s + \frac{1}{2}t^2. \end{aligned}$$

This completes the proof.

35.4.

(a) Let $f(t, x) = \exp\left(cx - \frac{1}{2}c^2 t\right)$. Then we have

$$\frac{\partial f}{\partial t} = -\frac{c^2}{2} \exp\left(cx - \frac{1}{2}c^2 t\right),$$

$$\frac{\partial f}{\partial x} = c \exp\left(cx - \frac{1}{2}c^2 t\right),$$

and

$$\frac{\partial^2 f}{\partial x^2} = c^2 \exp\left(cx - \frac{1}{2}c^2 t\right).$$

By Theorem 35.2, we get

$$M_t - 1 = \int_0^t c \exp\left(cB_s - \frac{1}{2}c^2 s\right) \mathrm{d}B_s.$$

Since $c \exp\left(cB_s - \frac{1}{2}c^2 s\right) \in L^2_{ad}([0,t] \times \Omega)$, M_t is a martingale by Theorem 32.2.

(b) Since $M_t^2 = e^{2cB_t - c^2 t}$, we consider the function $f(t,x) = e^{2cx - c^2 t}$. In this case, we have

$$\frac{\partial f}{\partial t} = -c^2 \exp\left(2cx - c^2 t\right),$$

$$\frac{\partial f}{\partial x} = 2c \exp\left(2cx - c^2 t\right),$$

and

$$\frac{\partial^2 f}{\partial x^2} = 4c^2 \exp\left(2cx - c^2 t\right).$$

By Theorem 35.2, we have

$$M_t^2 - 1 = 2c \int_0^t \exp\left(2cB_s - c^2 s\right) \mathrm{d}B_s + c^2 \int_0^t \exp\left(2cB_s - c^2 s\right) \mathrm{d}s.$$

By Theorem 32.2, $1 + 2c \int_0^t \exp\left(2cB_s - c^2 s\right) \mathrm{d}B_s$ is a martingale. Note that

$$c^2 \int_0^t \exp\left(2cB_s - c^2 s\right) \mathrm{d}s$$

is an increasing process. The compensator of M_t^2 is given by

$$\langle M \rangle_t = c^2 \int_0^t \exp\left(2cB_s - c^2 s\right) \mathrm{d}s.$$

35.5. Note that the X_t is an Itô process given by

$$X_t = x_0 + \int_0^t \alpha \mathrm{d}B_s + \int_0^t (-\beta X_s) \mathrm{d}s.$$

Let $\theta(t,x) = e^{\beta t} x$. Then we have $\frac{\partial \theta}{\partial t} = \beta e^{\beta t} x$, $\frac{\partial \theta}{\partial x} = e^{\beta t}$, and $\frac{\partial^2 \theta}{\partial x^2} = 0$. By Theorem 35.3, we have

$$\begin{aligned} e^{\beta t} X_t - x_0 &= \int_0^t \alpha e^{\beta s} \mathrm{d}B_s + \int_0^t \left[\beta e^{\beta s} X_s + (-\beta X_s) e^{\beta s} + 0\right] \mathrm{d}s \\ &= \alpha \int_0^t e^{\beta s} \mathrm{d}B_s, \end{aligned}$$

which gives

$$X_t = x_0 e^{-\beta t} + \alpha \int_0^t e^{-\beta(t-s)} \mathrm{d}B_s.$$

This completes the proof.

35.6. Let $\theta(t, x, y) = xy$. Then we have $\frac{\partial \theta}{\partial t} = 0$, $\frac{\partial \theta}{\partial x} = y$, $\frac{\partial \theta}{\partial y} = x$, $\frac{\partial^2 \theta}{\partial y^2} = 0$, and $\frac{\partial^2 \theta}{\partial x^2} = 0$. By Theorem 35.4, we have

$$\begin{aligned}
\mathrm{d}(X_t Y_t) &= Y_t \mathrm{d}X_t + X_t \mathrm{d}Y_t + \frac{1}{2}\mathrm{d}X_t \mathrm{d}Y_t + \frac{1}{2}\mathrm{d}Y_t \mathrm{d}X_t \\
&= Y_t \mathrm{d}X_t + X_t \mathrm{d}Y_t + \mathrm{d}X_t \mathrm{d}Y_t \\
&= Y_t f(t)\mathrm{d}B_t + Y_t \xi(t)\mathrm{d}t + X_t g(t)\mathrm{d}B_t + X_t \eta(t)\mathrm{d}t + f(t)g(t)\mathrm{d}t \\
&= [Y_t f(t) + X_t g(t)]\mathrm{d}B_t + [Y_t \xi(t) + X_t \eta(t) + f(t)g(t)]\mathrm{d}t.
\end{aligned}$$

The result follows from the above equation. This completes the proof.

35.7.

(a) Note that $(X - Y)^2 = (Y - X)^2$. By the uniqueness of the Doob-Meyer decomposition, we have $\langle X - Y \rangle_t = \langle Y - X \rangle_t$. Hence we have

$$\begin{aligned}
\langle X, Y \rangle_t &= \frac{1}{4}\left(\langle X + Y \rangle_t - \langle X - Y \rangle_t\right) \\
&= \frac{1}{4}\left(\langle Y + X \rangle_t - \langle Y - X \rangle_t\right) \\
&= \langle Y, X \rangle_t.
\end{aligned}$$

(b) By Theorem 35.8, $\langle \alpha X + \beta Y, Z \rangle_t$ is the only process of the form $A = A^{(1)} - A^{(2)}$ with $A^{(1)}$ and $A^{(2)}$ adapted and natural, such that $(\alpha X + \beta Y)Z - A$ is a martingale. However, $\alpha \langle X, Z \rangle_t + \beta \langle Y, Z \rangle_t$ is a process of the same form such that

$$(\alpha X + \beta Y)Z - \alpha \langle X, Z \rangle_t - \beta \langle Y, Z \rangle_t$$

is a martingale. Therefore, we have

$$\langle \alpha X + \beta Y, Z \rangle_t = \alpha \langle X, Z \rangle_t + \beta \langle Y, Z \rangle_t.$$

(c) By definition, we have

$$X_t^2 = M_t + \langle X \rangle_t,$$

where M_t is a martingale. Multiplying α^2 to both sides of the above equation gives

$$(\alpha X_t)^2 = \alpha^2 M_t + \alpha^2 \langle X \rangle_t.$$

Since $\alpha^2 \langle X \rangle_t$ is an adapted and natural process and $\alpha^2 M_t$ is a martingale, we have $\langle \alpha X \rangle_t = \alpha^2 \langle X \rangle_t$.

(d) By parts(b) and (c), we have

$$\langle X, X\rangle_t = \frac{1}{4}\langle 2X\rangle_t = \langle X\rangle_t.$$

(e) By part (b), we have

$$\langle X+Y, X+Y\rangle_t = \langle X\rangle_t + \langle Y\rangle_t + 2\langle X, Y\rangle_t$$

and

$$\langle X-Y, X-Y\rangle_t = \langle X\rangle_t + \langle Y\rangle_t - 2\langle X, Y\rangle_t.$$

Since

$$\langle X+Y, X+Y\rangle_t = \langle X+Y\rangle_t \geq 0$$

and

$$\langle X-Y, X-Y\rangle_t = \langle X-Y\rangle_t \geq 0,$$

we have

$$2|\langle X, Y\rangle_t| \leq \langle X\rangle_t + \langle Y\rangle_t.$$

Therefore, we have

$$4|\langle X, Y\rangle_t|^2 \leq (\langle X\rangle_t - \langle Y\rangle_t)^2 + 4\langle X\rangle_t\langle Y\rangle_t.$$

Replacing X by αX in the above equation and dividing α^2 in both sides, we get

$$4|\langle X, Y\rangle_t|^2 \leq \left(\alpha\langle X\rangle_t - \frac{1}{\alpha}\langle Y\rangle_t\right)^2 + 4\langle X\rangle_t\langle Y\rangle_t. \tag{35.4}$$

Suppose that there exist a ω_0 such that

$$|\langle X, Y\rangle_t(\omega_0)|^2 > \langle X\rangle_t(\omega_0)\cdot\langle Y\rangle_t(\omega_0). \tag{35.5}$$

But if we let $\alpha = \sqrt{\frac{\langle Y\rangle_t(\omega_0)}{\langle X\rangle_t(\omega_0)}}$ in Equation (35.4), we get

$$|\langle X, Y\rangle_t(\omega_0)|^2 \leq \langle X\rangle_t(\omega_0)\cdot\langle Y\rangle_t(\omega_0),$$

which contradicts Equation (35.5). Hence we have

$$4|\langle X, Y\rangle_t|^2 \leq 4\langle X\rangle_t\langle Y\rangle_t,$$

which gives $|\langle X, Y\rangle_t|^2 \leq \langle X\rangle_t\langle Y\rangle_t$.

This completes the proof.

35.8. Let us first calculate $\langle B^{(1)}, B^{(2)}\rangle_t$. To do that, let us consider

$$E\left(B_t^{(1)}B_t^{(2)}|\mathscr{F}_s\right)$$

for $s \le t$, where the $\mathscr{F}_t$ is a filtration under which $B_t^{(1)}$ and $B_t^{(2)}$ are martingales. By using the properties of conditional expectations and the assumption that $B_t^{(1)}$ and $B_t^{(2)}$ are independent, we have

$$\begin{aligned}
& E\left(B_t^{(1)} B_t^{(2)} | \mathscr{F}_s\right) \\
= \; & E\left((B_t^{(1)} - B_s^{(1)})(B_t^{(2)} - B_s^{(2)}) | \mathscr{F}_s\right) + E\left(B_s^{(1)}(B_t^{(2)} - B_s^{(2)}) | \mathscr{F}_s\right) \\
& + E\left(B_t^{(1)} B_s^{(2)} | \mathscr{F}_s\right) \\
= \; & E\left((B_t^{(1)} - B_s^{(1)})(B_t^{(2)} - B_s^{(2)})\right) + B_s^{(1)} E\left((B_t^{(2)} - B_s^{(2)}) | \mathscr{F}_s\right) \\
& + B_s^{(1)} B_s^{(2)} \\
= \; & B_s^{(1)} B_s^{(2)}.
\end{aligned}$$

Hence the $B_t^{(1)} B_t^{(2)}$ is a martingale. Therefore, we have

$$\langle B^{(1)}, B^{(2)} \rangle_t = 0.$$

Now by Problem 35.7, we have

$$\langle B^{(1)} + B^{(2)} \rangle_t = \langle B^{(1)} \rangle_t + \langle B^{(2)} \rangle_t + 2\langle B^{(1)}, B^{(2)} \rangle_t = 2t$$

and

$$\langle B^{(1)} - B^{(2)} \rangle_t = \langle B^{(1)} \rangle_t + \langle B^{(2)} \rangle_t - 2\langle B^{(1)}, B^{(2)} \rangle_t = 2t.$$

35.9. Since $f, g \in L^2_{ad}([a, b] \times \Omega)$, it follows from Theorems 32.1 and 32.2 that $X_t + Y_t = \int_a^t (f(s) + g(s))\mathrm{d}B_s$ and $X_t - Y_t = \int_a^t (f(s) - g(s))\mathrm{d}B_s$ are continuous, square-integrable martingales. By Problem 34.8 and Theorem 35.3, we have

$$\begin{aligned}
& \langle X + Y \rangle_t \\
= \; & (X_t + Y_t)^2 - (X_a + Y_a)^2 - 2\int_a^t (X_s + Y_s)(f(s) + g(s))\mathrm{d}B_s \\
= \; & \int_a^t (f(s) + g(s))^2 \mathrm{d}s.
\end{aligned}$$

Similarly, we have

$$\begin{aligned}
& \langle X - Y \rangle_t \\
= \; & (X_t - Y_t)^2 - (X_a - Y_a)^2 - 2\int_a^t (X_s - Y_s)(f(s) - g(s))\mathrm{d}B_s \\
= \; & \int_a^t (f(s) - g(s))^2 \mathrm{d}s.
\end{aligned}$$

Then by Definition 35.5, we have

$$\langle X, Y \rangle_t = \frac{1}{4}(\langle X + Y \rangle_t - \langle X - Y \rangle_t) = \int_a^t f(s)g(s)\mathrm{d}s.$$

This completes the proof.

35.10. Since $f, g \in L^2_{ad}([a, b] \times \Omega)$, it follows from Theorems 32.1 and 32.2 that X_t and Y_t are continuous, square-integrable martingales. Since

$$E\left[(X_t + Y_t)^2\right] \leq 2E\left[X_t^2\right] + 2E\left[Y_t^2\right],$$

$X_t + Y_t$ is also a continuous, square-integrable martingale. Similarly, $X_t - Y_t$ is also a continuous, square-integrable martingale. By Problem 34.8 and Theorem 35.4, we have

$$\begin{aligned} & \langle X + Y \rangle_t \\ = \ & (X_t + Y_t)^2 - (X_a + Y_a)^2 - 2\int_a^t (X_s + Y_s)(f(s)\mathrm{d}B_s^{(1)} + g(s)\mathrm{d}B_s^{(2)}) \\ = \ & \int_a^t [f^2(s) + g^2(s)]\mathrm{d}s. \end{aligned}$$

Similarly, we have

$$\begin{aligned} & \langle X - Y \rangle_t \\ = \ & (X_t - Y_t)^2 - (X_a - Y_a)^2 - 2\int_a^t (X_s - Y_s)(f(s)\mathrm{d}B_s^{(1)} - g(s)\mathrm{d}B_s^{(2)}) \\ = \ & \int_a^t [f^2(s) + g^2(s)]\mathrm{d}s. \end{aligned}$$

Then by Definition 35.5, we have

$$\langle X, Y \rangle_t = \frac{1}{4}(\langle X + Y \rangle_t - \langle X - Y \rangle_t) = 0.$$

This completes the proof.

35.11. Let

$$Y_t = \sum_{j=1}^m a_j B_t^{(j)}$$

and $\theta(t, x) = \exp(ct + x)$. Then we have $X_t = \theta(t, Y_t)$ and

$$\mathrm{d}Y_t = \sum_{j=1}^m \mathrm{d}B_t^{(j)}.$$

By Theorem 35.4, we have

$$\begin{aligned}
\mathrm{d}X_t &= \mathrm{d}\theta(t, Y_t) \\
&= c\theta(t, Y_t)\mathrm{d}t + \theta(t, Y_t)\mathrm{d}Y_t + \frac{1}{2}\theta(t, Y_t)\mathrm{d}Y_t\mathrm{d}Y_t \\
&= cX_t\mathrm{d}t + X_t\left(\sum_{j=1}^m \mathrm{d}B_t^{(j)}\right) + \frac{1}{2}X_t\sum_{j=1}^m a_j^2\mathrm{d}t \\
&= \left(c + \frac{1}{2}\sum_{j=1}^m a_j^2\right)X_t\mathrm{d}t + X_t\left(\sum_{j=1}^m \mathrm{d}B_t^{(j)}\right).
\end{aligned}$$

This completes the proof.

35.12. First let us assume that $|f| \le k$ for some constant k. Let

$$Y_t = \int_a^t f(s)\mathrm{d}B_s, \quad t \in [a, b].$$

Then by Problem 32.16, $\exp(\pm Y_t)$ is finite. Hence all moments of Y_t are finite. Therefore, we have

$$\int_a^b E\left[Y_s^{4n-2}f^2(s)\right]\mathrm{d}s \le k^2 E\left[\int_a^b Y_s^{4n-2}\mathrm{d}s\right] < \infty,$$

which shows that $Y_s^{2n-1}f(s) \in L^2_{ad}([a, b] \times \Omega)$. By Theorems 35.3 and 32.1, we have

$$\begin{aligned}
E\left[Y_t^{2n}\right] &= 2nE\left[\int_a^t Y_s^{2n-1}f(s)\mathrm{d}B_s\right] + n(2n-1)E\left[\int_a^t Y_s^{2n-2}f^2(s)\mathrm{d}s\right] \\
&= n(2n-1)E\left[\int_a^t Y_s^{2n-2}f^2(s)\mathrm{d}s\right]. \qquad (35.6)
\end{aligned}$$

By Theorem 8.1, we have

$$\begin{aligned}
&E\left[Y_t^{2n}\right] \\
\le\ & n(2n-1)\int_a^t \left(E\left[Y_s^{2n}\right]\right)^{\frac{n-1}{n}} \cdot \left(E\left[f^{2n}(s)\right]\right)^{\frac{1}{n}}\mathrm{d}s \\
\le\ & n(2n-1)\left(\int_a^t E\left[Y_s^{2n}\right]\mathrm{d}s\right)^{\frac{n-1}{n}} \cdot \left(\int_a^t E\left[f^{2n}(s)\right]\mathrm{d}s\right)^{\frac{1}{n}}. \qquad (35.7)
\end{aligned}$$

Since Y_t^{2n} is a submartingale (see Problem 22.8), we get

$$E\left[Y_s^{2n}\right] \le E\left[Y_t^{2n}\right], \quad s \le t.$$

Thus Equation (35.7) becomes

$$E\left[Y_t^{2n}\right] \le n(2n-1)\left((t-a)E\left[Y_t^{2n}\right]\right)^{\frac{n-1}{n}} \cdot \left(\int_a^t E\left[f^{2n}(s)\right] \mathrm{d}s\right)^{\frac{1}{n}}.$$

Taking powers of n on both sides of the above inequality gives the result with $C_n = [n(2n-1)]^n$.

Now we prove the general case. Let $f_m(s) = \min\{m, \max\{-m, f(s)\}\}$ for $m \ge 1$. By the assumption that $f^n \in L^2_{ad}([a,b]\times\Omega)$, we have $f \in L^2_{ad}([a,b]\times\Omega)$ (see Problem 8.5). Since

$$|f_m(s) - f(s)| \le |f_m(s)| + |f(s)| \le 2|f(s)|,$$

it follows from Theorem 6.4 that

$$\begin{aligned}
& \lim_{m\to\infty} E\left[\left(\int_0^t f_m(s)\mathrm{d}s - \int_0^t f(s)\mathrm{d}s\right)^2\right] \\
= & \lim_{m\to\infty} E\left[\int_0^t \left(f_m(s) - f(s)\right)^2 \mathrm{d}s\right] \\
= & E\left[\int_0^t \lim_{n\to\infty} \left(f_m(s) - f(s)\right)^2 \mathrm{d}s\right] = 0.
\end{aligned}$$

Hence

$$\int_0^t f_m(s)\mathrm{d}s \xrightarrow{L^2} \int_0^t f(s)\mathrm{d}s.$$

Then by Problem 9.1 and Problem 9.6, there exists a subsequence $f_{m_1}, f_{m_2}, \ldots$ such that

$$\int_0^t f_{m_i}(s)\mathrm{d}s \xrightarrow{\text{a.s.}} \int_0^t f(s)\mathrm{d}s.$$

Hence it follows from Fatou's lemma (Theorem 6.3) that

$$\begin{aligned}
& C_n(t-a)^{n-1} \int_a^t E\left[f^{2n}(s)\right] \mathrm{d}s \\
\ge & \liminf \left(C_n(t-a)^{n-1} \int_a^t E\left[f_{m_i}^{2n}(s)\right] \mathrm{d}s\right) \\
\ge & \liminf E\left[\left|\int_a^t f_{m_i}(s)\mathrm{d}B_s\right|^{2n}\right] \\
\ge & E\left[\liminf \left|\int_a^t f_{m_i}(s)\mathrm{d}B_s\right|^{2n}\right] \\
= & E\left[\left|\int_a^t f(s)\mathrm{d}B_s\right|^{2n}\right].
\end{aligned}$$

This completes the proof.

35.5 Bibliographic Notes

In this chapter, we introduced the simple Itô formula, the multidimensional Itô formula, and the Itô formula for martingales. For more information about Itô formula, readers are referred to Oksendal (1998) and Kuo (2006). Theorem 35.8 can be proved by using the same argument used to prove the uniqueness of the Doob-Meyer decomposition (Karatzas and Shreve, 1991, p31).

For some moment inequalities for Itô integrals, readers may read the papers by Zakai (1967) and Novikov (1973).

CHAPTER 36

MARTINGALE REPRESENTATION THEOREM

The Itô integral defined in Chapter 32 is a martingale with respect to the underlying filtration. The martingale representation theorem states that the converse is also true. In this chapter, we introduce the martingale representation theorem.

36.1 Basic Concepts and Facts

Theorem 36.1 (Martingale Representation Theorem). *Let* $\{\mathscr{F}_t^B : a \leq t \leq b\}$ *be the filtration generated by the Brownian motion* $\{B_t : a \leq t \leq b\}$*, that is,* $\mathscr{F}_t^B = \sigma(B_s : s \leq t)$*. Let* $\{M_t : a \leq t \leq b\}$ *be a square integrable martingale with respect to the filtration. Then* M_t *has a continuous version* $\tilde{M}_t$ *given by*

$$\tilde{M}_t = M_a + \int_a^t \theta(s)\mathrm{d}B_s, \quad a \leq t \leq b,$$

where $\theta \in L^2_{ad}([a, b] \times \Omega)$.

36.2 Problems

36.1. Let $\{B_t : t \geq 0\}$ be a Brownian motion with respect to a filtration $\{\mathscr{F}_t : t \geq 0\}$. Let $0 \leq a < b$ and X be a finite $\mathscr{F}_a$-measurable random variable. Show that there exists a stopping time τ with $a \leq \tau < b$ such that

(a)

$$X = \int_a^\tau \frac{1}{b-t} \mathrm{d}B_t.$$

(b) For any $\lambda > 0$ and $\beta \in \mathbf{R}$, we have

$$P\{h(\tau) \geq \lambda\} \leq \frac{2|\beta|}{\sqrt{\lambda}} + P\{|X| \geq |\beta|\},$$

where

$$h(\tau) = \int_a^\tau \frac{1}{(b-s)^2} \mathrm{d}s.$$

36.2 (Representation by Informative Increments). Let $\{B_t : 0 \leq t \leq T\}$ be a Brownian motion and $\mathscr{F}_T = \sigma(B_t : t \leq T)$. Let X be a $\mathscr{F}_T$-measurable random variable. Suppose that $\{a_n\}_{n\geq 1}$ and $\{b_n\}_{n\geq 1}$ are two summable sequences of decreasing real numbers. Show that there exists an increasing sequence $\{t_n\}_{n\geq 1}$ of real numbers such that $t_n \uparrow T$ and a sequence $\{\Delta_n\}_{n\geq 1}$ of random variables such that

(a) Δ_n is $\mathscr{F}_{t_n}$-measurable for all $n \geq 1$.

(b)

$$P\left\{X = \sum_{n=1}^{\infty} \Delta_n\right\} = 1.$$

(c) $P\{|\Delta_n| \geq a_n\} \leq b_n$ for all $n \geq 1$.

36.3 (Dudley's Representation Theorem). Let $\{B_t : 0 \leq t \leq T\}$ be a Brownian motion and $\mathscr{F}_T = \sigma(B_t : t \leq T)$. Let X be a $\mathscr{F}_T$-measurable random variable. Show that there exists a $\phi \in \mathcal{L}_{ad}(\Omega, L^2[0, T])$ (see Definition 33.1) such that

$$X = \int_0^T \phi(s, \omega) \mathrm{d}B_s \quad \text{a.s.}$$

36.4. Let $\{B_t : 0 \leq t \leq T\}$ be a Brownian motion and $\{\mathscr{F}_t : 0 \leq t \leq T\}$ the corresponding natural filtration. Let X and Y be two bounded random variables with the following representations:

$$X = x_0 + \int_0^T \phi(t, \omega) \mathrm{d}B_t,$$

$$Y = y_0 + \int_0^T \psi(t, \omega) \mathrm{d}B_t,$$

where ϕ and ψ are elements of $L^2_{ad}([0,T]\times\Omega)$ (see Definition 32.1). Suppose that

$$\int_0^T \phi(s,\omega)\psi(s,\omega)\mathrm{d}s = 0.$$

Show that

$$XY = x_0 y_0 + \int_0^T [X_s\psi(s,\omega) + Y_s\phi(s,\omega)]\mathrm{d}B_s,$$

where $\{X_t : 0 \le t \le T\}$ and $\{Y_t : 0 \le t \le T\}$ are bounded martingales defined as

$$X_t = E[X|\mathscr{F}_t], \quad Y_t = E[Y|\mathscr{F}_t].$$

36.3 Hints

36.1. To prove part (a), consider the process defined as

$$Y_t = \begin{cases} 0, & \text{if } 0 \le t \le a; \\ \int_a^t \frac{1}{b-u}\mathrm{d}B_u, & \text{if } a \le t < b, \end{cases}$$

and the τ defined as $\tau = \min\{t \ge a : Y_t = X\}$. To prove part (b), consider

$$\nu_\beta = \min\{t : Y_t = \beta, t \ge a\}$$

and then try to establish

$$\{h(\tau) \ge \lambda, |X| < |\beta|\} \le 2P\{h(\nu_\beta) \ge \lambda\}$$

and

$$h(\nu_\beta) = \min\{t : B_t = \beta, t \ge 0\}.$$

36.2. Consider the random variable $Y = \Phi(X)$ and its conditional expectations $Y_n = E(Y|\mathcal{G}_n)$, where $\Phi(\cdot)$ is the standard normal cumulative distribution function and

$$\mathcal{G}_n = \sigma\left(B_t : t \le T - \frac{1}{n}\right), \quad n \ge 1.$$

Then consider $X_n = \Phi^{-1}(Y_n)$ and use the result of Problem 25.5.

36.3. Use the results of Problems 36.1 and 36.2.

36.4. Use Theorem 32.1 and the results of Problems 14.16 and 35.6.

36.4 Solutions

36.1.

(a) Let $\{Y_t : 0 \leq t < b\}$ be a process defined by

$$Y_t = \begin{cases} 0, & \text{if } 0 \leq t \leq a; \\ \int_a^t \frac{1}{b-u} \mathrm{d}B_u, & \text{if } a \leq t < b, \end{cases}$$

Then we have $E[Y_s Y_t] = \min\{h(s), h(t)\}$, where

$$h(t) = \begin{cases} 0, & \text{if } 0 \leq t \leq a; \\ \int_a^t \frac{1}{(b-u)^2} \mathrm{d}u = \frac{1}{b-t} - \frac{1}{b-a}, & \text{if } a \leq t < b, \end{cases}$$

Hence the process $\{Y_t : 0 \leq t < b\}$ is equivalent to the time-changed Brownian motion $\{B_{h(t)} : 0 \leq t < b\}$. Now let τ be defined as

$$\tau = \min\{t \geq a : Y_t = X\}.$$

Then τ is a stopping time and, by Problem 28.12, $P\{\tau < b\} = 1$.

(b) Note that

$$\{h(\tau) \geq \lambda\} = \{h(\tau) \geq \lambda, |X| < |\beta|\} \cup \{h(\tau) \geq \lambda, |X| \geq |\beta|\}.$$

We have

$$\begin{aligned} P\{h(\tau) \geq \lambda\} &= P\{h(\tau) \geq \lambda, |X| < |\beta|\} + P\{h(\tau) \geq \lambda, |X| \geq |\beta|\} \\ &\leq P\{h(\tau) \geq \lambda, |X| < |\beta|\} + P\{|X| \geq |\beta|\}. \end{aligned}$$

Hence it is sufficient to show that

$$P\{h(\tau) \geq \lambda, |X| < |\beta|\} \leq \frac{2|\beta|}{\sqrt{\lambda}}. \tag{36.1}$$

To do that, let

$$\nu_\beta = \min\{t : Y_t = \beta, t \geq a\}.$$

Then

$$\{\tau \geq t, |X| < |\beta|\} \subseteq \{\nu_\beta \geq t\} \cup \{\nu_{-\beta} \geq t\}, \quad t \geq 0.$$

Since $h(\cdot)$ is monotone, we have

$$\{h(\tau) \geq \lambda, |X| < |\beta|\} \leq 2P\{h(\nu_\beta) \geq \lambda\}. \tag{36.2}$$

But noting that $\{Y_t : 0 \leq t < b\}$ and $\{B_{h(t)} : 0 \leq t < b\}$ are equivalent, we have

$$\begin{aligned} h(\nu_\beta) &= \min\{h(t) : Y_t = \beta, t \geq 0\} \\ &= \min\{h(t) : B_{h(t)} = \beta, t \geq 0\} \\ &= \min\{t : B_t = \beta\}. \end{aligned}$$

By Problem 28.15 and the fact that $\varphi(x) \leq \frac{1}{2}$, we have

$$P\{h(\nu_\beta) \geq \lambda\} = \int_\lambda^\infty |\beta| s^{-\frac{3}{2}} \varphi\left(\frac{|\beta|}{\sqrt{s}}\right) \mathrm{d}s \leq \frac{|\beta|}{\sqrt{\lambda}}. \tag{36.3}$$

Hence combining Equations (36.2) and (36.3) leads to the inequality given in Equation (36.1).

This completes the proof.

36.2. Let $Y = \Phi(X)$, where $\Phi(\cdot)$ is the standard normal cumulative distribution function. Let $Y_n = E(Y|\mathcal{G}_n)$, where

$$\mathcal{G}_n = \sigma\left(B_t : t \leq T - \frac{1}{n}\right), \quad n \geq 1.$$

Since $|Y| \leq 1$, by Problem 25.5, we have

$$Y_n \xrightarrow{\text{a.s.}} E(Y|\mathscr{F}_T) = \Phi(X).$$

Since $\Phi(X)$ is strictly monotone, we can define $X_n = \Phi^{-1}(Y_n)$. Then we have

$$X_n \xrightarrow{\text{a.s.}} X.$$

Note that $\{a_n\}_{n\geq 1}$ and $\{b_n\}_{n\geq 1}$ are summable sequences, we can choose a sequence $\{k_n\}_{n\geq 1}$ of integers such that

$$P\left\{|X - X_{k_n}| \geq \frac{a_{n+1}}{2}\right\} \leq \frac{b_{n+1}}{2}, \quad n \geq 1.$$

Now we let $\Delta_1 = X_{k_1}$ and

$$\Delta_n = X_{k_n} - X_{k_{n-1}}, \quad n \geq 2.$$

Then for $t_n = T - \frac{1}{k_n}$, we have $\Delta_n \in m\mathscr{F}_{t_n}$, that is, Δ_n is $\mathscr{F}_{t_n}$-measurable. Hence $\{\Delta_n\}_{n\geq}$ satisfies the first property. The second property follows from the fact that

$$\sum_{n=1}^N \Delta_n = X_{k_N}, \quad N \geq 1,$$

and that $X_{k_N} \xrightarrow{\text{a.s}} X$.

Note that

$$P\{|\Delta_n| \geq a_n\} \leq P\left\{|X - X_{k_n}| \geq \frac{a_n}{2}\right\} + P\left\{|X - X_{k_{n-1}}| \geq \frac{a_n}{2}\right\} \leq b_n, \quad n \geq 2.$$

Hence the third property is satisfied. This completes the proof.

36.3. Let $\{a_n\}_{n\geq 1}$ and $\{b_n\}_{n\geq 1}$ be two decreasing sequences that satisfy

$$\sum_{n=1}^{\infty} na_n < \infty$$

and

$$\sum_{n=1}^{\infty} b_n < \infty.$$

Then by Problem 36.2, we know that there exists an increasing sequence $\{t_n\}_{n\geq 1}$ of real numbers such that $t_n \uparrow T$ and a sequence $\{\Delta_n\}_{n\geq 1}$ of random variables such that

(a) Δ_n is $\mathscr{F}_{t_n}$-measurable for all $n \geq 1$.

(b) $P\{X = \sum_{n=1}^{\infty} \Delta_n\} = 1$ a.s.

(c) $P\{|\Delta_n| \geq a_n\} \leq b_n$ for all $n \geq 1$.

Now by Problem 36.1, we can find, for each $n \geq 1$, a stopping time $\tau_n \in [t_n, t_{n+1})$ such that

$$\Delta_n = \int_{t_n}^{\tau_n} \frac{1}{t_{n+1} - s} \mathrm{d}B_s = \int_0^T \phi_n(s,\omega)\mathrm{d}B_s,$$

where

$$\phi_n(s,\omega) = \begin{cases} \frac{1}{t_{n+1}-s} & \text{if } s \in [t_n, \tau_n); \\ 0, & \text{if otherwise.} \end{cases}$$

Let $\phi(s,\omega)$ be defined as

$$\phi(s,\omega) = \sum_{n=1}^{\infty} \phi_n(s,\omega).$$

We claim that $\phi \in \mathcal{L}_{ad}(\Omega, L^2[0,T])$. Since $\phi_1, \phi_2, \ldots$ have mutually disjoint supports, we have

$$\int_0^T \phi^2(s,\omega)\mathrm{d}s = \sum_{n=1}^{\infty} \int_0^T \phi_n^2(s,\omega)\mathrm{d}s.$$

Thus by Problem 36.1, we have

$$P\left\{\int_0^T \phi_n^2(s,\omega)\mathrm{d}s \geq \frac{1}{n}\right\} \leq 2na_n + P(|\Delta_n| \geq a_n) \leq 2a_n + b_n, \quad n \geq 1.$$

By Problem 9.8, we know that

$$\sum_{n=1}^{\infty} \int_0^T \phi_n^2(s,\omega)\mathrm{d}s$$

exists a.s. Hence $\phi \in \mathcal{L}_{ad}(\Omega, L^2[0,T])$.

Finally, we have

$$\sum_{n=1}^{N} \Delta_n = \sum_{n=1}^{N} \phi_n(s,\omega) = \int_0^{t_{N+1}} \phi(s,\omega)\mathrm{d}s, \quad N \geq 1. \tag{36.4}$$

By Theorem 33.3, we know that

$$t \to \int_0^t \phi(s,\omega)\mathrm{d}s$$

is continuous a.s. Hence letting $N \to \infty$ in Equation (36.4) gives

$$\sum_{n=1}^{\infty} \Delta_n = \sum_{n=1}^{\infty} \phi_n(s,\omega)\mathrm{d}s.$$

This completes the proof.

36.4. By Theorem 32.1 and Problem 14.16, we have

$$X_t = x_0 + \int_0^t \phi(s,\omega)\mathrm{d}B_s,$$

which can be written as $\mathrm{d}X_t = \phi(t,\omega)\mathrm{d}B_t$. Similarly, we have $\mathrm{d}Y_t = \psi(t,\omega)\mathrm{d}B_t$. Then by Problem 35.6, we have

$$X_tY_t = x_0y_0 + \int_0^t [X_s\psi(s,\omega) + Y_s\phi(s,\omega)]\mathrm{d}B_s.$$

This completes the proof.

36.5 Bibliographic Notes

In this chapter, we presented the martingale representation theorem. In mathematical finance, this theorem is used to establish the existence of a hedging strategy. For a proof of the martingale representation theorem, readers are referred to Oksendal (1998, p53), Steele (2003, Chapter 12), and Kuo (2006, p182).

CHAPTER 37

CHANGE OF MEASURE

The notion of change of measure has important applications in mathematical finance. In this chapter, we present some relevant concepts and theorems such as the Girsanov theorems.

37.1 Basic Concepts and Facts

Definition 37.1 (Exponential Process). Let $h \in \mathcal{L}_{ad}(\Omega, L^2[0,T])$. The exponential process given by h is defined to be

$$\mathcal{E}_{h,t} = \exp\left[\int_0^t h(s)\mathrm{d}B_s - \frac{1}{2}\int_0^t h(s)^2\mathrm{d}s\right], \quad 0 \le t \le T,$$

where the B_t is a standard Brownian motion.

Theorem 37.1 (Novikov's Theorem). *Let $\{X_t : 0 \le t \le T\}$ be a martingale, where $T \le \infty$. Let $\{M_t : 0 \le t \le T\}$ be defined as*

$$M_t = \exp\left(X_t - \frac{1}{2}\langle X\rangle_t\right),$$

where $\langle X \rangle_t$ is the compensator of X_t^2 (see Theorem 34.2). Suppose that

$$E\left[\exp\left(\frac{1}{2}\langle X \rangle_t\right)\right] < \infty, \quad \forall t \in [0, T].$$

Then $\{M_t : 0 \leq t \leq T\}$ is a continuous martingale.

Theorem 37.2 (Lévy Characterization of Brownian Motion). *Let $X = \{X_t = (X_t^{(1)}, \ldots, X_t^{(d)}) : t \geq 0\}$ be a continuous stochastic process on the probability space $(\Omega, \mathscr{H}, Q)$ with the state space $(\mathbf{R}^d, \mathcal{B}(\mathbf{R}^d))$, adapted to a filtration $\{\mathscr{H}_t : t \geq 0\}$. Suppose that for every $1 \leq k \leq d$, the process*

$$M_t^{(k)} = X_t^{(k)} - X_0^{(k)}, \quad t \geq 0,$$

is a continuous martingale with respect to $\{\mathscr{H}_t : t \geq 0\}$ under Q, and the cross-variations are given by

$$\left\langle M^{(k)}, M^{(j)} \right\rangle_t = \delta_{kj} t, \quad 1 \leq k, j \leq d,$$

where δ_{kj} is the Kronecker delta. Then X is a d-dimensional Brownian motion with respect to $\{\mathscr{H}_t : t \geq 0\}$ under Q.

37.2 Problems

37.1. Let μ and ν be two probability measures on a measurable space $(\Omega, \mathscr{F})$ such that

$$\frac{\mathrm{d}\nu}{\mathrm{d}\mu} = f$$

for some $f \in L^1(\Omega, \mathscr{F}, \mu)$. Let X be a random variable on $(\Omega, \mathscr{F})$ such that

$$E_\nu[|X|] = \int_\Omega |X(\omega)| f(\omega) \mathrm{d}\mu(\omega) < \infty.$$

Suppose that $\mathscr{H}$ is a sub-σ-algebra of $\mathscr{F}$. Show that

$$E_\nu[X|\mathscr{H}] \cdot E_\mu[f|\mathscr{H}] = E_\mu[Xf|\mathscr{H}], \quad \text{a.s.}$$

37.2 (Exponential Martingale). Let $B_t = (B_t^{(1)}, B_t^{(2)}, \ldots, B_t^{(m)})$ be an m-dimensional Brownian motion. Let $T \leq \infty$ be a constant. Let $\{M_t : 0 \leq t \leq T\}$ be defined as

$$M_t = \exp\left(-\sum_{j=1}^m \int_0^t f_j(s, \omega) \mathrm{d}B_s^{(j)} - \frac{1}{2}\sum_{j=1}^m \int_0^t f_j^2(s, \omega) \mathrm{d}s\right).$$

where $f_1, f_2, \ldots, f_m$ are elements of $L^2_{ad}([0,T] \times \Omega)$ and satisfy Novikov's condition:

$$E\left[\exp\left(\frac{1}{2}\sum_{j=1}^{m}\int_0^T f_j^2(s,\omega)\mathrm{d}s\right)\right] < \infty.$$

Show that $\{M_t : 0 \le t \le T\}$ is a martingale.

37.3. Let $\{M_t : 0 \le t \le T\}$ be a martingale under a filtered probability space $(\Omega, \mathscr{F}, \{\mathscr{F}_t : 0 \le t \le T\}, Q)$, where $T \le \infty$ is a constant. Let P be a probability measure on $(\Omega, \mathscr{F}_T)$ defined as

$$\mathrm{d}Q = M_T \mathrm{d}P.$$

Show that for $t \in [0, T]$,

$$\mathrm{d}Q = M_t \mathrm{d}P \quad \text{on } (\Omega, \mathscr{F}_t).$$

37.4 (The Girsanov Theorem I)**.** Let

$$B_t = (B_t^{(1)}, B_t^{(2)}, \ldots, B_t^{(m)})$$

be an m-dimensional Brownian motion on a filtered probability space $(\Omega, \mathscr{F}, \{\mathscr{F}_t : 0 \le t \le T\}, P)$, where $T \le \infty$ is a constant. Let $Y_t = (Y_t^{(1)}, Y_t^{(2)}, \ldots, Y_t^{(m)})$ be an m-dimensional Itô process given by

$$\mathrm{d}Y_t^{(j)} = f_j(t,\omega)\mathrm{d}t + \mathrm{d}B_t^{(j)}, \quad Y_0 = 0, \quad 0 \le t \le T,$$

where $f_1, f_2, \ldots, f_m$ satisfy Novikov's condition:

$$E_P\left[\exp\left(\frac{1}{2}\sum_{j=1}^{m}\int_0^T f_j^2(s,\omega)\mathrm{d}s\right)\right] < \infty.$$

Let $\{M_t : 0 \le t \le T\}$ be defined as

$$M_t = \exp\left(-\sum_{j=1}^{m}\int_0^t f_j(s,\omega)\mathrm{d}B_s^{(j)} - \frac{1}{2}\sum_{j=1}^{m}\int_0^t f_j^2(s,\omega)\mathrm{d}s\right).$$

Let Q be a probability measure on $(\Omega, \mathscr{F}_T)$ defined as

$$\mathrm{d}Q = M_T \mathrm{d}P.$$

Show that $\{Y_t : 0 \le t \le T\}$ is an m-dimensional Brownian motion with respect to the probability measure Q.

37.5 (The Girsanov Theorem II)**.** Let

$$B_t = (B_t^{(1)}, B_t^{(2)}, \ldots, B_t^{(m)})$$

be an m-dimensional Brownian motion on a filtered probability space $(\Omega, \mathscr{F}, \{\mathscr{F}_t : 0 \le t \le T\}, P)$, where $T \le \infty$ is a constant. Let $Y_t = (Y_t^{(1)}, Y_t^{(2)}, \ldots, Y_t^{(n)})$ be an n-dimensional Itô process given by

$$\mathrm{d}Y_t^{(j)} = \beta_j(t,\omega)\mathrm{d}t + \sum_{k=1}^{m} \theta_{jk}\mathrm{d}B_t^{(k)}, \quad 0 \le t \le T.$$

Suppose that there exist processes $u_1, u_2, \ldots, u_m$ and $v_1, v_2, \ldots, v_n$ such that

$$\begin{pmatrix} \theta_{11} & \theta_{12} & \cdots & \theta_{1m} \\ \theta_{11} & \theta_{22} & \cdots & \theta_{2m} \\ \vdots & \vdots & \ddots & \vdots \\ \theta_{n1} & \theta_{n2} & \cdots & \theta_{nm} \end{pmatrix} \cdot \begin{pmatrix} u_1 \\ u_2 \\ \vdots \\ u_m \end{pmatrix} = \begin{pmatrix} \beta_1 \\ \beta_2 \\ \vdots \\ \beta_n \end{pmatrix} - \begin{pmatrix} v_1 \\ v_2 \\ \vdots \\ v_n \end{pmatrix}.$$

Assume that $u_1, u_2, \ldots, u_m$ satisfy Novikov's condition:

$$E_P\left[\exp\left(\frac{1}{2}\sum_{j=1}^{m}\int_0^T u_j^2(s,\omega)\mathrm{d}s\right)\right] < \infty.$$

Let $\{M_t : 0 \le t \le T\}$ be defined as

$$M_t = \exp\left(-\sum_{j=1}^{m}\int_0^t u_j(s,\omega)\mathrm{d}B_s^{(j)} - \frac{1}{2}\sum_{j=1}^{m}\int_0^t u_j^2(s,\omega)\mathrm{d}s\right).$$

Let Q be a probability measure on $(\Omega, \mathscr{F}_T)$ defined as

$$\mathrm{d}Q = M_T \mathrm{d}P.$$

Let $\{\hat{B}_t : 0 \le t \le T\}$ be defined as

$$\hat{B}_t = \int_0^t u(s,\omega)\mathrm{d}s + B_t.$$

Show that

(a) $\{\hat{B}_t : 0 \le t \le T\}$ is an m-dimensional Brownian motion with respect to the probability measure Q.

(b) $\{Y_t : 0 \le t \le T\}$ has the following representation:

$$\mathrm{d}Y_t^{(j)} = v_j(t,\omega)\mathrm{d}t + \sum_{k=1}^{m}\theta_{jk}(t,\omega)\mathrm{d}\hat{B}_t^{(k)}, \quad j = 1, 2\ldots, n.$$

37.6. Let $B_t = \{(B_t^{(1)}, B_t^{(2)}) : 0 \le t \le T\}$ be a two-dimensional Brownian motion on a filtered probability space $(\Omega, \mathscr{F}, \{\mathscr{F}_t : 0 \le t \le T\}, P)$. Let $\tilde{B}_t = \{(\tilde{B}_t^{(1)}, \tilde{B}_t^{(2)}) : 0 \le t \le T\}$ be defined as

$$\tilde{B}_t^{(1)} = -3t + B_t^{(1)}, \quad \tilde{B}_t^{(2)} = t + B_t^{(2)}.$$

Let $Y_t = (Y_t^{(1)}, Y_t^{(2)})$ be defined as

$$\mathrm{d}Y_t^{(1)} = \mathrm{d}B_t^{(1)} + 3\mathrm{d}B_t^{(2)}, \quad \mathrm{d}Y_t^{(2)} = \mathrm{d}t - \mathrm{d}B_t^{(1)} - 2\mathrm{d}B_t^{(2)}.$$

Find a probability measure Q on $(\Omega, \mathscr{F}_T)$ such that

(a) $Q \sim P$ (see Definition 7.3);

(b) $\{Y_t : 0 \le t \le T\}$ can be represented as

$$\mathrm{d}Y_t^{(1)} = \mathrm{d}\tilde{B}_t^{(1)} + 3\mathrm{d}\tilde{B}_t^{(2)}, \quad \mathrm{d}Y_t^{(2)} = -\mathrm{d}\tilde{B}_t^{(1)} - 2\mathrm{d}\tilde{B}_t^{(2)}.$$

(c) $\{\tilde{B}_t : 0 \le t \le T\}$ is a Brownian motion under Q.

37.7 (Risk-Neutral Measure Q)**.** Let $\{B_t : 0 \le t \le T\}$ be a Brownian motion on a filtered probability space $(\Omega, \mathscr{F}, \{\mathscr{F}_t : 0 \le t \le T\}, P)$, where $T < \infty$. Let $\{W_t : t \ge 0\}$ be a process given by

$$W_t = B_t + \frac{(\mu - r)t}{\sigma},$$

where r, μ, and σ ($\sigma \ne 0$) are finite constants. Show that there exists a probability measure Q equivalent to P such that $\{W_t : 0 \le t \le T\}$ is a Brownian motion under Q.

37.8. Let $\{(B_t^{(1)}, B_t^{(2)}, \ldots, B_t^{(n)}) : 0 \le t \le T\}$ be an n-dimensional Brownian motion on some filtered probability space $(\Omega,\ \mathscr{F},\ \{\mathscr{F}_t : 0 \le t \le T\}, P)$. Let $\{\sigma_{ij}(t) : 0 \le t \le T\}$ be an adapted process for every $i = 1, 2, \ldots, d$ and $j = 1, 2, \ldots, n$. Let $\{(W_t^{(1)}, W_t^{(2)}, \ldots, W_t^{(n)}) : 0 \le t \le T\}$ be an n-dimensional stochastic process defined as

$$W_t^{(i)} = \sum_{j=1}^{n} \int_0^t \frac{\sigma_{ij}(u)}{\sigma_i(u)} \mathrm{d}B_u^{(j)},$$

where

$$\sigma_i(t) = \sqrt{\sum_{j=1}^{n} \sigma_{ij}^2(t)}, \quad i = 1, 2, \ldots, d.$$

Suppose that $\sigma_i(t) > 0$ for all $i = 1, 2, \ldots, d$ and $t \in [0, T]$. Show that for every $i = 1, 2, \ldots, d$, $\{W_t^{(i)} : 0 \le t \le T\}$ is a Brownian motion under the probability measure P.

37.3 Hints

37.1. Use the definition of conditional expectations (Definition 14.1) and the result of Problem 14.12.

37.2. Consider

$$X_t = \sum_{j=1}^{m} \int_0^t f_j(s, \omega) \mathrm{d}B_s^{(j)}$$

and use Theorem 34.4, Problem 34.4, and Theorem 37.1.

37.3. Use the tower property of conditional expectations (see Problem 14.10).

37.4. Use Lévy characterization of Brownian motion (see Theorem 37.2). To do that, use the result of Problem 37.2 to show that $Y_t = (Y_t^{(1)}, Y_t^{(2)}, \ldots, Y_t^{(m)})$ is a martingale with respect to Q and use Theorem 35.8 to calculate $\langle Y^{(i)}, Y^{(j)} \rangle_t$.

37.5. Part (a) follows from Problem 37.4. To prove part (b), consider $\mathrm{d}Y_t^{(j)}$.

37.6. Apply the result of Problem 37.5.

37.7. Use the Girsanov theorem I (see Problem 37.4).

37.8. Apply the Lévy characterization theorem (Theorem 37.2) and use the results of Problems 35.7, 35.9, and 35.10.

37.4 Solutions

37.1. By Definition 14.1, we have

$$\int_A E_\nu[X|\mathscr{H}] \mathrm{d}\nu = \int_A X \mathrm{d}\nu, \quad \forall A \in \mathscr{H}.$$

Since $\mathrm{d}\nu = f \mathrm{d}\mu$, this equation gives

$$\int_A E_\nu[X|\mathscr{H}] \cdot f \mathrm{d}\mu = \int_A X f \mathrm{d}\mu, \quad \forall A \in \mathscr{H}. \tag{37.1}$$

By the assumption that $E_\nu[|X|] < \infty$ and $E_\mu[|f|] < \infty$, by Problem 14.12, we have

$$E_\mu(E_\nu[X|\mathscr{H}] \cdot f|\mathscr{H}) = E_\nu[X|\mathscr{H}] \cdot E_\mu[f|\mathscr{H}].$$

Again by Definition 14.1, we have

$$\int_A E_\nu[X|\mathscr{H}] \cdot f \mathrm{d}\mu = \int_A E_\nu[X|\mathscr{H}] \cdot E_\mu[f|\mathscr{H}] \mathrm{d}\mu, \quad \forall A \in \mathscr{H}. \tag{37.2}$$

Combining Equations (37.1) and (37.2) gives

$$\int_A X f \mathrm{d}\mu = \int_A E_\nu[X|\mathscr{H}] \cdot E_\mu[f|\mathscr{H}]\mathrm{d}\mu, \quad \forall A \in \mathscr{H}.$$

Hence by Definition 14.1, we get

$$E_\nu[X|\mathscr{H}] \cdot E_\mu[f|\mathscr{H}] = E_\mu[Xf|\mathscr{H}], \quad a.s.$$

This completes the proof.

37.2. Let $\{X_t : 0 \le t \le T\}$ be defined as

$$X_t = \sum_{j=1}^{m} \int_0^t f_j(s,\omega)\mathrm{d}B_s^{(j)}.$$

Then by Theorems 32.1 and 32.2, $\{X_t : 0 \le t \le T\}$ is a square-integrable martingale. Note that by Itô's lemma (Theorem 35.4), we have

$$\mathrm{d}X_t^2 = 2X_t\mathrm{d}X_t + \sum_{j=1}^{m} f_j^2(t,\omega)\mathrm{d}t.$$

By Theorem 34.4, we know that $\int_0^t X_s \mathrm{d}X_s$ is a martingale. Therefore, the compensator of X_t is

$$\langle X\rangle_t = \sum_{j=1}^{m} \int_0^t f_j^2(s,\omega)\mathrm{d}s.$$

Thus by the assumption, we have

$$E\left[\exp\left(\frac{1}{2}\langle X\rangle_t\right)\right] < \infty.$$

It follows from Theorem 37.1 that $\{M_t : 0 \le t \le T\}$ is a martingale. This completes the proof.

37.3. By the definition of Q, we have

$$Q(A) = \int_A M_T \mathrm{d}P = E_P[M_T I_A], \quad \forall A \in \mathscr{F}_T. \tag{37.3}$$

Since $\mathscr{F}_t \subseteq \mathscr{F}_T$, Equation (37.3) leads to

$$Q(A) = \int_A M_T \mathrm{d}P = E_P[M_T I_A], \quad \forall A \in \mathscr{F}_t. \tag{37.4}$$

By Problem 14.10, we have

$$\begin{aligned} E_P[M_T I_A] &= E_P[E_P(M_T I_A|\mathscr{F}_s)] \\ &= E_P[I_A E(M_T|\mathscr{F}_t)] \\ &= E_P[I_A M_t], \quad \forall A \in \mathscr{F}_t. \end{aligned} \tag{37.5}$$

Combining Equation (37.4) and Equation (37.5) gives

$$Q(A) = E_P[M_t I_A] = \int_A M_t \mathrm{d}P, \quad \forall A \in \mathscr{F}_t.$$

Hence we have $\mathrm{d}Q = M_t \mathrm{d}P$ on $(\Omega, \mathscr{F}_t)$. This completes the proof.

37.4. First, let use show that $Y_t^{(i)}$ is a martingale under the probability measure Q. To do that, let $K_t^{(i)} = M_t Y_t^{(i)}$. Then by Theorem 35.4, we have

$$\begin{aligned}
\mathrm{d}K_t^{(i)} &= M_t \mathrm{d}Y_t^{(i)} + Y_t^{(i)} \mathrm{d}M_t + \mathrm{d}Y_t^{(i)} \mathrm{d}M_t \\
&= M_t \left[f_i(t,\omega)\mathrm{d}t + \mathrm{d}B_t^{(i)} \right] + Y_t^{(i)} M_t \left(-\sum_{k=1}^m f_k(t,\omega)\mathrm{d}B_t^{(k)} \right) \\
&\quad - M_t f_i(t,\omega)\mathrm{d}t \\
&= M_t \left(\mathrm{d}B_t^{(i)} - Y_t^{(i)} \sum_{k=1}^m f_k(t,\omega)\mathrm{d}B_t^{(k)} \right),
\end{aligned}$$

which shows that $K_t^{(i)}$ is a martingale under probability measure P. By Problem 37.2, M_t is also a martingale under probability measure P. Then by Problems 37.1 and 37.3, we have

$$\begin{aligned}
E_Q \left[Y_t^{(i)} | \mathscr{F}_s \right] &= \frac{E_P \left[M_t Y_t^{(i)} | \mathscr{F}_s \right]}{E_P \left[M_t | \mathscr{F}_s \right]} \\
&= \frac{K_s^{(i)}}{M_s} \\
&= Y_s^{(i)}.
\end{aligned}$$

Hence $Y_t^{(i)}$ is a martingale under probability measure Q.

Next, let us show that $\langle Y^{(i)}, Y^{(j)} \rangle_t = \delta_{ij} t$. Since

$$\begin{aligned}
\mathrm{d}\left(Y_t^{(i)} Y_t^{(j)} \right) &= Y_t^{(j)} \mathrm{d}Y_t^{(i)} + Y_t^{(i)} \mathrm{d}Y_t^{(j)} + \mathrm{d}Y_t^{(i)} \mathrm{d}Y_t^{(j)} \\
&= Y_t^{(j)} \mathrm{d}Y_t^{(i)} + Y_t^{(i)} \mathrm{d}Y_t^{(j)} + \delta_{ij} \mathrm{d}t,
\end{aligned}$$

we have

$$Y_t^{(i)} Y_t^{(j)} = \int_0^t Y_s^{(j)} \mathrm{d}Y_s^{(i)} + \int_0^t Y_s^{(i)} \mathrm{d}Y_s^{(j)} + \delta_{ij} t.$$

Since $Y_t^{(i)}$ and $Y_t^{(j)}$ are square integrable, it follows from Theorem 34.4 that

$$\int_0^t Y_s^{(j)} \mathrm{d}Y_s^{(i)} + \int_0^t Y_s^{(i)} \mathrm{d}Y_s^{(j)}$$

is a martingale under Q. Hence by Theorem 35.8, we have

$$\langle Y^{(i)}, Y^{(j)} \rangle_t = \delta_{ij} t.$$

Therefore it follows from Theorem 37.2 that $\{Y_t : 0 \leq t \leq T\}$ is an m-dimensional Brownian motion with respect to the probability measure Q. This completes the proof.

37.5.

(a) It follows from Problem 37.4 that $\{\hat{B}_t : 0 \leq t \leq T\}$ is an m-dimensional Brownian motion with respect to the probability measure Q.

(b) Since

$$\mathrm{d}\hat{B}_t^{(k)} = u_k(t,\omega)\mathrm{d}t + \mathrm{d}B_t^{(k)},$$

we have

$$\begin{aligned}
\mathrm{d}Y_t^{(j)} &= \beta_j(t,\omega)\mathrm{d}t + \sum_{k=1}^{m} \theta_{jk}\mathrm{d}B_t^{(k)} \\
&= \beta_j(t,\omega)\mathrm{d}t + \sum_{k=1}^{m} \theta_{jk}\left(\mathrm{d}\hat{B}_t^{(k)} - u_k(t,\omega)\mathrm{d}t\right) \\
&= \left(\beta_j(t,\omega) - \sum_{k=1}^{m} \theta_{jk}u_k(t,\omega)\right)\mathrm{d}t + \sum_{k=1}^{m} \theta_{jk}\mathrm{d}\hat{B}_t^{(k)} \\
&= v_j(t,\omega)\mathrm{d}t + \sum_{k=1}^{m} \theta_{jk}\mathrm{d}\hat{B}_t^{(k)}.
\end{aligned}$$

This completes the proof.

37.6. Using the symbols from Problem 37.5, we have

$$\begin{pmatrix}\beta_1\\ \beta_2\end{pmatrix} = \begin{pmatrix}0\\ 1\end{pmatrix}, \quad \begin{pmatrix}\theta_{11} & \theta_{12}\\ \theta_{21} & \theta_{22}\end{pmatrix} = \begin{pmatrix}1 & 3\\ -1 & -2\end{pmatrix}, \quad \begin{pmatrix}v_1\\ v_2\end{pmatrix} = \begin{pmatrix}0\\ 0\end{pmatrix}.$$

Then we can solve u_1 and u_2 to get $u_1 = -3$ and $u_2 = 1$. Let M_t be the martingale defined as

$$\begin{aligned}
M_t &= \exp\left(-\sum_{j=1}^{m}\int_0^t u_j(s,\omega)\mathrm{d}B_s^{(j)} - \frac{1}{2}\sum_{j=1}^{m}\int_0^t u_j^2(s,\omega)\mathrm{d}s\right) \\
&= \exp\left(3B_t^{(1)} - B_t^{(2)} - 5t\right).
\end{aligned}$$

The measure Q is defined as

$$\mathrm{d}Q = M_T\mathrm{d}P.$$

This finishes the solution.

37.7. Let $f(t,\omega) = \frac{\mu-r}{\sigma}$. Since r, μ, and σ are finite, f satisfies Novikov's condition. Let $\{M_t : 0 \le t \le T\}$ be a process given by

$$M_t = \exp\left(-\frac{\mu-r}{\sigma}B_t - \frac{(\mu-r)^2}{2\sigma^2}t\right), \quad 0 \le t \le T.$$

Then let Q be a probability measure on $(\Omega, \mathscr{F}_t)$ defined as

$$\mathrm{d}Q = M_T \mathrm{d}P.$$

By Problem 37.4, the process $\{W_t : 0 \le t \le T\}$ is a Brownian motion under Q.

Now we show that Q and P are equivalent. By Problem 7.6, we only need to show that

$$P(M_T > 0) = 1. \tag{37.6}$$

But

$$\begin{aligned} P(M_T > 0) &= P\left\{\exp\left(-\frac{\mu-r}{\sigma}B_T - \frac{(\mu-r)^2}{2\sigma^2}T\right) > 0\right\} \\ &= P\{-\infty < B_T < \infty\} \\ &= 1. \end{aligned}$$

Hence Equation (37.6) holds. This completes the proof.

37.8. For every $i = 1, 2, \ldots, d$ and $j = 1, 2, \ldots, n$, we have

$$\int_0^T E\left[\frac{\sigma_{ij}^2(u)}{\sigma_i^2(u)}\right] \mathrm{d}u \le \int_0^T E[1]\mathrm{d}u = T < \infty.$$

It follows from Theorems 32.2 and 32.3 that for every $i = 1, 2, \ldots, d$, we have

$$\{W_t^{(i)} : 0 \le t \le T\}$$

is a continuous martingale with respect to $\{\mathscr{F}_t : 0 \le t \le T\}$ under P. By Problems 35.7, 35.9, and 35.10, we have

$$\begin{aligned} \langle W_t^{(i)}, W_t^{(i)} \rangle_t &= \left\langle \sum_{j=1}^n \int_0^t \frac{\sigma_{ij}(u)}{\sigma_i(u)} \mathrm{d}B_u^{(j)}, \sum_{j=1}^n \int_0^t \frac{\sigma_{ij}(u)}{\sigma_i(u)} \mathrm{d}B_u^{(j)} \right\rangle_t \\ &= \sum_{j=1}^n \sum_{l=1}^n \left\langle \int_0^t \frac{\sigma_{ij}(u)}{\sigma_i(u)} \mathrm{d}B_u^{(j)}, \int_0^t \frac{\sigma_{il}(u)}{\sigma_i(u)} \mathrm{d}B_u^{(l)} \right\rangle_t \\ &= \sum_{j=1}^n \int_0^t \frac{\sigma_{ij}^2(u)}{\sigma_i^2(u)} \mathrm{d}u \\ &= t. \end{aligned}$$

It follows from Theorem 37.2 that $\{W_t^{(i)} : 0 \le t \le T\}$ is a Brownian motion under the probability measure P. This completes the proof.

37.5 Bibliographic Notes

In this chapter, we introduced some results relevant to the change of measure. For a good introduction to how the change of measure is applied, readers are referred to Baxter and Rennie (1996). For more information about this subject, readers can consult Hunt and Kennedy (2004), Elliott and Kopp (2005), Sondermann (2007), and Hull (2008).

Novikov's theorem (Theorem 37.1) pertains to exponential martingales constructed from martingales. A proof of this theorem can be found in Ikeda and Watanabe (1981, Theorem 5.3).

The Lévy characterization of Brownian motion theorem is very useful in that it allows us to check whether a stochastic process is a Brownian motion under a probability measure. A proof of this theorem can be found in Ikeda and Watanabe (1981, Chapter II) and Karatzas and Shreve (1991, Chapter 3).

CHAPTER 38

STOCHASTIC DIFFERENTIAL EQUATIONS

A stochastic differential equation is a differential equation that involves stochastic processes. The solution of a stochastic differential equation is also a stochastic process. In this chapter, we present stochastic differential equations and their solutions.

38.1 Basic Concepts and Facts

Definition 38.1 (Strong Solution of SDE). A stochastic process $\{X_t : a \le t \le b\}$ is called a *strong solution* of the following stochastic differential equation

$$\mathrm{d}X_t = \sigma(t, X_t)\mathrm{d}B_t + f(t, X_t)\mathrm{d}t, \quad X_a = \xi, \tag{38.1}$$

if it satisfies the following conditions:

(a) The stochastic process $\sigma(t, X_t)$ belongs to $\mathcal{L}_{ad}(\Omega, L^2[a, b])$ so that for each $t \in [a, b]$, $\int_a^t \sigma(s, X_s)\mathrm{d}B_s$ is an Itô integral.

(b) Almost all sample paths of $f(t, X_t)$ belong to $L^1[a, b]$.

(c) For each t, the following equation

$$X_t = \xi + \int_a^t \sigma(s, X_s)\mathrm{d}B_s + \int_a^t f(s, X_s)\mathrm{d}s, \quad a \le t \le b \tag{38.2}$$

holds almost surely.

Definition 38.2 (Weak Solution of SDE). If there exist a probability space with a filtration, a Brownian motion $\{\hat{B}_t : t \ge 0\}$, and a process $\{\hat{X}_t : a \le t \le b\}$ adapted to the filtration such that they satisfy the conditions given in Definition 38.1, then $\{\hat{X}_t : a \le t \le b\}$ is called a *weak solution* to the stochastic differential equation (38.1).

Definition 38.3 (Uniqueness of Solution). A stochastic differential equation is said to have a strongly unique (strong or weak) solution if any two solutions $\{X_t : a \le t \le b\}$ and $\{Y_t : a \le t \le b\}$ of the stochastic differential equation satisfy

$$P\{X_t = Y_t : a \le t \le b\} = 1.$$

A stochastic differential equation is said to have a weakly unique (strong or weak) solution if any two solutions of the stochastic differential equation have the same finite-dimensional distributions.

Definition 38.4 (Lipschitz Condition). A measurable function $f(t, x)$ on $[a, b] \times \mathbf{R}$ is said to satisfy the Lipschitz condition in x if there exists a constant $K > 0$ such that

$$|f(t, x) - f(t, y)| \le K|x - y|$$

for all $t \in [a, b]$ and $x, y \in \mathbf{R}$.

Definition 38.5 (Linear Growth Condition). A measurable function $f(t, x)$ on $[a, b]$ $\times\mathbf{R}$ is said to satisfy the linear growth condition in x if there exists a constant $K > 0$ such that

$$|f(t, x)| \le K(1 + |x|)$$

for all $t \in [a, b]$ and $x \in \mathbf{R}$.

Theorem 38.1 (Uniqueness and Existence of a Solution). *Let $f(\cdot, \cdot) : [a, b] \times \mathbf{R}^n \to \mathbf{R}^n$ and $\sigma(\cdot, \cdot) : [a, b] \times \mathbf{R}^n \to \mathbf{R}^{n\times m}$ be measurable functions satisfying*

$$|f(t, x)| + |\sigma(t, x)| \le C(1 + |x|), \quad x \in \mathbf{R}^n,\ t \in [a, b] \tag{38.3}$$

and

$$|f(t, x) - f(t, y)| + |\sigma(t, x) - \sigma(t, y)| \le D|x - y|, \quad x, y \in \mathbf{R}^n,\ t \in [a, b] \tag{38.4}$$

for some constants C and D, where $|v|^2$ is the sum of all the squared components of v. Let ξ be a random variable that is independent of the σ-algebra $\mathscr{F}_\infty = \sigma(B_s : s \ge 0)$ and satisfies

$$E\left[|\xi|^2\right] < \infty.$$

Then the stochastic differential equation

$$\mathrm{d}X_t = f(t, X_t)\mathrm{d}t + \sigma(t, X_t)\mathrm{d}B_t, \quad X_a = \xi \tag{38.5}$$

has a strongly unique continuous strong solution $\{X_t : a \le t \le b\}$ *such that*

$$E\left[\int_a^b |X_t|^2 \mathrm{d}t\right] < \infty$$

and $\{X_t : a \le t \le b\}$ *is adapted to the filtration* $\mathscr{F}_t^\xi = \sigma(\xi, B_s : s \le t)$.

Theorem 38.2. *Let* $f(t,x)$ *and* $\sigma(t,x)$ *satisfy the conditions given in Theorem 38.1. Then a (weak or strong) solution of Equation* (38.5) *is weakly unique.*

Theorem 38.3. *Let* $\{X_t : a \le t \le b\}$ *be the solution to the stochastic differential equation*

$$\mathrm{d}X_t = f(t, X_t)\mathrm{d}t + \sigma(t, X_t)\mathrm{d}B_t, \quad X_a = \xi,$$

where $f(t,x)$ *and* $\sigma(t,x)$ *satisfy the corresponding conditions given in Theorem 38.1. Suppose that*

$$E\left[\xi^{2n}\right] < \infty$$

for some integer $n \ge 1$. *Then for every* $t \in [a,b]$, *we have*

$$E\left[\sup_{0\le s\le t} |X_s|^{2n}\right] \le K_1\left(1 + E\left[|\xi|^{2n}\right]\right)e^{K_1(t-a)}, \quad t \in [a,b]$$

and

$$E\left[\sup_{0\le s\le t} |X_s - \xi|^{2n}\right] \le K_2\left(1 + E\left[|\xi|^{2n}\right]\right)(t-a)^n, \quad t \in [a,b],$$

where K_1 *and* K_2 *are constants depending only on* n, C, D, *and* $b - a$.

38.2 Problems

38.1 (Bellman-Gronwall Inequality). Let ϕ be a function in $L^1[a,b]$ satisfying the following condition

$$\phi(t) \le f(t) + \beta \int_a^t \phi(s)\mathrm{d}s, \quad \text{for all } t \in [a,b], \tag{38.6}$$

where $f \in L^1[a,b]$ and β is a positive constant. Show that

(a) For any $t \in [a,b]$,

$$\phi(t) \le f(t) + \beta \int_a^t f(s)e^{\beta(t-s)}\mathrm{d}s. \tag{38.7}$$

(b) If $f(t) = \alpha$ for some constant α, then for any $t \in [a, b]$,

$$\phi(t) \leq \alpha e^{\beta(t-a)}. \tag{38.8}$$

38.2. Let $\{\theta_n\}_{n\geq 1}$ be a sequence of functions in $L^1[a, b]$ such that

$$\theta_{n+1}(t) \leq f(t) + \beta \int_a^t \theta_n(s)\mathrm{d}s, \quad \text{for all } t \in [a, b],\ n \geq 1, \tag{38.9}$$

where $f \in L^1[a, b]$ and β is a positive constant. Show that

(a) For any $n \geq 1$ and any $t \in [a, b]$,

$$\theta_{n+1}(t) \leq f(t) + \beta \int_a^t f(s)e^{\beta(t-s)}\mathrm{d}s + \beta^n \int_a^t \frac{(t-s)^{n-1}}{(n-1)!}\theta_1(s)\mathrm{d}s.$$

(b) If $f(t) = \alpha$ and $\theta_1(t) = c$ for some constants α and c, then for any $n \geq 1$ and any $t \in [a, b]$,

$$\theta_{n+1}(t) \leq \alpha e^{\beta(t-a)} + c\frac{\beta^n(t-a)^n}{n!}.$$

38.3. Let $\{Z_t : a \leq t \leq b\}$ be a continuous stochastic process on $(\Omega, \mathscr{F}, P)$ such that for each $t \in [a, b]$, we have

$$Z_t = 0, \quad \text{a.s.}$$

Show that there exists a subset Ω_0 of Ω such that for each $\omega \in \Omega_0$, $Z_t(\omega) = 0$ for all $t \in [a, b]$.

38.4. Let $\sigma(t, x)$ and $f(t, x)$ be measurable functions on $[a, b] \times \mathbf{R}$ satisfying the Lipschitz condition in x. Let ξ be an $\mathscr{F}_a$-measurable random variable with $E(\xi^2) < \infty$. Then the stochastic differential equation (38.1) has at most one continuous solution X_t.

38.5. Let $f(t, x)$ be a measurable function satisfying the linear growth condition:

$$|f(t, x)| \leq K_1(1 + |x|), \quad t \in [0, T],\ x \in \mathbf{R}.$$

Let $\sigma(t, x)$ be a bounded measurable function:

$$|\sigma(t, x)| \leq K_2, \quad t \in [0, T],\ x \in \mathbf{R}.$$

Suppose that $\{X_t : 0 \leq t \leq T\}$ is a solution of the following stochastic differential equation

$$X_t = x_0 + \int_0^t \sigma(s, X_s)\mathrm{d}B_s + \int_0^t f(s, X_s)\mathrm{d}s,$$

where x_0 is some constant. Let $\tilde{X}_T = \sup_{0\leq t\leq T} |X_t|$. Show that

(a) There exist positive constants c_1, c_2, and c_3 such that

$$P\left\{\tilde{X}_T \geq \lambda\right\} \leq c_1 \exp\left(-c_2\lambda^2 + c_3\lambda\right);$$

(b) There exists a positive constant α such that

$$E\left[\exp\left(\alpha \tilde{X}_T^2\right)\right] < \infty.$$

38.6. Let $f(t,x)$ and $\sigma(t,x)$ be measurable functions satisfying the linear growth condition:

$$|f(t,x)| + |\sigma(t,x)| \leq K(1+|x|), \quad t \in [0,T],\ x \in \mathbf{R}.$$

Suppose that $\{X_t : 0 \leq t \leq T\}$ is a solution of the following stochastic differential equation

$$X_t = x_0 + \int_0^t \sigma(s, X_s)\mathrm{d}B_s + \int_0^t f(s, X_s)\mathrm{d}s,$$

where x_0 is some constant. For every $p \geq 1$, show that

$$E\left[\sup_{0\leq t\leq T} |X_t|^p\right] < \infty.$$

38.7. Let $\{B_t : 0 \leq t \leq T\}$ be a Brownian motion. Solve the following stochastic differential equation

$$\mathrm{d}X_t = X_t(\mu \mathrm{d}t + \sigma \mathrm{d}B_t), \quad t \in [0,T],$$

where μ and σ are constants.

38.8 (Stochastic Exponential)**.** Let $\{X_t : t \geq 0\}$ be a continuous Itô process (see Definition 35.3) given by

$$\mathrm{d}X_t = f(t)\mathrm{d}B_t + g(t)\mathrm{d}t,$$

where $f(t)$ and $g(t)$ are bounded. Let $\{U_t : t \geq 0\}$ be a stochastic process satisfying the following stochastic differential equation

$$\mathrm{d}U_t = U_t \mathrm{d}X_t, \quad U_0 = 1. \tag{38.10}$$

Suppose that $\{X_t :\geq 0\}$ has finite total variation. Show that the only solution of Equation (38.10) is given by

$$U_t = \exp\left(X_t - X_0 - \frac{1}{2}[X]_t\right),$$

where $[X]_t$ is the quadratic variation process of X (see Definition 34.6).

38.9 (Ornstein-Uhlenbeck Equation). Let $\{B_t : t \geq 0\}$ be a one-dimensional Brownian motion. Solve the following stochastic differential equation, where μ and σ are real constants:

$$\mathrm{d}X_t = \mu X_t \mathrm{d}t + \sigma \mathrm{d}B_t.$$

38.10 (Mean-Reverting Ornstein-Uhlenbeck Equation). Let

$$\{B_t : t \geq 0\}$$

be a one-dimensional Brownian motion. Solve the following stochastic differential equation, where m and σ are real constants:

$$\mathrm{d}X_t = (m - X_t)\mathrm{d}t + \sigma \mathrm{d}B_t.$$

38.11. Let $\{B_t : t \geq 0\}$ be a 1-dimensional Brownian motion. Solve the following stochastic differential equation, where r and α are real constants:

$$\mathrm{d}X_t = r\mathrm{d}t + \alpha X_t \mathrm{d}B_t.$$

38.12 (Brownian Bridge). Let $\{B_t : t \geq 0\}$ be a 1-dimensional Brownian motion. Solve the following stochastic differential equation

$$\mathrm{d}X_t = \frac{b - X_t}{1 - t}\mathrm{d}t + \mathrm{d}B_t, \quad 0 \leq t < 1, \quad X_0 = a,$$

where a and b are real constants.

38.13. Let $\{X_t : t \geq 0\}$ be a process satisfying the following nonlinear stochastic differential equation:

$$\mathrm{d}X_t = f(t, X_t)\mathrm{d}t + g(t)X_t \mathrm{d}B_t, \quad X_0 = x.$$

Here $f : \mathbf{R} \times \mathbf{R} \to \mathbf{R}$ and $g : \mathbf{R} \to \mathbf{R}$ are continuous deterministic functions. Let $\{Y_t : t \geq 0\}$ be a process defined as

$$Y_t = \exp\left(-\int_0^t g(s)\mathrm{d}B_s + \frac{1}{2}\int_0^t g^2(s)\mathrm{d}s\right).$$

(a) Show that

$$\mathrm{d}(X_t Y_t) = Y_t f(t, X_t)\mathrm{d}t.$$

(b) Solve the following stochastic differential equation

$$\mathrm{d}X_t = \frac{1}{X_t}\mathrm{d}t + \alpha X_t \mathrm{d}B_t, \quad X_0 = x > 0,$$

where α is a real constant.

38.3 Hints

38.1. To prove part (a), consider the function $g(t) = \int_a^t \phi(s)\mathrm{d}s$ and note that

$$\mathrm{d}[g(t)e^{-\beta t}] = e^{-\beta t}[g'(t) - \beta g(t)]\mathrm{d}t.$$

Part (b) is a special case of part (a).

38.2. Note that by integration by parts,

$$\int_a^t \int_a^s \theta_1(u)\mathrm{d}u\mathrm{d}s = \int_a^t (t-s)\theta_1(s)\mathrm{d}s$$

and use the induction method to establish that

$$\theta_{n+1}(t) \le f(t) + \sum_{k=0}^{n-2} \beta \int_a^t \frac{\beta^k (t-s)^k}{k!} f(s)\mathrm{d}s + \beta^n \int_a^t \frac{(t-s)^{n-1}}{(n-1)!}\theta_1(s)\mathrm{d}s.$$

38.3. Let $r_1, r_2, \ldots$ be an enumeration of all rational numbers in the interval $[a, b]$ and consider the sequence $\{A_n\}_{n\ge 1}$ of subsets of Ω, where $A_n = \{\omega \in \Omega : Z_{r_n}(\omega) = 0\}$.

38.4. Use the results of Problems 38.1 and 38.3.

38.5. Use the results of Problems 38.1 and 32.15 to prove part (a). Use the results of part (a) and Problem 13.12 to prove part (b).

38.6. Consider the stochastic differential of the process

$$Y_t = \ln\left(1 + X_t^2\right),$$

and use the result of Problem 38.5.

38.7. Apply Itô's lemma (Theorem 35.3) to $\ln X_t$.

38.8. The existence and uniqueness of a solution can be proved by Theorem 38.1. Note that $X_t - X_0$ is a martingale and

$$[X]_t = \int_0^t f(s)^2 \mathrm{d}s.$$

The exact formula of U can be obtained by applying Itô formula (Theorem 35.6).

38.9. Consider

$$\mathrm{d}\left(e^{-\mu t} X_t\right)$$

and use Itô's lemma (Theorem 35.3).

38.10. Consider

$$\mathrm{d}\left(e^{t} X_t\right)$$

and use Itô's lemma (Theorem 35.3).

38.11. Let

$$Y_t = \exp\left(-\alpha B_t + \frac{1}{2}\alpha^2 t\right).$$

Consider

$$\mathrm{d}\left(Y_t X_t\right)$$

and use the multidimensional Itô's lemma (Theorem 35.4).

38.12. Consider

$$\mathrm{d}\left(\frac{X_t}{1-t}\right)$$

and use Itô's lemma (Theorem 35.3).

38.13. Apply Itô's lemma (Theorems 35.3 and 35.4) to prove part (a). Then use part (a) to solve part (b).

38.4 Solutions

38.1.

(a) Let $g(t) = \int_a^t \phi(s)\mathrm{d}s$ for $t \in [a, b]$. Then we have $g'(t) = \phi(t)$ and $g(a) = 0$. By Equation (38.6), we get

$$g'(t) \leq f(t) + \beta g(t).$$

Note that $\mathrm{d}[g(t)e^{-\beta t}] = e^{-\beta t}[g'(t) - \beta g(t)]\mathrm{d}t$. Hence we have

$$\frac{\mathrm{d}}{\mathrm{d}t} g(t)e^{-\beta t} \leq f(t)e^{-\beta t},$$

which gives

$$g(t)e^{-\beta t} \leq \int_a^t f(s)e^{-\beta s}\mathrm{d}s.$$

Therefore by Equation (38.6), we get

$$\phi(t) \leq f(t) + \beta g(t) \leq f(t) + \beta \int_a^t f(s)e^{\beta(t-s)}\mathrm{d}s.$$

(b) This is implied from part (a).

This completes the proof.

38.2.

(a) When $n = 1$, by Equation (38.9), we have

$$\theta_2(t) \le f(t) + \beta \int_a^t \theta_1(s)\mathrm{d}s.$$

When $n = 2$, by Equation (38.9), the above equation, and integration by parts, we have

$$\begin{aligned}\theta_3(t) &\le f(t) + \beta \int_a^t \left[f(s) + \beta \int_a^s \theta_1(u)\mathrm{d}u \right] \mathrm{d}s \\ &= f(t) + \beta \int_a^t f(s)\mathrm{d}s + \beta^2 \int_a^t (t-s)\theta_1(s)\mathrm{d}s.\end{aligned}$$

Similarly, we can show that

$$\theta_{n+1}(t) \le f(t) + \sum_{k=0}^{n-2} \beta \int_a^t \frac{\beta^k (t-s)^k}{k!} f(s)\mathrm{d}s + \beta^n \int_a^t \frac{(t-s)^{n-1}}{(n-1)!}\theta_1(s)\mathrm{d}s.$$

Since

$$\sum_{k=0}^{n-2} \frac{\beta^k (t-s)^k}{k!} \le e^{\beta(t-s)},$$

we get

$$\theta_{n+1}(t) \le f(t) + \beta \int_a^t f(s) e^{\beta(t-s)}\mathrm{d}s + \beta^n \int_a^t \frac{(t-s)^{n-1}}{(n-1)!}\theta_1(s)\mathrm{d}s.$$

(b) This is a special case of part (a).

This completes the proof.

38.3. Let $r_1, r_2, \ldots$ be the sequence of all rational numbers in the interval $[a, b]$ and $A_n = \{\omega \in \Omega : Z_{r_n}(\omega) = 0\}$ for $n = 1, 2, \ldots$. By assumption, we have $P(A_n) = 1$ for all n. Let

$$A = \bigcap_{n=1}^{\infty} A_n.$$

Then by Problem 11.1, we have $P(A) = 1$. Now since the Z_t is a continuous process, there exists a subset B of Ω such that $P(B) = 1$ and for each $\omega \in B$, we see that $Z_t(\omega)$ is a continuous function of t on $[a, b]$. Let

$$\Omega_0 = A \cap B.$$

Then $P(\Omega_0) = 1$. Noting that for each $\omega \in \Omega_0$, we have $Z_{r_n}(\omega) = 0$ for all $n \ge 1$. Hence for each $\omega \in \Omega_0$, we have $Z_t(\omega) = 0$ for all $t \in [a, b]$ because each real number in $[a, b]$ can be approximated by rational numbers in $[a, b]$. This completes the proof.

38.4. Suppose that X_t and Y_t are two continuous solutions of the stochastic differential equation in Equation (38.1). Let $Z_t = X_t - Y_t$. Then Z_t is a continuous stochastic process and

$$Z_t = \int_a^t [\sigma(s, X_s) - \sigma(s, Y_s)]\mathrm{d}B_s + \int_a^t [f(s, X_s) - f(s, Y_s)]\mathrm{d}s.$$

Applying the inequality $(x+y)^2 \leq 2x^2 + 2y^2$ to the above equation gives

$$Z_t^2 \leq 2\left(\int_a^t [\sigma(s, X_s) - \sigma(s, Y_s)]\mathrm{d}B_s\right)^2 + 2\left(\int_a^t [f(s, X_s) - f(s, Y_s)]\mathrm{d}s\right)^2.$$

By the assumption that $\sigma(t, x)$ and $f(t, x)$ satisfy the Lipschitz condition in x, there exists a constant $K > 0$ such that

$$|\sigma(t, x) - \sigma(t, y)| \leq K|x - y|$$

and

$$|f(t, x) - f(t, y)| \leq K|x - y|$$

for all $t \in [a, b]$ and $x, y \in \mathbf{R}$. Hence we have

$$\begin{aligned} E\left[\left(\int_a^t [\sigma(s, X_s) - \sigma(s, Y_s)]\mathrm{d}B_s\right)^2\right] &= \int_a^t E\left[(\sigma(s, X_s) - \sigma(s, Y_s))^2\right] \mathrm{d}s \\ &\leq K^2 \int_a^t E(Z_s^2)\mathrm{d}s \end{aligned}$$

and by Schwarz's inequality (Theorem 8.3), we have

$$\begin{aligned} \left(\int_a^t [f(s, X_s) - f(s, Y_s)]\mathrm{d}s\right)^2 &\leq (t-a)\int_a^t [f(s, X_s) - f(s, Y_s)]^2\mathrm{d}s \\ &\leq (b-a)K^2 \int_a^t Z_s^2\mathrm{d}s. \end{aligned}$$

Therefore

$$Z_t^2 \leq 2K^2(1 + b - a)\int_a^t E(Z_s^2)\mathrm{d}s.$$

By Problem 38.1, we have $E(Z_t^2) = 0$ for all $t \in [a, b]$. Hence $Z_t = 0$ a.s. for each $t \in [a, b]$. By Problem 38.3, almost all sample paths of Z_t are zero. Hence X_t and Y_t are the same continuous solutions. This completes the proof.

38.5.

(a) Let

$$Y_t = \int_0^T \sigma(s, X_s)\mathrm{d}B_s.$$

Since

$$\begin{aligned} |X_t| &\le |x_0| + \int_0^t K_1(1+|X_s|)\mathrm{d}s + |Y_t| \\ &\le |x_0| + K_1T + \sup_{0\le t\le T} |Y_t| + K_1 \int_0^t |X_s|\mathrm{d}s, \end{aligned}$$

it follows from Problem 38.1 that

$$|X_t| \le \left(|x_0| + K_1T + \sup_{0\le t\le T} |Y_t| \right) e^{K_1T},$$

which implies that

$$P\left\{\tilde{X}_T \ge \lambda\right\} \le P\left\{ \sup_{0\le t\le T} |Y_t| \ge \lambda e^{-K_1T} - |x_0| - K_1T \right\}. \tag{38.11}$$

Then by Problem 32.15, we have

$$\begin{aligned} & P\left\{ \sup_{0\le t\le T} |Y_t| \ge \lambda e^{-K_1T} - |x_0| - K_1T \right\} \\ \le\ & 2\exp\left(-\frac{(\lambda e^{-K_1T} - |x_0| - K_1T)^2}{2K_2} \right). \end{aligned} \tag{38.12}$$

Now if we let

$$c_1 = 2\exp\left(-\frac{(|x_0| + K_1T)^2}{2K_2} \right),\ c_2 = \frac{e^{-2K_1T}}{2K_2},\ c_3 = \frac{e^{-K_1T}(|x_0| + K_1T)}{K_2},$$

then combining Equation (38.12) and Equation (38.11) gives

$$P\left\{\tilde{X}_T \ge \lambda\right\} \le c_1 \exp\left(-c_2\lambda^2 + c_3\lambda\right).$$

(b) Let $\alpha \in (0, c_2)$, where c_2 is given in part (a) of this proof. Then by Problem 13.12, we have

$$\begin{aligned} & E\left[\exp\left(\alpha \tilde{X}_T^2\right)\right] \\ =\ & 1 + \int_0^\infty 2\alpha\lambda \exp\left(\alpha\lambda^2\right) P\{\tilde{X}_T \ge \lambda\}\mathrm{d}\lambda \\ \le\ & 1 + \int_0^\infty 2\alpha\lambda \exp\left(\alpha\lambda^2\right) \cdot c_2 \exp\left(-c_2\lambda^2 + c_3\lambda\right)\mathrm{d}\lambda \\ <\ & \infty. \end{aligned}$$

This completes the proof.

38.6. Let $Y_t = \ln\left(1 + X_t^2\right)$. Then by Theorem 35.3, we have

$$Y_t = y_0 + \int_0^t \eta(s, Y_s)\mathrm{d}B_s + \int_0^t g(s, Y_s)\mathrm{d}s,$$

where $y_0 = \ln(1 + x_0^2)$,

$$\eta(t, Y_t) = \frac{2X_t\sigma(t, X_t)}{1 + X_t^2},$$

and

$$g(t, Y_t) = \frac{2X_t f(t, X_t)}{1 + X_t^2} + \frac{1 - X_t^2}{(1 + X_t^2)^2} f^2(t, X_t).$$

Since

$$|\eta(t, Y_t)| \le \frac{2|X_t| \cdot K(1 + |X_t|)}{1 + X_t^2} \le \frac{4K(1 + X_t^2)}{1 + X_t^2} = 4K$$

and

$$\begin{aligned} |g(t, Y_t)| &\le \frac{2|X_t| \cdot K(1 + |X_t|)}{1 + X_t^2} + \frac{|1 - X_t^2|}{(1 + X_t^2)^2} \cdot K^2(1 + |X_t|)^2 \\ &\le (1 + |X_t|) \cdot \left(K + K^2 \frac{1 + |X_t|}{1 + X_t^2}\right) \\ &\le \left(K + 2K^2\right)(1 + |X_t|), \end{aligned}$$

it follows from Problem 38.5 that for every $\lambda > 0$, we have

$$P\left\{\sup_{0 \le t \le T} Y_t \ge \lambda\right\} \le c_1 e^{-c_2\lambda^2 + c_3\lambda},$$

where c_1, c_2, and c_3 are some positive constants. Hence we have

$$\begin{aligned} P\left\{\sup_{0 \le t \le T} |X_t| \ge \lambda\right\} &= P\left\{\sup_{0 \le t \le T} Y_t \ge \ln(1 + \lambda^2)\right\} \\ &\le c_1 e^{-c_2 \ln^2(1+\lambda^2) + c_3 \ln(1+\lambda^2)} \\ &= c_1(1 + \lambda^2)^{-c_2 \ln(1+\lambda^2) + c_3}. \end{aligned}$$

Then by Problem 13.12, we have

$$\begin{aligned} E\left[\sup_{0 \le t \le T} |X_t|^p\right] &= \int_0^\infty p\lambda^{p-1} P\left\{\sup_{0 \le t \le T} |X_t| \ge \lambda\right\} \mathrm{d}\lambda \\ &\le pc_1 \int_0^\infty \lambda^{p-1}(1 + \lambda^2)^{-c_2 \ln(1+\lambda^2) + c_3} \mathrm{d}\lambda. \end{aligned}$$

Now let $\lambda_0 = \sqrt{\exp(\frac{c_3+p}{c_2}) - 1}$. Then we have

$$\begin{aligned}\int_{\lambda_0}^{\infty} \lambda^{p-1}(1+\lambda^2)^{-c_2\ln(1+\lambda^2)+c_3}\mathrm{d}\lambda &\le \int_{\lambda_0}^{\infty} \lambda^{p-1}(1+\lambda^2)^{-p}\mathrm{d}\lambda \\ &\le \int_{\lambda_0}^{\infty} \lambda^{-p-1}\mathrm{d}\lambda \\ &= \frac{1}{p}\lambda_0^{-p} \\ &< \infty.\end{aligned}$$

Hence we have

$$E\left[\sup_{0\le t\le T} |X_t|^p\right] < \infty.$$

This completes the proof.

38.7. By Theorem 35.3, we have

$$\begin{aligned}\mathrm{d}\ln X_t &= \frac{1}{X_t}\mathrm{d}X_t - \frac{1}{2}\sigma^2\mathrm{d}t \\ &= \left(\mu - \frac{1}{2}\sigma^2\right)\mathrm{d}t + \sigma\mathrm{d}B_t.\end{aligned}$$

Hence we have

$$\ln X_t - \ln X_0 = \left(\mu - \frac{1}{2}\sigma^2\right)t + \sigma B_t,$$

which gives

$$X_t = X_0\exp\left(\left(\mu - \frac{1}{2}\sigma^2\right)t + \sigma B_t\right).$$

This finishes the solution.

38.8. Since $f(t)$ and $g(t)$ are bounded, both $xf(t)$ and $xg(t)$ satisfy the Lipschitz condition in x and the linear growth condition in x. Hence by Theorem 38.1, the stochastic differential equation (38.10) has a unique solution.

By Theorems 34.2 and 34.4, we have

$$[X]_t = \langle X\rangle_t = \int_0^t f(s)^2\mathrm{d}\langle B\rangle_s = \int_0^t f(s)^2\mathrm{d}s.$$

Let U_t be defined as

$$U_t = \exp\left(X_t - X_0 - \frac{1}{2}\int_0^t f(s)\mathrm{d}s\right).$$

Then by Theorem 35.6, we have

$$\begin{aligned}\mathrm{d}U_t &= U_t[-f(t)]\mathrm{d}t + U_t\mathrm{d}X_t + \frac{1}{2}U_t\mathrm{d}\langle X\rangle_t \\ &= -U_tf(t)\mathrm{d}t + U_t\mathrm{d}X_t + U_tf(t)\mathrm{d}t \\ &= U_t\mathrm{d}X_t,\end{aligned}$$

which shows that $\{U_t : t \geq 0\}$ satisfies Equation (38.10). Hence $\{U_t : t \geq 0\}$ is the only solution of Equation (38.10). This completes the proof.

38.9. By Itô's lemma, we have

$$\begin{aligned} \mathrm{d}\left(e^{-\mu t} X_t\right) &= -\mu e^{-\mu t} X_t \mathrm{d}t + e^{-\mu t} \mathrm{d}X_t \\ &= e^{-\mu t}\left(-\mu X_t \mathrm{d}t + \mathrm{d}X_t\right) \\ &= e^{-\mu t} \sigma \mathrm{d}B_t. \end{aligned}$$

Integrating both sides of the above equation gives

$$e^{-\mu t} X_t = X_0 + \int_0^t e^{-\mu s} \sigma \mathrm{d}B_s.$$

Hence we have

$$X_t = X_0 e^{\mu t} + \int_0^t e^{\mu(t-s)} \sigma \mathrm{d}B_s.$$

38.10. By Itô's lemma, we have

$$\begin{aligned} \mathrm{d}\left(e^t X_t\right) &= e^t X_t \mathrm{d}t + e^t \mathrm{d}X_t \\ &= e^t\left(X_t \mathrm{d}t + \mathrm{d}X_t\right) \\ &= e^t (m \mathrm{d}t + \sigma \mathrm{d}B_t). \end{aligned}$$

Integrating both sides of the above equation gives

$$e^t X_t = X_0 + m(e^t - 1) + \sigma \int_0^t e^s \mathrm{d}B_s.$$

Hence we have

$$X_t = m + (X_0 - m)e^{-t} + \sigma \int_0^t e^{s-t} \mathrm{d}B_s.$$

This completes the solution.

38.11. Let

$$Y_t = \exp\left(-\alpha B_t + \frac{1}{2}\alpha^2 t\right).$$

Then by Theorem 35.4, we have

$$\begin{aligned} \mathrm{d}(Y_t X_t) &= Y_t \mathrm{d}X_t + X_t \mathrm{d}Y_t + \mathrm{d}X_t \mathrm{d}Y_t \\ &= Y_t(r\mathrm{d}t + \alpha X_t \mathrm{d}B_t) + X_t[\alpha Y_t(\alpha \mathrm{d}t - \mathrm{d}B_t)] + \alpha Y_t(-\alpha X_t \mathrm{d}t) \\ &= rY_t \mathrm{d}t. \end{aligned} \tag{38.13}$$

Integrating the two sides of Equation (38.13) gives

$$Y_t X_t = Y_0 X_0 + r \int_0^t Y_s \mathrm{d}s.$$

Hence we have

$$X_t = \exp\left(\alpha B_t - \frac{1}{2}\alpha^2 t\right)\left[X_0 + r\int_0^t \exp\left(-\alpha B_s + \frac{1}{2}\alpha^2 s\right) \mathrm{d}s\right].$$

This completes the solution.

38.12. By Theorem 35.3, we have

$$\begin{aligned}
\mathrm{d}\left(\frac{X_t}{1-t}\right) &= \frac{X_t}{(1-t)^2}\mathrm{d}t + \frac{\mathrm{d}X_t}{1-t} \\
&= \frac{1}{1-t}\left(\mathrm{d}X_t + \frac{X_t}{1-t}\mathrm{d}t\right) \\
&= \frac{1}{1-t}\left(\frac{b}{1-t}\mathrm{d}t + \mathrm{d}B_t\right). \qquad (38.14)
\end{aligned}$$

Integrating both sides of Equation (38.14) gives

$$\frac{X_t}{1-t} = a + \frac{bt}{1-t} + \int_0^t \frac{\mathrm{d}B_s}{1-s}.$$

We get the following result by multipling $1-t$ to both sides of the above equation:

$$X_t = a(1-t) + bt + (1-t)\int_0^t \frac{\mathrm{d}B_s}{1-s}.$$

This completes the solution.

38.13.

(a) By Theorem 35.3, we get

$$\mathrm{d}Y_t = Y_t\left(-g(t)\mathrm{d}B_t + \frac{1}{2}g^2(t)\mathrm{d}t\right). \qquad (38.15)$$

Now by Theorem 35.4, we have

$$\begin{aligned}
&\mathrm{d}(X_tY_t) \\
=\ & X_t\mathrm{d}Y_t + Y_t\mathrm{d}X_t + \mathrm{d}X_t\mathrm{d}Y_t \\
=\ & X_tY_t\left(-g(t)\mathrm{d}B_t + \frac{1}{2}g^2(t)\mathrm{d}t\right) + Y_t[f(t,X_t)\mathrm{d}t + g(t)X_t\mathrm{d}B_t] \\
& +\frac{1}{2}Y_t\left(-g(t)\mathrm{d}B_t + \frac{1}{2}g^2(t)\mathrm{d}t\right)(f(t,X_t)\mathrm{d}t + g(t)X_t\mathrm{d}B_t) \\
=\ & Y_t f(t,X_t)\mathrm{d}t. \qquad (38.16)
\end{aligned}$$

(b) For this particular stochastic differential equation, we have $f(t, X_t) = \frac{1}{X_t}$ and $g(t) = \alpha$. Then we have

$$Y_t = \exp\left(-\alpha B_t + \frac{1}{2}\alpha^2 t\right).$$

Now let $Z_t = X_t Y_t$. Then from Equation (38.15), we get

$$\mathrm{d}Z_t = \frac{Y_t^2}{Z_t}\mathrm{d}t$$

or

$$Z_t \mathrm{d}Z_t = Y_t^2 \mathrm{d}t. \tag{38.17}$$

Since Equation (38.17) is a deterministic equation, we can integrate both sides of the equation for each ω; that is, we have

$$\frac{1}{2}Z_t^2(\omega) = \frac{1}{2}Z_0^2(\omega) + \int_0^t Y_s^2(\omega)\mathrm{d}s,$$

which gives

$$X_t Y_t = x + 2\int_0^t Y_s^2 \mathrm{d}s.$$

Hence we have

$$X_t = \exp\left(\alpha B_t - \frac{1}{2}\alpha^2 t\right)\left[x + 2\int_0^t \exp\left(-2\alpha B_s + \alpha^2 s\right)\mathrm{d}s\right]^{\frac{1}{2}}.$$

This finishes the solution.

38.5 Bibliographic Notes

In this chapter, we presented stochastic differential equations. In fact, the stochastic differential equation in Equation (38.1) is just a symbolic expression and must be interpreted as the stochastic integral equation in Equation (38.2). Under certain conditions, the stochastic differential equation has a unique solution. A proof of the uniqueness is given in Problem 38.4. For a proof of the existence, readers are referred to Oksendal (1998) and Kuo (2006). For a proof of Theorem 38.2 and an example of a stochastic differential equation that has no strong solution but a weak solution, readers are referred to Oksendal (1998, p71).

Several texts on stochastic differential equations have been published since the mid-1990s. Oksendal (1998) is textbook on stochastic differential equations on the graduate level. Prevot and Rockner (2007) is a short book on stochastic partial differential equations. Allen (2007) introduced the theory, computation, and application of Itô stochastic differential equations. Other books that introduce stochastic differential equations include Friedman (1975), Friedman (1976), Ikeda and Watanabe (1981), Karatzas and Shreve (1988), and Protter (2003).

Numerical methods for stochastic differential equations are developed in Kloeden and Platen (1995) and Iacus (2008). For a proof of the moment inequalities given in Theorem 38.3, readers are referred to Kloeden and Platen (1995).

CHAPTER 39

DIFFUSION

A diffusion or a diffusion process is a solution to a stochastic differential equation. In this chapter, we will present some results about diffusion, in particular, Itô diffusion.

39.1 Basic Concepts and Facts

Definition 39.1 (Compact Set). A set $K \subseteq \mathbf{R}$ is called a *compact set* if every open cover of K contains a finite subcover; that is, if $\{V_i\}_{i \in I}$ is a collection of open sets such that

$$K \subseteq \bigcup_{i \in I} V_i,$$

then there exists a finite subcollection $\{V_{i_j}\}_{j=1,2,\ldots,n}$ such that

$$K \subseteq \bigcup_{j=1}^{n} V_{i_j}.$$

Definition 39.2 (Closure). Let $E \subseteq \mathbf{R}$. The closure of the set E, written as $\bar{E}$, is the smallest closed set in $\mathbf{R}$ that contains E.

Definition 39.3 (Support). The support of a function f on $\mathbf{R}$ is the closure of the set

$$\{x \in \mathbf{R} : f(x) \neq 0\}.$$

Definition 39.4 ($C^{1,2}$-Function). A function f on $\mathbf{R} \times \mathbf{R}$ is called a $C^{1,2}$*-function* if $f(t,x)$ is continuously differentiable on t (i.e., $\frac{\partial f}{\partial t}$ is continuous) and twice continuously differentiable on x (i.e., $\frac{\partial^2 f}{\partial x^2}$ is continuous).

Definition 39.5 (Diffusion). A diffusion is a solution to a stochastic differential equation of the form

$$\mathrm{d}X_t = \mu(t, X_t)\mathrm{d}t + \sigma(t, X_t)\mathrm{d}B_t, \quad t \geq s, X_s = x, \tag{39.1}$$

where $s \geq 0$ and x are some constants, μ and σ are functions satisfying the Lipschitz continuity condition and the linear growth condition given in Theorem 38.1. The functions μ and σ are called the *drift coefficient* and the *diffusion coefficient*, respectively.

The unique solution of Equation (39.1) is denoted by $X_t^{s,x}$, $t \geq s$. If $s = 0$, the solution is denoted by X_t^x.

Definition 39.6 (Itô Diffusion). An Itô diffusion is a stochastic process

$$X_t = X(t, \omega) : [0, \infty) \times \Omega \to \mathbf{R}$$

satisfying the following stochastic differential equation

$$\mathrm{d}X_t = b(X_t)\mathrm{d}t + \sigma(X_t)\mathrm{d}B_t, \quad t \geq s, \quad X_s = x, \tag{39.2}$$

where $s \geq 0$ and x are real constants, $\{B_t : t \geq 0\}$ is a standard Brownian motion, and $b : \mathbf{R} \to \mathbf{R}$ and $\sigma : \mathbf{R} \to \mathbf{R}$ are functions satisfying the Lipschitz continuity condition

$$|b(z) - b(y)| + |\sigma(z) - \sigma(y)| \leq c_1|z - y|, \quad z, y \in \mathbf{R}$$

and the linear growth condition

$$|b(z)| + |\sigma(z)| \leq c_2(1 + |z|), \quad z \in \mathbf{R}.$$

Definition 39.7 (Probability Measure Q^x). Let $\{X_t : t \geq 0\}$ be a diffusion as defined in Definition 39.5. The probability measure Q^x is the probability law (see Definition 21.2) of $\{X_t : t \geq 0\}$ assuming that $X_0 = x$. Mathematically, Q^x is given by

$$Q^x\{X_{t_1} \in E_1, \ldots, X_{t_k} \in E_k\} = P\{X_{t_1}^x \in E_1, \ldots, X_{t_k}^x \in E_k\},$$

where P is the probability law of the underlying Brownian motion, $t_1 < t_2 < \cdots < t_k$ are arbitrary times, and E_1, ..., E_k are arbitrary Borel sets.

The expectation under the probability measure Q^x is denoted by E^x.

Definition 39.8 (Transition Function). Let $\{X_t : t \geq 0\}$ be a diffusion as defined in Definition 39.5. The transition function of the diffusion is defined as

$$P_{s,t}(x, A) = P(X_t \in A | X_s = x), \quad 0 \leq s < t,\ x \in \mathbf{R},\ A \in \mathcal{B}.$$

Definition 39.9 (Differential Operator L_t). Let $\{X_t : t \geq 0\}$ be a diffusion as defined in Definition 39.5. Then the second order differential operator associated with the stochastic differential equation of X_t is defined as

$$\mathrm{L}_t f(x) = (\mathrm{L}_t f)(x) = \mu(t, x)\frac{\partial f}{\partial x}(x) + \frac{1}{2}\sigma^2(t, x)\frac{\partial^2 f}{\partial x^2}(x),$$

where $f(x)$ is a C^2 function, $\mu(t, x)$ is the drift coefficient, and $\sigma(t, x)$ is the diffusion coefficient.

If $\{X_t : t \geq 0\}$ is an Itô diffusion as defined in Definition 39.6, then the corresponding differential operator is defined as

$$\mathrm{L}f(x) = (\mathrm{L}f)(x) = b(x)\frac{\partial f}{\partial x}(x) + \frac{1}{2}\sigma^2(x)\frac{\partial^2 f}{\partial x^2}(x).$$

Definition 39.10 (Infinitesimal Generator for Itô Diffusions). Let $\{X_t : t \geq 0\}$ be an Itô diffusion as defined in Definition 39.6. Then the infinitesimal generator $\mathcal{A}$ of $\{X_t : t \geq 0\}$ is defined by

$$\mathcal{A}f(x) = \lim_{t \downarrow 0} \frac{E[f(X_t^x)] - f(x)}{t}, \quad x \in \mathbf{R},$$

where f is a function from $\mathbf{R}$ to $\mathbf{R}$.

Theorem 39.1 (The Markov Property for Itô Diffusions). *Let $\{X_t : t \geq 0\}$ be an Itô diffusion as defined in Definition 39.6. Then $\{X_t : t \geq 0\}$ is a Markov process with respect to the filtration $\{\mathscr{F}_t : t \geq 0\}$ generated by the Brownian motion $\{B_t : t \geq 0\}$; that is, for any bounded Borel function f, we have*

$$E[f(X_t)|\mathscr{F}_s] = E[f(X_t)|X_s] = \phi(X_s), \quad t \geq s,$$

where

$$\phi(x) = E[f(X_t^{s,x})].$$

Theorem 39.2 (The Strong Markov Property for Itô Diffusions). *Let $\{X_t : t \geq 0\}$ be an Itô diffusion as defined in Definition 39.6. Let f be a bounded Borel function on $\mathbf{R}$. Let τ be a stopping time with respect to the underlying filtration $\{\mathscr{F}_t : t \geq 0\}$. Suppose that $\tau < \infty$ a.s. Then*

$$E[f(X_{\tau+h})|\mathscr{F}_\tau] = E[f(X_{\tau+h})|X_\tau] = \phi(X_\tau), \quad h \geq 0,$$

where

$$\phi(x) = E[f(X_h^{0,x})].$$

39.2 Problems

39.1 (Time Homogeneity). Let $\{X_t : t \geq 0\}$ be an Itô diffusion as defined in Definition 39.6. Show that the process $\{X_t : t \geq 0\}$ is time-homogeneous; that is, the transition function $P_{s,t}$ of $\{X_t : t \geq 0\}$ depends only on $t - s$.

39.2. Let $\{X_t : t \geq 0\}$ be the solution of the following stochastic differential equation

$$\mathrm{d}X_t = \mu X_t \mathrm{d}t + \sigma X_t \mathrm{d}B_t, \quad t \geq 0,$$

where μ and σ are constants. Let $x > 0$ and $A = (-\infty, y]$, where $y > 0$. Calculate the transition function $P_{s,t}(x, A)$ of the process $\{X_t : t \geq 0\}$.

39.3 (Dynkin's Formula). Let f be a C^2 function on $\mathbf{R}$ with a compact support D. Let $\{Y_t : t \geq 0\}$ be an Itô process of the form

$$Y_t^x(\omega) = x + \int_0^t u(s,\omega)\mathrm{d}s + \int_0^t v(s,\omega)\mathrm{d}B_s(\omega).$$

Let τ be a stopping time with respect to the underlying filtration $\{\mathscr{F}_t : t \geq 0\}$. Suppose that $E^x[\tau] < \infty$ and that $u(t,\omega)$ and $v(t,\omega)$ are bounded on the set $\{(t,\omega) : Y_t^x(\omega) \in D\}$. Show that

$$f(x) = E^x[f(Y_\tau)] - E^x\left[\int_0^\tau \left(u(s,\omega)\frac{\partial f}{\partial x}(Y_s) + \frac{1}{2}v^2(s,\omega)\frac{\partial^2 f}{\partial x^2}(Y_s)\right)\mathrm{d}s\right].$$

39.4. Let $\{X_t : t \geq 0\}$ be an Itô diffusion as defined in Definition 39.6. Let f be a C^2 function on $\mathbf{R}$ with a compact support. Show that $\mathcal{A}f$ exists and that

$$\mathcal{A}f(x) = \mathrm{L}f(x),$$

where $\mathcal{A}$ is the infinitesimal generator for X_t and L is the differential operator for X_t.

39.5. Let $\{X_t : t \geq 0\}$ satisfy the stochastic differential equation

$$\mathrm{d}X_t = \mu(t, X_t)\mathrm{d}t + \sigma(t, X_t)\mathrm{d}B_t.$$

Let f be a $C^{1,2}$-function. Show that

$$\mathrm{d}f(t, X_t) = \left(\mathrm{L}_t f(t, X_t) + \frac{\partial f}{\partial t}(t, X_t)\right)\mathrm{d}t + \frac{\partial f}{\partial x}(t, X_t)\sigma(t, X_t)\mathrm{d}B_t,$$

where L_t is the second-order differential operator for X_t (see Definition 39.9).

39.6. Let $\{X_t : 0 \leq t \leq T\}$ solve the stochastic differential equation

$$\mathrm{d}X_t = \mu(t, X_t)\mathrm{d}t + \sigma(t, X_t)\mathrm{d}B_t,$$

where $\mu(t, x)$ and $\sigma(t, x)$ satisfy the corresponding conditions given in Theorem 38.1. Let $f(t, x)$ be a $C^{1,2}$-function with bounded first derivative in x. Show that

(a)
$$M_t = f(t, X_t) - \int_0^t \left(\mathrm{L}_u f(u, X_u) + \frac{\partial f}{\partial t}(u, X_u) \right) \mathrm{d}u$$
is a martingale.

(b) If $f(t, x)$ solves the backward equation
$$\mathrm{L}_t f(t, x) + \frac{\partial f}{\partial t}(t, x) = 0,$$
then $\{f(t, X_t) : 0 \le t \le T\}$ is a martingale.

39.7. Let $\{X_t : t \ge 0\}$ be an Itô diffusion as defined in Definition 39.6. Let $D \subset \mathbf{R}$ be a compact set. Let τ_D be the first exit time from the set D:
$$\tau_D = \inf\{t : X_t \notin D\}.$$
Suppose that there exists a C^2 function f on $\mathbf{R}$, nonnegative over D and such that
$$\mathrm{L} f(x) \le -1, \quad \forall x \in D,$$
where L is the differential operator defined in Definition 39.9. For every $x \in D$, show that
$$E^x[\tau_D] < \infty.$$

39.8. Let $\{X_t : t \ge 0\}$ be a geometric Brownian motion given by
$$X_t = x + \int_0^t \mu X_s \mathrm{d}s + \int_0^t \sigma X_s \mathrm{d}B_s, \quad t \ge 0,$$
where μ and $\sigma > 0$ are constants, and $\{B_t : t \ge 0\}$ is a Brownian motion. Suppose that $x \in (\frac{1}{n}, R)$, where R is a positive constant and n is a positive integer such that $nR > 1$. Let
$$\tau = \inf\{t > 0 : X_t \ge R\}$$
and
$$\tau_n = \inf\left\{t > 0 : X_t \le \frac{1}{n}\right\}.$$

Show that

(a) $E^x[\tau \wedge \tau_n] < \infty$.

(b) If $\mu \ne \frac{1}{2}\sigma^2$, then
$$Q^x\{\tau < \tau_n\} = \frac{n^{-\gamma} - x^\gamma}{n^{-\gamma} - R^\gamma},$$

where Q^x is as defined in Definition 39.7 and $\gamma = 1 - \frac{2\mu}{\sigma^2}$.

(c) If $\mu \neq \frac{1}{2}\sigma^2$, then

$$E^x[\tau \wedge \tau_n] = \frac{\ln \frac{R}{x} + (1 - P\{\tau < \tau_n\}) \ln \frac{1}{nR}}{\mu - \frac{1}{2}\sigma^2}.$$

(d)

$$Q^x\left\{\lim_{n\to\infty} \tau_n = \infty\right\} = 1.$$

(e) If $\mu < \frac{1}{2}\sigma^2$, then

$$Q^x\{\tau < \infty\} = \left(\frac{x}{R}\right)^\gamma,$$

where $\gamma = 1 - \frac{2\mu}{\sigma^2}$.

(f) If $\mu > \frac{1}{2}\sigma^2$, then

$$E^x[\tau] = \frac{\ln R - \ln x}{\mu - \frac{1}{2}\sigma^2}.$$

39.3 Hints

39.1. Show that $X_t^{0,x}$ and $X_{t-s}^{0,x}$ have the same distribution, and then use Theorem 39.1.

39.2. Use the result of Problem 38.7 and the Markov property of the process $\{X_t : t \geq 0\}$ (see Theorem 39.1).

39.3. Use Itô's lemma (Theorem 35.3) and the property of Itô integral (Theorem 32.1).

39.4. Use the results of Problems 39.3 and 9.10.

39.5. Use Itô's lemma (Theorem 35.3) and the definition of the generator (Definition 39.9).

39.6. To prove part (a), use the result of Problem 39.5 and Theorem 38.1 to show that

$$\frac{\partial f}{\partial x}(t, X_t)\sigma(t, X_t) \in L^2_{ad}([0, T] \times \Omega),$$

that is,

$$\int_0^T E\left[\left(\frac{\partial f}{\partial x}(u, X_u)\sigma(u, X_u)\right)^2\right] du < \infty,$$

and then apply Theorem 32.2. Use part (a) to prove part (b).

39.7. Apply Itô's formula (Theorem 35.3) to $f(X^x_{t\wedge\tau_D})$ and then use the monotone convergence theorem (Theorem 6.2).

39.8. To prove part (a), use the result of Problem 39.7. To prove part (b), consider the process X_t^γ and use Dynkin's formula (see Problem 39.3) and the result of Problem 14.18. To prove part (c), consider the process $Y_t = \ln X_t$ and use Dynkin's formula. To prove part (d), consider the sequence $\frac{1}{\tau_n}$ and use the result of Problem 9.3 and Doob's martingale inequality (Theorem 24.1). To prove part (e), use part (d) and the result of Problem 11.6 to show that

$$Q^x\{\tau < \infty\} = \lim_{k\to\infty} Q^x\{\tau < \tau_k\}$$

and then use part (b). To prove part (f), try to show that

$$E^x[\tau] = \lim_{k\to\infty} E^x[\tau \wedge \tau_k]$$

and then use parts (b) and (c).

39.4 Solutions

39.1. To show that $P_{s,t}$ depends only on $t-s$, it is sufficient to show that

$$P_{s,t} = P_{0,t-s}. \tag{39.3}$$

Since

$$\begin{aligned} X_t^{s,x} &= x + \int_s^t \mu\left(X_u^{s,x}\right) \mathrm{d}u + \int_s^t \sigma\left(X_u^{s,x}\right) \mathrm{d}B_u \\ &= x + \int_0^{t-s} \mu\left(X_{u+s}^{s,x}\right) \mathrm{d}u + \int_0^{t-s} \sigma\left(X_{u+s}^{s,x}\right) \mathrm{d}B_{u+s}, \end{aligned}$$

we have

$$Y_{t-s} = x + \int_0^{t-s} \mu\left(Y_u\right) \mathrm{d}u + \int_0^{t-s} \sigma(Y_u)\mathrm{d}\tilde{B}_u,$$

where $\tilde{B}_u = B_{u+s} - B_s$ and $Y_t = X^{s,x}_{t+s}$. Since $\{\tilde{B}_t : t \geq 0\}$ is also a Brownian motion, Y_{t-s} and $X^{0,x}_{t-s}$ satisfy the same stochastic differential equation. Hence Y_{t-s} and $X^{0,x}_{t-s}$ have the same distribution. It follows from Theorem 39.1 that

$$\begin{aligned} P_{s,t}(x,A) &= E[I_A(X_t^{s,x})] \\ &= E[I_A(Y_{t-s})] \\ &= E[I_A(X_{t-s}^{0,x})] \\ &= P_{0,t-s}(x,A). \end{aligned}$$

Hence Equation (39.3) holds. This completes the proof.

39.2. By Theorem 39.1, we have

$$P_{s,t}(x, A) = P(X_t \in A | X_s = x) = E[I_A(X_t) | X_s = x] = \phi(x),$$

where $\phi(x) = E[I_A(X_t^{s,x})]$. By Problem 38.7, we have

$$X_t^{s,x} = x \exp\left(\left(\mu - \frac{\sigma^2}{2}\right)(t-s) + \sigma(B_t - B_s)\right).$$

Then we have

$$\begin{aligned}
E[I_A(X_t^{s,x})] &= P\{X_t^{s,x} \in A\} \\
&= P\left\{x \exp\left(\left(\mu - \frac{\sigma^2}{2}\right)(t-s) + \sigma(B_t - B_s)\right) \le y\right\} \\
&= P\left\{\frac{(B_t - B_s)}{\sqrt{t-s}} \le \frac{\ln\left(\frac{y}{x}\right) - \left(\mu - \frac{\sigma^2}{2}\right)(t-s)}{\sigma\sqrt{t-s}}\right\} \\
&= \Phi\left(\frac{\ln\left(\frac{y}{x}\right) - \left(\mu - \frac{\sigma^2}{2}\right)(t-s)}{\sigma\sqrt{t-s}}\right),
\end{aligned}$$

where $\Phi(\cdot)$ is as defined in Equation (19.2). This finishes the solution.

39.3. By Theorem 35.3, we have

$$\begin{aligned}
f(Y_t) &= f(Y_0) + \int_0^t \left(u(s,\omega)\frac{\partial f}{\partial x}(Y_s) + \frac{1}{2}v^2(s,\omega)\frac{\partial^2 f}{\partial x^2}(Y_s)\right) \mathrm{d}s \\
&\quad + \int_0^t v(s,\omega)\frac{\partial f}{\partial x}(Y_s)\mathrm{d}B_s.
\end{aligned}$$

Hence we have

$$\begin{aligned}
E^x[f(Y_\tau)] &= f(x) + E^x\left[\int_0^\tau \left(u(s,\omega)\frac{\partial f}{\partial x}(Y_s) + \frac{1}{2}v^2(s,\omega)\frac{\partial^2 f}{\partial x^2}(Y_s)\right) \mathrm{d}s\right] \\
&\quad + E^x\left[\int_0^\tau v(s,\omega)\frac{\partial f}{\partial x}(Y_s)\mathrm{d}B_s\right]. \qquad (39.4)
\end{aligned}$$

By the assumption, $v(t,\omega)\frac{\partial f}{\partial x}(Y_t)$ is bounded. Then by Theorem 32.1, for every integer k, we have

$$E^x\left[\int_0^{\tau\wedge k} v(s,\omega)\frac{\partial f}{\partial x}(Y_s)\mathrm{d}B_s\right] = E^x\left[\int_0^k I_{\{s<\tau\}}v(s,\omega)\frac{\partial f}{\partial x}(Y_s)\mathrm{d}B_s\right] = 0.$$

Also by Theorem 32.1, we have

$$\begin{aligned}
&E^x\left[\left(\int_0^\tau v(s,\omega)\frac{\partial f}{\partial x}(Y_s)\mathrm{d}B_s - \int_0^{\tau\wedge k} v(s,\omega)\frac{\partial f}{\partial x}(Y_s)\mathrm{d}B_s\right)^2\right]\\
=\ & E^x\left[\int_{\tau\wedge k}^k \left(v(s,\omega)\frac{\partial f}{\partial x}(Y_s)\right)^2 \mathrm{d}s\right]\\
\leq\ & K^2E^x[\tau - \tau\wedge k],
\end{aligned}$$

where K is a bound of $v(t,\omega)\frac{\partial f}{\partial f}(Y_t)$. By the assumption that $E^x[\tau] < \infty$, we have $E^x[\tau - \tau\wedge k] \to 0$ as $k \to \infty$. Hence we have

$$\int_0^{\tau\wedge k} v(s,\omega)\frac{\partial f}{\partial x}(Y_s)\mathrm{d}B_s \xrightarrow{L^2} \int_0^\tau v(s,\omega)\frac{\partial f}{\partial x}(Y_s)\mathrm{d}B_s \text{ as } k\to\infty.$$

It follows from Problem 9.1 that

$$E^x\left[\int_0^\tau v(s,\omega)\frac{\partial f}{\partial x}(Y_s)\mathrm{d}B_s\right] = \lim_{k\to\infty} E^x\left[\int_0^{\tau\wedge k} v(s,\omega)\frac{\partial f}{\partial x}(Y_s)\mathrm{d}B_s\right] = 0.$$

The result follows by rearranging Equation (39.4). This completes the proof.

39.4. By letting $\tau = t$ in Problem 39.3, we get

$$\begin{aligned}
& E^x[f(X_t)] - f(x)\\
=\ & E^x\left[\int_0^t \mathrm{L}f(X_s)\mathrm{d}s\right]\\
=\ & E^x\left[\int_0^t \left(b(X_s)\frac{\partial f}{\partial x}(X_s) + \frac{1}{2}\sigma^2(X_s)\frac{\partial^2 f}{\partial x^2}(X_s)\right)\mathrm{d}s\right].
\end{aligned}$$

To prove the result, we only need to show that

$$\lim_{t\downarrow 0}\frac{E^x\left[\int_0^t \left(b(X_s)\frac{\partial f}{\partial x}(X_s) + \frac{1}{2}\sigma^2(X_s)\frac{\partial^2 f}{\partial x^2}(X_s)\right)\mathrm{d}s\right]}{t} = \mathrm{L}f(x). \tag{39.5}$$

Note that

$$\begin{aligned}
&\left|\frac{E^x\left[\int_0^t \left(b(X_s)\frac{\partial f}{\partial x}(X_s) + \frac{1}{2}\sigma^2(X_s)\frac{\partial^2 f}{\partial x^2}(X_s)\right)\mathrm{d}s\right]}{t} - \mathrm{L}f(x)\right|\\
=\ & \left|\frac{E^x\left[\int_0^t \left(b(X_s)\frac{\partial f}{\partial x}(X_s) + \frac{1}{2}\sigma^2(X_s)\frac{\partial^2 f}{\partial x^2}(X_s) - \mathrm{L}f(x)\right)\mathrm{d}s\right]}{t}\right|\\
\leq\ & E^x\left[\sup_{0\leq s\leq t}\left|b(X_s)\frac{\partial f}{\partial x}(X_s) + \frac{1}{2}\sigma^2(X_s)\frac{\partial^2 f}{\partial x^2}(X_s) - \mathrm{L}f(x)\right|\right].
\end{aligned}$$

Since $b(X_s)\frac{\partial f}{\partial x}(X_s) + \frac{1}{2}\sigma^2(X_s)\frac{\partial^2 f}{\partial x^2}(X_s)$ is continuous, we have

$$\lim_{t\downarrow 0}\sup_{0\le s\le t}\left|b(X_s)\frac{\partial f}{\partial x}(X_s) + \frac{1}{2}\sigma^2(X_s)\frac{\partial^2 f}{\partial x^2}(X_s) - \mathrm{L}f(x)\right| = 0.$$

It follows from Fatou's lemma (Theorem 6.3) that Equation (39.5) holds. This completes the proof.

39.5. By Theorem 35.3, we have

$$\begin{aligned}
\mathrm{d}f(t,X_t) &= \frac{\partial f}{\partial x}(t,X_t)\sigma(t,X_t)\mathrm{d}B_t \\
&\quad + \left(\frac{\partial f}{\partial t}(t,X_t) + \frac{\partial f}{\partial x}(t,X_t)\mu(t,X_t) + \frac{1}{2}\frac{\partial^2 f}{\partial x^2}(t,X_t)\sigma^2(t,X_t)\right)\mathrm{d}t \\
&= \left(\mathrm{L}_t f(t,X_t) + \frac{\partial f}{\partial t}(t,X_t)\right)\mathrm{d}t + \frac{\partial f}{\partial x}(t,X_t)\sigma(t,X_t)\mathrm{d}B_t.
\end{aligned}$$

This completes the proof.

39.6.

(a) By Problem 39.5, we have

$$M_t = \int_0^t \left(\frac{\partial f}{\partial x}(u,X_u)\sigma(u,X_u)\right)\mathrm{d}B_u.$$

By Theorem 32.2, it is sufficient to show that

$$\int_0^T E\left[\left(\frac{\partial f}{\partial x}(u,X_u)\sigma(u,X_u)\right)^2\right]\mathrm{d}u < \infty. \tag{39.6}$$

By the assumption that the first derivative of f in x is bounded, we have

$$\left|\frac{\partial f}{\partial x}(u,X_u)\right| < K_1$$

for some constant K_1. Also by the assumption that $\sigma(t,x))$ satisfies the linear growth condition, we have

$$|\sigma(t,x)| < K_2(1+|x|)$$

for some constant K_2. Hence we have

$$\begin{aligned}
&\int_0^T E\left[\left(\frac{\partial f}{\partial x}(u,X_u)\sigma(u,X_u)\right)^2\right]\mathrm{d}u \\
\le\ & K_1^2\int_0^T E\left[K_2^2(1+|X_u|)^2\right]\mathrm{d}u \\
\le\ & K_1^2K_2^2\int_0^T E\left[2+2|X_u|^2\right]\mathrm{d}u \\
=\ & K_1^2K_2^2\left(2T + 2\int_0^T E\left[|X_u|^2\right]\mathrm{d}u\right).
\end{aligned}$$

Since $\{X_t : 0 \leq t \leq T\}$ is the solution of the stochastic differential equation, it follows from Theorem 38.1 that

$$\int_0^T E\left[|X_u|^2\right] \mathrm{d}u < \infty.$$

Hence Equation (39.6) holds.

(b) By part (a) and the assumption that f solves the backward equation, we have

$$\begin{aligned} M_t &= f(t, X_t) - \int_0^t \left(\mathrm{L}_u f(u, X_u) + \frac{\partial f}{\partial t}(u, X_u) \right) \mathrm{d}u \\ &= f(t, X_t). \end{aligned}$$

By part (a) of this proof, the M_t is a martingale. Hence the $f(t, X_t)$ is a martingale.

This finishes the proof.

39.7. Let $t > 0$ be fixed and $x \in D$. By Theorem 35.3, we have

$$f(X^x_{t\wedge\tau_D}) = f(x) + \int_0^{t\wedge\tau_D} \mathrm{L}f(X^x_s)\mathrm{d}s + \int_0^{t\wedge\tau_D} \sigma(X^x_s)\frac{\partial f}{\partial x}(X^x_s)\mathrm{d}B_s. \quad (39.7)$$

Since D is a compact set in $\mathbf{R}$, D is bounded. Since $X^x_s \in D$ for $s \in [0, t \wedge \tau_d]$, $\sigma(X^x_s)\frac{\partial f}{\partial x}(X^x_s)$ is bounded for $s \in [0, t \wedge \tau_d]$. Hence it follows from Theorem 32.1 that

$$E\left[\int_0^{t\wedge\tau_D} \sigma(X^x_s)\frac{\partial f}{\partial x}(X^x_s)\mathrm{d}B_s\right] = 0.$$

Therefore, by Equation (39.7) and the assumption, we have

$$E\left[f(X^x_{t\wedge\tau_D})\right] \leq f(x) - E^x[t \wedge \tau_D].$$

Since f is nonnegative over D, we have

$$E^x[t \wedge \tau_D] \leq f(x) < \infty.$$

It follows from Theorem 6.2 that

$$E^x[\tau_D] = \lim_{t\to\infty} E^x[t \wedge \tau_D] < \infty.$$

This competes the proof.

39.8.

(a) By Problem 39.7, we only need to find a C^2 function, which is nonnegative over $D = [\frac{1}{n}, R]$, such that $\mathrm{L}f(x) \leq -1$ for all $x \in D$. In fact, if $\mu \neq 0$, then let $f(x) = \frac{n(R-x)}{\mu}$. In this case, we have

$$\mathrm{L}f(x) = \mu x \frac{-n}{\mu} = -nx \leq -1, \quad \forall x \in D.$$

If $\mu = 0$, then let $f(x) = \frac{n^2(R^2-x^2)}{\sigma^2}$. In this case, we have

$$\mathrm{L}f(x) = \frac{1}{2}\sigma^2 x^2 \frac{-2n^2}{\sigma^2} = -n^2x^2 \le -1, \quad \forall x \in D.$$

Thus $E^x[\tau \wedge \tau_n] < \infty$.

(b) Let $f(x) = x^\gamma$, where $\gamma = 1 - \frac{2\mu}{\sigma^2}$. Since $\mu \neq \frac{1}{2}\sigma^2$, we have $\gamma \neq 0$. Since $\mathrm{L}f(x) = 0$, by Problem 39.3, we get

$$\begin{aligned} & E^x[f(X_{\tau\wedge\tau_n})] \\ = \; & f(x) + E^x\left[\int_0^{\tau\wedge\tau_n} \mathrm{L}f(X_s)\mathrm{d}s\right] \\ = \; & x^\gamma. \end{aligned}$$

But by Problem 14.18, we have

$$\begin{aligned} & E^x[f(X_{\tau\wedge\tau_n})] \\ = \; & E^x[f(R)]P\{\tau < \tau_n\} + E^x\left[f\left(\frac{1}{n}\right)\right](1 - P\{\tau < \tau_n\}) \\ = \; & R^\gamma P\{\tau < \tau_n\} + \left(\frac{1}{n}\right)^\gamma (1 - P\{\tau < \tau_n\}). \end{aligned}$$

Combining the above two equations gives

$$P\{\tau < \tau_n\} = \frac{n^{-\gamma} - x^\gamma}{n^{-\gamma} - R^\gamma}.$$

(c) Let $f(x) = \ln x$. Then by Problem 39.3, we have

$$\begin{aligned} E^x[f(X_{\tau\wedge\tau_n})] &= f(x) + E^x\left[\int_0^{\tau\wedge\tau_n} \mathrm{L}f(X_s)\mathrm{d}s\right] \\ &= \ln x + \left(\mu - \frac{1}{2}\sigma^2\right) E^x[\tau \wedge \tau_n]. \end{aligned}$$

But by Problem 14.18, we have

$$\begin{aligned} & E^x[f(X_{\tau\wedge\tau_n})] \\ = \; & E^x[f(R)]P\{\tau < \tau_n\} + E^x\left[f\left(\frac{1}{n}\right)\right](1 - P\{\tau < \tau_n\}) \\ = \; & (\ln R)\, P\{\tau < \tau_n\} + \left(\ln \frac{1}{n}\right)(1 - P\{\tau < \tau_n\}). \end{aligned}$$

Solving the above two equations gives

$$E^x[\tau \wedge \tau_n] = \frac{\ln \frac{R}{x} + (1 - P\{\tau < \tau_n\}) \ln \frac{1}{nR}}{\mu - \frac{1}{2}\sigma^2}.$$

(d) Note that

$$\left\{\lim_{n\to\infty} \tau_n = \infty\right\} = \left\{\lim_{n\to\infty} \frac{1}{\tau_n} = 0\right\}.$$

By Problem 9.3, we only need to show that for every $\delta > 0$,

$$\lim_{m\to\infty} Q^x \left\{\bigcup_{k=m}^{\infty} \left\{\frac{1}{\tau_k} \geq \delta\right\}\right\} = 0$$

or

$$\lim_{m\to\infty} Q^x \left\{\bigcap_{k=m}^{\infty} \left\{\tau_k \geq \frac{1}{\delta}\right\}\right\} = 1. \tag{39.8}$$

Since $\tau_k \leq \tau_{k+1}$ for all $k > \frac{1}{x}$, we have

$$\bigcap_{k=m}^{\infty} \left\{\tau_k \geq \frac{1}{\delta}\right\} = \left\{\tau_m \geq \frac{1}{\delta}\right\}. \tag{39.9}$$

Since

$$\begin{aligned} Q^x \left\{\tau_m \geq \frac{1}{\delta}\right\} &= Q^x \left\{X_t \geq \frac{1}{m} : 0 \leq t \leq \frac{1}{\delta}\right\} \\ &= P\left\{x e^{(\mu - \frac{1}{2}\sigma^2)t + \sigma B_t} \geq \frac{1}{m} : 0 \leq t \leq \frac{1}{\delta}\right\} \\ &= P\left\{B_t \geq \frac{\ln\left(\frac{1}{mx}\right) - (\mu - \frac{1}{2}\sigma^2)t}{\sigma} : 0 \leq t \leq \frac{1}{\delta}\right\}, \end{aligned}$$

and by Theorem 24.1,

$$\begin{aligned} & P\left\{B_t \geq \frac{\ln\left(\frac{1}{mx}\right) - (\mu - \frac{1}{2}\sigma^2)t}{\sigma} : 0 \leq t \leq \frac{1}{\delta}\right\} \\ = \;& 1 - P\left\{B_t \geq \frac{\ln(mx) + (\mu - \frac{1}{2}\sigma^2)t}{\sigma} : 0 \leq t \leq \frac{1}{\delta}\right\} \\ \geq \;& 1 - P\left\{B_t \geq \frac{\ln(mx) - |\mu - \frac{1}{2}\sigma^2| R}{\sigma} : 0 \leq t \leq \frac{1}{\delta}\right\} \\ \geq \;& 1 - \frac{\sigma}{\ln(mx) - |\mu - \frac{1}{2}\sigma^2| R} E\left[B^+_{\frac{1}{\delta}}\right], \end{aligned}$$

we have

$$\lim_{m\to\infty} Q^x \left\{\tau_m \geq \frac{1}{\delta}\right\} = 1.$$

Hence Equation (39.8) holds.

(e) Since $\{\tau < \tau_k\} \subseteq \{\tau < \tau_{k+1}\}$ for all $k \geq \frac{1}{x}$, it follows from Problem 11.6 that

$$\lim_{k\to\infty} Q^x\{\tau < \tau_k\} = Q^x \left\{\lim_{k\to\infty} \{\tau < \tau_k\}\right\}. \tag{39.10}$$

Since

$$\{\tau < \infty\} \cap \left\{ \lim_{k\to\infty} \tau_k = \infty \right\} = \bigcup_{k>\frac{1}{x}} \{\tau < \tau_k\} = \lim_{k\to\infty} \{\tau < \tau_k\},$$

we have

$$Q^x\{\tau < \infty\} = Q^x \left\{ \lim_{k\to\infty} \{\tau < \tau_k\} \right\}. \tag{39.11}$$

Hence by part (b), and Equations (39.10) and (39.11), we get

$$Q^x\{\tau < \infty\} = \lim_{k\to\infty} Q^x\{\tau < \tau_k\} = \lim_{k\to\infty} \frac{k^{-\gamma} - x^{\gamma}}{k^{-\gamma} - R^{\gamma}}.$$

By the assumption that $\mu < \frac{1}{2}\sigma^2$, we have $\gamma > 0$. Therefore, we have

$$Q^x\{\tau < \infty\} = \left(\frac{x}{R}\right)^{\gamma}.$$

(f) By part (d), we have

$$Q^x \left\{ \lim_{k\to\infty} \tau \wedge \tau_k = \tau \right\} = 1.$$

Hence by Theorem 6.2 and Problem 6.9, we get

$$E^x[\tau] = E^x \left[\lim_{k\to\infty} \tau \wedge \tau_k \right] = \lim_{k\to\infty} E^x[\tau \wedge \tau_k].$$

Since $\mu > \frac{1}{2}\sigma^2$, we have

$$\lim_{k\to\infty} \frac{x^{\gamma} - R^{\gamma}}{k^{-\gamma} - R^{\gamma}} \ln \frac{1}{kR} = 0.$$

Hence it follows from parts (c) and (b) that

$$\begin{aligned} E^x[\tau] &= \lim_{k\to\infty} \frac{\ln \frac{R}{x} + (1 - P\{\tau < \tau_k\}) \ln \frac{1}{kR}}{\mu - \frac{1}{2}\sigma^2} \\ &= \lim_{k\to\infty} \frac{\ln \frac{R}{x} + \frac{x^{\gamma} - R^{\gamma}}{k^{-\gamma} - R^{\gamma}} \ln \frac{1}{kR}}{\mu - \frac{1}{2}\sigma^2} \\ &= \frac{\ln \frac{R}{x}}{\mu - \frac{1}{2}\sigma^2}. \end{aligned}$$

This completes the proof.

39.5 Bibliographic Notes

In this chapter, we presented diffusions, which are defined to be the solutions of stochastic differential equations. For general definition of diffusions or diffusion processes, readers are referred to Friedman (1975), Karlin and Taylor (1981), and Stirzaker (2005). Diffusions are also studied in several other books such as Stroock and Varadhan (1979), Durrett (1996), Oksendal (1998), Rogers and Williams (2000), Klebaner (2005), and Pinsky (1995). Revuz and Yor (1999) discusses properties of the generator of X_t.

Theorem 39.1 and Theorem 39.2 state that Itô diffusions have the Markov property and the strong Markov property, respectively. Proofs of these theorems can be found in Friedman (1975), Oksendal (1998), and Çinlar (2011).

Some concepts (e.g., compact and support) presented in this chapter are adopted from real analysis. For general definition of these concepts, readers are referred to Rudin (1970).

CHAPTER 40

THE FEYNMAN-KAC FORMULA

The solutions to certain second-order partial differential equations can be represented as expectations of stochastic functionals. In fact, the Feynman-Kac formula establishes a link between certain partial differential equations and stochastic processes. In this chapter, we present the Feynman-Kac formula.

40.1 Basic Concepts and Facts

Definition 40.1 (Dirichlet Problem)**.** Let $D = (a, b)$ be an interval. A Dirichlet problem is an ordinary differential equation of the form

$$b(x)\frac{\partial u}{\partial x}(x) + \frac{1}{2}\sigma^2(x)\frac{\partial^2 u}{\partial x^2}(x) - a(x)u(x) = f(x), \quad x \in D, \tag{40.1a}$$

$$u(x) = \varphi(x), \quad x \in \partial D = \{a, b\}, \tag{40.1b}$$

where b, σ, f, a, and φ are given functions, and ∂D is the boundary of of D. The function u is called a solution to the Dirichlet problem.

Definition 40.2 (Cauchy-Dirichlet Problem)**.** Let $Q = (0, T) \times D$, where $D = (a, b)$ is an interval and $T > 0$. Let $\partial_p Q = ([0, T] \times \{a, b\}) \cup (\{T\} \times D)$. A Cauchy-Dirichlet problem is a partial differential equation of the form

$$b\frac{\partial u}{\partial x} + \frac{1}{2}\sigma^2 \frac{\partial^2 u}{\partial x^2} - au + \frac{\partial u}{\partial t} = f, \quad (t, x) \in Q, \tag{40.2a}$$

$$u(t, x) = \varphi(t, x), \quad (t, x) \in \partial_p Q, \tag{40.2b}$$

where b, σ, f, a, and φ are given functions. The function u is called a solution to the Cauchy-Dirichlet problem.

Definition 40.3 (Cauchy Problem)**.** Let $Q = (0, T) \times \mathbf{R}$, where $T > 0$. A Cauchy problem is a partial differential equation of the form

$$b\frac{\partial u}{\partial x} + \frac{1}{2}\sigma^2 \frac{\partial^2 u}{\partial x^2} - au + \frac{\partial u}{\partial t} = f, \quad (t, x) \in Q, \tag{40.3a}$$

$$u(T, x) = \varphi(x), \quad x \in \mathbf{R}, \tag{40.3b}$$

where b, σ, f, a, and φ are given functions. The function u is called a solution to the Cauchy problem.

Theorem 40.1 (The Feynman-Kac Formula)**.** *Let f be a C^2 function on $\mathbf{R}$ with a compact support. Let q be a lower bounded continuous function on $\mathbf{R}$. Let $v(t, x)$ be a function on $\mathbf{R}^+ \times \mathbf{R}$ defined as*

$$v(t, x) = E^x \left[f(X_t) \exp\left(-\int_0^t q(X_s) \mathrm{d}s \right) \right],$$

where $\{X_t : t \geq 0\}$ is an Itô diffusion as defined in Definition 39.6. Then $v(t, x)$ is a solution to the following Cauchy problem

$$\frac{\partial v}{\partial t}(t, x) = \mathrm{L}v(t, x) - q(x)v(t, x), \quad t > 0, x \in \mathbf{R}, \tag{40.4a}$$

$$v(0, x) = f(x), \quad x \in \mathbf{R}, \tag{40.4b}$$

where L is the second order differential operator associated with the Itô diffusion (see Definition 39.9):

$$\mathrm{L}v(t, x) = b(x)\frac{\partial v}{\partial x}(t, x) + \frac{1}{2}\sigma^2(x)\frac{\partial^2 v}{\partial x^2}(t, x).$$

Moreover, if $w(t, x)$ is a $C^{1,2}$ function on $\mathbf{R}^+ \times \mathbf{R}$ that is bounded on $K \times \mathbf{R}^+$ for each compact $K \subseteq \mathbf{R}$ and $w(t, x)$ solves the initial-value problem given in Equation (40.4), then $w(t, x) = v(t, x)$.

40.2 Problems

40.1. Let $D = (a, b)$ be an interval. Let f be a bounded function on D, φ a function on ∂D, and a a continuous nonnegative function on D. Let $\{X_t : t \geq 0\}$ be an Itô diffusion as defined in Definition 39.6. Let τ_D be the first exit time of X_t^x from D. Suppose that $E^x[\tau_D] < \infty$ and that $u \in C^2(D) \cap C(\bar{D})$ (i.e., u is a C^2 function on D and is continuous on the closure of D) is a solution to the Dirichlet problem

$$\mathrm{L}u(x) - a(x)u(x) = f(x), \quad x \in D, \tag{40.5a}$$

$$u(x) = \varphi(x), \quad x \in \partial D, \tag{40.5b}$$

where L is the differential operator associated with X_t (see Definition 39.9). For every $x \in D$, show that

$$\begin{aligned} u(x) \quad &= \quad E^x\left[\exp\left(-\int_0^{\tau_D} a(X_t)\mathrm{d}t\right)\varphi(X_{\tau_D})\right] - \\ &\quad\quad E^x\left[\int_0^{\tau_D} \exp\left(-\int_0^t a(X_s)\mathrm{d}s\right) f(X_t)\mathrm{d}t\right]. \end{aligned}$$

40.2. Let $Q = (0, T) \times D$, where $D = (a, b)$ is an interval. Let $\partial_p Q = ([0, T] \times \{a, b\}) \cup (\{T\} \times D)$. Let f be a bounded function on Q, φ be a continuous function on $\partial_p Q$, and a be a lower bounded continuous function on Q. Let $\{X_t : t \geq 0\}$ be a diffusion as defined in Definition 39.5. If $u \in C^2(Q) \cap C(\bar{Q})$ is a solution of the following Cauchy-Dirichlet problem

$$\mathrm{L}_t u - au + \frac{\partial u}{\partial t} = f, \quad (t, x) \in Q, \tag{40.6a}$$

$$u(t, x) = \varphi(t, x), \quad (t, x) \in \partial_p Q, \tag{40.6b}$$

where f, a, and φ are given functions and L_t is the differential operator associated with X_t (see Definition 39.9). For every $(t, x) \in Q$, show that

$$\begin{aligned} u(t, x) \quad &= \quad E\left[\exp\left(-\int_t^{\tau_{t,x}} a(s, X_s^{t,x})\mathrm{d}s\right)\varphi(\tau_{t,x}, X_{\tau_{t,x}}^{t,x})\right] \\ &\quad\quad -E\left[\int_t^{\tau_{t,x}} \exp\left(-\int_t^s a(r, X_r^{t,x})\mathrm{d}r\right) f(s, X_s^{t,x})\mathrm{d}s\right], \end{aligned}$$

where $\tau_{t,x} = \inf\{s > t : X_s^{t,x} \notin D\} \wedge T$.

40.3. Let $\{X_t : 0 \leq t \leq T\}$ be a diffusion as defined in Definition 39.5. Let $Q = (0, T) \times \mathbf{R}$. Let $u \in C^2(Q) \cap C(\bar{Q})$ be a solution of the following Cauchy problem

$$\mathrm{L}_t u - au + \frac{\partial u}{\partial t} = f, \quad (t, x) \in Q, \tag{40.7a}$$

$$u(T, x) = \varphi(x), \quad x \in \mathbf{R}, \tag{40.7b}$$

where L_t is the differential operator associated with X_t (see Definition 39.9), a is a lower-bounded continuous function on Q, and f and φ are given functions. Suppose that

(a) The coefficients $\mu(t,x)$ and $\sigma(t,x)$ are measurable and have at most linear growth in x.

(b) For every $(t,x) \in Q$, there exists a solution $\{X_s^{t,x} : t \leq s \leq T\}$ of the stochastic differential equation

$$\mathrm{d}X_s = \mu(s, X_s)\mathrm{d}B_s + \sigma(s, X_s)\mathrm{d}s, \quad X_t = x, s \in [t,T];$$

(c) There exist two positive constants M and p such that

$$|u(t,x)| + |f(t,x)| \leq M\left(1 + |x|^p\right), \quad (t,x) \in Q;$$

For every $(t,x) \in Q$, show that

$$\begin{aligned} u(t,x) &= E\left[\exp\left(-\int_t^T a(s, X_s^{t,x})\mathrm{d}s\right)\varphi(X_T^{t,x})\right] \\ &\quad -E\left[\int_t^T \exp\left(-\int_t^s a(r, X_r^{t,x})\mathrm{d}r\right) f(s, X_s^{t,x})\mathrm{d}s\right]. \end{aligned}$$

40.4 (Kolmogorov's Backward Equation). Let f be a C^2 function on $\mathbf{R}$ with a compact support. Let $u(t,x)$ be a function on $\mathbf{R}^+ \times \mathbf{R}$ defined as

$$u(t,x) = E^x[f(X_t)],$$

where $\{X_t : t \geq 0\}$ is an Itô diffusion as defined in Definition 39.6. Show that $u(t,x)$ solves the following initial-value problem

$$\frac{\partial u}{\partial t}(t,x) = \mathrm{L}u(t,x), \quad t > 0, x \in \mathbf{R}, \tag{40.8a}$$

$$u(0,x) = f(x), \quad x \in \mathbf{R}, \tag{40.8b}$$

where $\mathrm{L}u(t,x)$ denotes that L is applied to the function $x \to u(t,x)$.

40.5. Let u be a $C^{1,2}$ function on $\mathbf{R}\times\mathbf{R}$ that satisfies the following partial differential equation

$$\frac{\partial u}{\partial t} = -qu + \mu x\frac{\partial u}{\partial x} + \frac{1}{2}\sigma^2 x^2 \frac{\partial^2 u}{\partial x^2}, \quad t > 0,\ x \in \mathbf{R}, \tag{40.9a}$$

$$u(0,x) = (x-K)^+, \quad x \in \mathbf{R}, \tag{40.9b}$$

where $q > 0$, μ, $\sigma > 0$, and $K > 0$ are constants and

$$(x-K)^+ = \max(x-K, 0).$$

Show that for every $t > 0$ and $x \in \mathbf{R}$,

$$u(t,x) = \frac{e^{-qt}}{\sqrt{2\pi t}} \int_{\mathbf{R}} \left[x \exp\left(\left(\mu - \frac{1}{2}\sigma^2 \right) t + \sigma y \right) - K \right]^+ \exp\left(-\frac{y^2}{2t} \right) \mathrm{d}y.$$

40.6 (Bivariate Laplace Transform). Let $\{B_t : t \geq 0\}$ be a Brownian motion. Show that

(a) For $0 \leq t \leq T$, we have

$$\begin{aligned} & E\left[e^{-w\int_t^T B_s \mathrm{d}s - uB_T} \,\middle|\, B_t = x \right] \\ = \quad & \exp\left(\frac{w^2}{6}(T-t)^3 + \frac{wu}{2}(T-t)^2 + \frac{u^2}{2}(T-t) - [w(T-t)+u]x \right). \end{aligned}$$

(b) For $0 \leq t \leq T \leq S$, we have

$$\begin{aligned} & E\left[e^{-w\int_t^S B_s \mathrm{d}s - uB_T} \,\middle|\, B_t = x \right] \\ = \quad & \exp\left(\frac{w^2}{6}(S-t)^3 + \frac{wu}{2}(S-t)^2 - \frac{wu}{2}(S-T)^2 + \frac{u^2}{2}(T-t) \right) \times \\ & \exp\left(-[w(S-t)+u]x \right). \end{aligned}$$

40.3 Hints

40.1. Apply Itô's lemma (Theorem 35.3) to the process $Y_t = u(X_t)Z_t$, where

$$Z_t = \exp\left(-\int_0^t a(X_s)\mathrm{d}s \right).$$

40.2. The proof is similar to that of Problem 40.1. That is, apply Itô's lemma (Theorem 35.3) to the process $Y_s = u(X_s)Z_s$, where

$$Z_s = \exp\left(-\int_t^s a(r, X_r)\mathrm{d}r \right).$$

40.3. Consider the exit time of X_t from $D = (-n, n)$, $n \geq 1$, and use the result of Problem 40.2 and the dominated convergence theorem (Theorem 6.4).

40.4. Use Theorem 39.2 and the result of Problem 39.4.

40.5. Consider the C^2 function $f_n(x)$ such that

$$f_n(x) = (x-K)^+, \quad x \in \left(K + \frac{1}{n}, K + n \right)$$

and

$$f_n(x) = 0, \quad x \in (-\infty, K) \cup (K + n + 1, \infty).$$

Then apply the Feynman-Kac formula (Theorem 40.1) and the dominated convergence theorem (Theorem 6.4).

40.6. Apply the Feynman-Kac formula (Theorem 40.1) to prove the first part. Use the first part to prove the second part.

40.4 Solutions

40.1. Let

$$Z_t = \exp\left(-\int_0^t a(X_s)\mathrm{d}s\right), \quad t \geq 0.$$

Then by Theorem 35.3, we have

$$\mathrm{d}Z_t = -a(X_t)Z_t\mathrm{d}t. \tag{40.10}$$

Also by Theorem 35.3 and the assumption, we have

$$\begin{aligned} \mathrm{d}u(X_t) &= \sigma(X_t)\frac{\partial u}{\partial x}(X_t)\mathrm{d}B_t + \mathrm{L}u(X_t)\mathrm{d}t \\ &= \sigma(X_t)\frac{\partial u}{\partial x}(X_t)\mathrm{d}B_t + [a(X_t)u(X_t) + f(X_t)]\mathrm{d}t. \end{aligned} \tag{40.11}$$

Then it follows from Problem 35.6, and Equations (40.10) and (40.11) that

$$Z_t^x u(X_t^x) = u(x) + \int_0^t Z_s^x \sigma(X_s^x)\frac{\partial u}{\partial x}(X_s^x)\mathrm{d}B_s + \int_0^t Z_s^x f(X_s^x)\mathrm{d}s,$$

where $Z_s^x = \exp\left(-\int_0^t a(X_s^x)\mathrm{d}s\right)$. Hence we have

$$Z_{\tau_D}^x u(X_t^x) = u(x) + \int_0^{\tau_D} Z_s^x \sigma(X_s^x)\frac{\partial u}{\partial x}(X_s^x)\mathrm{d}B_s + \int_0^{\tau_D} Z_s^x f(X_s^x)\mathrm{d}s.$$

By the assumption, we have $Z_s^x \in [0, 1]$. Hence $Z_s^x \sigma(X_s^x)\frac{\partial u}{\partial x}(X_s^x)$ and $Z_s^x f(X_s^x)$ are bounded on $\{(t, \omega) : X_t^x(\omega) \in D\}$. Therefore, by Theorem 32.1 we have

$$E\left[\int_0^{\tau_D} Z_s^x \sigma(X_s^x)\frac{\partial u}{\partial x}(X_s^x)\mathrm{d}B_s\right] = 0.$$

Hence we have

$$u(x) = E^x\left[Z_{\tau_D} u(X_{\tau_D})\right] - E^x\left[\int_0^{\tau_D} Z_t f(X_t)\mathrm{d}t\right].$$

The result follows by noting that $u(X_{\tau_D}^x) = \varphi(X_{\tau_D}^x)$. This completes the proof.

40.2. Let

$$Z_s = \exp\left(-\int_t^s a(r, X_r)\mathrm{d}r\right), \quad s \geq t.$$

Then by Theorem 35.3, we have

$$\mathrm{d}Z_s = -a(s, X_s)Z_s\mathrm{d}s, \quad s > t.$$

Also by Theorem 35.3 and the assumption, we have

$$\begin{aligned} \mathrm{d}u(s, X_s) &= \sigma\frac{\partial u}{\partial x}\mathrm{d}B_s + \left(\frac{\partial u}{\partial t} + \mathrm{L}_s u\right)\mathrm{d}s \\ &= \sigma\frac{\partial u}{\partial x}\mathrm{d}B_s + (au + f)\,\mathrm{d}s. \end{aligned}$$

Then by Problem 35.6, we have

$$\mathrm{d}Z_s u(s, X_s) = Z_s\sigma(s, X_s)\frac{\partial u}{\partial x}(s, X_s)\mathrm{d}B_s + Z_s f(s, X_s)\mathrm{d}s.$$

Hence we have

$$\begin{aligned} Z^{t,x}_{\tau_{t,x}} u(\tau_{t,x}, X^{t,x}_{\tau_{t,x}}) &= Z_t u(t, X^{t,x}_t) + \int_t^{\tau_{t,x}} Z^{t,x}_s f(s, X^{t,x}_s)\mathrm{d}s \\ &\quad + \int_t^{\tau_{t,x}} Z^{t,x}_s \sigma(s, X^{t,x}_s)\frac{\partial u}{\partial x}(s, X^{t,x}_s)\mathrm{d}B_s, \quad (40.12) \end{aligned}$$

where

$$Z^{t,x}_s = \exp\left(-\int_t^s a(r, X^{t,x}_r)\mathrm{d}r\right).$$

By the assumption that $a(s, X_s)$ is lower-bounded, σ satisfies the linear growth condition, and u is a C^2 function on Q, $Z^{t,x}_s\sigma(s, X^{t,x}_s)\frac{\partial u}{\partial x}(s, X^{t,x}_s)$ is bounded on $\{(s, \omega) : X^{t,x}_s(\omega) \in D\}$. Hence we have

$$E\left[\int_t^{\tau_{t,x}} Z^{t,x}_s \sigma(s, X^{t,x}_s)\frac{\partial u}{\partial x}(s, X^{t,x}_s)\mathrm{d}B_s\right] = 0.$$

Therefore, taking expectation in both sides of Equation (40.12) gives

$$\begin{aligned} u(t, x) &= E\left[Z^{t,x}_{\tau_{t,x}} u(\tau_{t,x}, X^{t,x}_{\tau_{t,x}})\right] - E\left[\int_t^{\tau_{t,x}} Z^{t,x}_s f(s, X^{t,x}_s)\mathrm{d}s\right] \\ &= E\left[Z^{t,x}_{\tau_{t,x}} \varphi(\tau_{t,x}, X^{t,x}_{\tau_{t,x}})\right] - E\left[\int_t^{\tau_{t,x}} Z^{t,x}_s f(s, X^{t,x}_s)\mathrm{d}s\right]. \end{aligned}$$

This completes the proof.

40.3. Let $n \geq 1$ and $\tau_n = \inf\{s \geq t : X^{t,x}_s \notin (-n, n)\}$. Then by Problem 40.2, we have

$$\begin{aligned} u(t, x) &= E\left[\exp\left(-\int_t^{T\wedge\tau_n} a(s, X^{t,x}_s)\mathrm{d}s\right) u(T\wedge\tau_n, X^{t,x}_{T\wedge\tau_n})\right] \\ &\quad - E\left[\int_t^{T\wedge\tau_n} \exp\left(-\int_t^s a(r, X^{t,x}_r)\mathrm{d}r\right) f(s, X^{t,x}_s)\mathrm{d}s\right] \quad (40.13) \end{aligned}$$

By the assumption, we have

$$\left|\exp\left(-\int_t^{T\wedge\tau_n} a(s, X_s^{t,x})\mathrm{d}s\right) u(T\wedge\tau_n, X_{T\wedge\tau_n}^{t,x})\right| \le Me^{|a_0|T}\left(1+\tilde{X}_T^p\right),$$

$$\left|\int_t^{T\wedge\tau_n} \exp\left(-\int_t^s a(r, X_r^{t,x})\mathrm{d}r\right) f(s, X_s^{t,x})\mathrm{d}s\right| \le Te^{|a_0|T}\left(1+\tilde{X}_T^p\right),$$

where a_0 is the lower bound of a and

$$\tilde{X}_T = \sup_{t\le s\le T} |X_s^{t,x}|.$$

By Problem 38.6, $\tilde{X}_T$ is integrable. Since $\tau_n \wedge T \to T$ as $n\to\infty$, it follows from Theorem 6.4 that

$$\begin{aligned} u(t,x) &= E\left[\exp\left(-\int_t^T a(s, X_s^{t,x})\mathrm{d}s\right) u(T, X_T^{t,x})\right] \\ &\quad -E\left[\int_t^T \exp\left(-\int_t^s a(r, X_r^{t,x})\mathrm{d}r\right) f(s, X_s^{t,x})\mathrm{d}s\right]. \end{aligned}$$

This completes the proof.

40.4. Let $t \ge 0$ be fixed. Let $g(x) = u(t,x)$. By Theorem 39.2, $g(X_r) = E[f(X_{t+r})|\mathscr{F}_r]$ for every $r > 0$ (r is a constant and hence a stopping time). Then we have

$$\begin{aligned} \frac{E^x[g(X_r)] - g(x)}{r} &= \frac{E^x(E[f(X_{t+r})|\mathscr{F}_r]) - E^x[f(X_t)]}{r} \\ &= \frac{E^x[f(X_{t+r})] - E^x[f(X_t)]}{r} \\ &= \frac{u(t+r,x) - u(t,x)}{r}. \end{aligned}$$

By Problem 39.4, we have

$$\lim_{r\downarrow 0} \frac{E^x[g(X_r)] - g(x)}{r} = \mathrm{L}u(t,x).$$

Combining the above two equations gives

$$\frac{\partial u}{\partial t}(t,x) = \mathrm{L}u(t,x).$$

This completes the proof.

40.5. Let $\{X_t : t \ge 0\}$ be the Itô diffusion given by

$$\mathrm{d}X_t = \mu X_t \mathrm{d}t + \sigma X_t \mathrm{d}B_t, \quad X_0 = x,\ t \ge 0.$$

Then the second-order differential operator associated with X_t is given by

$$\mathrm{L}u = \mu x \frac{\partial u}{\partial x} + \frac{1}{2}\sigma^2 x^2 \frac{\partial^2 u}{\partial x^2}.$$

For every $n \geq 1$, let $f_n(x)$ be a function on $\mathbf{R}$ given by

$$f_n(x) = \begin{cases} 0, & \text{if } x \in (-\infty, K); \\ g_n(x), & \text{if } x \in [K, K + \frac{1}{n}]; \\ x - K, & \text{if } x \in \left(K + \frac{1}{n}, K + n\right); \\ h_n(x), & \text{if } x \in [K + n, K + n + 1]; \\ 0, & \text{if } x \in (K + n + 1, \infty), \end{cases}$$

where

$$g_n(x) = 6n^2(x-K)^3 - 8n^3(x-K)^4 + 3n^4(x-K)^5$$

and

$$\begin{aligned} h_n(x) &= n + (x - K - n) - (10n+6)(x-K-n)^3 \\ &\quad + (15n+8)(x-K-n)^4 - (6n+3)(x-K-n)^5. \end{aligned}$$

Then $f_n(x)$ is a C^2 function with a compact support. It follows from Theorem 40.1 that

$$u_n(t, x) = E^x\left[f_n(X_t)e^{-qt}\right] \tag{40.14}$$

solves the following partial differential equation

$$\frac{\partial u_n}{\partial t} = -qu_n + \mu x \frac{\partial u_n}{\partial x} + \frac{1}{2}\sigma^2 x^2 \frac{\partial^2 u_n}{\partial x^2}, \quad t > 0,\ x \in \mathbf{R},$$

$$u_n(0, x) = f_n(x), \quad x \in \mathbf{R}.$$

Since for every $x \in [K, K + \frac{1}{n}]$,

$$\begin{aligned} |g_n(x)| &\leq 6n^2(x-K)^3 + 8n^3(x-K)^4 + 3n^4(x-K)^5 \\ &\leq (x-K)(6+8+3) \\ &\leq 17 \end{aligned}$$

and for every $x \in [K + n, K + n + 1]$,

$$\begin{aligned} |h_n(x)| &\leq n + (x-K-n) + (10n+6)(x-K-n)^3 \\ &\quad + (15n+8)(x-K+n)^4 + (6n+3)(x-K-n)^5 \\ &\leq n + 1 + (10n+6) + (15n+8) + (6n+3) \\ &\leq 32x + 8, \end{aligned}$$

we have

$$|f_n(X_t^x)| \leq 32|X_t^x| + 8 + K \leq 32 \sup_{0 \leq s \leq t} |X_s^x| + 17 + K.$$

Since $f_n(t, x) \to (x - K)^+$ as $n \to \infty$ and $\sup_{0\leq s\leq t} |X_s^x|$ is integrable (see Problem 38.6), it follows from Theorem 6.4 that

$$\lim_{n\to\infty} u_n(t, x) = u(t, x).$$

Hence we have

$$u(t, x) = e^{-qt} E^x[(X_t - K)^+].$$

The result follows by Problem 38.7 and integrating out the expectation. This completes the proof.

40.6.

(a) Let $X_r = B_{r+t}$ for all $r \geq 0$, $f(x) = e^{-ux}$, and $q(x) = wx$. Then by Theorem 40.1, the function

$$v(r, x) = E\left[e^{-w\int_0^r X_s \mathrm{d}s - uX_r}\middle| X_0 = x\right]$$

satisfies the following Cauchy problem

$$\frac{\partial v}{\partial r}(r, x) = \frac{1}{2}\frac{\partial^2 v}{\partial x^2}(r, x) - wxv(r, x), \quad r > 0, x \in \mathbf{R}, \tag{40.15a}$$

$$v(0, x) = f(x) = e^{-ux}, \quad x \in \mathbf{R}, \tag{40.15b}$$

But the function

$$\tilde{v}(r, x) = \exp\left(\frac{w^2}{6}r^3 + \frac{wu}{2}r^2 + \frac{u^2}{2}r - [wr + u]x\right)$$

also satisfies the Cauchy problem given in Equation (40.15). Therefore, it follows from Theorem 40.1 that

$$\begin{aligned}
&E\left[e^{-w\int_0^r X_s \mathrm{d}s - uX_r}\middle| X_0 = x\right] \\
= \; & \exp\left(\frac{w^2}{6}r^3 + \frac{wu}{2}r^2 + \frac{u^2}{2}r - [wr + u]x\right).
\end{aligned}$$

The result follows by letting $r = T - t$.

(b) By the first part and the tower property of conditional expectations (see Problem 14.10), we have

$$
\begin{aligned}
& E\left[e^{-w\int_t^S B_s ds - uB_T}\,\middle|\, B_t = x\right] \\
= \; & E\left[e^{-w\int_t^T B_s ds - uB_T} E\left[e^{-w\int_T^S B_s ds}\,\middle|\, B_T\right]\,\middle|\, B_t = x\right] \\
= \; & E\left[e^{-w\int_t^T B_s ds - uB_T} \exp\left(\frac{w^2}{6}(S-T)^3 - w(S-T)B_T\right)\,\middle|\, B_t = x\right] \\
= \; & \exp\left(\frac{w^2}{6}(S-T)^3\right) E\left[e^{-w\int_t^T B_s ds - [w(S-T)+u]B_T}\,\middle|\, B_t = x\right] \\
= \; & \exp\left(\frac{w^2}{6}(S-T)^3\right) \times \\
& \exp\left(\frac{w^2}{6}(T-t)^3 + \frac{w\tilde{u}}{2}(T-t)^2 + \frac{\tilde{u}^2}{2}(T-t) - [w(T-t)+\tilde{u}]x\right) \\
= \; & \exp\left(\frac{w^2}{6}(S-t)^3 + \frac{wu}{2}(S-t)^2 - \frac{wu}{2}(S-T)^2 + \frac{u^2}{2}(T-t)\right) \times \\
& \exp\left(-[w(S-t)+u]x\right),
\end{aligned}
$$

where $\tilde{u} = w(S-T) + u$.

This completes the proof.

40.5 Bibliographic Notes

In this chapter, we presented several versions of the Feynman-Kac formula, each version of which involves some technical conditions on underlying coefficients and initial data. Although these conditions are different, the idea behind the Feynman-Kac formula is the same. For a discussion of the technical conditions for the Feynman-Kac formula, readers are referred to Duffie (2001).

The Feynman-Kac formula is studied in many texts, for example, Karatzas and Shreve (1991), Durrett (1996), Bass (1998), Oksendal (1998), Steele (2003), and Pascucci (2011). In this chapter, we presented only the one-dimensional Feynman-Kac formula. For the multidimensional Feynman-Kac formula, readers are referred to Karatzas and Shreve (1991), Oksendal (1998), Shreve (2004), Pinsky (1995), Pham (2009), and Pascucci (2011).

The Dirichlet problem, the Cauchy problem, and the heat equation are subjects studied in partial differential equations. Steele (2003, Chapter 11) gives a very good introduction to the derivation of the heat equation. For an introduction to the heat equation and Cauchy problem, readers are referred to Cannon (1984). The connection between the heat equation and the Brownian motion is discussed in Karatzas and Shreve (1991) and Lawler (2006).

PART V

STOCHASTIC FINANCIAL MODELS

CHAPTER 41

DISCRETE-TIME MODELS

In discrete-time models, it is assumed that trading takes place at discrete time points. In this chapter, we present discrete-time models of financial markets. In particular, we will introduce the binomial model.

41.1 Basic Concepts and Facts

Definition 41.1 (Discrete Market). Let T be a positive real number and N be a positive integer. Let $0 = t_0 < t_1 < \cdots < t_N = T$. Let $(\Omega, \mathscr{F}, P)$ be a probability space, where Ω is finite, $\mathscr{F} = 2^\Omega$, and $P\{\omega\} > 0$ for every $\omega \in \Omega$. A discrete market is a $(d+1)$-dimensional stochastic process

$$S = \{(S_n^{(0)}, S_n^{(1)}, \ldots, S_n^{(d)}) : n = 0, 1, \ldots, N\}$$

defined on the probability space $(\Omega, \mathscr{F}, P)$.

The first entry $S_n^{(0)}$ is the price at time t_n of a riskless asset and is deterministic, that is,

$$S_0^{(0)} = 1, \quad S_n^{(0)} = S_{n-1}^{(0)} (1 + r_n), \quad n = 1, 2, \ldots, N,$$

where $r_n > -1$ is the risk-free interest rate in the nth period $[t_{n-1}, t_n]$.

For $j \geq 1$, the jth entry $S_n^{(j)}$ is the price at time t_n of the jth risky asset and follows the following stochastic dynamics:

$$S_0^{(j)} > 0, \quad S_n^{(j)} = S_{n-1}^{(j)} \left(1 + \mu_n^{(j)}\right), \quad n = 1, 2, \ldots, N.$$

Here $S_0^{(j)}$ is a real number and $\mu_n^{(j)} > -1$ is a real random variable.

The amount of information available in the market at time t_n is represented by the σ-algebra $\mathscr{F}_n$ given by

$$\mathscr{F}_0 = \{\emptyset, \Omega\}$$

and

$$\mathscr{F}_n = \sigma\left(\left(\mu_i^{(1)}, \mu_i^{(2)}, \ldots, \mu_i^{(d)}\right) : i = 1, 2, \ldots, n\right), \quad n = 1, 2, \ldots, N.$$

We assume that $\mathscr{F}_N = \mathscr{F}$.

Definition 41.2 (Normalized Market)**.** Let S be a discrete market and $\{\mathscr{F}_n : n = 0, \ldots, N\}$ be the filtration as defined in Definition 41.1. Let $Y = \{Y_n : n = 0, 1, \ldots, N\}$ be a positive $\mathscr{F}_n$-adapted price process. The normalized market of S with respect to Y is defined as

$$\bar{S} = \{(\bar{S}_n^{(0)}, \bar{S}_n^{(1)}, \ldots, \bar{S}_n^{(d)}) : n = 0, 1, \ldots, N\},$$

where

$$\bar{S}_n^{(j)} = \frac{S_n^{(j)}}{Y_n}, \quad j = 0, \ldots, d, \; n = 0, \ldots, N.$$

The process Y is called a *numeraire.*

Definition 41.3 (Portfolio)**.** A portfolio or strategy is a $(d+1)$-dimensional stochastic process

$$\alpha = \{(\alpha_n^{(0)}, \alpha_n^{(1)}, \ldots, \alpha_n^{(d)}) : n = 0, 1, \ldots, N\}.$$

The value of the portfolio α at time t_n is

$$V_n(\alpha) = \sum_{j=0}^{d} \alpha_n^{(j)} S_n^{(j)}.$$

Definition 41.4 (Self-Financing Portfolio)**.** A portfolio α is called self-financing if for every n in $\{1, 2, \ldots, N\}$, we have

$$V_{n-1}(\alpha) = \sum_{j=0}^{d} \alpha_n^{(j)} S_{n-1}^{(j)}.$$

Definition 41.5 (Predictable Portfolio)**.** A portfolio α is called predictable if for every n in $\{1, 2, \ldots, N\}$, $(\alpha_n^{(0)}, \alpha_n^{(1)}, \ldots, \alpha_n^{(d)})$ is $\mathscr{F}_{n-1}$-measurable, where $\mathscr{F}_n$ is as given in Definition 41.1.

Definition 41.6 (Arbitrage). Let $\mathcal{A}$ be the family of all self-financing, predictable portfolios of the discrete market. A portfolio $\alpha \in \mathcal{A}$ is called an *arbitrage* if $V_0(\alpha) = 0$ and there exists $n \geq 1$ such that $P\{V_n(\alpha) \geq 0\} = 1$ and $P\{V_n(\alpha) > 0\} > 0$.

The discrete market is called arbitrage-free if $\mathcal{A}$ does not contain arbitrage portfolios.

Definition 41.7 (Admissible Portfolio). A portfolio α is called admissible if for every n in $\{1, 2, \ldots, N\}$, we have

$$P\{V_n(\alpha) \geq 0\} = 1.$$

Definition 41.8 (Equivalent Martingale Measure). Let Y be an numeraire. An equivalent martingale measure with Y is a probability measure Q on $(\Omega, \mathscr{F})$ such that

(a) $Q \sim P$ (see Definition 7.3).

(b) the Y-normalized price processes are Q-martingales:

$$\frac{S_{n-1}^{(j)}}{Y_{n-1}} = E_Q\left[\frac{S_n^{(j)}}{Y_n}|\mathscr{F}_{n-1}\right], \quad n = 1, 2, \ldots, N, \; j = 0, 1, \ldots, d,$$

where E_Q denotes the expectation under the probability measure Q.

Definition 41.9 (European Derivative). A European derivative is an $\mathscr{F}_N$-measurable random variable on $(\Omega, \mathscr{F}, P)$, where $\mathscr{F}_n$ is given in Definition 41.1.

Definition 41.10 (Replicating Portfolio for European Derivatives). Let X be a European derivative. A portfolio $\alpha \in \mathcal{A}$ is said to be a replicating portfolio of X if

$$V_N(\alpha) = X \quad P-\text{a.s.}$$

If such a portfolio exists, the derivative is said to be replicable.

A portfolio $\alpha \in \mathcal{A}$ is said to be a superreplicating portfolio for X if

$$V_N(\alpha) \geq X \quad P-\text{a.s.}$$

The family of all super-replicating portfolio for X is denoted by $\mathcal{A}_X^+$:

$$\mathcal{A}_X^+ = \{\alpha \in \mathcal{A} : P\{V_N(\alpha) \geq X\} = 1\}.$$

A portfolio $\alpha \in \mathcal{A}$ is said to be a subreplicating portfolio for X if

$$V_N(\alpha) \leq X \quad P-\text{a.s.}$$

The family of all sub-replicating portfolio for X is denoted by $\mathcal{A}_X^-$:

$$\mathcal{A}_X^- = \{\alpha \in \mathcal{A} : P\{V_N(\alpha) \leq X\} = 1\}.$$

Definition 41.11 (American Derivative). An American derivative is a nonnegative discrete stochastic process $X = \{X_n : n = 0, 1, \ldots, N\}$ adapted to the filtration $\{\mathscr{F}_n : n = 0, 1, \ldots, N\}$, where the filtration is as given in Definition 41.1.

Definition 41.12 (Exercise Strategy). An exercise strategy or exercise time of an American derivative is a stopping time $\nu : \Omega \to \{0, 1, \ldots, N\}$. The set of all exercise strategies is denoted by $\mathcal{T}_0$.

Definition 41.13 (Payoff of American Derivatives). Let X be an American derivative and ν an exercise strategy. The payoff of X relative to the strategy ν is the random variable X_ν defined by

$$X_\nu(\omega) = X_{\nu(\omega)}(\omega), \quad \forall \omega \in \Omega.$$

Let Q be an equivalent martingale measure with numeraire Y. An exercise strategy ν_0 is called optimal under Q if

$$E_Q\left[\frac{X_{\nu_0}}{Y_{\nu_0}}\right] = \sup_{\nu \in \mathcal{T}_0} E_Q\left[\frac{X_\nu}{Y_\nu}\right].$$

Definition 41.14 (Replicating Portfolio for American Derivatives). Let X be an American derivative. A portfolio $\alpha \in \mathcal{A}$ is said to be a superreplicating portfolio for X if

$$V_n(\alpha) \geq X_n \quad P-\text{a.s.}, \quad \forall n = 0, 1, \ldots, N.$$

The family of all superreplicating portfolios for X is denoted by $\mathcal{A}_X^+$:

$$\mathcal{A}_X^+ = \{\alpha \in \mathcal{A} : P\{V_n(\alpha) \geq X_n\} = 1,\ n = 0, 1, \ldots, N\}.$$

The family $\mathcal{A}_X^-$ is defined as

$$\mathcal{A}_X^- = \{\alpha \in \mathcal{A} : \text{there exists } \nu \in \mathcal{T}_0 \text{ such that } P\{X_\nu \geq V_\nu(\alpha)\} = 1\}.$$

Definition 41.15 (Complete Market). A market is said to be complete if every European derivative is replicable.

Definition 41.16 (Arbitrage Price). Let X be a replicable European derivative in a market that is free of arbitrage. Then arbitrage price or risk-neutral price of X is $V(\alpha)$, where α is a replicating portfolio of X (see Problem 41.16).

Definition 41.17 (Binomial Model). In the binomial model, the market is composed of two assets: a riskless asset and a risky asset. The time intervals have the same length:

$$t_n - t_{n-1} = \frac{T}{N}, \quad n = 1, 2, \ldots, N.$$

The interest rate is assumed to be constant, namely, $r_n = r$ for every n. The dynamics of the riskless asset is given by

$$S_n^{(0)} = (1 + r)^n, \quad n = 0, 1, \ldots, N.$$

The random variables $\mu_1, \mu_2, \ldots, \mu_N$ are assumed to be independent and identically distributed on the probability space $(\Omega, \mathscr{F}, P)$:

$$P\{1+\mu_n = u\} = p, \quad P\{1+\mu_n = d\} = 1-p,$$

where $0 < d < u$ and $p \in (0,1)$. The sample space Ω is assumed to be

$$\{(e_1, e_2, \ldots, e_N) : e_k = u \text{ or } e_k = d\},$$

which contains 2^N elements. Each element in Ω is a path in the binomial tree.

Theorem 41.1 (First Fundamental Theorem of Asset Pricing)**.** *Let S be a discrete market as defined in Definition 41.1. Then S is arbitrage-free if and only if there exists at least one equivalent martingale measure.*

Theorem 41.2 (Second Fundamental Theorem of Asset Pricing)**.** *Let S be a discrete market as defined in Definition 41.1. Suppose that S is arbitrage-free. Then S is complete if and only if there exists a unique equivalent martingale measure with numeraire $\{S_n^{(0)} : n = 0, 1, \ldots, N\}$.*

Theorem 41.3 (European Derivative Pricing in Binomial Models)**.** *Let S be the binomial market as defined in Definition 41.17. Suppose that $d < 1+r < u$. Then*

(a) The binomial market is arbitrage-free and complete.

(b) The arbitrage price at time t_n of a European derivative X is given by

$$H_n = \frac{1}{(1+r)^{N-n}} E_Q[X|\mathscr{F}_n], \tag{41.1}$$

where Q is the equivalent martingale measure given in Problem 41.8.

(c) If a European derivative $X = f(S_N^{(1)})$, then the initial price of X is given by

$$\begin{aligned} H_0 &= \frac{1}{(1+r)^N} E_Q[f(S_N^{(1)})] \\ &= \frac{1}{(1+r)^N} \sum_{k=0}^{N} \binom{n}{k} q^k (1-q)^{N-k} f\left(u^k d^{N-k} S_0^{(1)}\right), \end{aligned} \tag{41.2}$$

where q is as given in Problem 41.8.

41.2 Problems

41.1. Let S be a discrete market as defined in Definition 41.1. Show that S is arbitrage-free if and only if there exist no admissible arbitrage portfolios.

41.2. Let Q be an equivalent martingale measure with numeraire Y. Let $\alpha \in \mathcal{A}$ (see Definition 41.6). Define $\bar{V}(\alpha)$ as

$$\bar{V}_n(\alpha) = \frac{V_n(\alpha)}{Y_n}, \quad n = 0, 1, \ldots, N.$$

Show that

(a) $\bar{V}(\alpha)$ is a Q-martingale:

$$\bar{V}_{n-1}(\alpha) = E_Q\left[\bar{V}_n(\alpha)|\mathscr{F}_{n-1}\right], \quad n = 1, 2, \ldots, N.$$

(b) The following risk-neutral pricing formula holds:

$$\bar{V}_0(\alpha) = E_Q\left[\bar{V}_n(\alpha)\right], \quad \forall n \leq N.$$

41.3 (No-Arbitrage Principle)**.** Let S be an arbitrage-free discrete market and $\mathcal{A}$ the family of all self-financing, predictable portfolios in S. Suppose that $\alpha, \beta \in \mathcal{A}$ and

$$V_N(\alpha) = V_N(\beta) \quad P-\text{a.s.}$$

Show that for every n in $\{0, 1, \ldots, N\}$,

$$V_n(\alpha) = V_n(\beta) \quad P-\text{a.s.}$$

41.4 (Change of Numeraire)**.** Let S be a discrete market and $\{\mathscr{F}_n : n = 0, \ldots, N\}$ be the filtration as defined in Definition 41.1. Let Q be an equivalent martingale measure with numeraire Y. Let X be a positive $\mathscr{F}_n$-adapted process such that

$$\left\{\frac{X_n}{Y_n} : n = 0, 1, \ldots, N\right\}$$

is a Q-martingale. Let Q_X be a probability measure defined as

$$\mathrm{d}Q_X = \frac{X_N Y_0}{X_0 Y_N}\mathrm{d}Q.$$

Show that

(a) For every random variable Z, we have

$$Y_n E_Q\left[\frac{Z}{Y_N}|\mathscr{F}_n\right] = X_n E_{Q_X}\left[\frac{Z}{X_N}|\mathscr{F}_n\right], \quad \forall n \leq N.$$

(b) Q_X is an equivalent martingale measure with numeraire X.

41.5. Let X be a European derivative and Q be an equivalent martingale measure with numeraire Y. Show that for every n in $\{0, 1, \ldots, N\}$, we have

$$\sup_{\alpha \in \mathcal{A}_X^-} \bar{V}(\alpha) \leq E_Q\left[\frac{X}{Y_N}|\mathscr{F}_n\right] \leq \inf_{\alpha \in \mathcal{A}_X^+} \bar{V}(\alpha),$$

where $\mathcal{A}_X^-$ and $\mathcal{A}_X^+$ are as defined in Definition 41.10.

41.6 (Arbitrage Price). Let X be a replicable European derivative in an arbitrage-free market. For every replicating portfolio $\alpha \in \mathcal{A}$ and for every equivalent martingale measure Q with numeraire Y, show that

$$E_Q\left[\frac{X}{Y_N}|\mathscr{F}_n\right] = \frac{V_n(\alpha)}{Y_n}, \quad n = 0, 1, \ldots, N.$$

41.7 (Markov Property). Let S be a discrete market as defined in Definition 41.1. Let $\mu_n = (\mu_n^{(1)}, \mu_n^{(2)}, \ldots, \mu_n^{(d)})$ and $S_n = (S_n^{(1)}, S_n^{(2)}, \ldots, S_n^{(d)})$ for $n = 1, 2, \ldots, N$. Suppose that μ_1, μ_2, ..., μ_N are independent. Show that $\{S_n : n = 0, 1, \ldots, N\}$ is a Markov process; that is, for every positive function $f \in m\mathcal{B}(\mathbf{R}^d)$, that

$$E[f(S_n)|\mathscr{F}_{n-1}] = E[f(S_n)|S_{n-1}], \quad n = 1, 2, \ldots, N.$$

41.8. Let S be the binomial market given in Definition 41.17. Suppose that $d < 1 + r < u$. Show that

(a) There exists a unique equivalent martingale measure Q with numeraire $\{S_n^{(0)} : n = 0, 1, \ldots, N\}$, which is given by

$$Q\{1 + \mu_n = u\} = q, \quad Q\{1 + \mu_n = d\} = 1 - q,$$

where $q = \frac{1+r-d}{u-d}$.

(b) The random variables $\mu_1, \mu_2, \ldots, \mu_N$ are independent under the probability measure Q and

$$Q\left\{S_n^{(1)} = u^k d^{n-k} S_0^{(1)}\right\} = \binom{n}{k} q^k (1-q)^{n-k}, \quad 0 \le k \le n \le N.$$

(c) The process $\{S_n^{(1)} : n = 0, 1, \ldots, N\}$ is a Markov process; that is, for every positive Borel function φ, we have

$$E_Q[\varphi(S_n^{(1)})|\mathscr{F}_{n-1}] = E_Q[\varphi(S_n^{(1)})|S_{n-1}^{(1)}] = q\varphi(uS_{n-1}^{(1)}) + (1-q)\varphi(dS_{n-1}^{(1)}).$$

41.9 (Snell Envelope). Let S be the discrete market as defined in Definition 41.1. Let X be an American derivative and Q be an equivalent martingale measure with numeraire Y. Let $\bar{H} = \{\bar{H}_n : n = 0, 1, \ldots, N\}$ be a stochastic process defined by

$$\bar{H}_n = \begin{cases} \bar{X}_N, & \text{if } n = N; \\ \max\left\{\bar{X}_n, E_Q\left[\bar{H}_{n+1}|\mathscr{F}_n\right]\right\}, & \text{if } n = 0, 1, \ldots, N-1, \end{cases}$$

where $\bar{X}_n = \frac{X_n}{Y_n}$ for $n = 0, 1, \ldots, N$.

Show that $\bar{H}$ is the smallest Q-supermartingale greater than or equal to $\bar{X}$:

$$\bar{H}_n \ge \bar{X}_n, \quad \forall n = 0, 1, \ldots, N.$$

41.10. Let X be an American derivative and Q be an equivalent martingale measure with numeraire Y. Show that

$$\sup_{\alpha \in \mathcal{A}_X^-} \bar{V}_0(\alpha) \leq \sup_{\nu \in \mathcal{T}_0} E_Q[\bar{X}_\nu] \leq \inf_{\alpha \in \mathcal{A}_X^+} \bar{V}_0(\alpha),$$

where $\mathcal{A}_X^-$ and $\mathcal{A}_X^+$ are as defined in Definition 41.14, $\bar{X}_\nu = \frac{X_\nu}{Y_\nu}$, and $\bar{V}_0(\alpha) = \frac{V_0(\alpha)}{Y_0}$.

41.11 (Arbitrage Price for American Derivatives)**.** Let X be an American derivative in the discrete market S defined in Definition 41.1. Let $\mathcal{A}_X^+$ and $\mathcal{A}_X^-$ be defined as in Definition 41.14. Suppose that S is arbitrage-free and complete and Q is the equivalent martingale measure in the market with numeraire Y. Show that there exists a portfolio $\alpha_0 \in \mathcal{A}_X^+ \cap \mathcal{A}_X^-$ such that

(a) $V_n(\alpha_0) \geq X_n$ for all $n = 0, 1, \ldots, N$.

(b) There exists an exercise strategy ν_0 such that $V_{\nu_0}(\alpha_0) = X_{\nu_0}$.

(c) The initial arbitrage price of X is given by

$$V_0(\alpha_0) = Y_0 E_Q\left[\frac{X_{\nu_0}}{Y_{\nu_0}}\right] = Y_0 \sup_{\nu \in \mathcal{T}_0} E_Q\left[\frac{X_\nu}{Y_\nu}\right].$$

41.3 Hints

41.1. The sufficiency ("only if") part is obvious. To prove the necessity part, use the method of contradiction.

41.2. To prove the first part, follow the definitions of equivalent martingale measure, self-financing portfolio, and predictable portfolio. The second part follows from the first part.

41.3. Apply the first fundamental theorem of asset pricing (Theorem 41.1) and use the result of Problem 41.2.

41.4. Use the result of Problem 14.20 to prove the first part. To prove the second part, first use the first part to show that

$$\frac{S_n^{(j)}}{X_n} = E_{Q^X}\left[\frac{S_N^{(j)}}{X_N} | \mathscr{F}_n\right], \quad j = 0, \ldots, d,\ n = 0, \ldots, N,$$

and then apply the tower property (see Problem 14.10).

41.5. Use the results of Problems 14.5 and 14.6.

41.6. Use the results of Problems 41.3 and 41.5.

41.7. Use the result of Problem 14.13.

41.8. Follow the definition of equivalent martingale measures (Definition 41.8) and use the result of Problem 14.19 to prove part (a). Follow the definition of independence (Definition 12.2) and use the result of Problem 14.10 to prove part (b). Use the result of Problem 41.7 and part (a) to prove part (c).

41.9. Use the induction method to show that $\bar{H}$ is the smallest Q-supermartingale; that is, assume $\bar{H}_n \leq Y_n$ and then show that $\bar{H}_{n-1} \leq Y_{n-1}$.

41.10. Note that $\bar{V}(\alpha)$ is a Q-martingale and use the optimal stopping theorem (Theorem 23.1).

41.11. Use the results of Problems 41.9 and 22.11.

41.4 Solutions

41.1. The sufficiency part is clear. If S is arbitrage-free, then there exist no arbitrage portfolios. Hence there exist no admissible arbitrage portfolios.

Now we prove the necessity part. Assume that there exist no admissible arbitrage portfolios. We show that S is arbitrage-free by contradiction. If there exists an arbitrage portfolio α, then by Definition 41.6, we have

$$V_0(\alpha) = \sum_{j=0}^{d} \alpha_1^{(j)} S_0^{(0)} = 0,$$

and there exists an integer $k \geq 1$ such that

$$P\{V_k(\alpha) \geq 0\} = 1 \text{ and } P\{V_k(\alpha) > 0\} > 0.$$

By the hypothesis, α is not admissible. Hence there exist an integer $m \leq N$ and a set $F \in \mathscr{F}_m$ with $P(F) > 0$ such that

$$\sum_{j=0}^{d} \alpha_m^{(j)}(\omega) S_m^{(j)}(\omega) < 0, \quad \forall \omega \in F$$

and

$$P\left\{\omega : \sum_{j=0}^{d} \alpha_n^{(j)}(\omega) S_n^{(j)}(\omega) \geq 0\right\} = 1, \quad \forall m < n \leq N.$$

If $m = N$, we let $\beta_n^{(j)} = 0$ on $\Omega \backslash F$ for all $j = 0, 1, \ldots, d$ and $n = 0, 1, \ldots, N$, and

$$\beta_n^{(j)} = \begin{cases} -V_N(\alpha), & \text{if } n = N \text{ and } j = 0; \\ 0, & \text{otherwise.} \end{cases} \qquad j = 0, \ldots, d, \ n = 0, \ldots, N.$$

Hence we have $V_n(\beta) = 0$ on $\Omega\backslash F$, $V_n(\beta) = 0$ on F for $n < N$, and $V_N(\beta) = -V_N(\alpha)S_N^{(0)} > 0$ on F. Therefore, β is an admissible portfolio. This contradicts the assumption.

If $m < N$, we can also find an admissible portfolio, which contradicts the assumption. This completes the proof.

41.2.

(a) By the assumption that $\alpha \in \mathcal{A}$, we know that α is self-financing and predictable. Then by Problem 14.12, for every $n \in \{1, 2, \ldots, N\}$, we have

$$\begin{aligned} E_Q\left[\bar{V}_n(\alpha)|\mathscr{F}_{n-1}\right] &= E_Q\left[\frac{1}{Y_n}\sum_{j=0}^{d}\alpha_n^{(j)}S_n^{(j)}|\mathscr{F}_{n-1}\right] \\ &= \sum_{j=0}^{d}\alpha_n^{(j)}E_Q\left[\frac{S_n^{(j)}}{Y_n}|\mathscr{F}_{n-1}\right] \\ &= \sum_{j=0}^{d}\alpha_n^{(j)}\frac{S_{n-1}^{(j)}}{Y_{n-1}} \\ &= \bar{V}_{n-1}(\alpha), \end{aligned}$$

which shows that $\bar{V}(\alpha)$ is a Q-martingale.

(b) By the first part and Problem 14.10, we have

$$E_Q\left[\bar{V}_n(\alpha)\right] = E_Q\left(E_Q\left[\bar{V}_n(\alpha)|\mathscr{F}_{n-1}\right]\right) = E_Q\left[\bar{V}_{n-1}(\alpha)\right].$$

Repeating the above calculation gives

$$E_Q\left[\bar{V}_n(\alpha)\right] = E_Q\left[\bar{V}_0(\alpha)\right] = \bar{V}_0(\alpha),$$

because $\bar{V}_0(\alpha)$ is a real number.

This completes the proof.

41.3. Since S is arbitrage-free, it follows from Theorem 41.1 that there exists an equivalent martingale measure Q with numeraire Y. Let $\bar{V}(\alpha)$ and $\bar{V}(\beta)$ be the normalized portfolios with respect to Y. By Problem 41.2, $\bar{V}(\alpha)$ and $\bar{V}(\beta)$ are Q-martingales. By the assumption that $P\{V_N(\alpha) = V_N(\beta)\} = 1$ and the fact that $P \sim Q$, we have $Q\{V_N(\alpha) = V_N(\beta)\} = 1$. Hence we have for every n in $\{0, 1, \ldots, N\}$,

$$\bar{V}_n(\alpha) = E_Q\left[\bar{V}_N(\alpha)|\mathscr{F}_n\right] = E_Q\left[\bar{V}_N(\beta)|\mathscr{F}_n\right] = \bar{V}_n(\beta) \quad Q-\text{a.s.}$$

Since $P \sim Q$, we get

$$V_n(\alpha) = V_n(\beta) \quad P-\text{a.s.}$$

This completes the proof.

41.4.

(a) Let $L = \frac{dQ_X}{dQ}$ be the Radon-Nikodym derivative. Then by the assumption and by Problem 14.20, we have

$$\begin{aligned} Y_n E_Q\left[\frac{Z}{Y_N}|\mathscr{F}_n\right] &= Y_n E_Q\left[\frac{ZX_0}{X_N Y_0} L|\mathscr{F}_n\right] \\ &= Y_n E_{Q_X}\left[\frac{ZX_0}{X_N Y_0}|\mathscr{F}_n\right] \cdot E_Q\left[\frac{X_N Y_0}{Y_N X_0}|\mathscr{F}_n\right] \\ &= Y_n E_{Q_X}\left[\frac{ZX_0}{X_N Y_0}|\mathscr{F}_n\right] \cdot \frac{X_n Y_0}{Y_n X_0} \\ &= X_n E_{Q_X}\left[\frac{Z}{X_N}|\mathscr{F}_n\right]. \end{aligned}$$

(b) By the assumption and the first part, we have for every $j = 0, 1, \ldots, d$ and every $n = 0, 1, \ldots, N$,

$$\begin{aligned} \frac{S_n^{(j)}}{X_n} &= \frac{Y_n}{X_n}\frac{S_n^{(j)}}{Y_n} \\ &= \frac{Y_n}{X_n} E_Q\left[\frac{S_N^{(j)}}{Y_N}|\mathscr{F}_n\right] \\ &= E_{Q_X}\left[\frac{S_N^{(j)}}{X_N}|\mathscr{F}_n\right]. \end{aligned}$$

Then by Problem 14.10, we have

$$\begin{aligned} E_{Q_X}\left[\frac{S_n^{(j)}}{X_n}|\mathscr{F}_{n-1}\right] &= E_{Q_X}\left[E_{Q_X}\left[\frac{S_N^{(j)}}{X_N}|\mathscr{F}_n\right]|\mathscr{F}_{n-1}\right] \\ &= E_{Q_X}\left[\frac{S_N^{(j)}}{X_N}|\mathscr{F}_{n-1}\right] \\ &= \frac{S_{n-1}^{(j)}}{X_{n-1}}. \end{aligned}$$

This finishes the proof.

41.5. Let $\alpha \in \mathcal{A}_X^-$. Then we have

$$V_N(\alpha) \leq X \quad P-a.s.$$

By Problems 14.5 and 14.6, we have

$$\bar{V}_n(\alpha) = E_Q\left[\bar{V}_N(\alpha)|\mathscr{F}_n\right] \leq E_Q\left[\frac{X}{Y_N}|\mathscr{F}_n\right].$$

Since α is arbitrary, we have

$$\sup_{\alpha \in \mathcal{A}_X^-} \bar{V}(\alpha) \leq E_Q\left[\frac{X}{Y_N} | \mathscr{F}_n\right].$$

Similarly, we can show that

$$E_Q\left[\frac{X}{Y_N} | \mathscr{F}_n\right] \leq \inf_{\alpha \in \mathcal{A}_X^+} \bar{V}(\alpha).$$

This completes the proof.

41.6. Let $\alpha, \beta \in \mathcal{A}$ be replicating portfolios of X. Then by Problem 41.3, we have

$$V_n(\alpha) = V_n(\beta) \quad P-\text{a.s.}, \quad n = 0, 1, \ldots, N.$$

In addition, we have $\alpha \in \mathcal{A}_X^- \cap \mathcal{A}_X^+$. It follows from Problem 41.5 that

$$E_Q\left[\frac{X}{Y_N} | \mathscr{F}_n\right] = \frac{V_n(\alpha)}{Y_n}$$

for every equivalent martingale measure Q with numeraire Y. This completes the proof.

41.7. Let $n \in \{1, 2, \ldots, N\}$. Since

$$\begin{aligned} S_n &= \left(S_n^{(1)}, S_n^{(2)}, \ldots, S_n^{(d)}\right) \\ &= \left(S_{n-1}^{(1)}(1 + \mu_n^{(1)}), S_n^{(2)}(1 + \mu_n^{(2)}), \ldots, S_n^{(d)}(1 + \mu_n^{(d)})\right), \end{aligned}$$

we can write $f(S_n) = \varphi(S_{n-1}, \mu_n)$ for some function φ. By assumption that μ_1, μ_2, ..., μ_N are independent, μ_n is independent of $\mathscr{F}_{n-1}$. Since S_{n-1} is $\mathscr{F}_{n-1}$-measurable, it follows from Problem 14.13 that

$$E[f(S_n)|\mathscr{F}_{n-1}] = E[\varphi(S_{n-1}, \mu_n)|\mathscr{F}_{n-1}] = g(S_{n-1}),$$

where $g(s) = E[\varphi(s, \mu_n)]$. Similarly, we have

$$E[f(S_n)|S_{n-1}] = E[\varphi(S_{n-1}, \mu_n)|S_{n-1}] = g(S_{n-1}).$$

Combining the above two equations gives

$$E[f(S_n)|\mathscr{F}_{n-1}] = E[f(S_n)|S_{n-1}].$$

This completes the proof.

41.8.

(a) If Q is a equivalent martingale measure with numeraire $\{S_n^{(0)} : n = 0, 1, \ldots, N\}$, then by Definition 41.8 we have

$$\frac{S_{n-1}^{(1)}}{S_{n-1}^{(0)}} = E_Q\left[\frac{S_n^{(1)}}{S_n^{(0)}}|\mathscr{F}_{n-1}\right], \quad n = 1, 2, \ldots, N.$$

Since $S_n^{(0)} = (1+r)^n$ and $S_n^{(1)} = S_{n-1}^{(1)}(1+\mu_n)$, the above equation gives

$$(1+r)S_{n-1}^{(1)} = S_{n-1}^{(1)} E_Q\left[1+\mu_n|\mathscr{F}_{n-1}\right].$$

Since $S_{n-1}^{(1)} > 0$, we have

$$1+r = E_Q\left[1+\mu_n|\mathscr{F}_{n-1}\right].$$

Let $q = Q(1+\mu_n = u|\mathscr{F}_{n-1})$. Then by Problem 14.19 we get

$$1+r = uq + d(1-q),$$

which gives a unique solution $q = \frac{1+r-d}{u-d}$. Hence we have

$$Q\{1+\mu_n = u\} = E_Q[Q(1+\mu_n = u|\mathscr{F}_{n-1})] = q.$$

(b) Let $E_n \in \sigma(\mu_n)$ for $n = 1, 2, \ldots, N$. Since $\sigma(\mu_n) = \{\emptyset, \{1+\mu_n = u\}, \{1+\mu_n = d\}, \Omega\}$, by the proof of part (a) we know that $E_Q[I_{E_n}|\mathscr{F}_{n-1}] = Q(E_n|\mathscr{F}_{n-1}) = Q(E_n)$ is a constant for all $n = 1, 2, \ldots, N$. Therefore by Problem 14.10 we have

$$\begin{aligned}
Q(E_1 \cap E_2 \cap \cdots \cap E_N) &= E_Q[I_{E_1} I_{E_2} \cdots I_{E_N}] \\
&= E_Q(E_Q[I_{E_1} I_{E_2} \cdots I_{E_N}|\mathscr{F}_{N-1}]) \\
&= E_Q(I_{E_1} I_{E_2} \cdots I_{E_{N-1}} E_Q[I_{E_N}|\mathscr{F}_{N-1}]) \\
&= E_Q(I_{E_1} I_{E_2} \cdots I_{E_{N-1}}) E_Q[I_{E_N}|\mathscr{F}_{N-1}] \\
&= Q(E_N) E_Q(I_{E_1} I_{E_2} \cdots I_{E_{N-1}}).
\end{aligned}$$

Repeating the above calculation gives

$$Q(E_1 \cap E_2 \cap \cdots \cap E_N) = Q(E_1)Q(E_2)\cdots Q(E_N).$$

Hence $\mu_1, \mu_2, \ldots, \mu_N$ are Q-independent.

Since $S_n^{(1)} = S_0^{(1)} \prod_{i=1}^n (1+\mu_i)$ and $S_0^{(1)} > 0$, we have

$$Q\left\{S_n^{(1)} = u^k d^{n-k} S_0^{(1)}\right\} = Q\left\{\prod_{i=1}^n (1+\mu_i) = u^k d^{n-k}\right\}.$$

Since $\mu_1, \mu_2, \ldots, \mu_N$ are Q-independent and identically distributed, we have

$$Q\left\{\prod_{i=1}^n (1+\mu_i) = u^k d^{n-k}\right\} = \binom{n}{k} q^k (1-q)^{n-k}.$$

(c) By part (b) and Problem 41.7, we have

$$E_Q[\varphi(S_n^{(1)})|\mathscr{F}_{n-1}] = E_Q[\varphi(S_n^{(1)})|S_{n-1}^{(1)}].$$

By Problem 14.13, we have

$$E_Q[\varphi(S_n^{(1)})|\mathscr{F}_{n-1}] = E_Q[\varphi(S_{n-1}^{(1)}(1+\mu_n))|\mathscr{F}_{n-1}] = g(S_{n-1}^{(1)}),$$

where

$$g(x) = E_Q[\varphi(x(1+\mu_n))] = q\varphi(ux) + (1-q)\varphi(dx).$$

This completes the proof.

41.9. By the definition of $\bar{H}$, $\bar{H}$ is adapted to the filtration and

$$\bar{H}_n \geq E_Q\left[\bar{H}_{n+1}|\mathscr{F}_n\right]$$

for all $n = 0, 1, \ldots, N-1$. Hence $\bar{H}$ is a Q-supermartingale. Also by the definition of $\bar{H}$, we have

$$\bar{H}_n \geq \bar{X}_n, \quad n = 0, 1, \ldots, N.$$

Now let $Y = \{Y_n : n = 0, 1, \ldots, N\}$ be a Q-supermartingale such that

$$Y_n \geq \bar{X}_n, \quad n = 0, 1, \ldots, N.$$

Then $Y_n \geq \bar{X}_N = \bar{H}_N$. Suppose that $Y_n \geq \bar{H}_n$. Then

$$\begin{aligned}
\bar{H}_{n-1} &= \max\{\bar{X}_{n-1}, E_Q[\bar{H}_n|\mathscr{F}_{n-1}]\} \\
&\leq \max\{\bar{X}_{n-1}, E_Q[Y_n|\mathscr{F}_{n-1}]\} \\
&\leq \max\{\bar{X}_{n-1}, Y_{n-1}\} \\
&\leq Y_{n-1}.
\end{aligned}$$

Therefore, we have

$$\bar{H}_n \leq Y_n, \quad n = 0, 1, \ldots, N.$$

Thus $\bar{H}$ is the smallest Q-supermartingale that dominates $\bar{X}$. This completes the proof.

41.10. Let $\alpha \in \mathcal{A}_X^-$. Then by Definition 41.14, there exists an exercise strategy ν_0 such that

$$\bar{V}_{\nu_0}(\alpha) \leq \bar{X}_{\nu_0} \quad P-\text{a.s.}$$

Since $P \sim Q$, we have

$$\bar{V}_{\nu_0}(\alpha) \leq \bar{X}_{\nu_0} \quad Q-\text{a.s.}$$

Then by Theorem 23.1 and Problem 41.2, we get

$$\bar{V}_0(\alpha) = E_Q[\bar{V}_{\nu_0}(\alpha)|\mathscr{F}_0] = E_Q[\bar{V}_{\nu_0}(\alpha)] \leq E_Q[\bar{X}_{\nu_0}] \leq \sup_{\nu\in\mathcal{T}_0} E_Q[\bar{X}_\nu].$$

Since α is arbitrary, we have

$$\sup_{\alpha\in\mathcal{A}_X^-} \bar{V}_0(\alpha) \leq \sup_{\nu\in\mathcal{T}_0} E_Q[\bar{X}_\nu].$$

Now let $\alpha \in \mathcal{A}_X^+$. Then by Definition 41.14 and the equivalence of P and Q, we have

$$\bar{V}_n(\alpha) \geq \bar{X}_n \quad Q-\text{a.s.}, \ n = 0, 1, \ldots, N.$$

Hence for every $\nu \in \mathcal{A}_X^+$, we have

$$\bar{V}_\nu(\alpha) \geq \bar{X}_\nu \quad Q-\text{a.s.},$$

which gives

$$\bar{V}_0(\alpha) = E_Q[\bar{V}_\nu(\alpha)] \geq E_Q[\bar{X}_\nu].$$

Since the above inequality holds for every $\nu \in \mathcal{A}_X^+$, we have

$$\bar{V}_0(\alpha) \geq \sup_{\nu\in\mathcal{T}_0} E_Q[\bar{X}_\nu].$$

Since α is arbitrary, we get

$$\inf_{\alpha\in\mathcal{A}_X^+} \bar{V}_0(\alpha) \geq \sup_{\nu\in\mathcal{T}_0} E_Q[\bar{X}_\nu].$$

This completes the proof.

41.11. Let $\bar{H}$ be the stochastic process as defined in Problem 41.9. Since $\bar{H}$ is a supermartingale, $-\bar{H}$ is a submartingale. Then by Problem 22.11, there exist a Q-martingale M and an increasing predictable process A with $A_0 = 0$ such that

$$-\bar{H}_n = M_n + A_n, \quad n = 0, 1, \ldots, N$$

or

$$\bar{H}_n = -M_n - A_n, \quad n = 0, 1, \ldots, N.$$

By the assumption that the market is complete, there exists a replicating portfolio α_0 for the European derivative $-Y_N M_N$. Then we have

$$\bar{V}_N(\alpha_0) = \frac{V_N(\alpha_0)}{Y_N} = -M_N$$

and

$$\bar{V}_n(\alpha_0) = E_Q[\bar{V}_N|\mathscr{F}_n] = -E_Q[M_N|\mathscr{F}_n] = -M_n \geq \bar{H}_n \geq \bar{X}_n$$

for all $n = 0, 1, \ldots, N$. In addition, since $A_0 = 0$, we have $\bar{H}_0 = V_0(\alpha_0)$. Therefore,

$$\alpha_0 \in \mathcal{A}_X^+.$$

Let ν_0 be a random variable defined by

$$\nu_0(\omega) = \min\{n \in \{0, 1, \ldots, N\} : \bar{H}_n(\omega) = \bar{X}_n(\omega)\}, \quad \omega \in \Omega.$$

For every $n \in \{0, 1, \ldots, N\}$, since

$$\{\nu_0 = n\} = \{\bar{H}_n = \bar{X}_n\} \cap \left(\bigcap_{k=0}^{n-1} \{\bar{H}_k > \bar{X}_k\} \right) \in \mathscr{F}_n,$$

ν_0 is a stopping time. By the definition of $\bar{H}$, we have

$$\bar{H}_k = E_Q[\bar{H}_{k+1} | \mathscr{F}_k] \text{ over } \{k < \nu_0\}.$$

By the proof of Problem 22.11, we have

$$M_n = -\bar{H}_n - \sum_{k=0}^{n-1} (E_Q[\bar{H}_{k+1} | \mathscr{F}_k] - \bar{H}_k).$$

Hence we have

$$M_{\nu_0} = -\bar{H}_{\nu_0}.$$

Therefore,

$$\bar{V}_{\nu_0}(\alpha_0) = -M_{\nu_0} = \bar{H}_{\nu_0} = \bar{X}_{\nu_0}.$$

Hence

$$\alpha_0 \in \mathcal{A}_X^-.$$

In addition, by Theorem 23.1 we have

$$V_0(\alpha_0) = Y_0 \bar{V}_0(\alpha_0) = Y_0 E_Q[\bar{V}_{\nu_0}(\alpha_0)] = Y_0 E_Q[\bar{X}_{\nu_0}].$$

Now we show that ν_0 is an optimal exercise strategy. In fact, by Problem 41.10 and the fact that $\alpha_0 \in \mathcal{A}_X^+ \cap \mathcal{A}_X^-$, we get

$$\bar{V}_0(\alpha_0) = \sup_{\nu \in \mathcal{T}_0} E_Q[\bar{X}_\nu].$$

This completes the proof.

41.5 Bibliographic Notes

In this chapter, we presented discrete-time models and how European derivatives and American derivatives are priced. For proofs of the fundamental theorems of asset pricing (Theorems 41.1 and 41.2), readers are referred to Pascucci (2011, pp31–34).

The binomial model was introduced by Cox et al. (1979). Hoek and Elliott (2006) is a textbook on binomial models. Pliska (1997) and Shreve (2005) are two textbooks devoted to discrete-time models. Follmer and Schied (2004) introduced discrete-time probabilistic methods in finance at the graduate level. Discrete-time financial models are also studied in Lamberton and Lapeyre (1996), Kallianpur and Karandikar (1999), Capinski and Zastawniak (2003), Franke et al. (2004), Bhar and Hamori (2005), Elliott and Kopp (2005), Dokuchaev (2007), Dana and Jeanblanc (2007), and Björk (2009). Borak et al. (2010) presented many exercises related to binomial models.

For numerical implementation of binomial models, readers are referred to James (2003), Higham (2004), Brandimarte (2006), and Rouah and Vainberg (2007). As the number of time steps increases, the binomial price of an European option converges to the Black-Scholes price (Follmer and Schied, 2004).

CHAPTER 42

BLACK-SCHOLES OPTION PRICING MODELS

The Black-Scholes model for pricing stock options is a revolutionary result in mathematical finance. Many of the pricing techniques and models used today in finance are rooted in the methods and ideas from the Black-Scholes model. In this chapter, we present the Black-Scholes model and relevant results.

42.1 Basic Concepts and Facts

Definition 42.1 (Continuous Market). Let $\{(B_t^{(1)}, B_t^{(2)}, \ldots, B_t^{(n)}) : 0 \le t \le T\}$ be an n-dimensional Brownian motion on some filtered probability space $(\Omega, \mathscr{F}, \{\mathscr{F}_t : 0 \le t \le T\}, P)$. Let $\{r_t : 0 \le t \le T\}$ be an adapted interest rate process. A continuous market is a $(d+1)$-dimensional stochastic process

$$S = \{(S_t^{(0)}, S_t^{(1)}, \ldots, S_t^{(d)}) : 0 \le t \le T\},$$

where

$$S_t^{(i)} = \alpha_i(t) S_t^{(i)} \mathrm{d}t + S_t^{(i)} \sum_{j=1}^{n} \sigma_{ij}(t) \mathrm{d}B_t^{(j)}, \quad i = 1, 2, \ldots, d,$$

and

$$S_t^{(0)} = \exp\left(\int_0^t r_u du\right).$$

Here the mean rate of return vector $\{(\alpha_1(t), \alpha_2(t), \ldots, \alpha_d(t)) : 0 \leq t \leq T\}$ and the volatility matrix $\{(\sigma_{ij}(t))_{i=1,2,\ldots,d;j=1,2,\ldots,n} : 0 \leq t \leq T\}$ are also adapted processes.

Definition 42.2 (Black-Scholes Market). Let $\{B_t : t \geq 0\}$ be a Brownian motion on some probability space $(\Omega, \mathscr{F}, P)$ and let $\{\mathscr{F}_t : t \geq 0\}$ be its natural filtration. A market is an $\mathscr{F}_t$-adapted two-dimensional Itô process $\{(X_t^{(0)}, X_t) : 0 \leq t \leq T\}$, where $X^{(0)}$ is the price of a riskless asset given by

$$\mathrm{d}X_t^{(0)} = rX_t^{(0)}\mathrm{d}t, \quad X^{(0)}(0) = 1,$$

and X_t is the price of a risky asset given by

$$\mathrm{d}X_t = X_t(\mu\mathrm{d}t + \sigma\mathrm{d}B_t).$$

Here r, μ, and σ are constants.

Definition 42.3 (Portfolio). A portfolio in the Black-Scholes market $\{(X_t^{(0}, X_t) : 0 \leq t \leq T\}$ is an two-dimensional, jointly measurable (with respect to t and ω), and $\mathscr{F}_t$-adapted stochastic process

$$\phi = \{(\theta_t^{(0)}, \theta_t) : 0 \leq t \leq T\}.$$

The value at time t of the portfolio is defined as

$$V_t(\phi) = \theta_t^{(0)} X_t^{(0)} + \theta_t X_t.$$

A portfolio is also referred to as a *trading strategy*.

Definition 42.4 (Normalized Black-Scholes Market). Let $\{(X_t^{(0)}, X_t) : 0 \leq t \leq T\}$ be a Black-Scholes market given in Definition 42.2. Then $X_t^{(0)} = e^{rt}$. The normalized Black-Scholes market is the process given by $\{(1, \bar{X}_t) : 0 \leq t \leq T\}$, where $\bar{X}_t = e^{-rt}X_t$.

The value at time t of a portfolio $\phi = \{(\theta_t^{(0)}, \theta_t) : 0 \leq t \leq T\}$ for the normalized market is given by

$$\bar{V}_t(\phi) = \theta_t^{(0)} + \theta_t \bar{X}_t.$$

Definition 42.5 (Self-Financing Portfolio). A portfolio $\phi = \{(\theta_t^{(0)}, \theta_t) : 0 \leq t \leq T\}$ is said to be self-financing if it satisfies the following properties:

(a)

$$\int_0^T \left|\theta_t^{(0)}\right| \mathrm{d}t + \int_0^T \theta_t^2 \mathrm{d}t < \infty \quad \text{a.s.},$$

(b) For every $t \in [0, T]$, we have

$$\mathrm{d}V_t(\phi) = \theta_t^{(0)} \mathrm{d}X_t^{(0)} + \theta_t \mathrm{d}X_t \quad \text{a.s.},$$

specifically,

$$V_t(\phi) - V_0(\phi) = \int_0^t \theta_u^{(0)} \mathrm{d}X_u^{(0)} + \int_0^t \theta_u \mathrm{d}X_u \quad \text{a.s.},$$

where $V_t(\phi) = \theta_t^{(0)} X_t^{(0)} + \theta_t X_t$.

Definition 42.6 (Admissible Portfolio). A portfolio $\phi = \{(\theta_t^{(0)}, \theta_t) : 0 \leq t \leq T\}$ is said to be admissible if it satisfies the following conditions:

(a) ϕ is self-financing.

(b) The value $\bar{V}_t(\phi) = \theta_t^{(0)} + \theta_t \bar{X}_t$ is nonnegative for all $t \in [0, T]$ and

$$\sup_{0 \leq t \leq T} \bar{V}_t(\phi)$$

is square-integrable under Q, where Q is the probability measure defined in Problem 37.7.

Definition 42.7 (European Derivative). Let $\{\mathscr{F}_t : 0 \leq t \leq T\}$ be the natural filtration given in Definition 42.2. An European derivative h with maturity T is a $\mathscr{F}_T$-measurable random variable such that $h \in L^p(\Omega, \mathscr{F}, P)$ for some $p > 1$.

Definition 42.8 (Replicable European Derivative). An European derivative h is said to be replicable if there is an admissible portfolio ϕ such that

$$h = V_T(\phi), \quad P - \text{a.s.},$$

i.e.,

$$P\{h = V_T(\phi)\} = 1.$$

The admissible portfolio that replicates h is called a *replicating portfolio* for h.

Definition 42.9 (Arbitrage). An admissible portfolio ϕ is called an *arbitrage* if the corresponding value process $\{V_t(\phi) : 0 \leq t \leq T\}$ satisfies

(a) $V_0(\phi) = 0$.

(b) $V_T(\phi) \geq 0$ a.s.

(c) $P\{V_T(\phi) > 0\} > 0$.

Definition 42.10 (Risk-Neutral Measure). Let $(\Omega, \mathscr{F}, \{\mathscr{F}_t : 0 \leq t \leq T\}, P)$ be the filtered probability space and $S = \{(S_t^{(0)}, S_t^{(1)}, \ldots, S_t^{(d)}) : 0 \leq t \leq T\}$ be the continuous market as defined in Definition 42.1. Let $\{Y_t : 0 \leq t \leq T\}$ be a positive price process adapted to $\{\mathscr{F}_t : 0 \leq t \leq T\}$. A probability measure Q on $(\Omega, \mathscr{F})$ is said to be risk-neutral with respect to $\{Y_t : 0 \leq t \leq T\}$ if

(a) $Q \sim P$, that is, Q and P are equivalent.

(b) under Q, the discounted price process

$$\left\{ \frac{S_t^{(i)}}{Y_t} : 0 \leq t \leq T \right\}$$

is a martingale for every $i = 0, 1, \ldots, d$.

The price process $\{Y_t : 0 \leq t \leq T\}$ is called a *numeraire*.

Theorem 42.1 (First Fundamental Theorem of Asset Pricing). *Let S be a continuous market as defined in Definition 42.1. Then S is arbitrage-free if and only if there exists at least one risk-neutral measure.*

Theorem 42.2 (Second Fundamental Theorem of Asset Pricing). *Let S be a continuous market as defined in Definition 42.1. Suppose that S is arbitrage-free. Then S is complete (see Definition 41.15) if and only if there exists a unique risk-neutral measure with numeraire $\{S_n^{(0)} : n = 0, 1, \ldots, N\}$.*

Theorem 42.3 (Black-Scholar Option Price). *Let $\{(X_t^{(0)}, X_t) : 0 \leq t \leq T\}$ be the Black-Scholes market given in Definition 42.2. Let*

$$d_1 = \frac{\ln \frac{X_T}{K} + (r + \frac{1}{2}\sigma^2)(T - t)}{\sigma\sqrt{T - t}}, \tag{42.1a}$$

$$d_2 = d_1 - \sigma\sqrt{T - t}, \tag{42.1b}$$

and

$$N(d) = \frac{1}{\sqrt{2\pi}} \int_{-\infty}^{d} e^{-\frac{x^2}{2}} \mathrm{d}x. \tag{42.1c}$$

Then the price at time t of a European call option that matures at time T with strike K is given by

$$C_t = X_t N(d_1) - K e^{-r(T-t)} N(d_2), \tag{42.2}$$

and the price at time t of a European put option that matures at time T with strike K is given by

$$P_t = K e^{-r(T-t)} N(-d_2) - X_t N(-d_1), \tag{42.3}$$

where d_1, d_2, $N(\cdot)$ are as defined in Equation (42.1).

42.2 Problems

42.1. Let $\{(X_t^{(0)}, X_t) : 0 \leq t \leq T\}$ be a Black-Scholes market given in Definition 42.2. Let $\phi = \{(\theta_t^{(0)}, \theta_t) : 0 \leq t \leq T\}$ be a portfolio for the market. Let $\bar{X}_t = e^{-rt} X_t$ be the normalized price. Show that ϕ is self-financing for market $\{(X_t^{(0)}, X_t) : 0 \leq t \leq T\}$ if and only if it is self-financing for the normalized market $\{(1, \bar{X}_t) : 0 \leq t \leq T\}$.

42.2. Let $\{(X_t^{(0)}, X_t) : 0 \leq t \leq T\}$ be a Black-Scholes market given in Definition 42.2. Let $\phi = \{(\theta_t^{(0)}, \theta_t) : 0 \leq t \leq T\}$ be an admissible portfolio for the market. Let $\bar{X}_t = e^{-rt} X_t$ be the normalized price. Show that $\{\bar{V}_t : 0 \leq t \leq T\}$ is a square-integrable martingale under Q, where $\bar{V}_t = \theta_t^{(0)} + \theta_t \bar{X}_t$ is the value at time t of the portfolio for the normalized market and Q is the probability measure defined in Problem 37.7.

42.3 (No-Arbitrage Principle)**.** Let ϕ and ψ be two admissible portfolios in the Black-Scholes market given in Definition 42.2. Suppose that $V_T(\phi) = V_T(\psi)$, P-a.s. For every $t \in [0, T]$, show that

$$V_t(\phi) = V_t(\psi), \quad P - \text{a.s.}$$

42.4. Let $\{B_t : 0 \leq t \leq T\}$, $\{\mathscr{F}_t : 0 \leq t \leq T\}$, and $\{(X_t^{(0)}, X_t) : 0 \leq t \leq T\}$ be a Brownian motion, the natural filtration, and the Black-Scholes market given in Definition 42.2, respectively. Let $\{W_t : 0 \leq t \leq T\}$ be a process defined by

$$W_t = B_t + \frac{(\mu - r)t}{\sigma}, \quad t \in [0, T].$$

Let Q be the probability measure under which $\{W_t : 0 \leq t \leq T\}$ is a Brownian motion (see Problem 37.7). Let h be a nonnegative random variable that is $\mathscr{F}_T$-measurable and square-integrable under Q. Suppose that there are no arbitrages in the market. Show that

(a) There exists an admissible portfolio $\phi = \{(\theta_t^{(0)}, \theta_t) : 0 \leq t \leq T\}$ such that for every $t \in [0, T]$, we have

$$\theta_t^{(0)} X_t^{(0)} + \theta_t X_t = E_Q \left[e^{-r(T-t)} h | \mathscr{F}_t \right],$$

where E_Q denotes the expectation under the probability measure Q.

(b) If the payoff of an European option that matures at time T is defined by h, then the value at time t of the option is given by

$$V_t = E_Q \left[e^{-r(T-t)} h | \mathscr{F}_t \right], \quad \text{a.s.}, \quad 0 \leq t \leq T.$$

42.5. Let $\{(X_t^{(0)}, X_t) : 0 \le t \le T\}$ be the Black-Scholes market given in Definition 42.2. Let $h = f(X_T)$, where f is a Borel function. Suppose that h is nonnegative and square-integrable under Q, where Q is the probability measure given in Problem 37.7. Show that the value at time t of an European derivative that matures at T with payoff h is given by $F(t, X_t)$, where

$$F(t,x) = E_Q\left[e^{-r(T-t)} f\left(xe^{(r-\frac{1}{2}\sigma^2)(T-t)+\sigma(W_T-W_t)}\right)\right].$$

Here $\{W_t : 0 \le t \le T\}$ is the Q-Brownian motion given in Problem 37.7.

42.6 **(Black-Scholes Option Price).** Let $\{(X_t^{(0)}, X_t) : 0 \le t \le T\}$ be the Black-Scholes market given in Definition 42.2. Let h be the payoff function of an European derivative that matures at T. Show that

(a) If $h = \max\{X_T - K, 0\}$ be the payoff of a call option, then the call option value at time t is given by

$$C_t = X_t N(d_1) - Ke^{-r(T-t)} N(d_2),$$

where

$$d_1 = \frac{\ln \frac{X_T}{K} + (r + \frac{1}{2}\sigma^2)(T-t)}{\sigma\sqrt{T-t}}, \tag{42.4a}$$

$$d_2 = d_1 - \sigma\sqrt{T-t}, \tag{42.4b}$$

and

$$N(d) = \frac{1}{\sqrt{2\pi}} \int_{-\infty}^{d} e^{-\frac{x^2}{2}} dx. \tag{42.4c}$$

(b) If $h = \max\{K - X_T, 0\}$ is the payoff of a put option, then the put option value at time t is given by

$$P_t = Ke^{-r(T-t)} N(-d_2) - X_t N(-d_1),$$

where d_1, d_2, $N(\cdot)$ are as defined in Equation (42.4).

42.7 **(Put-Call Parity).** Let $\{(X_t^{(0)}, X_t) : 0 \le t \le T\}$ be the Black-Scholes market given in Definition 42.2. Let C_t and P_t be the values at time t of a European call option and a European put option, respectively. Suppose that both options mature at time T and have a strike of K. Show that

$$C_t - P_t = X_t - e^{-r(T-t)} K.$$

42.8 **(Delta of European Options).** Let $\{(X_t^{(0)}, X_t) : 0 \le t \le T\}$ be the Black-Scholes market given in Definition 42.2 and $K > 0$. Let Q be the probability measure and $\{W_t : 0 \le t \le T\}$ the Q-Brownian motion given in Problem 37.7. Show that

(a) For any sequence $\{s_n\}_{n\geq 1}$ of nonzero real numbers such that $s_n \to 0$ as $n \to \infty$,

$$\lim_{n\to\infty} \frac{E_Q[g(x+s_n)] - E_Q[g(x)]}{s_n} = E_Q\left[\lim_{n\to\infty} \frac{g(x+s_n)-g(x)}{s_n}\right] = N(d_1),$$

where

$$g(x) = e^{-r(T-t)}\left(xe^{(r-\frac{1}{2}\sigma^2)(T-t)+\sigma(W_T-W_t)} - K\right)^+,$$

and d_1 and $N(\cdot)$ are as defined in Equation (42.4);

(b) The delta of a European call option that matures at T with strike K is

$$\frac{\partial C_t}{\partial X_t} = N(d_1),$$

where d_1 and $N(\cdot)$ are as defined in Equation (42.4);

(c) The delta of a European put option that matures at T with strike K is

$$\frac{\partial P_t}{\partial X_t} = N(d_1) - 1,$$

where d_1 and $N(\cdot)$ are as defined in Equation (42.4).

42.3 Hints

42.1. Note that $X_t^{(0)} = e^{rt}$ and use Itô's lemma (Theorem 35.3) to show that $\mathrm{d}\bar{V}_t(\phi) = \theta_t \mathrm{d}\bar{X}_t$, where $\bar{V}_t = e^{-rt}V_t$.

42.2. Use the result of Problem 33.7.

42.3. Use the result of Problem 42.2 and note that P and Q are equivalent.

42.4. Use the martingale representation theorem (Theorem 36.1) with

$$M_t = E_Q\left[e^{-rT}h|\mathscr{F}_t\right]$$

to prove the first part. Use the first part and the result of Problem 42.3 to prove the second part.

42.5. Use the results of Problems 38.7, 42.4, and 14.13.

42.6. Apply the result of Problem 42.5.

42.7. Use the result of Problem 42.6.

42.8. To prove part (a), use the results of Problems9.9 and 9.10. Use part (a) to prove part (b). To prove part (c), use part (b) and the put-call parity (see Problem 42.7).

42.4 Solutions

42.1. Let $V_t(\phi)$ be the value of the portfolio at time t. Then the value at time t of the portfolio for the normalized market is $\bar{V}_t(\phi) = e^{-rt}V_t(\phi)$.

First, let us prove the sufficiency ("if") part. Suppose that ϕ is self-financing for the market $\{(X_t^{(0)}, X_t) : 0 \leq t \leq T\}$. To show that ϕ is self-financing for the normalized market, we only need to show that for every $t \in [0, T]$, we have

$$\mathrm{d}\bar{V}_t(\phi) = \theta_t \mathrm{d}\bar{X}_t, \quad \text{a.s.} \tag{42.5}$$

By Theorem 35.3, we get

$$\mathrm{d}\bar{X}_t = -re^{-rt}X_t + e^{-rt}\mathrm{d}X_t. \tag{42.6}$$

Similarly, we have

$$\mathrm{d}\bar{V}_t(\phi) = -re^{-rt}V_t(\phi) + e^{-rt}\mathrm{d}V_t(\phi). \tag{42.7}$$

Since $X_t^{(0)} = e^{rt}$ and ϕ is self-financing for the market, we have $V_t(\phi) = \theta_t^{(0)}e^{rt} + \theta_t X_t$ and $\mathrm{d}V_t(\phi) = re^{rt}\theta_t^{(0)} + \theta_t \mathrm{d}X_t$ a.s. Replacing $V_t(\phi)$ and $\mathrm{d}V_t(\phi)$ in Equation (42.7) gives

$$\mathrm{d}\bar{V}_t(\phi) = \theta_t e^{-rt}(-rX_t + \mathrm{d}X_t), \quad \text{a.s.} \tag{42.8}$$

Combining Equation (42.6) and Equation (42.8) leads to Equation (42.5).

Next, let us prove the necessity ("only if") part. Suppose that ϕ is self-financing for the normalized market [i.e., Equation (42.5) holds]. Since $V_t(\phi) = e^{rt}\bar{V}_t(\phi)$, we have

$$\mathrm{d}V_t(\phi) = re^{rt}\bar{V}_t(\phi) + e^{rt}\mathrm{d}\bar{V}_t(\phi) = rV_t(\phi) + e^{rt}\theta_t \mathrm{d}\bar{X}_t. \tag{42.9}$$

Combining Equations (42.6) and (42.9) gives

$$\mathrm{d}V_t(\phi) = rV_t(\phi) - r\theta_t X_t + \theta_t \mathrm{d}X_t = \theta_t^{(0)}\mathrm{d}X_t^{(0)} + \theta_t \mathrm{d}X_t.$$

Hence ϕ is self-financing for $\{(X_t^{(0)}, X_t) : 0 \leq t \leq T\}$.

This finishes the proof.

42.2. Since ϕ is self-financing for the Black-Scholes market, we have

$$\mathrm{d}\bar{V}_t = \theta_t \mathrm{d}\bar{X}_t. \tag{42.10}$$

Let $\{W_t : 0 \leq t \leq T\}$ be the Brownian motion under Q that was defined in Problem 37.7. Then $\mathrm{d}\bar{X}_t = \sigma \bar{X}_t \mathrm{d}W_t$. Hence from Equation (42.10), we have

$$\bar{V}_t = \bar{V}_0 + \int_0^t \theta_s \sigma \bar{X}_s \mathrm{d}W_s, \quad t \in [0, T].$$

By the assumption that ϕ is admissible, $\sup_{0 \leq t \leq T} \bar{V}_t$ is square-integrable. By Problem 33.7 and Theorem 32.1, $\bar{V}_t$ is a square-integrable martingale. This finishes the proof.

42.3. By Problem 42.2, the value processes $\bar{V}_T(\phi)$ and $\bar{V}_T(\psi)$ are martingales under Q. Hence we have

$$\bar{V}_t(\phi) = E_Q[\bar{V}_T(\phi)|\mathscr{F}_t], \quad Q-\text{a.s.} \tag{42.11}$$

and

$$\bar{V}_t(\psi) = E_Q[\bar{V}_T(\psi)|\mathscr{F}_t], \quad Q-\text{a.s.} \tag{42.12}$$

By the assumption, we have $P\{V_T(\phi) \neq V_T(\psi)\} = 0$. By Problem 37.7, $P \sim Q$. Hence we have $Q\{V_T(\phi) \neq V_T(\psi)\} = 0$. Therefore, we have

$$E_Q[\bar{V}_T(\phi)|\mathscr{F}_t] = E_Q[\bar{V}_T(\psi)|\mathscr{F}_t], \quad Q-\text{a.s.} \tag{42.13}$$

Combining Equations (42.11), (42.12), and (42.13) gives $\bar{V}_t(\phi) = \bar{V}_t(\psi)$, Q-a.s. The result follows from the fact that Q and P are equivalent. This completes the proof.

42.4.

(a) Let $\{(1, \bar{X}_t) : 0 \leq t \leq T\}$ be the normalized market. By Problem 42.1, we only need to show that there exists an admissible portfolio $\phi = \{(\theta_t^{(0)}, \theta_t) : 0 \leq t \leq T\}$ such that for every $t \in [0, T]$, we get

$$\theta_t^{(0)} + \theta_t \bar{X}_t = E_Q\left[e^{-rT}h|\mathscr{F}_t\right]. \tag{42.14}$$

To do that, let $M_t = E_Q\left[e^{-rT}h|\mathscr{F}_t\right]$. Then $\{M_t : 0 \leq t \leq T\}$ is a square-integrable martingale adapted to the $\mathscr{F}_t$. Since $\{\mathscr{F}_t : 0 \leq t \leq T\}$ is also the natural filtration of the Brownian motion $\{W_t : 0 \leq t \leq T\}$ under Q, by Theorem 36.1, there exists an adapted process $\{K_t : 0 \leq t \leq T\}$ such that

$$\int_0^T E_Q\left[K_s^2\right] \mathrm{d}s < \infty$$

and for every $t \in [0, T]$, we get

$$M_t = M_0 + \int_0^t K_s \mathrm{d}W_s, \quad \text{a.s.} \tag{42.15}$$

Now we define

$$\theta_t = \frac{K_t}{\sigma \bar{X}_t}, \quad \theta_t^{(0)} = M_t - \frac{K_t}{\sigma}, \quad t \in [0, T]. \tag{42.16}$$

Then we have

$$\mathrm{d}\bar{V}_t(\phi) = \mathrm{d}M_t = K_t \mathrm{d}W_t = \theta_t \sigma \bar{X}_t \mathrm{d}W_t = \theta_t \mathrm{d}\bar{X}_t,$$

which shows that ϕ is self-financing. Since h is square-integrable under Q, it follows from Problem 24.5 that $\sup_{0 \leq t \leq T} \bar{V}_t(\phi)$ is square-integrable under Q. Thus ϕ is an admissible portfolio.

(b) Let $\phi = \{(\theta_t^{(0)}, \theta_t) : 0 \leq t \leq T\}$ be the admissible portfolio found in the first part. Since h is $\mathscr{F}_T$-measurable, by Problem 14.4, the value at time T of the portfolio is

$$V_T(\phi) = E_Q[h|\mathscr{F}_T] = h.$$

Therefore, the portfolio ϕ has the same payoff as the option at maturity. By Problem 42.3, the value at time t of the option is

$$V_t(\phi) = E_Q\left[e^{-r(T-t)}h|\mathscr{F}_t\right] \quad \text{a.s.}$$

This completes the proof.

42.5. By Problem 38.7 and the definition of W_t in Problem 37.7, we have

$$X_T = X_t e^{(r-\frac{1}{2}\sigma^2)(T-t)+\sigma(W_T-W_t)}, \quad t \in [0, T]. \tag{42.17}$$

Then by Problem 42.4 and Equation (42.17), we have

$$\begin{aligned} & E_Q\left[e^{-r(T-t)}h|\mathscr{F}_t\right] \\ = \;& E_Q\left[e^{-r(T-t)} f\left(X_t e^{(r-\frac{1}{2}\sigma^2)(T-t)+\sigma(W_T-W_t)}\right)|\mathscr{F}_t\right]. \end{aligned}$$

Note that $W_T - W_t$ is independent of $\mathscr{F}_t$ and that X_t is $\mathscr{F}_t$-measurable. It follows from Problem 14.13 that the value of the European derivative is $F(t, X_t)$, where where

$$F(t, x) = E_Q\left[e^{-r(T-t)} f\left(x e^{(r-\frac{1}{2}\sigma^2)(T-t)+\sigma(W_T-W_t)}\right)\right].$$

This completes the proof.

42.6.

(a) If $h = \max\{X_T - K, 0\}$, then

$$F(t, X_t) = \int_A e^{-r(T-t)}\left(X_t e^{(r-\frac{1}{2}\sigma^2)(T-t)+\sigma(W_T-W_t)} - K\right) \mathrm{d}Q, \tag{42.18}$$

where $A = \{X_t e^{(r-\frac{1}{2}\sigma^2)(T-t)+\sigma(W_T-W_t)} > K\}$. Let $Z = \frac{W_T-W_t}{\sqrt{T-t}}$. Then $Z \sim N(0, 1)$, that is, Z is a standard normal random variable. After some algebraic calculation, we can get $A = \{Z > -d_2\}$, where d_2 is as defined in Equation (42.4). Then from Equation (42.18) and the fact that $N(d) = 1 - N(-d)$ (see Problem 19.8), we have

$$\begin{aligned} & F(t, X_t) \\ = \;& \int_{-d_2}^{\infty} e^{-r(T-t)}\left(X_t e^{(r-\frac{1}{2}\sigma^2)(T-t)+\sigma\sqrt{T-t}y} - K\right) \frac{1}{\sqrt{2\pi}} e^{-\frac{y^2}{2}} \mathrm{d}y \\ = \;& -e^{-r(T-t)} K N(d_2) + X_t \int_{-d_2}^{\infty} \frac{1}{\sqrt{2\pi}} e^{-\frac{(y-\sigma\sqrt{T-t})^2}{2}} \mathrm{d}y \\ = \;& X_t N(d_1) - e^{-r(T-t)} K N(d_2). \end{aligned}$$

(b) If $h = \max\{K - X_T, 0\}$, then

$$F(t, X_t) = \int_B e^{-r(T-t)} \left(K - X_t e^{(r-\frac{1}{2}\sigma^2)(T-t)+\sigma(W_T - W_t)}\right) \mathrm{d}Q, \quad (42.19)$$

where $B = \{X_t e^{(r-\frac{1}{2}\sigma^2)(T-t)+\sigma(W_T-W_t)} < K\}$. Similarly, we have $B = \{Z < -d_2\}$ and

$$\begin{aligned}
& F(t, X_t) \\
= & \int_{-\infty}^{-d_2} e^{-r(T-t)} \left(K - X_t e^{(r-\frac{1}{2}\sigma^2)(T-t)+\sigma\sqrt{T-t}y}\right) \frac{1}{\sqrt{2\pi}} e^{-\frac{y^2}{2}} \mathrm{d}y \\
= & e^{-r(T-t)} K N(-d_2) - X_t \int_{-\infty}^{-d_2} \frac{1}{\sqrt{2\pi}} e^{-\frac{(y-\sigma\sqrt{T-t})^2}{2}} \mathrm{d}y \\
= & e^{-r(T-t)} K N(-d_2) - X_t N(-d_1).
\end{aligned}$$

This completes the proof.

42.7. By Problem 42.6, we have

$$\begin{aligned}
& C_t - P_t \\
= & X_t N(d_1) - K e^{-r(T-t)} N(d_2) - K e^{-r(T-t)} N(-d_2) + X_t N(-d_1) \\
= & X_t (N(d_1) + N(-d_1)) - K e^{-r(T-t)} (N(d_2) + N(-d_2)) \\
= & X_t - e^{-r(T-t)} K,
\end{aligned}$$

where $d_1, d_2, N(\cdot)$ are as defined in Equation (42.4). This completes the proof.

42.8.

(a) Let

$$f_n = \frac{g(x + s_n) - g(x)}{s_n}, \quad n \geq 1.$$

Let $A = \{\omega \in \Omega : x e^{(r-\frac{1}{2}\sigma^2)(T-t)+\sigma(W_T(\omega)-W_t(\omega))} > K\}$ and $B = \{\omega \in \Omega : x e^{(r-\frac{1}{2}\sigma^2)(T-t)+\sigma(W_T(\omega)-W_t(\omega))} < K\}$. Since

$$\lim_{n\to\infty} f_n = \frac{\partial g(x)}{\partial x} = \begin{cases} 0, & \text{if } \omega \in B; \\ e^{-\frac{1}{2}\sigma^2(T-t)+\sigma(W_T-W_t)}, & \text{if } \omega \in A, \end{cases}$$

we have

$$\begin{aligned}
E_Q\left[\lim_{n\to\infty} f_n\right] &= \int_A e^{-\frac{1}{2}\sigma^2(T-t)+\sigma(W_T-W_t)} \mathrm{d}Q \\
&= N(d_1). \qquad (42.20)
\end{aligned}$$

Now we show that

$$\lim_{n\to\infty} E_Q[f_n] = E_Q\left[\lim_{n\to\infty} f_n\right]. \qquad (42.21)$$

To do that, let

$$A_n = \{\omega \in \Omega : (x + s_n)e^{(r-\frac{1}{2}\sigma^2)(T-t)+\sigma(W_T(\omega)-W_t(\omega))} > K\}.$$

On the event $A_n \cap A$, we have

$$|f_n|^2 = \left|\frac{g(x+s_n) - g(x)}{s_n}\right|^2 = e^{-\sigma^2(T-t)+2\sigma(W_T-W_t)}.$$

On the event $A_n^c \cap A^c$, we have $|f_n|^2 = 0$. On the event $A_n \cap A^c$, we have

$$(x + s_n)e^{(r-\frac{1}{2}\sigma^2)(T-t)+\sigma(W_T-W_t)} > K$$

and

$$xe^{(r-\frac{1}{2}\sigma^2)(T-t)+\sigma(W_T-W_t)} \leq K,$$

which give us

$$s_n > \frac{K - xe^{(r-\frac{1}{2}\sigma^2)(T-t)+\sigma(W_T-W_t)}}{e^{(r-\frac{1}{2}\sigma^2)(T-t)+\sigma(W_T-W_t)}} \geq 0.$$

Hence on the event $A_n \cap A^c$ we have

$$|f_n|^2 = \left|\frac{g(x+s_n)}{s_n}\right|^2 < 4e^{-\sigma^2(T-t)+2\sigma(W_T-W_t)}.$$

Similarly, on the event $A_n^c \cap A$ we have

$$|f_n|^2 = \left|\frac{g(x)}{s_n}\right|^2 < e^{-\sigma^2(T-t)+2\sigma(W_T-W_t)}.$$

Combining the above estimations gives

$$\begin{aligned}
& \int_\Omega |f_n|^2 \mathrm{d}Q \\
= & \int_{A_n \cap A} |f_n|^2 \mathrm{d}Q + \int_{A_n \cap A^c} |f_n|^2 \mathrm{d}Q + \int_{A_n^c \cap A} |f_n|^2 \mathrm{d}Q \\
& + \int_{A_n^c \cap A^c} |f_n|^2 \mathrm{d}Q \\
\leq & \ 6 \int_\Omega e^{-\sigma^2(T-t)+2\sigma(W_T-W_t)} \mathrm{d}Q \\
= & \ 6e^{\sigma^2(T-t)} \\
< & \ \infty.
\end{aligned}$$

It follows from Problem 9.9 that $\{f_n\}_{n\geq 1}$ is uniformly integrable. Hence Equation (42.21) follows immediately from Problem 9.10.

(b) By Problem 42.5 and part (a), we have

$$\frac{\partial C_t}{\partial X_t} = \frac{\partial E_Q[g(X_t)]}{\partial X_t} = E_Q\left[\frac{\partial g(X_t)}{\partial X_t}\right] = N(d_1).$$

(c) By part (b) and Problem 42.7, we have

$$\frac{\partial P_t}{\partial X_t} = \frac{\partial C_t}{\partial X_t} - 1 = N(d_1) - 1.$$

This completes the proof.

42.5 Bibliographic Notes

In this chapter, we presented the Black-Scholes option pricing model, which was first discovered by Fischer Black and Myron Scholes (Black and Scholes, 1973) and further developed by Robert Merton (Merton, 1973). There are several approaches to deriving the model (Andreasen et al., 1998).

The Black-Scholes model is discussed in many books such as Merton (1990), Wilmott et al. (1995), Lamberton and Lapeyre (1996), Shiryaev (1999), Ross (1999), Neftci (2000), Duffie (2001), Karatzas and Shreve (2001), Etheridge (2002), Steele (2003), Shreve (2004), Bingham and Kiesel (2004), Dineen (2005), Jiang (2005), Elliott and Kopp (2005), Back (2005), Klebaner (2005), Musiela and Rutkowski (2005), Björk (2009), and Pascucci (2011), to name only a few.

The option pricing models presented in this chapter are based on one risky asset, whose price follows a geometric Brownian motion. Cherubini et al. (2004) and Allen (2007) discussed models based on two risky assets. Jondeau et al. (2007) discussed option pricing under non-Gaussian distributions. Delbaen and Schachermayer (2006) presented the mathematics of arbitrage and several original papers on asset pricing. Back et al. (2004) introduced models for incomplete markets, where European options may not be replicable. Katz and McCormick (2005) presented other models (e.g., neural networks) for option pricing.

Smith (1976) presented a comprehensive discussion of the analytic properties of the option pricing functions. Hobson (2004) is a survey paper on mathematical finance. For introduction to various financial derivatives, readers can consult Kolb (1993), Wilmott et al. (1995), Houthakker and Williamson (1996), Hull (2001), Wilmott (2001), Eales and Choudhry (2003), Franke et al. (2004), Back (2005), McDonald (2005), Wilmott (2006), Kwok (2008), Neftci (2008), and Hull (2008).

Theorems 42.1 and 42.2 are the two fundamental asset pricing theorems for continuous-time markets. For proofs of these theorems, readers are referred to Shreve (2004, Section 5.4).

CHAPTER 43

PATH-DEPENDENT OPTIONS

Path-dependent options are also referred to as "exotic options", whose payoffs depend on the path of the underlying asset. In this chapter, we present pricing formulas for some path-dependent options within the Black-Scholes framework.

43.1 Basic Concepts and Facts

Theorem 43.1 (European Barrier Option Price). *Let $\{(X_t^{(0)}, X_t) : 0 \leq t \leq T\}$ be the Black-Scholes market given in Definition 42.2. Let*

$$A(\phi) = \phi S N(\phi x_1) - \phi K e^{-rT} N(\phi x_1 - \phi\sigma\sqrt{T}), \tag{43.1a}$$

$$B(\phi) = \phi S N(\phi x_2) - \phi K e^{-rT} N(\phi x_2 - \phi\sigma\sqrt{T}), \tag{43.1b}$$

$$C(\eta, \phi) = \phi H \left(\frac{H}{S}\right)^{\frac{2r}{\sigma^2}} N(\eta y_1) - \phi K e^{-rT} \left(\frac{H}{S}\right)^{\frac{2r}{\sigma^2}-1} N(\eta y_1 - \eta\sigma\sqrt{T}), \tag{43.1c}$$

Measure, Probability, and Mathematical Finance.
By Guojun Gan, Chaoqun Ma, and Hong Xie

and

$$D(\eta, \phi) = \phi H \left(\frac{H}{S}\right)^{\frac{2r}{\sigma^2}} N(\eta y_2) - \phi K e^{-rT} \left(\frac{H}{S}\right)^{\frac{2r}{\sigma^2}-1} N(\eta y_2 - \eta\sigma\sqrt{T}), \quad (43.1d)$$

where $S = X_0$ is the initial price of the risky asset, K is the strike, T is the expiration time, H is the barrier, $N(\cdot)$ is as defined in Equation (42.1c), *and*

$$x_1 = \frac{\ln \frac{S}{K} + (r + \frac{1}{2}\sigma^2)T}{\sigma\sqrt{T}}, \quad x_2 = \frac{\ln \frac{S}{H} + (r + \frac{1}{2}\sigma^2)T}{\sigma\sqrt{T}},$$

$$y_1 = \frac{\ln \frac{H^2}{SK} + (r + \frac{1}{2}\sigma^2)T}{\sigma\sqrt{T}}, \quad y_2 = \frac{\ln \frac{H}{S} + (r + \frac{1}{2}\sigma^2)T}{\sigma\sqrt{T}}.$$

Then

(a) *If $S > H$ and the payoff of a down-and-in call is*

$$\begin{cases} \max(X_T - K, 0), & \text{if } \min_{0 \le t \le T} X_t \le H; \\ 0, & \text{if otherwise,} \end{cases}$$

then the price at time 0 of the down-and-in call is

$$C_{di} = \begin{cases} C(1,1), & \text{if } K \ge H; \\ A(1) - B(1) + D(1,1), & \text{if } K < H. \end{cases} \quad (43.2)$$

(b) *If $S < H$ and the payoff of an up-and-in call is*

$$\begin{cases} \max(X_T - K, 0), & \text{if } \max_{0 \le t \le T} X_t \ge H; \\ 0, & \text{if otherwise,} \end{cases}$$

then the price at time 0 of the up-and-in call is

$$C_{ui} = \begin{cases} A(1), & \text{if } K \ge H; \\ B(1) - C(-1,1) + D(-1,1), & \text{if } K < H. \end{cases} \quad (43.3)$$

(c) *If $S > H$ and the payoff of a down-and-in put is*

$$\begin{cases} \max(K - X_T, 0), & \text{if } \min_{0 \le t \le T} X_t \le H; \\ 0, & \text{if otherwise,} \end{cases}$$

then the price at time 0 of the down-and-in put is

$$P_{di} = \begin{cases} B(-1) - C(1,-1) + D(1,-1), & \text{if } K \ge H; \\ A(-1), & \text{if } K < H. \end{cases} \quad (43.4)$$

(d) *If* $S < H$ *and the payoff of an up-and-in put is*

$$\begin{cases} \max(K - X_T, 0), & \textit{if } \max_{0 \le t \le T} X_t \ge H; \\ 0, & \textit{if otherwise,} \end{cases}$$

then the price at time 0 of the up-and-in put is

$$P_{ui} = \begin{cases} A(-1) - B(-1) + D(-1,-1), & \textit{if } K \ge H; \\ C(-1,-1), & \textit{if } K < H. \end{cases} \tag{43.5}$$

(e) *If* $S > H$ *and the payoff of a down-and-out call is*

$$\begin{cases} \max(X_T - K, 0), & \textit{if } \min_{0 \le t \le T} X_t > H; \\ 0, & \textit{if otherwise,} \end{cases}$$

then the price at time 0 of the down-and-out call is

$$C_{do} = \begin{cases} A(1) - C(1,1), & \textit{if } K \ge H; \\ B(1) - D(1,1), & \textit{if } K < H. \end{cases} \tag{43.6}$$

(f) *If* $S < H$ *and the payoff of an up-and-out call is*

$$\begin{cases} \max(X_T - K, 0), & \textit{if } \max_{0 \le t \le T} X_t < H; \\ 0, & \textit{if otherwise,} \end{cases}$$

then the price at time 0 of the up-and-out call is

$$C_{uo} = \begin{cases} 0, & \textit{if } K \ge H; \\ A(1) - B(1) + C(-1,1) - D(-1,1), & \textit{if } K < H. \end{cases} \tag{43.7}$$

(g) *If* $S > H$ *and the payoff of a down-and-out put is*

$$\begin{cases} \max(K - X_T, 0), & \textit{if } \min_{0 \le t \le T} X_t > H; \\ 0, & \textit{if otherwise,} \end{cases}$$

then the price at time 0 of the down-and-out put is

$$P_{do} = \begin{cases} A(-1) - B(-1) + C(1,-1) - D(1,-1), & \textit{if } K \ge H; \\ 0, & \textit{if } K < H. \end{cases} \tag{43.8}$$

(h) *If* $S < H$ *and the payoff of a up-and-out put is*

$$\begin{cases} \max(K - X_T, 0), & \textit{if } \max_{0 \le t \le T} X_t < H; \\ 0, & \textit{if otherwise,} \end{cases}$$

then the price at time 0 of the up-and-out put is

$$P_{uo} = \begin{cases} B(-1) - D(-1,-1), & \text{if } K \geq H; \\ A(-1) - C(-1,-1), & \text{if } K < H. \end{cases} \tag{43.9}$$

Theorem 43.2 (Floating-Strike Lookback European Option Price). *Let*

$$\{(X_t^{(0)}, X_t) : 0 \leq t \leq T\}$$

be the Black-Scholes market given in Definition 42.2. Let

$$H_t = \max_{0 \leq u \leq t} X_u, \quad L_t = \min_{0 \leq u \leq t} X_u, \quad 0 \leq t \leq T.$$

(a) The payoff of a floating-strike lookback European call option is given by

$$\max(X_T - L_T, 0) = X_T - L_T.$$

The value at time t $(0 \leq t < T)$ of the floating-strike lookback European call option is

$$\begin{aligned} & C_{floating}(t) \\ = \; & X_t N(a_1) - L_t e^{-r(T-t)} N(a_2) - X_t \frac{\sigma^2}{2r} N(-a_1) \\ & + X_t e^{-r(T-t)} \frac{\sigma^2}{2r} \left(\frac{X_t}{L_t}\right)^{-\frac{2r}{\sigma^2}} N\left(-a_1 + \frac{2r}{\sigma}\sqrt{T-t}\right), \end{aligned} \tag{43.10}$$

where $N(\cdot)$ is as defined in Equation (42.1c) *and*

$$a_1 = \frac{\ln \frac{X_t}{L_t} + (r + \frac{\sigma^2}{2})(T-t)}{\sigma\sqrt{T-t}}, \quad a_2 = a_1 - \sigma\sqrt{T-t}.$$

(b) The payoff of a floating-strike lookback European put option is given by

$$\max(H_T - X_T, 0) = H_T - X_T.$$

The value at time t $(0 \leq t < T)$ of the floating-strike lookback European put option is

$$\begin{aligned} & P_{floating}(t) \\ = \; & H_t e^{-r(T-t)} N(-b_2) - X_t N(-b_1) + X_t \frac{\sigma^2}{2r} N(b_1) \\ & - X_t e^{-r(T-t)} \frac{\sigma^2}{2r} \left(\frac{X_t}{H_t}\right)^{-\frac{2r}{\sigma^2}} N\left(b_1 - \frac{2r}{\sigma}\sqrt{T-t}\right), \end{aligned} \tag{43.11}$$

where $N(\cdot)$ is as defined in Equation (42.1c) *and*

$$b_1 = \frac{\ln \frac{X_t}{H_t} + (r + \frac{\sigma^2}{2})(T-t)}{\sigma\sqrt{T-t}}, \quad b_2 = b_1 - \sigma\sqrt{T-t}.$$

Theorem 43.3 (Fixed-Strike Lookback Option Price). *Let* $\{(X_t^{(0)}, X_t) : 0 \leq t \leq T\}$ *be the Black-Scholes market given in Definition 42.2. Let*

$$H_t = \max_{0 \leq u \leq t} X_u, \quad L_t = \min_{0 \leq u \leq t} X_u, \quad 0 \leq t \leq T.$$

Let K *be the strike and*

$$d_1 = \frac{\ln \frac{X_t}{K} + (r + \frac{\sigma^2}{2})(T-t)}{\sigma\sqrt{T-t}}, \quad d_2 = d_1 - \sigma\sqrt{T-t}.$$

(a) The payoff of a fixed-strike lookback European call option is given by

$$\max(H_T - K, 0).$$

When $K > H_t$*, the value at time* t $(0 \leq t < T)$ *of the fixed-strike lookback European call option is*

$$\begin{aligned} & C_{fixed}(t) \\ = \; & X_t \left(1 + \frac{\sigma^2}{2r}\right) N(d_1) - Ke^{-r(T-t)} N(d_2) \\ & -X_t e^{-r(T-t)} \frac{\sigma^2}{2r} \left(\frac{X_t}{K}\right)^{-\frac{2r}{\sigma^2}} N\left(d_1 - \frac{2r}{\sigma}\sqrt{T-t}\right), \end{aligned} \quad (43.12)$$

where $N(\cdot)$ *is as defined in Equation* (42.1c).

When $K \leq H_t$*, the value at time* t $(0 \leq t < T)$ *of the fixed-strike lookback European call option is*

$$\begin{aligned} & C_{fixed}(t) \\ = \; & e^{-r(T-t)}(H_t - K) + X_t \left(1 + \frac{\sigma^2}{2r}\right) N(e_1) - H_t e^{-r(T-t)} N(e_2) \\ & -X_t e^{-r(T-t)} \frac{\sigma^2}{2r} \left(\frac{X_t}{H_t}\right)^{-\frac{2r}{\sigma^2}} N\left(e_1 - \frac{2r}{\sigma}\sqrt{T-t}\right), \end{aligned} \quad (43.13)$$

where $N(\cdot)$ *is as defined in Equation* (42.1c) *and*

$$e_1 = \frac{\ln \frac{X_t}{H_t} + (r + \frac{\sigma^2}{2})(T-t)}{\sigma\sqrt{T-t}}, \quad e_2 = e_1 - \sigma\sqrt{T-t}.$$

(b) The payoff of a fixed-strike lookback European put option is given by

$$\max(K - L_T, 0).$$

When $K < L_t$, the value at time t $(0 \leq t < T)$ of the fixed-strike lookback European put option is

$$\begin{aligned} & P_{fixed}(t) \\ = \; & Ke^{-r(T-t)}N(-d_2) - X_t\left(1+\frac{\sigma^2}{2r}\right)N(-d_1) \\ & + X_t e^{-r(T-t)}\frac{\sigma^2}{2r}\left(\frac{X_t}{K}\right)^{-\frac{2r}{\sigma^2}} N\left(-d_1 + \frac{2r}{\sigma}\sqrt{T-t}\right), \quad (43.14) \end{aligned}$$

where $N(\cdot)$ is as defined in Equation (42.1c).

When $K \geq L_t$, the value at time t $(0 \leq t < T)$ of the fixed-strike lookback European put option is

$$\begin{aligned} & P_{fixed}(t) \\ = \; & e^{-r(T-t)}(K - L_t) - X_t\left(1+\frac{\sigma^2}{2r}\right)N(-f_1) + L_t e^{-r(T-t)}N(-f_2) \\ & + X_t e^{-r(T-t)}\frac{\sigma^2}{2r}\left(\frac{X_t}{L_t}\right)^{-\frac{2r}{\sigma^2}} N\left(-f_1 + \frac{2r}{\sigma}\sqrt{T-t}\right), \quad (43.15) \end{aligned}$$

where $N(\cdot)$ is as defined in Equation (42.1c) *and*

$$f_1 = \frac{\ln\frac{X_t}{L_t} + (r + \frac{\sigma^2}{2})(T-t)}{\sigma\sqrt{T-t}}, \quad f_2 = f_1 - \sigma\sqrt{T-t}.$$

43.2 Problems

43.1. Let $\{B_t : 0 \leq t \leq T\}$ be a Brownian motion on a probability space $(\Omega, \mathscr{F}, P)$. Let $\{W_t : 0 \leq t \leq T\}$ be defined as

$$W_t = \alpha t + B_t, \quad 0 \leq t \leq T,$$

where $\alpha \in \mathbf{R}$. Let $M_T = \max_{0 \leq t \leq T} W_t$. Show that under P, the joint density of the pair (M_T, W_T) is

$$f_{M_T,W_T}(x,y) = \frac{2(2x-y)}{T\sqrt{2\pi T}}\exp\left(\alpha y - \frac{1}{2}\alpha^2 T - \frac{1}{2T}(2x-y)^2\right), \quad y \leq x, x \geq 0,$$

and is zero for other values of x and y.

43.2. Let $\{B_t : 0 \leq t \leq T\}$ be a Brownian motion on a probability space $(\Omega, \mathscr{F}, P)$. Let $\{W_t : 0 \leq t \leq T\}$ be defined as

$$W_t = \alpha t + B_t, \quad 0 \leq t \leq T,$$

where $\alpha \in \mathbf{R}$. Let $L_T = \min_{0\leq t\leq T} W_t$. Show that under P, the joint density of the pair (L_T, W_T) is

$$f_{L_T,W_T}(x,y) = \frac{-2(2x-y)}{T\sqrt{2\pi T}} \exp\left(\alpha y - \frac{1}{2}\alpha^2 T - \frac{1}{2T}(2x-y)^2\right), \quad y \geq x, x \leq 0,$$

and is zero for other values of x and y.

43.3. Let $\{B_t : 0 \leq t \leq T\}$ be a Brownian motion on a probability space $(\Omega, \mathscr{F}, P)$. Let M_T be a random variable defined as

$$M_T = \max_{0\leq t\leq T} (\alpha t + B_t),$$

where $\alpha \in \mathbf{R}$. Let $N(\cdot)$ be the function defined in Problem 19.8. Show that

(a)

$$P\{M_T \leq x\} = N\left(\frac{x-\alpha T}{\sqrt{T}}\right) - e^{2\alpha x} N\left(\frac{-x-\alpha T}{\sqrt{T}}\right), \quad x \geq 0.$$

(b) The density under P of the random variable M_T is

$$f_{M_T}(x) = \frac{2}{\sqrt{2\pi T}} \exp\left(-\frac{1}{2T}(x-\alpha T)^2\right) - 2\alpha e^{2\alpha x} N\left(\frac{-x-\alpha T}{\sqrt{T}}\right)$$

for $x \geq 0$ and is zero for $x < 0$.

43.4 (Down-and-In European Call Price)**.** Let $\{(X_t^{(0)}, X_t) : 0 \leq t \leq T\}$ be the Black-Scholes market given in Definition 42.2. Let $S = X_0$ be the initial price of the risky asset and H a barrier such that $H < S$. Suppose that a down-and-in European call option has the following payoff function:

$$\begin{cases} \max(X_T - K, 0), & \text{if } \min_{0\leq t\leq T} X_t \leq H; \\ 0, & \text{if otherwise,} \end{cases}$$

where T is the maturity time and K is the strike. Show that the time-zero price of the down-and-in European call option is given by

$$C_{di} = \begin{cases} C(1,1), & \text{if } K \geq H; \\ A(1) - B(1) + D(1,1), & \text{if } K < H, \end{cases}$$

where A, B, C, and D are defined in Equation (43.1).

43.5 (Up-and-Out European Call Price)**.** Let $\{(X_t^{(0)}, X_t) : 0 \leq t \leq T\}$ be the Black-Scholes market given in Definition 42.2. Let $S = X_0$ be the initial price of the risky asset and H be a barrier such that $H > S$. Suppose that an up-and-out European call option has the following payoff function:

$$\begin{cases} \max(X_T - K, 0), & \text{if } \max_{0\leq t\leq T} X_t < H; \\ 0, & \text{if otherwise,} \end{cases}$$

where T is the maturity time and K is the strike. Show that the time-zero price of the up-and-out European call option is given by

$$C_{uo} = \begin{cases} 0, & \text{if } K \geq H; \\ A(1) - B(1) + C(-1,1) - D(-1,1), & \text{if } K < H, \end{cases}$$

where A, B, C, and D are as defined in Equation (43.1).

43.6 (Floating-Strike Lookback European Put Price)**.** Let $\{(X_t^{(0)}, X_t) : 0 \leq t \leq T\}$ be the Black-Scholes market given in Definition 42.2. Let

$$H_t = \max_{0 \leq u \leq t} X_u, \quad 0 \leq t \leq T.$$

A standard floating-strike lookback European put option has the following payoff function:

$$\max(H_T - X_T, 0) = H_T - X_T.$$

Show that the value at time t $(0 \leq t < T)$ of the floating-strike lookback European put option is given by

$$\begin{aligned} P_{floating}(t) &= H_t e^{-r(T-t)} N(-b_2) - X_t N(-b_1) + X_t \frac{\sigma^2}{2r} N(b_1) \\ &\quad - X_t e^{-r(T-t)} \frac{\sigma^2}{2r} \left(\frac{X_t}{H_t}\right)^{-\frac{2r}{\sigma^2}} N\left(b_1 - \frac{2r}{\sigma}\sqrt{T-t}\right), \end{aligned}$$

where $N(\cdot)$ is as defined in Equation (42.1c) and

$$b_1 = \frac{\ln \frac{X_t}{H_t} + (r + \frac{\sigma^2}{2})(T-t)}{\sigma\sqrt{T-t}}, \qquad b_2 = b_1 - \sigma\sqrt{T-t}.$$

43.3 Hints

43.1. First use the Girsanov Theorem I (see Problem 37.4) to find a probability measure Q such that $\{W_t : 0 \leq t \leq T\}$ is a Brownian motion under Q. Then use the result of Problem 28.16.

43.2. Apply the result of Problem 43.1 and note that the two processes $\{\alpha t + B_t : t \geq 0\}$ and $\{\alpha t - B_t : t \geq 0\}$ have the same distribution.

43.3. To prove the first part, use the result of Problem 43.1 and the fact that

$$\begin{aligned} P\{M_T \leq x\} &= \int_0^x \int_{-\infty}^z f_{M_T,W_T}(z,y)\mathrm{d}y\mathrm{d}z \\ &= \int_0^x \int_0^z f_{M_T,W_T}(z,y)\mathrm{d}y\mathrm{d}z + \int_0^x \int_{-\infty}^0 f_{M_T,W_T}(z,y)\mathrm{d}y\mathrm{d}z \\ &= \int_0^x \int_y^x f_{M_T,W_T}(z,y)\mathrm{d}z\mathrm{d}y + \int_{-\infty}^0 \int_0^x f_{M_T,W_T}(z,y)\mathrm{d}z\mathrm{d}y \end{aligned}$$

Use the first part to prove the second part.

43.4. Apply the results of Problems 42.4, 43.2, and 19.9.

43.5. Apply the results of Problems 42.4, 43.1, and 19.9.

43.6. Use the results of Problems 42.4, 14.15, 14.13, and 43.3.

43.4 Solutions

43.1. Let Q be a probability measure on $(\Omega, \mathscr{F})$ defined as

$$\mathrm{d}Q = Z_T \mathrm{d}P,$$

where

$$Z_t = \exp\left(-\alpha B_t - \frac{1}{2}\alpha^2 t\right) = \exp\left(-\alpha W_t + \frac{1}{2}\alpha^2 t\right), \quad 0 \le t \le T.$$

Then it follows from Problem 37.4 that $\{W_t : 0 \le t \le T\}$ is a Brownian motion under Q. Then by Problem 28.16, the joint density of (M_T, W_T) under Q is

$$g_{M_T,W_T}(x,y) = \frac{2(2x-y)}{T\sqrt{2\pi T}} \exp\left(-\frac{(2x-y)^2}{2T}\right), \quad y \le x,\ x > 0$$

and is zero for other values of x and y. Therefore, we have

$$\begin{aligned}
& P\{M_T \le x, W_T \le y\} \\
= & \int_\Omega I_{\{M_T \le x, W_T \le y\}} \mathrm{d}P \\
= & \int_\Omega \frac{1}{Z_T} I_{\{M_T \le x, W_T \le y\}} \mathrm{d}Q \\
= & \int_{-\infty}^{x} \int_{-\infty}^{y} e^{\alpha v - \frac{1}{2}\alpha^2 T} g_{M_T,W_T}(u,v) \mathrm{d}v \mathrm{d}u.
\end{aligned}$$

Differentiating the above equation with respect to x and y gives the following joint density of (M_T, W_T) under P:

$$\begin{aligned}
f_{M_T,W_T}(x,y) &= g_{M_T,W_T}(x,y) e^{\alpha y - \frac{1}{2}\alpha^2 T} \\
&= \frac{2(2x-y)}{T\sqrt{2\pi T}} \exp\left(-\frac{(2x-y)^2}{2T} + \alpha y - \frac{1}{2}\alpha^2 T\right).
\end{aligned}$$

This completes the proof.

43.2. Let $x \le 0$ and $y \ge x$. Then we have

$$P\{L_T \le x, W_T \le y\} = \int_{-\infty}^{x} \int_{-\infty}^{y} f_{L_T,W_T}(r,s) \mathrm{d}s \mathrm{d}r. \tag{43.16}$$

Since $\{-\alpha t - B_t : t \geq 0\}$ and $\{-\alpha t + B_t : t \geq 0\}$ have the same distribution, we have

$$\begin{aligned} & P\{L_T \leq x, W_T \leq y\} \\ = \ & P\{-L_T \geq -x, -W_T \geq -y\} \\ = \ & P\left\{ \max_{0\leq t\leq T} (-\alpha t - B_t) \geq -x, -\alpha T - B_T \geq -y \right\} \\ = \ & P\left\{ \max_{0\leq t\leq T} (-\alpha t + B_t) \geq -x, -\alpha T + B_T \geq -y \right\}. \end{aligned} \tag{43.17}$$

But by Problem 43.1, we have

$$\begin{aligned} & P\left\{ \max_{0\leq t\leq T} (-\alpha t + B_t) \geq -x, -\alpha T + B_T \geq -y \right\} \\ = \ & \int_{-x}^{\infty} \int_{-y}^{\infty} \frac{2(2r-s)}{T\sqrt{2\pi T}} \exp\left(-\frac{(2r-s)^2}{2T} - \alpha s - \frac{1}{2}\alpha^2 T \right) \mathrm{d}s\mathrm{d}r. \end{aligned} \tag{43.18}$$

Combining Equations (43.16), (43.17), and (43.18) gives

$$\begin{aligned} & \int_{-\infty}^{x} \int_{-\infty}^{y} f_{L_T,W_T}(r,s)\mathrm{d}s\mathrm{d}r \\ = \ & \int_{-x}^{\infty} \int_{-y}^{\infty} \frac{2(2r-s)}{T\sqrt{2\pi T}} \exp\left(-\frac{(2r-s)^2}{2T} - \alpha s - \frac{1}{2}\alpha^2 T \right) \mathrm{d}s\mathrm{d}r. \end{aligned}$$

Differentiating both sides of the above equation with respect to x and y gives the following joint density of (L_T, W_T) under P:

$$f_{L_T,W_T}(x,y) = \frac{-2(2x-y)}{T\sqrt{2\pi T}} \exp\left(-\frac{(2x-y)^2}{2T} + \alpha y - \frac{1}{2}\alpha^2 T \right).$$

This completes the proof.

43.3.

(a) By Problem 43.1 and Theorem 10.1, we have

$$\begin{aligned} & P\{M_T \leq x\} \\ = \ & \int_0^x \int_{-\infty}^{z} f_{M_T,W_T}(z,y)\mathrm{d}y\mathrm{d}z \\ = \ & \int_0^x \int_0^z f_{M_T,W_T}(z,y)\mathrm{d}y\mathrm{d}z + \int_0^x \int_{-\infty}^{0} f_{M_T,W_T}(z,y)\mathrm{d}y\mathrm{d}z \\ = \ & \int_0^x \int_y^x f_{M_T,W_T}(z,y)\mathrm{d}z\mathrm{d}y \\ & + \int_{-\infty}^{0} \int_0^x f_{M_T,W_T}(z,y)\mathrm{d}z\mathrm{d}y. \end{aligned} \tag{43.19}$$

Note that

$$
\begin{aligned}
& \int_v^x f_{M_T,W_T}(z,w)\mathrm{d}z \\
= & \int_v^x \frac{2(2z-w)}{T\sqrt{2\pi T}} \exp\left(\alpha w - \frac{1}{2}\alpha^2 T - \frac{1}{2T}(2z-w)^2\right) \mathrm{d}z \\
= & -\frac{1}{\sqrt{2\pi T}} \exp\left(\alpha w - \frac{1}{2}\alpha^2 - \frac{1}{2T}(2z-w)^2\right)\Bigg|_v^x \\
= & \frac{1}{\sqrt{2\pi T}} \exp\left(\alpha w - \frac{1}{2}\alpha^2 - \frac{1}{2T}(2v-w)^2\right) - \\
& \frac{1}{\sqrt{2\pi T}} \exp\left(\alpha w - \frac{1}{2}\alpha^2 - \frac{1}{2T}(2x-w)^2\right).
\end{aligned} \tag{43.20}
$$

Combining Equations (43.19) and (43.20) gives

$$
\begin{aligned}
P\{M_T \le x\} &= \int_{-\infty}^x \frac{1}{\sqrt{2\pi T}} \exp\left(\alpha y - \frac{1}{2}\alpha^2 - \frac{1}{2T}y^2\right) \mathrm{d}y - \\
& \quad \int_{-\infty}^x \frac{1}{\sqrt{2\pi T}} \exp\left(\alpha y - \frac{1}{2}\alpha^2 - \frac{1}{2T}(2x-y)^2\right) \mathrm{d}y \\
&= N\left(\frac{x-\alpha T}{\sqrt{T}}\right) - e^{2\alpha x} N\left(\frac{-x-\alpha T}{\sqrt{T}}\right).
\end{aligned}
$$

(b) Note that

$$f_{M_T}(x) = \frac{\mathrm{d}P\{M_T \le x\}}{\mathrm{d}x}.$$

By the first part of this problem, we have

$$
\begin{aligned}
f_{M_T}(x) &= \frac{\mathrm{d}}{\mathrm{d}x} N\left(\frac{x-\alpha T}{\sqrt{T}}\right) - \frac{\mathrm{d}}{\mathrm{d}x} e^{2\alpha x} N\left(\frac{-x-\alpha T}{\sqrt{T}}\right) \\
&= \varphi\left(0,1,\frac{x-\alpha T}{\sqrt{T}}\right)\frac{1}{\sqrt{T}} - 2\alpha e^{2\alpha x} N\left(\frac{-x-\alpha T}{\sqrt{T}}\right) \\
& \quad + e^{2\alpha x}\varphi\left(0,1,\frac{-x-\alpha T}{\sqrt{T}}\right)\frac{1}{\sqrt{T}} \\
&= \varphi\left(0,1,\frac{x-\alpha T}{\sqrt{T}}\right)\frac{2}{\sqrt{T}} - 2\alpha e^{2\alpha x} N\left(\frac{-x-\alpha T}{\sqrt{T}}\right) \\
&= \frac{2}{\sqrt{2\pi T}} \exp\left(-\frac{1}{2T}(x-\alpha T)^2\right) - 2\alpha e^{2\alpha x} N\left(\frac{-x-\alpha T}{\sqrt{T}}\right),
\end{aligned}
$$

where $\varphi(0,1,\cdot)$ is defined in Equation (19.1).

This completes the proof.

43.4. Let $W_t = \frac{\mu - r}{\sigma}t + B_t$ for $0 \le t \le T$. Let Q be the risk neutral measure Q defined in Problem 37.7. Then $\{W_t : 0 \le t \le T\}$ is a Brownian motion under Q.

By Problem 42.4, the time-zero price of the down-and-in European call option is

$$C_{di} = E_Q\left[e^{-rT}(X_T - K)^+ I_{\{\min_{0\le t\le T} X_t \le H\}}\right]. \tag{43.21}$$

The stochastic differential equation of X_t can be written as

$$\mathrm{d}X_t = rX_t\mathrm{d}t + \sigma X_t \mathrm{d}W_t, \quad 0 \le t \le T.$$

Solving the above stochastic differential equation (see Problem 38.7) gives

$$X_t = S\exp(\sigma\alpha t + \sigma W_t) = Se^{\sigma Z_t}, \quad 0 \le t \le T,$$

where $\alpha = (r - \frac{1}{2}\sigma^2)/\sigma$ and $Z_t = \alpha t + W_t$. Let $L_T = \min_{0\le t\le T} Z_t$. Then

$$\min_{0\le t\le T} X_t = Se^{\sigma L_T}.$$

Therefore, we can rewrite Equation (43.21) as

$$\begin{aligned} C_{di} &= E_Q\left[e^{-rT}(X_T - K)I_{\{X_T>K, S\exp(\sigma L_T)\le H\}}\right] \\ &= E_Q\left[e^{-rT}(S\exp(\sigma Z_T) - K)I_{\{Z_T > \frac{1}{\sigma}\ln\frac{K}{S}, L_T \le \frac{1}{\sigma}\ln\frac{H}{S}\}}\right]. \end{aligned} \tag{43.22}$$

If $K \ge H$, then by Problem 43.2, we can calculate Equation (43.22) as follows:

$$\begin{aligned} C_{di} &= \int_{\frac{1}{\sigma}\ln\frac{K}{S}}^{\infty}\int_{-\infty}^{\frac{1}{\sigma}\ln\frac{H}{S}} e^{-rT}\left(Se^{\sigma y} - K\right) \\ &\quad \frac{-2(2x-y)}{T\sqrt{2\pi T}}\exp\left(\alpha y - \frac{1}{2}\alpha^2 T - \frac{1}{2T}(2x-y)^2\right)\mathrm{d}x\mathrm{d}y \\ &= \int_{\frac{1}{\sigma}\ln\frac{K}{S}}^{\infty} \frac{e^{-rT}(Se^{\sigma y} - K)}{\sqrt{2\pi T}}\exp\left(\alpha y - \frac{1}{2}\alpha^2 T - \frac{(\frac{2}{\sigma}\ln\frac{H}{S} - y)^2}{2T}\right)\mathrm{d}y. \end{aligned}$$

It follows from Problem 19.9 that

$$C_{di} = H\left(\frac{H}{S}\right)^{\frac{2r}{\sigma^2}} N(y_1) - Ke^{-rT}\left(\frac{H}{S}\right)^{\frac{2r}{\sigma^2}-1} N(y_1 - \sigma\sqrt{T}) = C(1,1).$$

Similarly, if $K < H$, we have

$$\begin{aligned} C_{di} &= \int_{\frac{1}{\sigma}\ln\frac{H}{S}}^{\infty}\int_{-\infty}^{\frac{1}{\sigma}\ln\frac{H}{S}} e^{-rT}\left(Se^{\sigma y} - K\right) \\ &\quad \frac{-2(2x-y)}{T\sqrt{2\pi T}}\exp\left(\alpha y - \frac{1}{2}\alpha^2 T - \frac{1}{2T}(2x-y)^2\right)\mathrm{d}x\mathrm{d}y \\ &\quad + \int_{\frac{1}{\sigma}\ln\frac{K}{S}}^{\frac{1}{\sigma}\ln\frac{H}{S}}\int_{-\infty}^{y} e^{-rT}\left(Se^{\sigma y} - K\right) \\ &\quad \frac{-2(2x-y)}{T\sqrt{2\pi T}}\exp\left(\alpha y - \frac{1}{2}\alpha^2 T - \frac{1}{2T}(2x-y)^2\right)\mathrm{d}x\mathrm{d}y \\ &= A(1) - B(1) + D(1,1). \end{aligned}$$

This completes the proof.

43.5. The proof is similar to that of Problem 43.4. Let $W_t = \frac{\mu - r}{\sigma} t + B_t$ for $0 \le t \le T$ and Q the risk-neutral measure Q defined in Problem 37.7. Then $\{W_t : 0 \le t \le T\}$ is a Brownian motion under Q. By Problem 42.4, the time-zero price of the up-and-out European call option is

$$C_{uo} = E_Q \left[e^{-rT} (X_T - K)^+ I_{\{\max_{0 \le t \le T} X_t < H\}} \right]. \tag{43.23}$$

The stochastic differential equation of X_t can be written as

$$\mathrm{d}X_t = rX_t \mathrm{d}t + \sigma X_t \mathrm{d}W_t, \quad 0 \le t \le T.$$

Solving the above stochastic differential equation (see Problem 38.7) gives

$$X_t = S \exp(\sigma \alpha t + \sigma W_t) = S e^{\sigma Z_t}, \quad 0 \le t \le T,$$

where $\alpha = (r - \frac{1}{2}\sigma^2)/\sigma$ and $Z_t = \alpha t + W_t$. Let $M_T = \max_{0 \le t \le T} Z_t$. Then

$$\max_{0 \le t \le T} X_t = S e^{\sigma M_T}.$$

Hence, we can rewrite Equation (43.23) as

$$\begin{aligned} C_{uo} &= E_Q \left[e^{-rT} (X_T - K) I_{\{X_T > K, S \exp(\sigma M_T) < H\}} \right] \\ &= E_Q \left[e^{-rT} (S \exp(\sigma Z_T) - K) I_{\{Z_T > \frac{1}{\sigma} \ln \frac{K}{S}, M_T < \frac{1}{\sigma} \ln \frac{H}{S}\}} \right]. \end{aligned} \tag{43.24}$$

If $K \ge H$, the up-and-out call option is knocked out before being in the money. Hence the payoff is zero and the option price is also zero.

If $K < H$, then by Problem 43.1, we have

$$\begin{aligned} C_{uo} &= \int_{\frac{1}{\sigma} \ln \frac{K}{S}}^{\frac{1}{\sigma} \ln \frac{H}{S}} \int_{\max(y,0)}^{\frac{1}{\sigma} \ln \frac{H}{S}} e^{-rT} \left(S e^{\sigma y} - K \right) \\ &\quad \frac{2(2x - y)}{T\sqrt{2\pi T}} \exp\left(\alpha y - \frac{1}{2}\alpha^2 T - \frac{1}{2T}(2x - y)^2 \right) \mathrm{d}x \mathrm{d}y \\ &= -\int_{\frac{1}{\sigma} \ln \frac{K}{S}}^{\frac{1}{\sigma} \ln \frac{H}{S}} \frac{e^{-rT} \left(S e^{\sigma y} - K \right)}{\sqrt{2\pi T}} \exp\left(\alpha y - \frac{1}{2}\alpha^2 T - \frac{(\frac{2}{\sigma} \ln \frac{H}{S} - y)^2}{2T} \right) dy \\ &\quad + \int_{\frac{1}{\sigma} \ln \frac{K}{S}}^{\frac{1}{\sigma} \ln \frac{H}{S}} \frac{e^{-rT} \left(S e^{\sigma y} - K \right)}{\sqrt{2\pi T}} \exp\left(\alpha y - \frac{1}{2}\alpha^2 T - \frac{y^2}{2T} \right) dy. \end{aligned}$$

By Problem 19.9, we have

$$\begin{aligned} &\int_{\frac{1}{\sigma} \ln \frac{K}{S}}^{\frac{1}{\sigma} \ln \frac{H}{S}} \frac{e^{-rT} \left(S e^{\sigma y} - K \right)}{\sqrt{2\pi T}} \exp\left(\alpha y - \frac{1}{2}\alpha^2 T - \frac{(\frac{2}{\sigma} \ln \frac{H}{S} - y)^2}{2T} \right) dy \\ = \; & D(-1, 1) - C(-1, 1), \end{aligned}$$

and

$$\begin{aligned} & \int_{\frac{1}{\sigma}\ln\frac{K}{S}}^{\frac{1}{\sigma}\ln\frac{H}{S}} \frac{e^{-rT}\left(Se^{\sigma y}-K\right)}{\sqrt{2\pi T}} \exp\left(\alpha y - \frac{1}{2}\alpha^2 T - \frac{y^2}{2T}\right) \mathrm{d}y \\ = \; & A(1) - B(1). \end{aligned}$$

Hence we have

$$C_{uo} = A(1) - B(1) + C(-1,1) - D(-1,1).$$

This completes the proof.

43.6. Let $W_t = \frac{\mu - r}{\sigma} t + B_t$ for $0 \le t \le T$ and Q the risk neutral measure Q defined in Problem 37.7. Then $\{W_t : 0 \le t \le T\}$ is a Brownian motion under Q. By Problem 42.4, the price at time t of the floating-strike lookback European put option is

$$\begin{aligned} P_{floating}(t) &= E_Q\left[e^{-r(T-t)}(H_T - X_T)|\mathscr{F}_t\right] \\ &= e^{-r(T-t)}\left(E_Q\left[H_T|\mathscr{F}_t\right] - E_Q\left[X_T|\mathscr{F}_t\right]\right). \end{aligned} \tag{43.25}$$

Since $\{e^{-rt}X_t : 0 \le t \le T\}$ is a Q-martingale, we have

$$E_Q\left[e^{-rT}X_T|\mathscr{F}_t\right] = e^{-rt}X_t.$$

Hence we can write Equation (43.25) as

$$P_{floating}(t) = e^{-r(T-t)}E_Q\left[H_T|\mathscr{F}_t\right] - X_t. \tag{43.26}$$

To compute $E_Q[H_T|\mathscr{F}_t]$, we let $\alpha = (r - \frac{1}{2}\sigma^2)/\sigma$ and $Z_t = \alpha t + W_t$. Then we can write X_t as follows (see Problem 38.7):

$$X_t = X_0 \exp\left(\sigma\alpha t + \sigma W_t\right) = X_0 e^{\sigma Z_t}, \quad 0 \le t \le T,$$

Let $M_t = \max_{0 \le u \le t} Z_u$ for $t \in [0, T]$. Then we have

$$H_T = X_0 e^{\sigma M_t} e^{\sigma(M_T - M_t)} = H_t e^{\sigma(M_T - M_t)}. \tag{43.27}$$

Since

$$M_T - M_t = \left[\max_{t \le u \le T} Z_u - M_t\right]^+ = \left[\max_{t \le u \le T}(Z_u - Z_t) - (M_t - Z_t)\right]^+$$

and

$$M_t - Z_t = \frac{1}{\sigma}\ln\frac{H_t}{X_t},$$

we have

$$\sigma(M_T - M_t) = \left[\max_{t \le u \le T}\sigma(Z_u - Z_t) - \ln\frac{H_t}{X_t}\right]^+. \tag{43.28}$$

Combining Equations (43.27), (43.28), and (43.26) gives

$$E_Q[H_T|\mathscr{F}_t] = E_Q\left[H_t \exp\left(\left[\max_{t\le u\le T}\sigma(Z_u - Z_t) - \ln\frac{H_t}{X_t}\right]^+\right)|\mathscr{F}_t\right].$$

Let $V_1 = \max_{t\le u\le T}\sigma(Z_u - Z_t)$ and $V_2 = \ln\frac{H_t}{X_t}$. Since H_t and V_2 are $\mathscr{F}_t$-measurable and V_1 is independent of $\mathscr{F}_t$, it follows from Problems 14.15 and 14.13 that

$$E_Q[H_T|\mathscr{F}_t] = H_t\varphi(V_2), \tag{43.29}$$

where

$$\varphi(x) = E_Q\left[\exp\left(\left[\max_{t\le u\le T}\sigma(Z_u - Z_t) - x\right]^+\right)\right].$$

Since $Z_u - Z_t = \alpha(u-t) + W_u - W_t$ and $Z_{u-t} - Z_0 = \alpha(u-t) + W_{u-t}$ have the same distribution under Q, $\max_{t\le u\le T}\sigma(Z_u - Z_t)$ and $\max_{0\le u\le T-t}\sigma(Z_u - Z_0) = \sigma M_{T-t}$ also have the same distribution under Q. Hence we can write $\varphi(x)$ as

$$\begin{aligned}\varphi(x) &= E_Q\left[\exp\left([\sigma M_{T-t} - x]^+\right)\right]\\ &= Q\left\{M_{T-t} \le \frac{x}{\sigma}\right\} + e^{-x}E_Q\left[e^{\sigma M_{T-t}}I_{\{M_{T-t}>\frac{x}{\sigma}\}}\right]. \end{aligned}\tag{43.30}$$

By Problem 43.3, we have

$$\begin{aligned}Q\left\{M_{T-t} \le \frac{x}{\sigma}\right\} &= N\left(\frac{x - \sigma\alpha(T-t)}{\sigma\sqrt{T-t}}\right) - e^{\frac{2\alpha x}{\sigma}}N\left(\frac{-x-\sigma\alpha(T-t)}{\sigma\sqrt{T-t}}\right)\\ &= N\left(\frac{x - (r-\frac{1}{2}\sigma^2)(T-t)}{\sigma\sqrt{T-t}}\right)\\ &\quad -e^{\frac{2r-\sigma^2}{\sigma^2}x}N\left(\frac{-x-(r-\frac{1}{2}\sigma^2)(T-t)}{\sigma\sqrt{T-t}}\right), \end{aligned}\tag{43.31}$$

and

$$\begin{aligned}&E_Q\left[e^{\sigma M_{T-t}}I_{\{M_{T-t}>\frac{x}{\sigma}\}}\right]\\ &= \int_{\frac{x}{\sigma}}^{\infty} e^{\sigma y}\frac{2}{\sqrt{2\pi(T-t)}}\exp\left(-\frac{[y-\alpha(T-t)]^2}{2(T-t)}\right)\mathrm{d}y\\ &\quad - \int_{\frac{x}{\sigma}}^{\infty} e^{\sigma y}2\alpha e^{2\alpha y}N\left(\frac{-y-\alpha(T-t)}{\sqrt{T-t}}\right)\mathrm{d}y. \end{aligned}\tag{43.32}$$

By Problem 19.9, the first integral in Equation (43.32) can be written as

$$\begin{aligned}&\int_{\frac{x}{\sigma}}^{\infty} e^{\sigma y}\frac{2}{\sqrt{2\pi(T-t)}}\exp\left(-\frac{[y-\alpha(T-t)]^2}{2(T-t)}\right)\mathrm{d}y\\ &= 2e^{r(T-t)}N\left(\frac{-x+(r+\frac{1}{2}\sigma^2)(T-t)}{\sigma\sqrt{T-t}}\right). \end{aligned}\tag{43.33}$$

By Problem 19.10, the second integral in Equation (43.32) can be calculated as

$$\begin{aligned}
&\int_{\frac{x}{\sigma}}^{\infty} e^{\sigma y} 2\alpha e^{2\alpha y} N\left(\frac{-y-\alpha(T-t)}{\sqrt{T-t}}\right) dy \\
= & -\left(1-\frac{\sigma^2}{2r}\right) \exp\left(\frac{2rx}{\sigma^2}\right) N\left(\frac{-x-(r-\frac{1}{2}\sigma^2)(T-t)}{\sigma\sqrt{T-t}}\right) \\
& +\left(1-\frac{\sigma^2}{2r}\right) e^{r(T-t)} N\left(\frac{-x+(r+\frac{1}{2}\sigma^2)(T-t)}{\sigma\sqrt{T-t}}\right). \qquad (43.34)
\end{aligned}$$

Then it follows from Equations (43.30) – (43.34) that

$$\begin{aligned}
\varphi(x) = & N\left(\frac{x-(r-\frac{1}{2}\sigma^2)(T-t)}{\sigma\sqrt{T-t}}\right) - \\
& \frac{\sigma^2}{2r} e^{\left(\frac{2r}{\sigma^2}-1\right)x} N\left(\frac{-x-(r-\frac{1}{2}\sigma^2)(T-t)}{\sigma\sqrt{T-t}}\right) + \\
& \left(1+\frac{\sigma^2}{2r}\right) e^{-x} e^{r(T-t)} N\left(\frac{-x+(r+\frac{1}{2}\sigma^2)(T-t)}{\sigma\sqrt{T-t}}\right). \qquad (43.35)
\end{aligned}$$

Now, combining Equations (43.26), (43.29), and (43.35) leads to

$$\begin{aligned}
P_{floating}(t) = & e^{-r(T-t)} H_t \varphi\left(\ln \frac{H_t}{X_t}\right) - X_t \\
= & H_t e^{-r(T-t)} N(-b_2) - X_t N(-b_1) + X_t \frac{\sigma^2}{2r} N(b_1) \\
& - X_t e^{-r(T-t)} \frac{\sigma^2}{2r} \left(\frac{X_t}{H_t}\right)^{-\frac{2r}{\sigma^2}} N\left(b_1 - \frac{2r}{\sigma}\sqrt{T-t}\right).
\end{aligned}$$

This completes the proof.

43.5 Bibliographic Notes

In this chapter, we presented some closed-form pricing formulas for certain path-dependent options. The pricing formulas for the European barrier options were introduced by Merton (1973), Rubinstein (1991), and Rubinstein and Reiner (1991). General versions of these formulas can be found in Haug (2006).

Like the European barrier options, the lookback options also have closed form pricing formulas. The pricing formulas for floating-strike lookback options and fixed-strike lookback options were introduced by Goldman et al. (1979) and Conze and Viswanathan (1991), respectively. These formulas are also presented in Haug (2006).

Asian options are also path-dependent options. But unlike the barrier options and lookback options, Asian options do not have closed form pricing formulas. The price of Asian options is obtained by solving partial differential equations numerically. For more details about Asian option pricing, readers are referred to Shreve (2004) and Haug (2006).

CHAPTER 44

AMERICAN OPTIONS

Unlike European options, American options can be exercised at any time up to and including the expiration date. This early exercise feature makes American options more difficult to price than European ones. In this chapter, we present risk-neutral pricing of American options.

44.1 Basic Concepts and Facts

Definition 44.1 (Reward Function). A reward function is any function $g : (0, \infty) \times [0, T] \to \mathbf{R}$ that is continuous and satisfies the linear growth condition

$$|g(x, t)| \leq K_1 + K_2 x$$

for some constants K_1 and K_2.

Definition 44.2 (American Option). In a Black-Scholes market $\{(X_t^{(0)}, X_t) : t \geq 0\}$, an American option with reward function g and maturity T is a derivative that pays the amount $g(X_t, t)$ when exercised at any time $t \in [0, T]$.

The reward function for an American call option with strike K is $g(X_t, t) = (X_t - K)^+$. The reward function for an American put option with strike K is $g(X_t, t) = (K - X_t)^+$.

Definition 44.3 (Consumption Strategy). Let $\{(X_t^{(0)}, X_t) : t \geq 0\}$ be the Black-Scholes market given in Definition 42.2. A consumption strategy is a stochastic process $\{A_t : 0 \leq t \leq T\}$ satisfying the following conditions:

(a) It is adapted to the underlying filtration $\{\mathscr{F}_t : 0 \leq t \leq T\}$.

(b) Almost all of its sample paths are nondecreasing and right-continuous with left-hand limits.

(c) $A_0 = 0$ a.s.

Definition 44.4 (Trading and Consumption Strategy). A trading and consumption strategy is a process $(\phi, A) = \{(\theta_t^{(0)}, \theta_t, A_t) : 0 \leq t \leq T\}$, where $\phi = \{(\theta_t^{(0)}, \theta_t) : 0 \leq t \leq T\}$ is a trading strategy (see Definition 42.3) and $A = \{A_t : 0 \leq t \leq T\}$ is a consumption strategy.

Definition 44.5 (Self-Financing Trading and Consumption Strategy). In a Black-Scholes market $\{(X_t^{(0)}, X_t) : t \geq 0\}$, a trading and consumption strategy (ϕ, A) is said to be self-financing on $[0, T]$ if it satisfies the following properties:

(a)

$$\int_0^T \left|\theta_t^{(0)}\right| \mathrm{d}t + \int_0^T \theta_t^2 \mathrm{d}t < \infty \quad \text{a.s.,}$$

(b) For every $t \in [0, T]$,

$$V_t(\phi, A) - V_0(\phi, A) = \int_0^t \theta_u^{(0)} \mathrm{d}X_u^{(0)} + \int_0^t \theta_u \mathrm{d}X_u - A_t \quad \text{a.s.,}$$

where $V_t(\phi, A)$ is its wealth process defined as

$$V_t(\phi, A) = \theta_t^{(0)} X_t^{(0)} + \theta_t X_t, \quad \forall t \in [0, T].$$

Definition 44.6 (Admissible Trading and Consumption Strategy). A trading and consumption strategy (ϕ, A) in a Black-Scholes market is said to be admissible if it is self-financing and

$$E_Q\left[\int_0^T (\theta_u X_u)^2 \mathrm{d}u\right] < \infty,$$

where Q is the risk-neutral measure defined in Problem 37.7.

Definition 44.7 (Buy-and-Hold Strategy). Let $\mathcal{T}_{0,T}$ be the set of all stopping times of the underlying filtration $\{\mathscr{F}_t : 0 \leq t \leq T\}$ that take values in $[0, T]$ a.s. A buy-and-hold strategy associated with an American option $X^{(a)}$ is a pair (c, τ), where

$c \in \mathbf{R}$ and $\tau \in \mathcal{T}_{0,T}$; that is, c units of the American option are bought (or sold if $c < 0$) at time 0 and held until time τ.

Definition 44.8 (Extended Admissible Trading and Consumption Strategy). Let

$$\{(X_t^{(0)}, X_t) : 0 \leq t \leq T\}$$

be a Black-Scholes market and $X^{(a)}$ an American option. An extended admissible trading and consumption strategy in the market is a collection (ϕ, A, c, τ) such that

(a) (ϕ, A) is an admissible trading and consumption strategy in the Black-Scholes market.

(b) (c, τ) is a buy-and-hold strategy in $X^{(a)}$.

(c) For any $t \in (\tau, T]$, we have

$$\theta_t^{(0)} = \theta_\tau^{(0)} + \theta_\tau \frac{X_\tau}{X_\tau^{(0)}} + \frac{cg(X_\tau, \tau)}{X_\tau^{(0)}}, \quad \theta_t = 0,$$

where $g(\cdot, \cdot)$ is the reward function of the American option.

(d) For any $t \in (\tau, T]$, we have $A_t = A_\tau$.

The set of all extended admissible trading and consumption strategies is denoted by $\mathscr{A}$. Let $\mathscr{A}_0$ denote the set of strategies in $\mathscr{A}$ that satisfy the following conditions:

$$\theta_0^{(0)} + \theta_0 X_0 + cU_0 < 0$$

and

$$\theta_T^{(0)} X_T^{(0)} \geq 0, \quad \text{a.s.}$$

Definition 44.9 (Arbitrage). In a Black-Scholes market $\{(X_t^{(0)}, X_t) : 0 \leq t \leq T\}$ with trading in the American option $X^{(a)}$ with initial price U_0, there is arbitrage if either

(a) there is long arbitrage, that is, there exists a buy-and-hold strategy (c, τ) with $c \geq 0$ such that for some trading and consumption strategy (ϕ, A), the strategy $(\phi, A, c, \tau) \in \mathscr{A}_0$, or

(b) there is short arbitrage, that is, there exists a trading and consumption strategy (ϕ, A) such that for any buy-and-hold strategy (c, τ) with $c < 0$, the strategy $(\phi, A, c, \tau) \in \mathscr{A}_0$.

Definition 44.10 (Essential Supremum). Let $\mathcal{X}$ be a nonempty family of nonnegative random variables defined on a probability space $(\Omega, \mathscr{F}, P)$. The essential supremum of $\mathcal{X}$, denoted by $\operatorname{ess\,sup} \mathcal{X}$, is a random variable X_* satisfying the following conditions:

(a) For every $X \in \mathcal{X}$, $X \leq X_*$ a.s.

(b) If Y is a random variable such that $X \leq Y$ a.s. for all $X \in \mathcal{X}$, then $X_* \leq Y$ a.s.

Theorem 44.1 (Existence of Essential Supremum). *Let $\mathcal{X}$ be a nonempty family of nonnegative random variables. Then*

(a) The essential supremum of $\mathcal{X}$ exists.

(b) If $\mathcal{X}$ is closed under pairwise maximization (i.e., $X, Y \in \mathcal{X}$ implies $X \vee Y \in \mathcal{X}$), then there exists a nondecreasing sequence $\{Z_n\}_{n\geq 1}$ of random variables in $\mathcal{X}$ such that

$$\operatorname{ess\,sup} \mathcal{X} = \lim_{n\to\infty} Z_n, \quad a.s.$$

Theorem 44.2 (Snell Envelope). *Let $T \in (0, \infty]$. Let $\{\mathscr{F}_t : 0 \leq t \leq T\}$ be a right-continuous filtration such that $\mathscr{F}_0$ contains only sets of probability zero or one, that $\mathscr{F}_\infty = \sigma(\bigcup_{0\leq t<\infty} \mathscr{F}_t)$ if $T = \infty$, and that for every $t \in [0, T]$, $\mathscr{F}_t$ contains the null sets of $\mathscr{F}_T$. Let $\{Y_t : 0 \leq t \leq T\}$ be a nonnegative right-continuous stochastic process adapted to $\{\mathscr{F}_t : 0 \leq t \leq T\}$ such that*

$$E\left[\sup_{0\leq t\leq T} Y_t\right] < \infty.$$

Let $\{U_t : 0 \leq t \leq T\}$ be a process defined as

$$U_t = \operatorname*{ess\,sup}_{\rho\in\mathcal{T}_{t,T}} E[Y_\rho|\mathscr{F}_t],$$

where $\mathcal{T}_{t,T}$ is the set of $\{\mathscr{F}_t : 0 \leq t \leq T\}$-stopping times with values in $[t, T]$ a.s. Then

(a) The process $\{U_t : 0 \leq t \leq T\}$ is a supermartingale with respect to the filtration $\{\mathscr{F}_t : 0 \leq t \leq T\}$.

(b) For every $t \in [0, T]$, we have

$$E[U_t] = \sup_{\rho\in\mathcal{T}_{t,T}} E[Y_\rho].$$

(c) The process $\{U_t : 0 \leq t \leq T\}$ has a Càdlàg (see Definition 21.4) modification, which is called the Snell envelope of $\{Y_t : 0 \leq t \leq T\}$.

(d) The Snell envelope of $\{Y_t : 0 \leq t \leq T\}$ is the smallest supermartingale that dominates $\{Y_t : 0 \leq t \leq T\}$ (see Definition 21.12).

Theorem 44.3 (Optimal Stopping Time). *Let $T \in (0, \infty]$. Let $\{Y_t : 0 \leq t \leq T\}$ be a nonnegative stochastic process adapted to some filtration $\{\mathscr{F}_t : 0 \leq t \leq T\}$. Suppose that $\{Y_t : 0 \leq t \leq T\}$ and $\{\mathscr{F}_t : 0 \leq t \leq T\}$ satisfy the assumptions given in Theorem 44.2 and that almost all the sample paths of $\{Y_t : 0 \leq t \leq T\}$ are continuous. Let*

$$\rho_*(t) = \inf\{s \in [t, T] : U_s = Y_s\}, \quad t \in [0, T],$$

where $\{U_t : 0 \leq t \leq T\}$ *is the Snell envelope of* $\{Y_t : 0 \leq t \leq T\}$. *Then*

$$E[Y_{\rho_*(t)}|\mathscr{F}_t] = U_t = \operatorname*{ess\,sup}_{\rho\in\mathcal{T}_{t,T}} E[Y_\rho|\mathscr{F}_t], \quad a.s.$$

Theorem 44.4 (Snell Envelope of the American Put Option). *Let* $\{(X_t^{(0)}, X_t) : 0 \leq t \leq T\}$ *be the Black-Scholes market given in Definition 42.2, and let* Q *be the risk-neutral measure defined in Problem 37.7. Let* $\{J_t : 0 \leq t \leq T\}$ *be the Snell envelope of the discount reward of an American put option with strike* K *and maturity* T*:*

$$J_t = \operatorname*{ess\,sup}_{\rho\in\mathcal{T}_{t,T}} E_Q\left[e^{-r\rho}(K - X_\rho)^+|\mathscr{F}_t\right], \quad t \in [0, T], \ a.s.$$

Let $P(x, t)$ *be a function on* $(0, \infty) \times [0, T)$ *defined as*

$$P(x,t) = \sup_{\rho\in\mathcal{T}_{t,T}} E_Q\left[e^{-r(\rho-t)}(K - X_\rho)^+|X_t = x\right]. \tag{44.1}$$

Then the Snell envelope admits the following decomposition

$$J_t = E_Q\left[e^{-rT}(K - X_T)^+|\mathscr{F}_t\right] + E_Q\left[\int_t^T e^{-rs} rKI_{\{X_s<X_s^*\}}\mathrm{d}s|\mathscr{F}_t\right],$$

where $X_t^* = \sup\{x : (x,t) \in \mathcal{D}\}$ *with*

$$\mathcal{D} = \left\{(x,t) \in (0,\infty) \times [0,T) : P(x,t) = (K-x)^+\right\}.$$

44.2 Problems

44.1. Let $\{Y_t : 0 \leq t \leq T\}$ be a nonnegative stochastic process adapted to some filtration $\{\mathscr{F}_t : 0 \leq t \leq T\}$, where $T \in (0, \infty]$. Suppose that $\{Y_t : 0 \leq t \leq T\}$ and $\{\mathscr{F}_t : 0 \leq t \leq T\}$ satisfy the assumptions given in Theorem 44.2. Let $v \in \mathcal{T}_{0,T}$ and $\tau \in \mathcal{T}_{v,T}$, where $\mathcal{T}_{v,T} = \{u \in \mathcal{T}_{0,T} : u \geq v \ a.s.\}$. Show that

(a) The set $\{E[Y_\rho|\mathscr{F}_v] : \rho \in \mathcal{T}_{\tau,T}\}$ is closed under pairwise maximization, where $\mathcal{T}_{\tau,T} = \{u \in \mathcal{T}_{0,T} : u \geq \tau \ a.s.\}$.

(b) There exists a sequence $\{\rho_n\}_{n\geq1}$ of stopping times in $\mathcal{T}_{\tau,T}$ such that $E[Y_{\rho_n}|\mathscr{F}_v]$ is nondecreasing and

$$\operatorname*{ess\,sup}_{\rho\in\mathcal{T}_{\tau,T}} E[Y_\rho|\mathscr{F}_v] = \lim_{n\to\infty} E[Y_{\rho_n}|\mathscr{F}_v], \quad \text{a.s.}$$

44.2. Let $\{Y_t : 0 \leq t \leq T\}$ be a nonnegative stochastic process adapted to some filtration $\{\mathscr{F}_t : 0 \leq t \leq T\}$, where $T \in (0, \infty]$. Suppose that $\{Y_t : 0 \leq t \leq T\}$ and $\{\mathscr{F}_t : 0 \leq t \leq T\}$ satisfy the assumptions given in Theorem 44.2. For any $w \in \mathcal{T}_{0,T}$, define

$$U_w = \operatorname*{ess\,sup}_{\rho\in\mathcal{T}_{w,T}} E[Y_\rho|\mathscr{F}_w].$$

Let $v, u \in \mathcal{T}_{0,T}$ and $\tau \in \mathcal{T}_{v,T}$. Let $A = \{v = u\}$. Show that

(a)

$$I_A E[Y_{\tau_A}|\mathscr{F}_v] = I_A E[Y_{\tau_A}|\mathscr{F}_{v \wedge u}] = I_A E[Y_{\tau_A}|\mathscr{F}_u],$$

where $\tau_A = \tau I_A + T I_{A^c}$;

(b) $U_v = U_u$ a.s. on $\{v = u\}$.

(c) $E[U_\tau|\mathscr{F}_v] = \operatorname{ess\,sup}_{\rho \in \mathcal{T}_{\tau,T}} E[Y_\rho|\mathscr{F}_v]$ a.s.

(d) $E[U_\tau|\mathscr{F}_v] \leq U_v$ a.s.

(e) $E[U_\tau] = \sup_{\rho \in \mathcal{T}_{\tau,T}} E[Y_\rho] \leq U_0 < \infty$.

44.3. Let $\{Y_t : 0 \leq t \leq T\}$ be a nonnegative stochastic process adapted to some filtration $\{\mathscr{F}_t : 0 \leq t \leq T\}$, where $T \in (0, \infty]$. Suppose that $\{Y_t : 0 \leq t \leq T\}$ and $\{\mathscr{F}_t : 0 \leq t \leq T\}$ satisfy the assumptions given in Theorem 44.2. Let $\{U_t : 0 \leq t \leq T\}$ be the Snell envelope of $\{Y_t : 0 \leq t \leq T\}$. Show that a stopping time $\tau_* \in \mathcal{T}_{0,T}$ is optimal, specifically,

$$E[Y_{\tau_*}] = U_0 = \sup_{\tau \in \mathcal{T}_{0,T}} E[Y_\tau],$$

if and only if

(a) $U_{\tau_*} = Y_{\tau_*}$ a.s., and

(b) The stopped process $\{U_{t \wedge \tau_*} : 0 \leq t \leq T\}$ is a martingale.

44.4. Let $\{(X_t^{(0)}, X_t) : 0 \leq t \leq T\}$ be the Black-Scholes market given in Definition 42.2, and let Q be the risk-neutral measure defined in Problem 37.7. Let $\{V_t : 0 \leq t \leq T\}$ be a process defined as

$$V_t = \operatorname*{ess\,sup}_{\rho \in \mathcal{T}_{t,T}} E_Q\left[e^{-r(\rho - t)}(K - X_\rho)^+|\mathscr{F}_t\right], \quad t \in [0, T],$$

where $K > 0$ is a constant. Show that $\{V_t : 0 \leq t \leq T\}$ is a wealth process, that is, there exists an admissible trading and consumption strategy (ϕ, A) corresponding to V_t.

44.5 **(Arbitrage-Free Price of the American Put Option).** Let $\{(X_t^{(0)}, X_t) : 0 \leq t \leq T\}$ be the Black-Scholes market given in Definition 42.2 and let Q be the risk-neutral measure defined in Problem 37.7. Let $X^{(a)}$ be an American put option with strike K and maturity T. Then the arbitrage-free price at time zero of $X^{(a)}$ is

$$V_0 = \operatorname*{ess\,sup}_{\rho \in \mathcal{T}_{0,T}} E_Q\left[e^{-r\rho}(K - X_\rho)^+\right].$$

44.6. Let $\{B_t : t \geq 0\}$ be a standard Brownian motion under a probability measure P. Let $\{W_t : t \geq 0\}$ be a process defined as

$$W_t = \alpha t + B_t,$$

where $\alpha \in \mathbf{R}$. Let $\beta > 0$ and τ be a stopping time defined as

$$\tau = \begin{cases} \min\{t \geq 0 : W_t = \beta\}, & \text{if } \{t \geq 0 : W_t = \beta\} \neq \emptyset; \\ \infty, & \text{if } \{t \geq 0 : W_t = \beta\} = \emptyset. \end{cases}$$

For every $\lambda > 0$, show that

$$E_P\left[e^{-\lambda\tau}\right] = \exp\left(\alpha\beta - \beta\sqrt{\alpha^2 + 2\lambda}\right).$$

44.7 (Perpetual American Put)**.** Let $\{(X_t^{(0)}, X_t) : 0 \leq t \leq T\}$ be the Black-Scholes market given in Definition 42.2 with $r > 0$. A person bought a perpetual American put option with strike $K > 0$. The person sets a positive level $L \in (0, K)$ and will exercise the option at the stopping time τ_L given by

$$\tau_L = \begin{cases} 0, & \text{if } X_0 \leq L; \\ \min\{t \geq 0 : X_t = L\}, & \text{if } X_0 > L. \end{cases}$$

Suppose that under this exercise strategy, the value of the perpetual American put is defined as

$$v_L(X_0) = E_Q\left[e^{-r\tau_L}(K - X_{\tau_L})\right] = \begin{cases} K - X_0, & \text{if } X_0 \leq L; \\ (K - L)E_Q\left[e^{-r\tau_L}\right], & \text{if } X_0 > L, \end{cases}$$

where Q is the risk-neutral measure defined in Problem 37.7. Show that

(a) The function $v_L(x)$ is given by

$$v_L(x) = \begin{cases} K - x, & \text{if } 0 < x \leq L; \\ (K - L)\left(\frac{x}{L}\right)^{-\frac{2r}{\sigma^2}}, & \text{if } x > L. \end{cases}$$

(b) Given $x > 0$, the function $v_L(x)$ is maximized at $L_* = \frac{2r}{\sigma^2 + 2r}K$:

$$v_{L_*}(x) \geq v_L(x), \quad \forall L \in (0, K).$$

(c) Let $L_* = \frac{2r}{\sigma^2+2r}K$. show that $v_{L_*}(x)$ satisfies the following linear complementarity conditions:

$$v(x) \geq (K - x)^+, \quad \forall x \geq 0, \tag{44.2a}$$

$$rv(x) - rxv'(x) - \frac{1}{2}\sigma^2 x^2 v''(x) \geq 0, \quad \forall x \geq 0, \tag{44.2b}$$

and for each $x \geq 0$, equality holds in either Equation (44.2a) or Equation (44.2b). Here $v'(x) = \frac{dv(x)}{dx}$ and $v''(x) = \frac{dv'(x)}{dx}$. If $v''(x)$ is undefined, then $v''(x)$ is replaced by its left-hand or right-hand limit.

(d) The process $\{Y_t : t \geq 0\}$ is a supermartingale under Q, where

$$Y_t = e^{-rt} v_{L_*}(X_t), \quad t \geq 0.$$

Moreover, the stopped process $\{Y_{t \wedge \tau_{L_*}} : t \geq 0\}$ is a martingale under Q.

(e) The price of the perpetual American option is $v_{L_*}(X_0)$, where $\mathcal{T}$ is the set of all stopping times:

$$v_{L_*}(X_0) = \max_{\tau \in \mathcal{T}} E_Q\left[e^{-r\tau}(K - X_\tau)\right].$$

44.8 (American Call Option). Let $\{(X_t^{(0)}, X_t) : 0 \leq t \leq T\}$ be the Black-Scholes market given in Definition 42.2, and let Q be the risk-neutral measure defined in Problem 37.7. Let $Y^{(a)}$ be an American call option with strike K and maturity T. Show that

(a) The discounted intrinsic value process $\{e^{-rt}(X_t - K)^+ : 0 \leq t \leq T\}$ of the American call option is a submartingale under Q.

(b) The price of the American option is the same as the price of the European call option with strike K and maturity T.

44.3 Hints

44.1. To prove the first part, let $\rho_1, \rho_2 \in \mathcal{T}_{\tau,T}$ and consider $\rho_3 = \rho_1 I_A + \rho_2 I_{A^c}$, where

$$A = \{E[Y_{\rho_1}|\mathscr{F}_v] \geq E[Y_{\rho_2}|\mathscr{F}_v]\}.$$

The second part follows from the first part and Theorem 44.1.

44.2. To prove part (a), use the definition of conditional expectations (Definition 14.1) and the result of Problem 23.9. To prove part (b), use part (a) to show that

$$I_{\{v=u\}} E[Y_\rho|\mathscr{F}_v] \leq I_{\{v=u\}} U_u, \quad \text{a.s.} \quad \forall \rho \in \mathcal{T}_{v,T}$$

and

$$I_{\{v=u\}} E[Y_\rho|\mathscr{F}_u] \leq I_{\{v=u\}} U_v, \quad \text{a.s.} \quad \forall \rho \in \mathcal{T}_{u,T}.$$

Use the results of Problems 44.1 and 14.7 to prove part (c). Part (d) follows from part (c). Part (e) follows from parts (c) and (d).

44.3. To prove the "only if" part, note that $\{U_t : 0 \leq t \leq T\}$ dominates $\{Y_t : 0 \leq t \leq T\}$ (see Theorem 44.2) and $E[U_{\tau_*}] = E[Y_{\tau_*}]$.

44.4. Use Theorem 44.4 and the martingale representation theorem (Theorem 36.1).

44.5. Follow the definition of arbitrage and use the result of Problem 44.4 and Theorems 44.2 and 44.3.

44.6. Note that $\{\tau \geq y\} = \{\max_{0\leq t\leq y} W_t \leq \beta\}$ and then apply the result of Problem 43.3.

44.7. To prove part (a), use the result of Problem 44.6. To prove part (b), consider $\ln v_L(x)$ and the cases when $x \in (0, L_*)$ and $x \in (L_*, \infty)$. Part (c) can be proved by direct calculation. To prove part (d), apply Itô's formula (Theorem 35.3) and use part (c). To prove part (e), use the optimal stopping theorem (Theorem 23.1) and part (d).

44.8. To prove the first part, note that $(x-K)^+$ is a convex function of x (Definition 15.1) and use the result of Problem 22.7. The second part follows from the first part.

44.4 Solutions

44.1.

(a) Let $\rho_1, \rho_2 \in \mathcal{T}_{\tau,T}$ and

$$A = \{E[Y_{\rho_1}|\mathscr{F}_v] \geq E[Y_{\rho_2}|\mathscr{F}_v]\}.$$

Then $A \in \mathscr{F}_v$ (see Definition 14.1). Let $\rho_3 = \rho_1 I_A + \rho_2 I_{A^c}$. For every $s \in [0,T]$, since

$$\begin{aligned}
& \{\rho_1 I_A + \rho_2 I_{A^c} \leq s\} \\
= \; & (\{\rho_1 \leq s\} \cap A) \cup (\{\rho_2 \leq s\} \cap A^c) \\
= \; & (\{\rho_1 \leq s\} \cap [(A \cap \{v \leq s\}) \cup (A \cap \{v > s\})]) \cup \\
& (\{\rho_2 \leq s\} \cap [(A^c \cap \{v \leq s\}) \cup (A^c \cap \{v > s\})]) \\
\in \; & \mathscr{F}_s
\end{aligned}$$

and $\tau \leq \rho_3 \leq T$ a.s., we have $\rho_3 \in \mathcal{T}_{\tau,T}$. Since

$$\begin{aligned}
E[Y_{\rho_3}|\mathscr{F}_v] &= E[Y_{\rho_1} I_A + Y_{\rho_2} I_{A^c}|\mathscr{F}_v] \\
&= I_A E[Y_{\rho_1}|\mathscr{F}_v] + I_{A^c} E[Y_{\rho_2}|\mathscr{F}_v] \\
&= E[Y_{\rho_1}|\mathscr{F}_v] \vee E[Y_{\rho_2}|\mathscr{F}_v],
\end{aligned}$$

we know that the set $\{E[Y_\rho|\mathscr{F}_v] : \rho \in \mathcal{T}_{\tau,T}\}$ is closed under pairwise maximization.

(b) This follows from the first part and Theorem 44.1.

This completes the proof.

44.2.

(a) Let $X = E[Y_{\tau_A}|\mathscr{F}_{v\wedge u}]$. Then by Definition 14.1, we know that X is $\mathscr{F}_{v\wedge u}$-measurable and

$$\int_B X\mathrm{d}P = \int_B Y_{\tau_A}\mathrm{d}P, \quad \forall B \in \mathscr{F}_{v\wedge u}. \tag{44.3}$$

By Problem 23.9, X is also $\mathscr{F}_v$-measurable. Now let $C \in \mathscr{F}_v$. Since

$$A\cap C\cap\{u\le t\} = \{v=u\}\cap C\cap\{v\le t\}\in\mathscr{F}_t, \quad \forall t\in[0,T],$$

we have $A\cap C\in\mathscr{F}_u$. Hence $A\cap C\in\mathscr{F}_{v\wedge u}$. By Equation (44.3), we get

$$\int_{A\cap C} X\mathrm{d}P = \int_{A\cap C} Y_{\tau_A}\mathrm{d}P,$$

or

$$\int_C I_A X\mathrm{d}P = \int_C I_A Y_{\tau_A}\mathrm{d}P.$$

Since $I_A X$ is $\mathscr{F}_v$-measurable and C is an arbitrary element of $\mathscr{F}_v$, the above equation implies that

$$I_A X = E[I_A Y_{\tau_A}|\mathscr{F}_v] = I_A E[Y_{\tau_A}|\mathscr{F}_v],$$

which gives

$$I_A E[Y_{\tau_A}|\mathscr{F}_{v\wedge u}] = I_A E[Y_{\tau_A}|\mathscr{F}_v].$$

Similarly, we have $I_A E[Y_{\tau_A}|\mathscr{F}_{v\wedge u}] = I_A E[Y_{\tau_A}|\mathscr{F}_u]$.

(b) By part (a), we have for every $\rho\in\mathcal{T}_{v,T}$,

$$\begin{aligned} I_A E[Y_\rho|\mathscr{F}_v] &= I_A E[Y_{\rho_A}|\mathscr{F}_v] \\ &= I_A E[Y_{\rho_A}|\mathscr{F}_u] \\ &= I_A E[Y_\rho|\mathscr{F}_u] \\ &\le I_A U_u, \end{aligned}$$

where $\rho_A = \rho I_A + T I_{A^c}$. The above equation implies that $I_A U_v \le I_A U_u$. Similarly, we have $I_A U_u \le I_A U_v$. Hence $U_v = U_u$ a.s. on A.

(c) By Problem 44.1, there exists a sequence $\{\rho_n\}_{n\ge1}$ of stopping times in $\mathcal{T}_{\tau,T}$ such that $\{E[Y_{\rho_n}|\mathscr{F}_\tau]\}_{n\ge1}$ is nondecreasing and

$$U_\tau = \lim_{n\to\infty} E[Y_{\rho_n}|\mathscr{F}_\tau].$$

Then by Problem 14.13, we have

$$\begin{aligned} E[U_\tau|\mathscr{F}_v] &= \lim_{n\to\infty} E(E[Y_{\rho_n}|\mathscr{F}_\tau]|\mathscr{F}_v) \\ &= \lim_{n\to\infty} E[Y_{\rho_n}|\mathscr{F}_v] \\ &\le \operatorname*{ess\,sup}_{\rho\in\mathcal{T}_{\tau,T}} E[Y_\rho|\mathscr{F}_v]. \end{aligned} \tag{44.4}$$

On the other hand, we have $U_\tau \geq E[Y_\rho|\mathscr{F}_\tau]$ for every $\rho \in \mathcal{T}_{\tau,T}$. Hence

$$E[U_\tau|\mathscr{F}_v] \geq E(E[Y_\rho|\mathscr{F}_\tau]|\mathscr{F}_v) = E[Y_\rho|\mathscr{F}_v], \quad \forall \rho \in \mathcal{T}_{\tau,T},$$

which implies

$$E[U_\tau|\mathscr{F}_v] \geq \operatorname*{ess\,sup}_{\rho\in\mathcal{T}_{\tau,T}} E[Y_\rho|\mathscr{F}_v]. \tag{44.5}$$

The result follows from Equations (44.4) and (44.5).

(d) Since $\mathcal{T}_{\tau,T} \subseteq \mathcal{T}_{v,T}$, we have

$$\operatorname*{ess\,sup}_{\rho\in\mathcal{T}_{\tau,T}} E[Y_\rho|\mathscr{F}_v] \leq U_v.$$

Hence it follows from part (c) that

$$E[U_\tau|\mathscr{F}_v] \leq U_v.$$

(e) This follows from parts (c) and (d) by letting $v = 0$ and noting that $E[Y_\rho]$ is a constant for every $\rho \in \mathcal{T}_{\tau,T}$.

This completes the proof.

44.3. First let us prove the "only if" part. Suppose that τ_* is optimal. Then by Problem 44.2, we have

$$E[U_{\tau_*}] \leq U_0 = E[Y_{\tau_*}].$$

But by Theorem 44.2, $U_{\tau_*} \geq Y_{\tau_*}$ a.s. The above equation implies that $U_{\tau_*} = Y_{\tau_*}$ a.s. Also by Theorem 44.2, $\{U_t : 0 \leq t \leq T\}$ is a supermartingale. It follows from Theorem 23.1 that $\{U_{t\wedge\tau_*} : 0 \leq t \leq T\}$ is also a supermartingale. Since

$$U_0 = E[Y_{\tau_*}] = E[U_{\tau_*}],$$

we have

$$E[U_{T\wedge\tau_*}] = E[U_{\tau_*}] = E[E[U_{\tau_*}]] = E[U_0] = E[U_{0\wedge\tau_*}].$$

By Problem 22.12, $\{U_{t\wedge\tau_*} : 0 \leq t \leq T\}$ is a martingale.

Now we prove the "if" part. Suppose that $U_{\tau_*} = Y_{\tau_*}$ a.s. and the stopped process $\{U_{t\wedge\tau_*} : 0 \leq t \leq T\}$ is a martingale. Then we have

$$U_0 = E[U_{T\wedge\tau_*}] = E[U_{\tau_*}] = E[Y_{\tau_*}],$$

which shows that τ_* is optimal. This completes the proof.

44.4. Let $\{J_t : 0 \leq t \leq T\}$ be defined as

$$J_t = \operatorname*{ess\,sup}_{\rho\in\mathcal{T}_{t,T}} E_Q\left[e^{-r\rho}(K - X_\rho)^+|\mathscr{F}_t\right], \quad t \in [0,T], \text{ a.s.}$$

Then by Theorem 44.4, we have

$$J_t = E_Q\left[e^{-rT}(K - X_T)^+|\mathscr{F}_t\right] + E_Q\left[\int_t^T e^{-rs} rKI_{\{X_s < X_s^*\}}\mathrm{d}s|\mathscr{F}_t\right],$$

where X_t^* is as defined in Theorem 44.4. For every $t \in [0, T]$, let

$$M_t = E_Q\left[e^{-rT}(K - X_T)^+ + \int_0^T e^{-rs} rKI_{\{X_s < X_s^*\}}\mathrm{d}s|\mathscr{F}_t\right]$$

and

$$C_t = E_Q\left[\int_0^t e^{-rs} rKI_{\{X_s < X_s^*\}}\mathrm{d}s|\mathscr{F}_t\right] = \int_0^t e^{-rs} rKI_{\{X_s < X_s^*\}}\mathrm{d}s.$$

Then we have $J_t = M_t - C_t$. Since $\{M_t : 0 \le t \le T\}$ a bounded martingale under Q, it is a square-integrable martingale under Q. It follows from Theorem 36.1 that

$$\mathrm{d}M_t = \eta_t \mathrm{d}W_t, \quad t \in [0, T] \text{ a.s.},$$

where $W_t = \frac{\mu - r}{\sigma}t + B_t$ and $\{\eta_t : 0 \le t \le T\}$ is some progressively measurable process with

$$E\left[\int_0^T \eta_t^2 \mathrm{d}t\right] < \infty.$$

Now let (ϕ, A) be a trading and consumption strategy defined as

$$\theta_t^{(0)} = J_t - \frac{\eta_t}{\sigma}, \quad \theta_t = \frac{e^{rt}\eta_t}{\sigma X_t}, \quad A_t = \int_0^t e^{rs}\mathrm{d}C_s = \int_0^t rKI_{\{X_s < X_s^*\}}\mathrm{d}s.$$

Then we have

$$\theta_t^{(0)} X_t^{(0)} + \theta_t X_t = \left(J_t - \frac{\eta_t}{\sigma}\right)e^{rt} + \frac{e^{rt}\eta_t}{\sigma X_t}X_t = e^{rt}J_t = V_t$$

and

$$\theta_t^{(0)}\mathrm{d}X_t^{(0)} + \theta_t \mathrm{d}X_t - \mathrm{d}A_t = re^{rt}J_t\mathrm{d}t + e^{rt}\mathrm{d}M_t - e^{rt}\mathrm{d}C_t = \mathrm{d}(e^{rt}J_t) = \mathrm{d}V_t.$$

Hence (ϕ, A) is an admissible trading and consumption strategy and $\{V_t : 0 \le t \le T\}$ is a wealth process. This completes the proof.

44.5. Let V_t be defined as

$$V_t = \operatorname*{ess\,sup}_{\rho \in \mathcal{T}_{t,T}} E_Q\left[e^{-r(\rho - t)}(K - X_\rho)^+|\mathscr{F}_t\right], \quad t \in [0, T].$$

Then by Problem 44.4, $\{V_t : 0 \le t \le T\}$ is a wealth process and there exists an admissible trading and consumption strategy $(\phi, A) = (\theta^{(0)}, \theta, A)$ corresponding to it. Let $X_0^{(a)}$ be the price at time zero of the American put option. If $X_0^{(a)} > V_0$, we show that there is short arbitrage. To do that, let $\tau \in \mathcal{T}_{0,T}$ be an exercise policy selected by the buyer of the American put option. We consider the following trading and consumption strategy $(\beta, C) = (\gamma^{(0)}, \gamma, C)$:

$$\gamma_t^{(0)} = \begin{cases} \theta_t^{(0)}, & \text{if } t \in [0,\tau]; \\ \theta_\tau^{(0)} + e^{-r\tau}\theta_\tau X_\tau - e^{-r\tau}(K - X_\tau)^+, & \text{if } t \in (\tau, T], \end{cases}$$

$$\gamma_t = \theta_t I_{[0,\tau]}(t), \quad C_t = A_{t\wedge\tau}.$$

Since $\{e^{-rt}V_t : 0 \le t \le T\}$ dominates $\{e^{-rt}(K - X_t)^+ : 0 \le t \le T\}$ (see Theorem 44.2), we have

$$\gamma_T^{(0)} X_T^{(0)} = (V_\tau e^{-r\tau} - e^{-r\tau}(K - X_\tau)^+)e^{rT} \ge 0 \quad \text{a.s.}$$

Also we have

$$\gamma_0^{(0)} + \gamma_0 X_0 - X_0^{(a)} = V_0 - X_0^{(0)} < 0.$$

Hence the extended strategy $(\beta, C, -1, \tau) \in \mathscr{A}_0$.

If $X_0^{(a)} < V_0$, we show that there is long arbitrage. Let ρ_0 be the optimal stopping time (see Theorem 44.3). We consider the following trading and consumption strategy $(\beta, C) = (\gamma^{(0)}, \gamma, C)$:

$$\gamma_t^{(0)} = \begin{cases} -\theta_t^{(0)}, & \text{if } t \in [0,\rho_0]; \\ -\theta_{\rho_0}^{(0)} - e^{-r\rho_0}\theta_{\rho_0} X_{\rho_0} + e^{-r\rho_0}(K - X_{\rho_0})^+, & \text{if } t \in (\rho_0, T], \end{cases}$$

$$\gamma_t = -\theta_t I_{[0,\rho_0]}(t), \quad C_t = -A_{t\wedge\rho_0}.$$

By Theorem 44.3, we have

$$\gamma_T^{(0)} X_T^{(0)} = (V_{\rho_0} e^{-r\rho_0} - e^{-r\rho_0}(K - X_{\rho_0})^+)e^{rT} = 0 \quad \text{a.s.}$$

Also we have

$$\gamma_0^{(0)} + \gamma_0 X_0 - X_0^{(a)} = -V_0 + X_0^{(0)} < 0.$$

Hence the extended strategy $(\beta, C, 1, \rho_0) \in \mathscr{A}_0$. This completes the proof.

44.6. Since $\{\tau \ge y\} = \{\max_{0\le t\le y} W_t \le \beta\}$, we have

$$P\{\tau < y\} = 1 - P\{\tau \ge y\} = 1 - P\left\{\max_{0\le t\le y} W_t \le \beta\right\}.$$

By Problem 43.3, we have

$$P\{\tau < y\} = 1 - N\left(\frac{\beta - \alpha y}{\sqrt{y}}\right) + e^{2\alpha\beta} N\left(\frac{-\beta - \alpha y}{\sqrt{y}}\right),$$

which gives the density of τ:

$$f_\tau(y) = \frac{\mathrm{d}P\{\tau < y\}}{\mathrm{d}y} = \frac{\beta}{\sqrt{2\pi}} \exp\left(-\frac{1}{2}\left(\beta y^{-\frac{1}{2}} - \alpha y^{\frac{1}{2}}\right)^2\right) y^{-\frac{3}{2}}.$$

Then we have

$$\begin{aligned}
& E_P\left[e^{-\lambda\tau}\right] \\
= & \int_0^\infty e^{-\lambda y} f_\tau(y)\mathrm{d}y \\
= & \int_0^\infty \frac{\beta}{\sqrt{2\pi}} \exp\left(-\frac{1}{2}\left(\beta y^{-\frac{1}{2}} - \alpha y^{\frac{1}{2}}\right)^2 - \lambda y\right) y^{-\frac{3}{2}}\mathrm{d}y \\
= & e^{\alpha\beta - \beta\sqrt{\alpha^2+2\lambda}} \int_0^\infty \frac{\beta}{\sqrt{2\pi}} e^{-\frac{1}{2}\left(\beta y^{-\frac{1}{2}} - \sqrt{\alpha^2+2\lambda}y^{\frac{1}{2}}\right)^2} y^{-\frac{3}{2}}\mathrm{d}y. \qquad (44.6)
\end{aligned}$$

To evaluate the integral, we let

$$g_1 = \int_0^\infty \frac{1}{\sqrt{2\pi}} e^{-\frac{1}{2}\left(\beta y^{-\frac{1}{2}} - \sqrt{\alpha^2+2\lambda}y^{\frac{1}{2}}\right)^2} y^{-\frac{3}{2}}\mathrm{d}y$$

and

$$g_0 = \int_0^\infty \frac{1}{\sqrt{2\pi}} e^{-\frac{1}{2}\left(\beta y^{-\frac{1}{2}} - \sqrt{\alpha^2+2\lambda}y^{\frac{1}{2}}\right)^2} y^{-\frac{1}{2}}\mathrm{d}y.$$

Then we have

$$\begin{aligned}
& -\frac{\beta}{2}g_1 - \frac{\sqrt{\alpha^2+2\lambda}}{2}g_0 \\
= & \int_0^\infty \frac{1}{\sqrt{2\pi}} e^{-\frac{1}{2}\left(\beta y^{-\frac{1}{2}} - \sqrt{\alpha^2+2\lambda}y^{\frac{1}{2}}\right)^2} \mathrm{d}\left(\beta y^{-\frac{1}{2}} - \sqrt{\alpha^2+2\lambda}y^{\frac{1}{2}}\right).
\end{aligned}$$

We use a change of variable $z = \beta y^{-\frac{1}{2}} - \sqrt{\alpha^2+2\lambda}y^{\frac{1}{2}}$. Since

$$\frac{\mathrm{d}z}{\mathrm{d}y} = -\frac{\beta}{2}y^{-\frac{3}{2}} - \frac{1}{2}\sqrt{\alpha^2+2\lambda}y^{-\frac{1}{2}} < 0,$$

we know that z is a monotonically decreasing function of y. Hence the above equation becomes

$$-\frac{\beta}{2}g_1 - \frac{\sqrt{\alpha^2+2\lambda}}{2}g_0 = \int_\infty^{-\infty} \frac{1}{\sqrt{2\pi}} e^{-\frac{1}{2}z^2}\mathrm{d}z = -1. \qquad (44.7)$$

Also, we have

$$\begin{aligned}
& e^{2\beta\sqrt{\alpha^2+2\lambda}}\left(-\frac{\beta}{2}g_1 + \frac{\sqrt{\alpha^2+2\lambda}}{2}g_0\right) \\
= & \int_0^\infty \frac{1}{\sqrt{2\pi}} e^{-\frac{1}{2}\left(\beta y^{-\frac{1}{2}} + \sqrt{\alpha^2+2\lambda}y^{\frac{1}{2}}\right)^2} \mathrm{d}\left(\beta y^{-\frac{1}{2}} + \sqrt{\alpha^2+2\lambda}y^{\frac{1}{2}}\right).
\end{aligned}$$

We also use a change of variable $z = \beta y^{-\frac{1}{2}} + \sqrt{\alpha^2+2\lambda}y^{\frac{1}{2}}$. In this case, z is a monotonically decreasing function of y when $y \in (0, \frac{\beta}{\sqrt{\alpha^2+2\lambda}})$ and z is a monotonically increasing function of y when $y \in (\frac{\beta}{\sqrt{\alpha^2+2\lambda}}, \infty)$. Therefore the above equation becomes

$$
\begin{aligned}
& e^{2\beta\sqrt{\alpha^2+2\lambda}}\left(-\frac{\beta}{2}g_1 + \frac{\sqrt{\alpha^2+2\lambda}}{2}g_0\right) \\
= & \int_{\infty}^{\frac{\beta}{\sqrt{\alpha^2+2\lambda}}} \frac{1}{\sqrt{2\pi}}e^{-\frac{1}{2}z^2}\mathrm{d}z + \int_{\frac{\beta}{\sqrt{\alpha^2+2\lambda}}}^{\infty} \frac{1}{\sqrt{2\pi}}e^{-\frac{1}{2}z^2}\mathrm{d}z \\
= & \ 0. \qquad (44.8)
\end{aligned}
$$

Combining Equations (44.7) and (44.8) gives $\beta g_1 = 1$. The result follows by plugging this into Equation (44.6). This completes the proof.

44.7.

(a) We only need to show that for $X_0 > L$,

$$E_Q\left[e^{-r\tau_L}\right] = \left(\frac{X_0}{L}\right)^{-\frac{2r}{\sigma^2}}. \qquad (44.9)$$

Let $W_t = \frac{\mu - r}{\sigma}t + B_t$ for $t \geq 0$. Then $\{W_t : t \geq 0\}$ be a Brownian motion under Q and $X_t = X_0 \exp\left((r - \frac{1}{2}\sigma^2)t + \sigma W_t\right)$. Hence we have

$$\tau_L = \min\left\{t \geq 0 : -\alpha t - W_t = \frac{1}{\sigma}\ln\frac{X_0}{L}\right\},$$

where $\alpha = \frac{1}{\sigma}(r - \frac{1}{2}\sigma^2)$. Note that $-W_t$ is also a Brownian motion under Q. By Problem 44.6, we have

$$E_Q\left[e^{-r\tau_L}\right] = e^{\frac{1}{\sigma}\ln\frac{X_0}{L}(-\alpha - \sqrt{\alpha^2+2r})} = \left(\frac{X_0}{L}\right)^{-\frac{2r}{\sigma^2}},$$

which verifies Equation (44.9).

(b) Let $f(L) = \ln v_L(x)$. By part (a), we have

$$f'(L) = \frac{\mathrm{d}f(L)}{\mathrm{d}L} = \begin{cases} 0, & \text{if } 0 < x < L; \\ \frac{2rK - (\sigma^2+2r)L}{(K-L)L}, & \text{if } x > L. \end{cases}$$

If $x \in (0, L_*)$, then we have $f'(L) > 0$ for $L < x$; that is, $f(L)$ is increasing in the interval $(0, x)$. In this case, $f(L)$ attains its maximum at x. Hence

$$v_{L_*}(x) = f(L_*) = K - x = f(x) \geq f(L) = v_L(x), \quad \forall L \in (0, K).$$

If $x \in (L_*, \infty)$, then $f(L)$ is increasing in the interval $(0, L_*)$ and decreasing in the interval (L_*, x). In this case, we also have

$$v_{L_*}(x) = f(L_*) \geq f(L) = v_L(x), \quad \forall L \in (0, K).$$

(c) By part (a), we have

$$v'_{L_*}(x) = \begin{cases} -1, & \text{if } 0 < x \le L_*; \\ -\frac{2r}{\sigma^2 x}(K - L_*)\left(\frac{x}{L_*}\right)^{-\frac{2r}{\sigma^2}}, & \text{if } x > L_*. \end{cases}$$

The first derivative $v'(x)$ is continuous at L_* since both the left-hand and right-hand limits of $v(x)$ at L_* are equal to -1. Differentiating the above equation gives

$$v''_{L_*}(x) = \begin{cases} 0, & \text{if } 0 < x < L_*; \\ \frac{2r(\sigma^2+2r)}{\sigma^4 x^2}(K - L_*)\left(\frac{x}{L_*}\right)^{-\frac{2r}{\sigma^2}}, & \text{if } x > L_*. \end{cases}$$

The second derivative $v''(x)$ is undefined at L_* as $v''(L_*-) = 0$ and $v''(L_*+) > 0$.

When $x \in (0, L_*]$, we have $v_{L_*}(x) = K - x = (K - x)^+$ and

$$\begin{aligned} & rv_{L_*}(x) - rxv'_{L_*}(x) - \frac{1}{2}\sigma^2 x^2 v''_{L_*}(x) \\ = \; & r(K - x) + rx \\ = \; & rK. \end{aligned}$$

When $x \in (L_*, \infty)$, we have $v_{L_*}(x) = (K - L_*)\left(\frac{x}{L_*}\right)^{-\frac{2r}{\sigma^2}} > 0$ and

$$\begin{aligned} & rv_{L_*}(x) - rxv'_{L_*}(x) - \frac{1}{2}\sigma^2 x^2 v''_{L_*}(x) \\ = \; & r(K - L_*)\left(\frac{x}{L_*}\right)^{-\frac{2r}{\sigma^2}} + rx\frac{2r}{\sigma^2 x}(K - L_*)\left(\frac{x}{L_*}\right)^{-\frac{2r}{\sigma^2}} - \\ & \frac{1}{2}\sigma^2 x^2 \frac{2r(\sigma^2 + 2r)}{\sigma^4 x^2}(K - L_*)\left(\frac{x}{L_*}\right)^{-\frac{2r}{\sigma^2}} \\ = \; & 0. \end{aligned}$$

Hence $v_{L_*}(x)$ satisfies the linear complementarity conditions.

(d) Note that $\mathrm{d}X_t = rX_t\mathrm{d}t + \sigma X_t \mathrm{d}W_t$. By Theorem 35.3 and part (c), we have

$$\begin{aligned} \mathrm{d}Y_t &= \mathrm{d}e^{-rt}v_{L_*}(X_t) \\ &= e^{-rt}\sigma X_t v'_{L_*}(X_t)\mathrm{d}W_t - \\ & \quad e^{-rt}\left[rv_{L_*}(X_t) - rX_t v'_{L_*}(X_t) - \frac{1}{2}\sigma^2 X_t^2 v''_{L_*}(X_t)\right]\mathrm{d}t \\ &= e^{-rt}\sigma X_t v'_{L_*}(X_t)\mathrm{d}W_t - rKe^{-rt}I_{(0,L_*)}(X_t)\mathrm{d}t, \end{aligned}$$

or

$$Y_t = Y_0 + \int_0^t e^{-rs}\sigma X_s v'_{L_*}(X_s)\mathrm{d}W_s - rK\int_0^t e^{-rs}I_{(0,L_*)}(X_s)\mathrm{d}s. \quad (44.10)$$

By Theorem 32.2, we have

$$\left\{ \int_0^t e^{-rs}\sigma X_s v'_{L_*}(X_s)\mathrm{d}W_s : t \geq 0 \right\}$$

is a martingale, and since $rK > 0$, that $\{Y_t : t \geq 0\}$ is a supermartingale.

If $X_0 > L_*$, then $X_s > L_*$ during the period $(0, t \wedge \tau_{L_*})$. Hence Equation (44.10) gives

$$Y_{t\wedge\tau_{L_*}} = Y_0 + \int_0^{t\wedge\tau_{L_*}} e^{-rs}\sigma X_s v'_{L_*}(X_s)\mathrm{d}W_s. \tag{44.11}$$

If $X_0 \leq L_*$, we have $\tau_{L_*} = 0$. The above equation still holds. It follows from the optimal sampling theorem (Theorem 23.2) that $\{Y_{t\wedge\tau_{L_*}} : t \geq 0\}$ is a martingale.

(e) By part (d), the process $\{e^{-rt}v_{L_*}(X_t) : t \geq 0\}$ is a supermartingale. For any stopping time τ, by Theorem 23.1 we have

$$v_{L_*}(X_0) \geq E\left[e^{-r\tau}v_{L_*}(X_\tau)\right] \geq E\left[e^{-r\tau}(K - X_\tau)\right].$$

But by the definition of $v_L(X_0)$, we have

$$v_{L_*}(X_0) = E\left[e^{-r\tau_{L*}}(K - X_{\tau_{L_*}})\right].$$

Hence $v_{L_*}(X_0)$ is the price of the perpetual American put option.

This completes the proof.

44.8.

(a) Let $0 \leq s \leq t \leq T$. We need to show that

$$E_Q\left[e^{-rt}(X_t - K)^+ | \mathscr{F}_s\right] \geq e^{-rs}(X_s - K)^+. \tag{44.12}$$

Since $h(x) = (x - K)^+$ is a convex function of x, for $\lambda \in [0, 1]$ and $0 \leq x_1 \leq x_2$, we have

$$h((1-\lambda)x_1 + \lambda x_2) \leq (1-\lambda)h(x_1) + \lambda h(x_2).$$

Letting $x_1 = 0$, $x_2 = X_t$, and $\lambda = e^{-r(t-s)}$ in the above equation gives

$$\left(e^{-r(t-s)}X_t - K\right)^+ \leq e^{-r(t-s)}(X_t - K)^+$$

or

$$\left(e^{-rt}X_t - Ke^{-rs}\right)^+ \leq e^{-rt}(X_t - K)^+$$

Note that $\{e^{-rt}X_t : 0 \le t \le T\}$ is a martingale under Q. From the above equation and Problem 22.7, we have

$$\begin{aligned} E_Q\left[e^{-rt}(X_t - K)^+|\mathscr{F}_s\right] &\ge E_Q\left[\left(e^{-rt}X_t - Ke^{-rs}\right)^+|\mathscr{F}_s\right] \\ &\ge \left(e^{-rs}X_s - Ke^{-rs}\right)^+ \\ &= e^{-rs}(X_s - K)^+, \end{aligned}$$

from which Equation (44.12) follows.

(b) By Problem 42.4, the price at time t of the European call option with strike K and maturity T is given by

$$E_Q\left[e^{-r(T-t)}(X_T - K)^+|\mathscr{F}_t\right].$$

But by the first item, we have

$$E_Q\left[e^{-r(T-t)}(X_T - K)^+|\mathscr{F}_t\right] \ge e^{rt}e^{-rt}(X_t - K)^+ = (X_t - K)^+,$$

which shows that the price of the European call option always dominates the intrinsic value of the American call option. Hence the option to exercise early is worthless. Thus the price of the American call option agrees with the price of the European call option.

This completes the proof.

44.5 Bibliographic Notes

The pricing problem of the American option can be formulated in different ways (Barone-Adesi, 2005; Wilmott et al., 1995). In this chapter, we presented the pricing problem in terms of optimal stopping. For more information about optimal stopping and American options, readers are referred to Bensoussan (1984), Karatzas (1988), Jacka (1991), and Myneni (1992). For a survey of the results on the pricing of American options, readers are referred to Barone-Adesi (2005) and Ahn et al. (2011). The American option is also studied in Briys et al. (1998), Oksendal (1998), Karatzas and Shreve (2001), Elliott and Kopp (2005), and Musiela and Rutkowski (2005).

Analytic solutions of the pricing models for American options are not available. Numerical methods for pricing American options can be found in Broadie and Glasserman (1997), Longstaff and Schwartz (2001), Topper (2005), Chan and Wong (2006), Seydel (2006), Duffy (2006), and Korn et al. (2010). Amin and Khanna (1994) proved the convergence of the binomial prices of the American option.

Although there exist no analytic pricing formulas for American options, approximate analytic pricing formulas for American options have been developed. For such approximate pricing formulas, readers are referred to Barone-Adesi and Whaley (1987), Kim (1990), and Haug (2006).

For more information about the essential supremum of a family of random variables, readers are referred to Karatzas and Shreve (2001, Appendix A), where a proof of Theorem 44.1 can be found. Proofs of Theorems 44.2 and 44.3 can be found in Karatzas and Shreve (2001, Appendix D). A proof of Theorem 44.4 can be found in Myneni (1992, Theorem 3.2).

CHAPTER 45

SHORT RATE MODELS

Short rate models are the earliest stochastic interest rate models. In this chapter, we present several short rate models and relevant concepts.

45.1 Basic Concepts and Facts

Definition 45.1 (Zero-Coupon Bond)**.** A zero-coupon bond with maturity date T is a contract that guarantees the holder a cash payment of one unit on the date T. Zero-coupon bonds are also referred to as *T-bonds*. The price at time t of a zero-coupon bond maturing at T is denoted by $P(t, T)$.

Definition 45.2 (Spot Rate and Short Rate)**.** Let $0 \leq t < T$. The simple spot rate for $[t, T]$ is defined to be

$$F(t,T) = \frac{1}{T-t}\left(\frac{1}{P(t,T)} - 1\right).$$

The continuously compounded spot rate for $[t, T]$ is defined to be

$$R(t,T) = \frac{-\ln P(t,T)}{T-t}.$$

The function $T \to R(t,T)$ is called the *zero-coupon yield curve* or *yield curve*. The instantaneous short rate at time t is defined as

$$r(t) = \lim_{T \downarrow t} R(t,T).$$

Definition 45.3 (Bank Account). A bank account or money-market account represents a risk-free investment. The value $\beta(t)$ at time t of a bank account is defined as

$$\beta(t) = e^{\int_0^t r(s)\mathrm{d}s},$$

where $r(s)$ is the short rate at time s.

Definition 45.4 (Discount Factor). Let $0 \le t \le T$. The (stochastic) discount factor $D(t,T)$ between time t and time T is defined as

$$D(t,T) = \frac{\beta(t)}{\beta(T)} = e^{-\int_t^T r(s)\mathrm{d}s}.$$

Definition 45.5 (Short Rate Model). In general, a short rate model has the following form

$$\mathrm{d}r(t) = b(t)\mathrm{d}t + \sum_{j=1}^{n} \sigma_j(t)\mathrm{d}B_t^{(j)},$$

where $\{(B_t^{(1)}, B_t^{(2)}, \ldots, B_t^{(n)}) : t \ge 0\}$ is an n-dimensional Brownian motion on some filtered probability space $(\Omega, \mathscr{F}, \{\mathscr{F}_t : t \ge 0\}, P)$, and $\{b(t) : t \ge 0\}$ and $\{\sigma_j(t) : t \ge 0\}$ are adapted processes that satisfy standard conditions (see Theorem 38.1) required to guarantee the existence of the short rate process.

Definition 45.6 (Equivalent Martingale Measure). Let $(\Omega, \mathscr{F}, P)$ be the real world probability space for the bond market. A probability measure Q defined on $(\Omega, \mathscr{F})$ is called an *equivalent martingale measure* if

(a) $Q \sim P$.

(b) For every $T > 0$, the discounted price process $\left\{ \frac{P(t,T)}{\beta(t)} : 0 \le t \le T \right\}$ is a martingale.

In this definition, we can restrict $T \in [0, T^*]$ for some upper bound T^*. An equivalent martingale measure is also referred to as a *risk-neutral measure*.

Definition 45.7 (T-Contingent Claim). Let $(\Omega, \mathscr{F}, \{\mathscr{F}_t : t \ge 0\}, Q)$ be a filtered risk-neutral probability space for the bond market, where Q is an equivalent martingale measure. A T-contingent claim X is a $\mathscr{F}_T$-measurable random variable such that

$$\frac{X}{\beta(T)} \in L^1(\Omega, \mathscr{F}_T, Q).$$

Theorem 45.1 (Risk-Neutral Price). *Let $(\Omega, \mathscr{F}, \{\mathscr{F}_t : t \geq 0\}, Q)$ be a filtered risk-neutral probability space for the bond market, where Q is an equivalent martingale measure. Then the price process of a T-contingent claim X is given by*

$$V_t(X) = \beta(t) E_Q\left[\frac{X}{\beta(T)}\bigg| \mathscr{F}_t\right] = E_Q\left[X e^{-\int_t^T r(s)\mathrm{d}s}\Big| \mathscr{F}_t\right], \quad 0 \leq t \leq T.$$

In particular, the price process of a zero-coupon bond maturing at time T is given by

$$P(t,T) = E_Q\left[e^{-\int_t^T r(s)\mathrm{d}s}\Big| \mathscr{F}_t\right], \quad 0 \leq t \leq T.$$

45.2 Problems

45.1. Let $0 < T < S$. A gentleman entered into a contract such that he will pay a price K at time T in return for a repayment of 1 at time S. Show that at time $0 \leq t < T$, the contract is worth

$$P(t,S) - KP(t,T).$$

45.2. Let $\{r(t) : 0 \leq t \leq T\}$ be a short rate process satisfying the following stochastic differential equation

$$\mathrm{d}r(t) = a(t)\mathrm{d}t + b(t)\mathrm{d}B_t,$$

where $\{B_t : 0 \leq t \leq T\}$ is a Brownian motion on some filtered probability space $(\Omega, \mathscr{F}, \{\mathscr{F}_t : t \geq 0\}, P)$, and $\{a(t) : t \geq 0\}$ and $\{b(t) : t \geq 0\}$ are adapted processes that satisfy standard conditions (see Theorem 38.1). Suppose that there exists an equivalent martingale measure on $(\Omega, \mathscr{F})$.

(a) Show that $P(t,T)$ is a function of $r(t)$.

(b) Show that

$$\mathrm{d}P(t,T) = P(t,T)[m(t,T)\mathrm{d}t + v(t,T)\mathrm{d}B_t] \tag{45.1}$$

for some adapted processes $m(t,T)$ and $v(t,T)$.

(c) Suppose that the processes $m(t,T)$ and $v(t,T)$ in Equation (45.1) are bounded and that

$$E_P\left[\exp\left(\frac{1}{2}\int_0^T \gamma(u)^2 \mathrm{d}u\right)\right] < \infty,$$

where $\gamma(t)$ is the market price of risk defined as

$$\gamma(t) = \frac{m(t,T) - r(t)}{v(t,T)}, \quad 0 \leq t \leq T.$$

Let $\{W_t : 0 \leq t \leq T\}$ be defined as

$$W_t = B_t + \int_0^t \gamma(u)\mathrm{d}u$$

and Q be a measure defined as

$$\frac{\mathrm{d}Q}{\mathrm{d}P} = \exp\left(-\int_0^T \gamma(u)\mathrm{d}B_u - \frac{1}{2}\int_0^T \gamma(u)^2 \mathrm{d}u \right).$$

Show that $\{W_t : 0 \le t \le T\}$ is a Brownian motion under Q and that the discounted price process $\{P(t,T)e^{-\int_0^t r(u)\mathrm{d}u} : 0 \le t \le T\}$ is a martingale under Q.

45.3 (Term Structure Equation)**.** Let $\{r(t) : 0 \le t \le T\}$ be a short rate process satisfying the following stochastic differential equation

$$\mathrm{d}r(t) = a(t)\mathrm{d}t + b(t)\mathrm{d}B_t,$$

where $\{B_t : 0 \le t \le T\}$ is a Brownian motion on some filtered probability space $(\Omega, \mathscr{F}, \{\mathscr{F}_t : t \ge 0\}, P)$, and $\{a(t) : t \ge 0\}$ and $\{b(t) : t \ge 0\}$ are adapted processes that satisfy standard conditions (see Theorem 38.1). Suppose that $P(t,T) = P(r(t), t, T)$; that is, the time t price of a zero-coupon bond with maturity T is a function of the current short rate $r(t)$. Suppose that the market is efficient. Let

$$\gamma(t,T) = \frac{m(t,T) - r(t)}{v(t,T)},$$

where

$$m(t,T) = \frac{1}{P}\left(\frac{\partial P}{\partial t} + a(t)\frac{\partial P}{\partial r} + \frac{1}{2}b(t)^2\frac{\partial^2 P}{\partial r^2} \right)$$

and

$$v(t,T) = \frac{b(t)}{P}\frac{\partial P}{\partial r}.$$

Show that

(a) $\gamma(t,T)$ is independent of T.

(b) $P(t,T)$ satisfies the following partial differential equation, where $\gamma(t) = \gamma(t,T)$:

$$\frac{\partial P}{\partial t} + [a(t) - \gamma(t)b(t)]\frac{\partial P}{\partial r} + \frac{1}{2}b(t)^2\frac{\partial^2 P}{\partial r^2} - r(t)P(t,T) = 0.$$

45.4 (Term Structure Equation Under Q)**.** Let $\{r(t) : 0 \le t \le T\}$ be a short rate process following the Itô diffusion (see Definition 39.6)

$$\mathrm{d}r(t) = a(r(t))\mathrm{d}t + b(r(t))\mathrm{d}W_t,$$

where $\{W_t : 0 \le t \le T\}$ is a Brownian motion under the risk-neutral measure Q. Show that the zero-coupon bond price satisfies the following partial differential equation:

$$\frac{1}{2}b(r(t))^2\frac{\partial^2 P}{\partial r^2} + a(r(t))\frac{\partial P}{\partial r} - \frac{\partial P}{\partial t} - r(t)P = 0.$$

45.5 (Affine Term Structure). Let $\{r(t) : 0 \leq t \leq T\}$ be a short rate process following the Itô diffusion

$$\mathrm{d}r(t) = [a(t) + b(t)r(t)]\mathrm{d}t + \sqrt{\sigma(t) + \eta(t)r(t)}\mathrm{d}W_t,$$

where $a(t), b(t), \sigma(t)$, and $\eta(t)$ are some continuous functions, and $\{W_t : 0 \leq t \leq T\}$ is a Brownian motion under the risk neutral measure Q. Suppose that the price of T-bonds has the following form

$$P(t,T) = \exp\left(-A(t,T) - B(t,T)r(t)\right)$$

for some smooth functions $A(t,T)$ and $B(t,T)$. Show that $A(t,T)$ and $B(t,T)$ satisfy the following system of ordinary differential equations

$$\frac{\partial A(t,T)}{\partial t} = \frac{1}{2}\sigma(t)B^2(t,T) - a(t)B(t,T), \quad A(T,T) = 0, \tag{45.2a}$$

$$\frac{\partial B(t,T)}{\partial t} = \frac{1}{2}\eta(t)B^2(t,T) - b(t)B(t,T) - 1, \quad B(T,T) = 0. \tag{45.2b}$$

45.6 (Vasicek Model). Let $\{r(t) : 0 \leq t \leq T\}$ be a short rate process satisfying the following stochastic differential equation

$$\mathrm{d}r(t) = \alpha[\mu - r(t)]\mathrm{d}t + \sigma\mathrm{d}W_t,$$

where $\alpha > 0$, $\mu > 0$, $\sigma > 0$, and $\{W_t : 0 \leq t \leq T\}$ is a Brownian motion under the risk neutral measure Q. Show that

(a) For every $t \in [0,T]$,

$$r(t) = r(0)e^{-\alpha t} + \mu\left(1 - e^{-\alpha t}\right) + \sigma e^{-\alpha t}\int_0^t e^{\alpha u}\mathrm{d}W_u.$$

(b) For every $t \in [0,T]$,

$$E_Q[r(t)] = r(0)e^{-\alpha t} - \mu\left(e^{-\alpha t} - 1\right)$$

and

$$\mathrm{Var}_Q[r(t)] = \frac{\sigma^2}{2\alpha}\left(1 - e^{-2\alpha t}\right),$$

where $\mathrm{Var}_Q[X] = E_Q[X^2] - E_Q[X]^2$.

(c) The time t price of the T-bond is

$$P(t,T) = \exp\left(-A(t,T) - B(t,T)r(t)\right),$$

where

$$A(t,T) = \frac{\mu\left(e^{-\alpha(T-t)} + \alpha(T-t) - 1\right)}{\alpha} - \frac{\sigma^2\left(4e^{-\alpha(T-t)} - e^{-2\alpha(T-t)} + 2\alpha(T-t) - 3\right)}{4\alpha^3}$$

and

$$B(t,T) = \frac{1 - e^{-\alpha(T-t)}}{\alpha}.$$

45.7 (CIR Model). Let $\{r(t) : 0 \leq t \leq T\}$ be a short rate process satisfying the following stochastic differential equation

$$\mathrm{d}r(t) = \alpha[\mu - r(t)]\mathrm{d}t + \sigma\sqrt{r(t)}\mathrm{d}W_t,$$

where $\alpha > 0$, $\mu > 0$, $\sigma > 0$, and $\{W_t : 0 \leq t \leq T\}$ is a Brownian motion under the risk-neutral measure Q. Suppose that the time t price of the T-bond has the following form:

$$P(t,T) = \exp(-A(t,T) - B(t,T)r(t)), \quad t \in [0,T].$$

Show that

(a) The mean of the short rate at time t is

$$E_Q[r(t)] = \mu - [\mu - r(0)]e^{-\alpha t}.$$

(b) The variance of the short rate at time t is

$$\mathrm{Var}_Q[r(t)] = \frac{\mu\sigma^2}{2\alpha} + \frac{[r(0) - \mu]\sigma^2}{\alpha} e^{-\alpha t} - \frac{[2r(0) - \mu]\sigma^2}{2\alpha} e^{-2\alpha t}.$$

(c) The functions $A(t,T)$ and $B(t,T)$ are given by

$$A(t,T) = -\frac{2\alpha\mu}{\sigma^2} \ln\left(\frac{2\theta e^{\frac{1}{2}(\theta+\alpha)(T-t)}}{(\theta + \alpha)\left[e^{\theta(T-t)} - 1\right] + 2\theta} \right)$$

and

$$B(t,T) = \frac{2\left[e^{\theta(T-t)} - 1\right]}{(\theta + \alpha)\left[e^{\theta(T-t)} - 1\right] + 2\theta},$$

where $\theta = \sqrt{\alpha^2 + 2\sigma^2}$.

45.8 (Hull-White Model). Let $\{r(t) : 0 \leq t \leq T\}$ be a short rate process satisfying the following stochastic differential equation

$$\mathrm{d}r(t) = [b(t) + \alpha r(t)]\mathrm{d}t + \sigma \mathrm{d}W_t,$$

where α and σ are constants and $\{W_t : 0 \leq t \leq T\}$ is a Brownian motion under the risk-neutral measure Q. Suppose that the time t price of the T-bond has the following form

$$P(t,T) = \exp(-A(t,T) - B(t,T)r(t)), \quad t \in [0,T].$$

Show that

$$B(t,T) = \frac{e^{\alpha(T-t)}}{\alpha}$$

and

$$A(t,T) = -\frac{\sigma^2}{2}\int_t^T B^2(s,T)\mathrm{d}s + \int_s^T b(s)B(s,T)\mathrm{d}s.$$

45.3 Hints

45.1. Apply Theorem 45.1 and the tower property of conditional expectations (see Problem 14.10).

45.2. To prove part (a), use Theorems 45.1 and 39.1, the result of Problem 29.2, and Theorem 11.2. To prove part (b), use part (a) and Itô's lemma (Theorem 35.3). To prove part (c), apply the Girsanov Theorem (see Problem 37.4) and the martingale property of Itô integrals (see Theorem 32.2).

45.3. To prove part (a), consider at time t a portfolio of $-V_1(t)$ amount in T_1-bond and $V_2(t)$ amount in T_2-bond and use the arbitrage argument to establish

$$\frac{m(t,T_1)-r(t)}{v(t,T_1)} = \frac{m(t,T_2)-r(t)}{v(t,T_2)}.$$

Use part (a) to prove part (b).

45.4. Apply Theorem 45.1 and the Feynman-Kac formula (see Theorem 40.1).

45.5. Use the result of Problem 45.4.

45.6. To prove part (a), consider the process $e^{\alpha t}r(t)$ and use Itô's lemma (Theorem 35.3). Part (b) follows from part (a) and Theorem 32.1. To prove part (c), use Theorem 45.1 and note that $\{r(u)-r(t): t \le u \le T\}$ is independent of $\mathscr{F}_t$.

45.7. To prove part (a), use Theorem 32.1 and solve $E_Q[r(t)]$ from an ordinary differential equation. To prove part (b), use Itô's lemma (Theorem 35.3) to calculate $E_Q[r(t)^2]$ and then use the formula $\text{Var}_Q[r(t)] = E_Q[r(t)^2] - E_Q[r(t)]^2$ and the result of part (a). Use the result of Problem 45.5 to prove part (c).

45.8. Use the result of Problem 45.5.

45.4 Solutions

45.1. Let V_t be the value at time t of the contract. Then $V_T = P(T,S) - K$. It follows from Theorem 45.1 that

$$\begin{aligned} V_t &= E_Q\left[(P(T,S)-K)e^{-\int_t^T r(u)\mathrm{d}u}\middle|\mathscr{F}_t\right] \\ &= E_Q\left[P(T,S)e^{-\int_t^T r(u)\mathrm{d}u}\middle|\mathscr{F}_t\right] - KE_Q\left[e^{-\int_t^T r(u)\mathrm{d}u}\middle|\mathscr{F}_t\right] \\ &= E_Q\left[E_Q\left[e^{-\int_T^S r(u)\mathrm{d}u}\middle|\mathscr{F}_T\right]e^{-\int_t^T r(u)\mathrm{d}u}\middle|\mathscr{F}_t\right] - KP(t,T). \quad (45.3) \end{aligned}$$

From Problem 14.10, we have

$$\begin{aligned} & E_Q\left[E_Q\left[e^{-\int_T^S r(u)\mathrm{d}u}\middle|\mathscr{F}_T\right]e^{-\int_t^T r(u)\mathrm{d}u}\middle|\mathscr{F}_t\right] \\ =\ & E_Q\left[e^{-\int_T^S r(u)\mathrm{d}u}e^{-\int_t^T r(u)\mathrm{d}u}\middle|\mathscr{F}_t\right] \\ =\ & P(t,S). \end{aligned} \quad (45.4)$$

Combining Equations (45.3) and (45.4) gives

$$V_t = P(t, S) - KP(t, T).$$

This completes the proof.

45.2.

(a) Let Q be the equivalent martingale measure on $(\Omega, \mathscr{F})$. Then by Theorem 45.1, we have

$$P(t, T) = E_Q\left[e^{-\int_t^T r(s)\mathrm{d}s}\middle| \mathscr{F}_t\right].$$

By Theorem 39.1, the short rate process $\{r(t) : 0 \le t \le T\}$ is Markovian. Hence it follows from Problem 29.2 that

$$P(t, T) = E_Q\left[e^{-\int_t^T r(s)\mathrm{d}s}\middle| r(t)\right].$$

By Theorem 11.2, $P(t, T)$ is a function of $r(t)$.

(b) By part (a), $P(t, T)$ is a function of $r(t)$. Let $P(t, T) = P(r(t), t, T)$. Then by Theorem 35.3, we have

$$\mathrm{d}P(t, T) = \frac{\partial P}{\partial r} b(t)\mathrm{d}B_t + \left(\frac{\partial P}{\partial t} + a(t)\frac{\partial P}{\partial r} + \frac{1}{2}b(t)^2\frac{\partial^2 P}{\partial r^2}\right)\mathrm{d}t.$$

Now let

$$m(t, T) = \frac{1}{P}\left(\frac{\partial P}{\partial t} + a(t)\frac{\partial P}{\partial r} + \frac{1}{2}b(t)^2\frac{\partial^2 P}{\partial r^2}\right)$$

and

$$v(t, T) = \frac{b(t)}{P}\frac{\partial P}{\partial r}.$$

Then we have

$$\mathrm{d}P(t, T) = P(t, T)[m(t, T)\mathrm{d}t + v(t, T)\mathrm{d}B_t].$$

(c) By the Girsanov Theorem (see Problem 37.4), $\{W_t : 0 \le t \le T\}$ is a Brownian motion under Q. Now we show that the discounted price process $\{P(t, T)e^{-\int_0^t r(u)\mathrm{d}u}$ $0 \le t \le T\}$ is a martingale under Q. Let $Z(t, T) = P(t, T)e^{-\int_0^t r(u)\mathrm{d}u}$. By Theorem 35.4, we have

$$\begin{aligned}
& \mathrm{d}Z(t, T) \\
= \; & e^{-\int_0^t r(u)\mathrm{d}u}\mathrm{d}P(t, T) + P(t, T)\mathrm{d}e^{-\int_0^t r(u)\mathrm{d}u} + \mathrm{d}P(t, T)\mathrm{d}e^{-\int_0^t r(u)\mathrm{d}u} \\
= \; & Z(t, T)[m(t, T)\mathrm{d}t + v(t, T)\mathrm{d}B_t] - Z(t, T)r(t)\mathrm{d}t \\
= \; & Z(t, T)v(t, T)\mathrm{d}W_t.
\end{aligned}$$

By the assumption that $v(t,T)$ and $m(t,T)$ are bounded and Theorem 38.1, we have

$$E_Q\left[\int_0^T Z(t,T)^2 \mathrm{d}t\right] < \infty.$$

Hence we have

$$E_Q\left[\int_0^T Z(t,T)^2 v(t,T)^2 \mathrm{d}t\right] < \infty.$$

It follows from Theorem 32.2 that $Z(t,T)$ is a martingale.

This completes the proof.

45.3.

(a) Let $T_1 < T_2$. Let V be a portfolio constructed by shorting V_1 amount in T_1-bond and purchasing V_2 amount in T_2-bond. Then the value at time t of the portfolio is $V(t) = V_2(t) - V_1(t)$. The instantaneous investment gain from t to $t + \mathrm{d}t$ is

$$\mathrm{d}V(t) = \mathrm{d}V_2(t) - \mathrm{d}V_1(t) = \frac{V_2(t)}{P(t,T_2)}\mathrm{d}P(t,T_2) - \frac{V_1(t)}{P(t,T_1)}\mathrm{d}P(t,T_1).$$

By Itô's lemma (Theorem 35.3), we have

$$\mathrm{d}P(t,T_i) = P(t,T_i)[m(t,T_i)\mathrm{d}t + v(t,T_i)\mathrm{d}B_t], \quad i = 1,2.$$

Then we have

$$\begin{aligned}\mathrm{d}V(t) &= [V_2(t)m(t,T_2) - V_1(t)m(t,T_1)]\mathrm{d}t \\ &\quad +[V_2(t)v(t,T_2) - V_1(t)v(t,T_1)]\mathrm{d}B_t. \end{aligned} \tag{45.5}$$

Suppose that we choose $V_1(t)$ and $V_2(t)$ in such a way that

$$V_2(t)v(t,T_2) - V_1(t)v(t,T_1) = 0.$$

Note that $V(t) = V_2(t) - V_1(t)$. Solving the above equations gives

$$V_1(t) = \frac{V(t)v(t,T_2)}{v(t,T_1) - v(t,T_2)}, \quad V_2(t) = \frac{V(t)v(t,T_1)}{v(t,T_1) - v(t,T_2)}.$$

Then Equation (45.5) becomes

$$\mathrm{d}V(t) = V(t)\frac{m(t,T_2)v(t,T_1) - m(t,T_1)v(t,T_2)}{v(t,T_1) - v(t,T_2)}\mathrm{d}t, \tag{45.6}$$

which shows that the portfolio V has deterministic growth. Hence the growth rate must be the risk-free rate $r(t)$:

$$\frac{m(t,T_2)v(t,T_1) - m(t,T_1)v(t,T_2)}{v(t,T_1) - v(t,T_2)} = r(t).$$

Manipulating the above equation gives

$$\frac{m(t,T_1) - r(t)}{v(t,T_1)} = \frac{m(t,T_2) - r(t)}{v(t,T_2)}.$$

Hence $\gamma(t,T_1) = \gamma(t,T_2)$.

(b) By part (a), we have

$$m(t,T) = \gamma(t)v(t,T) + r(t) = \gamma(t)\frac{b(t)}{P}\frac{\partial P}{\partial r} + r(t).$$

But by definition we have

$$m(t,T) = \frac{1}{P}\left(\frac{\partial P}{\partial t} + a(t)\frac{\partial P}{\partial r} + \frac{1}{2}b(t)^2\frac{\partial^2 P}{\partial r^2}\right).$$

Combining the above two equations gives

$$\frac{\partial P}{\partial t} + [a(t) - \gamma(t)b(t)]\frac{\partial P}{\partial r} + \frac{1}{2}b(t)^2\frac{\partial^2 P}{\partial r^2} - r(t)P(t,T) = 0.$$

This completes the proof.

45.4. By Theorem 45.1 and Problem 29.2, we have

$$P(r(t),t,T) = E_Q\left[e^{-\int_t^T r(s)\mathrm{d}s}\,\middle|\,\mathscr{F}_t\right] = E_Q\left[e^{-\int_t^T r(s)\mathrm{d}s}\,\middle|\,r(t)\right].$$

Let

$$v(t,x) = P(x, T-t, T) = E_Q\left[e^{-\int_0^t r(s+T-t)\mathrm{d}s}\,\middle|\,r(T-t) = x\right].$$

Then by Theorem 40.1, we have

$$\frac{\partial v}{\partial t}(t,x) = a(x)\frac{\partial v}{\partial x}(t,x) + \frac{1}{2}b(x)^2\frac{\partial^2 v}{\partial x^2}(t,x) - xv(t,x)$$

or

$$a(x)\frac{\partial v}{\partial x}(t,x) + \frac{1}{2}b(x)^2\frac{\partial^2 v}{\partial x^2}(t,x) - \frac{\partial v}{\partial t}(t,x) - xv(t,x) = 0.$$

Since

$$\frac{\partial P(r(t),t,T)}{\partial t} = -\frac{\partial v(T-t, r(t))}{\partial t},$$

we have

$$a(r(t))\frac{\partial P}{\partial r} + \frac{1}{2}b(r(t))^2\frac{\partial^2 P}{\partial r^2} + \frac{\partial P}{\partial t} - r(t)P = 0.$$

This completes the proof.

45.5. By the assumption that $P(t,T) = \exp(-A(t,T) - B(t,T)r(t))$, we have

$$\frac{\partial P}{\partial r} = -B(t,T)P(t,T), \quad \frac{\partial^2 P}{\partial r^2} = B^2(t,T)P(t,T),$$

and

$$\frac{\partial P}{\partial t} = -P(t,T)\left(\frac{\partial A(t,T)}{\partial t} + r(t)\frac{\partial B(t,T)}{\partial t}\right).$$

Then by Problem 45.4, we have

$$\frac{1}{2}[\sigma(t) + \eta(t)r(t)]B^2(t,T)P(t,T) - [a(t) + b(t)r(t)]B(t,T)P(t,T) -$$
$$P(t,T)\left(\frac{\partial A(t,T)}{\partial t} + r(t)\frac{\partial B(t,T)}{\partial t}\right) - r(t)P(t,T) = 0.$$

Reorganizing the above equation gives

$$\begin{aligned} &\left(\frac{1}{2}\eta(t)B^2(t,T) - b(t)B(t,T) - \frac{\partial B(t,T)}{\partial t} - 1\right) r(t) \\ = \quad &\frac{\partial A(t,T)}{\partial t} - \frac{1}{2}\sigma(t)B^2(t,T) + a(t)B(t,T) \end{aligned}$$

Since the above equation holds for arbitrary $r(t)$, the following two equations must be satisfied:

$$\frac{1}{2}\eta(t)B^2(t,T) - b(t)B(t,T) - \frac{\partial B(t,T)}{\partial t} - 1 = 0,$$

$$\frac{\partial A(t,T)}{\partial t} - \frac{1}{2}\sigma(t)B^2(t,T) + a(t)B(t,T) = 0.$$

This completes the proof.

45.6.

(a) By Itô's lemma (Theorem 35.3), we have

$$\begin{aligned} \mathrm{d}\left(e^{\alpha t}r(t)\right) &= e^{\alpha t}\mathrm{d}r(t) + r(t)\mathrm{d}e^{\alpha t} \\ &= e^{\alpha t}(\alpha[\mu - r(t)]\mathrm{d}t + \sigma \mathrm{d}W_t) + r(t)\alpha e^{\alpha t}\mathrm{d}t \\ &= \alpha\mu e^{\alpha t}\mathrm{d}t + \sigma e^{\alpha t}\mathrm{d}W_t. \end{aligned}$$

Hence we have

$$e^{\alpha t}r(t) - r(t) = \alpha\mu\int_0^t e^{\alpha u}\mathrm{d}u + \sigma\int_0^t e^{\alpha u}\mathrm{d}W_u$$

or

$$r(t) = r(0)e^{-\alpha t} + \mu\left(1 - e^{-\alpha t}\right) + \sigma e^{-\alpha t}\int_0^t e^{\alpha u}\mathrm{d}W_u.$$

(b) By part (a) and Theorem 32.1, we have

$$\begin{aligned} E_Q[r(t)] &= r(0)e^{-\alpha t} + \mu\left(1 - e^{-\alpha t}\right) + E_Q\left[\sigma e^{-\alpha t}\int_0^t e^{\alpha u}\mathrm{d}W_u\right] \\ &= r(0)e^{-\alpha t} + \mu\left(1 - e^{-\alpha t}\right). \end{aligned}$$

and

$$\mathrm{Var}_Q[r(t)] = \int_0^t \sigma^2 e^{-2\alpha t}e^{2\alpha u}\mathrm{d}u = \frac{\sigma^2}{2\alpha}\left(1 - e^{-2\alpha t}\right).$$

(c) By part (a), we have for $u \in [t, T]$,

$$r(u) - r(t) = [\mu - r(t)]\left(1 - e^{\alpha(t-u)}\right) + \sigma e^{-\alpha u}\int_t^u e^{\alpha s}\mathrm{d}W_s.$$

Since $\{r(u) - r(t) : t \le u \le T\}$ is independent of $\mathscr{F}_t$, it follows from Theorem 45.1 and Problem 14.16 that

$$\begin{aligned} P(t,T) &= E_Q\left[e^{-\int_t^T r(u)\mathrm{d}u}\middle|\mathscr{F}_t\right] \\ &= e^{-r(t)(T-t)}E_Q\left[e^{-\int_t^T (r(u)-r(t))\mathrm{d}u}\middle|\mathscr{F}_t\right] \\ &= e^{-r(t)(T-t)}E_Q\left[e^{-\int_t^T (r(u)-r(t))\mathrm{d}u}\right]. \end{aligned} \tag{45.7}$$

Note that

$$E_Q\left[e^{-\int_t^T (r(u)-r(t))\mathrm{d}u}\right] = \int_\Omega e^{-\int_t^T (r(u,\omega)-r(t,\omega))\mathrm{d}u}\mathrm{d}P. \tag{45.8}$$

For a fixed $\omega \in \Omega$, we have

$$\begin{aligned} &\int_t^T (r(u,\omega) - r(t,\omega))\mathrm{d}u \\ = &\int_t^T [\mu - r(t,\omega)]\left(1 - e^{\alpha(t-u)}\right)\mathrm{d}u + \sigma\int_t^T e^{-\alpha u}\int_t^u e^{\alpha s}\mathrm{d}W_s(\omega)\mathrm{d}u \\ = &[\mu - r(t,\omega)]\frac{e^{-\alpha(T-t)} + \alpha(T-t) - 1}{\alpha} - \\ &\frac{\sigma}{\alpha}\left[e^{-\alpha u}\int_t^u e^{\alpha s}\mathrm{d}W_s(\omega)\Big|_t^T - \int_t^T \mathrm{d}W_u(\omega)\right] \\ = &[\mu - r(t,\omega)]\frac{e^{-\alpha(T-t)} + \alpha(T-t) - 1}{\alpha} - \\ &\frac{\sigma}{\alpha}\int_t^T \left(e^{\alpha(s-T)} - 1\right)\mathrm{d}W_s(\omega). \end{aligned} \tag{45.9}$$

Combining Equations (45.8) and (45.9) gives

$$\begin{aligned} &E_Q\left[e^{-\int_t^T (r(u)-r(t))\mathrm{d}u}\right] \\ = &E_Q\left[\exp\left([r(t) - \mu]\frac{e^{-\alpha(T-t)} + \alpha(T-t) - 1}{\alpha} + Z\right)\right], \end{aligned} \tag{45.10}$$

where

$$Z = \frac{\sigma}{\alpha} \int_t^T \left(e^{\alpha(s-T)} - 1 \right) \mathrm{d}W_s.$$

By Theorem 32.1, Z is a normal random variable with mean 0 and variance

$$\begin{aligned} \mathrm{Var}_Q(Z) &= \frac{\sigma^2}{\alpha^2} \int_t^T \left(e^{\alpha(s-T)} - 1 \right)^2 \mathrm{d}s \\ &= \frac{4e^{-\alpha(T-t)} - e^{-2\alpha(T-t)} + 2\alpha(T-t) - 3}{2\alpha^3} \sigma^2. \end{aligned}$$

It follows from Equation (45.10) and Problem 19.1 that

$$\begin{aligned} & E_Q\left[e^{-\int_t^T (r(u)-r(t))\mathrm{d}u} \right] \\ = \;& \exp\left([r(t) - \mu] \frac{e^{-\alpha(T-t)} + \alpha(T-t) - 1}{\alpha} \right) E_Q\left[e^Z \right] \\ = \;& \exp\left([r(t) - \mu] \frac{e^{-\alpha(T-t)} + \alpha(T-t) - 1}{\alpha} + \frac{1}{2} \mathrm{Var}_Q(Z) \right) \quad (45.11) \end{aligned}$$

The result follows by combining Equation (45.7) and the above equation.

This completes the proof.

45.7.

(a) Let $g(t) = E_Q[r(t)]$. Then by Theorem 32.1, we have

$$g(t) = r(0) + E_Q\left[\int_0^t \alpha[\mu - r(s)]\mathrm{d}s \right] = r(0) + \int_0^t \alpha[\mu - g(s)].$$

Differentiating both sides of the above equation gives

$$\frac{\mathrm{d}g(t)}{\mathrm{d}t} = \alpha\mu - \alpha g(t).$$

Hence we have

$$\mathrm{d}\left(g(t)e^{\alpha t} \right) = e^{\alpha t} \left[\mathrm{d}g(t) + \alpha g(t)\mathrm{d}t \right] = e^{\alpha t} \alpha\mu \mathrm{d}t,$$

which leads to

$$g(t)e^{\alpha t} = g(0) + \alpha\mu \int_0^t e^{\alpha s}\mathrm{d}s = g(0) + \mu\left[e^{\alpha t} - 1 \right].$$

Note that $g(0) = E_Q[r(0)] = r(0)$. The above equation gives

$$E_Q[r(t)] = \mu + [r(0) - \mu]e^{-\alpha t}.$$

(b) Let us first calculate $g(t) = E_Q[r(t)^2]$. By Theorem 35.3, we have

$$r(t)^2 = r(0)^2 + \int_0^t 2\sigma r(s)^{\frac{3}{2}} \mathrm{d}W_s + \int_0^t \left[(2\alpha\mu + \sigma^2) r(s) - 2\alpha r(s)^2\right] \mathrm{d}s.$$

Hence we have

$$g(t) = r(0)^2 + \int_0^t \left[(2\alpha\mu + \sigma^2) E_Q[r(s)] - 2\alpha g(s)\right] \mathrm{d}s,$$

which gives

$$\begin{aligned} \frac{\mathrm{d}g(t)}{\mathrm{d}t} &= (2\alpha\mu + \sigma^2) E_Q[r(t)] - 2\alpha g(t) \\ &= (2\alpha\mu + \sigma^2)(\mu + [r(0) - \mu]e^{-\alpha t}) - 2\alpha g(t). \end{aligned}$$

Now we have

$$\begin{aligned} \frac{\mathrm{d}g(t)e^{2\alpha t}}{\mathrm{d}t} &= e^{2\alpha t}\left(\frac{\mathrm{d}g(t)}{\mathrm{d}t} + 2\alpha g(t)\right) \\ &= (2\alpha\mu + \sigma^2)\left(\mu e^{2\alpha t} + [r(0) - \mu]e^{\alpha t}\right). \end{aligned}$$

Integrating both sides of the above equation gives

$$g(t)e^{2\alpha t} - g(0) = (2\alpha\mu + \sigma^2)\left(\frac{\mu}{2\alpha}\left[e^{2\alpha t} - 1\right] + \frac{r(0) - \mu}{\alpha}\left[e^{\alpha t} - 1\right]\right).$$

Therefore,

$$\begin{aligned} &\mathrm{Var}_Q[r(t)] \\ = \;& g(t) - E_Q[r(t)]^2 \\ = \;& r(0)e^{-2\alpha t} + \frac{(2\alpha\mu + \sigma^2)\mu}{2\alpha}\left[1 - e^{-2\alpha t}\right] + \\ & \frac{(2\alpha\mu + \sigma^2)(r(0) - \mu)}{\alpha}\left[e^{-\alpha t} - e^{-2\alpha t}\right] - (\mu + [r(0) - \mu]e^{-\alpha t})^2 \\ = \;& \frac{\mu\sigma^2}{2\alpha} + \frac{[r(0) - \mu]\sigma^2}{\alpha}e^{-\alpha t} - \frac{[2r(0) - \mu]\sigma^2}{2\alpha}e^{-2\alpha t}. \end{aligned}$$

(c) By Problem 45.5, the functions $A(t, T)$ and $B(t, T)$ satisfy the following system of ordinary differential equations

$$\frac{\partial A(t, T)}{\partial t} = -\alpha\mu B(t, T), \quad A(T, T) = 0, \tag{45.12a}$$

$$\frac{\partial B(t, T)}{\partial t} = \frac{1}{2}\sigma^2 B^2(t, T) + \alpha B(t, T) - 1, \quad B(T, T) = 0. \tag{45.12b}$$

Let us first solve $B(t,T)$. To do that, let $g(t) = \sigma B(t,T) + \frac{\alpha}{\sigma}$. Then Equation (45.12b) becomes

$$\frac{\partial g(t)}{\partial t} = \frac{1}{2}\sigma g^2(t) - \frac{\theta^2}{2\sigma}, \quad g(T) = \frac{\alpha}{\sigma}. \tag{45.13}$$

Now let $f(t) = \ln[g(t) + \frac{\theta}{\sigma}]$. Then Equation (45.13) becomes

$$\frac{\partial f(t)}{\partial t} = \frac{1}{2}\sigma e^{f(t)} - \theta, \quad f(T) = \ln\frac{\alpha+\theta}{\sigma}$$

or

$$\mathrm{d}e^{-f(t)-\theta t} = -\frac{\sigma}{2}e^{-\theta t}\mathrm{d}t, \quad f(T) = \ln\frac{\alpha+\theta}{\sigma}.$$

Integrating both sides of the above equation from t to T leads to

$$e^{-f(T)-\theta T} - e^{-f(t)-\theta t} = \frac{\sigma}{2\theta}\left[e^{-\theta T} - e^{-\theta t}\right].$$

Reorganizing the above equation and using the initial conditions, we can get

$$B(t,T) = \frac{2\left[e^{\theta(T-t)} - 1\right]}{(\theta+\alpha)\left[e^{\theta(T-t)} - 1\right] + 2\theta}.$$

From Equation (45.12a), we have

$$\begin{aligned} A(t,T) &= A(T,T) + \int_t^T \alpha\mu B(s,T)\mathrm{d}s \\ &= \alpha\mu\int_t^T \frac{2\left[e^{\theta(T-s)} - 1\right]}{(\theta+\alpha)\left[e^{\theta(T-s)} - 1\right] + 2\theta}\mathrm{d}s \\ &= 2\alpha\mu\int_{e^{\theta(T-t)}}^1 \frac{y-1}{(\theta+\alpha)(y-1)+2\theta}\frac{-\mathrm{d}y}{\theta y} \\ &= 2\alpha\mu\int_{e^{\theta(T-t)}}^1 \left(\frac{2}{\alpha-\theta}\frac{1}{(\theta+\alpha)y+\theta-\alpha} + \frac{1}{\theta-\alpha}\frac{1}{\theta y}\right)\mathrm{d}y. \end{aligned}$$

Since

$$\int_{e^{\theta(T-t)}}^1 \frac{2}{\alpha-\theta}\frac{1}{(\theta+\alpha)y+\theta-\alpha}\mathrm{d}y = -\frac{1}{\sigma^2}\ln\frac{2\theta}{(\theta+\alpha)[e^{\theta(T-t)}-1]+2\theta}$$

and

$$\int_{e^{\theta(T-t)}}^1 \frac{1}{\theta-\alpha}\frac{1}{\theta y}\mathrm{d}y = -\frac{1}{\sigma^2}\frac{(\theta+\alpha)(T-t)}{2},$$

we have

$$A(t,T) = -\frac{2\alpha\mu}{\sigma^2}\ln\left(\frac{2\theta e^{\frac{1}{2}(\theta+\alpha)(T-t)}}{(\theta+\alpha)\left[e^{\theta(T-t)} - 1\right] + 2\theta}\right).$$

This completes the proof.

45.8. By Problem 45.5, $B(t,T)$ satisfies the following ordinary differential equation

$$\frac{\partial B(t,T)}{t} = -\alpha B(t,T) - 1, \quad B(T,T) = 0.$$

Hence

$$\mathrm{d}\left(B(t,T)e^{\alpha t}\right) = e^{\alpha t}[\mathrm{d}B(t,T) + \alpha B(t,T)] = -e^{\alpha t},$$

which gives

$$B(t,T) = \frac{e^{\alpha(T-t)-1}}{\alpha}.$$

Also by Problem 45.5, we have

$$\frac{\partial A(t,T)}{t} = \frac{1}{2}\sigma^2 B^2(t,T) - b(t)B(t,T), \quad A(T,T) = 0.$$

Integrating both sides of the above equation from t to T leads to

$$A(T,T) - A(t,T) = \frac{1}{2}\sigma^2 \int_t^T B^2(s,T)\mathrm{d}s - \int_t^T b(s)B(s,T)\mathrm{d}s$$

or

$$A(t,T) = -\frac{1}{2}\sigma^2 \int_t^T B^2(s,T)\mathrm{d}s + \int_t^T b(s)B(s,T)\mathrm{d}s.$$

This completes the proof.

45.5 Bibliographic Notes

In this chapter, we introduced some continuous-time short rate models for interest rates. In particular, we presented the Vasicek model and the CIR model.

Table 45.1 Some short rate models.

Model	Reference
Vasiček	Vasiček (1977), Hull and White (1990)
Dothan	Dothan (1978)
CIR	Cox et al. (1985), Hull and White (1990)
Ho-Lee	Ho and Lee (1986)
Black-Derman-Toy	Black et al. (1990)
Black-Karasinski	Black and Karasinski (1991)

There are several other short rate models. For example, Table 45.1 lists some short rate models and the corresponding references. Assumptions of various short rate models are discussed in Svoboda (2004). For discrete interest rate models, readers are referred to Cairns (2004) and Pascucci and Runggaldier (2012).

Short rate models are also studied in many other books, such as Hunt and Kennedy (2004), Shreve (2004), Musiela and Rutkowski (2005), Brigo and Mercurio (2006), Björk (2009), Filipović (2009), Ekstrand (2011), and Lyuu (2001), to name only a few. For general term structure models, readers are referred to Baxter (1997), Björk et al. (1997), and Jin and Glasserman (2001). For a review on interest rate models, readers are referred to Rebonato (2004).

Theorem 45.1 states the risk-neutral price of interest rate derivatives, which include zero-coupon bonds. A proof of this theorem can be found in Cairns (2004, Theorem 4.1).

CHAPTER 46

INSTANTANEOUS FORWARD RATE MODELS

The short rate models presented in the previous chapter are not flexible for calibration. As a consequence, Heath et al. (1992) introduced a new uniform framework, called the HJM framework, for modeling the interest rates. In the Heath-Jarrow-Morton (HJM) framework, the entire forward curve is modeled directly. These interest rate models are called *HJM models* or *instantaneous forward rate* models. In this chapter, we introduce some concepts related to the HJM framework.

46.1 Basic Concepts and Facts

Definition 46.1 (Forward Rate). Let $0 \le t < T < S$. The simple (or simply compounded) forward rate for $[T, S]$ prevailing at t is defined to be

$$F(t; T, S) = \frac{1}{S - T}\left(\frac{P(t, T)}{P(t, S)} - 1\right),$$

where $P(t,T)$ is the value at time t of a dollar at time T. The continuously compounded forward rate for $[T,S]$ prevailing at t is defined to be

$$R(t;T,S) = \frac{\ln P(t,T) - \ln P(t,S)}{S-T}.$$

The instantaneous forward rate with maturity T prevailing at t is defined as

$$f(t,T) = \lim_{S \downarrow T} R(t;T,S) = -\frac{\partial \ln P(t,T)}{\partial T}.$$

The function $T \to f(t,T)$ is called the forward curve at time t.

Definition 46.2 (Forward Contract). Let $t < T$. A forward contract on an underlying $\mathcal{Y}$, entered at time t, with maturity T is defined as follows:

- At time T, the contract holder at the long side pays $f(t,T,\mathcal{Y})$ and receives $\mathcal{Y}$ from the contract holder at the short side, where $f(t,T,\mathcal{Y})$ is the forward price of $\mathcal{Y}$ at time t.
- At time t, the forward price is chosen such that the value of the contract is zero, where $r(u)$ is the short rate and Q is the equivalent martingale measure for the bond market:

$$f(t,T,\mathcal{Y}) = \frac{1}{P(t,T)} E_Q\left[\mathcal{Y} e^{-\int_t^T r(u)\mathrm{d}u} | \mathscr{F}_t\right].$$

Definition 46.3 (Futures Contract). Let $t < T$. A futures contract on an underlying $\mathcal{Y}$, entered at time t, with maturity T is defined as follows:

- At time T, the contract holder at the long side pays $F(t,T,\mathcal{Y})$ and receives $\mathcal{Y}$ from the contract holder at the short side, where $F(t,T,\mathcal{Y})$ is the futures price quoted at time t for the futures contract.
- At time t, the futures price is chosen such that the value of the contract is zero.
- During any infinitesimal time interval $(t, t+\delta]$, the contract holder at the long side receives $F(t+\delta,T,\mathcal{Y}) - F(t,T,\mathcal{Y})$ from the contract holder at the short side.

Definition 46.4 (HJM Framework). Let $T^* > 0$. Let $\{(B_t^{(1)}, B_t^{(2)}, \ldots, B_t^{(n)}) : 0 \le t \le T^*\}$ be an n-dimensional Brownian motion on some filtered probability space $(\Omega, \mathscr{F}, \{\mathscr{F}_t : 0 \le t \le T^*\}, P)$. For every $T \in [0, T^*]$, the forward rate process $\{f(t,T) : 0 \le t \le T\}$ satisfies the following stochastic differential equation:

$$f(t,T) = f(0,T) + \int_0^t \alpha(\omega,s,T)\mathrm{d}s + \sum_{i=1}^n \int_0^t \sigma_i(\omega,s,T)\mathrm{d}B_s^{(i)}. \tag{46.1}$$

Here $f(0,T)$, α, and σ_i $(i = 1, 2, \ldots, n)$ satisfy the following technical conditions:

(a) $\{f(0,T) : 0 \leq T \leq T^*\}$ is a fixed and deterministic initial forward curve. The map $f(0,\cdot) : [0,T^*] \to \mathbf{R}$ is $\mathcal{B}[0,T^*]$-measurable and

$$\int_0^{T^*} |f(0,v)|\mathrm{d}v < \infty.$$

(b) $\alpha : \Omega \times \{(s,t) : 0 \leq s \leq t \leq T^*\} \to \mathbf{R}$ is a family of drift functions and is $\mathscr{F} \otimes \mathcal{B}\{(s,t) : 0 \leq s \leq t \leq T^*\}$-measurable. For every $T \in [0,T^*]$, the process $\{\alpha(\cdot,t,T) : 0 \leq t \leq T\}$ is adapted to the filtration $\{\mathscr{F}_t : 0 \leq t \leq T\}$ and

$$\int_0^{T^*}\int_0^{v} |\alpha(\cdot,t,v)|\mathrm{d}t\mathrm{d}v < \infty, \quad P-\text{a.s.}$$

(c) For every $i = 1,2,\ldots,n$, $\sigma_i : \Omega \times \{(s,t) : 0 \leq s \leq t \leq T^*\} \to \mathbf{R}$ is a family of drift functions and is $\mathscr{F} \otimes \mathcal{B}\{(s,t) : 0 \leq s \leq t \leq T^*\}$-measurable. For for every $i = 1,2,\ldots,n$ and every $T \in [0,T^*]$, the process $\{\sigma_i(\cdot,t,T) : 0 \leq t \leq T\}$ is adapted to the filtration $\{\mathscr{F}_t : 0 \leq t \leq T\}$ and satisfies the following conditions:

$$\int_0^T \left(\int_s^T \sigma_i(\cdot,s,t)\mathrm{d}t\right)^2 \mathrm{d}s < \infty, \quad P-\text{a.s.},$$

$$\int_0^v \left(\int_v^T \sigma_i(\cdot,s,t)\mathrm{d}t\right)^2 \mathrm{d}s < \infty, \quad P-\text{a.s.}, \quad \forall v \in [0,T],$$

and

$$v \to \int_v^T \left(\int_0^v \sigma_i(\cdot,s,t)\mathrm{d}B_s^{(i)}\right)^2 \mathrm{d}t \quad \text{is continuous P-a.s.}$$

Theorem 46.1 (Stochastic Fubini Theorem). *Let $\{B_t : 0 \leq t \leq T\}$ be a Brownian motion on some filtered probability space $(\Omega, \mathscr{F}, \{\mathscr{F}_t : 0 \leq t \leq T\}, P)$. Let $\{\phi(\omega,s,t) : 0 \leq s,t \leq T\}$ be a family of random variables such that*

(a) The map $((\omega,s),t) \in (\Omega \times [0,T]) \times [0,T] \to \phi(\omega,s,t)$ is $\mathcal{L} \otimes \mathcal{B}[0,T]$-measurable, where $\mathcal{L}$ is the smallest σ-field to which all left-continuous $\{\mathscr{F}_t : 0 \leq t \leq T\}$-adapted stochastic processes $\{X_t = X(\omega,t) : 0 \leq t \leq T\}$ are measurable (see Definition 34.2).

(b) For every $v \in [0,T]$ and every $t \in [0,T]$, we have

$$\int_0^v \phi^2(\cdot,s,t)\mathrm{d}s < \infty, \quad P-a.s.$$

(c) For every $v \in [0,T]$, we have

$$\int_0^v \left(\int_0^T \phi(\cdot,s,t)\mathrm{d}t\right)^2 \mathrm{d}s < \infty.$$

(d) The map

$$v \to \int_0^T \left(\int_0^v \phi(\cdot, s, t) \mathrm{d}B_s \right)^2 \mathrm{d}t \quad \textit{is continuous } P - a.s.$$

Then for every $v \in [0, T]$,

$$\int_0^v \left(\int_0^T \phi(\cdot, s, t) \mathrm{d}t \right) \mathrm{d}B_s = \int_0^T \left(\int_0^v \phi(\cdot, s, t) \mathrm{d}B_s \right) \mathrm{d}t.$$

Theorem 46.2 (HJM Drift Condition). *Let Q be an equivalent martingale measure for the bond market and $\{W_t : 0 \le t \le T^*\}$ a Brownian motion with respect to Q. Suppose that the forward rate satisfies the following stochastic differential equation*

$$\mathrm{d}f(t, T) = \alpha(t, T)\mathrm{d}t + \sigma(t, T)\mathrm{d}W_t, \quad 0 \le t \le T \le T^*.$$

Then

$$\alpha(t, T) = \sigma(t, T) \int_t^T \sigma(t, s)\mathrm{d}s, \quad Q - a.s., \quad 0 \le t \le T \le T^*.$$

The above equation is referred to as the HJM drift condition.

Theorem 46.3 (Futures Price). *Let $t < T$. For a futures contract with maturity T and underlying $\mathcal{Y}$, the futures price at time t is given by*

$$F(t, T, \mathcal{Y}) = E_Q[\mathcal{Y} | \mathscr{F}_t],$$

where Q is the equivalent martingale measure for the bond market.

46.2 Problems

46.1. Let $P(t, T)$ be the price at time t of a zero-coupon bond maturing at time T. Let $f(t, T)$ be the instantaneous forward rate with maturity T prevailing at t. Show that for $t \le S \le T$, we have

$$P(t, T) = P(t, S) \exp\left(- \int_S^T f(t, u)\mathrm{d}u \right).$$

46.2. Let $T > 0$. Suppose that the price process $\{P(t, T) : 0 \le t \le T\}$ of the T-bond satisfies the following stochastic differential equation:

$$\mathrm{d}P(t, T) = P(t, T)[m(t, T)\mathrm{d}t + v(t, T)\mathrm{d}B_t].$$

Here $\{B_t : 0 \le t \le T\}$ is a Brownian motion on some filtered probability space (Ω, $\mathscr{F}$, $\{\mathscr{F}_t : 0 \le t \le T\}, P$), and $\{m(t, T) : 0 \le t \le T\}$ and $\{v(t, T) : 0 \le t \le T\}$

are adapted processes that satisfy standard conditions (see Theorem 38.1). Show that the forward rate process $\{f(t,T) : 0 \leq t \leq T\}$ satisfies the following stochastic differential equation

$$\mathrm{d}f(t,T) = \alpha(t,T)\mathrm{d}t + \sigma(t,T)\mathrm{d}B_t,$$

where

$$\alpha(t,T) = v(t,T)\frac{\partial v}{\partial T}(t,T) - \frac{\partial m}{\partial T}(t,T)$$

and

$$\sigma(t,T) = -\frac{\partial v}{\partial T}(t,T).$$

46.3. Let $T > 0$. Suppose that the forward rate process $\{f(t,T) : 0 \leq t \leq T\}$ satisfies the following stochastic differential equation

$$\mathrm{d}f(t,T) = \alpha(t,T)\mathrm{d}t + \sigma(t,T)\mathrm{d}B_t,$$

where $\{B_t : 0 \leq t \leq T\}$ is a Brownian motion on some filtered probability space $(\Omega, \mathscr{F}, \{\mathscr{F}_t : 0 \leq t \leq T\}, P)$, and $\{\alpha(t,T) : 0 \leq t \leq T\}$ and $\{\sigma(t,T) : 0 \leq t \leq T\}$ satisfy the conditions given in Definition 46.4. Show that

(a) The short rate process satisfies the following stochastic differential equation

$$\mathrm{d}r(t) = \left(\alpha(t,t) + \frac{\partial f}{\partial T}(t,t)\right)\mathrm{d}t + \sigma(t,t)\mathrm{d}B_t.$$

(b) For every $t \in [0,T]$, we have

$$\begin{aligned}\int_t^T \int_0^t \alpha(s,u)\mathrm{d}s\mathrm{d}u &= \int_0^t \int_t^T \alpha(s,u)\mathrm{d}u\mathrm{d}s \\ &= \int_0^t \int_s^T \alpha(s,u)\mathrm{d}u\mathrm{d}s - \int_0^t \int_0^u \alpha(s,u)\mathrm{d}s\mathrm{d}u.\end{aligned}$$

(c) For every $t \in [0,T]$, we have

$$\begin{aligned}\int_t^T \int_0^t \sigma(s,u)\mathrm{d}B_s\mathrm{d}u &= \int_0^t \int_t^T \sigma(s,u)\mathrm{d}u\mathrm{d}B_s \\ &= \int_0^t \int_s^T \sigma(s,u)\mathrm{d}u\mathrm{d}B_s - \int_0^t \int_0^u \sigma(s,u)\mathrm{d}B_s\mathrm{d}u.\end{aligned}$$

(d) The T-bond price process satisfies the following stochastic differential equation:

$$\mathrm{d}P(t,T) = P(t,T)\left[m(t,T)\mathrm{d}t + v(t,T)\mathrm{d}B_t\right].$$

Here

$$m(t,T) = r(t) + \frac{1}{2}\left(\int_t^T \sigma(t,s)\mathrm{d}s\right)^2 - \int_t^T \alpha(t,s)\mathrm{d}s$$

and

$$v(t, T) = -\int_t^T \sigma(t, s)\mathrm{d}s.$$

46.4 (Risk-Neutral Measure). Let $T^* > 0$. Suppose that for every $T \in [0, T^*]$, the forward rate process $\{f(t, T) : 0 \leq t \leq T\}$ satisfies the following stochastic differential equation

$$\mathrm{d}f(t, T) = \alpha(t, T)\mathrm{d}t + \sigma(t, T)\mathrm{d}B_t,$$

where $\{B_t : 0 \leq t \leq T\}$ is a Brownian motion on some filtered probability space $(\Omega, \mathcal{F}, \{\mathcal{F}_t : 0 \leq t \leq T\}, P)$, and $\{\alpha(t, T) : 0 \leq t \leq T\}$ and $\{\sigma(t, T) : 0 \leq t \leq T\}$ satisfy the conditions given in Definition 46.4. Suppose that there exists an adapted process $\{\lambda(t) : 0 \leq t \leq T^*\}$ with the following properties:

- For every $T \in [0, T^*]$, we have

$$\int_0^T \lambda^2(t)\mathrm{d}t < \infty, \quad P-\text{a.s.}$$

- For every $T \in [0, T^*]$, $E_P[L(T)] = 1$, where

$$L(t) = \exp\left(-\int_0^t \lambda(u)\mathrm{d}B_u - \frac{1}{2}\int_0^t \lambda^2(u)\mathrm{d}u\right); \tag{46.2}$$

- For every $T \in [0, T^*]$ and every $t \in [0, T]$, we have

$$\alpha(t, T) - \sigma(t, T)\int_t^T \sigma(t, s)\mathrm{d}s = \sigma(t, T)\lambda(t).$$

Show that

(a) For any $T, S \in [0, T^*]$ with $T < S$, $Q_T = Q_S$ on $(\Omega, \mathcal{F}_T)$, where

$$\frac{\mathrm{d}Q_T}{\mathrm{d}P} = L(T), \quad \frac{\mathrm{d}Q_S}{\mathrm{d}P} = L(S).$$

(b) There exists an equivalent martingale measure Q (see Definition 45.6).

46.5 (General HJM Model). Let $T^* > 0$. Suppose that for every $T \in [0, T^*]$, the forward rate process $\{f(t, T) : 0 \leq t \leq T\}$ satisfies the following stochastic differential equation

$$\mathrm{d}f(t, T) = \sigma(t, T)\int_t^T \sigma(t, s)\mathrm{d}s\mathrm{d}t + \sigma(t, T)\mathrm{d}W_t,$$

where $\{W_t : 0 \leq t \leq T\}$ is a Brownian motion under the risk-neutral probability Q and the coefficients of $\mathrm{d}t$ and $\mathrm{d}W_t$ satisfy the technical conditions given in Definition 46.4. Let

$$\eta(t, T) = -\int_t^T \sigma(t, u)\mathrm{d}u.$$

Show that

(a) For $0 \le t \le T \le T^*$, we have

$$\frac{P(t,T)}{\beta(t)} = P(0,T)\exp\left(-\frac{1}{2}\int_0^t \eta^2(s,T)\mathrm{d}s + \int_0^t \eta(s,T)\mathrm{d}W_s\right).$$

(b) For $0 \le t \le T \le S \le T^*$, the price at time T of a zero-coupon bond with maturity S is given by

$$P(T,S) = \frac{P(t,S)}{P(t,T)}e^Z,$$

where

$$Z = -\frac{1}{2}\int_t^T \left[\eta^2(u,S) - \eta^2(u,T)\right]\mathrm{d}u + \int_t^T [\eta(u,S) - \eta(u,T)]\mathrm{d}W_u.$$

(c) For $0 \le t \le T \le S \le T^*$, the futures price, quoted at time t for a futures contract with maturity T on a zero-coupon bond with maturity S, is given by

$$V(t,T,S) = \frac{P(t,S)}{P(t,T)}\exp\left(\int_t^T \eta(u,T)[\eta(u,T) - \eta(u,S)]\mathrm{d}u\right).$$

46.6 (Gaussian HJM Model). Let $T^* > 0$. Suppose that for every $T \in [0,T^*]$, the forward rate process $\{f(t,T) : 0 \le t \le T\}$ satisfies the following stochastic differential equation

$$\mathrm{d}f(t,T) = \sigma^2(T-t)\mathrm{d}t + \sigma\mathrm{d}W_t,$$

where $\{W_t : 0 \le t \le T\}$ is a Brownian motion under the risk neutral probability Q and σ is a nonnegative constant. Let $0 \le t \le T \le S \le T^*$. Show that the price at time t of a European call option with strike K and maturity T on a zero-coupon bond with maturity S is given by

$$C(t,T,S,K) = P(t,S)N(d_1) - KP(t,T)N(d_2),$$

where $N(\cdot)$ is as defined in Definition 19.1, specifically

$$d_1 = \frac{\ln\frac{P(t,S)}{P(t,T)K} + \frac{1}{2}\sigma^2(S-T)^2(T-t)}{\sigma(S-T)\sqrt{T-t}},$$

and

$$d_2 = d_1 - \sigma(S-T)\sqrt{T-t}.$$

46.3 Hints

46.1. Follow the definition of instantaneous forward rates (Definition 46.1).

46.2. Use the result of Problem 46.1 and Itô's lemma (Theorem 35.3).

46.3. To prove part (a), note that $r(t) = f(t,t)$,

$$\int_0^t \alpha(s,t)\mathrm{d}s = \int_0^t \alpha(s,s)\mathrm{d}s + \int_0^t \int_0^u \frac{\partial \alpha}{\partial T}(s,u)\mathrm{d}s\mathrm{d}u,$$

and

$$\int_0^t \sigma(s,t)\mathrm{d}B_s = \int_0^t \sigma(s,s)\mathrm{d}B_s + \int_0^t \int_0^u \frac{\partial \sigma}{\partial T}(s,u)\mathrm{d}B_s\mathrm{d}u.$$

To prove parts (b) and (c), use the classical Fubini theorem (Theorem 10.2) and the stochastic Fubini theorem (Theorem 46.1), respectively. Part (d) can be proved by parts (b) and (c).

46.4. To prove the first part, use the results of Problems 14.10 and 14.12 to show that $Q_T(A) = Q_S(A)$ for every $A \in \mathscr{F}_T$. To prove the second part, follow the definition of equivalent martingale measure (Definition 45.6) and consider the probability measure defined as $\frac{\mathrm{d}Q}{\mathrm{d}P} = L(T^*)$.

46.5. To prove part (a), use the result of Problem 46.1 and Fubini's theorem (Theorems 10.2 and 46.1). To prove part (b), use Theorem 45.1 and Fubini's theorem (Theorems 10.2 and 46.1). Use part (b) and Theorem 46.3 to prove part (c).

46.6. Use Theorem 45.1 and the results of Problems 40.6 and 46.5.

46.4 Solutions

46.1. By Definition 46.1, we have

$$\mathrm{d}\ln P(t,u) = -f(t,u)\mathrm{d}u.$$

Integrating both sides of the above equation from S to T gives

$$\ln P(t,T) - \ln P(t,S) = -\int_S^T f(t,u)\mathrm{d}u,$$

or

$$P(t,T) = P(t,S)\exp\left(-\int_S^T f(t,u)\mathrm{d}u\right).$$

This completes the proof.

46.2. By Problem 46.1 and the fact that $P(t,t) = 1$, we have

$$\ln P(t,T) = -\int_t^T f(t,u)\mathrm{d}u.$$

Differentiating both sides of the above equation with respect to t gives

$$\frac{\partial \ln P(t,T)}{\partial t} = -\frac{\partial}{\partial t}\int_t^T f(t,u)\mathrm{d}u = -f(t,t) - \int_t^T \frac{\partial f}{\partial t}(t,u)\mathrm{d}u.$$

Now differentiating the above equation with respect to T gives

$$\frac{\partial^2 \ln P(t,T)}{\partial T \partial t} = -\frac{\partial f}{\partial t}(t,T).$$

But by Theorem 35.3, we get

$$\ln P(t,T) = \int_t^T \left(m(s,T) - \frac{1}{2}v^2(s,T)\right)\mathrm{d}s + \int_t^T v(s,T)\mathrm{d}B_s$$

or

$$\frac{\partial \ln P(t,T)}{\partial t} = \left(m(t,T) - \frac{1}{2}v^2(t,T)\right)\mathrm{d}t + v(t,T)\mathrm{d}B_t.$$

Combining the above equations gives

$$\frac{\partial f}{\partial t}(t,T) = \left(v(t,T)\frac{\partial v}{\partial T}(t,T) - \frac{\partial m}{\partial T}(t,T)\right)\mathrm{d}t - \frac{\partial v}{\partial T}(t,T)\mathrm{d}B_t.$$

This completes the proof.

46.3.

(a) By Definitions 45.2 and 46.1, we have

$$r(t) = f(t,t) = f(0,t) + \int_0^t \alpha(s,t)\mathrm{d}s + \int_0^t \sigma(s,t)\mathrm{d}B_s.$$

Since

$$\alpha(s,t) = \alpha(s,s) + \int_s^t \frac{\partial \alpha}{\partial T}(s,u)\mathrm{d}u$$

and

$$\sigma(s,t) = \sigma(s,s) + \int_s^t \frac{\partial \sigma}{\partial T}(s,u)\mathrm{d}u,$$

we have

$$\begin{aligned}
r(t) &= f(0,t) + \int_0^t \alpha(s,s)\mathrm{d}s + \int_0^t\int_s^t \frac{\partial \alpha}{\partial T}(s,u)\mathrm{d}u\mathrm{d}s \\
&\quad + \int_0^t \sigma(s,s)\mathrm{d}B_s + \int_0^t\int_s^t \frac{\partial \sigma}{\partial T}(s,u)\mathrm{d}u\mathrm{d}B_s \\
&= f(0,t) + \int_0^t \alpha(s,s)\mathrm{d}s + \int_0^t\int_0^u \frac{\partial \alpha}{\partial T}(s,u)\mathrm{d}s\mathrm{d}u \\
&\quad + \int_0^t \sigma(s,s)\mathrm{d}B_s + \int_0^t\int_0^u \frac{\partial \sigma}{\partial T}(s,u)\mathrm{d}B_s\mathrm{d}u,
\end{aligned}$$

or

$$\begin{aligned} \mathrm{d}r(t) &= \frac{\partial f}{\partial T}(0,t)\mathrm{d}t + \alpha(t,t)\mathrm{d}t + \int_0^t \frac{\partial \alpha}{\partial T}(s,t)\mathrm{d}s\mathrm{d}t \\ & \sigma(t,t)\mathrm{d}B_t + \int_0^t \frac{\partial \sigma}{\partial T}(s,t)\mathrm{d}B_s\mathrm{d}t. \end{aligned} \tag{46.3}$$

Note also that

$$f(t,T) = f(0,T) + \int_0^t \alpha(s,T)\mathrm{d}s + \int_0^t \sigma(s,T)\mathrm{d}B_s.$$

Differentiating the above equation with respect to T gives

$$\frac{\partial f}{T}(t,T) = \frac{\partial f}{\partial T}(0,T) + \int_0^t \frac{\partial \alpha}{\partial T}(s,T)\mathrm{d}s + \int_0^t \frac{\partial \sigma}{\partial T}(s,T)\mathrm{d}B_s.$$

Now let $T = t$ in the above equation, we get

$$\frac{\partial f}{T}(t,t) = \frac{\partial f}{\partial T}(0,t) + \int_0^t \frac{\partial \alpha}{\partial T}(s,t)\mathrm{d}s + \int_0^t \frac{\partial \sigma}{\partial T}(s,t)\mathrm{d}B_s. \tag{46.4}$$

Combining Equations (46.3) and (46.4) gives

$$\mathrm{d}r(t) = \left(\frac{\partial f}{T}(t,t) + \alpha(t,t)\right)\mathrm{d}t + \sigma(t,t)\mathrm{d}B_t.$$

(b) By assumption and Theorem 10.2, we have

$$\begin{aligned} \int_t^T \int_0^t \alpha(s,u)\mathrm{d}s\mathrm{d}u &= \int_0^t \int_t^T \alpha(s,u)\mathrm{d}u\mathrm{d}s \\ &= \int_0^t \int_s^T \alpha(s,u)\mathrm{d}u\mathrm{d}s - \int_0^t \int_s^t \alpha(s,u)\mathrm{d}u\mathrm{d}s \\ &= \int_0^t \int_s^T \alpha(s,u)\mathrm{d}u\mathrm{d}s - \int_0^t \int_0^u \alpha(s,u)\mathrm{d}s\mathrm{d}u. \end{aligned}$$

(c) Let $\phi(\omega, s, u)$ be defined as

$$\phi(\omega,s,u) = \begin{cases} 0, & \text{if } (s,u) \notin [0,t]\times[t,T]; \\ \sigma(s,u)(\omega), & \text{if } (s,u) \in [0,t]\times[t,T]. \end{cases}$$

Then by assumption and Theorem 46.1, we have

$$\begin{aligned}
\int_t^T \int_0^t \sigma(s,u)\mathrm{d}B_s\mathrm{d}u &= \int_0^T \int_0^t \phi(\cdot,s,u)\mathrm{d}B_s\mathrm{d}u \\
&= \int_0^t \int_0^T \phi(\cdot,s,u)\mathrm{d}u\mathrm{d}B_s \\
&= \int_0^t \int_t^T \sigma(s,u)\mathrm{d}u\mathrm{d}B_s \\
&= \int_0^t \int_s^T \sigma(s,u)\mathrm{d}u\mathrm{d}B_s - \int_0^t \int_s^t \sigma(s,u)\mathrm{d}u\mathrm{d}B_s \\
&= \int_0^t \int_s^T \sigma(s,u)\mathrm{d}u\mathrm{d}B_s - \int_0^t \int_0^u \sigma(s,u)\mathrm{d}B_s\mathrm{d}u.
\end{aligned}$$

(d) By Itô's lemma (Theorem 35.3), the stochastic differential equation

$$\mathrm{d}P(t,T) = P(t,T)\left[m(t,T)\mathrm{d}t + v(t,T)\mathrm{d}B_t\right]$$

is equivalent to the stochastic differential equation

$$\mathrm{d}\ln P(t,T) = \left(m(t,T) - \frac{1}{2}v^2(t,T)\right)\mathrm{d}t + v(t,T)\mathrm{d}B_t. \tag{46.5}$$

Now by assumption we have

$$f(t,u) = f(0,u) + \int_0^t \alpha(s,u)\mathrm{d}s + \int_0^t \sigma(s,u)\mathrm{d}s.$$

Since $\ln P(t,T) = -\int_t^T f(t,u)\mathrm{d}u$, we have

$$\begin{aligned}
&\ln P(t,T) \\
= &-\int_t^T f(0,u)\mathrm{d}u - \int_t^T \int_0^t \alpha(s,u)\mathrm{d}s\mathrm{d}u - \int_t^T \int_0^t \sigma(s,u)\mathrm{d}s\mathrm{d}u.
\end{aligned}$$

By parts (b) and (c), the above equation becomes

$$\begin{aligned}
&\ln P(t,T) \\
= &-\int_t^T f(0,u)\mathrm{d}u - \int_0^t \int_s^T \alpha(s,u)\mathrm{d}u\mathrm{d}s + \int_0^t \int_0^u \alpha(s,u)\mathrm{d}s\mathrm{d}u \\
&-\int_0^t \int_s^T \sigma(s,u)\mathrm{d}u\mathrm{d}B_s + \int_0^t \int_0^u \sigma(s,u)\mathrm{d}B_s\mathrm{d}u
\end{aligned}$$

or

$$
\begin{aligned}
& \mathrm{d}\ln P(t,T) \\
= \; & f(0,t)\mathrm{d}t - \int_t^T \alpha(t,u)\mathrm{d}u\mathrm{d}t + \int_0^t \alpha(s,t)\mathrm{d}s\mathrm{d}t \\
& - \int_t^T \sigma(t,u)\mathrm{d}u\mathrm{d}B_t + \int_0^t \sigma(s,t)\mathrm{d}B_s\mathrm{d}t \\
= \; & f(t,t)\mathrm{d}t - \int_t^T \alpha(t,u)\mathrm{d}u\mathrm{d}t - \int_t^T \sigma(t,u)\mathrm{d}u\mathrm{d}B_t. \qquad (46.6)
\end{aligned}
$$

Note that $r(t) = f(t,t)$. Comparing Equations (46.5) and (46.6) gives

$$
m(t,T) = r(t) + \frac{1}{2}\left(\int_t^T \sigma(t,s)\mathrm{d}s\right)^2 - \int_t^T \alpha(t,s)\mathrm{d}s
$$

and

$$
v(t,T) = -\int_t^T \sigma(t,s)\mathrm{d}s.
$$

This completes the proof.

46.4.

(a) Let $A \in \mathscr{F}_T$. Since $Q_S(A) = E_{Q_S}[I_A] = E_P[L(S)I_A]$, it follows from Problems 14.10 and 14.12 that

$$
\begin{aligned}
Q_S(A) &= E_P[L(S)I_A] \\
&= E_P[E_P[L(S)I_A]|\mathscr{F}_T] \\
&= E_P[I_A E_P[L(S)|\mathscr{F}_T]].
\end{aligned}
$$

But

$$
\begin{aligned}
& E_P[L(S)|\mathscr{F}_T] \\
= \; & E_P\left[L(T)\exp\left(-\int_T^S \lambda(u)\mathrm{d}B_u - \frac{1}{2}\int_T^S \lambda^2(u)\mathrm{d}u\right) \middle| \mathscr{F}_T\right] \\
= \; & L(T)E_P\left[\exp\left(-\int_T^S \lambda(u)\mathrm{d}B_u - \frac{1}{2}\int_T^S \lambda^2(u)\mathrm{d}u\right)\right] \\
= \; & L(T).
\end{aligned}
$$

Hence we have

$$
Q_S(A) = E_P[I_A L(T)] = E_{Q_T}[I_A] = Q_T(A).
$$

(b) Let $Z(t,t) = \frac{P(t,T)}{\beta(t)}$. Then it follows from Theorem 35.3 that

$$\mathrm{d}Z(t,T) = Z(t,T)\left[m(t,T)\mathrm{d}t + v(t,T)\mathrm{d}B_t\right], \tag{46.7}$$

where

$$m(t,T) = \frac{1}{2}\left(\int_t^T \sigma(t,s)\mathrm{d}s\right)^2 - \int_t^T \alpha(t,s)\mathrm{d}s$$

and

$$v(t,T) = -\int_t^T \sigma(t,s)\mathrm{d}s.$$

Let Q be a probability measure on $(\Omega, \mathscr{F}_{T^*})$ defined as $\frac{\mathrm{d}Q}{\mathrm{d}P} = L(T^*)$. Let $\{W_t : 0 \le t \le T^*\}$ be a process defined as

$$W_t = B_t + \int_0^t \lambda(s)\mathrm{d}s.$$

Then by Problem 37.4, $\{W_t : 0 \le t \le T^*\}$ is a Brownian motion under Q. Equation (46.7) becomes

$$\mathrm{d}Z(t,T) = Z(t,T)\left[(m(t,T) - v(t,T)\lambda(t))\mathrm{d}t + v(t,T)\mathrm{d}W_t\right]. \tag{46.8}$$

Since

$$\begin{aligned}\int_t^T \alpha(t,s)\mathrm{d}s &= \int_t^T \left(\sigma(t,s)\int_t^s \sigma(t,u)\mathrm{d}u + \sigma(t,s)\lambda(t)\right)\mathrm{d}s \\ &= \frac{1}{2}\left(\int_t^T \sigma(t,s)\mathrm{d}s\right)^2 - \lambda(t)v(t,T),\end{aligned}$$

we have

$$m(t,T) - v(t,T)\lambda(t) = 0.$$

Hence Equation (46.8) becomes

$$\mathrm{d}Z(t,T) = Z(t,T)v(t,T)\mathrm{d}W_t.$$

By Theorem 32.2, $\{Z(t,T) : 0 \le t \le T\}$ is a martingale.

This completes the proof.

46.5.

(a) Since

$$f(u,u) = f(t,u) - \int_t^u \sigma(s,u)\eta(s,u)\mathrm{d}s + \int_t^u \sigma(s,u)\mathrm{d}W_s,$$

we have

$$\begin{aligned}\int_t^T r(u)\mathrm{d}u &= \int_t^T f(t,u)\mathrm{d}u - \int_t^T \int_t^u \sigma(s,u)\eta(s,u)\mathrm{d}s\mathrm{d}u \\ &\quad + \int_t^T \int_t^u \sigma(s,u)\mathrm{d}W_s\mathrm{d}u.\end{aligned}$$

By Theorems 10.2 and 46.1, the above equation becomes

$$\begin{aligned}\int_t^T r(u)\mathrm{d}u &= \int_t^T f(t,u)\mathrm{d}u - \int_t^T \int_t^u \sigma(s,u)\eta(s,u)\mathrm{d}s\mathrm{d}u \\ &\quad + \int_t^T \int_t^u \sigma(s,u)\mathrm{d}W_s\mathrm{d}u \\ &= \int_t^T f(t,u)\mathrm{d}u - \int_t^T \int_s^T \sigma(s,u)\eta(s,u)\mathrm{d}u\mathrm{d}s \\ &\quad + \int_t^T \int_s^T \sigma(s,u)\mathrm{d}u\mathrm{d}W_s \\ &= \int_t^T f(t,u)\mathrm{d}u + \frac{1}{2}\int_t^T \eta^2(s,T)\mathrm{d}s \\ &\quad - \int_t^T \eta(s,T)\mathrm{d}W_s.\end{aligned}$$

By Problem 46.1, we have

$$P(t,T) = \exp\left(-\int_t^T f(t,u)\mathrm{d}u\right).$$

Combining the above two equations, we get

$$\frac{\beta(T)}{\beta(t)} = \frac{1}{P(t,T)} \exp\left(\frac{1}{2}\int_t^T \eta^2(s,T)\mathrm{d}s - \int_t^T \eta(s,T)\mathrm{d}W_s\right). \tag{46.9}$$

Letting $t = 0$ in the above equation gives

$$\beta(T) = \frac{1}{P(0,T)} \exp\left(\frac{1}{2}\int_0^T \eta^2(s,T)\mathrm{d}s - \int_0^T \eta(s,T)\mathrm{d}W_s\right).$$

Plugging $\beta(T)$ in Equation (46.9) and rearranging the terms leads to

$$\frac{P(t,T)}{\beta(t)} = P(0,T)\exp\left(-\frac{1}{2}\int_0^t \eta^2(s,T)\mathrm{d}s + \int_0^t \eta(s,T)\mathrm{d}W_s\right).$$

(b) By Theorem 45.1, the price at time T of a zero-coupon bond with maturity S is

$$P(T,S) = E_Q\left[e^{-\int_T^S r(u)\mathrm{d}u}\middle| \mathscr{F}_T\right]. \tag{46.10}$$

Since

$$r(u) = f(u, u) = f(t, u) - \int_t^u \sigma(s, u)\eta(s, u)\mathrm{d}s + \int_t^u \sigma(s, u)\mathrm{d}W_s,$$

we have

$$\begin{aligned}\int_T^S r(u)\mathrm{d}u &= \int_T^S f(t, u)\mathrm{d}u - \int_T^S \int_t^u \sigma(s, u)\eta(s, u)\mathrm{d}s\mathrm{d}u \\ &\quad + \int_T^S \int_t^u \sigma(s, u)\mathrm{d}W_s\mathrm{d}u. \qquad (46.11)\end{aligned}$$

By Theorem 10.2, we have

$$\begin{aligned}&\int_T^S \int_t^u \sigma(s, u)\eta(s, u)\mathrm{d}s\mathrm{d}u \\ =&\int_t^T \int_T^S \sigma(s, u)\eta(s, u)\mathrm{d}u\mathrm{d}s + \int_T^S \int_s^S \sigma(s, u)\eta(s, u)\mathrm{d}u\mathrm{d}s \\ =&\frac{1}{2}\int_t^T \left[\eta^2(s, T) - \eta^2(s, S)\right]\mathrm{d}s - \frac{1}{2}\int_T^S \eta^2(s, S)\mathrm{d}s. \qquad (46.12)\end{aligned}$$

By Theorem 46.1, we have

$$\begin{aligned}&\int_T^S \int_t^u \sigma(s, u)\mathrm{d}W_s\mathrm{d}u \\ =&\int_t^T \int_T^S \sigma(s, u)\mathrm{d}u\mathrm{d}W_s + \int_T^S \int_s^S \sigma(s, u)\mathrm{d}u\mathrm{d}W_s \\ =&\int_t^T [\eta(s, T) - \eta(s, S)]\,\mathrm{d}W_s - \int_T^S \eta(s, S)\mathrm{d}W_s. \qquad (46.13)\end{aligned}$$

Plugging Equations (46.12) and (46.13) into Equation (46.11) gives

$$\begin{aligned}\int_T^S r(u)\mathrm{d}u &= \int_T^S f(t, u)\mathrm{d}u - Z + \frac{1}{2}\int_T^S \eta^2(s, S)\mathrm{d}s \\ &\quad - \int_T^S \eta(s, S)\mathrm{d}W_s. \qquad (46.14)\end{aligned}$$

Since $\int_T^S \eta(s, S)\mathrm{d}W_s$ is a normal random variable independent of $\mathscr{F}_T$, combining Equations (46.10) and (46.14) gives

$$P(T, S) = \frac{P(t, S)}{P(t, T)}e^Z.$$

(c) By Theorem 46.3 and the first part, we have

$$\begin{aligned}
V(t,T,S) &= F(t,T,P(T,S)) \\
&= E_Q\left[\frac{P(t,S)}{P(t,T)}e^Z \middle| \mathscr{F}_t\right] \\
&= \frac{P(t,S)}{P(t,T)}E_Q[e^Z] \\
&= \frac{P(t,S)}{P(t,T)}\exp\left(\int_t^T \eta(u,T)[\eta(u,T)-\eta(u,S)]\mathrm{d}u\right).
\end{aligned}$$

This completes the proof.

46.6. By Theorem 45.1, we have

$$\begin{aligned}
C(t,T,S,K) &= E_Q\left[(P(T,S)-K)^+e^{-\int_t^T r(u)\mathrm{d}u}\middle| \mathscr{F}_t\right] \\
&= E_Q\left[I_A P(T,S)e^{-\int_t^T r(u)\mathrm{d}u}\middle| \mathscr{F}_t\right] \\
&\quad -KE_Q\left[I_A e^{-\int_t^T r(u)\mathrm{d}u}\middle| \mathscr{F}_t\right],
\end{aligned} \tag{46.15}$$

where $A = \{P(T,S) > K\}$. Also by Theorem 45.1, we have

$$P(T,S) = E_Q\left[e^{-\int_T^S r(u)\mathrm{d}u}\middle| \mathscr{F}_T\right].$$

Hence it follows from Problem 14.10 that

$$\begin{aligned}
&E_Q\left[I_A P(T,S)e^{-\int_t^T r(u)\mathrm{d}u}\middle| \mathscr{F}_t\right] \\
=\ &E_Q\left[I_A E_Q\left[e^{-\int_T^S r(u)\mathrm{d}u}\middle| \mathscr{F}_T\right]e^{-\int_t^T r(u)\mathrm{d}u}\middle| \mathscr{F}_t\right] \\
=\ &E_Q\left[E_Q\left[I_A e^{-\int_t^T r(u)\mathrm{d}u}e^{-\int_T^S r(u)\mathrm{d}u}\middle| \mathscr{F}_T\right]\middle| \mathscr{F}_t\right] \\
=\ &E_Q\left[I_A e^{-\int_t^S r(u)\mathrm{d}u}\middle| \mathscr{F}_t\right].
\end{aligned}$$

Plugging the above equation into Equation (46.15) gives

$$\begin{aligned}
C(t,T,S,K) &= E_Q\left[I_A e^{-\int_t^S r(u)\mathrm{d}u}\middle| \mathscr{F}_t\right] \\
&\quad -KE_Q\left[I_A e^{-\int_t^T r(u)\mathrm{d}u}\middle| \mathscr{F}_t\right].
\end{aligned} \tag{46.16}$$

Now from the assumption, we have

$$r(u) = f(u,u) = f(t,u) + \frac{\sigma^2}{2}(u-t)^2 + \sigma(W_u - W_t).$$

Hence by Theorem 46.1, we have

$$\begin{aligned}\int_t^S r(u)\mathrm{d}u &= \int_t^S f(t,u)\mathrm{d}u + \int_t^S \frac{\sigma^2}{2}(u-t)^2\mathrm{d}u + \sigma\int_t^S (W_u - W_t)\mathrm{d}u \\ &= \int_t^S f(t,u)\mathrm{d}u + \frac{\sigma^2}{6}(S-t)^3 + \sigma\int_t^S (W_u - W_t)\mathrm{d}u \\ &= \int_t^S f(t,u)\mathrm{d}u + \frac{\sigma^2}{6}(S-t)^3 + \sigma\int_0^{S-t} \tilde{W}_u\mathrm{d}u, \qquad (46.17)\end{aligned}$$

where

$$\tilde{W}_u = W_{t+u} - W_t.$$

By Problem 46.5, we have

$$P(T,S) = \frac{P(t,S)}{P(t,T)}e^Z,$$

where

$$Z = -\frac{\sigma^2}{2}(S-T)(S-t)(T-t) - (S-T)(W_T - W_t).$$

Hence we have

$$A = \left\{ \tilde{W}_{T-t} < \frac{\ln\frac{P(t,S)}{KP(t,T)} - \frac{\sigma^2}{2}(S-T)(S-t)(T-t)}{\sigma(S-T)} \right\}.$$

Since $\tilde{W}_u$ is independent of $\mathscr{F}_t$ for all $u \geq 0$, we have

$$\begin{aligned}& E_Q\left[I_A e^{-\int_t^S r(u)\mathrm{d}u} \middle| \mathscr{F}_t \right] \\ =\ & E_Q\left[I_A \exp\left(-\int_t^S f(t,u)\mathrm{d}u - \frac{\sigma^2}{6}(S-t)^3 - \sigma\int_0^{S-t}\tilde{W}_u\mathrm{d}u \right) \middle| \mathscr{F}_t \right] \\ =\ & E_Q\left[I_A \exp\left(-\int_t^S f(t,u)\mathrm{d}u - \frac{\sigma^2}{6}(S-t)^3 - \sigma\int_0^{S-t}\tilde{W}_u\mathrm{d}u \right) \right] \\ =\ & P(t,S)\exp\left(-\frac{\sigma^2}{6}(S-t)^3\right) E_Q\left[I_A \exp\left(-\sigma\int_0^{S-t}\tilde{W}_u\mathrm{d}u \right)\right]. \qquad (46.18)\end{aligned}$$

By Problem 40.6, we have

$$E_Q\left[\exp\left(-\sigma\int_0^{S-t}\tilde{W}_u\mathrm{d}u\right)\right] = \exp\left(\frac{\sigma^2}{6}(S-t)^3\right).$$

In order to evaluate the expectation in Equation (46.18), we consider the probability measure P_1 defined by

$$\frac{\mathrm{d}P_1}{\mathrm{d}Q} = \frac{\exp\left(-\sigma\int_0^{S-t}\tilde{W}_u\mathrm{d}u\right)}{E_Q\left[\exp\left(-\sigma\int_0^{S-t}\tilde{W}_u\mathrm{d}u\right)\right]} = \frac{\exp\left(-\sigma\int_0^{S-t}\tilde{W}_u\mathrm{d}u\right)}{\exp\left(\frac{\sigma^2}{6}(S-t)^3\right)}.$$

Then we have

$$\begin{aligned}
& E_Q\left[I_A \exp\left(-\sigma \int_0^{S-t} \tilde{W}_u \mathrm{d}u\right)\right] \\
= \; & \exp\left(\frac{\sigma^2}{6}(S-t)^3\right) E_Q\left[I_A \frac{\mathrm{d}P_1}{\mathrm{d}Q}\right] \\
= \; & \exp\left(\frac{\sigma^2}{6}(S-t)^3\right) E_{P_1}\left[I_A\right] \\
= \; & \exp\left(\frac{\sigma^2}{6}(S-t)^3\right) P_1(A). \qquad (46.19)
\end{aligned}$$

To calculate $P_1(A)$, we need to know the distribution of $\tilde{W}_{T-t}$ under P_1. To do that, we consider

$$\begin{aligned}
& E_{P_1}\left[e^{-w\tilde{W}_{T-t}}\right] \\
= \; & E_Q\left[e^{-w\tilde{W}_{T-t}} \frac{\exp\left(-\sigma \int_0^{S-t} \tilde{W}_u \mathrm{d}u\right)}{\exp\left(\frac{\sigma^2}{6}(S-t)^3\right)}\right] \\
= \; & \exp\left(-\frac{\sigma^2}{6}(S-t)^3\right) E_Q\left[\exp\left(-\sigma \int_0^{S-t} \tilde{W}_u \mathrm{d}u - w\tilde{W}_{T-t}\right)\right].
\end{aligned}$$

By Problem 40.6, the above equation becomes

$$E_{P_1}\left[e^{-w\tilde{W}_{T-t}}\right] = \exp\left(\frac{\sigma w}{2}(S-t)^2 - \frac{\sigma w}{2}(S-T)^2 + \frac{w^2}{2}(T-t)\right).$$

By Problem 19.1, $\tilde{W}_{T-t}$ is a normal random variable under P_1 with mean

$$\mu_1 = -\frac{\sigma}{2}(S-t)^2 + \frac{\sigma}{2}(S-T)^2$$

and variance $T-t$. Then we have

$$\begin{aligned}
P_1(A) &= P_1\left\{\tilde{W}_{T-t} < \frac{\ln \frac{P(t,S)}{KP(t,T)} - \frac{\sigma^2}{2}(S-T)(S-t)(T-t)}{\sigma(S-T)}\right\} \\
&= P_1\left\{\frac{\tilde{W}_{T-t} - \mu_1}{\sqrt{T-t}} < \frac{\ln \frac{P(t,S)}{KP(t,T)} + \frac{\sigma^2}{2}(S-T)^2(T-t)}{\sigma(S-T)\sqrt{T-t}}\right\} \\
&= N(d_1). \qquad (46.20)
\end{aligned}$$

Now combining Equations (46.18), (46.19), and (46.20) gives

$$E_Q\left[I_A e^{-\int_t^S r(u)\mathrm{d}u} \,\middle|\, \mathscr{F}_t\right] = P(t,S)N(d_1).$$

In a similar way we can show that

$$E_Q\left[I_A e^{-\int_t^T r(u)\mathrm{d}u} \middle| \mathscr{F}_t \right] = P(t,T)N(d_2).$$

Plugging the above two equations into Equation (46.16) completes the proof.

46.5 Bibliographic Notes

In this chapter, we presented the HJM framework and the HJM drift condition. For more information about HJM framework, readers are referred to Bingham and Kiesel (2004), Svoboda (2004), Musiela and Rutkowski (2005), Carmona (2006), Brigo and Mercurio (2006), and Filipović (2009).

Theorem 46.1 is a version of Fubini's theorem for stochastic integrals. A proof of this theorem can be found in Svoboda (2004, p166). For a multidimensional version of the stochastic Fubini theorem, readers are referred to Filipović (2009, Theorem 6.2). A proof of Theorem 46.2 can be found in Bingham and Kiesel (2004). For the rational behind the futures price (see Theorem 46.3), readers are referred to Filipović (2009, Section 8.2).

CHAPTER 47

LIBOR MARKET MODELS

The short rate models and the HJM models introduced in the previous chapters concentrate on unobservable rates, including short rates and instantaneous forward rates. Brace et al. (1997) introduced LIBOR (London interbank offered rate) market models, which model the observable LIBOR rates directly. In this chapter, we introduce the LIBOR model and relevant concepts.

47.1 Basic Concepts and Facts

Definition 47.1 (LIBOR). Let $\delta > 0$ and $0 \le t < T$. The δ-period forward LIBOR for the future date T prevailing at time t is defined as

$$L(t,T) = F(t;T,T+\delta) = \frac{1}{\delta}\left(\frac{P(t,T)}{P(t,T+\delta)} - 1\right).$$

Definition 47.2 (T-Forward Measure). Let Q be an equivalent martingale measure for the bond market. Let $T > 0$. Then the T-forward measure on $(\Omega, \mathscr{F}_T)$ is defined as

$$\frac{\mathrm{d}Q^T}{\mathrm{d}Q} = \frac{1}{P(0,T)\beta(T)},$$

where $\beta(\cdot)$ is defined in Definition 45.3.

For $t \leq T$, the conditional expectation of $\frac{\mathrm{d}Q^T}{\mathrm{d}Q}$ given $\mathscr{F}_t$, written as $\left.\frac{\mathrm{d}Q^T}{\mathrm{d}Q}\right|_{\mathscr{F}_t}$, is defined as

$$\left.\frac{\mathrm{d}Q^T}{\mathrm{d}Q}\right|_{\mathscr{F}_t} = E_Q\left[\left.\frac{\mathrm{d}Q^T}{\mathrm{d}Q}\right|\mathscr{F}_t\right].$$

Definition 47.3 (LIBOR Market Model)**.** Let $\delta > 0$ and M be a positive integer. Let $T_M = M\delta$ be the finite time horizon. Let $(\Omega, \mathscr{F}, \{\mathscr{F}_t : 0 \leq t \leq T_M\}, Q^{T_M})$ be the underlying filtered probability space, where Q^{T_M} is the T_M-forward measure. For every integer $m = 0, 1, \ldots, M-1$, there is a deterministic bounded measurable function $\lambda(t, T_m)$, where $T_m = m\delta$. For every integer $m = 0, 1, \ldots, M-1$, the initial forward LIBOR rate is given by

$$L(0, T_m) = \frac{1}{\delta}\left(\frac{P(0, T_m)}{P(0, T_{m+1})} - 1\right).$$

A LIBOR market model is defined recursively as follows. For $i = M, M-2, \ldots, 1$ and $t \in [0, T_{i-1}]$, we have

$$L(t, T_{i-1}) = L(0, T_{i-1}) \exp\left(-\frac{1}{2}\int_0^t \lambda^2(s, T_{i-1})\mathrm{d}s + \int_0^t \lambda(s, T_{i-1})\mathrm{d}W_s^{T_i}\right),$$

$$\sigma_{T_{i-1},T_i}(t) = \frac{\delta L(t, T_{i-1})}{\delta L(t, T_{i-1}) + 1}\lambda(t, T_{i-1}),$$

and

$$W_t^{T_{i-1}} = W_t^{T_i} - \int_0^T \sigma_{T_{i-1},T_i}(s)\mathrm{d}s,$$

where the process $\{W_t^{T_{i-1}} : 0 \leq t \leq T_{i-1}\}$ is a Brownian motion under $Q^{T_{i-1}}$, which is defined as

$$\frac{\mathrm{d}Q^{T_{i-1}}}{\mathrm{d}Q^{T_i}} = \exp\left(-\frac{1}{2}\int_0^{T_{i-1}} \sigma^2_{T_{i-1},T_i}(s)\mathrm{d}s + \int_0^{T_{i-1}} \sigma_{T_{i-1},T_i}(s)\mathrm{d}W_s^{T_i}\right).$$

47.2 Problems

47.1. Let Q be an equivalent martingale measure for the bond market and Q^T be the T-forward measure. For $t \leq T$, show that

$$\left.\frac{\mathrm{d}Q^T}{\mathrm{d}Q}\right|_{\mathscr{F}_t} = \frac{P(t, T)}{P(0, T)\beta(t)}.$$

47.2. Let $T^* > 0$. Suppose that for every $T \in [0, T^*]$, the forward rate process $\{f(t, T) : 0 \leq t \leq T\}$ satisfies the HJM model

$$\mathrm{d}f(t, T) = \sigma(t, T)\int_t^T \sigma(t, s)\mathrm{d}s\mathrm{d}t + \sigma(t, T)\mathrm{d}W_t,$$

where $\{W_t : 0 \le t \le T\}$ is a Brownian motion under the risk-neutral probability Q and the coefficients of $\mathrm{d}t$ and $\mathrm{d}W_t$ satisfy the technical conditions given in Definition 46.4. Let

$$\eta(t,T) = -\int_t^T \sigma(t,u)\mathrm{d}u.$$

Show that

(a) For $0 \le t \le T \le T^*$,

$$\left.\frac{\mathrm{d}Q^T}{\mathrm{d}Q}\right|_{\mathscr{F}_t} = \exp\left(-\frac{1}{2}\int_0^t \eta^2(s,T)\mathrm{d}s + \int_0^t \eta(s,T)\mathrm{d}W_s\right).$$

(b) The process $\{W_t^T : 0 \le t \le T\}$ given by

$$W_t^T = W_t - \int_0^t \eta(s,T)\mathrm{d}s \tag{47.1}$$

is a Brownian motion under Q^T.

(c) For $0 \le t \le T \le T^*$ and $0 \le t \le S \le T^*$,

$$\frac{P(t,S)}{P(t,T)} = \frac{P(0,S)}{P(0,T)} \times$$
$$\exp\left(-\frac{1}{2}\int_0^t [\eta(s,S) - \eta(s,T)]^2\mathrm{d}s + \int_0^t [\eta(s,S) - \eta(s,T)]\mathrm{d}W_s^T\right)$$

and the T-bond discounted S-bond price process $\{\frac{P(t,S)}{P(t,T)} : 0 \le t \le T\}$ is a Q^T-martingale, where W_s^T is as defined in Equation (47.1).

(d) For $0 \le t \le T \le T^*$ and $0 \le t \le S \le T^*$, the T-forward measure and S-forward measure satisfy

$$\left.\frac{\mathrm{d}Q^S}{\mathrm{d}Q^T}\right|_{\mathscr{F}_t} = \frac{P(t,S)P(0,T)}{P(t,T)P(0,S)}.$$

47.3 (Arbitrage Price)**.** Let Q be an equivalent martingale measure for the bond market. Let X be a T-contingent claim such that

$$E_Q\left[\frac{|X|}{\beta(T)}\right] < \infty.$$

Show that

$$E_{Q^T}[|X|] < \infty$$

and for $0 \le t \le T$,

$$\beta(t)E_Q\left[\left.\frac{X}{\beta(T)}\right| \mathscr{F}_t\right] = P(t,T)E_{Q^T}[X|\mathscr{F}_t].$$

47.4 (Expectation Hypothesis). Let $T^* > 0$. Suppose that for every $T \in [0, T^*]$, the forward rate process $\{f(t,T) : 0 \le t \le T\}$ satisfies the HJM model

$$df(t,T) = \sigma(t,T)\int_t^T \sigma(t,s)\mathrm{d}s\mathrm{d}t + \sigma(t,T)\mathrm{d}W_t,$$

where $\{W_t : 0 \le t \le T\}$ is a Brownian motion under the risk-neutral probability Q and the coefficients of $\mathrm{d}t$ and $\mathrm{d}W_t$ satisfy the technical conditions given in Definition 46.4. Suppose that $\sigma(t,T) \in L^2_{ad}([0,T] \times \Omega)$. Show that for $t \le T$,

$$f(t,T) = E_{Q^T}[r(T)|\mathscr{F}_t].$$

47.5. Let $L(t,T_i)$, $i = 0, 1, \ldots, M$ be LIBOR rates that satisfy the LIBOR market model given in Definition 47.3. Let m, n be integers in $\{0, 1, \ldots, M\}$. Show that for $t \in [0, T_m \wedge T_n]$,

$$\mathrm{d}W_t^{T_m} = \begin{cases} \mathrm{d}W_t^{T_n} - \sum_{k=m}^{n-1} \sigma_{T_k,T_{k+1}}(t)\mathrm{d}t, & \text{if } m < n; \\ \mathrm{d}W_t^{T_n}, & \text{if } m = n; \\ \mathrm{d}W_t^{T_n} + \sum_{k=n}^{m-1} \sigma_{T_k,T_{k+1}}(t)\mathrm{d}t, & \text{if } m > n, \end{cases}$$

and for $t \in [0, T_m \wedge T_{n+1}]$,

$$\frac{\mathrm{d}L(t,T_m)}{L(t,T_m)} = \begin{cases} \lambda(t,T_m)\mathrm{d}W_t^{T_{n+1}} - \lambda(t,T_m)\sum_{k=m+1}^{n} \sigma_{T_k,T_{k+1}}(t)\mathrm{d}t, & \text{if } m < n; \\ \lambda(t,T_m)\mathrm{d}W_t^{T_{n+1}}, & \text{if } m = n; \\ \lambda(t,T_m)\mathrm{d}W_t^{T_{n+1}} + \lambda(t,T_m)\sum_{k=n+1}^{m} \sigma_{T_k,T_{k+1}}(t)\mathrm{d}t, & \text{if } m > n. \end{cases}$$

47.6. Consider the LIBOR market model defined in Definition 47.3. Let $1 \le k \ne m \le M$. Show that

(a) For $t \in [0, T_k \wedge T_m]$,

$$\frac{P(t,T_k)}{P(t,T_m)} = \frac{P(0,T_k)}{P(0,T_m)} \exp\left(-\frac{1}{2}\int_0^t \sigma^2_{T_k,T_m}(s)\mathrm{d}s + \int_0^t \sigma_{T_k,T_m}(s)\mathrm{d}W_s^{T_m}\right),$$

where

$$\sigma_{T_k,T_m}(t) = \begin{cases} \sum_{i=k}^{m-1} \sigma_{T_i,T_{i+1}}(t), & \text{if } k < m; \\ -\sum_{i=m}^{k-1} \sigma_{T_i,T_{i+1}}(t), & \text{if } k > m. \end{cases}$$

(b) The forward price process

$$\left\{\frac{P(t,T_k)}{P(t,T_m)} : 0 \le t \le T_k \wedge T_m\right\}$$

is a positive Q^{T_m}-martingale.

47.7. Consider the LIBOR market model defined in Definition 47.3. Let $1 \leq m < n \leq M$. Show that the price at time t of a T_m-contingent claim X is given by

$$P(t, T_n) E_{Q^{T_n}} \left[\frac{X}{P(T_m, T_n)} \middle| \mathscr{F}_t \right].$$

47.8 (Caplet)**.** Consider the LIBOR market model defined in Definition 47.3. Let $m + 1 < n \leq M$. Let X be a caplet with reset date T_{n-1} and settlement date T_n. At the settlement date, the caplet pays the holder the difference between the LIBOR rate $L(T_{n-1}, T_n)$ and the strike rate κ; that is, the caplet pays the holder at time T_n the following amount

$$\delta(L(T_{n-1}, T_n) - \kappa)^+.$$

Show that the price at time T_m of the caplet is given by

$$V(T_m, T_{n-1}, T_n) = \delta P(T_m, T_n)[L(T_m, T_{n-1}) N(d_1) - \kappa N(d_2)],$$

where

$$d_1 = \frac{\ln \frac{L(T_m, T_{n-1})}{\kappa} + \frac{1}{2} \int_{T_m}^{T_{n-1}} \lambda^2(s, T_{n-1}) \mathrm{d}s}{\sqrt{\int_{T_m}^{T_{n-1}} \lambda^2(s, T_{n-1}) \mathrm{d}s}}$$

and

$$d_2 = d_1 - \sqrt{\int_{T_m}^{T_{n-1}} \lambda^2(s, T_{n-1}) \mathrm{d}s}.$$

47.9 (Implied Money-Market Account)**.** Consider the LIBOR market model defined in Definition 47.3. Let $\{\beta^*(T_i) : i = 0, 1, \ldots, M\}$ be a discrete-time process defined as

$$\beta^*(T_i) = \begin{cases} 1, & \text{if } i = 0; \\ [1 + \delta L(T_{i-1}, T_{i-1})] \, \beta^*(T_{i-1}), & \text{if } i = 1, 2, \ldots, M. \end{cases}$$

Show that

(a) For every $t \in [0, T_{M-1}]$,

$$\begin{aligned} & E_{Q^{T_M}} \left[\beta^*(T_M) P(0, T_M) | \mathscr{F}_t \right] \\ = \; & \exp\left(-\frac{1}{2} \int_0^t \Sigma^2(s) \mathrm{d}s + \int_0^t \Sigma(s) \mathrm{d}W_s^{T_M} \right), \end{aligned}$$

where

$$\Sigma(t) = \sum_{i=\eta(t)}^{M-1} \sigma_{T_i, T_{i+1}}(t)$$

with $\eta(t) = k$ if $t \in [T_{k-1}, T_k)$.

(b) For every $m = 0, 1, \ldots, M$,

$$E_{Q^{T_M}} \left[\beta^*(T_M) | \mathscr{F}_{T_m} \right] = \frac{\beta^*(T_m)}{P(T_m, T_M)}.$$

47.10 (Sport LIBOR Measure). Consider the LIBOR market model defined in Definition 47.3. Let $\{\beta^*(T_i) : i = 0, 1, \ldots, M\}$ be a discrete-time process defined as

$$\beta^*(T_i) = \begin{cases} 1, & \text{if } i = 0; \\ [1 + \delta L(T_{i-1}, T_{i-1})]\, \beta^*(T_{i-1}), & \text{if } i = 1, 2, \ldots, M. \end{cases}$$

Let Q^* be a probability measure on $(\Omega, \mathscr{F}_{T_M})$ defined as

$$\frac{\mathrm{d}Q^*}{\mathrm{d}Q^{T_M}} = \beta^*(T_M) P(0, T_M).$$

Show that

(a) For every $t \in [0, T_{M-1}]$,

$$\left.\frac{\mathrm{d}Q^*}{\mathrm{d}Q^{T_M}}\right|_{\mathscr{F}_t} = \exp\left(-\frac{1}{2}\int_0^t \Sigma^2(s)\mathrm{d}s + \int_0^t \Sigma(s)\mathrm{d}W_s^{T_M}\right),$$

where

$$\Sigma(t) = \sum_{i=\eta(t)}^{M-1} \sigma_{T_i, T_{i+1}}(t)$$

with $\eta(t) = k$ if $t \in [T_{k-1}, T_k)$.

(b) For $m \le n$, the price at time T_m of a T_n-contingent claim X is given by

$$V_{T_m}(X) = \beta^*(T_m) E_{Q^*}\left[\left.\frac{X}{\beta^*(T_n)}\right| \mathscr{F}_{T_m}\right].$$

47.3 Hints

47.1. Follow the definition of equivalent martingale measures (Definition 45.6) and the definition of $\left.\frac{\mathrm{d}Q^T}{\mathrm{d}Q}\right|_{\mathscr{F}_t}$ (Definition 47.2).

47.2. Part (a) follows from Problems 47.1 and 46.5. Part (b) follows from part (a) and Problem 37.4. Use the result of Problem 47.1 and part (a) to prove part (c). Use the result of Problem 47.1 to prove part (d).

47.3. Use the result of Problem 47.1 and Bayes' formula (see Problem 14.20).

47.4. Consider the forward rate process under the forward measure Q^T and apply Theorem 32.2.

47.5. Follow the definition of LIBOR market models (Definition 47.3).

47.6. To prove the first part, use the definitions of $L(t, T_{m-1})$ and $\sigma_{T_{m-1}, T_m}(t)$ (see Definition 47.3). Use the first part to prove the second part.

47.7. Use the results of Problems 47.2 and 47.3.

47.8. Use the result of Problem 47.3.

47.9. To prove the first part, use the result of Problem 47.6 and consider

$$\tilde{\sigma}_{T_{i-1},T_i}(s) = \begin{cases} \sigma_{T_{i-1},T_i}(s), & \text{if } s \in [0, T_{i-1}); \\ 0, & \text{if } s \in [T_{i-1}, T_M], \end{cases}$$

and note that

$$\sum_{i=0}^{M-1} \tilde{\sigma}_{T_i,T_{i+1}}(s) = \sum_{i=\eta(s)}^{M-1} \sigma_{T_i,T_{i+1}}(s).$$

The second part can be proved similarly.

47.10. To prove the first part, use the result of Problem 47.9 and note that

$$\left.\frac{\mathrm{d}Q^*}{\mathrm{d}Q^{T_M}}\right|_{\mathscr{F}_t} = E_{Q^{T_M}}\left[\left.\frac{\mathrm{d}Q^*}{\mathrm{d}Q^{T_M}}\right| \mathscr{F}_t\right].$$

To prove the second part, use the result of Problems 47.7 and 14.20.

47.4 Solutions

47.1. By the definition of equivalent martingale measures, the process $\{\frac{P(t,T)}{\beta(t)} : 0 \leq t \leq T\}$ is a martingale under Q. Hence we have

$$\left.\frac{\mathrm{d}Q^T}{\mathrm{d}Q}\right|_{\mathscr{F}_t} = E_Q\left[\left.\frac{\mathrm{d}Q^T}{\mathrm{d}Q}\right| \mathscr{F}_t\right] = E_Q\left[\left.\frac{P(T,T)}{P(0,T)\beta(T)}\right| \mathscr{F}_t\right] = \frac{P(t,T)}{P(0,T)\beta(t)}.$$

This completes the proof.

47.2.

(a) This result follows from Problems 47.1 and 46.5 directly.

(b) This result follows from part (a) and Problem 37.4 directly.

(c) By Problem 47.1 and part (a), we have

$$
\begin{aligned}
&\frac{P(t,S)}{P(t,T)} \\
=\ & \frac{P(0,S)}{P(0,T)} \frac{\left.\frac{\mathrm{d}Q^S}{\mathrm{d}Q}\right|_{\mathscr{F}_t}}{\left.\frac{\mathrm{d}Q^T}{\mathrm{d}Q}\right|_{\mathscr{F}_t}} \\
=\ & \frac{P(0,S)}{P(0,T)} \times \\
& \exp\left(-\frac{1}{2}\int_0^t [\eta^2(s,S)-\eta^2(s,T)]\mathrm{d}s + \int_0^t [\eta(s,S)-\eta(s,T)]\mathrm{d}W_s\right) \\
=\ & \frac{P(0,S)}{P(0,T)} \times \\
& \exp\left(-\frac{1}{2}\int_0^t [\eta(s,S)-\eta(s,T)]^2\mathrm{d}s + \int_0^t [\eta(s,S)-\eta(s,T)]\mathrm{d}W_s^T\right).
\end{aligned}
$$

(d) By Problem 47.1, we have

$$
\left.\frac{\mathrm{d}Q^S}{\mathrm{d}Q^T}\right|_{\mathscr{F}_t} = \left.\frac{\mathrm{d}Q^S}{\mathrm{d}Q}\right|_{\mathscr{F}_t} \left.\frac{\mathrm{d}Q}{\mathrm{d}Q^T}\right|_{\mathscr{F}_t} = \frac{P(t,S)P(0,T)}{P(t,T)P(0,S)}.
$$

This completes the proof.

47.3. By the assumption, we have

$$
E_{Q^T}[|X|] = E_Q\left[\frac{|X|}{P(0,T)\beta(T)}\right] < \infty.
$$

By Problems 14.20 and 47.1, we have

$$
\begin{aligned}
E_{Q^T}[X|\mathscr{F}_t] &= \frac{E_Q\left[\left.\frac{X}{P(0,T)\beta(T)}\right|\mathscr{F}_t\right]}{E_Q\left[\left.\frac{\mathrm{d}Q^T}{\mathrm{d}Q}\right|\mathscr{F}_t\right]} \\
&= \frac{\frac{1}{P(0,T)}E_Q\left[\left.\frac{X}{\beta(T)}\right|\mathscr{F}_t\right]}{\frac{P(t,T)}{P(0,T)\beta(t)}} \\
&= \frac{\beta(t)E_Q\left[\left.\frac{X}{\beta(T)}\right|\mathscr{F}_t\right]}{P(t,T)}.
\end{aligned}
$$

This completes the proof.

47.4. By part (b) of Problem 47.2, the process $\{W_t^T : 0 \le t \le T\}$ given by

$$
W_t^T = W_t - \int_0^t \eta(s,T)\mathrm{d}s
$$

is a Brownian motion under Q^T. Then we have

$$f(t,T) = f(0,T) + \int_0^t \sigma(s,T)\mathrm{d}W_s^T, \quad 0 \le t \le T.$$

By the assumption that $\sigma(t,T) \in L^2_{ad}([0,T] \times \Omega)$ and Theorem 32.2, the forward rate process $\{f(t,T) : 0 \le t \le T\}$ is a martingale under Q^T. Hence

$$E_{Q^T}[r(T)|\mathscr{F}_t] = E_{Q^T}[f(T,T)|\mathscr{F}_t] = f(t,T).$$

This completes the proof.

47.5. By the definition of LIBOR market models, we have for $i = 1, 2, \ldots, M$,

$$\mathrm{d}W_t^{T_{i-1}} = \mathrm{d}W_t^{T_i} - \sigma_{T_{i-1},T_i}(t)\mathrm{d}t, \quad t \in [0, T_{i-1}]. \tag{47.2}$$

If $m < n$, summing both sides of the above equation from $m+1$ to n gives

$$\sum_{i=m+1}^{n} \mathrm{d}W_t^{T_{i-1}} = \sum_{i=m+1}^{n} \mathrm{d}W_t^{T_i} - \sum_{i=m+1}^{n} \sigma_{T_{i-1},T_i}(t)\mathrm{d}t$$

or

$$\mathrm{d}W_t^{T_m} = \mathrm{d}W_t^{T_n} - \sum_{i=m+1}^{n} \sigma_{T_{i-1},T_i}(t)\mathrm{d}t, \quad t \in [0, T_m].$$

If $m > n$, summing both sides of Equation (47.2) from $n+1$ to m gives

$$\sum_{i=n+1}^{m} \mathrm{d}W_t^{T_{i-1}} = \sum_{i=n+1}^{m} \mathrm{d}W_t^{T_i} - \sum_{i=n+1}^{m} \sigma_{T_{i-1},T_i}(t)\mathrm{d}t$$

or

$$\mathrm{d}W_t^{T_n} = \mathrm{d}W_t^{T_m} - \sum_{i=n+1}^{m} \sigma_{T_{i-1},T_i}(t)\mathrm{d}t, \quad t \in [0, T_n].$$

Hence we have

$$\mathrm{d}W_t^{T_m} = \begin{cases} \mathrm{d}W_t^{T_n} - \sum_{k=m}^{n-1} \sigma_{T_k,T_{k+1}}(t)\mathrm{d}t, & \text{if } m < n; \\ \mathrm{d}W_t^{T_n}, & \text{if } m = n; \\ \mathrm{d}W_t^{T_n} + \sum_{k=n}^{m-1} \sigma_{T_k,T_{k+1}}(t)\mathrm{d}t, & \text{if } m > n, \end{cases} \tag{47.3}$$

Also by the definition of LIBOR market models, for $i = 1, 2, \ldots, M$, we have

$$\mathrm{d}L(t, T_{i-1}) = \lambda(t, T_{i-1})L(t, T_{i-1})\mathrm{d}W_t^{T_i}, \quad t \in [0, T_{i-1}]$$

or

$$\frac{\mathrm{d}L(t, T_{i-1})}{L(t, T_{i-1})} = \lambda(t, T_{i-1})\mathrm{d}W_t^{T_i}, \quad t \in [0, T_{i-1}]. \tag{47.4}$$

Combining Equations (47.3) and (47.4) gives

$$\frac{\mathrm{d}L(t,T_m)}{L(t,T_m)} = \begin{cases} \lambda(t,T_m)\mathrm{d}W_t^{T_{n+1}} - \lambda(t,T_m)\sum_{k=m+1}^{n}\sigma_{T_k,T_{k+1}}(t)\mathrm{d}t, & \text{if } m < n; \\ \lambda(t,T_m)\mathrm{d}W_t^{T_{n+1}}, & \text{if } m = n; \\ \lambda(t,T_m)\mathrm{d}W_t^{T_{n+1}} + \lambda(t,T_m)\sum_{k=n+1}^{m}\sigma_{T_k,T_{k+1}}(t)\mathrm{d}t, & \text{if } m > n. \end{cases}$$

This completes the proof.

47.6.

(a) By the definition of $L(t, T_{m-1})$, we have

$$\frac{P(t,T_{m-1})}{P(t,T_m)} = \delta L(t,T_{m-1}) + 1, \quad t \in [0, T_{m-1}],$$

which gives

$$\mathrm{d}\left(\frac{P(t,T_{m-1})}{P(t,T_m)}\right) = \delta \mathrm{d}L(t,T_{m-1}).$$

Since

$$\mathrm{d}L(t,T_{m-1}) = \lambda(t,T_{m-1})L(t,T_{m-1})\mathrm{d}W_t^{T_m}$$

and

$$\begin{aligned} \sigma_{T_{m-1},T_m}(t) &= \frac{\delta L(t,T_{m-1})}{\delta L(t,T_{m-1})+1}\lambda(t,T_{m-1}) \\ &= \frac{P(t,T_m)}{P(t,T_{m-1})}\delta L(t,T_{m-1})\lambda(t,T_{m-1}), \end{aligned}$$

we have

$$\mathrm{d}\left(\frac{P(t,T_{m-1})}{P(t,T_m)}\right) = \frac{P(t,T_{m-1})}{P(t,T_m)}\sigma_{T_{m-1},T_m}(t)\mathrm{d}W_t^{T_m}.$$

Solving the above stochastic differential equation gives

$$\frac{P(t,T_{m-1})}{P(t,T_m)} = \frac{P(0,T_{m-1})}{P(0,T_m)} \times \exp\left(-\frac{1}{2}\int_0^t \sigma^2_{T_{m-1},T_m}(s)\mathrm{d}s + \int_0^t \sigma_{T_{m-1},T_m}(s)\mathrm{d}W_s^{T_m}\right). \quad (47.5)$$

If $k < m$, multiplying both sides of Equation (47.5) from $k+1$ to m gives

$$\prod_{i=k+1}^{m}\frac{P(t,T_{i-1})}{P(t,T_i)} = \prod_{i=k+1}^{m}\frac{P(0,T_{i-1})}{P(0,T_i)} \times \exp\left(-\frac{1}{2}\int_0^t \sum_{i=k+1}^{m}\sigma^2_{T_{i-1},T_i}(s)\mathrm{d}s + \int_0^t \sum_{i=k+1}^{m}\sigma_{T_{i-1},T_i}(s)\mathrm{d}W_s^{T_i}\right)$$

or

$$\frac{P(t,T_k)}{P(t,T_m)} = \frac{P(0,T_k)}{P(0,T_m)} \times$$
$$\exp\left(-\frac{1}{2}\int_0^t \sum_{i=k+1}^{m} \sigma^2_{T_{i-1},T_i}(s)\mathrm{d}s + \int_0^t \sum_{i=k+1}^{m} \sigma_{T_{i-1},T_i}(s)\mathrm{d}W_s^{T_i}\right).$$

But by Problem 47.5, we have for every $i < m$,

$$\mathrm{d}W_t^{T_i} = \mathrm{d}W_t^{T_m} - \sum_{j=i}^{m-1} \sigma_{T_j,T_{j+1}}(t)\mathrm{d}t.$$

We have

$$\begin{aligned}
& -\frac{1}{2}\int_0^t \sum_{i=k+1}^{m} \sigma^2_{T_{i-1},T_i}(s)\mathrm{d}s + \int_0^t \sum_{i=k+1}^{m} \sigma_{T_{i-1},T_i}(s)\mathrm{d}W_s^{T_i} \\
= & -\frac{1}{2}\int_0^t \sum_{i=k+1}^{m} \sigma^2_{T_{i-1},T_i}(s)\mathrm{d}s + \\
& \int_0^t \sum_{i=k+1}^{m} \sigma_{T_{i-1},T_i}(s)\left(\mathrm{d}W_t^{T_m} - \sum_{j=i}^{m-1} \sigma_{T_j,T_{j+1}}(t)\mathrm{d}t\right) \\
= & -\frac{1}{2}\int_0^t \sigma^2_{T_k,T_m}(s)\mathrm{d}s + \int_0^t \sigma_{T_k,T_m}(s)\mathrm{d}W_s^{T_m}.
\end{aligned}$$

Therefore,

$$\frac{P(t,T_k)}{P(t,T_m)} = \frac{P(0,T_k)}{P(0,T_m)} \exp\left(-\frac{1}{2}\int_0^t \sigma^2_{T_k,T_m}(s)\mathrm{d}s + \int_0^t \sigma_{T_k,T_m}(s)\mathrm{d}W_s^{T_m}\right).$$

The case when $k > m$ can be proved similarly.

(b) By the assumption that $\lambda(t, T_i)$ is bounded, we know that $\sigma(T_k, T_m)$ satisfies the Novikov's condition. Hence it follows from Problem 37.2 that the forward price process

$$\left\{\frac{P(t,T_k)}{P(t,T_m)} : 0 \le t \le T_k \wedge T_m\right\}$$

is a positive Q^{T_m}-martingale.

This completes the proof.

47.7. By Problem 47.3, the price at time t of the T_m-contingent claim X is given by

$$P(t,T_m)E_{Q^{T_m}}[X|\mathscr{F}_t]. \tag{47.6}$$

By Problem 47.2, we have

$$\left.\frac{\mathrm{d}Q^{T_m}}{\mathrm{d}Q^{T_n}}\right|_{\mathscr{F}_t} = \frac{P(t,T_m)P(0,T_n)}{P(t,T_n)P(0,T_m)}, \quad t \in [0, T_m].$$

Hence we have

$$\frac{\mathrm{d}Q^{T_m}}{\mathrm{d}Q^{T_n}} = \left.\frac{\mathrm{d}Q^{T_m}}{\mathrm{d}Q^{T_n}}\right|_{\mathscr{F}_{T_m}} = \frac{P(0,T_n)}{P(T_m,T_n)P(0,T_m)}.$$

It follows from Problem 14.20 that

$$\begin{aligned} E_{Q^{T_m}}[X|\mathscr{F}_t] &= \frac{E_{Q^{T_n}}\left[\left.\frac{\mathrm{d}Q^{T_m}}{\mathrm{d}Q^{T_n}}X\right|\mathscr{F}_t\right]}{E_{Q^{T_n}}\left[\left.\frac{\mathrm{d}Q^{T_m}}{\mathrm{d}Q^{T_n}}\right|\mathscr{F}_t\right]} \\ &= \frac{E_{Q^{T_n}}\left[\left.\frac{P(0,T_n)}{P(T_m,T_n)P(0,T_m)}X\right|\mathscr{F}_t\right]}{\frac{P(t,T_m)P(0,T_n)}{P(t,T_n)P(0,T_m)}} \\ &= \frac{P(t,T_n)}{P(t,T_m)}E_{Q^{T_n}}\left[\left.\frac{X}{P(T_m,T_n)}\right|\mathscr{F}_t\right]. \end{aligned} \tag{47.7}$$

Combining Equations (47.6) and (47.7) gives the price

$$P(t,T_n)E_{Q^{T_n}}\left[\left.\frac{X}{P(T_m,T_n)}\right|\mathscr{F}_t\right].$$

This completes the proof.

47.8. By Problem 47.3, the price at time T_m of the caplet is calculated as

$$\begin{aligned} V(T_m,T_{n-1},T_n) &= P(T_m,T_n)E_{Q^{T_n}}\left[\delta(L(T_{n-1},T_n)-\kappa)^+\middle|\mathscr{F}_{T_m}\right] \\ &= \delta P(T_m,T_n)E_{Q^{T_n}}\left[(L(T_{n-1},T_n)-\kappa)^+\middle|\mathscr{F}_{T_m}\right]. \end{aligned}$$

By Definition 47.3, we have

$$\begin{aligned} &L(T_{n-1},T_{n-1}) \\ = \; &L(0,T_{n-1})\exp\left(-\frac{1}{2}\int_0^{T_{n-1}}\lambda^2(s,T_{n-1})\mathrm{d}s + \int_0^{T_{n-1}}\lambda(s,T_{n-1})\mathrm{d}W_s^{T_n}\right) \end{aligned}$$

and

$$\begin{aligned} &L(T_m,T_{n-1}) \\ = \; &L(0,T_{n-1})\exp\left(-\frac{1}{2}\int_0^{T_m}\lambda^2(s,T_{n-1})\mathrm{d}s + \int_0^{T_m}\lambda(s,T_{n-1})\mathrm{d}W_s^{T_n}\right), \end{aligned}$$

which give

$$\begin{aligned} & L(T_{n-1}, T_{n-1}) \\ = \; & L(T_m, T_{n-1}) \cdot \\ & \exp\left(-\frac{1}{2} \int_{T_m}^{T_{n-1}} \lambda^2(s, T_{n-1}) \mathrm{d}s + \int_{T_m}^{T_{n-1}} \lambda(s, T_{n-1}) \mathrm{d}W_s^{T_n} \right). \end{aligned}$$

Since

$$\int_{T_m}^{T_{n-1}} \lambda(s, T_{n-1}) \mathrm{d}W_s^{T_n} = \int_{T_m}^{T_{n-1}} \lambda(s, T_{n-1}) \mathrm{d}\left(W_s^{T_n} - W_{T_m}^{T_n} \right)$$

and $\{W_s^{T_n} - W_{T_m}^{T_n} : T_m \le s \le T_{n-1}\}$ is independent of $\mathscr{F}_{T_m}$, we know that $L(T_{n-1}, T_{n-1})$ is independent of $\mathscr{F}_{T_m}$. Hence

$$\begin{aligned} & V(T_m, T_{n-1}, T_n) \\ = \; & \delta P(T_m, T_n) E_{Q^{T_n}} \left[(L(T_{n-1}, T_n) - \kappa)^+ \right] \\ = \; & \delta P(T_m, T_n) \times E_{Q^{T_n}} \\ & \left[\left(L(T_m, T_{n-1}) e^{-\frac{1}{2} \int_{T_m}^{T_{n-1}} \lambda^2(s, T_{n-1}) \mathrm{d}s + \int_{T_m}^{T_{n-1}} \lambda(s, T_{n-1}) \mathrm{d}W_s^{T_n}} - \kappa \right)^+ \right]. \end{aligned}$$

Now let

$$\begin{aligned} A &= \left\{ L(T_m, T_{n-1}) e^{-\frac{1}{2} \int_{T_m}^{T_{n-1}} \lambda^2(s, T_{n-1}) \mathrm{d}s + \int_{T_m}^{T_{n-1}} \lambda(s, T_{n-1}) \mathrm{d}W_s^{T_n}} > \kappa \right\} \\ &= \left\{ \frac{\int_{T_m}^{T_{n-1}} \lambda(s, T_{n-1}) \mathrm{d}W_s^{T_n}}{\sqrt{\int_{T_m}^{T_{n-1}} \lambda^2(s, T_{n-1}) \mathrm{d}s}} > -d_2 \right\}. \end{aligned}$$

Since

$$Z = \frac{\int_{T_m}^{T_{n-1}} \lambda(s, T_{n-1}) \mathrm{d}W_s^{T_n}}{\sqrt{\int_{T_m}^{T_{n-1}} \lambda^2(s, T_{n-1}) \mathrm{d}s}}$$

is a normal random variable with mean 0 and variance 1 under Q^{T_n}, we have

$$E_{Q^{T_n}}[I_A] = Q^{T_n}(A) = N(d_2)$$

and

$$\begin{aligned} E_{Q^{T_n}}[L(T_{n-1}, T_{n-1}) I_A] &= E_{Q^{T_n}} \left[L(T_m, T_{n-1}) e^{-\frac{1}{2}\sigma^2 + \sigma Z} I_{\{Z > -d_2\}} \right] \\ &= L(T_m, T_{n-1}) \int_{-d_2}^{\infty} e^{-\frac{1}{2}\sigma^2 + \sigma x} \frac{1}{\sqrt{2\pi}} e^{-\frac{1}{2}x^2} \mathrm{d}x \\ &= L(T_m, T_{n-1}) \int_{-d_2}^{\infty} \frac{1}{\sqrt{2\pi}} e^{-\frac{1}{2}(x - \sigma)^2} \mathrm{d}x \\ &= L(T_m, T_{n-1}) N(d_1), \end{aligned}$$

where

$$\sigma = \sqrt{\int_{T_m}^{T_{n-1}} \lambda^2(s, T_{n-1})\mathrm{d}s}.$$

Hence

$$\begin{aligned} V(T_m, T_{n-1}, T_n) &= \delta P(T_m, T_n)[E_{Q^{T_n}}[L(T_{n-1}, T_{n-1})I_A] - \kappa E_{Q^{T_n}}[I_A]] \\ &= \delta P(T_m, T_n)[L(T_m, T_{n-1})N(d_1) - \kappa N(d_2)]. \end{aligned}$$

This completes the proof.

47.9.

(a) By Definition 47.1, we have

$$1 + \delta L(T_{i-1}, T_{i-1}) = \frac{1}{P(T_{i-1}, T_i)}, \quad i = 1, 2, \ldots, M,$$

leads to

$$\beta^*(T_M) = \prod_{i=1}^{M} \frac{1}{P(T_{i-1}, T_i)}. \tag{47.8}$$

By Problem 47.6, we have for every $i = 1, 2, \ldots, M$,

$$\frac{1}{P(T_{i-1}, T_i)} = \frac{P(0, T_{i-1})}{P(0, T_i)} \times$$
$$\exp\left(-\frac{1}{2}\int_0^{T_{i-1}} \sigma^2_{T_{i-1},T_i}(s)\mathrm{d}s + \int_0^{T_{i-1}} \sigma_{T_{i-1},T_i}(s)\mathrm{d}W_s^{T_i}\right).$$

Plugging the above equation in Equation (47.8) gives

$$\beta^*(T_M) = \frac{1}{P(0, T_M)} \times$$
$$\exp\left(-\frac{1}{2}\sum_{i=1}^{M}\int_0^{T_{i-1}} \sigma^2_{T_{i-1},T_i}(s)\mathrm{d}s + \sum_{i=1}^{M}\int_0^{T_{i-1}} \sigma_{T_{i-1},T_i}(s)\mathrm{d}W_s^{T_i}\right).$$

By Problem 47.5, we have for every $i = 0, 1, \ldots, M-1$,

$$\mathrm{d}W_s^{T_i} = \mathrm{d}W_s^{T_M} - \sum_{k=i}^{M-1} \sigma_{T_k,T_{k+1}}(s)\mathrm{d}s.$$

Then we have

$$\begin{aligned} \beta^*(T_M) &= \frac{1}{P(0, T_M)} \exp\left(-\frac{1}{2}\sum_{i=1}^{M}\int_0^{T_{i-1}} \sigma^2_{T_{i-1},T_i}(s)\mathrm{d}s - \right. \\ &\quad \sum_{i=1}^{M}\sum_{k=i+1}^{M}\int_0^{T_{i-1}} \sigma_{T_{i-1},T_i}(s)\sigma_{T_{k-1},T_k}(s)\mathrm{d}s + \\ &\quad \left. \sum_{i=1}^{M}\int_0^{T_{i-1}} \sigma_{T_{i-1},T_i}(s)\mathrm{d}W_s^{T_M}\right). \end{aligned} \tag{47.9}$$

Now for every $s \in [0, T_M]$, let

$$\tilde{\sigma}_{T_{i-1},T_i}(s) = \begin{cases} \sigma_{T_{i-1},T_i}(s), & \text{if } s \in [0, T_{i-1}); \\ 0, & \text{if } s \in [T_{i-1}, T_M], \end{cases}$$

Then we have

$$\int_0^{T_{i-1}} \sigma^2_{T_{i-1},T_i}(s)\mathrm{d}s = \int_0^{T_M} \tilde{\sigma}^2_{T_{i-1},T_i}(s)\mathrm{d}s,$$

$$\int_0^{T_{i-1}} \sigma_{T_{i-1},T_i}(s)\sigma_{T_{k-1},T_k}(s)\mathrm{d}s = \int_0^{T_M} \tilde{\sigma}_{T_{i-1},T_i}(s)\tilde{\sigma}_{T_{k-1},T_k}(s)\mathrm{d}s,$$

and

$$\int_0^{T_{i-1}} \sigma_{T_{i-1},T_i}(s)\mathrm{d}W_s^{T_i} = \int_0^{T_M} \tilde{\sigma}_{T_{i-1},T_i}(s)\mathrm{d}W_s^{T_M}.$$

Plugging the above three equations into Equation (47.9) gives

$$\begin{aligned}\beta^*(T_M) &= \frac{1}{P(0,T_M)} \exp\left(-\frac{1}{2}\sum_{i=1}^{M}\int_0^{T_M} \tilde{\sigma}^2_{T_{i-1},T_i}(s)\mathrm{d}s - \right.\\ &\quad \sum_{i=1}^{M}\sum_{k=i+1}^{M}\int_0^{T_M} \tilde{\sigma}_{T_{i-1},T_i}(s)\tilde{\sigma}_{T_{k-1},T_k}(s)\mathrm{d}s + \\ &\quad \left.\sum_{i=1}^{M}\int_0^{T_M} \tilde{\sigma}_{T_{i-1},T_i}(s)\mathrm{d}W_s^{T_M}\right)\end{aligned}$$

or

$$\begin{aligned}\beta^*(T_M) &= \frac{1}{P(0,T_M)} \exp\left(-\frac{1}{2}\int_0^{T_M} \left(\sum_{i=1}^{M} \tilde{\sigma}_{T_{i-1},T_i}(s)\right)^2 \mathrm{d}s + \right.\\ &\quad \left.\int_0^{T_M}\sum_{i=1}^{M} \tilde{\sigma}_{T_{i-1},T_i}(s)\mathrm{d}W_s^{T_M}\right). \end{aligned} \tag{47.10}$$

Since $T_{\eta(s)-1} \le s < T_{\eta(s)}$, we have

$$\sum_{i=1}^{M} \tilde{\sigma}_{T_{i-1},T_i}(s) = \sum_{i=\eta(s)+1}^{M} \sigma_{T_{i-1},T_i}(s) = \sum_{i=\eta(s)}^{M-1} \sigma_{T_i,T_{i+1}}(s) = \Sigma(s).$$

Thus from Equation (47.10) and Problem 37.2, we get

$$\begin{aligned}& E_{Q^{T_M}}\left[\beta^*(T_M)P(0,T_M)|\mathscr{F}_t\right] \\ =\;& E_{Q^{T_M}}\left[\exp\left(-\frac{1}{2}\int_0^{T_M} \Sigma^2(s)\mathrm{d}s + \int_0^{T_M} \Sigma(s)\mathrm{d}W_s^{T_M}\right)\middle|\mathscr{F}_t\right] \\ =\;& \exp\left(-\frac{1}{2}\int_0^{t} \Sigma^2(s)\mathrm{d}s + \int_0^{t} \Sigma(s)\mathrm{d}W_s^{T_M}\right).\end{aligned}$$

(b) By a similar argument used in the proof of the first part, we have

$$\begin{aligned}\beta^*(T_m) &= \frac{1}{P(0,T_m)}\exp\left(-\frac{1}{2}\int_0^{T_m}\left(\sum_{i=1}^{m}\tilde{\sigma}_{T_{i-1},T_i}(s)\right)^2 \mathrm{d}s+\right.\\ &\quad \left.\int_0^{T_m}\sum_{i=1}^{m}\tilde{\sigma}_{T_{i-1},T_i}(s)\mathrm{d}W_s^{T_m}\right)\\ &= \frac{1}{P(0,T_m)}\exp\left(-\frac{1}{2}\int_0^{T_m}\left(\sum_{i=1}^{m}\tilde{\sigma}_{T_{i-1},T_i}(s)\right)^2 \mathrm{d}s-\right.\\ &\quad \int_0^{T_m}\sum_{i=1}^{m}\sum_{k=m+1}^{M}\tilde{\sigma}_{T_{i-1},T_i}(s)\tilde{\sigma}_{T_{k-1},T_k}(s)\mathrm{d}s+\\ &\quad \left.\int_0^{T_m}\sum_{i=1}^{m}\tilde{\sigma}_{T_{i-1},T_i}(s)\mathrm{d}W_s^{T_M}\right). \qquad (47.11)\end{aligned}$$

Note that

$$\begin{aligned}&-\frac{1}{2}\int_0^{T_m}\left(\sum_{i=1}^{m}\tilde{\sigma}_{T_{i-1},T_i}(s)\right)^2 \mathrm{d}s-\\ &\int_0^{T_m}\sum_{i=1}^{m}\sum_{k=m+1}^{M}\tilde{\sigma}_{T_{i-1},T_i}(s)\tilde{\sigma}_{T_{k-1},T_k}(s)\mathrm{d}s\\ =& -\frac{1}{2}\int_0^{T_m}\left(\sum_{i=1}^{M}\tilde{\sigma}_{T_{i-1},T_i}(s)\right)^2 \mathrm{d}s+\frac{1}{2}\int_0^{T_m}\left(\sum_{i=m+1}^{M}\tilde{\sigma}_{T_{i-1},T_i}(s)\right)^2 \mathrm{d}s\\ =& -\frac{1}{2}\int_0^{T_m}\Sigma^2(s)\mathrm{d}s+\frac{1}{2}\int_0^{T_m}\sigma^2_{T_m,T_M}(s)\mathrm{d}s,\end{aligned}$$

where

$$\sigma_{T_m,T_M}(s)=\sum_{i=m}^{M-1}\sigma_{T_i,T_{i+1}}(s)=\sum_{i=m+1}^{M}\tilde{\sigma}_{T_{i-1},T_i}(s),\quad s\in[0,T_m].$$

Equation (47.11) becomes

$$\begin{aligned}\beta^*(T_m) &= \frac{1}{P(0,T_m)}\exp\left(-\frac{1}{2}\int_0^{T_m}\Sigma^2(s)\mathrm{d}s+\frac{1}{2}\int_0^{T_m}\sigma^2_{T_m,T_M}(s)\mathrm{d}s+\right.\\ &\quad \left.\int_0^{T_m}\sum_{i=1}^{m}\tilde{\sigma}_{T_{i-1},T_i}(s)\mathrm{d}W_s^{T_M}\right). \qquad (47.12)\end{aligned}$$

But by the first item, we have

$$\begin{aligned} & E_{Q^{T_M}}\left[\beta^*(T_M)P(0,T_M)|\mathscr{F}_{T_m}\right] \\ = & \exp\left(-\frac{1}{2}\int_0^{T_m}\Sigma^2(s)\mathrm{d}s + \int_0^{T_m}\Sigma(s)\mathrm{d}W_s^{T_M}\right), \end{aligned}$$

which gives

$$\begin{aligned} & E_{Q^{T_M}}\left[\beta^*(T_M)|\mathscr{F}_{T_m}\right] \\ = & \frac{\exp\left(-\frac{1}{2}\int_0^{T_m}\Sigma^2(s)\mathrm{d}s + \int_0^{T_m}\Sigma(s)\mathrm{d}W_s^{T_M}\right)}{P(0,T_M)}. \end{aligned} \tag{47.13}$$

Also by Problem 47.6, we have

$$\begin{aligned} & \frac{1}{P(T_m,T_M)} = \frac{P(0,T_m)}{P(0,T_M)} \times \\ & \exp\left(-\frac{1}{2}\int_0^{T_m}\sigma^2_{T_m,T_M}(s)\mathrm{d}s + \int_0^{T_m}\sigma_{T_m,T_M}(s)\mathrm{d}W_s^{T_M}\right) \end{aligned} \tag{47.14}$$

Combining Equation (47.13) and Equation (47.14) gives

$$\begin{aligned} & E_{Q^{T_M}}\left[\beta^*(T_M)|\mathscr{F}_{T_m}\right] \\ = & \frac{1}{P(T_m,T_M)P(0,T_m)}\exp\left(-\frac{1}{2}\int_0^{T_m}\Sigma^2(s)\mathrm{d}s + \int_0^{T_m}\Sigma(s)\mathrm{d}W_s^{T_M}\right. \\ & \left.+\frac{1}{2}\int_0^{T_m}\sigma^2_{T_m,T_M}(s)\mathrm{d}s - \int_0^{T_m}\sigma_{T_m,T_M}(s)\mathrm{d}W_s^{T_M}\right). \end{aligned} \tag{47.15}$$

Note that for every $s \in [0, T_m]$,

$$\begin{aligned} \Sigma(s) - \sigma_{T_m,T_M}(s) &= \sum_{i=1}^{M}\tilde{\sigma}_{T_{i-1},T_i}(s) - \sum_{i=m+1}^{M}\tilde{\sigma}_{T_{i-1},T_i}(s) \\ &= \sum_{i=1}^{m}\tilde{\sigma}_{T_{i-1},T_i}(s). \end{aligned} \tag{47.16}$$

Combining Equations (47.12), (47.15), and (47.16), we get

$$E_{Q^{T_M}}\left[\beta^*(T_M)|\mathscr{F}_{T_m}\right] = \frac{\beta^*(T_m)}{P(T_m,T_M)}.$$

This completes the proof.

47.10.

(a) Since

$$\left.\frac{\mathrm{d}Q^*}{\mathrm{d}Q^{T_M}}\right|_{\mathscr{F}_t} = E_{Q^{T_M}}\left[\left.\frac{\mathrm{d}Q^*}{\mathrm{d}Q^{T_M}}\right| \mathscr{F}_t\right] = E_{Q^{T_M}}\left[\beta^*(T_M)P(0,T_M)\right| \mathscr{F}_t\right],$$

it follows from Problem 47.9 that

$$\left.\frac{\mathrm{d}Q^*}{\mathrm{d}Q^{T_M}}\right|_{\mathscr{F}_t} = \exp\left(-\frac{1}{2}\int_0^t \Sigma^2(s)\mathrm{d}s + \int_0^t \Sigma(s)\mathrm{d}W_s^{T_M}\right).$$

(b) By Problem 47.7, the price at time T_m of a T_n-contingent claim X is given by

$$P(T_m, T_M)E_{Q^{T_M}}\left[\left.\frac{X}{P(T_n,T_M)}\right| \mathscr{F}_{T_m}\right].$$

Hence we only need to show that

$$E_{Q^*}\left[\left.\frac{X}{\beta^*(T_n)}\right| \mathscr{F}_{T_m}\right] = \frac{P(T_m,T_M)}{\beta^*(T_m)}E_{Q^{T_M}}\left[\left.\frac{X}{P(T_n,T_M)}\right| \mathscr{F}_{T_m}\right]. \quad (47.17)$$

But by Problems 14.20 and 47.9, we have

$$\begin{aligned}
&E_{Q^*}\left[\left.\frac{X}{\beta^*(T_n)}\right| \mathscr{F}_{T_m}\right] \\
=\ &\frac{E_{Q^{T_M}}\left[\left.\frac{X}{\beta^*(T_n)}\frac{\mathrm{d}Q^*}{\mathrm{d}Q^{T_M}}\right| \mathscr{F}_{T_m}\right]}{E_{Q^{T_M}}\left[\left.\frac{\mathrm{d}Q^*}{\mathrm{d}Q^{T_M}}\right| \mathscr{F}_{T_m}\right]} \\
=\ &\frac{E_{Q^{T_M}}\left[\left.\frac{X}{\beta^*(T_n)}\beta^*(T_M)P(0,T_M)\right| \mathscr{F}_{T_m}\right]}{\frac{\beta^*(T_m)P(0,T_M)}{P(T_m,T_M)}} \\
=\ &\frac{P(T_m,T_M)E_{Q^{T_M}}\left[\left.X\frac{\beta^*(T_M)}{\beta^*(T_n)}\right| \mathscr{F}_{T_m}\right]}{\beta^*(T_m)}. \quad (47.18)
\end{aligned}$$

Since X is a T_n-contingent claim, then X is $\mathscr{F}_{T_n}$-measurable. Then by Problem 14.10, we have

$$\begin{aligned}
&E_{Q^{T_M}}\left[\left.X\frac{\beta^*(T_M)}{\beta^*(T_n)}\right| \mathscr{F}_{T_m}\right] \\
=\ &E_{Q^{T_M}}\left[\left.\frac{X}{\beta^*(T_n)}E_{Q^{T_M}}\left(\beta^*(T_M)\right| \mathscr{F}_{T_n}\right)\right| \mathscr{F}_{T_m}\right].
\end{aligned}$$

But by Problem 47.9, we have

$$E_{Q^{T_M}}\left(\beta^*(T_M)\right| \mathscr{F}_{T_n}\right) = \frac{\beta^*(T_n)}{P(T_n,T_M)}.$$

Combining the above two equations gives

$$E_{Q^{T_M}}\left[X\frac{\beta^*(T_M)}{\beta^*(T_n)}\middle|\mathscr{F}_{T_m}\right]=E_{Q^{T_M}}\left[\frac{X}{P(T_n,T_M)}\middle|\mathscr{F}_{T_m}\right]. \tag{47.19}$$

Equation (47.17) follows from Equations (47.18) and (47.19).

This completes the proof.

47.5 Bibliographic Notes

In this chapter, we introduced the LIBOR market model, which is also referred to as the BGM model. The model is a collection of dynamics for the forward LIBOR rates that traders are used to working with. For more information about LIBOR market models, readers are referred to Rebonato (2002), Hunt and Kennedy (2004), Bingham and Kiesel (2004), Musiela and Rutkowski (2005), Hull (2008), and Filipović (2009).

References

Ahn, S., Bae, H.-O., Koo, H., and Lee, K. (2011). A survey on american options: Old approaches and new trends. *Bulletin of the Korean Mathematical Society*, 48:791–812.

Allen, E. (2007). *Modeling with Itô Stochastic Differential Equations*. Springer, New York.

Amin, K. and Khanna, A. (1994). Convergence of american option values from discrete- to continuous-time financial models. *Mathematical Finance*, 4(4):289–304.

Andreasen, J., Jensen, B., and Poulsen, R. (1998). Eight valuation methods in financial mathematics: The Black-Scholes formula as an example. *Mathematical Scientist*, 23(1):18–40.

Applebaum, D. (2004). Lévy processes – from probability to finance and quantum groups. *Notices of the American Mathematical Society*, 51(11):1336–1347.

Applebaum, D. (2009). *Lévy Processes and Stochastic Calculus*. Cambridge University Press, Cambridge, UK.

Ash, R. and Doleans-Dade, C. (1999). *Probability and Measure Theory*. Academic Press, New York.

Athreya, K. and Lahiri, S. (2006). *Measure Theory and Probability Theory*. Springer, New York.

Back, K. (2005). *A Course in Derivative Securities: Introduction to Theory and Computation*. Springer Finance. Springer, New York.

Back, K., Bielecki, T., Hipp, C., Peng, S., and Schachermayer, W. (2004). *Stochastic Methods in Finance*. Springer, New York.

Bailey, N. (1964). *The Elements of Stochastic Processes with Applications to the Natural Sciences*. Wiley, New York.

Balakrishnan, N. and Nevzorov, V. (2003). *A Primer on Statistical Distributions*. Wiley, New York.

Baldi, P., Mazliak, L., and Priouret, P. (2002). *Martingales and Markov Chains: Solved Exercises and Elements of Theory*. Chapman & Hall/CRC, Boca Raton, FL.

Barone-Adesi, G. (2005). The saga of the american put. *Journal of Banking & Finance*, 29(11):2909–2918.

Barone-Adesi, G. and Whaley, R. (1987). Efficient analytic approximation of American option values. *Journal of Finance*, 42(2):301–320.

Bartoszynski, R. and Niewiadomska-Bugaj, M. (2007). *Probability and Statistical Inference*. Wiley, Hoboken, NJ.

Bass, R. (1998). *Diffusions and Elliptic Operators*. Springer, New York.

Baxter, M. (1997). General interest-rate models and the universality of HJM. In *Mathematics of Derivative Securities*. Cambridge University Press, Cambridge, UK.

Baxter, M. and Rennie, A. (1996). *Financial Calculus: An Introduction to Derivative Pricing*. Cambridge University Press, Cambridge, UK.

Beichelt, F. (2006). *Stochastic Processes in Science, Engineering and Finance*. Chapman & Hall/CRC, Boca Raton, FL.

Bensoussan, A. (1984). On the theory of option pricing. *Acta Applicandae Mathematicae*, 2:139–158.

Bertoin, J. (1996). *Lévy Processes*. Cambridge University Press, Cambridge, UK.

Bhar, R. and Hamori, S. (2005). *Empirical Techniques in Finance*. Springer, New York.

Bhattacharya, R. and Waymire, E. (2007). *A Basic Course in Probability Theory*. Springer, New York.

Bhattacharya, R. and Waymire, E. (2009). *Stochastic Processes with Applications*. Society for Industrial and Applied Mathematics, Philadelphia.

Bichteler, K. (2002). *Stochastic Integration with Jumps*. Cambridge University Press, Cambridge, UK.

Billingsley, P. (1986). *Probability and measure*. Wiley, 2nd edition.

Billingsley, P. (1995). *Probability and measure*. Wiley, New York, 3rd edition.

Billingsley, P. (1999). *Convergence of Probability Measures*. Wiley, New York, 2nd edition.

Bingham, N. and Kiesel, R. (2004). *Risk-Neutral Valuation: Pricing and Hedging of Financial Derivatives*. Springer, New York, NY, 2nd edition.

Björk, T. (2009). *Arbitrage Theory in Continuous Time*. Oxford University Press, Oxford, UK, 3rd edition.

Björk, T., di Masi, G., Kabanov, Y., and Runggaldier, W. (1997). Towards a general theory of bond markets. *Finance and Stochastics*, 1:141–174.

Black, F., Derman, E., and Toy, W. (1990). A one-factor model of interest rates and its application to treasury bond options. *Financial Analysts Journal*, 46(1):33–39.

Black, F. and Karasinski, P. (1991). Bond and option pricing when short rates are lognormal. *Financial Analysts Journal*, 47(4):52–59.

Black, F. and Scholes, M. (1973). The pricing of options and corporate liabilities. *Journal of Political Economy*, 81(3):637–654.

Borak, S., Härdle, W., and Cabrera, B. L. (2010). *Statistics of Financial Markets: Exercises and Solutions*. Springer, New York.

Borodin, A. and Salminen, P. (2002). *Handbook of Brownian Motion – Facts and Formulae*. Birkhäuser, Boston, 2nd edition.

Borokov, A. (1999). *Mathematical Statistics*. CRC Press, Boca Raton, FL.

Brace, A., Gatarek, D., and Musiela, M. (1997). The market model of interest rate dynamics. *Mathematical Finance*, 7(2):127–154.

Brandimarte, P. (2006). *Numerical Methods in Finance and Economics: A MATLAB-Based Introduction*. Wiley, New York, 2nd edition.

Brigo, D. and Mercurio, F. (2006). *Interest Rate Models – Theory and Practice*. Springer, New York, 2nd edition.

Briys, E., Bellalah, M., Mai, H., and de Varenne, F. (1998). *Options, Futures and Exotic Derivatives: Theory, Application and Practice*. Wiley, New York.

Broadie, M. and Glasserman, P. (1997). Pricing american-style securities using simulation. *Journal of Economic Dynamics and Control*, 21:1323–1352.

Brzezniak, Z. and Zastawniak, T. (1999). *Basic Stochastic Processes: A Course Through Exercises*. Springer, New York.

Buchanan, J. (2006). *An Undergraduate Introduction to Financial Mathematics*. World Scientific Publishing, Singapore.

Cairns, A. (2004). *Interest Rate Models: An Introduction*. Princeton University Press, Princeton, NJ.

Cannon, J. (1984). *The One-Dimensional Heat Equation*, volume 23 of *Encyclopedia of Mathematics and Its Applications*. Cambridge University Press, Cambridge, UK.

Capasso, V. and Bakstein, D. (2005). *An Introduction to Continuous-Time Stochastic Processes: Theory, Models, and Applications to Finance, Biology, and Medicine*. Springer, New York.

Capinski, M. and Kopp, P. (2007). *Measure, Integral and Probability*. Springer, New York, 2nd edition.

Capinski, M. and Zastawniak, T. (2000). *Probability through Problems*. Springer, New York.

Capinski, M. and Zastawniak, T. (2003). *Mathematics for Finance: An Introduction to Financial Engineering*. Springer, New York.

Carmona, R. (2006). *Interest Rate Models: an Infinite Dimensional Stochastic Analysis Perspective*. Springer, New York, NY.

Çinlar, E. (1974). *Introduction to Stochastic Processes*. Prentice Hall, New York, NY.

Çinlar, E. (2011). *Probability and Stochastics*. Springer, New York, NY.

Chan, N. and Wong, H.-Y. (2006). *Simulation Techniques in Financial Risk Management*. Wiley, New York.

Chaumont, L. and Yor, M. (2003). *Exercises in Probability: A Guided Tour from Measure Theory to Random Processes, via Conditioning*. Cambridge University Press, Cambridge, UK.

Cherubini, U., Luciano, E., and Vecchiato, W. (2004). *Copula Methods in Finance*. Wiley, New York.

Chung, F. and Lu, L. (2006). Concentration inequalities and martingale inequalities: A survey. *Internet Mathematics*, 3(1):79–127.

Chung, K. (2000). *A Course in Probability Theory*. Academic Press, Singapore, 3rd edition.

Chung, K. and Aitsahlia, F. (2007). *Elementary Probability Theory: With Stochastic Processes and an Introduction to Mathematical Finance*. Springer, New York.

Chung, K. and Williams, R. (1990). *Introduction to Stochastic Integration*. Birkhäuser, Boston, 2nd edition.

Cont, R. and Tankov, P. (2003). *Financial Modelling with Jump Processes*. Chapman & Hall/CRC Financial Mathematics Series. Chapman & Hall/CRC, Boca Raton, FL.

Conze, A. and Viswanathan (1991). Path dependent options: The case of lookback options. *The Journal of Finance*, 46(5):1893–1907.

Cox, J., Ingersoll, Jr., J., and Ross, S. (1985). A theory of the term structure of interest rates. *Econometrica*, 53(2):385–407.

Cox, J., Ross, S., and Rubinstein, M. (1979). Option pricing: A simplified approach. *Journal of Financial Economics*, 7(3):229–263.

Dana, R.-A. and Jeanblanc, M. (2007). *Financial Markets in Continuous Time*. Springer, New York, 2nd edition.

DasGupta, A. (2010). *Fundamentals of Probability: A First Course*. Springer, New York.

Dekking, F., Kraaikamp, C., Lopuhaa, H., and Meester, L. (2005). *A Modern Introduction to Probability and Statistics*. Springer, New York.

Delbaen, F. and Schachermayer, W. (2006). *The Mathematics of Arbitrage*. Springer, New York.

Dineen, S. (2005). *Probability Theory in Finance: A Mathematical Guide to the Black-Scholes Formula*. American Mathematical Society, Providence, RI.

Dokuchaev, N. (2007). *Mathematical Finance: Core Theory, Problems and Statistical Algorithms*. Routledge, London.

Doob, J. (1990). *Stochastic Processes*. Wiley, New York.

Dothan, L. (1978). On the term structure of interest rates. *Journal of Financial Economics*, 6(1):59–69.

Dudley, R. (2002). *Real Analysis and Probability*. Cambridge University Press, Cambridge, UK, 2nd edition.

Duffie, D. (2001). *Dynamic Asset Pricing Theory*. Princeton University Press, Princeton, NJ, 3rd edition.

Duffy, D. (2006). *Finite Difference Methods in Financial Engineering: A Partial Differential Equation Approach*. Wiley, New York.

Durrett, R. (1996). *Stochastic Calculus: A Practical Introduction*. CRC Press, Boca Raton, FL.

Durrett, R. (2010). *Probability: Theory and Examples*. Cambridge University Press, Cambridge, UK, 4th edition.

Eales, B. and Choudhry, M. (2003). *Derivative Instruments: A Guide to Theory and Practice*. Butterworth-Heinemann, Oxford, UK.

Ekstrand, C. (2011). *Financial Derivatives Modeling*. Springer, New York.

Elliott, R. and Kopp, P. (2005). *Mathematics of Financial Markets*. Springer, New York.

Etheridge, A. (2002). *A Course in Financial Calculus*. Cambridge University Press, Cambridge, UK.

Ethier, S. and Kurtz, T. (1986). *Markov Processes: Characterization and Convergence*. Wiley, New York.

Evans, M., Hastings, N., and Peacock, B. (2000). *Statistical Distributions*. Wiley, Hoboken, NJ, 3rd edition.

Feller, W. (1968). *An Introduction to Probability Theory and Its Applications, Volume 1*. Wiley, New York.

Filipović, D. (2009). *Term-Structure Models: A Graduate Course*. Springer, New York.

Follmer, H. and Schied, A. (2004). *Stochastic Finance: An Introduction in Discrete Time 2*. Walter de Gruyter, Berlin, 2nd edition.

Forbes, C., Evans, M., Hastings, N., and Peacock, B. (2011). *Statistical Distributions*. Wiley, Hoboken, NJ, 4th edition.

Franke, J., Hardle, W., and Hafner, C. (2004). *Statistics of Financial Markets: An Introduction*. Springer, New York.

Friedman, A. (1975). *Stochastic Differential Equations and Applications, Volume 1*. Academic Press, New York.

Friedman, A. (1976). *Stochastic Differential Equations and Applications, Volume 2*. Academic Press, New York.

Goldman, M., Sosin, H., and Gatto, M. (1979). Path-dependent options: Buy at the low, sell at the high. *Journal of Finance*, 34(5):1111–1127.

Grimmett, G. and Stirzaker, D. (2001). *One Thousand Exercises in Probability*. Oxford University Press, Oxford, UK.

Grinstead, C. and Snell, J. (1997). *Introduction to Probability*. American Mathematical Society, Providence, RI, 2nd edition.

Gusak, D., Kukush, A., Kulik, A., Mishura, Y., and Pilipenko, A. (2010). *Theory of Stochastic Processes: With Applications to Financial Mathematics and Risk Theory*. Springer, New York.

Gut, A. (2007). *Probability: A Graduate Course*. Springer, New York.

Gut, A. (2009). *An Immediate Course in Probability*. Springer, New York, 2nd edition.

Haas, M. and Pigorsch, C. (2011). Financial economics: Fat-tailed distributions. In Meyers, R. A., editor, *Complex Systems in Finance and Econometrics*, pages 308–339. Springer, New York.

Hall, P. and Heyde, C. (1980). *Martingale Limit Theory and Its Application*. Academic Press, Singapore.

Haug, E. (2006). *The Complete Guide to Option Pricing Formulas*. McGraw-Hill, New York, 2nd edition.

Heath, D., Jarrow, R., and Morton, A. (1992). Bond pricing and the term structure of interest rates: A new methodology for contingent claims valuation. *Econometrica*, 60(1):77–105.

Herrlich, H. (2006). *Axiom of Choice*. Springer-Verlag, New York.

Higham, D. (2004). *An Introduction to Financial Option Valuation: Mathematics, Stochastics and Computation*. Cambridge University Press, Cambridge, UK.

Ho, T. and Lee, S. (1986). Term structure movements and pricing interest rate contingent claims. *Journal of Finance*, 41(5):1011–1029.

Hobson, D. (2004). A survey of mathematical finance. *Proceedings of the Royal Society of London. Series A, Mathematical and Physical Sciences*, 460(2052):3369–3401.

Hoek, J. and Elliott, R. (2006). *Binomial Models in Finance*. Springer, New York.

Hoel, P., Port, S., and Stone, C. (1972). *Introduction to Stochastic Processes*. Houghton-Mifflin, Boston.

Houthakker, H. and Williamson, P. (1996). *The Economics of Financial Markets*. Oxford University Press, Oxford, UK.

Hsu, H. (1996). *Schaum's Outline of Probability, Random Variables, and Random Processes*. McGraw-Hill, New York.

Hull, J. (2001). *Fundamentals of Futures and Options Markets*. Prentice Hall, New York, 4th edition.

Hull, J. (2008). *Options, Futures, and Other Derivatives*. Prentice Hall, New York, 7th edition.

Hull, J. and White, A. (1990). Pricing interest-rate-derivative securities. *The Review of Financial Studies*, 3(4):573–592.

Hunt, G. (1956). Some theorems concerning brownian motion. *Transactions of the American Mathematical Society*, 81(2):294–319.

Hunt, P. and Kennedy, J. (2004). *Financial Derivatives in Theory and Practice*. Wiley, New York, revised edition.

Iacus, S. (2008). *Simulation and Inference for Stochastic Differential Equations: with R Examples*. Springer, New York.

Ikeda, N. and Watanabe, S. (1981). *Stochastic Differential Equations and Diffusion Processes*. North Holland, Amsterdam.

Itô, K. (1944). Stochastic integral. *Proceedings of the Imperial Academy*, 20(8):519–524.

Jacka, S. (1991). Optimal stopping and the american put. *Mathematical Finance*, 1(2):1–14.

Jacod, J. and Protter, P. (2004). *Probability Essentials*. Springer, New York, 2nd edition.

James, P. (2003). *Option Theory*. Wiley, New York.

Jeanblanc, M., Yor, M., and Chesney, M. (2009). *Mathematical Methods for Financial Markets*. Springer, New York.

Jiang, L. (2005). *Mathematical Modeling and Methods of Option Pricing*. World Scientific Publishing, Singapore.

Jin, Y. and Glasserman, P. (2001). Equilibrium positive interest rates: A unified view. *The Review of Financial Studies*, 14(1):187–214.

Johnson, N., Kemp, A., and Kotz, S. (2005). *Univariate Discrete Distributions*. Wiley, Hoboken, NJ, 3rd edition.

Johnson, N., Kotz, S., and Balakrishnan, N. (1994). *Continuous Univariate Distributions, Volumn 1*. Wiley, Hoboken, NJ, 2nd edition.

Johnson, N., Kotz, S., and Balakrishnan, N. (1995). *Continuous Univariate Distributions, Volumn 2*. Wiley, Hoboken, NJ, 2nd edition.

Jondeau, E., Poon, S.-H., and Rockinger, M. (2007). *Financial Modeling under Non-Gaussian Distributions*. Springer, New York.

Kallenberg, O. (1997). *Foundations of Modern Probability*. Springer, New York.

Kallianpur, G. and Karandikar, R. (1999). *Introduction to Option Pricing Theory*. Birkhäuser Boston, Boston.

Kannan, D. (1979). *Introduction to Stochastic Processes*. Elsevier, New York.

Kannan, D. and Lakshmikantham, V. (2001). *Handbook of Stochastic Analysis & Applications*. CRC Press, Boca Raton, FL.

Kao, E. (1997). *An Introduction to Stochastic Processes*. Duxbury Press, Pacific Grove, CA.

Karatzas, I. (1988). On the pricing of american options. *Applied Mathematics and Optimization*, 17:37–60.

Karatzas, I. and Shreve, S. (1988). *Brownian Motion and Stochastic Calculus*. Springer, New York.

Karatzas, I. and Shreve, S. (1991). *Brownian Motion and Stochastic Calculus*. Springer, New York, 2nd edition.

Karatzas, I. and Shreve, S. (2001). *Methods of Mathematical Finance*. Springer, New York.

Karlin, S. and Taylor, H. (1975). *A First Course in Stochastic Processes*. Academic Press, Singapore, 2nd edition.

Karlin, S. and Taylor, H. (1981). *A Second Course in Stochastic Processes*. Academic Press, Singapore.

Katz, J. and McCormick, D. (2005). *Advanced Option Pricing Models*. McGraw-Hill, New York.

Kim, I. (1990). The analytic valuation of american options. *Review of Financial Studies*, 3(4):547–572.

Kinney, J. (2009). *A Probability and Statistics Companion*. Wiley, Hoboken, NJ.

Klebaner, F. (2005). *Introduction to Stochastic Calculus with Applications*. World Scientific Publishing, Singapore, 2nd edition.

Klenke, A. (2006). *Probability Theory: A Comprehensive Course*. Springer, New York.

Kloeden, P. and Platen, E. (1995). *Numerical Solution to Stochastic Differential Equations*. Springer, New York, 2nd (corrected) edition.

Kolb, R. (1993). *Financial Derivatives*. Prentice Hall Press, New York.

Kopp, E. (2011). *From Measures to Itô Integrals*. Cambridge University Press, Cambridge, UK.

Koralov, L. and Sinai, Y. (2007). *Theory of Probability and Random Processes*. Springer, New York, 2nd edition.

Korn, R., Korn, E., and Kroisandt, G. (2010). *Monte Carlo Methods and Models in Finance and Insurance*. Chapman & Hall/CRC, Boca Raton, FL.

Krishnan, V. (2006). *Probability and Random Processes*. Wiley, Hoboken, NJ.

Kuo, H.-H. (2006). *Introduction to Stochastic Integration*. Springer, New York.

Kwok, Y.-K. (2008). *Mathematical Models of Financial Derivatives*. Springer, New York, 2nd edition.

Kyprianou, A. (2006). *Introductory Lectures on Fluctuations of Lévy Processes with Applications*. Springer, New York.

Lamberton, D. and Lapeyre, B. (1996). *Introduction to Stochastic Calculus Applied to Finance*. Chapman & Hall/CRC, Boca Raton, FL. translated by N. Rabeu and F. Mantion.

Lawler, G. (2006). *Introduction to Stochastic Processes*. Chapman & Hall/CRC, Boca Raton, FL, 2nd edition.

Lawler, G. and Limic, V. (2010). *Random Walk: A Modern Introduction*. Cambridge University Press, Cambridge, UK.

Lefebvre, M. (2006). *Applied Probability and Statistics*. Springer, New York.

Lin, X. (2006). *Introductory Stochastic Analysis for Finance and Insurance*. Wiley, Hoboken, NJ.

Lin, Z. and Bai, Z. (2010). *Probability Inequalities*. Springer, New York.

Liptser, R. and Shiryayev, A. (1977). *Statistics of Stochastic Processes I: General Theory*. Springer, New York, NY.

Loeve, M. (1977). *Probability Theory I*. Springer, New York, 4th edition.

Loeve, M. (1978). *Probability Theory II*. Springer, New York, 4th edition.

Longstaff, F. and Schwartz, E. (2001). Valuing American options by simulation: A simple least-squares approach. *The Review of Financial Studies*, 14(1):113–147.

Lukacs, E. (1970). *Characteristic Functions*. Hafner Publishing, New York, 2nd edition.

Lyuu, Y.-D. (2001). *Financial Engineering and Computation: Principles, Mathematics, and Algorithms*. Cambridge University Press, Cambridge, UK.

Malliavin, P. and Thalmaier, A. (2006). *Stochastic Calculus of Variations in Mathematical Finance*. Springer, New York.

McDonald, R. (2005). *Derivatives Markets*. Addison-Wesley, New York.

Medvegyev, P. (2007). *Stochastic Integration Theory*. Oxford University Press, Oxford, UK.

Merton, R. (1973). Theory of rational option pricing. *Bell Journal of Economics and Management Science*, 4(1):141–183.

Merton, R. (1990). *Continuous-Time Finance*. Basil Blackwell, Cambridge, MA.

Meucci, A. (2005). *Risk and Asset Allocation*. Springer, New York.

Meyer, M. (2000). *Continuous Stochastic Calculus with Applications to Finance*. Chapman & Hall/CRC, Boca Raton, FL.

Mikosch, T. (1999). *Elementary Stochastic Calculus with Finance in View*, volume 6 of *Advanced Series on Statistical Science and Applied Probability*. World Scientific Publishing, Singapore.

Mills, T. (2001). *Problems in Probability*. World Scientific Publishing, Singapore.

Mörters, P. and Peres, Y. (2010). *Brownian Motion*. Cambridge University Press, Cambridge, UK.

Moschovakis, Y. (2006). *Notes on Set Theory*. Springer, New York, 2nd edition.

Mosteller, F. (1987). *Fifty Challenging Problems in Probability with Solutions*. Dover Publications, Mineola, NY.

Musiela, M. and Rutkowski, M. (2005). *Martingale Methods in Financial Modelling*. Springer, New York, 2nd edition.

Myneni, R. (1992). The pricing of the american option. *Annals of Applied Probability*, 2(1):1–23.

Neftci, S. (2000). *Introduction to the Mathematics of Financial Derivatives*. Academic Press, Singapore, 2nd edition.

Neftci, S. (2008). *Principles of Financial Engineering*. Academic Press, Singapore, 2nd edition.

Novikov, A. (1973). On moment inequalities and identities for stochastic integrals. In *Proceedings of the Second Japan-USSR Symposium on Probability Theory*, volume 330 of *Lecture Notes in Mathematics*, pages 333–339.

Oksendal, B. (1998). *Stochastic Differential Equations: An Introduction with Applications*. Springer, New York, 5th edition.

Pap, E. (2002). *Handbook of Measure Theory*. North Holland, Amsterdam.

Papoulis, A. (1991). *Probability, Random Variables and Stochastic Processes*. McGraw-Hill, New York, 3rd edition.

Pascucci, A. (2011). *PDE and Martingale Methods in Option Pricing*. Springer, New York.

Pascucci, A. and Runggaldier, W. (2012). *Financial Mathematics: Theory and Problems for Multi-period Models*. Springer, New York.

Paul, W. and Baschnagel, J. (1999). *Stochastic Processes: From Physics to Finance*. Springer, Berlin.

Pham, H. (2009). *Continuous-Time Stochastic Control and Optimisation with financial applications*. Springer, New York.

Pinsky, R. (1995). *Positive Harmonic Functions and Diffusion*. Cambridge University Press, Cambridge, UK.

Pliska, S. (1997). *Introduction to Mathematical Finance: Discrete Time Models*. Wiley, New York.

Pollard, D. (2001). *A User's Guide to Measure Theoretic Probability*. Cambridge University Press, Cambridge, UK.

Prevot, C. and Rockner, M. (2007). *A Concise Course on Stochastic Partial Differential Equations*. Springer, New York.

Protter, P. (2003). *Stochastic Integration and Differential Equations*. Springer, New York, 2nd edition.

Rao, M. and Swift, R. (2006). *Probability Theory with Applications*. Springer, New York, 2nd edition.

Rebonato, R. (2002). *Modern Pricing of Interest-Rate Derivatives: The LIBOR Market Model and Beyond*. Princeton University Press, Princeton, NJ.

Rebonato, R. (2004). Interest-rate term-structure pricing models: A review. *Proceedings: Mathematical, Physical and Engineering Sciences*, 460(2043):667–728.

Reitano, R. (2010). *Introduction to Quantitative Finance: A Math Tool Kit*. The MIT Press, Cambridge, MA.

Resnick, S. (1992). *Adventures in Stochastic Processes*. Springer, New York.

Resnick, S. (1999). *A Probability Path*. Birkhäuser, Boston.

Revuz, D. and Yor, M. (1999). *Continuous Martingales and Brownian Motion.* Springer, New York, 3rd edition.

Roberts, A. (2009). *Elementary Calculus of Financial Mathematics.* SIAM, Philadelphia.

Rogers, L. and Williams, D. (2000). *Diffusions, Markov Processes and Martingales: Volume 2, It Calculus.* Cambridge University Press, Cambridge, UK, 2nd edition.

Rosenthal, J. (2006). *A First Look at Rigorous Probability Theory.* World Scientific Publishing, Singapore, 2nd edition.

Ross, S. (1995). *Stochastic Processes.* Wiley, 2nd edition.

Ross, S. (1999). *An Introduction to Mathematical Finance: Options and Other Topics.* Cambridge University Press, Cambridge, UK.

Ross, S. (2010). *Introduction to Probability Models.* Academic Press, Burlington, MA, 6th edition.

Ross, S. and Pekoz, E. (2007). *A Second Course in Probability.* ProbabilityBookstore.com, Boston.

Rouah, F. and Vainberg, G. (2007). *Option Pricing Models and Volatility Using Excel-VBA.* Wiley, Hoboken, NJ.

Roussas, G. (1997). *A Course in Mathematical Statistics.* Academic Press, Singapore, 2nd edition.

Roussas, G. (2005). *An Introduction to Measure-Theoretic Probability.* Academic Press, Burlington, MA.

Royden, H. (1988). *Real Analysis.* Prentice Hall, New York, 3rd edition.

Rubinstein, M. (1991). Double trouble. *Risk Magazine*, 5:53–56.

Rubinstein, M. and Reiner, E. (1991). Breaking down the barriers. *Risk Magazine*, 4(9):28–35.

Rudin, W. (1970). *Real and Complex Analysis.* McGraw-Hills, New York.

Rudnick, J. and Gaspari, G. (2004). *Elements of the Random Walk: An Introduction for Advanced Students and Researchers.* Cambridge University Press, Cambridge, UK.

Sato, K.-I. (1999). *Lévy Processes and Infinitely Divisible Distributions.* Cambridge University Press, Cambridge, UK.

Schilling, R. (2005). *Measures, Integrals and Martingales.* Cambridge University Press, Cambridge, UK.

Schoutens, W. (2003). *Lévy Processes in Finance: Pricing Financial Derivatives*. Wiley, Hoboken, NJ.

Schoutens, W. and Cariboni, C. (2009). *Lévy Processes in Credit Risk*. Wiley, Hoboken, NJ.

Seydel, R. (2006). *Tools for Computational Finance*. Springer, New York, 3rd edition.

Shafer, G. and Vovk, V. (2001). *Probability and Finance: It's only a Game!* Wiley, Hoboken, NJ.

Shao, J. (2003). *Mathematical Statistics*. Springer, New York, 2nd edition.

Shiryaev, A. (1999). *Essentials of Stochastic Finance: Facts, Models, Theory*. World Scientific Publishing, Singapore.

Shiryaev, A. and Boas, R. (1995). *Probability*. Springer, New York, 2nd edition.

Shorack, G. (2000). *Probability for Statisticians*. Springer, New York.

Shreve, S. (2004). *Stochastic Calculus for Finance II: Continuous-Time Models*. Springer, New York.

Shreve, S. (2005). *Stochastic Calculus for Finance I: The Binomial Asset Pricing Model*. Springer, New York.

Smith, C. (1976). Option pricing: A review. *Journal of Financial Economics*, 3(1-2):3–51.

Sondermann, D. (2007). *Introduction to Stochastic Calculus for Finance: A New Didactic Approach*. Springer, New York.

Spitzer, F. (2001). *Principles of Random Walks*, volume 34 of *Graduate Texts in Mathematics*. Springer, New York, 2nd edition.

Steele, J. (2003). *Stochastic Calculus and Financial Applications*. Springer, New York, NY.

Stirzaker, D. (2003). *Elementary Probability*. Cambridge University Press, Cambridge, UK, 2nd edition.

Stirzaker, D. (2005). *Stochastic Processes and Models*. Oxford University Press, Oxford, UK.

Stoyanov, J. (1997). *Counterexamples in Probability*. Wiley, New York, 2nd edition.

Stroock, D. (2005). *An Introduction to Markov Processes*. Springer, New York.

Stroock, D. (2011). *Probability Theory: An Analytic View*. Cambridge University Press, Cambridge, UK, 2nd edition.

Stroock, D. and Varadhan, S. (1979). *Multidimensional Diffusion Processes*. Springer, New York, NY.

Svoboda, S. (2004). *Interest Rate Modelling*. Palgrave Macmillan, Gordonsville, VA.

Takacs, L. (1960). *Stochastic Processes: Problems and Solutions*. Wiley, New York.

Taylor, S. (1973). *Introduction to Measure and Integration*. Cambridge University Press, Cambridge, UK.

Tijms, H. (2003). *A First Course in Stochastic Models*. Wiley, Hoboken, NJ.

Topper, J. (2005). *Financial Engineering with Finite Elements*. Wiley, Hoboken, NJ.

Vasiček, O. (1977). An equilibrium characterization of the term structure. *Journal of Financial Economics*, 5(2):177–188.

Vaught, R. (1995). *Set Theory: An Introduction*. Birkhäuser, Boston, 2nd edition.

Vestrup, E. (2003). *The Theory of Measures and Integration*. Wiley, Hoboken, NJ.

Watsham, T. and Parramore, K. (1997). *Quantitative Methods in Finance*. Thomson, London, UK.

Weir, A. (1979). *General integration and measure*. Cambridge University Press, Cambridge, UK.

Wheeden, R. and Zygmund, A. (1977). *Measure and Integral: An Introduction to Real Analysis*. CRC Press, Boca Raton, FL.

Wiersema, U. (2008). *Brownian Motion Calculus*. Wiley, Hoboken, NJ.

Williams, D. (1991). *Probability with Martingales*. Cambridge University Press, Cambridge, UK.

Wilmott, P. (2001). *Paul Wilmott Introduces Quantitative Finance*. Wiley, Hoboken, NJ, 2nd edition.

Wilmott, P. (2006). *Paul Wilmott on Quantitative Finance 3 Volume Set*. Wiley, Hoboken, NJ, 2nd edition.

Wilmott, P., Howison, S., and Dewynne, J. (1995). *The Mathematics of Financial Derivatives: A Student Introduction*. Cambridge University Press, Cambridge, UK.

Wimmer, G. and Altmann, G. (1999). *Thesaurus of Univariate Discrete Probability Distributions*. STAMM Verlag, Essen, Germany.

Wise, G. and Hall, E. (1993). *Counterexamples in Probability and Real Analysis*. Oxford University Press, Oxford, UK.

Yeh, J. (2006). *Real Analysis: Theory of Measure And Integration*. World Scientific Publishing, Singapore, 2nd edition.

Zakai, M. (1967). Some moment inequalities for stochastic integrals and for solutions of stochastic differential equations. *Israel Journal of Mathematics*, 5(3):170–176.

List of Symbols

Subject Index

E

F